"We see great peril if governments and societies do not take action now to render nuclear weapons obsolete and to prevent further climate change".

Stephen Hawking, "Brief Answers to the Big Questions", 2018.

CLIMATE CRISIS, CLIMATE GENOCIDE & SOLUTIONS

To my dear late wife Zareena née Lateef.

KP

TABLE OF CONTENTS

PREFACE

"Climate Crisis, Climate Genocide & Solutions" has been written to set out the acute seriousness of the worsening Climate Emergency, and the existential threat to Humanity and the Biosphere from man-made climate change. The animal extinction rate in our present Anthropocene Era is 100-1,000 times greater than normal [1], and there is widespread speciescide and ecocide pushing the Earth towards omnicide and terracide [2]. Presently each year about 1 million people die from climate change, and 9 million die from the effects of pollutants from burning carbon fuels [3], but it is estimated that an average of 100 million people will die each year this century due to a worsening Climate Genocide en route to a sustainable population in 2100 of only about 1 billion people – a near-terminal Climate Genocide involving the avoidable deaths of 10 billion people and the destruction of much of the Biosphere [2].

Irish chemist John Tyndall discovered the greenhouse effect 150 years ago (indeed I have a copy of one of his books that describes this phenomenon in detail [4]). The Swedish physical chemist Svante Arrhenius predicted man-made global warming due to greenhouse gases (GHGs) 120 years ago. The Intergovernmental Panel on Climate Change (IPCC) was set up 30 years ago. Scientists have been individually and collectively warning about the seriousness of man-made global warming for 30 years but the worsening impacts of global warming-exacerbated droughts, floods, tropical storms, sea level rise, storm surges, ecosystem loss, biodiversity loss and catastrophic forest fires have now lent a new desperation to their pleas for action.

Thus an Open Letter by 255 members of the US National Academy of Sciences, including 11 Nobel Laureates, concluded (2010): "Delay is not an option" [5]. The Synthesis Report of the 2,500-delegate March 2009 Copenhagen Scientific Climate Change Conference concluded that "Inaction is inexcusable" [6, 7]. One of the world's greatest minds, Stephen Hawking, in his last book (2018) reiterated that "We see great peril if

governments and societies do not take action now to render
nuclear weapons obsolete and to prevent further climate
change" [8].

 Unfortunately the endlessly greedy One Percenters running the
World are committed to endless growth and their obscene
acquisition of 50% of the World's wealth means that One
Percenter-owned and One Percenter-beholden Mainstream
media can spread a message of outright climate change denial or
of effective climate change denial through climate change
inaction [9-11]. Neoliberal greed, egregious inequity and
ruthless mendacity are at the heart of this deadly problem.
Inequity is bad for the economy (the poor cannot purchase the
goods and services they produce), and bad for democracy (Big
Money determines public perception of reality and hence buys
votes) [12, 13].

The worsening Climate Crisis and Biodiversity Crisis demand
urgent constraint on both population and resource utilization,
but the deadly neoliberal message of endless "jobs and growth"
continues to dominate the world. Neoliberalism demands
maximal freedom for the smart and advantaged to exploit
natural and human resources for private profit. In contrast,
social humanism (socialism, eco-socialism, democratic
socialism, human rights-observant communism, the welfare
state, Universal Basic Income) seeks to maximize happiness,
dignity and opportunity for everyone through culturally
cognizant and evolving intra-national and international social
contracts [14-19].

It is estimated that if everyone had an American lifestyle we
would need 7 planets [20], and at the World's present level of
resource utilization 1 planet is clearly already insufficient, as
evidenced by mass species extinction. There is no Planet B. The
worsening Climate Crisis needs a population decrease of about
50%, a corresponding economic de-growth, and sustainable and
equitable sharing of scarce resources (eco-socialism) to enable a
decent life for everyone on Spaceship Earth [21]. This will not
happen unless there is a Climate Revolution (non-violent of
course) led by today's young people who with future

generations will have to bear the cost (Carbon Debt) of past, present and remorselessly continuing neoliberal profligacy [22-28].

Science-trained Pope Francis has demanded that the environmental and social cost of pollution be "fully borne" by the polluters [29, 30]. Eminent economist Lord Nicholas Stern has described the non-application of a Carbon Price as "the greatest market failure in history" [31]. Numerous scientists and science-informed activists say that we must return the atmospheric CO_2 from the present disastrous 415 ppm CO_2 to a safe and sustainable, pre-Industrial Revolution level of about 300 ppm CO_2 for a safe and sustainable environment for all peoples and all species [32, 33]. However, at a damage-related Carbon Price of $200 per tonne CO_2 the World has a Carbon Debt of $200-250 trillion that is increasing at $16 trillion per year [34-37].

The task is daunting. Indeed it is now apparent that the Paris target of plus 1.5C will be reached within this coming decade. Given the global inaction and remorselessly increasing atmospheric CO_2 it is increasingly unlikely that a catastrophic plus 2C will be avoided. Indeed national commitments made at the 2015 Paris Climate Change Conference amount to a disastrous temperature rise of 3.2C [38]. Nevertheless, decent people are obliged to do everything they can to make the future "less bad" for our children and future generations.

Indeed "Climate Crisis, Climate Genocide & Solutions" has been written to set out both the dimensions of the problem and the Solutions that can and must be implemented as soon as possible (ASAP). This book is based on articles and books I have published over the last 25 years, from 1995 [39] to October 2020 [40]. The Introduction provides a rapid introduction for non-scientists to the chemistry and biochemistry of climate change in note form. It is an updated version of a 2011 climate change course I gave to final year university students at a big university and to the Banyule (Melbourne) branch of the University of the Third Age. The book is divided into 8 Sections A-H, and within each Section (excepting Section C and Chapter 11) I have reproduced as Chapters articles published in the progressive web magazine Countercurrents. Each Section commences with a detailed summary of the present situation and of my writings on the subject, mostly over the last decade. A final Epilogue provides a summary of actions that must be taken.

In his novel "Bliss" [41] Australian author Peter Carey has a character who creeps into work on Saturday and writes and files indignant un-sent letters to wicked corporate polluters in the hope that he will be spared when the Revolution comes. This book records how I have carefully researched, written and published for public perusal my concerns about the worsening Climate Crisis over the last 25 years. Decent people are obliged to "bear witness" especially about events involving mass suffering and mass mortality.

Because of how the book was constructed the numbers quoted (e.g. of atmospheric CO_2 in parts per million) changed over the last decade. Occasionally I have inserted amplifying comments or updates in square brackets (e.g. [415 ppm CO_2 in 2020]). Further, key quotations from top scientific experts are repeated in the various chapters because I sought to rigorously reiterate the expert basis for my assessments in the various contexts of the book sections.

Peace is the only way but silence kills and silence is complicity. To that end I have published a number of carefully documented large books "bearing witness" to atrocities that the endlessly mendacious Mainstream resolutely ignores, namely "Jane Austen and the Black Hole of British History. Colonial rapacity, holocaust denial and the crisis in biological sustainability" (1998, 2008) [42], "Body Count. Global avoidable mortality since 1950" (2007) [43], and "US-Imposed Post-9/11 Muslim Holocaust & Muslim Genocide" (that includes Chapter 21, "Climate Genocide & War on Terra") [44, 45]. Indeed, space permitting, I was able to insert many brief, several word and pertinent references to such events into the scientific Tables in my huge pharmacological text "Biochemical Targets of Plant

Bioactive Compounds. A pharmacological reference guide to sites of action and biological effects" (2003) [46].

This present book has had a long genesis over 60 years of my life as a student and thence a researcher of biological chemistry. During that time the "monthly average atmospheric CO_2", measured as parts per million (ppm) CO_2 at Mauna Loa, Hawaii, has steadily increased as noted in brackets in the following key steps in my personal intellectual history re biodiversity, climate change and sustainability:

 (a) in 1962 I read Rachel Carson's "Silent Spring" [47] and immediately regarded all species and ecosystems as priceless i.e. we cannot destroy what we cannot replace (318 ppm CO_2);

(b) in 1968 I was made aware of the serious population problem through Paul Ehrlich's (and Anne Ehrlich's) "The Population Bomb" [48] (321 ppm CO_2);

(c) in 1972 I was made aware of the finiteness of exploitable resources by "The Limits to Growth" report to the Club of Rome [49] (328 ppm CO_2);

(d) in the early 1980s I was made aware of the fragility of the atmosphere through the discovery of the chlorofluorocarbon (CFC)-generated Ozone Hole, and the subsequent Montreal Protocol to ban CFC refrigerant gases of 1987 [50] (348 ppm CO_2);

(e) I became aware of the seriousness of the potential impact of atmospheric GHGs in the 1980s and especially with the founding of the Intergovernmental Panel on Climate Change (IPCC) in 1988 [51](351 ppm CO_2);

(f) in 1995 I was aware of the serious threat of global warming-driven sea level rise to Bengal, sent a detailed account to MPs, university VCs and media (that was subject to a speech and tabling in the Australian Federal Parliament by Greens Senator Christobal Chamarette) (see [42]), and published a detailed account in the International Network on Holocaust and Genocide [39] (360 ppm CO_2);

(g) while holding all species and ecosystems as priceless, I gained insight into the economic value of wild nature through reading a key 2002 paper in Science by Andrew Balmford et al. [52] (373 ppm CO_2):

(h) in 2007 I got a great insight into planetary biological and non-biological interactions in reading James Lovelock's "Revenge of Gaia"[53] (383 ppm CO_2);

(i) I was greatly influenced by the continuing cogent advocacy of Australian climate activists David Spratt and Philip Sutton as presented at meetings and in their 2008 book "Climate Code Red" [54](386 ppm CO_2);

(j) in 2009 I was a founding member of the Yarra Valley Climate Action Group [55] (that transmuted into the Banyule Climate Action Now or BCAN in 2019) [56], and was involved in the activities of the Climate Emergency Network (387 ppm CO_2);

 (k) I was greatly influenced by climate scientists, most notably by Dr James Hansen of NASA and thence of Columbia University as expressed in numerous articles and his 2009 book "Storms of My Grandchildren: The Truth About the Coming Climate Catastrophe and Our Last Chance to Save Humanity"[57] (387 ppm CO_2);

(l) I have been informed by IPCC reports, reports from the admirable Australian Climate Council, and papers from climate scientists, most recently a 2019 paper by variously eminent climate scientists in the top journal Nature stating "Act now. In our view, the evidence from tipping points alone suggests that we are in a state of planetary emergency: both the risk and urgency of the situation are acute" [58, 59] (410 ppm CO_2).

Some hope has been taken from remarkable teenage activist Greta Thunberg, her now global School Strike movement, her

2019 book "No One Is Too Small to Make a Difference" [60], and especially her impassioned and rivetting: "How dare you!" at the United Nations [61]. The monthly average atmospheric CO_2 at Mauna Loa, Hawaii, is now about 415 ppm CO_2.

I have dedicated this book to my dear late wife Zareena née Lateef. She was from Fiji but had her high school education in Hobart, capital of the Australian island state of Tasmania, and obtained a Bachelor of Science degree from the University of Tasmania, later acquiring a Diploma of Education and a Diploma of Librarianship from "Mainland" universities. She introduced me to bush walking, the Tasmania Wilderness, and native Australian plants. A science teacher and teacher-librarian, she was the president of the Yarra Valley Climate Action Group for many years. She was endlessly cheerful and sociable, and was always very amused when in handing out how to vote cards for the Greens she was called a watermelon (green on the outside and red on the inside). Zareena was of Bengali and Bihari origin, all of her grandparents having been indentured Indian labourers ("5-year slaves") brought to Fiji by the British to slave on British and Australian sugar cane plantations [62].

Her British colonial and Australian education left Zareena (and likewise her family) utterly ignorant of the WW2 Bengali Holocaust (the 1942-1945 Bengal Famine in which the British with Australian complicity deliberately starved 6-7 million Indians to death for strategic reasons in Bengal and the adjoining provinces of Bihar, Assam and Orissa). Having discovered this assiduously "forgotten" atrocity I have spent a lot of time in the last 25 years bringing it to the attention of people around the world through books, articles and broadcasts [42-45, 63-65].

Indeed my message as articulated back in 1995 was acutely related to man-made climate change: "History may yet repeat itself with respect to the Bengalis. Australia is per capita one of the most profligate contributors to "greenhouse" gas emission and hence global warming. There is unfortunately a current bipartisan political consensus in Australia opposed to effective short-term reduction of such emission. The recent Berlin

Climate Change Conference, heavily influenced by the U.S. and similarly-inclined Australia and Canada, concluded with no agreed targets on greenhouse gas emission and produced merely a motherhood commitment to continuing assessment. While global warming may have advantages for some, there is a widespread concern that global warming and consequent sea level rises will have a major impact on low-lying regions such as densely populated and impoverished West Bengal and Bangladesh. It is quite conceivable that continuing global greed and irresponsibility in relation to greenhouse gas emissions, desertification and forest destruction will have an even more devastating impact on Bengal in the 21st century than that of ruthless imperialism in past centuries. We must learn from the appalling consequences of the blindness, complacency and insensitivity with which the world responded at the time to the Jewish Holocaust and to the now-forgotten Bengali Holocaust. Never again" [39]. In 2020 monsoon floods left one third of Bangladesh under water and displaced more than 1.5 million people [66].

Zareena's home country of Fiji is acutely threatened by man-made climate change as are other tropical Island Nations, and is a world leader in protesting the callous GHG pollution and climate change inaction by the industrialized world that is set to see some Island Nations totally inundated [67]. This is "Climate Genocide" with "genocide" defined by Article 2 of the UN Genocide Convention as "acts committed with intent to destroy a national, ethnic, racial or religious group" [2, 68]. The Island Nations will be devastated first, followed by the most vulnerable mega-delta countries like Bangladesh. However all nations will be impacted by a worsening Climate Genocide en route to a sustainable human population of only about 1 billion by 2100 [2]. One notes that the South Pacific Island Continent nation of Australia is fervently pro-fossil fuels, and is among world leaders in 16 areas of climate criminality in an extraordinary, continuing process of cognitive dissonance, myopia, greed and mendacity common to all significantly climate criminal countries. This book has been written to counter the deadly Mainstream media silencing of the worsening Climate Crisis and Climate Genocide [70].

Finally, I must express my deep gratitude to Søren Korsgaard for his enthusiasm, expertise, dedication and commitment to this important, truth-telling, and humanitarian project.

References.

[1]. Phillip Levin, Donald Levin, "The real biodiversity crisis", American Scientist, January-February 2002: http://www.americanscientist.org/issues/pub/the-real-biodiversity-crisis.
[2]. Gideon Polya, editor, "Climate Genocide": https://sites.google.com/site/climategenocide/.
[3]. Gideon Polya, editor, "Stop air pollution deaths": https://sites.google.com/site/300orgsite/stop-air-pollution-deaths.
[4]. John Tyndall, "Fragments of Science: a series of detached essays, addresses and reviews", fifth edition, D. Appleton, New York, 1886, pages 27-70.
[5]. 2010 Open Letter by 255 members of the US National Academy of Sciences, "Open Letter: climate change and the integrity of science", Guardian, 6 May 2010: http://www.guardian.co.uk/environment/2010/may/06/climate-science-open-letter.
[6]. Synthesis Report from the March 2009 Copenhagen Climate Change Conference, Climate Change, Global risks, challenges & decisions", Copenhagen 10-12 March, 2009, University of Copenhagen, Denmark: http://lyceum.anu.edu.au/wp-content/blogs/3/uploads//Synthesis%20Report%20Web.pdf.
[7]. Summary of Synthesis Report from 2009 Copenhagen Climate Change Conference, YVCAG: https://sites.google.com/site/yarravalleyclimateactiongroup/synthesis-report-of-the-2009-copenhagen-climate-change-conference.
[8] (Stephen Hawking, "Brief Answers to the Big Questions", John Murray, 2018, Chapter 7).
[9]. Gideon Polya, editor, "Boycott Murdoch media": https://sites.google.com/site/boycottmurdochmedia/.
[10]. Gideon Polya, editor, "Mainstream media censorship": https://sites.google.com/site/mainstreammediacensorship/home.
[11]. Gideon Polya, editor, "Mainstream media lying": https://sites.google.com/site/mainstreammedialying/home.
[12]. Thomas Piketty, "Capital in the Twenty-First Century", Harvard University Press, 2014.
[13]. Gideon Polya, "Key Book Review: "Capital In The Twenty-First Century" By Thomas Piketty", Countercurrents, 1 July, 2014: https://countercurrents.org/polya010714.htm.
[14]. Brian Ellis, "The New Enlightenment. On Steven Pinker & beyond", Australian Scholarly Publishing, Melbourne, 2019.

[15]. Gideon Polya, "Review: "The New Enlightenment" by Brian Ellis – World Government & Social Humanism to save Planet", Countercurrents, 7 October 2019: https://countercurrents.org/2019/10/review-the-new-enlightenment-by-brian-ellis-world-government-social-humanism-to-save-humanity/.
[16]. Brian Ellis,"Social Humanism. A New Metaphysics", Routledge, UK, 2012.
[17]. Gideon Polya, "Book Review: "Social Humanism. A New Metaphysics" By Brian Ellis – Last Chance To Save Planet?", Countercurrents, 19 August, 2012: https://countercurrents.org/polya190812.htm.
[18]. Brian Ellis, "Rationalism. A critique of pure theory", Australian Scholarly, Melbourne, 2017.
[19]. Gideon Polya, "Review: "Rationalism" by Brian Ellis, Countercurrents, 14 August 2017: https://countercurrents.org/2017/08/review-rationalism-by-brian-ellis.
[20]. Irene Banos Ruiz, "China's new love affair with dogs – as pets, not food – presents environmental problems", DW, 21 June 2016: https://www.dw.com/en/chinas-new-love-affair-with-dogs-as-pets-not-food-presents-environmental-problems/a-19197523.
[21]. Gideon Polya, "How much negative carbon emissions, negative population growth & negative economic growth is needed to save Planet?", Countercurrents, 28 November 2018: https://countercurrents.org/2018/11/how-much-negative-carbon-emissions-negative-population-growth-negative-economic-growth-is-needed-to-save-planet/.
[22]. Gideon Polya, editor, "Carbon Debt Carbon Credit": https://sites.google.com/site/carbondebtcarboncredit/.
[23]. Gideon Polya, editor, "Climate Revolution Now": https://sites.google.com/site/300orgsite/climate-revolution.
[24]. Gideon Polya, editor, "Stop climate crime": https://sites.google.com/site/300orgsite/stop-climate-crime.
[25]. Gideon Polya, editor, "Stop state terrorism": https://sites.google.com/site/stopstateterrorism/.
[26]. Gideon Polya, editor, "State crime and non-state terrorism": https://sites.google.com/site/statecrimeandnonstateterrorism/.
[27]. "Climate terrorism: 400,000 climate change-related deaths globally annually versus an average of 4 US deaths from political terrorism annually since 9-11": https://sites.google.com/site/statecrimeandnonstateterrorism/climate-terrorism.
[28]. "Carbon terrorism: 3 million US air pollution deaths versus 53 US political terrorism deaths since 9-11 (2001-2015)":

https://sites.google.com/site/statecrimeandnonstateterrorism/carbon-terrorism.

[29]. Pope Francis, Encyclical Letter "Laudato si'", 2015: http://w2.vatican.va/content/francesco/en/encyclicals/documents/papa-francesco_20150524_enciclica-laudato-si.html.

[30]. Gideon Polya, "Green Left Pope Francis Demands Climate Action "Without Delay" To Prevent Climate "Catastrophe"", Countercurrents, 10 August 2015: https://countercurrents.org/polya100815.htm.

[31]. Alison Benjamin, "Stern: climate change a "market failure"", Guardian, 29 November 2007: https://www.theguardian.com/environment/2007/nov/29/climatechange.carbonemissions.

[32]. Gideon Polya, "300.org Urges Reduction Of Atmospheric CO_2 To 300 ppm", Countercurrents, 28 May 2009: https://www.countercurrents.org/polya280509.htm.

[33]. Gideon Polya, "300.org Message To World – Act Urgently Over Climate Emergency And Climate Genocide", Countercurrents, 23 September, 2009: https://countercurrents.org/polya230909.htm.

[34]. Chris Hope, "How high should climate change taxes be?", Working Paper Series, Judge Business School, University of Cambridge, 9, 2011: http://www.jbs.cam.ac.uk/media/assets/wp1109.pdf.

[35]. James Hansen, "Climate change in a nutshell: the gathering storm", Columbia University, 18 December 2018: http://www.columbia.edu/~jeh1/mailings/2018/20181206_Nutshell.pdf.

[36]. Gideon Polya, "Inescapable $200-250 trillion global carbon debt increasing by $16 trillion annually", Countercurrents, 27 April 2019: https://countercurrents.org/2019/04/inescapable-200-250-trillion-global-carbon-debt-increasing-by-16-trillion-annually-gideon-polya.

[37]. Gideon Polya, "Australia rejects IMF Carbon Tax & preventing 4 million pollution deaths by 2030", Countercurrents, 15 October 2019: https://countercurrents.org/2019/10/australia-rejects-imf-carbon-tax-preventing-4-million-pollution-deaths-by-2030.

[38]. "UN emissions report: World on course for more than 3 degree spike, even if climate commitments are met", UN News, 26 November 2019: https://news.un.org/en/story/2019/11/1052171.

[39]. Gideon Polya (1995) "The Forgotten Holocaust - The 1943/44 Bengal Famine": http://globalavoidablemortality.blogspot.com.au/2005/07/forgotten-holocaust-194344-bengal.html (an edited version of this account has been published as G.M. Polya, "The famine of history: Bengal 1943.

International Network on Holocaust and Genocide vol.10, pages 10-15, 1995).
[40]. Gideon Polya, "The lie that "gas is cleaner"", Green Left, 7 October 2020: https://www.greenleft.org.au/content/lie-gas-cleaner.
[41]. Peter Carey, "Bliss", University of Queensland Press, 1981.
[42]. Gideon Polya, "Jane Austen and the Black Hole of British History. Colonial rapacity, holocaust denial and the crisis in biological sustainability", G.M. Polya, Melbourne, 1998, 2008, Chapter 17: http://janeaustenand.blogspot.com.
[43]. Gideon Polya, "Body Count. Global avoidable mortality since 1950" that includes an avoidable mortality-related history of every country since Neolithic times and is now available for free perusal on the web: http://globalbodycount.blogspot.com/.
[44]. Gideon Polya, "US-Imposed Post-9/11 Muslim Holocaust & Muslim Genocide", 400 pages, Korsgaard Publishing, Germany, 4 June 2020: https://www.amazon.com/US-Imposed-Post-9-Muslim-Holocaust-Genocide/dp/8793987056.
[45]. Gideon Polya, "Racist Mainstream ignores "US-Imposed Post-9/11 Muslim Holocaust & Muslim Genocide"", Countercurrents, 17 July 2020: https://countercurrents.org/2020/07/racist-mainstream-ignores-us-imposed-post-9-11-muslim-holocaust-muslim-genocide/.
[46]. Gideon Polya, "Biochemical Targets of Plant Bioactive Compounds. A pharmacological reference guide to sites of action and biological effects", Taylor & Francis and CRC Press, London & New York, 2003.
[47]. Rachel Carson, "Silent Spring", Houghton Mifflin, 1962.
[48]. Paul Ehrlich (and Anne Ehrlich) "The Population Bomb", Sierra Club / Ballantine Books, 1968.
[49]. Dennis Meadows, Donella Meadows, Jørgen Randers, William W. Behrens III, "The Limits to Growth" report to the Club of Rome, 1972: http://www.donellameadows.org/wp-content/userfiles/Limits-to-Growth-digital-scan-version.pdf.
[50]. "Montreal Protocol", Wikipedia: https://en.wikipedia.org/wiki/Montreal_Protocol.
[51]. Intergovernmental Panel on Climate Change (IPCC): https://www.ipcc.ch/.
[52]. Andrew Balmford et al. "Economic Reasons for Conserving Wild Nature", Science Vol. 297, pp. 950 – 953, 9 August 2002: https://www.yumpu.com/en/document/read/40893022/economic-reasons-for-conserving-wild-nature-science.
[53]. James Lovelock, "Revenge of Gaia", Penguin, 2007.
[54]. David Spratt and Philip Sutton, "Climate Code Red", Scribe, 2008.

[55]. "Yarra Valley Climate Action Group":
https://sites.google.com/site/yarravalleyclimateactiongroup/Home.
[56]. "Banyule Climate Action Now":
https://sites.google.com/site/banyuleclimateactionnow/.
[57]. James Hansen, "Storms of My Grandchildren: The Truth About the Coming Climate Catastrophe and Our Last Chance to Save Humanity", Bloomsbury Publishing, 2009.
[58]. Timothy Lenton, Johan Rockstrom, Owen Gaffney, Stefan Rahmsdorf, Katherine Richardson, Will Steffen and Hans Joachim Schellnhuber, "Climate tipping points – too risky to bet against", Nature 575, 592-595, 27 November 2019:
https://www.nature.com/articles/d41586-019-03595-0.
[59]. Gideon Polya, "Climate scientists: Planetary Emergency, Planet in Peril, Act Now", Countercurrents, 3 December 2019:
https://countercurrents.org/2019/12/climate-scientists-planetary-emergency-planet-in-peril-act-now/.
[60]. Greta Thunberg, "No One Is Too Small to Make a Difference", Penguin, 2019.
[61]. Greta Thunberg, ""How dare you!": transcript of Greta Thunberg's [23 September 2019] UN climate speech", Nikkei Asia, 25 September 2019:
https://asia.nikkei.com/Spotlight/Environment/How-dare-you-Transcript-of-Greta-Thunberg-s-UN-climate-speech.
[62]. Gideon Polya, "Review: "Tears In Paradise. Suffering and Struggle Of Indians In Fiji 1879-2004" by Rajendra Prasad – Britain's Indentured Indian "5 Year Slaves"", Countercurrents, 4 March, 2015: https://countercurrents.org/polya040315.htm.
[63]. "Bengali Holocaust (WW2 Bengal Famine) writings of Gideon Polya", Gideon Polya:
https://sites.google.com/site/drgideonpolya/bengali-holocaust.
[64]. Gideon Polya (2011), "Australia And Britain Killed 6-7 Million Indians In WW2 Bengal Famine", Countercurrents, 29 September, 2011: http://www.countercurrents.org/polya290911.htm.
[65]. Gideon Polya, "Economist Mahima Khanna, Cambridge Stevenson Prize And Dire Indian Poverty", Countercurrents, 20 November, 2011: https://countercurrents.org/polya201111.htm.
[66]. "Floods leave one third of Bangladesh under water and displace 1.5 million, raising coronavirus risk", ABC News, 13 August 2020: https://www.abc.net.au/news/2020-08-13/bangladesh-floods-sees-a-third-of-nation-underwater-coronavirus/12555448.
[67]. Pacific Islands Development Forum 4 September 2015 "Suva Declaration on Climate Change": http://pacificidf.org/wp-content/uploads/2013/06/PACIFIC-ISLAND-DEVELOPMENT-FORUM-SUVA-DECLARATION-ON-CLIMATE-CHANGE.v2.pdf.

[68]. Convention on the Prevention and Punishment of the Crime of Genocide, Adopted by Resolution 260 (III) A of the United Nations General Assembly on 9 December 1948: http://www.hrweb.org/legal/genocide.html.
[69]. Richard Hil and Gideon Polya, "The silencing of the climate emergency and what we should do about it", New Matilda, 5 March 2019: https://newmatilda.com/2019/03/05/silencing-climate-emergency/.

"We would then have some right to indulge in the pleasant belief that our descendants, albeit after many generations, might live under a milder sky and in less barren surroundings than is our lot at present." Svante Arrhenius (Swedish physical chemist and chemistry Nobel laureate) on the greenhouse effect from increasing atmospheric carbon dioxide, 1896.

"On the influence of carbonic acid [carbon dioxide] in the air on the temperature of the ground." Title of a key paper by Svante Arrhenius, 1896.

 "This result suggests that stabilization at today's greenhouse gas levels may already commit Earth to an eventual total warming of 5 degrees Celsius (range 3 to 7 degrees Celsius, 95 per cent credible interval) over the next few millennia as ice sheets, vegetation and atmospheric dust continue to respond to global warming." Dr Carolyn Snyder, "Evolution of global temperature over the last two million years", Nature, 2016.

"The Holocene has ended. The Garden of Eden is no more. We have changed the world so much that scientists say we are in a new geological age: the Anthropocene, the age of humans." Sir David Attenborough (famed naturalist), 2019.

"What becomes of the wave-motion thus intercepted? It is applied to the heating of the absorbing gas." John Tyndall (discoverer of the greenhouse effect), "Radiation", 1865.

INTRODUCTION: CLIMATE CHANGE, BIOCHEMISTRY & SOLUTIONS COURSE NOTES

[First published as "2011 Climate Change Course": https://sites.google.com/site/300orgsite/2011-climate-change-course.This started out as course notes for a mini-course on Climate Change I gave for final year Animal Science students at a big Australian university, and for an 8 session course entitled "Climate change, biochemistry and solutions" I gave for the Yarra Valley University of the Third Age (U3A) in Melbourne, Victoria, Australia (now the Banyule U3A). Including its appended Glossary of Terms and References, this section provides a very concise but detailed introduction for people with just a basic High School science background. Only very few updating changes have been made for the purposes of the book.]

A. Man-made global warming and GHGs

1. Earth's atmosphere: troposphere (surface to 9 km at poles, 17 km at equator); stratosphere (from tropopause boundary to 50 km; UV-absorbing layer); mesosphere (from stratopause boundary out to 80-85 km; where most meteors burn up); ionosphere (from 50 km out to 1,000 km; solar radiation ionizes molecules).

2. Dry air composition: 78% nitrogen (N_2), 21% oxygen (O_2), 0.9% argon (Ar), 0.04% CO_2. Air typically has about 1% water (H_2O) which increases with temperature in the range 0.01% (dry, polar) to 20% (humid, tropical).

3. Greenhouse effect: thermal radiation from sun absorbed by surface and air; re-emitted and reflected light absorbed and re-radiated by air molecules, notably carbon dioxide (CO_2), nitrous oxide (N_2O), methane (CH_4), H_2O, man-made greenhouse gases (GHGs). Greenhouse effect (John Tyndall, 1858, UK) keeps planet warm. CO_2-equivalent (CO_2-e): total GHGs including CO_2 and other GHGs but not H_2O.

4. Radiative forcing and Global Warming Potential (GWP) of GHGs. Radiative forcing measures warming (positive) effects of e.g. GHGs and carbon particles and cooling (negative) effects of e.g. sulphate aerosols and reflectivity (albedo) of ice, snow and clouds. Relative GWP based on infra-red (IR) absorbance properties and half-life in atmosphere is directly proportional to radiative forcing and inversely proportional to half-life in the atmosphere.

GWP relative to same mass of CO_2 on a 20 year time frame: CO_2 (1.0), CH_4 (79; 105 if aerosol impacts are considered), N_2O (289), chlorofluorohydrocarbon CFC-12, CCl_2F_2 (11,000), hydrochlorofluorohydrocarbon HFC-22, $CHClF_2$ (5,160), hydrofluorohydrocarbon HFC-23, CHF_3 (12,000), sulphur hexafluoride SF_6 (16,300), nitrogen trifluoride NF_3 (12,300).

GWP relative to same mass of CO_2 (1.0) on a 100 year time frame: CO_2 (1.0), CH_4 (21), N_2O (298), chlorofluorohydrocarbon CFC-12, CCl_2F_2 (10,900), hydrochlorofluorohydrocarbon HFC-22, $CHClF_2$ (1,810), hydrofluorohydrocarbon HFC-23, CHF_3 (14,800), sulphur hexafluoride SF_6 (22,800), nitrogen trifluoride NF_3 (17,200).

5. O_3-destroying chlorofluorohydrocarbons (CFCs) such as chlorofluorohydrocarbon CFC-12, CCl_2F_2 were replaced under the Montreal Convention (1987) by hydrofluorohydrocarbon refrigerants and propellants such as hydrofluorohydrocarbon HFC-23 (CHF_3) but the HFCs are now posing an increasing threat because of their increasing use and high GWP.

6. CO_2 is the major contributor to anthropogenic global warming (AGW), deriving from aerobic respiration involving oxidation of carbohydrate ($(CH_2O)_n + O_2 \rightarrow n\ CO_2 + nH_2O$), lime (CaO) generation in cement production ($CaCO_3 \rightarrow CaO + CO_2$), and the combustion of fossil fuels such as coal ($C + O_2 \rightarrow CO_2$), oil ($CH_3(CH_2)nH + (n + 1)O_2 + (n +2)O_2 \rightarrow (n+1)CO_2 +$

(n+2)H_2O) and natural gas, mainly methane ($CH_4 + 2O_2$ -> CO_2 + 2 H_2O).

Major CO_2 sinks include photosynthetic bacterial photosynthesis and plant photosynthesis yielding cellulose and related carbohydrates of wood and thence of soil carbonaceous compounds of humus. Photosynthesis yields carbohydrate: $nCO_2 + nH_2O$ + solar energy -> $(CH_2O)_n + nO_2$. Most of photosynthesis is reversed by carbohydrate oxidation by fires or aerobic organisms: $(CH_2O)_n + nO_2$ -> $nCO_2 + nH_2O$. Some CO_2 dissolves in ocean water, this resulting in biologically deleterious acidification: $CO_2 + H_2O$ -> H_2CO_3 -> $HCO_3^- + H^+$; HCO_3^- -> $CO_3^{2-} + H^+$ (see later: threat to coral from ocean acidification as well as from warming).

Coal derives from anaerobic geologic conversion of cellulosic carbohydrates to carbon (($CH_2O)_n$ + heat, pressure -> $nC + nH_2O$). Subterranean **oil** derives from anaerobic decarboxylation of biologically-derived fatty acids (CH_3-$(CH_2)_n$-COOH + heat, pressure -> $CH_3(CH_2)_nH + CO_2$. Subterranean **methane** derives from anaerobic reduction of carbohydrates by anaerobic bacteria (reduction being addition of electrons (e-), addition of hydrogen atoms (H) or removal of oxygen (O)): $(CH_2O)_n$ + 4nH (derived from catabolism and reduced coenzymes) -> $nCH_4 + nH_2O$.

Biochar. Atmospheric CO_2 can be reduced from the current 415 ppm CO_2 [2020] back to a safe and sustainable 300 ppm CO_2 by fixing CO_2 as cellulose via solar-energy-driven photosynthesis ($nCO_2 + nH_2O$ -> $(CH_2O)_n + nO_2$) with subsequent anaerobic pyrolysis of cellulosic material (e.g. waste wood and straw) at 500-700C to yield carbon (C, charcoal) (($CH_2O)_n$ -> $nC + n$ H_2O) which can then be added to soil or buried in holes in the ground (e.g. used coal mines). Thus p224, Progress in Thermochemical Biomass Conversion, volume 1, IAE Bioenergy, ed. A,V, Bridgewater (Blackwell Science) informs us that we could obtain 1.7 GtC/yr (straw from agriculture) + 4.2 GtC/yr (total grass upgrowth from grasslands upgrowth) + 6 GtC/yr (possible sustainable wood harvest) = 11.9 GtC/yr. From this one can see why biochar expert Professor Johannes

Lehmann of Cornell University is correct calculating that it is realistically possible to fix 9.5bn tonnes of carbon per year using biochar, noting that global annual production of carbon from fossil fuels is about 9 bn tonnes. Unfortunately biochar is expensive to produce and thus for every \$1 derived from coal burning, \$1-\$2 will have to be spent converting the consequent CO_2 back to carbon (biochar) (based on US mid-West corn stalk-derived biochar; multiply this by 4 for the cost in the UK).

7. Methane derives from anaerobic degradation of biological material e.g. in swamps, waste dumps, livestock digestion: $(CH_2O)_n$ + 4H (derived from reduced coenzymes) -> nCH_4 + nH_2O or, overall: $2(CH_2O)n$ -> nCH_4 + nCO_2. Global warming is already releasing CH_4 from H_2O-CH_4 clathrates in tundra and in shallow parts of the Arctic Ocean. Fugitive CH_4 emissions occur from coal mines, coal seam gas (CSG) extraction, conventional natural gas extraction, from coal seam and shale fracking (hydrological fracturing), and from systemic gas reticulation leakage. It is estimated that 50 Gt CH_4 will be released from the Arctic Ocean sea bed in coming decades.

8. Nitrous oxide (N_2O) derives from agricultural use of nitrogenous fertilizers and from fossil fuel (coal, gas and oil) combustion.

9. CO_2 concentration. As determined from ice cores the atmospheric CO_2 concentration has been 180-280 parts per million (ppm) for the last 800,000 years (excluding the last century), during which time *Homo sapiens* finally evolved (glaciation at low CO_2 and inter-glacial at high CO_2). Indeed these circa 100,000 year cycle oscillations (determined by the earth's orientation towards the sun and the ellipticity of its orbit) crucially contributed to the final evolution of man (repeated severe selection pressures). Atmospheric CO_2, now 400 ppm and increasing at over 2.4 ppm per year (3.05 ppm per year in February 2016 [415 ppm in 2020]; seasonally oscillating atmospheric CO_2 at Mauna Loa Observatory, Hawaii, is reported as a dry air mole fraction defined as the number of molecules of CO_2 divided by the number of all molecules in air, including CO_2 itself, after water vapour has been removed.

Man-made from burning fossil fuels (decreasing ^{14}C; fossil CO_2 deficient in ^{14}C) and deforestation (initially non-tropical, now mostly tropical). Atmospheric CO_2 reached 400 ppm in 2013 and the peak in 2015 was 400 ppm CO_2. In February 2016 it reached 404 ppm CO_2 [415 ppm CO_2 in 2020].

10. Atmospheric CH_4 concentration 1774 parts per billion (ppb) (2011) versus 700 ppb in 1750 (i.e. pre-industrial). Atmospheric CH_4 increased to 1843 ppb CH_4 by December 2015 as compared to a pre-Industrial Revolution level of 700 ppb CH_4 [1,880 ppb in 2020]. **Atmospheric N_2O concentration** is 334 ppb [2020] as compared to 270 ppb in 1750.

11. Temperature change correlates with GHG change. Modelled temperature change from GHG forcing fits observed pattern in nearly all zones (IPCC; key evidence for Anthropogenic Global Warming, AGW).

12. Forcing of man-made GHG and absorbing particles 30x that of change in solar input effect. 1750-2005 heating change in watts/m^2: air CO_2 (+1.7), CH_4, N_2O, CFCs (+1.0), net O_3 (troposphere up, stratosphere down; +0.3), soot (+0.3), reflective particles e.g. sulphate aerosols (-0.7), indirect, cloud-forming particle effects (-0.7), human land-use increasing reflectivity (-0.2), solar input change (+ 0.1).

13. Photosynthesis and re-oxidation carbon cycle. Terrestrial carbon fixation of 121.3 GtC/y (x 44/12 = 3.67 -> 445 Gt CO_2 = 445 billion tonnes of CO_2) of which about half returns annually to the atmosphere through plant and animal respiration and most of the remaining half returns to the air through the action of soil fungi and bacteria. About 2 Gt C (7 Gt CO_2) ends up stored as soil carbon or cellulosic biomass (wood). Net terrestrial biome production of fixed CO_2 is 0.7 GtC/year (2.6 Gt CO_2/year). Ocean photosynthesis (prokaryotic cyanobacteria and eukaryotic algae) fixes 50 GtC /year (184 Gt CO_2/year) of which all but circa 2 Gt C is thence re-oxidized back to CO_2.

14. Biochar from anaerobic pyrolysis (at 400-700C) conversion of cellulosic material to carbon (C, charcoal, biochar, Amazonian Indian terra preta): $(CH_2O)_n$ + heat -> nC + nH_2O. Current potential: 1.7 GtC/yr (straw from agriculture) + 4.2 GtC/yr (total grass upgrowth from grasslands upgrowth) + 6 GtC/yr (possible sustainable wood harvest) = 11.9 GtC/yr. Professor Johannes Lehmann (Cornell University) could fix 9.5bn tonnes of carbon per year using biochar, noting global annual production of carbon from fossil fuels is 8.5bn tonnes C (= 31 Gt CO_2) (see later: biochar is a major means of returning atmospheric CO_2 to 300 ppm from current dangerous and damaging 400 ppm CO_2). Alternative geoengineering abatement proposals of increasing albedo, fertilizing the oceans to increase ocean photosynthesis, sulphur in aviation fuel gas to increase global dimming atmospheric sulphur oxide particulates, shades in space).

15. Carbon storage. 700- 750 GtC in atmosphere (mostly as 750 x 3.67 = circa 2,800 Gt CO_2; half due to historical fossil fuel combustion); 700 GtC in biomass (mostly wood); 1,600 GtC in soil; 36,000 GtC in ocean as bicarbonate ion (HCO_3^-); no net CO_2 from vulcanism and weathering (time scale < 100,000 years).

16. World Bank analysts have recently re-assessed annual man-made (anthropogenic) global greenhouse gas (GHG) pollution as 64 Gt CO_2-e (17.4 GtC), about 50% bigger than the 42 Gt CO_2-e (11.4 GtC) hitherto thought, and that the livestock contribution is over 51% of the bigger figure (major element: 20 year time frame considered for CH_4 GWP of 72, noting that the GWP for CH_4 on a 20 year time frame with aerosol impacts considered is 105 times greater than that of the same mass of CO_2).

17. The Paleocene-Eocene Thermal Maximum (PETM) 55 million years ago was associated with extremely rapid 5C warming and mass extinctions - following a doubling in carbon dioxide levels, the surface of the ocean turned acidic over a period of weeks or months and global temperatures rose by 5 degrees Centigrade – all in the space of about 13 years. Methane release may have been involved. The 50 Gt (billion tonnes) CH_4 in the East Siberian Arctic Shelf and predicted to be released in coming decades is equivalent to 50 billion tonnes CH_4 x 105 tonnes CO_2-equivalent/tonne CH_4 = 5,250 tonnes CO_2-e or about nine (9) times more than the world's terminal greenhouse gas (GHG) pollution budget. We are doomed unless we can stop this Arctic CH4 release. We are evidently facing a possible PETM scenario in coming decades.

B. Already observed climatic disruption consequences of global warming.

1. Climate is the pattern of weather. Average surface temperature + 0.8C in 2010, +1.1C (2020); +1.5C by 2030 (IPCC); +2C already but for the cooling effects of sulphate aerosols; +2C inevitable on current trends; +3.2C (2015 Paris Climate Conference national promises); may reach +4C by 2100.

2. Non-uniform temperature increase e.g. +2C (Indian Ocean), +4C (Arctic), average about +1C; thermal inertia, >90% extra heat in oceans.

3. Uneven heating changes wind patterns e.g. East Asia monsoon weakening.

4. Glaciers are shrinking world-wide. No net deposition of ice in Himalayas (no net deposition of atmospheric radioactivity from nuclear tests) – feeds mega-rivers from Pakistan to China.

5. Permafrost thawing (Fairbanks, Alaska: average circa 0C, +2C over last 50 years). Permafrost melts at T> 0C (noting CH_4 release and positive feedback of warming -> methane release -> more warming ->). It is estimated that 50 Gt CH_4 will be released from the Arctic Ocean sea bed in coming decades.

6. Arctic summer sea ice disappearing (80% decrease of total mass; NIDC data: half surface area gone already, the rest may go in coming decades, NW Passage already open in summer).

7. Surface melting in Greenland expanding (+ 7 metres sea level if it all goes).

8. China: increased floods (south), increased drought (north); same pattern in Australia of increased floods (closer to equator), increased drought (south).

9. A 4-10-fold increase in major flood events per decade around the world (1950-2000) (increased sea temperature means increased humidity, increased precipitation). Statistically proven AGW (Anthropogenic Global Warming) cause for recent Welsh floods but hard to prove in general because of weather variability (cf cannot prove that an individual smoker's lung cancer is due to smoking).

10. Consensus prediction of an increased number of the more intense storms as AGW increases (arguable doubling of tropical hurricane intensity 1950-2000; tropical cyclone power increase parallels increased sea temperature).

11. Melting land ice and thermal expansion increasing sea level (3.0 mm/yr, 1993-2003; 1.5 mm/yr, 1910-1990; circa 20 cm). Major problem for India, Bangladesh, Myanmar, other tropical megadelta regions (storm surges).

12. Ocean acidification due to CO_2 dissolution and ionization (0.1 pH decrease; see below: coral threat; some CO_2 dissolves in ocean water, this resulting in biologically deleterious acidification: $CO_2 + H_2O \rightarrow H_2CO_3 \rightarrow HCO_3^- + H^+$; $HCO_3^- \rightarrow CO_3^{2-} + H^+$).

13. Loss of major CO_2 uptake by Southern Ocean (increased storm intensity impact).

14. Over 90% of extra heat in oceans.

<u>**C. Already observed biological impacts related to
Anthropogenic Global Warming (AGW).**</u>

1. Ecosystem migration towards poles with AGW (e.g. ocean phytoplankton, deciduous trees in Canada).

2. Heat waves 2x more frequent in Europe (2003 heat wave killed 35,000-50,000 people). January 2009 heat wave prior to 7 February 2009 Victoria Black Saturday killed 500 in SE Australia (elderly more frail, decreased stress signalling).

3. Drought and heat causing increased forest fires (4-10-fold increase in forest area burned in W USA, 1970-2000). Amazon forests threatened by drought and burning positive feedback cycle. Threat to Australia, US, Russia, Southern Europe, Indonesia.

4. Mountain pine beetle (MPB) *Dendroctonus ponderosae* blight (warmer winters, increased survival; larval feeding; fungus infection prevents tree resin defence; devastation of North American boreal conifer forests, US, Canada).

5. Animal, plant and disease migration towards poles e.g. northerly deciduous tree migration in Canada; fish polar migration away from equator; dengue fever and malaria spread through mosquito vector migration away from equator.

6. Increased drought impact in Southern US, Central America, Brazil, Europe, Russia, Mediterranean, Sub-Saharan Africa, East Africa, Siberia, Central Asia, Northern China, Southern Australia, South East Asia (+1C -> 10% decrease in sub-tropical grain yield).

7. Increased temperature, droughts, floods and coastal loss coupled with increased population (9.5 billion by 2050) means greater impact on ecosystems.

8. Coral loss started at 320 ppm CO_2 (inevitable general death above 450 ppm CO_2 due to warming-induced expulsion of photosynthetic zooanthellae symbionts and difficulty making calcareous exoskeleton due to lowered pH: some CO_2 dissolves in ocean water, this resulting in biologically deleterious acidification: $CO_2 + H_2O \rightarrow H_2CO_3 \rightarrow HCO_3^- + H^+$; $HCO_3^- \rightarrow CO_3^{2-} + H^+$; ecosystem disaster.

9. Sex ratio changes in reptilian species: feminization due to increased average temperature.

10 Species extinction rate is now 100-1,000 times greater than previously, impacted by AGW, increasing human population and land use (hence term Anthropocene Era to describe the current mass extinction event); flora and fauna threatened.

11. 6 major previous mass extinction events in Paleaozoic (490-250 Mya; successively Ordovician, Silurian, Devonian, Carboniferous, Permian): **(1) End-Ordovician, Ordovician - Silurian Mass Extinction, 434 Mya**, **(2) End-Devonian Mass Extinction, 354 Mya**, and **(3) End-Permian Mass Extinction, 251 Mya**; in Mesozoic (250-65 Mya; successively Triassic, Jurassic, Cretaceous): **(4) End-Triassic, Triassic-Jurassic Mass Extinction 205 Mya**); and Cenozoic (65 Mya to present): **(5) Cretaceous-Tertiary (K/T) Mass Extinction or Cretaceous-Paleogene boundary (KPB) mass extinction, 66 Mya**, and **(6) Paleocene-Eocene Thermal Maximum (PETM), 55 Mya** (dinosaurs disappeared around this time, likely due to a meteor strike).

12. The most recent mass extinction event occurred in the Cenozoic 55 Mya and is called the Paleocene-Eocene Thermal Maximum (PETM). PETM was associated with rapid global warming, profound changes in ecosystems, and major perturbations in the carbon cycle (increase in CO_2 as well as CH_4 perhaps due to vulcanism). Many ocean foraminifera and land mammals became extinct, but numerous modern mammalian groups emerged. Massive dissolution of ocean deposited carbonate occurred detected as a big entry of ^{13}C-depleted (old) carbon into the hydrosphere or atmosphere at the start of the PETM during which global temperatures rose by about 6 °C over a period of approximately 20,000 years i.e.

0.0003 °C per year as compared the current Anthropocene Mass Extinction (extinction rate 100-1,000 times greater than normal) and a temperature increase of 0.8C over 100 years i.e. 0.009°C per year or possibly as much as 6°C over 100 years or 0.06°C per year.

13. Successive loss of Antarctic sea ice, sea ice-associated phytoplankton, krill and thence krill-eating animals e.g. fish, penguins, seals, whales (massive decline already in krill for various reasons).

D. Projected biological impacts.

1. Increasing temperature, drought, floods, sea level rise (possibly 2 metre by 2100), increased high intensity storms and storm surges (see (C) above) resulting in further loss of arable land, ecosystems, pressure on remaining ecosystems.

2. Coral loss due to ocean warming (expulsion of zooxanthellae photosynthetic algae symbionts and coral bleaching) and ocean acidification from CO_2 dissolution impacts on calcareous exoskeleton formation. Major coral death above 450 ppm CO_2. IPCC: 70-90% coral loss at +1.5C, 99% coral loss at 2C.

3. Ocean acidification impacting all ocean organisms with calcareous exoskeletons (about one-third of the carbon dioxide (CO_2) released into the atmosphere as a result of human activity has been absorbed by the oceans, where it partitions into the constituent ions of carbonic acid, this leading to ocean acidification, one of the major threats to marine ecosystems and particularly to calcifying organisms such as corals (animals with photosynthetic symbionts), foraminifera (single-celled eukaryote amoeboid protists with a calcareous external shell), coccolithophores (photosynthetic algae that generate calcareous plates) and lobster, crab, shrimp and krill crustaceans with food chain implications).

4. Above 500 ppm CO_2 major loss of phytoplankton (bottom of the ocean food chain) and dimethyl sulphide (DMS) production (critically involved in cloud seeding); complete loss of Greenland ice sheet (long-term loss yielding 7 m sea level rise); loss of terrestrial plants as CO_2 sinks (i.e. net CO_2 emission).

5. Complete loss of Arctic summer sea ice in coming decades (loss of ecosystems, positive feedback effect to get even more GHG pollution (Albedo Flip: warming -> loss of light-reflecting ice, increased light-absorbing black sea -> more warming ->…), increased risk of oil pollution in Arctic from increased shipping and mineral extraction).

6. Successive loss of Antarctic sea ice, sea ice-associated phytoplankton, krill and thence krill-eating animals e.g. fish, penguins, seals, whales.

7. Water stress in particular regions with agricultural, ecosystem, peace impacts.

8. Concerns that ocean Coccolithophore algae that make calcareous plates would be damaged by warming and decreasing pH may not necessarily be borne out (during the Paleocene–Eocene Thermal Maximum (PETM; 55 Million Years Ago), rapid release of isotopically light C to the ocean–atmosphere system elevated the greenhouse effect and warmed temperatures by 5–7 °C for 10^5 yr. However despite acidification of the ocean (some CO_2 dissolves in ocean water, this resulting in biologically deleterious acidification: $CO_2 + H_2O$ -> H_2CO_3 -> $HCO_3^- + H^+$; HCO_3^- -> $CO_3^{2-} + H^+$) there was not a productivity crisis among calcifying phytoplankton (negative effects could have been balanced by increased weathering i.e. increased nutrient inventories could have resulted from climatically enhanced weathering). Similarly, a heavily calcified *Emiliania huxleyi* morphotype has been found in modern waters with low pH.

8. Climate genocide. Both Dr James Lovelock FRS (Gaia hypothesis) and Professor Kevin Anderson (Tyndall Centre, Manchester University, UK) have estimated that only about 0.5 billion people will survive this century due to unaddressed, man-made global warming. Noting that the world population is expected to reach 9.5 billion by 2050, these estimates translate to a Climate Genocide en route to a sustainable population in 2100 of only about 1 billion people and involving deaths of 10 billion people this century, this including roughly 2 times the present populations of various non-European groups, specifically 6 billion under-5 year old infants, 3 billion Muslims, 2 billion Indians, 1.3 billion non-Arab Africans, 0.5 billion Bengalis, 0.3 billion Pakistanis and 0.3 billion Bangladeshis. **Biofuel genocide** (food for fuel, price increase, volatility). It is estimated that 15 million people die avoidably each year, from deprivation (2020). 9 million people die each year from the long-term effects of carbon burning pollutants. About 1 million die annually from climate change. These horrendous projections raise the further moral issues of climate change justice, climate justice, climate change injustice, climate injustice, climate change terrorism, climate terrorism, climate change denial, climate denial.

<u>E. Urgency of required action.</u>

1. Just as we turn to top medical specialists for advice on life-threatening disease, so we turn to the opinions of top scientists and in particular top biological and climate scientists for Climate Change risk assessment. Thus some opinions: **(a) Professor James Hansen** (top US climate scientist, head, NASA's Goddard Institute for Space Studies): "We face a climate emergency"; **(b) Nobel Laureate Professor Peter Doherty**: "We are in real danger"; **(c) Professor David de Kretser AC** (eminent medical scientist and former Governor of Victoria, Australia): "There is no doubt in my mind that this is the greatest problem confronting mankind at this time and that it has reached the level of a state of emergency"; **(d) Dr Andrew Glikson** (palaeo-climate scientist, ANU): "The continuing use of the atmosphere as an open sewer for industrial pollution has … raised CO_2 levels to 387 ppm CO_2 to date, leading toward conditions which existed on Earth about 3 million years (Ma) ago (mid-Pliocene), when CO_2 levels rose to about 400 ppm, temperatures to about 2–3 degrees C and sea levels by about 25 +/- 12 metres"; **(e) Synthesis Report of the March 2009**

Copenhagen Scientific Climate Change Conference: "Inaction is inexcusable"; and **(f) 2010 Open Letter by 255 members of the US National Academy of Sciences: "Delay is not an option".**

2. Climate emergency actions urgently required: (a) Change of societal philosophy to one of scientific risk management and biological sustainability with complete cessation of species extinctions and zero tolerance for lying; **(b) Urgent reduction of atmospheric CO_2 to a safe level of about 300 ppm** (as recommended by leading climate and biological scientists to address the current mass species extinction event and to permit return and sustainability of Arctic sea ice); **(c) Rapid switch to the best non-carbon and renewable energy** (solar, wind, geothermal, wave, tide and hydro options that presently [2020] are cheaper than coal burning-based power and have a 4 times cheaper "true price" taking environmental and human impacts into account) and to **energy efficiency, public transport, needs-based production, re-afforestation and return of carbon as biochar to soils** coupled with correspondingly **rapid cessation of fossil fuel burning, deforestation, methanogenic livestock production and population growth.**

3. Budget approach to last remaining permissible GHG pollution.

(a). Professor Hans Joachim Schellnhuber (Director, Potsdam Institute for Climate Impact Research, Germany) stated that for a 67% chance of avoiding a catastrophic 2C temperature rise the World must cease CO_2 emissions by 2050 and top per capita greenhouse gas (GHG) polluters such as the US and Australia must get to zero CO_2 emissions before 2020.

(b) Australian Climate Commission's "The Critical Decade" reports, 2011, 2012, 2013 stated that for a 75% chance of avoiding a disastrous 2 degree Centigrade temperature rise the World can emit no more than 1 trillion tonnes of CO_2 before reaching zero emissions in about 2050. Australia's high Domestic plus Exported GHG pollution rate means that relative to 2010 it must get to zero emissions in 1.9 years (or in 4.6

years ignoring Exported GHG pollution). In the 2013 Report the Climate Commission estimates a present terminal budget of 600 Gt CO_2 and 16 years at current rates of pollution before this is exceeded.

(c) WBGU (that advises German Government on climate change), "Solving the climate dilemma: the budget approach" (2009) estimated that for a 75% chance of avoiding a disastrous 2 degree Centigrade (2C) temperature rise the World can emit no more than 0.6 trillion tonnes (600 Gt) of CO_2 before reaching zero emissions in about 2050. Australia's high Domestic plus Exported GHG pollution rate means that by August 2011 it had already used up its "fair share". Australia's huge resources and bipartisan policy of unlimited coal, gas and iron ore exports means that it is set to exceed the whole world's terminal budget of 600 Gt CO_2 by a factor of three (3). Global fossil fuel corporations intend to exceed the world's terminal budget by a factor of about five (5).

(d) Using latest data from the US EIA, World Bank and NASA one estimates 2 years left relative to 2016 before the terminal budget (Terminal Carbon Pollution Budget) is exceeded i.e. the Budget has now been used up (2020).

4. Climate Genocide. Estimates by Dr. James Lovelock FRS and Professor Kevin Anderson (Deputy Director of the Tyndall Centre for Climate Change Research, University of Manchester) that only 0.5 billion will survive unaddressed climate change this century translate to a Climate Genocide involving deaths of 10 billion people this century, this including roughly twice the present population of particular mainly non-European groups, specifically 6 billion under-5 year old infants, 3 billion Muslims in a terminal Muslim Holocaust, 2 billion Indians, 1.3 billion non-Arab Africans, 0.5 billion Bengalis, 0.3 billion Pakistanis and 0.3 billion Bangladeshis.

5. Humanitarian organisation DARA report (2012) commissioned by 20 countries says that more than 100 million people will die and global economic growth will be cut by 3.2 percent of gross domestic product (GDP) by 2030 if the world fails to tackle climate change; DARA estimated that about 5 million people die annually from climate change (0.5 million) or carbon burning (4.5 million).

6. Bangladesh's Prime Minister Sheikh Hasina on DARA report (2012): "One degree Celsius rise in temperature is associated with 10 percent productivity loss in farming. For us, it means losing about 4 million metric tonnes of food grain, amounting to about $2.5 billion. That is about 2 percent of our GDP".

7. Threat of 50Gt methane from East Siberian Arctic Shelf. Gail Whiteman, Chris Hope and Peter Wadhams (2013): "Economic time bomb. As the amount of Arctic sea ice declines at an unprecedented rate, the thawing of offshore permafrost releases methane. A 50-gigatonne (Gt) reservoir of methane, stored in the form of hydrates, exists on the East Siberian Arctic Shelf. It is likely to be emitted as the seabed warms, either steadily over 50 years or suddenly. Higher methane concentrations in the atmosphere will accelerate global warming and hasten local changes in the Arctic, speeding up sea-ice retreat, reducing the reflection of solar energy and accelerating the melting of the Greenland ice sheet. The ramifications will be felt far from the poles… To quantify the effects of Arctic methane release on the global economy, we used PAGE09. This integrated assessment model calculates the impacts of climate change and the costs of mitigation and adaptation measures… The methane pulse will bring forward by 15–35 years the average date at which the global mean temperature rise exceeds 2°C above pre-industrial levels — to 2035 for the business-as-usual scenario and to 2040 for the low-emissions case (see 'Arctic methane'). This will lead to an extra $60 trillion (net present value) of mean climate-change impacts for the scenario with no mitigation, or 15% of the mean total predicted cost of climate-change impacts (about $400 trillion)."

8. Further to point #7, a methane GWP of 105 means that 50Gt CH_4 from the Arctic sea bed amounts to 50 x 105 =5,250 Gt CO_2-e i.e. 5,250/600 = 8.8 or roughly 9 times greater than the world's terminal GHG budget that must

not be exceeded if we are to avoid a catastrophic 2C temperature rise.

9. Arctic CH_4, cost of returning to 300 ppm CO_2 and are we doomed? Failure of the world to tackle man-made climate change massively violates intergenerational equity and intergenerational justice. The cost of conversion of cellulosic waste to biochar in the US mid-West is about \$49-\$74 per tonne CO_2 as compared to \$210-\$303 per tonne CO_2 in the UK. There is 700 billion tonnes of CO_2 in the atmosphere and the cost of reversing 2 centuries of fossil fuel burning by means of biochar-based return of the atmospheric CO_2 concentration to the pre-industrial 300 ppm from the current 400 ppm by removing 0.25 x 700 = 175 billion tonnes CO_2 has been estimated at \$13 trillion to \$53 trillion (US dollars) or 15-62% of the current world annual GDP of \$85 trillion. These numbers, while horrifying, nevertheless show that it is still conceivable (albeit very unlikely) that we could save the world from climate disaster.

However methane (CH_4) has a Global Warming Potential (GWP) 105 times greater than that of carbon dioxide (CO_2) on a 20 year time frame and taking aerosol impacts into account. Accordingly the predicted entry of 50 billion tonnes of methane (CH_4) into the atmosphere from the Arctic in the coming decades amounts to 50 x 105 = 5,250 billion tonnes CO_2-equivalent (CO_2-e), and the cost of \$13 trillion to \$53 trillion for removing 175 billion tonnes CO_2 as biochar translates conservatively to \$390 trillion to \$1,590 trillion for conversion of 5,250 billion tonnes of CO_2-e via CO_2 to biochar. This estimate represents about 5 to 20 years of world GDP and tells even the most optimistic person that we are doomed unless radical action is taken now to stop Arctic warming immediately.

10. Market-based asserted "solutions".

(a) Carbon Price, Carbon Tax. A suitably high Carbon Price can help shift to 100% renewable energy. Dr James Hansen and others argue for a "Fee and Dividend" system in which fossil fuels are taxed at the mine gate and the receipts are handed directly to the citizens. Climate economist Dr Chris Hope, University of Cambridge, and Dr James Hansen, Columbia University, have estimated a damage-related Carbon Price of about $200 per tonne CO_2-e. The true Carbon Price in Australia based on existing fossil fuel subsidies ($12 billion pa) and 10,000 annual carbon-burning-related deaths is about $150 per tonne CO_2-e. However the EU ETS price is about $6 per tonne CO_2-e, and in 2013 the Australian Government abolished a short-lived and flawed Carbon Tax approach (that involved a "futile cycle" of raising $10 billion pa in Carbon Taxes while making fugitive emissions effectively tax free, and fast-tracking to a 2014 ETS with an EU Carbon Price of about $6 per tonne CO_2-e.)

(b) Cap-and -Trade Emissions Trading Scheme (ETS). The ETS approach involves setting a Cap on pollution and establishing a "market" for licences to pollute. However this is seen as an immense political confidence trick because (1) the ETS approach has not been empirically shown to reduce GHG pollution, (2) this approach is accordingly counterproductive, and (3) it involves a national government, e.g. the Australian Government, fraudulently selling licences to pollute the one common atmosphere of all countries in the world.

(c) Direct Action. "Direct Action" is what is needed to install 100% renewable energy ASAP **(for details see section H)**. In Australia the Coalition Opposition has a policy of "Direct Action" that is too little too late (20% rather than 100% renewable energy by 2020) and which counterproductively involves providing subsidies to polluters to pollute less.

(d) Divest from fossil fuels. Divestment from fossil fuel corporations is an important way individuals, religious organizations, social organizations in general (e.g. societies, unions), institutions (e.g. universities), superannuation funds, pension funds, companies and indeed local, state and federal governments can take concrete action against man-made climate change that is threatening the Biosphere and Humanity. The world's terminal GHG pollution budget, that must not be exceeded if we are to have a 75% chance of avoiding a catastrophic 2C temperature rise, is 600 Gt CO_2. The fossil fuel industry intends to exploit its stock exchange-listed reserves of 762 $GtCO_2$ that is approximately a quarter of the world's total reserves of 2,795 Gt CO_2 on combustion. However historical GHG pollution since the start of the Industrial Revolution has now brought us to the point at which an estimated 50 Gt of methane (CH_4) is predicted to be released from the East Siberian Arctic Shelf in coming decades, GHG pollution of 5,250 Gt CO_2-equivalent, 9 times greater than the world's terminal GHG pollution budget.

11. For alphabetically-organized compendia of expert views on the seriousness of the situation see "Are we doomed?" and "Too late to avoid climate catastrophe".

12. Neoliberal, corporatist Mainstream media lying and Mainstream media censorship hides the worsening climate emergency. Proper public perception of the gravity of the worsening climate emergency is due to corporatist, neoliberal Mainstream media lying and Mainstream media censorship in Western Murdochracies Lobbyocracies and Corporatocracies (Kleptocracies, Plutocracies, Dollarocracies) in which Big Money buys politicians, parties, policies, public perception of reality, votes and political power.

Polya's 3 Laws of Economics mirror the 3 Laws of Thermodynamics of science and are (1) Price minus COP (Cost of Production) equals profit; (2) Deception about COP strives to a maximum; and (3) No work, price or profit on a dead planet. These fundamental laws help expose the failure of neoliberal capitalism in relation to wealth inequality, massive tax evasion by multinational corporations, and horrendous avoidable deaths from poverty and pollution culminating in general ecocide, speciescide, climate genocide, omnicide and terracide (see Gideon Polya. "Polya's 3 Laws Of Economics Expose Deadly, Dishonest And Terminal Neoliberal Capitalism", Countercurrents, 17 October, 2015: http://www.countercurrents.org/polya171015.htm).

F. Further key points.

1. Gas is not clean, it is dirty, 1 tonne of methane (CH_4) generates 2.75 tonnes CO_2 on combustion (burning 1 tonne carbon generates 3.67 tonne CO_2). Gas burning is cleaner than coal burning in terms of twice the MWh/tonne CO_2 emitted and less health damaging pollutants, but gas is not necessarily cleaner than coal burning GHG-wise. Thus methane (CH_4) leaks (circa 3% in the US; 7.9% from fracking shale deposits) and is 105 times worse than CO_2 as a GHG on a 20 year time frame taking aerosol impacts into account, this meaning that at 2.6% leakage the Global Warming Potential of the leaked gas is the same as that from burning the remaining gas. A Carbon Tax-driven coal to gas transition could double electric power industry-derived GHG pollution (if shale gas is used). A coal-to-gas transition is contraindicated because (a) gas is dirty, and (b) gas can be dirtier than coal on a per mass basis and even on a per energy yield basis depending upon the degree of gas leakage.

2. Climate change is damaging and destroying ecosystems (ecocide), and the species extinction rate is now 100-1,000 times greater than normal (Australia is a world leader). We must not destroy what we cannot replace. Every ecosystem and species is priceless.

3. Leading scientists, economists and analysts slam the Carbon Trading Emissions Trading Scheme (ETS) approach as empirically ineffective, dangerously counterproductive and inherently fraudulent because it involves governments selling licences to pollute the one common atmosphere of all countries.

4. 2.6 % leakage of CH_4 yields the same greenhouse effect as burning the remaining 97.4% (noting that 1g CH_4 has 105 times the GWP of 1 g CO_2) – ergo, stop gas exploitation, aquifer-poisoning and aquifer-depleting fracking of shale and coal seams.

5. Many countries (e.g. EU countries and Australia) support a 450 ppm CO_2 -e and 2C temperature rise "cap". However the Synthesis Report of the 2,500-delegate March 2009 scientific Copenhagen Climate Conference indicates that we have already exceeded 450 ppm CO_2-e and over 90% of delegates polled thought 2C was inevitable this century [500 ppm CO_2-e already in 2020].

6. Atmospheric CO_2 must be urgently returned to about 300 ppm CO_2 for a safe Planet for all peoples and all species. Circa 320 ppm CO_2 is required for restoration of the Arctic sea ice and for coral sustainability. However atmospheric CO_2 concentration is currently 415 ppm [2020] and is increasing at about 2 ppm per year. **Less not more!**

7. Stop shale oil and tar sands exploitation (e.g. Canada-US keystone oil pipeline) means "game over" for Planet. Dr James Hansen (NASA): "The tar sands are estimated (e.g. see IPCC Fourth Assessment Report) to contain at least 400 GtC (equivalent to about 200 ppm CO_2). Easily available reserves of conventional oil and gas are enough to take atmospheric CO_2 well above 400 ppm, which is unsafe for life on earth. However, if emissions from coal are phased out over the next few decades and if unconventional fossil fuels including tar sands are left in the ground, it is conceivable to stabilize earth's climate. Phasing out emissions from coal is itself an enormous challenge. However, if the tar sands are thrown into the mix, it is essentially game over." 70% recovery of oil from the Alberta tar sands can be estimated to yield about 600 Gt CO_2 -e on combustion (i.e. equivalent to the WBGU's terminal GHG pollution budget that must not be exceeded if we are to have a 75% chance of avoiding a catastrophic 2C temperature rise).

8. "Annual per capita greenhouse gas (GHG) pollution" in units of "tonnes CO_2-equivalent per person per year" (2005-2008 data) is 0.9 (Bangladesh), 0.9 (Pakistan), 2.2 (India), less than 3 (many African and Island countries), 3.2 (the Developing World), 5.5 (China), 6.7 (the World), 11 (Europe), 16 (the Developed World), 27 (the US) and 30 (Australia; or 74 if Australia's huge Exported CO_2 pollution is included).

9. Biofuel from crops such as sugar cane or corn carries an enormous Carbon Debt. Biofuel from photosynthetic algae (e.g. in CO_2-enriched ponds close to power stations) is not carbon neutral in a carbon economy. Legislatively mandated biofuel in the US, UK and the EU is unconscionable use of food for fuel in a hungry world, and drives up world food prices with consequent malnourishment and starvation (Biofuel Genocide).

10. Nuclear options have been suggested but in the context of a carbon economy, the overall nuclear cycle produces CO_2 (from mining, transport, processing and enrichment, cement production for power station construction, disposal of waste and ultimately decommissioning of power stations). Further, when limited high quality uranium oxide reserves are used up, use of low grade ores may mean that a new nuclear power station could release the same CO_2/MWh as a new gas-fired power station. Thorium-based nuclear power is cleaner and safer than uranium-based power but thorium ores are limited. Fast breeder reactors have been advocated as a highly efficient option but there are major fears relating to expense, security, nuclear terrorism and predicted severe human rights impacts of a plutonium economy. A former proponent, Dr James Lovelock FRS, thence argued that we have run out of time for the nuclear stop-gap solution, and advocated the biochar option for drawing down atmospheric CO_2.

11. Rail transport is vastly more energy efficient than road transport (e.g. a tram can be 20 times more energy efficient per passenger than a 4WD vehicle.

12. Carbon burning pollutants are deadly and kill about 9 million people each year worldwide (2020).

G. Australia.

1. Carbon burning pollutants have been estimated from Canadian and New Zealand data to kill about 10,000 Australians yearly. Australians dying each year from the effects of pollutants from vehicles, coal burning for electricity and other carbon burning total about 2,200, 4,600 and 2,800, respectively (as of 2020 about 9 million people die annually worldwide from carbon burning pollutants of whom about 75,000 die from the burning of Australian coal exports).

2. Australia has about 0.3% of the World's population but its Domestic plus Exported GHG pollution is about 3% of the World total (climate exceptionalism, climate racism, and climate injustice in addition to horrific intergenerational inequity) [2020 revision: 5.4%].

3. Australia already has a huge negative carbon tax of $11 billion annually (2013) to subsidize carbon burning.

4. It is estimated that an Australian carbon tax of circa $23/tonne CO_2-e will encourage gas-fired power, $70/tonne CO_2-e will encourage wind and about $200/tonne CO_2-e will encourage concentrated solar thermal installation (indeed the former Australian Labor Government hoped for a Carbon Tax-driven coal to gas transition but (a) gas can be dirtier than coal GHG-wise depending upon the degree of CH_4 leakage, and (b) the Australian Labor Government decided to slash the Carbon Price in 2013 from $25 per tonne CO_2-e to about $6 per tonne CO_2-e.

5. True carbon price. A risk avoidance-based estimate of $7.6 million for the risk-avoidance-based Value of a Statistical Life, Australian annual subsidies of $11 billion for fossil fuel burning, 9,600 annual Australian carbon pollution-related deaths, and Australia's annual Domestic GHG pollution (2009) of 552 million tonnes CO_2-e yields a Carbon Price of $7.6 million x 9,600 annual deaths = $73 billion + $11 billion = $84 billion / 552 million tonnes CO_2-e = $152/tonne CO_2-e (= $562 /tonne C).

6. The Australian Bureau of Agricultural and Resource Economics (ABARE) has projected that Australia's black coal exports will increase at an average rate of 2.6% per year over the next 20 years, and that liquid natural gas (LNG) exports will increase at 9% per year over the same period. Further, it is estimated that Australian exports of dried brown coal will reach

20 million tonnes by 2020, this corresponding to about 59 million tonnes CO$_2$-e after combustion.

7. The initial Carbon Tax-ETS-Ignore Agriculture (CTETSIA) policy of the Australian Labor Government failed comprehensively in 3 key areas, specifically (1) it **entrenched climate change inaction** for decades by promoting a Carbon Tax-driven coal to gas transition (that will double electricity generation-derived GHG pollution if shale gas used) and scuppered science-demanded 100% renewable energy by 2020; (2) it adopted an **empirically ineffective, disastrously counterproductive and inherently fraudulent ETS** approach and (3) **ignored major GHG sources** of petrol, diesel, biofuel, fossil fuel exports (apart from fugitive emissions, extraction and transport costs), soil, forestry and agriculture (agriculture is responsible for over 50% of GHG pollution).

8. Success in "tackling climate change" is surely measured in terms of GHG pollution reduction but Australia's Domestic plus Exported GHG pollution is huge and ever-increasing.

From Treasury, ABARE, and US EIA data and assuming an 11% annual growth in iron ore exports, 2.4% annual growth in coal exports and 9% annual growth in gas exports, Australia's Domestic and Exported greenhouse gas (GHG) pollution is as follows (million tonnes CO$_2$-e or Mt CO$_2$-e). The data are based on fossil fuel combustion and ignore fugitive emissions due leakage of gas (mainly methane, CH$_4$) from coal mines and in coal-seam gas (CSG) production.

2000: 565 (Domestic) + 505 (coal exports) + 17 (LNG exports) + 105 (iron ore exports) = 1,192.

[2005: 572 (Domestic) + 654 (coal exports) + 26 (LNG exports) + 199 (iron ore exports) = 1,451; 2005 data estimated as the mean of the 2000 and 2010 data].

2009: 600 (Domestic) + 784 (coal exports) + 31 (LNG exports) + 97 (iron ore exports) = 1,512.

2010: 578 (Domestic) + 803 (coal exports) + 34 (LNG exports) + 293 (iron ore exports) = 1,708.

[2015: 600 (Domestic) +921 (coal exports) + 57 (LNG exports) + [30 (brown coal exports) notional, unrealized] + 533 (iron ore exports) = 2,141 - 30 = 2,111; 2015 data estimated as the mean of the 2010 and 2020 data].

2020: 621 (Domestic) + 1,039 (black coal exports) + 80 (LNG exports) + [59 (brown coal exports) notional] + 772 (iron ore exports) = 2,571 Mt CO_2-e (2.6 Gt CO_2-e).

9. Australia used up its "fair share" of the world's terminal greenhouse gas (GHG) pollution budget of 600 billion tonnes CO_2-e in 2011. The German WBGU and the Australian Climate Commission have estimated that no more than 600 billion tonnes of CO_2 can be emitted between 2010 and zero emissions in 2050 if the world is to have a 75% chance of avoiding a catastrophic 2C temperature rise (Australia's "fair share" [2020 population data] is 25.5M x 600 Gt CO_2/7,600M = 2 Gt CO_2-e). Inspection of item #8 reveals that Australia had already exceeded its fair share of this terminal greenhouse gas pollution budget of 600 billion tonnes CO_2-e in 2011.

10. The Australian bipartisan policy of unlimited coal, gas and iron ore exports means that, given Australia's huge resources, Australia is set to exceed the WBGU's terminal GHG pollution budget by a factor of three (3) (the World is set to exceed this budget by a factor of 5).

11. In August 2015 the Coalition Government offered a 2030 "target" of "26% off 2005 GHG pollution by 2030" i.e. 423 Mt CO_2-e of Domestic GHG pollution. If due to international pressure Australia's 2030 Exported GHG pollution remains at the level predicted for 2020 (1,950 Mt CO_2-e) then Australia's 2030 Domestic plus Exported GHG pollution will be 423 + 1,950 = 2,373 Mt CO_2-e or 112% of the 2015 value and 164% of the 2005 value i.e. Australia was promising the World that its 2030 Domestic plus Exported GHG pollution would be 1.1 times bigger than that in 2015.

In November 2015 the Labor Opposition proposed a notional 2030 "target" of "45% off 2005 GHG pollution by 2030" i.e. 315 Mt CO_2-e of Domestic GHG pollution. If Australia's 2030 Exported GHG pollution remains at the level predicted for 2020 (1,950 Mt CO_2-e) then Australia's 2030 Domestic plus Exported GHG pollution would be 315 + 1,950 = 2,265 or 107% of the 2015 value and 156% of the 2005 value i.e. both Australia's Labor Opposition and it pro-coal COALition Government are promising the World that Australia's Domestic plus Exported GHG pollution in 2030 will be about 1.1 times bigger than that in 2015.

H. 100% renewable energy, cessation of GHG pollution, re-afforestation, biochar & bicarbonate

1. The Beyond Zero Emissions (BZE) plan for 100% renewable stationary energy for Australia by 2020 (Zero Carbon Australia by 2020, ZCA 2020) involves 40% wind energy, 60% Concentrated Solar Thermal (CST) with molten salts energy storage for 24/7 baseload power, biomass and hydroelectric backup (for days of no wind and low sunshine), and an efficient HV DC and HV AC national power grid. The BZE scheme was costed at $370 billion over 10 years, with roughly half spent on CST, one quarter on wind and one quarter on the national electricity grid.

2. Seligman scheme. A scheme for 100% renewable energy for Australia has been set out by top electrical engineer Professor Peter Seligman (a major player in development of the bionic ear). Professor Seligman's scheme involves wind, solar thermal, other energy sources, hydrological energy storage (in dams on the Nullabor Plain in Southern Australia), an efficient HV AC and HV DC electricity transmission grid, and a cost over 20 years of $253 billion. [Battery storage is an alternative to gas compression, hydrological and molten salts energy storage].

3. Wind power. Ignoring cost-increasing energy storage and transmission grid costs and cost-decreasing economies of scale for a 2- to 10-fold size increase, here are 2 similar cost estimates for installation of wind power for 80% of Australia's projected 325,000 GWh of annual electrical energy by 2020: (1) 90,000 MW capacity, 260,000 GWh/year, $200 billion/10 years (10-fold scale-up from GL Garrad Hassan), and (2) 96,000 MW, 260,000 GWh/year, $144 billion (2-fold scale up from BZE).

4. Science-demanded reduction of atmospheric CO_2 from current 415 ppm [2020] to 300 ppm requires "negative GHG emissions" achieved by cessation of GHG pollution ASAP and atmospheric CO_2 reduction though re-afforestation, renewable energy driven CO_2 trapping in alkaline solutions, and biochar (as much as 12 billion tonnes carbon as biochar can be fixed annually globally from renewable energy-driven anaerobic pyrolysis of existing agricultural and forestry cellulosic waste).

5. Re-afforestation (SE Australian native forests are World's best forest carbon sinks; 14 M ha, 25.5 Gt CO_2, 460 Mt CO_2/yr avoided for next 100 years if retained. Sir Nicholas Stern: only $20 billion pa to halve annual global deforestation.

6. Livestock production inefficient, methanogenic and requires compensating carbon sinks – however, we are all in this together (dairy farmers feed methanogenic cows, but most urban adults drive cars that presently are overwhelmingly fossil fuel-based rather than electric and based on renewable energy).

7. Biochar. Atmospheric CO_2 can be reduced from the current 415 ppm CO_2 [2020] back to a safe and sustainable 300 ppm CO_2 by fixing CO_2 as cellulose via solar-energy-driven photosynthesis ($nCO_2 + nH_2O \rightarrow (CH_2O)_n + O_2$) with subsequent anaerobic pyrolysis of cellulosic material (e.g. waste wood and straw) to yield carbon (C, charcoal) ($(CH_2O)_n \rightarrow nC + n H_2O$).which can then be added to soil or buried in holes in the ground (e.g. used coal mines). Thus p224, Progress in Thermochemical Biomass Conversion, volume 1, IAE Bioenergy, ed. A.V. Bridgewater (Blackwell Science) informs us that we could obtain 1.7 GtC/yr (straw from agriculture) + 4.2 GtC/yr (total grass upgrowth from grasslands upgrowth) + 6 GtC/yr (possible sustainable wood harvest) = 11.9 GtC/yr. From this one can see why biochar expert Professor Johannes

Lehmann of Cornell University was correct in calculating that it is realistically possible to fix 9.5bn tonnes of carbon per year as biochar, noting that global annual production of carbon pollution from fossil fuels is about 9 bn tonnes.

Unfortunately, biochar is expensive to produce, and thus for every \$1 derived from coal burning, \$2-\$3 will have to be spent at today's prices by future generations converting the consequent CO_2 back to carbon (biochar) (based on US mid-West corn stalk-derived biochar; multiply this by 4 for the cost in the UK) i.e. a major issue of intergenerational equity.

8. CO_2 trapping as carbonate (CO_3^{2-}). Theoretically one can use solar energy or indeed other renewable energy to generate alkaline solutions (high OH^- concentration) from electrolysis of saline solutions (i.e. NaCl solutions e.g. sea water) and thence trap CO_2 thus: $2Na^+ + CO_2 + 2OH^- -> CO_3^{2-} + H_2O + 2\,Na^+$ (solid Na_2CO_3 can be stored underground). Unfortunately, while electrolysis of sea water (NaCl solution) yields useful and combustible hydrogen (H_2), the other gaseous product is chlorine (Cl_2) that is toxic, as is hypochorite (ClO^-) produced when chlorine reacts with hydroxyl ions (OH^-) (see "Process for preparing sodium hydroxide, chlorine and hydrogen from aqueous salt solution using solar energy", US Patent: https://patents.google.com/patent/US20090107850). However, the Direct Air Capture (DAC) system captures as water-insoluble calcium carbonate ($CaCO_3$) all the CO_2 in the incoming air stream (circa 80% nitrogen, N_2; 20% oxygen, O_2; 0.04% CO_2) by passage through a lime solution of calcium ions (Ca^{2+}) and hydroxyl ions (OH^-).The calcium carbonate is then heated to generate lime (calcium oxide, CaO) and a stream of circa 100% carbon dioxide (CO_2) which can then be compressed and hopefully permanently sequestered (e.g. in deep ocean, underground in spaces from former coal or gas extraction, or by underground reactions to form carbonates with basalt rocks).

(a) CO_2 scrubbed out of air by passage through lime solution ($Ca^{2+} + 2\,OH^-$) in water (H_2O) to form carbonate ions (CO_3^{2-}): $CO_2 + 2OH^- -> CO_3^{2-} + H_2O$

(b) Carbonate (CO_3^{2-}) precipitated as calcium carbonate ($CaCO_3$): $Ca^{2+} + CO_3^{2-} \rightarrow CaCO_3$

(c) Lime (calcium oxide, CaO) regenerated by heating calcium carbonate ($CaCO_3$) with the resultant CO_2 being sequestered: $CaCO_3 \rightarrow CaO + CO_2 \rightarrow CO_2$ sequestered.

(d) Lime (CaO) dissolved in water (H_2O): $CaO + H_2O \rightarrow Ca^{2+} + 2\ OH^-$

A DAC pilot plant has been constructed that sequesters 1 Mt CO_2 per year (equivalent to annual emissions of 250,000 average cars) at a cost of \$100-150 per tonne of CO_2 captured, purified, and compressed to 150 bar.

CO_2 pumped underground can react with basaltic minerals to yield carbonates.

9. Accelerated Weathering of Limestone (AWL). The waste gas from burning coal or gas is passed through a sea water-limestone ($CaCO_3$) scrubber with the following reaction: CO_2 (gas) + $CaCO_3$ (solid) + $H_2O \rightarrow Ca^{2+}$ (aqueous) + 2 HCO_3^- (aqueous). The scrubbing solution is then piped to the sea. Carbon in the oceans as bicarbonate is 10 times that in all recoverable fossil fuel reserves and about 60 times that of the CO_2 in the atmosphere. The carbon in carbonate minerals is about 4,000 times greater than the carbon in oil and coal fossil fuel reserves, and the AWL process would in part reverse the deleterious acidification of the oceans due to the massive CO_2 pollution of the atmosphere.

10. Terrestrial enhanced weathering of silicate rock flour. The carbonation of silicate rocks through enhancement of natural weathering is another route for atmospheric CO_2 draw-down.

I. Biochemistry, climate change and solutions.

This section ambitiously summarizes the essential chemistry and biochemistry underlying the biological basis of fossil fuels, climate change and solutions in just a few pages. For a similar but much more extensive summary see Chapter 2, "Biochemistry – the chemistry of life" in Gideon Polya, "Biochemical Targets of Plant Bioactive Compounds. A pharmacological guide to sites of action and biological effects", Taylor & Francis/CRC Press, 2003. For a vastly lengthier, vastly less specific and much more readable attempt at such simplification see Bill Bryson, "The Body. A guide for occupants", Penguin, 2019. For a detailed, circa 1,200-page exposition of Biochemistry and Molecular Biology for undergraduate university students see the classic D.L. Nelson and Michael M. Cox, "Lehninger Principles of Biochemistry", Worth, 2000.

1. Biochemistry (Biological Chemistry) is the Chemistry of Life.

2. Atoms. Atoms are composed of subatomic particles Neutrons (n), Protons (p^+), Electrons (e^-). Atomic Number of atom = no. of protons. Atomic Mass = mass of neutrons + protons + electrons. In element (uncharged); sum of (Σ) electrons = Σ protons. H (hydrogen, $1p^+ + 1e^-$). H ionizes -> p^+ (proton, H^+) + e^- i.e. $H^+ = p^+$ = proton (later: more H^+ in solution -> more acidic, lower ($\downarrow$) pH).

3. Atomic structure. Hydrogen atom: 1 p^+ in Nucleus (most of mass, extremely tiny volume), 1 e^- in 1s shell. Larger atoms: more ns, p^+s in Nucleus plus e^-s. For stability want to fill up shells (orbitals) i.e. 2 e^-s in 1s, 2 e^-s in 2s, 6 e^-s in 2p orbitals etc. Particle versus wave duality: imagine e^- as a particle or as a cloud of electron density (described by wave equations).

4. Molecules. Molecules form when atoms share electrons to form Covalent Bonds (Single -, Double = or Triple ≡). Thus H• + Hx (monovalent H atoms, 1s (1)) -> stable H•xH (hydrogen molecule, diatomic H_2 molecule, H-H) i.e. 1s filled for both Hs & 1 covalent bond (a single bond) links the 2 Hs. Other key examples: 2H (1s(1)) + O (divalent oxygen, 1s(2)2s(2)2p(4)) -> H_2O = H-O-H = water (1s(2)2s(2) 2p(6)) (stable 1s, 2s and 6p orbitals);

N (trivalent nitrogen, 1s (2)2s(2)2p(3)) + 3H (1s(1)) -> NH_3 (ammonia, N now: 1s(2)2s(2)2p(6), 3 single bonds)

C (tetravalent carbon, 1s(2)2s(2)2p(2)) plus 4H (1s(1)) -> CH_4 = methane (C now: 1s(2)2s(2)2p(6), 4 single bonds) [Jane Austen, "Pride and Prejudice", first sentence: "It is a truth universally acknowledged, that a single man in possession of a good fortune must be in want of a wife."].

5. Valency. H monovalent, oxygen (O) divalent (has 2 friends), nitrogen (N) trivalent (has 3 friends), carbon (C) tetravalent (has 4 friends), sulphur (S) divalent or tetravalent, phosphorus (P) pentavalent e.g.

Hydrogen (H) univalent: H- (e.g. H-O-H);

Oxygen (O) divalent: -O- (e.g. H-O-H, water) or O= (e.g. O=O, O_2, oxygen molecule & O=C=O, carbon dioxide);

Nitrogen (N) trivalent: e.g. NH_3 (ammonia), -N= (e.g. the free radical nitric oxide, •N=O) & N≡ (e.g. N≡N i.e. N_2, nitrogen molecule & -C≡N, nitrile e.g. acetonitrile CH_3-C≡N);

Carbon (C) tetravalent: C (e.g. CH_4, methane), =C= (e.g. O=C=O, carbon dioxide (CO_2) & ethylene H_2C=CH_2) & -C≡C- (e.g. H-C≡C-H, C_2H_2, acetylene & $^{\delta-}$C≡O$^{\delta+}$, carbon monoxide).

Sulphur (S) divalent: -S- (e.g, H-S-H, H_2S, hydrogen sulphide) and Sulphur (S) tetravalent: =S= (e.g. O=S=O, SO_2, sulphur dioxide);

Phosphorus (P) pentavalent: P (e.g. H_3PO_4, phosphoric acid).

6. Atomic size. Van der Waals radii represents the "space" occupied by an atom and from which intrusion by other atoms is repelled by a huge energy barrier (in nm = nanometers = 10^{-9} metres): H (0.1), O (0.14), F (0.14), N (0.15), C (0.17), S (0.18), Cl (0.19), P (0.19).

7. Electronegativity of atomic constituents of molecules i.e. some atoms in a molecule get more of the shared electron cloud than others e.g. $F > O > Cl > N > Br > S > C > I > Se > P$. Thus unequal sharing of e^-s in H_2O so that O is more negative than H in H_2O: $H^{\delta+} - O^{\delta-} - H^{\delta+}$.

8. Hydrogen bonding. Differential charge on H and O atoms of H_2O means that an H atom can be "shared" by electronegative O atoms in adjacent molecules: $H^{\delta+} - O^{\delta-} - H^{\delta+}$. This means that H-O-H molecules can be linked via Hydrogen Bonds i.e. $H^{\delta+} - O^{\delta-} \ldots H^{\delta+} \ldots O^{\delta-}(-H^{\delta+})_2$. Hydrogen bonding can occur between other O-containing molecules of biochemical importance. Hydrogen bonds (H bonds) are very weak and readily broken at room temperature (RT). Hence the solid, liquid & gas phases of H_2O in which molecules are tightly and exactly arrayed in ice (solid; below 0C), there is some short-range order of "flickering clusters' in liquid water (0-100C), and H bonding ceases above 100C (gas phase).

9. Ionization & protonation-deprotonation in aqueous solution.

H-O-H (water) $\leftrightarrow$ OH$^-$ (hydroxyl ion) + H$^+$ (proton)

$H_2O + H^+ \leftrightarrow H_3O^+$ (hydronium ion)

R-C(=O)-OH (R-COOH, a carboxylic acid) $\leftrightarrow$ R-C(=O)-O$^-$ (carboxylate ion) + H$^+$ (proton) (acids donate protons so [H$^+$] (H$^+$ concentration) increases i.e. $pH = -\log_{10}$ [H$^+$] decreases e.g. vinegar (acetic acid) is acidic, low pH).

CH_3-COOH (acetic acid) $\leftrightarrow$ CH_3-COO$^-$ (acetate ion) + H$^+$ (vinegar is acetic acid $CH_3COOH = CH_3C(=O)OH$). Acids donate protons through dissociation.

NH_3 (ammonia) + H$^+$ (proton) $\leftrightarrow$ NH_4^+ (ammonium ion, [NH$_4$]$^+$). CH_3-NH_2 (methylamine) + H$^+$ $\leftrightarrow$ CH_3-NH_3^+ (protonated methylamine, [CH$_3$-NH$_3$]$^+$). Bases bind protons i.e. $pH = -\log_{10}$ [H$^+$] increases i.e. [OH$^-$] increases e.g. ammonia is basic or alkaline).

10. Isotopes. Increasing neutrons (ns) in the nucleus increases mass and can increase instability (basket of eggs analogy) while not impacting chemical reactivity properties. Radioactive or non-radioactive isotopes of major elements found in biological systems (e.g. of H, C, O, N, P, K, Na, S, Ca, Cl, I) have been very useful in paleoclimatology and also in biochemical research as "tracers" for the normal, abundant isotope. Genetic damage is directly proportional to dose => minimize ingestion (e.g. gloves) and minimize exposure (e.g. to high energy emitters such as P-32) (e.g. lead jacket, perspex screens).

H-1 (^{1}H) abundant (normal) hydrogen ($1p^+ + 1e^-$)

H-2 (^{2}H) deuterium (e.g. in heavy water D_2O) ($1p^+ + n + 1e^-$)

H-3 (^{3}H) tritium (radioactivity, disintegrates, half-life = time for half to disintegrate = $t_{1/2}$ = 12.32 years, low energy 0.019 Mev ß emitter) ($1p^+ + 2n + 1e^-$)

C-12 (^{12}C) abundant carbon ($6n + 6p^+ + 6e^-$)

C-13 (^{13}C) very low abundance, C-12 plus 1 n, non-radioactive

C-14 (^{14}C) very low abundance (C-12 plus 2n, radioactive, half-life = $t_{1/2}$ = 5,730 years, useful in carbon dating; relatively low energy 0.15 Mev ß emitter)

P-31 (^{31}P) non-radioactive abundant P ($15p^+ + 16n + 15e^-$)

P-32 (^{32}P) radioactive P isotope (half-life 14.29 days, relatively high energy 1.71 Mev ß emitter) ($15p^+ + 17n + 15e^-$)

11. Covalent bonds & light absorption.

Single X-Y e.g. –O-H (rotation and vibration e.g. from light absorption).

Double X=Y e.g. –C=O (keto group), $H_2C=CH_2$ (ethylene) (vibration but no rotation about double bond; cf electric plug).

Triple bond X≡Y e.g. H-C≡C-H (acetylene) and N≡N (N_2, nitrogen molecule), vibration but no rotation about triple bond.

Light photon wave-particle duality. Electromagnetic spectrum from highest energy (smallest wavelength) to lowest energy (highest wavelength): Gamma rays, X-rays, Ultra-violet (UV), Visible, Infra-Red, Microwave, Radio. Prisms separate (e.g. rainbow).

Red things reflect red light, absorb other visible wavelengths.

12. Non-covalent H bonds & weak interactions. H bonds (X…H…Y); Van der Waals (weak inter-nuclear attraction); hydrophobic (very important water-repelling interactions between non-polar, hydrophobic, "water-fearing" entities e.g. olive oil in water); ionic (electrostatic) interactions (e.g. $X\text{-}O^-\ldots H_3N^+\text{-}Y$).

13. Living organisms are self-replicating systems. On earth inanimate systems (very likely near undersea vulcanism) evolved into living systems that are (1) water-based; (2) having membrane-bounded cells; (3) self-repairing; and (4) self-replicating. Living systems are highly ordered cells in a randomizing universe. Living cells feed on available free energy (Gibbs free energy, G) → energy-requiring synthesis, maintenance & replication

14. Life and the Laws of Thermodynamics. 1. The energy of the universe is constant. 2. The disorder (or entropy, S) of the universe increases. Living systems extract energy from a chaotic universe to repair and replicate highly ordered systems. [Indeed for secular Humanists there is no "purpose" in existence but they can extract a sense of "meaning" by defying the Second Law through increasing order via increasing truth and beauty as in John Keats' poem "Ode on a Grecian Urn": "Beauty is truth, truth beauty, that is all/ Ye know on earth, and all ye need to know"].

15. Cells are the minimum units of autonomous, self-replicating life (self-aware, self-evolving, self-repairing and self-replicating robots are "living" in this sense and hopefully might conceivably be generated in coming decades before man destroys himself through nuclear weapons or climate change).

Bulk constituent water (H_2O)

Cell membrane = plasma membrane (PM) encloses cell

PM = phospholipid (PL) bilayer + proteins, oligosaccharides

PM 0.01 micrometre (μm, 10^{-6} metre) = 10 nanometre (nm) thick

Bimolecular layer of hydrophobic (water fearing, water repelling, non-polar, fatty) phospholipid (PL) molecules (O, the polar, hydrophilic "head" and = the hydrophobic tail of a PL molecule =O):

$$O= =O$$

Outside cell, aqueous   O= =O Inside cell, aqueous

 (O_2, oxidizing)   O= =O (XH_2, reducing)

where = is the hydrophobic (water fearing, water repelling, non-polar) fatty acyl chain (CH_3- $(CH_2)_n$-) part of fatty acids (FAs, R-COOH) that are esterified to glycerol and make up the hydrophobic interior of biological membranes; O is the polar head group, C-O-P(=O, -OH) -O-X that interacts with polar water molecules.

[Esterification is the reaction of an organic acid (X-COOH) with an alcohol (Y-OH) to yield an ester: X-COOH + Y-OH -> X(CO)-OY + H_2O. Hydrolysis is the reverse reaction].

16. Subcellular structure and organelles (city and economy analogy). Cells are mostly water, membranes are mostly phospholipid bilayers (plus cholesterol, proteins, and some carbohydrates), and proteins (amino acid polymers) have structural, signalling and catalytic (enzyme) functions.

Prokaryotes, anaerobic and aerobic bacteria, have no nucleus. Eukaryote cells have a membrane-bounded, DNA-containing nucleus (plants, fungi, animals). PL bilayer membranes enclose specialized organelles within the cytosol of the cell listed below with functions in parentheses):

Smooth endoplasmic reticulum (ER) (tubular network: lipid synthesis, detoxification of xenobiotics);

Rough ER (tubular network; decorated with ribosomes involved in the synthesis of proteins for export);

Rough ER-linked Golgi vesicles (protein modification and export);

Lysosomes (breaking down proteins, polynucleotides, fats);

Vacuoles (storage and transport of molecules);

Peroxisomes (getting rid of toxic hydrogen peroxide, H_2O_2, and superoxide, $\cdot O_2^-$, generated through oxidative metabolism);

Nucleus (contains double-stranded DNA (dsDNA) constituting genes encoding proteins. DNA is "transcribed" to yield messenger RNA (mRNA) that is "translated" to yield proteins on ribosomes in the cytosol or on the rough ER; also contains a nucleolus where ribosomal RNAs, rRNAs, are made by transcription of rRNA genes);

Mitochondria (DNA-containing, double membrane-enclosed powerhouses of the cell; derived evolutionarily from prokaryote symbionts; generate ATP, the energy currency of the cell, from Oxidative Phosphorylation (ox phos) whereby reduced coenzymes, XH_2, from oxidation of Glucose by Glycolysis and the Tricarboxylic Acid (TCA) Cycle are oxidized via an Electron Transport Chain (ETC) that ultimately passes electrons from XH_2 to O_2 as the Terminal Electron Acceptor and in the process generates a Proton Gradient across the Inner Mitochondrial Membrane that drives the rotation of the membrane-associated ATP Synthase (ATPase) that makes the High Energy Compound or HEC ATP);

Chloroplasts (in plant cells) are double-membrane-bounded organelles containing thylakoid membrane-bounded compartments that are involved in the Light Reactions of Photosynthesis: sun-derived light energy from photon capture by the Light Harvesting Assemblies of photosystems I and II (PSI and PSII) drive a process of electron flow through the Photosynthetic Electron Transport Chain in which electrons flow from the photolysis of water (generating O_2) to Q (the Primary Acceptor of PSII), thence "downhill" via cytochromes and plastoquinone to PSI (associated with proton gradient generation and coupled ATP generation by Photophosphorylation via a rotating photosynthetic ATP Synthase (ATPase)) and thence via Z (the Primary Acceptor of PSI) to reduce (add H or electrons to) the coenzyme $NADP^+$ to form NADPH. A transmembrane proton gradient is used to make ATP in both Cyclic Photophosphorylation and Non-Cyclic Photophosphorylation. The intermembrane stroma contains enzymes of the Dark Reactions of Photosynthesis catalyzing the formation of Glucose (C6) in reactions of the Calvin Cycle of which some crucially involve use of ATP and NADPH generated in the Light Reactions.

The Plasma Membrane (PM) is a phospholipid bilayer that contains solute transporters, hormone receptors, clathrin (for receptor internalization), and glycoproteins (glycosylated with sugars that are oriented towards the outside; the basis of blood groups, immunological incompatibility, and tissue rejection).

The Cytosol contains the organelles, proteins and metabolites plus complex structures such as ribosomes (complex of rRNAs and proteins; involved in protein synthesis), proteasomes (macromolecular assemblies involved in degradation of proteins tagged for destruction with the protein ubiquitin), cytoskeletal elements (actin and myosin that make up contractile microfilaments involved in cell shape and motility, and tubulin-composed microtubules involved in intracellular transport and separation of chromosomes at mitosis, or cell division).

17. The Cytosol also contains enzymes for catalyzing the following key reaction sequences (numbers of carbon atoms in the key molecules are given in parentheses):

Anaerobic Glycolysis: glucose (C6) oxidation -> -> phosphoenolpyruvate (PEP, C3, a HEC) -> pyruvate (C3) <-> lactate (C3) and ATP by substrate level phosphorylation (in emergency anaerobic conditions e.g. running from a predator or for a bus, lactate accumulates but in aerobic conditions the pyruvate (C3) enters the mitochondria (singular: mitochondrion) and is oxidized to CO_2 (C1) in the TCA Cycle with resultant reduced coenzymes (general structure XH_2) feeding electrons into the inner membrane-associated Electron Transport Chain (ETC) (and thence to O_2 with concomitant energy storage through proton gradient generation) for consequent Oxidative Phosphorylation that yields 36 ATP per glucose (C6) as compared to only 2 ATP per glucose for Anaerobic Glycolysis.

Anaerobic glycolysis (running from a predator or for the bus) (cytosol): glycogen (animal starch, $(CH_2O)_n$) -> glucose (C6) (C6) -> -> using 2 NAD^+ and 2 ATP -> 2 NADH + 4 ATP + 2 CH_3COOH (pyruvate, C3) -> 2 CH_3CH_2OH (lactate = reduced pyruvate) + 2 NAD^+ (used to keep the process going) -> net 2 ATP per glucose and accumulation of lactate (muscles feel sore) (in brewing beer, yeast does the same sort of thing in anaerobic conditions but pyruvate (C3) is decarboxylated to yield $2CO_2$ (C1) and 2 acetaldehyde (CH_3CHO, C2) which is reduced to 2 ethanol (C2; CH_3CH_2OH) and yielding $2NAD^+$ to permit the process to continue). [It is speculated that pleasure-giving ethanol from anaerobic fermentation of grain may have been important in the Agrarian Revolution of 10,000 years ago that has resulted in the Anthropocene Era with Humanity and the Biosphere presently facing catastrophe in the worsening Climate Crisis.]

Aerobic glycolysis (cytosol): glycogen (animal starch, $(CH_2O)_n$) -> glucose (C6) -> -> 2 NADH + 2 ATP + 2 CH_3COOH (pyruvate, C3) -> 2 pyruvate (C3) **enters mitochondrial matrix (inside the mitochondrion)** -> oxidized via TCA Cycle (= Tricarboxylic Acid Cycle = Krebs Cycle =

Citric Acid Cycle) -> 6 CO_2 (C1), 2GTP (= 2 ATP equivalent), 8 NADH, 2 $FADH_2$ (reduced coenzymes) -> reduced coenzymes (XH_2) oxidized via the ETC so that oxidative phosphorylation (ox phos for short) yields 30 ATP (from 10 NADH) & 4 ATP (from 2 $FADH_2$) -> 38 ATP/glucose in total (cf only 2 ATP/glucose via "emergency" anaerobic glycolysis).

Fatty acid (FA) synthesis: synthesis, in times of Plenty, of Fatty Acids (FAs) from acetyl(C2)Coenzyme A via malonate (C3), and critically involving the enzymes (protein catalysts) acetyl coenzyme A carboxylase (ACC) and FA synthase. Conversely **FA oxidation, or β-oxidation of FAs** to form acetate (C2) and the mobile ketone bodies acetoacetate (C4) and its reduced XH2 form, 3-hydroxybutyrate (C4), is confined to the mitochondrial matrix, and entry of FAs into the matrix as Fatty acyl-carnitine is prevented by malonyl(C3)CoenzymeA inhibition of the enzyme catalyzing the formation of this, namely Carnitine Acyl Transferase I, CATI, this preventing a "futile cycle" of concomitant FA synthesis and destruction by FA oxidation.

Pentose phosphate pathway: uses excess glucose in times of Plenty to generate the reduced coenzyme NADPH used for biosynthetic reactions such as FA synthesis and DNA synthesis, formation of ribose (the precursor for nucleotides, DNA and RNA), **and synthesis of enzymes (protein catalysts) e.g. for:**

Gluconeogenesis (G'neo) (liver and kidney) required to maintain blood glucose at circa 4 mmoles per litre (4 millimolar, 4 mM, 4×10^{-3} M) during Fasting (e.g. when sleeping): [pyruvate (C3) + CO_2 (C1)-> oxaloacetate (C4; reduced form malate) via Pyruvate Carboxylase in mitochondria] $\rightarrow$ in the cytosol via PEP Carboxykinase, oxaloacaetate (C4) $\rightarrow CO_2$ (C1) + phosphoenolpyruvate (PEP, C3) -> via the enzymes of Glycolysis -> -> glucose-6-phosphate (C6) -> glucose (C6) in the blood (liver and kidney cells) (G'neo converts pyruvate (C3), its reduced form lactate (C3) and amino acids from protein breakdown to glucose; through a Glyoxylate Cycle and G'neo plants but not animals can make glucose from FAs and acetyl(C2) coenzyme A). Protein phosphatase-reversed phosphorylation and inactivation by Fasting-activated cyclic AMP-dependent protein kinase, PKA, of liver Pyruvate Kinase (catalyzes PEP + ADP -> pyruvate + ATP, a key substrate level phosphorylation in Glycolysis) prevents a "futile cycle" of simultaneously making and destroying PEP. Cyclic AMP is a key cellular "hunger signal" in animal cells.

Urea cycle (liver and small intestine) is required to remove excess ammonia (NH_3, N1) from amino acids from protein degradation by [ammonia (NH_3, N1) plus CO_2 (C1) -> carbamoyl phosphate (C1, N1) via Carbamoylphosphate Synthase in the mitochondrion -> citrulline formation by reaction with ornithine ($X-NH_2$) catalyzed by Ornithine Transcarbamoylase in mitochondrial matrix]. In the cytosol citrulline -> arginosuccinate -> fumarate plus arginine, which finally (via Arginase-catalyzed reaction) yields ornithine + urea (C1N2; $H_2N-CO-NH_2$) in the cytosol (liver and kidney cells) -> urea excreted via kidneys (in man; different mechanism for N excretion in birds).

18. Overview of protein-, DNA- and RNA-based life:

a. Life: cell-based; self-replicating, self-repairing, liquid H_2O-based (operates between 0-100°C).

b. Information flow:

DNA -> DNA - DNA replication (in eukaryote nucleus, mitochondria or chloroplasts). DNA -> DNA (DNA replication via DNA polymerase, DNA ligase; DNA polynucleotide with bases (A, T, G, C) attached via N-glycosidic links to the 1'-position of (2')-deoxyribose of a 5'-deoxyribose-3'-phosphate-5'-deoxyribose- 3'-phosphate backbone, noting that there is a 5' end and a 3' end; plectonemically coiled, base-paired, complementary, anti-parallel sense and anti-sense strands of double-stranded dsDNA, A=T, G≡C base-pairing via hydrogen bonding; Genetic Code of 64 triplet codons, codon degeneracy (61 codons for 20 common amino acids with >1 codon for all except for Methionine and Tryptophan, 3 stop codons)

 DNA -> mRNA (messenger RNA) - transcription (in eukaryote nucleus, mitochondrion or chloroplast). DNA (e.g. sense strand 5'-ATG GGG … 3' in this example, noting that there is a 5' end and a 3' end) -> mRNA (transcription catalyzed by RNA polymerases; RNA polynucleotide with bases (A, U, G, C) attached via N-glycosidic links to the 1'-position of ribose on a 5'-ribose-3'-phosphate-5' ribose backbone; fidelity yields in this example 5'- AUG-GGG).

 mRNA -> protein - translation (on ribosomes in eukaryote cytosol, mitochondria or chloroplasts). Protein synthesis (translation) from post-transcriptionally-modified mRNA on ribosomes to yield a polypeptide with the sequence in this example N-terminal-Methionine- Glycine… C-terminus -> post-translational protein processing through covalent modification (by additions or snipping). Fidelity of information flow is achieved via mRNA codon hydrogen bonding to corresponding tRNA (transfer RNA) anticodon (A=U, G≡C). Post-translational processing by polypeptide cleavage, chemical additions (e.g. glycosylation), and correct protein folding to form the 3-D structure of the protein.

c. Need for catalysts.

(1). Need thousands of specific catalysts to "speed up" (catalyze) specific metabolite conversions by lowering the "activation energy barrier" for reactions.

(2). Need catalysts operating in H_2O & stable in 0-100°C range for liquid H_2O.

(3). Needed an evolutionary mechanism to select better catalysts and macromolecule structural "scaffolds" (mutation, and thence Darwinian natural selection).

d. Why proteins as catalysts?

 (1). Strings of 20 different amino acids (aas) linked by peptide bonds gives a huge diversity (20^{100} different sequences 100-aas long – but utility in a mostly aqueous

environment limits the useful proportion) plus "post-translational modification -> a huge molecular diversity.

(2). Proteins can be soluble in H_2O (solvated by hydrogen bonded H_2O).

(3). Gene mutation -> variability, evolutionary selection mechanism -> evolution of "better" proteins for catalytic and other functions.

19. Overview of protein structure.

Primary structure - amino acid sequence (20 common Genetic Code-encoded amino acids - general formula R-CH (-NH_2, -COOH), different R groups that have different polarity e.g. positively charged, negatively charged, uncharged polar, hydrophobic, the alphabet of life - that are inserted in translation; however some of these can be modified post-translationally e.g, proline -> 4-hydroxyproline in the very abundant protein collagen).

Secondary structure - 3D structure of sequence elements; alpha-helices (right handed helical and compact, favoured by large amino acid R groups that point away from the structure; spring analogy), beta-strands (stretched, favoured by small amino acids like glycine, serine and alanine), beta-turns (ordered turn structure involving hydrogen bonds), random coils (no evident structure).

Tertiary structure - overall 3D structure, unique structure (some "breathing"), compact (very little internal space for simple molecules like H_2O), disulphide (S-S) links, very important hydrophobic interactions in aqueous solution, ionic, H-bonding, dipole-dipole, and weak Van der Waals interactions.

Quaternary structure - subunit complexity & arrangement e.g. protein monomers, homodimers A-A, heterodimers A-B (e.g. oxygen-carrying Haemoglobin, $\alpha_2\beta_2$), multimeric complexes (e.g. mitochondrial and chloroplast ATPase), macromolecular assemblies (mitochondrial and chloroplast ETCs, ribosomes).

20. Enzymes (Es), kinetics and thermodynamics: enzymes are proteins and catalysts; active site, substrate (S), transition state (E-S*), product (P) release, lock & key versus induced fit models (for specificity of E for S, stereospecificity (e.g. only L-amino acids and not the mirror-image D-amino acids occur in proteins on this planet), other molecular interactions as for protein tertiary structure): E + S $\leftrightarrow$ E-S $\leftrightarrow$ E-S* (transition state) $\leftrightarrow$ E–P $\leftrightarrow$ E + P (Product). Enzymes are protein catalysts that speed up specific chemical reactions without themselves undergoing net change. In the kitchen or the laboratory we speed up reactions by heating the reagents – this lowers the Activation Free Energy Barrier (ΔG_{act}) that must be overcome for an exergonic (free energy releasing, $\Delta G<0$) reaction to proceed. Thus a piece of paper (composed of cellulose; a linear polysaccharide polymer, $(CH_2O)_n$) will survive and may only brown slightly for thousands of years but briefly apply a flame (overcome the ΔG_{act}) and it will burn to completion in seconds: $(CH_2O)_n + O_2 \rightarrow nCO_2 + nH_2O$. An enzyme changes the kinetics (speed) of reaction but does not change the overall thermodynamic feasibility of a reaction ($\Delta G<0$). [Ditto a largely cellulosic forest at high temperature and critically low humidity in our present "Pyrocene Era" happily exists intact until flame initiated by a lightning strike lowers the Activation Free Energy Barrier and initiates catastrophic conflagration.]

A useful analogy for biological thermodynamics I used in lectures: a stick man Jack at a fence on top of a cliff. Before he can jump off he has to use energy to jump onto the fence (ΔG_{act}). If he jumps off he releases free energy in the exergonic process of falling to the ground ($\Delta G<0$) and when he splatters to his death at the bottom, disorder (entropy S) is increased (he is dead and order cannot be restored in this instance). However if he abseils down via a pulley he can couple the free energy loss from falling (free energy out, $\Delta G<0$) to lift Jill from the ground to the top of the cliff (an energy-requiring, endergonic reaction; $\Delta G> 0$ [one is reminded here of the "The Bricklayer's Lament" told by Gerald Hoffnung at the Oxford Union, 4 December 1958: http://bluegum.com/Humour/Assorted/bricks.html].

Thus plants use solar energy to drive the endergonic reaction (free energy to be put in, $\Delta G > 0$): $nCO_2 + nH_2O + \text{solar energy} -> (CH_2O)_n$ (glucose, starch) $+ nO_2$.

Plant-eaters oxidize carbohydrates like glucose (C6) in an exergonic (G releasing) reaction ($\Delta G < 0$): $(CH_2O)_n + O_2 -> nCO_2 + nH_2O$ but couple this exergonic reaction to the endergonic formation ($\Delta G > 0$) of the High Energy Compound (HEC) ATP (adenosine-5'-triphosphate) from ADP (adenosine-5'-diphosphate) and inorganic phosphate (HPO_4^{2-}, P_i): $ADP + Pi -> ATP + H_2O$.

ATP is called the "Energy Currency of Cells" because the exergonic hydrolyis of ATP: $ATP + H_2O -> ADP + Pi$ ($\Delta G < 0$; $\Delta G° = -30.5$ kJ/mol (-7.3 kcal/mol)) can be "coupled" to energy requiring, endergonic reactions ($\Delta G > 0$) such as biosynthesis, walking, thinking. Note that 1 cal (calorie) raises the Temperature (T) of 1 ml of H_2O by 1C; an active adult needs about 2,000 kcal per day.

21. Overview of metabolism:

(1). Primary solar energy-driven plant photosynthesis in chloroplasts (or photosynthetic bacteria) involves CO_2 fixation via Ribulose-bis-phosphate carboxylase (Rubisco; about 50% of leaf protein) to form 3-phosphoglycerate (3-PGA)(C3) and thence reduction and interconversions involving Light Reactions-derived ATP and NADPH into carbohydrate $(CH_2O)_n$ e.g. glucose (C6), starch or cellulose $((CH_2O)_n)$. Energy for plant-eating organisms comes from exergonic ($\Delta G < 0$) oxidation of photosynthetically-derived glucose (C6) - $(CH_2O)_n + O_2 -> nCO_2 + nH_2O$ - coupled to endergonic ($\Delta G > 0$) ATP formation; endergonic (free energy requiring, $\Delta G > 0$) biosynthetic reactions (e.g. protein, DNA, RNA, lipid synthesis) are coupled to exergonic ($\Delta G < 0$) ATP hydrolysis; in general exergonic reactions (free energy releasing, $\Delta G < 0$) are coupled to endergonic ($\Delta G > 0$) ATP synthesis. G is released in small steps in aerobic and anaerobic metabolism e.g. in the TCA cycle reactions pyruvate (C3) from the glycolytic oxidation of glucose (C6) is oxidized to CO_2 (C1) in successive steps involving the coupled formation of reduced coenzymes XH_2 from their respective oxidized forms, X; thence electrons from XH_2 go to O_2 as the Terminal Electron Acceptor via the ETC, and in the process an energy-conserving Proton Gradient is generated across the Inner Mitochondrial Membrane that drives the rotation of the membrane-associated ATP Synthase (ATPase) that makes ATP in Oxidative Phosphorylation. [In the kitchen or the chemical laboratory we can use high temperatures to drive reactions but living things only operate in the range 0-100C; heat-labile proteins are "denatured" within and above this range].

(2). The High Energy Compound (HEC) ATP is the energy currency for energy-requiring motion (involving actin and myosin of muscle), solute transport (e.g. PM-located transporters to get solutes across the hydrophobic PM), biosynthesis (e.g. of proteins, DNA, RNA, FAs, PLs, and of glucose in G'neo) and signalling (e.g. neuronal and muscle cell excitability involves making transmembrane Ca^{2+} gradients via PM and ER Ca^{2+} ATPases, and Na^+ and K^+ gradients via the $Na^+ K^+$ ATPase).

(3). The reduced coenzyme NADPH is used in reductive biosynthesis and is generated in Plenty conditions, notably by the cytosolic Pentose Phosphate Pathway for making NADPH (the reduced form of $NADP^+$) and the monosaccharide ribose for Anabolic (building up) reactions in "Plenty" situations (e.g. FA synthesis, nucleotide synthesis). In contrast, the coenzyme NAD^+ (oxidized form; NADH the reduced form) are used in Catabolic (breaking down) reactions (e.g. Glycolysis, FA oxidation, TCA Cycle aka Krebs Cycle, Citric Acid Cycle). Note that oxidation-reduction (e.g. $Y + XH_2 <-> YH_2 + X$) is like a buyer and seller - Oxidation variously involves the addition to a reductant of an O, removal of a H or removal of an electron (e^-) whereas Reduction involves with the oxidized reaction partner the corresponding removal of an O, addition of a H or addition of an electron.

(4). Monomer-polymer conversions involve amino acids (monomers of polypeptides or proteins), monosaccharides (the sugar monomers of polysaccharides e.g. the monosaccharide glucose (C6) is the monomer of the polysaccharides $((CH_2O)_n)$ cellulose (linear, water-insoluble) and the non-linear and more water-soluble starch, and the sugar-base-phosphate mononucleotides (monomers of the polynucleotides DNA and RNA). Notably adenosine monophosphate, AMP, is adenine-ribose-phosphate; the HEC adenosine diphosphate, ADP, is adenine-ribose-phosphate-phosphate; and the HEC adenosine triphosphate, ATP, is adenine-ribose-phosphate-phosphate-phosphate).

(5). There are distinct degradative and biosynthetic pathways, Anabolic (building up) and Catabolic (breaking down) pathways with metabolic compartmentation (e.g. FA synthesis in the cytosol, FA catabolism in mitochondria) that maintains homeostasis or balance and avoids "futile cycles" in different states (notably Plenty and Fasting).

(6). Overall homeostasis, cell division versus Apoptosis (programmed cell death), and enzyme activity regulated by enzyme amount and by regulation of the activity of existing enzymes by allosteric modifiers (Feed Forward Activation (FF+) and Feedback Inhibition (FB-), signalling-controlled, reversible covalent modification, including second messenger-mediated hormonal control by switching enzymes on or off by phosphorylation and dephosphorylation (important second messengers being Ca^{2+}, cyclic AMP, cyclic GMP and particular inositol phospholipids that activate particular second messenger-activated protein kinases that put phosphoryl (phosphate) groups on particular proteins, thereby functionally switching them on or off. Thus, for example, a lion in the street -> fear -> increased adrenaline release from adrenal medulla -> adrenalin (epinephrine) binds to a specific PM beta-adrenergic receptor -> via trimeric G protein complex activation -> Adenylate Cyclase activated -> cyclic AMP made from ATP -> activation of cyclic AMP-dependent protein kinase -> proteins phosphorylated e.g. phosphotriglyceride lipase active -> triglyceride breakdown to glycerol plus FAs -> catabolism -> increased ATP -> increased energy and ability to run away (i.e. stimulus-response coupling). The breakdown and new synthesis of enzymes can be hormonally regulated in a stress-responsive way e.g. sustained stress -> corticosteroid hormone cortisol released from adrenal cortex -> binds to a cytosolic receptor which translocates to the nucleus -> binds to PEPCK gene promoter -> transcription of PEPCK -> PEPCK made -> increased Gluconeogenesis -> increased [glucose] (glucose concentration) in the blood -> better ability to cope with sustained stress.

22. Regulation of animal (human, rat, mouse etc) metabolism in Plenty and in Fasting by Insulin/Glucagon control & stimulus-response coupling (a lesson from Mother Nature for a world engaged in a non-sustainable and terracidal carbon burning economy):

Plenty: increased [blood glucose] -> increased insulin (peptide hormone from pancreas beta cells) -> binds to insulin-Receptor Tyrosine Kinase (RTK ; on PM) -> PKB activated by PL second messengers -> specific protein phosphorylation (and indirectly insulin activated dephosphorylation by protein phosphatases) -> increased glucose transport, glucose-dependent gene expression, glycolysis, and Anabolic reactions (increased glycogen synthesis, protein synthesis, FA synthesis) and decreased Catabolic reactions (glycogenolysis, FA oxidation, protein degradation, gluconeogenesis) -> ultimately lowered blood [glucose] (should be about 4 mM).

Fasting: lowered blood [glucose] -> glucagon (another peptide hormone; from pancreas alpha cells) -> glucagon binds to glucagon receptor on PM -> Adenylate Cyclase activated -> increased cyclic AMP (second messenger hunger signal) (and lowered Fructose-2,6-bisphosphate, F26BP, Plenty signal) -> activated cyclic AMP-dependent protein kinase (PKA) -> specific protein phosphorylation (reversed by dephosphorylation by protein phosphatases) -> increased Catabolism (increased fatty acid oxidation, gluconeogenesis, glycogenolysis) and decreased Anabolism (decreased glycolysis, fatty acid synthesis) -> ultimately increased blood [glucose].

Brain function obligately depends on glucose catabolism and organs maintain steady blood [glucose] of about 4 mM in fasting per: 1. use of glycogen (muscle and liver); 2. Gluconeogenesis & glucose release by liver; 3. use of fatty acids from adipocytes, ketone bodies from liver; 4. shift of adipocytes, muscle and liver from glucose to fatty acids (and eventually ketone bodies).

23. Biochemistry, climate change problem, and biological solutions.

Major CO_2 sinks include photosynthetic bacterial photosynthesis and plant photosynthesis yielding cellulose and related carbohydrates of wood and thence of soil carbonaceous compounds of humus (roughly half of forest carbon in tree wood and half in soil carbon; SE Australian *Eucalyptus regnans* (Mountain ash) forests are the best forest carbon sinks in the world). Photosynthesis yields carbohydrate: $nCO_2 + nH_2O +$ solar energy $-> (CH_2O)_n$ (relatively stable, water-insoluble cellulose) $+ nO_2$. Most of photosynthesis is reversed by carbohydrate oxidation by fires or aerobic organisms: $(CH_2O)_n + O_2 -> nCO_2 + nH_2O$. Some CO_2 dissolves in ocean water, this resulting in acidification (biologically deleterious to organism with calcareous exoskeletons e.g. coral, crustaceans): $CO_2 + H_2O -> H_2CO_3 -> HCO_3^- + H^+$; $HCO_3^- -> CO_3^{2-} + H^+$ (see later: threat to coral from ocean acidification as well as warming). Fixation of CO_2 by ocean photosynthetic algae and prokaryote phytoplankton could be increased by adding fertilizer to oceans (a major geoengineering proposal).

Coal derives from anaerobic geologic conversion of cellulosic carbohydrates to carbon $((CH_2O)_n$ + heat, pressure $-> nC + n\ H_2O)$. Subterranean oil drives from anaerobic decarboxylation of biologically-derived fatty acids $(CH_3(CH_2)_n\text{-}COOH$ + heat, pressure $-> CH_3(CH_2)_nH + CO_2$. Subterranean methane (CH_4) derives from anaerobic reduction of carbohydrates by anaerobic bacteria (reduction being addition of electrons (e-), addition of hydrogen atoms (H) or removal of oxygen (O)): $(CH_2O)_n + 4H$ (derived from catabolism and reduced coenzymes) $-> nCH_4 + nH_2O$. We must keep coal, oil and gas in the ground, and get energy from renewable and geothermal sources.

Oil derives from decarboxylation of biological fatty acids (FAs). Oil is a vital precursor for plastics and pharmaceuticals yet we are burning this vital feedstock as well as polluting the atmosphere and oceans with CO_2 from oil combustion. Biofuel production carries a huge carbon pollution debt (Carbon Debt) as well as being an obscene use of food for fuel, driving up food prices and condemning 2 billion to malnourishment, 1 billion to hunger and millions to starvation (Biofuel Genocide).

Methane derives from anaerobic degradation of biological material e.g. in swamps, waste dumps, livestock digestion: $(CH_2O)_n + 4nH$ (derived from reduced coenzymes) $\rightarrow nCH_4 + nH_2O$ or, overall: $2(CH_2O)_n \rightarrow n\,CH_4 + nCO_2$. Global warming is already releasing CH_4 from H_2O-CH_4 clathrates in the Arctic tundra and in shallow parts of the Arctic Ocean. Fugitive CH_4 emissions occur from coal mines, coal seam gas (CSG) extraction, conventional natural gas extraction, from coal seam and shale fracking, and from systemic gas reticulation leakage. World Bank re-assessment of the livestock contribution to GHG pollution (e.g. deforestation, cattle belching) indicates that annual global greenhouse gas (GHG) pollution is 64 Gt CO_2-e, about 50% bigger than the 42 Gt CO_2-e hitherto thought, and that the livestock contribution is over 51% of the bigger figure (a major element of this analysis being a 20 year time frame considered for the CH_4 GWP rather than a 100 year time frame). Solutions include keeping gas in the ground, decreasing livestock production (different bacterial flora, vaccination against methanogens, shifting to kangaroo meat, fish aquaculture involving 2 kg grain per kg meat versus 7 for beef, vegetarian or vegan culture, plant-derived protein, soy milk).

Nitrous oxide (N_2O) derives from agricultural use of nitrogenous fertilizers and from fossil fuel (coal, gas and oil) combustion - ergo, use renewable energy and apply smarter use of nitrogenous fertilizers.

Albedo (light reflectivity) decreased by disappearance of Arctic sea ice - restore albedo by painting roads and roof tops white, and biologically by appropriate plant ground cover.

Biochar. Atmospheric CO_2 can be reduced from the current 415 ppm CO_2 back to a safe and sustainable 300 ppm CO_2 by fixing CO_2 as cellulose via solar-energy-driven photosynthesis ($nCO_2 + nH_2O \rightarrow (CH_2O)_n + O_2$) with subsequent anaerobic pyrolysis of cellulosic material (e.g. waste wood and straw) to yield carbon (C, charcoal) ($(CH_2O)_n \rightarrow nC + nH_2O$) which can then be added to soil or buried in holes in the ground (e.g. used coal mines). Thus p224, Progress in Thermochemical Biomass Conversion, volume 1, IAE Bioenergy, ed. A.V. Bridgewater (Blackwell Science) informs us that we could obtain 1.7 GtC/yr (straw from agriculture) + 4.2 GtC/yr (total grass upgrowth from grasslands upgrowth) + 6 GtC/yr (possible sustainable woodharvest) = 11.9 GtC/yr. From this one can see why biochar expert Professor Johannes Lehmann of Cornell University is correct calculating that it is realistically possible to fix 9.5bn tonnes of carbon per year using biochar, noting that global annual production of carbon from fossil fuels is about 9 bn tonnes.

Introduced plant pests (e.g. introduced "mimosa", *Mimosa pigra* and Gamba grass, *Andropogon gayanus*, in the Northern Territory of Australia) are good candidates for disposal via biochar production.

Unfortunately biochar is expensive to produce and thus for every \$1 derived from coal or gas burning, \$1-\$2 will have to be spent converting the consequent CO_2 back to carbon (biochar) (based on US mid-West corn stalk-derived biochar; multiply this by 4 for the cost in the UK).

The cost of conversion of cellulosic waste to biochar in the US mid-West is about \$49-\$74 per tonne CO_2 as compared to \$210-\$303 per tonne CO_2 in the UK. There is 2,800 billion tonnes of CO_2 in the atmosphere, and the cost of reversing 2 centuries of fossil fuel burning by means of biochar-based return of the atmospheric CO_2 concentration to the pre-industrial 300 ppm from the current 415 ppm [2020] through removing 1,400 billion tonnes CO_2 at \$200 per tonne is \$280 trillion (US dollars) (cf the current world annual GDP of \$85 trillion). The corresponding cost of conversion of 5,250 billion tonnes of CO_2-e via CO_2 to biochar is \$1,050 trillion. This estimate represents about 12 years of world GDP and tells even the most optimistic person that we are doomed unless radical action is taken now to stop Arctic warming immediately. Ergo, keep coal and other fossil fuels in the ground. It is still conceivable (albeit very unlikely) that we could save the world from climate disaster.

Various methods have been proposed for removing atmospheric CO_2 from past or continuing pollution (e.g. back to 300 ppm CO_2) including (with critiques in parenthesis):

(a) Carbon Capture and Storage or Carbon Capture and Sequestration (CCS) underground of CO_2 from power stations (widely touted but very expensive and not presently commercially feasible [4]);

(b) Ocean storage of power station-derived CO_2 in the deep ocean (similarly costly in relation to compression and also contraindicated by associated ocean acidification deadly to calcifying photosynthetic and animal organisms with calcium carbonate-based exteriors e.g. coccolithophore algae, corals, foraminifera, echinoderms, crustaceans and molluscs);

(c) Algal photosynthesis use to generate biomass through photosynthesis in power plant-associated algal ponds (huge expense and incomplete CO_2 removal);

(d) Ocean algal photosynthesis per fertilization of oceans to promote photosynthetic fixation of CO_2 into biomass (huge cost and massive global ecosystem disruption with unknown consequences);

(e) Re-afforestation - biological removal of CO_2 as cellulose in tree trunks and as carbon compounds in the soil (clearly a major part of the solution but there are limits to re-afforestation and soil carbon storage; there is presently about 700 Gt C in biomass (mostly wood) and 1,600 Gt C in soil as humus; SE Australian *Eucalyptus regnans* forests are the best forest carbon sinks in the world);

(f) Accelerated Weathering of Limestone (AWL) - relatively cheap CO_2 sequestration as bicarbonate in oceans through Accelerated Weathering of Limestone (AWL) by scrubbing power plant emissions in sea water plus limestone ($CaCO_3$) (compression costs would make more general application to removing CO_2 from the atmosphere even more expensive – however in the context of a presently existing and unfortunately entrenched and dominant carbon economy this mechanism is important);

(g). Terrestrial enhanced weathering of pulverized silicate rocks through carbonation;

(h) Biochar (carbon, charcoal) production and storage through anaerobic pyrolysis at 400-700C of waste biomass cellulosic materials - the annual biochar production using existing agricultural and forestry waste would be similar to the amount of carbon pollution produced annually by industry.

NB. Other so-called geo-engineering approaches to global warming variously with severe ecological and biological implications include (i) shades in space, (ii) adding sulphate aerosols to the stratosphere or troposphere to promote global dimming, and (iii) changing albedo by large-scale plant monocultures or a-biotically (e.g. by painting roofs or roads white).

However methane (CH_4) has a Global Warming Potential (GWP) 105 times greater than that of carbon dioxide (CO_2) on a 20 year time frame and taking aerosol impacts into account. Accordingly the predicted entry of 50 billion tonnes of methane (CH_4) into the atmosphere from the Arctic in the coming decades is equivalent to $50 \times 105 = 5,250$ billion tonnes CO_2-equivalent (CO_2-e).

The cost at \$200 per tonne CO_2 for removing 1,400 billion tonnes CO_2 (half atmospheric CO_2) as biochar is \$280 trillion. The corresponding cost for conversion of 5,250 billion tonnes of CO_2 to biochar is \$1,400 trillion, this representing about 3 years of world GDP, and telling even the most optimistic person that we are doomed unless radical action is taken now to stop warming immediately. Ergo, keep coal and other fossil fuels in the ground to avoid what biochemists call a "futile cycle".

24. Alphabetically-organized, detailed explanations of key terms relating to climate change and to advocacy and achievement of Climate Justice, Intergenerational Justice, and Intergenerational Equity.

ABC Strategy. Successful –isms and –ists have an ABC strategy involving Accountability (negative feedback for non-adherence), Badge (symbol) and Credo (brief statement of position). Environmental-ism requires (A) accountability e.g. holding people, politicians, parties, candidates, countries and corporations accountable by boycotts, divestment, sanctions, exposure and scorn, (B) wearing a badge at all times (e.g. "300 ppm CO_2"), and (C) a simple credo e.g. "For a safe planet for all peoples and all species we must urgently return the atmospheric CO_2 from the present 400 ppm CO_2 to the safe and sustainable pre-industrial 300 ppm CO_2".

Accelerated Weathering of Limestone (AWL). Proposed but not yet commercial Accelerated Weathering of Limestone (AWL) involves waste gas from burning coal or gas being passed through a sea water-limestone ($CaCO_3$) scrubber with the following reaction: CO_2 (gas) + $CaCO_3$ (solid) + H_2O <-> Ca^{2+} (aqueous) + 2 HCO_3^- (aqueous).The scrubbing solution is then piped to the sea. Carbon in the oceans as bicarbonate is 10 times that in all recoverable fossil fuel reserves and about 60 times that in the CO_2 in the atmosphere. The carbon in carbonate minerals is about 4,000 times greater than the carbon in oil and coal fossil fuel reserves, and the AWL process would in part reverse the deleterious acidification of the oceans due to the massive CO_2 pollution of the atmosphere [1].

Anthropogenic Global Warming (AGW) (see Climate Change and Man-made Climate Change). Anthropogenic Climate Change refers to the change in the Earth's climate due to Man's pollution of the atmosphere with greenhouse gases (GHGs), principally carbon dioxide (CO_2) and methane (CH_4), since the start of the Industrial Revolution in the mid-18th century [1]. CO_2 is the major contributor to Anthropogenic Global Warming (AGW), deriving from aerobic respiration involving oxidation of carbohydrate ($(CH_2O)_n + nO_2 \rightarrow nCO_2 + nH_2O$), lime (CaO) in cement production ($CaCO_3 \rightarrow CaO + CO_2$), and the combustion of fossil fuels such as coal ($C + O_2 \rightarrow CO_2$), oil ($CH_3(CH_2)nH + (n+1)O_2 + (n+2)/2O_2 \rightarrow (n+1)CO_2 + (n+2)H_2O$) and natural gas, mainly methane ($CH_4 + 2O_2 \rightarrow CO_2 + 2H_2O$). Other GHGs (Global Warming Potential, GWP, relative to CO_2 on a 20 year time frame in parentheses) include methane (CH_4; 79, or 105 with aerosol impacts included) (from natural gas and anaerobic biomass degradation), nitrous oxide (N_2O, 289) (from power plants and agriculture) and man-made GHGs e.g. chlorofluorohydrocarbon CFC-12, CCl_2F_2 (11,000), hydrochlorofluorohydrocarbon HFC-22, $CHClF_2$ (5,160), hydrofluorohydrocarbon HFC-23, CHF_3 (12,000), sulphur hexafluoride SF_6 (16,300), nitrogen trifluoride NF_3 (12,300).

Avoidable Death (see Avoidable Mortality, Excess Death, Excess Mortality). Avoidable Mortality (Avoidable Death, Excess Death, Excess Mortality, Premature Death, Untimely Death) is the difference between the actual mortality in a country and the mortality expected in a peaceful, decently-run country with the same demographics. By 1950 all the World potentially had access to the requisites for the very low avoidable mortality obtaining in European countries, namely clean water, sanitation, proper nutrition, literacy (especially female literacy), primary health care, antibiotics and major preventive medicine programs including public health education, prophylactics (such as insecticides, antiseptics, mosquito netting, soap and condoms) and major vaccinations. However such benefits took decades to arrive in many countries and are still variously lacking in African countries. Nevertheless, in most countries outside Africa the annual mortality rate (expressed as deaths per 1,000 people per year)

typically declined to a minimum and in the best countries
(typically European and East Asian countries) eventually began
to rise, with this reflecting aging populations. Using United
Nations Population Division demographic data going back to
1950 (the 2002 Revision), it was possible to make base-line
estimates of expected mortality for various demographically
distinct groups. For many high birth-rate Third World countries
the base-line mortality estimates clustered about 4 deaths per
1,000 of population [2].

**Avoidable Mortality (see Avoidable Death, Excess Death,
Excess Mortality).**

Biochar. Biochar is the charcoal (i.e. carbon, C) derived from
the anaerobic (minus oxygen) heating of cellulosic waste to
about 400-700C [3]. This represents one of the major ways of
reducing the atmospheric CO_2 from the present circa 415 ppm
(2020) to a safe and sustainable maximum of circa 300 ppm
CO_2 that had obtained until the last century for about 1 million
years (as advocated by 300.org [4, 5], as compared to the
politically pragmatic but insufficient 350 ppm CO_2 advocated
by 350.org [6]).

Biofuel Genocide. The legislatively mandated "food for fuel"
perversion is grossly wasteful, involves a huge Carbon Debt,
and drives up grain price in a hungry world with the potential
for famine as in the WW2 Bengal Famine in which 6-7 million
people died because of a 4-fold increase in the price of rice [2]
(see the "Biofuel Genocide" website).

Boycott (see Boycott, Divestment and Sanctions (BDS)).
Peace is the only way but silence kills and silence is complicity.
Decent people are obliged to (a) inform everyone they can about
the worsening Climate Crisis, and (b) urge and apply Boycotts
as appropriate against all people, politicians, parties, policies,
companies, corporations and countries complicit in the
worsening Climate Emergency.e.g. Boycott the climate change
denialist Murdoch Media [7].

Boycott, Divestment and Sanctions (BDS) (see Boycott). BDS was successfully applied against Apartheid South Africa, and is now widely applied against Apartheid Israel in the interests of 1-man-1-vote democracy, justice, all human rights for all, and reconciliation in Palestine [8]. Many people, notably anti-racist Jewish humanitarians, demand BDS in support of equal human rights and indeed any human rights for 10 million human rights-deprived Palestinians [8], and BDS is surely required against the climate criminals threatening to extinguish the very lives of 10 billion mostly Third World people this century through unaddressed climate change. Some advocate Boycott of Murdoch media and other Mainstream media (MSM) which have variously promoted Climate Change Denialism (Climate Change Scepticism) and/or Effective Climate Change Denialism (Effective Climate Change Scepticism) through climate change inaction [7].

Cap and Trade (see Emissions Trading Scheme (ETS)). This is basically a fraudulent device to allow particular countries to set a notional cap on greenhouse gas (GHG) pollution and then trade limited and dodgy Carbon Credits and limited and dodgy licences to pollute the one common atmosphere of all countries. This might be a bit more plausible if the licences were sold by acutely global warming-threatened countries like mega-delta nations (e.g. Bangladesh and Myanmar) and Island States (e.g. Tuvalu and Kiribati)

Carbon Burning. Carbon burning is the burning of carbon-containing substances e.g. wood, coal, gas and oil to yield carbon dioxide (CO_2) and other substances (see Carbon Burning Pollutants) [2, 5]. Thus, for example, the compete combustion of carbon-containing fuels is as follows: carbohydrate (e.g. wood cellulose) ($(CH_2O)_n + O_2 \rightarrow n\ CO_2 + nH_2O$), coal ($C + O_2 \rightarrow CO_2$), oil ($CH_3(CH_2)nH + (n + 1)O_2 + (n + 2)/2\ O_2 \rightarrow (n+1)CO_2 + (n+2)H_2O$) and natural gas, mainly methane ($CH_4 + 2O_2 \rightarrow CO_2 + 2\ H_2O$) [5].

Carbon Burning-related Deaths. About 10,000 Australians die each year from the effects of pollutants from vehicles, coal burning for electricity and other carbon burning, the estimates being 2,200, 4,600 and 2,800, respectively. Given Australia's annual Domestic GHG pollution of about 552 Mt CO_2-e (2011) and given a risk avoidance-based "value of a statistical life" (VOSL) of $7.6 million per person ($73 billion pa for Australian carbon burning-related deaths) and $10 billion pa in fossil fuel subsidies, the minimum Carbon Price to cover carbon burning-derived deaths and carbon burning subsidies is $7.6 million x 9,600 = $73 billion plus $10 billion = $83,000 million/552 Mt CO_2-e = or $150 per tonne CO_2-e (A$555 per tonne C) as compared to the best political offer yet in Australia of $23 per tonne CO_2-e [9]. It is estimated that about 5 million people die each year worldwide from climate change (0.5 million) and 4.5 million (carbon burning pollutants), it being estimated that 100 million people will have died thus by 2030 if climate change is not properly addressed [10, 11].

Carbon Burning Pollutants. Gas-fired power plants (GFPPs) are clean-er than coal-fired power plants (CFPPs) in the sense that per unit of electrical energy they emit lower amounts of pollutants such as the greenhouse gas carbon dioxide (CO_2), and the very toxic pollutants carbon monoxide (CO), nitrogen oxides (NOx), sulphur dioxide (SO_2), particulates (fine soot to ultrafine particles), radioactivity, and heavy metals such as mercury (Hg) [12]. Thus according to Naturalgas.org [13], the fossil fuel-derived pollutant emissions levels (in pounds per Btu of energy input) are as follows (noting that this is for average gas burning and not specifically for gas-turbine power stations) - for CO_2: 117,000 (gas), 164,000 (oil) and 208,000 (coal); for CO: 40 (gas), 33 (oil) and 208 (coal); for NOx: 92 (gas), 448 (oil) and 457 (coal); for SO_2: 1 (gas), 1,122 (oil), and 2,591 (coal); for particulates: 7 (gas), 84 (oil) and 2,744 (coal); and for Hg: 0.000 (gas), 0.007 (oil) and 0.016 (coal). In addition, radioactive material is a significant pollutant from coal burning but not from gas burning. However, while Hg and radioactivity are negligible pollutants for gas burning and SO_2 emission is very low, CO_2, CO, NOx and particulate matter (PM) are significant pollutants from GFPPs [12, 13]. Particulate matter (PM) is generated by GFPPs and CFPPs but gas-fired power plants disproportionately produce fine PM that is of particular concern because it can penetrate deep into the lungs [12]. Thus according to the World Health Organization (WHO) "no threshold for PM has been identified below which no damage to health is observed." [14].

Carbon Credit (see Carbon Debt). The Historical Carbon Debt (aka Historical Climate Debt) of a country can be measured by the amount of greenhouse gas (GHG) it has introduced into the atmosphere since the start of the Industrial revolution in the mid-18th century. The Carbon Credit of a country can be measured by the amount of GHG it can generate as its "fair share" of the 600 billion tonnes of CO_2 the World is permitted to generate between 2010 and zero emissions in 2050 if it is to have a 75% chance of avoiding a catastrophic 2C temperature rise (EU policy). Historical Carbon Debt minus Carbon Credit equals Net Carbon Debt - and when a negative number can be called Net Carbon Credit (aka Net Climate Credit) [15].

Carbon Debt (see Carbon Credit). NASA's Dr James Hansen has provided a breakdown of global responsibility for fossil fuel-derived CO_2 pollution between 1751 and 2006 as a percentage (%) of the Historical Carbon Debt (1751-2006) of 346 Gt C [16]. A US dollar value can be placed on that Carbon Debt from the estimate of a requisite Carbon Price for effective climate change action of $150 per tonne CO_2-e [15]. Thus $150 per tonne CO_2-e x 346 billion tonnes C x 3.7 tonne CO_2-e per tonne C = $192 trillion or about 2.3 times the World's GDP. Annual GHG pollution is 64 billion tonnes CO_2-e, and thus Carbon Debt for future generations is increasing each year by $150 per tonne CO_2-e x 0.064 trillion tonnes CO_2-e = $9.6 trillion i.e. by about $10 trillion annually.

Carbon dioxide (CO_2). Carbon dioxide (CO_2) is the major contributor to anthropogenic global warming (AGW), and derives from aerobic respiration involving oxidation of carbohydrate (e.g. wood cellulose) ($(CH_2O)_n + O_2 \rightarrow n\ CO_2 + nH_2O$), lime (CaO) generation in cement production ($CaCO_3 \rightarrow CaO + CO_2$), and the combustion of fossil fuels such as coal ($C + O_2 \rightarrow CO_2$), oil ($CH_3(CH_2)nH + (n + 1)O_2 + (n + 2)/2\ O_2 \rightarrow$

$(n+1)CO_2 + (n+2)H_2O)$ and natural gas, mainly methane $(CH_4 + 2O_2 \rightarrow CO_2 + 2H_2O)$ [1].

Carbon Dioxide-equivalent (CO_2-e). CO_2-equivalent (CO_2-e) is the sum of all the GHGs (excluding water, H_2O) expressed as CO_2-equivalents). Water is excluded on the assumption that the hydrological cycle means that there no net effect of water [1].

Carbon Pollution. Carbon pollution refers to the increased atmospheric CO_2 and CO_2-e [1]. Note that carbon (C) has an atomic weight of 12 and carbon dioxide (CO_2) has a molecular weight of 44 so one converts from "atmospheric C" to "atmospheric CO_2" by multiplying by $44/12 = 3.67$ e.g. 750 Gt atmospheric C = 750 GtC x 3.7 t CO_2/t C = circa 2,800 Gt atmospheric CO_2. More specifically, Carbon Pollution refers to extra carbon dioxide (CO_2), extra methane (CH_4) or extra CO_2-equivalent (CO_2-e, this including all GHGs excluding water (H_2O) expressed as CO_2-equivalents). Until the mid-20th century, CO_2 in the atmosphere had been no higher than 300 ppm for about 1 million years, but has now reached 400 ppm (at Mauna Loa, Hawaii) and is increasing at about 2 ppm CO_2 each year. CO_2 also dissolves in the ocean, and about 90% of the extra heat from global warming is in the ocean [1].

Carbon Pollution-related Deaths (also see Carbon Burning-related Deaths). About 9 million people die each year from the long-term effects of pollutants from carbon fuel burning. About 1 million people die annually world-wide due to climate change, it presently being estimated (2020) that 300 million people will have died thus by 2030 if climate change is not properly addressed [10, 11]. Toxic pollutants from the burning of carbon-containing fuels (e.g. in wood burning, vehicles, industry and in gas-fired and coal-fired power stations) variously include carbon monoxide (CO), nitrogen oxides (NOx e.g. N_2O, NO_2), heavy metals (e.g. mercury), radioactivity, fine Particulate Matter (e.g. PM_{10}, $PM_{2.5}$), Volatile Organic Compounds (some carcinogenic) and sulphur dioxide (SO_2) [12-14].

Carbon Price (see Carbon Tax). Man-made climate change is happening because GHG polluters are not held financially responsible for the health and environmental consequences i.e. governments are hugely subsidizing the killing of the planet. Thus, for example, in Australia (one of the world's worst annual per capita GHG polluters), the Coalition Federal Government insists on a Carbon Price of $0 (zero dollars) per tonne CO_2-e i.e. it allows climate criminal polluters to dangerously pollute the one common atmosphere and ocean of all countries on earth for free. In contrast, leading climate change economist Dr Chris Hope from 89-Nobel-Laureate Cambridge University estimates that a Carbon Price of about $200 per tonne CO_2-e is required for effective climate change action [17]. Australia's current annual Domestic greenhouse gas (GHG) pollution is about 2 Gt CO_2-e (2 billion CO_2-e) [1] and hence the climate criminal Australian Coalition Government is subsidizing this deadly, terracidal activity to the tune of $200 per tonne CO_2-e x 2 billion tonnes CO_2-e per year = $400 billion per year [18].

Carbon Sequestration (also see Carbon Storage). Carbon Sequestration can refer to the so far non-commercially-proven scheme for sequestering CO_2 from power stations underground or at the bottom of the oceans. Carbon sequestration can also be achieved by generating Biochar (charcoal, carbon) and burying it safe from combustion. Proposed but not yet commercial Accelerated Weathering of Limestone (AWL) involves waste gas from burning coal or gas being passed through a sea water-limestone ($CaCO_3$) scrubber with the following reaction: CO_2 (gas) + $CaCO_3$ (solid) + H_2O <-> Ca^{2+} (aqueous) + 2 HCO_3^- (aqueous). The scrubbing solution is then piped to the sea. Carbon in the oceans as bicarbonate is 10 times that in all recoverable fossil fuel reserves and about 60 times that as the CO_2 in the atmosphere. The carbon in carbonate minerals is about 4,000 times greater than the carbon in oil and coal fossil fuel reserves. The AWL process would in part reverse the deleterious acidification of the oceans due to the massive CO_2 pollution of the atmosphere [1].

Carbon Storage (also see Carbon Sequestration). There is presently: 700- 750 GtC (billion tonne carbon) in atmosphere (mostly as 750 GtC x 3.7 t CO_2/t C = circa 2,800 Gt CO_2, and about half due to historical fossil fuel combustion); 700 GtC in biomass (mostly wood); 1,600 GtC in soil; 36,000 GtC in ocean as bicarbonate ion (HCO_3^-); there is no net CO_2 from vulcanism and weathering (on a time scale of less than 100,000 years) [1].

Carbon Tax (see Carbon Price). A Carbon Price can be established fraudulently via a Cap and Trade (see Emissions Trading Scheme (ETS)). However it can be honestly set by a direct Carbon Tax analogous to an excise tax on alcohol. Thus leading climate scientist Dr James Hansen has suggested the following: "A price on emissions that cause harm is essential. Yes, a carbon tax. Carbon tax with 100 percent dividend is needed to wean us off fossil fuel addiction. Tax and dividend allows the marketplace, not politicians, to make investment decisions. Carbon tax on coal, oil and gas is simple, applied at the first point of sale or port of entry. The entire tax must be returned to the public, an equal amount to each adult, a half-share for children. This dividend can be deposited monthly in an individual's bank account. Carbon tax with 100 percent dividend is non-regressive. On the contrary, you can bet that low and middle income people will find ways to limit their carbon tax and come out ahead. Profligate energy users will have to pay for their emissions. Demand for low-carbon high-efficiency products will spur innovation, making our products more competitive on international markets. Carbon emissions will plummet as energy efficiency and renewable energies grow rapidly. Black soot, mercury and other fossil fuel emissions will decline. A brighter, cleaner future, with energy independence, is possible" [19].

Climate Change (see Man-made Climate Change). Climate Change refers to the change in the Earth's climate due to Man's pollution of the atmosphere with greenhouse gases (GHGs), principally carbon dioxide (CO_2) and methane (CH_4), since the start of the Industrial Revolution in the mid-18th century [1]. The Greenhouse Effect due to absorption of (infrared) radiation by CO_2 was discovered by UK scientist John Tyndall in the mid-19th century [1].

Climate Change Action (see Climate Change Inaction). The World is rapidly running out of time to deal with man-made climate change. Thus according to the German WBGU and Australian Climate Commission (sacked by the effective climate change denialist Coalition Australian Federal Government and now surviving unfunded as the Australian Climate Council), relative to 2014 we have only 15 years left before we exceed the Terminal Carbon Budget of 600 Gt CO_2 that must not be exceeded if we are to have a 75% chance of avoiding a catastrophic 2C temperature rise [20, 21]. However a more stringent analysis taking methane CO_2-e contributions into account estimates that relative to 2014 we have only 4 years left before we exceed 600 Gt CO_2-e [22, 23]. Business As Usual (BAU) and exploitation of all fossil fuel reserves means that the world will exceed the Terminal Carbon Budget by a factor of 5 [24] and Australian exploitation of all its fossil fuel and iron ore reserves means that it will exceed the terminal Carbon Budget by a factor of 3 [25]. An estimate of "years left to zero emissions" has been determined for all countries relative to mid-2011 - countries that must cease GHG pollution within 10 years relative to mid-2011 are Belize (1.3 years), Qatar (2.3), Guyana (2.4), Malaysia (3.4), United Arab Emirates (3.4), Kuwait (4.1), Papua New Guinea (4.3), Brunei (4.8), Australia (4.8), Antigua & Barbuda (4.9), Zambia (5.1), Canada (5.1), Bahrain (5.2), United States (5.5), Trinidad & Tobago (6.4), Luxembourg (5.9), Panama (6.3), New Zealand (6.5), Estonia (6.9), Botswana (7.0), Ireland (7.4), Saudi Arabia (7.6), Venezuela (7.9), Indonesia (8.4), Equatorial Guinea (8.6), Belgium (8.7), Turkmenistan (8.8), Singapore (8.9), Czech Republic (9.0), Liberia (9.0), Netherlands (9.3), Russia (9.3), Nicaragua (9.3), Finland (9.5), Oman (9.7), Palau (9.8), Brazil (9.8), Uruguay (9.8), Denmark (10.0). If one considers Australia's Domestic plus Exported GHG pollution, Australia used up its "fair share" of the World's Terminal Carbon Budget in 2011 (3 years ago) [26]. Indeed as of 2020 the whole World has used up this Terminal Carbon Pollution Budget.

Climate Change Cost (see Climate Cost). (A). Leading climate change economist Dr Chris Hope from 89-Nobel-Laureate Cambridge University estimates that a Carbon Price of about $200 per tonne CO_2-e is required for effective climate change action [17]. World Bank analysts have revised upwards the annual global GHG pollution from 42 Gt CO_2-e to 64 Gt CO_2-e [27]. One can accordingly estimate the annual cost of climate change inaction (i.e. a Carbon Price of $0 per tonne CO_2-e) at $200 per tonne CO_2-e x 64 billion tonnes CO_2-e per year = $13 trillion or about 15% of the Global GDP of $85 trillion. (B). Another way is to consider the biochar-based cost of reducing the atmospheric CO_2 from the present dangerous 400 ppm to a safe and sustainable 300 ppm CO_2 [3-4]. The cost of conversion of cellulosic waste to biochar in the US mid-West is about $49-$74 per tonne CO_2 as compared to $210-$303 per tonne CO_2 in the UK. There are 2,800 billion tonnes of CO_2 in the atmosphere and at $200 per tonne CO_2 the cost of reversing 2 centuries of fossil fuel burning by means of biochar-based return of the atmospheric CO_2 concentration to the pre-industrial 300 ppm from the current 400 ppm by removing 0.5 x 2,800 = 1,400 billion tonnes CO_2 is $280 trillion or 3.3 times the current world annual GDP of $85 trillion. (C). A further way is to consider the US EPA recently estimating the Value of a Statistical Life (VOSL, VSL) at $7.4 million per person. All men being equal, we can apply this estimate to the 10 million people who die annually [2020] from climate change and carbon burning [10, 11] to get a Climate Change Cost estimate of $74 trillion per year. These numbers, while horrifying, nevertheless show that it is still conceivable (albeit very unlikely) that we could save the world from climate disaster [1, 28, 29].

Climate Change Denialism (see Climate Change Scepticism). Science is about the critical testing of potentially falsifiable hypotheses and thus is inherently sceptical. Climate Change Denialism (Climate Change Scepticism) of a tiny minority of independent or fossil fuel corporation-backed scientists is overly sceptical to the point of radical and dangerous divergence from an overwhelming consensus of circa 97% of scientists. Big Money (notably fossil fuel corporations and the Murdoch media Empire) has ensured substantial layperson Climate Change Denialism (Climate Change Scepticism) in the neoliberal Western Murdochracies, Lobbyocracies and Corporatocracies with this leading to Effective Climate Change Denialism (Effective Climate Change Scepticism) and dangerous Climate Change Inaction in countries like Canada, Australia and the

USA. However US Secretary of State John Kerry has signalled a change in the US Democrat position in a recent speech in Djakarta, Indonesia: "But because of climate change, it is no secret that today, Indonesia is also one of the most vulnerable countries on Earth… First and foremost, we should not allow a tiny minority of shoddy scientists and science and extreme ideologues to compete with scientific fact… The science is unequivocal. And those who refuse to believe it are simply burying their heads in the sand" [30].

Climate Change Inaction (see Climate Change Action, Climate Change Denialism and Climate Change Scepticism). The Synthesis Report from the 2,500-delegate, March 2009 Copenhagen Climate Change Scientific Conference concluded "Inaction is inexcusable" [31]. A 2010 Open Letter by 255 members of the prestigious US National Academy of Sciences (including 11 Nobel Laureates) concluded "Delay is not an option" [32]. The World is rapidly running out of time and money to deal with the worsening climate crisis - thanks to climate criminals the Carbon Debt cost for future generations to restore a safe planet is now as much as $250 trillion and increasing by about $13 trillion each year [2020 estimates].

Climate Change-related Deaths. DARA has estimated 0.5 million climate change-related deaths each year [10, 11]. However this may be a considerable under-estimate because it can be readily determined from UN Population Division data [2020] that 15 million people die avoidably each year in a Developing World that is increasingly impacted by climate change (it was estimated that total global annual avoidable deaths in 2003 totalled 16 million [1]).

Climate Change Scepticism (see Climate Change Denialism).

Climate Corruption. Worsening GHG pollution means continuing pollution of the one common atmosphere and one common ocean of all countries i.e. it is theft of common resources. This dangerous theft is compounded by the absence of a Carbon Price due to fossil fuel lobbying in the neoliberal Murdochracies, Lobbyocracies and Corporatocracies in which

Big Money buys people, politicians, parties, policies, public perception of reality and political power in utter perversion of public life and democracy i.e. egregious corruption that also involves murderous depraved indifference toward the 10 million people who presently perish each year due to climate change (1 million) and carbon burning pollutants (9 million) [10, 11]. Thanks to climate criminals the Carbon Debt cost for future generations to restore a safe planet is now $250 trillion and increasing by about $13 trillion each year.

Climate Cost (see Climate Change Cost). The annual cost of climate change can be estimated at (A) $13 trillion (from non-application of a Carbon Price of $200 per tonne CO_2), and (B) $74 trillion (from a US VOSL of $7.4 million per person applied to 10 million annual deaths from carbon burning and climate change and assuming that "all men are created equal". Thanks to climate criminals the Carbon Debt cost for future generations to restore a safe planet is now $250 trillion and increasing by about $13 trillion each year [2020].

Climate Crimes. The current Climate Crimes involve (1) illegal, remorseless, corrupt, charge-free pollution of the one common atmosphere and ocean of all countries; (2) depraved indifference to 10 million annual deaths from climate change and carbon burning; (3) threat to vulnerable Humanity (climate racism, climate terrorism); (4) worsening damage to the Biosphere (the extinction rate is already 100-1,000 times greater than normal [33], most coral and coral reefs have only 20 years to go before the tipping point of irreversible decline [34]; (5) all species and ecosystems are priceless; and (6) worsening Climate Genocide that will see 10 billion mostly Third World people perish this century if climate change is not properly addressed [35]. Thanks to climate criminals the Carbon Debt cost for future generations to restore a safe planet is now $250 trillion and increasing by about $13 trillion each year.

Climate Criminal. Climate criminals are those who commit climate crimes (see Climate Crimes) and include corporate leaders and complicit politicians "persuaded" into climate change inaction and complicity in climate criminality. Thanks to climate criminals the Carbon Debt cost for future generations to restore a safe planet is now $250 trillion and increasing by about $13 trillion each year.

Climate Crisis. With zero years left at current rates of GHG pollution before the World exceeds the Terminal Carbon Budget of 600 Gt CO_2 [22] (as perceived in 2020), energy-related GHG pollution at a record high [36], the World running at the worst-case scenario of the IPCC [37, 38], the Arctic summer sea ice set to disappear in decades [39], over 5,025 Gt CO_2-e of methane set to be released in coming decades [40], atmospheric CO_2 at 415 ppm (2020) and increasing at a record rate [41], we have a Climate Crisis and a Climate Emergency that are made worse by World governments not in effect recognizing the crisis and emergency [42, 43]. Thanks to climate criminals the Carbon Debt cost for future generations to restore a safe planet is now a stunning $250 trillion and increasing by about $13 trillion each year.

Climate Denialism (see Climate Change Denialism, Climate Change Scepticism, and Climate Scepticism).

Climate Emergency (see Climate Crisis).

Climate Genocide. Under International Law the definition of "genocide" is by Article 2 of the UN Genocide Convention: it critically involves Assessment of "intent" by those responsible, and states "In the present Convention, genocide means any of the following acts committed with intent to destroy, in whole or in part, a national, ethnic, racial or religious group, as such: a) Killing members of the group; b) Causing serious bodily or mental harm to members of the group; c) Deliberately inflicting on the group conditions of life calculated to bring about its physical destruction in whole or in part; d) Imposing measures intended to prevent births within the group; e) Forcibly transferring children of the group to another group" [44]. "Intent" to commit mass murder is only rarely explicitly expressed but can be readily established through the horrendous, ongoing reality. Javier Sethness-Castro comments in relation to "intent" in relation to the related phenomenon of "climate genocide" that similarly arises from racist First World greed and depraved indifference: "Dominant relations can hence be characterized as governed by what Chomsky calls "depraved indifference" to human life. Australian scientist Gideon Polya has termed the current situation "climate genocide", while Bangladeshi climatologist Atiq Rahman similarly labels it "climatic genocide". The phrases are accurate if the word genocide is to be understood as murder of persons belonging to particular classes and social groups, as originally formulated by Raphael Lemkin, the concept's inventor. If the definition is extended to membership or residence in particular geographic regions – a collective of sorts – the term fits better, even if the question of intent for such eventualities is left unresolved: Under the internationally accepted definition, acts of genocide occur only if governed by conscious intent. Against this view, Chomsky is right to suggest that those concerned with such problems focus on "predictable outcome as evidence for intent". Not to work to undermine global capitalism is effectively to be complicit with the genocide of southern peoples. Jean-Paul Sartre put it well in a statement he issued as president of the International War Crimes Tribunal on Vietnam: "The genocidal intent is implicit in the facts. It is not necessarily premeditated." [45]. I have made a Formal Complaint to the International Criminal Court over Australian involvement in Climate Genocide and other genocidal atrocities (2008) [46].

Climate Holocaust. Holocaust simply means death of a huge number of people_as exampled by the WW2 Jewish Holocaust (5-6 million people killed by violence or deprivation), the WW2 Holocaust in general (30 million Slavs, Jews and Gypsies killed), the WW2 Bengali Holocaust (6-7 million Indians deliberately starved to death by the British with Australian complicity) and the 35 million Chinese killed associated with the Japanese occupation of China in the1930s and 1940s [2]. A predicted Climate Genocide involving the avoidable deaths of 10 billion people this century due to unaddressed climate change certainly represents a Climate Holocaust as well as a Climate Genocide (see Climate Genocide).

Climate Injustice (see Climate Justice, Intergenerational Injustice, and Intergenerational Justice). Unaddressed man-made climate change represents an enormous injustice to the vulnerable of the World, in particular those of the Developing World, about 10 billion of whom will die this century under Business As Usual (BAU). However there is also an immense intergenerational injustice as future generations in both the North and the South are bequeathed a mounting Carbon Debt of about \$13 trillion each year (as outlined above, a Carbon Price of about \$200 per tonne CO_2-e is required for effective climate change action [17]. World Bank analysts have revised upwards the annual global GHG pollution from 42 Gt CO_2-e to 64 Gt CO_2-e [27]. One can accordingly estimate the annual cost of climate change inaction (i.e. a Carbon Price of \$0 per tonne CO_2-e) at \$200 per tonne CO_2-e x 64 billion tonnes CO_2-e per year = \$13 trillion or about 15% of Global GDP of \$85 trillion). Dr James Hansen et al.: "We assess climate impacts of global warming using ongoing observations and paleoclimate data. We use Earth's measured energy imbalance, paleoclimate data, and simple representations of the global carbon cycle and temperature to define emission reductions needed to stabilize climate and avoid potentially disastrous impacts on today's young people, future generations, and nature. A cumulative industrial-era limit of ~500 GtC fossil fuel emissions and 100 GtC storage in the biosphere and soil would keep climate close to the Holocene range to which humanity and other species are adapted. Cumulative emissions of ~1000 GtC, sometimes associated with 2°C global warming, would spur "slow" feedbacks and eventual warming of 3–4°C with disastrous consequences. Rapid emissions reduction is required to restore Earth's energy balance and avoid ocean heat uptake that would practically guarantee irreversible effects. Continuation of high fossil fuel emissions, given current knowledge of the consequences, would be an act of extraordinary witting intergenerational injustice. Responsible policymaking requires a rising price on carbon emissions that would preclude emissions from most remaining coal and unconventional fossil fuels and phase down emissions from conventional fossil fuels" [47].

Climate Justice (see Climate Justice, Intergenerational Injustice, and Intergenerational Justice).

Climate Racism. The worsening Climate Genocide will overwhelmingly severely impact the non-European South rather than the European North and there is thus a climate racism dimension to the worsening climate genocide. Between 1950 and 2005 global avoidable deaths from deprivation totalled 1.3 billion on Spaceship Earth with the First World in charge of the flight deck and 1.2 billion of these deaths were associated with the non-European World and 0.6 billion with the Muslim World [2]. Today, about 15 million people die avoidably from deprivation every year in the Developing World (minus China, a nation for which the annual avoidable mortality is essentially zero) [2]. Unaddressed global warming will worsen this Developing World avoidable mortality holocaust with a worst case scenario of an annual average of 100 million avoidable deaths [35].

Climate Scepticism (see Climate Change Denialism, Climate Change Scepticism, and Climate Denialism).

Climate Terrorism. Terrorism is the ruthless killing of innocents in support of an ideological or political agenda. The agenda of the capitalist Establishment is maximum profit from a Carbon economy at the expense of the Biosphere and the bulk of Humanity as a whole that depends upon it - knowing, remorseless mass murder arising from a ruthless economic agenda i.e. climate terrorism.

Climate War. Climate war is a predictable outcome of the worsening resource depletion and worsening Climate Genocide. Indeed the violence associated with the Arab Spring can be seen as driven by economic desperation as well as by ideology.

Coal. Coal derives from anaerobic geologic conversion of cellulosic carbohydrates to carbon ($(CH_2O)_n$ + heat, pressure -> nC + $n\ H_2O$). Complete burning of coal (C) generates CO_2: $C + O_2$ -> CO_2.

Divestment (see Boycott, Divestment and Sanctions, BDS).

Divestment From Deforestation. Deforestation in the non-European World and land use in general (including food price-elevating, Carbon Debt-incurring and food-for-fuel agriculture, the Biofuel Genocide [48- 50)]) are now major contributors to GHG pollution [27, 51, 52].

Divestment From Fossil Fuels. The Western democracies have become Kleptocracies, Plutocracies, Murdochracies, Lobbyocracies, Corporatocracies and Dollarocracies in which Big Money buys people, politicians, parties, policies, public perception of reality and political power. This transformation has crippled effective action against climate change. Some key strategies that should be adopted by decent people and NGOs around the world are (a) to urge and apply Boycotts, Divestment and Sanctions (BDS) against all people, politicians, parties, candidates, countries and corporations involved in man-made climate change, and (b) to inform everyone they can about the worsening climate emergency and the need for urgent action. Top climate change economist Professor Lord Stern reporting about risks to investment in fossil fuels: "Smart investors can see that investing in companies that rely solely or heavily on constantly replenishing reserves of fossil fuels is becoming a very risky decision. The report raises serious questions as to the ability of the financial system to act on industry-wide long term risk, since currently the only measure of risk is performance against industry benchmarks" [53]. According to Carbon Tracker: "Between 60-80% of coal, oil and gas reserves of publicly listed companies are unburnable if the world is to have a chance of not exceeding global warming of $2^{o}C$. The total coal, oil and gas reserves listed on the world's stock exchanges equals 762Gt CO_2 – approximately a quarter of the world's total reserves [of 3,000 Gt CO_2]" [53]. Note that the 2010-2050 Terminal Carbon Budget that must not be exceeded if we are to have a 75% chance of avoiding a catastrophic 2C temperature rise is 600 Gt CO_2 [20, 21] – and has finally been used up in 2020. It gets worse – thus, the Global Warming Potential (GWP) of CH_4 on a 20 year time frame and with aerosol impacts considered is 105 times that of CO_2 [55, 56]. The 50 Gt (billion tonnes) CH_4 to be released from the East Siberian Arctic Shelf in coming decades [40] is thus equivalent to 50 billion tonnes CH_4 x 105 tonnes CO_2-equivalent/tonne CH_4 = 5,250 tonnes CO_2-e or about nine (9) times more than the world's terminal greenhouse gas (GHG) pollution budget. We are doomed unless we can stop this Arctic CH_4 release [28]. "Divest from fossil fuels" [54] details towns, cities, countries, colleges, universities, religious organizations, other organizations and individuals committed to divestment from fossil fuel companies.

Divestment from Greenhouse Gas Pollution (see Divestment From Fossil Fuels, Vegan and Vegetarian).

Effective Climate Change Denialism (see Effective Climate Change Scepticism). This term is well illustrated by climate criminal Australia where the major political parties, the governing Liberal Party-National Party Coalition and the opposition Labor Party (collectively also known as the Lib-Labs or Liberal-Laborals) variously pay lip service to climate change but have a common policy of climate change inaction involving a derisory 5% off 2000 GHG pollution by 2020 coupled with unlimited coal, gas and iron ore exports that will see Australia exceed the whole World's Terminal Carbon Budget of 600 Gt CO_2 by a factor of 3 [25].

Effective Climate Change Scepticism (see Effective Climate Change Denialism).

Emissions Trading Scheme (ETS) (see Cap and Trade). This is basically a fraudulent device to allow particular countries to set a notional cap on greenhouse gas (GHG) pollution, and then trade limited and dodgy Carbon Credits and limited and dodgy licences to pollute the one common atmosphere and one common ocean of all countries. The ETS approach (a) is empirically unsuccessful, (b) is accordingly counterproductive and (c) is fraudulent because it involves the government of a particular country selling its corporations licences to pollute the one common atmosphere and ocean of all countries on earth. Numerous science and economics experts dismiss the ETS approach and demand an effective Carbon Price through a Carbon Tax [57]. Thus Dr Chris Hope from 90-Nobel-Laureate Cambridge University demands a Carbon Tax of about \$200 per tonne CO_2-e [17], and Dr James Hansen (former head of NASA's Goddard Institute for Space Studies and Adjunct Professor at 101-Nobel-Laureate Columbia University) argues cogently for an upfront, transparent and effective Carbon Tax of about \$200 per tonne CO_2 with the receipts going to all citizens [19, 58].

Excess Death (see Excess Mortality, Avoidable Death, Avoidable Mortality). Excess Death (Excess Mortality, Avoidable Death, Avoidable Mortality) is the difference between the actual mortality in a country and the mortality expected in a peaceful, decently-run country with the same demographics. By 1950 all the World potentially had access to the requisites for the very low excess death obtaining in European countries, namely clean water, sanitation, proper nutrition, literacy (especially female literacy), primary health care, antibiotics and major preventive medicine programs including public health education, prophylactics (such as insecticides, antiseptics, mosquito netting, soap and condoms) and major vaccinations. However such benefits took decades to arrive in many countries and are still variously lacking in African countries. Nevertheless, in most countries outside Africa the annual mortality rate (expressed as deaths per 1,000 people per year) typically declined to a minimum and in the best countries (typically European and East Asian countries) eventually began to rise, with this reflecting aging populations. Using United Nations Population Division demographic data going back to 1950 (the 2002 Revision), it was possible to make base-line estimates of expected mortality for various demographically distinct groups. For many high birth-rate Third World countries the base-line mortality estimates clustered about 4 deaths per 1,000 of population [2].

Excess Mortality (see Excess Death, Avoidable Death, Avoidable Mortality).

Gas (see Natural Gas). Subterranean methane derives from anaerobic reduction of carbohydrates by anaerobic bacteria (reduction being addition of electrons (e-), addition of hydrogen atoms (H) or removal of oxygen (O)): $(CH_2O)_n + 4H$ (derived from catabolism and reduced coenzymes) $->$ $nCH_4 + nH_2O$ [1]. Methane (CH_4) (about 85% of natural gas and deriving from anaerobic bacterial action on biomass) is 105 times worse than CO_2 as a greenhouse gas (GHG) on a 20 year time frame and taking aerosol impacts into account. Methane leaks (3% in the US based on the latest US EPA data and as high as 7.9% for methane from "fracking" or hydrologically fracturing geological seams). Using this information one can determine that gas burning for electricity can be much dirtier than coal burning greenhouse gas-wise (GHG-wise). While gas burning for power generates twice as much electrical energy per tonne of CO_2 produced (MWh/tonne CO_2) than coal burning, and the health-adverse pollution from gas burning is lower than for coal burning, 3% gas leakage in the system actually means that gas burning for power is worse GHG-wise than coal burning either on a per mass basis or a per energy yield basis. At 2.6% gas leakage, the global warming effect from the leaked gas is that same as that from burning the remaining gas (see the website "Gas is not clean energy") [1].

Global Warming Potential (GWP). Relative GWP based on infra-red (IR) absorbance properties and half-life in atmosphere is directly proportional to radiative forcing and inversely proportional to half-life in the atmosphere. GWP relative to same mass of CO_2 on a 20 year time frame: CO_2 (1.0), CH_4 (79; 105 if aerosol impacts are considered [55, 56]), N_2O (289), chlorofluorohydrocarbon CFC-12, CCl_2F_2 (11,000), hydrochlorofluorohydrocarbon HFC-22, $CHClF_2$ (5,160), hydrofluorohydrocarbon HFC-23, CHF_3 (12,000), sulphur hexafluoride SF_6 (16,300), and nitrogen trifluoride NF_3 (12,300). GWP relative to same mass of CO_2 (1.0) on a 100 year time frame: CO_2 (1.0), CH_4 (21), N_2O (298), chlorofluorohydrocarbon CFC-12, CCl_2F_2 (10,900), hydrochlrofluorohydrocarbon HFC-22, $CHClF_2$ (1,810), hydrofluorohydrocarbon HFC-23, CHF_3 (14,800), sulphur hexafluoride SF_6 (22,800), and nitrogen trifluoride NF_3 (17,200) [1].

Greenhouse Gas (GHG). The greenhouse effect (discovered by UK chemist John Tyndall, 1858) involves thermal radiation from the sun being absorbed by surface and air with re-emitted and reflected light being absorbed and re-radiated by air molecules, notably carbon dioxide (CO_2), nitrous oxide (N_2O), methane (CH_4), H_2O, and man-made greenhouse gases (GHGs). The greenhouse effect keeps planet warm. Note that the term CO_2-equivalent (CO_2-e) refers to total GHGs including CO_2 and other GHGs (excepting H_2O) expressed as CO_2-equivalent.

Greenhouse Gas Pollution (GHG Pollution). World Bank analysts have re-assessed the contribution of livestock to World annual GHG pollution. Their re-assessment is that annual World GHG pollution is 64 Gt CO_2-e, 50% higher than the previous estimate of 42 Gt CO_2-e and with methanogenic livestock raising and attendant land use contributing over 51% of the bigger figure [27]. CO_2 is the major contributor to Anthropogenic Global Warming (AGW), deriving from aerobic respiration involving oxidation of carbohydrate ($(CH_2O)_n + O_2$ -> n CO_2 + nH_2O), lime (CaO) in cement production ($CaCO_3$ -> $CaO + CO_2$), and the combustion of fossil fuels such as coal (C + O_2 -> CO_2), oil ($CH_3(CH_2)nH + (n + 1)O_2 + (n + 2)/2\ O_2$-> $(n+1)CO_2 + (n+2)H_2O$) and natural gas, mainly methane ($CH_4 + 2O_2$ -> $CO_2 + 2\ H_2O$) [1].

Intergenerational Equity (see Climate Injustice, Climate Justice, Intergenerational Inequity, Intergenerational Injustice, Intergenerational Justice).

Intergenerational Inequity (see Climate Injustice, Climate Justice, Intergenerational Equity, Intergenerational Injustice, Intergenerational Justice). Profiting from pollution and leaving the mess for future generations to clean up is Climate Injustice and Intergenerational Inequity. Dr James Hansen: "Cap and trade with offsets, in contrast, is astoundingly ineffective. Global emissions rose rapidly in response to Kyoto, as expected, because fossil fuels remained the cheapest energy. Cap and trade is an inefficient compromise, paying off numerous special interests. It must be replaced with an honest approach, raising the price of carbon emissions and leaving the dirtiest fossil fuels in the ground. Are we going to stand up and give global politicians a hard slap in the face, to make them face the truth? It will take a lot of us – probably in the streets. Or are we going to let them continue to kid themselves and us and cheat our children and grandchildren? Intergenerational inequity is a moral issue. Just as when Abraham Lincoln faced slavery and when Winston Churchill faced Nazism, the time for compromises and half-measures is over. Can we find a leader who understands the core issue and will lead?"[58].

Intergenerational Injustice (see Climate Injustice, Climate Justice, Intergenerational Equity, Intergenerational Inequity, Intergenerational Justice). The present generations are drastically diminishing the resources of the planet – notably the forests, fisheries, atmosphere and ocean of the Earth – to the detriment of future generations. Climate change inaction means that in the absence of a carbon price (circa $150 per tonne CO_2-e) we are generating a Carbon Debt of future generations of $9.6 trillion each year. Indeed the Carbon Debt inherent in the extra CO_2 in the atmosphere is $250 trillion or about 3 times the World's GDP [1]. It can be estimated that for every $1 received for coal by the Australian coal industry our children, grandchildren and further generations will have to spend $2 to $3 at present prices converting the consequent CO_2 back to biochar to save the Planet. One can accordingly similarly determine (for Australia) that the biochar production-related debt for future generations for every $1 received by the gas-based electricity industry will be about $0.6-$1.0 at present prices (these estimates must be multiplied by 4 for the cost of biochar production in the UK) - Intergenerational Injustice indeed [29].

Intergenerational Justice (see Climate Injustice, Climate Justice, Intergenerational Equity, Intergenerational Inequity, Intergenerational Injustice).

Natural Gas (see Gas).

Ocean Acidification. Dissolving CO_2 in the ocean results in acidification. The ocean pH has already dropped by 0.1 pH unit, with this threatening all ocean organisms with calcareous exoskeletons. About one-third of the carbon dioxide (CO_2) released into the atmosphere as a result of human activity has been absorbed by the oceans, where it partitions into the constituent ions of carbonic acid, this leading to ocean acidification, one of the major threats to marine ecosystems and particularly to calcifying organisms such as corals (animals with photosynthetic symbionts), foraminifera (single-celled eukaryote amoeboid protists with a calcareous external shell), coccolithophores (photosynthetic algae that generate calcareous plates) and lobster, crab, shrimp and krill crustaceans with major food chain implications) [1].

Ocean Warming. While average surface temperature has increased by +1.1C [2020] relative to 1900, 90% of the extra heat has gone into the ocean. At the current 415ppm CO_2 in the atmosphere most coral is facing irreversible decline from ocean warming and acidification [1].

Oil Oil derives from decarboxylation of biological fatty acids (FAs) over geological time spans. Subterranean oil derives from anaerobic decarboxylation of biologically-derived fatty acids thus: $CH_3(CH_2)_n\text{-}COOH$ + heat, pressure -> $CH_3(CH_2)_nH$ + CO_2 [1].

Per Capita Greenhouse Gas Pollution. Collective, national responsibility for the worsening Climate Holocaust and Climate Genocide is in direct proportion to per capita national pollution of the atmosphere with greenhouse gases (GHGs). Indeed, fundamental to any international agreement on national rights to pollute our common atmosphere and oceans should be the belief that "all men are created equal". However neoliberal-imposed reality is otherwise: "annual per capita greenhouse gas (GHG) pollution" in units of "tonnes CO_2-equivalent per person per year" (2005-2008 data) is 0.9 (Bangladesh), 0.9 (Pakistan), 2.2 (India), less than 3 (many African and Island countries), 3.2 (the Developing World), 5.5 (China), 6.7 (the World), 11 (Europe), 16 (the Developed World), 27 (the US) and 25 (Australia; or 74 in 2010 if Australia's huge Exported CO_2 pollution is included) [1, 35, 59] [these estimates ignoring land use and the GWP of CH_4 on a 20 year time frame].

Preventable Deaths. Avoidable Mortality (Avoidable Mortality, Excess Death, Excess Mortality) is the difference between the actual mortality in a country and the mortality expected in a peaceful, decently-run country with the same demographics. However it is useful to use the term "Preventable Deaths" for intra-national avoidable deaths in prosperous, First World countries such as Australia and the US that have effectively zero avoidable mortality on the international scale of comparison as defined above but in which annual preventable deaths from "lifestyle choice" and "political choice" reasons total [in 2020] about 85,000 [60, 61] and 1.7 million [62, 63], respectively. Thus eminently preventable carbon burning-related deaths in Australia and the US annually total 10,000 [9, 60] and 70,000 [62], respectively.

Radiative Forcing. Radiative forcing measures warming (positive) effects of e.g. GHGs and carbon particles, and cooling (negative) effects of e.g. sulphate aerosols and the reflectivity (albedo) of ice, snow and clouds [1].

Renewable Energy. 100% Renewable Energy (including geothermal energy) is crucial for effectively dealing with the climate emergency and returning atmospheric CO_2 from the current dangerous 415 ppm CO_2 (2020) to a safe and sustainable 300 ppm CO_2. Crucial are the following: (1) Change of societal philosophy to one of scientific risk management and biological sustainability with complete cessation of species extinctions and zero tolerance for lying; (2) Urgent reduction of atmospheric CO_2 to a safe level of about 300 ppm as recommended by leading climate and biological scientists; and

(3) a rapid switch to the best non-carbon and renewable energy
(solar PV, solar thermal, wind, geothermal, wave, tide and
hydro options that in 2020 are cheaper than coal burning-based
power) and to energy efficiency, public transport, needs-based
production, re-afforestation and return of carbon as biochar to
soils coupled with correspondingly rapid cessation of fossil fuel
burning, deforestation, methanogenic livestock production and
population growth [1, 28]. The goal of "100% renewable energy
by 2020" and the related goal of "Cut carbon emissions 80% by
2020" have already been embraced by various Island nations,
states, cities and corporations [64, 65].

**Sanctions (see Boycott, Boycott, Divestment and
Sanctions (BDS), Divestment, Divestment From
Deforestation, Divestment From Fossil Fuels).** Peace is the
only way but inaction is complicity. Sanctions and indeed
Boycott, Divestment and Sanctions (BDS) are required to stop
the climate criminals destroying the planet [54].

**Terminal Carbon Budget (see Terminal Carbon Pollution
Budget).** The World is rapidly running out of time to deal with
man-made climate change. Thus according to the German
WBGU and Australian Climate Commission (sacked by the
effective climate change denialist Coalition Federal
Government and now surviving privately funded as the
Australian Climate Council), relative to 2014 we have only 15
years left before we exceed the Terminal Carbon Budget of 600
Gt CO_2 that must not be exceeded if we are to have a 75%
chance of avoiding a catastrophic 2C temperature rise [20, 21].
As of 2020 the World's Terminal Carbon Pollution Budget has
been used up.

**Terminal Carbon Pollution Budget (see Terminal Carbon
Budget).**

Tipping Points. The World is rapidly approaching crucial
tipping points for irreversible damage to key ecosystems [19,
66]. Thus scientists say that the Arctic summer sea ice may be
gone in coming decades [39]. Whiteman and colleagues at 121-
Nobel-Laureate Cambridge University say that 50 Gt methane

will be released from the East Siberian Arctic sea bed in coming decades, this translating to a catastrophic 5,250 Gt CO_2-e, 9 times greater than the Whole World's Terminal Carbon Budget [40]. In 2020 the World finally exceeded the Terminal Carbon Budget of 600 Gt CO_2 that must not be exceeded if we are to have a 75% chance of avoiding a catastrophic 2C temperature rise [20 - 22]. The atmospheric CO_2 has reached 415 ppm (2020) and is increasing at about 2 ppm per year i.e. within 20 years it will reach 450 ppm CO_2 at which point most coral and coral reefs are doomed – indeed top coral experts of the UK Royal Society Coral Working Group say that at about 400 ppm CO_2 most coral is facing irreversible decline from ocean warming and acidification [67]. The species extinction rate is already 100-1,000 times greater than normal [33].

Total Economic Value of Wild Nature. The Total Economic Value (TEV) of Wild Nature may be about 50% of World GDP, and Andrew Balmford and colleagues have estimated that: "Loss and degradation of remaining natural habitats has continued largely unabated. However, evidence has been accumulating that such systems generate marked economic benefits, which the available data suggest exceed those obtained from continued habitat conversion. We estimate that the overall benefit:cost ratio of an effective global program for the conservation of remaining wild nature is at least 100:1" [68].

Unknown (see X).

Value Of a Statistical Life (VOSL, VSL). Climate change and carbon burning already kill about 10 million people annually. Assuming that "all men are created equal" we can use the US EPA's estimate of a Value Of a Statistical Life (VOSL, VSL) of $7.4 million to estimate a risk avoidance based cost of $74 trillion each year, nearly half the World's annual GDP.

Vegan (see Divestment from Greenhouse Gas Pollution, Vegetarian). World Bank analysts have re-assessed the contribution of livestock to World annual GHG pollution. Their re-assessment is that annual World GHG pollution is 64 Gt CO_2-e, 50% higher than the previous estimate of 42 Gt CO_2-e and with methanogenic livestock raising and attendant land use contributing over 51% of the bigger figure [27]. The conversion efficiency (kg grain to produce 1 kg gain in live weight) is: herbivorous farmed fish (e.g. carp, tilapia, catfish; less than 2), chicken (2), pork (4), and beef (7). In 2003, 37 percent of the world grain harvest, or nearly 700 million tons, was used to produce animal protein. Meat use is contraindicated because of GHG pollution and inefficient diversion of scarce food resources in addition to the "food for fuel" Biofuel Genocide obscenity that drives up grain price [1, 69, 70].

Vegetarian (see Divestment from Greenhouse Gas Pollution, Vegan).

World's Worst for GHG Pollution. "Annual per capita greenhouse gas (GHG) pollution" in units of "tonnes CO_2-equivalent per person per year" (2005-2008 data) is 0.9 (Bangladesh), 0.9 (Pakistan), 2.2 (India), less than 3 (many African and Island countries), 3.2 (the Developing World), 5.5 (China), 6.7 (the World), 11 (Europe), 16 (the Developed World), 27 (the US) and 25 (Australia; or 74 in 2010 if Australia's huge Exported CO_2 pollution is included) [1, 35, 59]. An estimate of "years left to zero emissions" has been determined for all countries relative to mid-2011 – the worst polluting countries that must cease GHG pollution within 10 years relative to mid-2011 are Belize (1.3 years), Qatar (2.3), Guyana (2.4), Malaysia (3.4), United Arab Emirates (3.4), Kuwait (4.1), Papua New Guinea (4.3), Brunei (4.8), Australia (4.8), Antigua & Barbuda (4.9), Zambia (5.1), Canada (5.1), Bahrain (5.2), United States (5.5), Trinidad & Tobago (6.4), Luxembourg (5.9), Panama (6.3), New Zealand (6.5), Estonia (6.9), Botswana (7.0), Ireland (7.4), Saudi Arabia (7.6), Venezuela (7.9), Indonesia (8.4), Equatorial Guinea (8.6), Belgium (8.7), Turkmenistan (8.8), Singapore (8.9), Czech Republic (9.0), Liberia (9.0), Netherlands (9.3), Russia (9.3), Nicaragua (9.3), Finland (9.5), Oman (9.7), Palau (9.8), Brazil (9.8), Uruguay (9.8), Denmark (10.0). If one considers Australia's Domestic plus Exported GHG pollution, Australia used up its "fair share" of the World's Terminal Carbon Budget in 2011 (9 years ago) [26].

World's Worst for Climate Inaction. The Climate Change Performance Index – Results 2014 [71] shows that of 61 countries examined the following were the worst: Australia (57), Canada (58), Iran (59), Kazakhstan (60), and Saudi Arabia (61). Australia and Canada were the worst OECD countries. In 2020 the US was worst for performance and Australia was the worst for climate policy [72].

X (see Unknown). X stands for an unknown. The danger of being so close to various tipping points is that some unpredictable but not unexpected disaster (e.g. huge volcanic eruptions, catastrophic forest fires, changes in ocean currents or atmosphere jet stream patterns) might suddenly inject a lot more GHGs into the atmosphere and bring the world past particular tipping points. Thus, for example, the Global Warming Potential (GWP) of CH_4 on a 20 year time frame and with aerosol impacts considered is 105 times that of CO_2 [55, 56]. The 50 Gt (billion tonnes) CH_4 to be released from the East Siberian Arctic Shelf in coming decades [40] is thus equivalent to 50 billion tonnes CH_4 x 105 tonnes CO_2-equivalent/tonne CH_4 = 5,250 tonnes CO_2-e or about nine (9) times more than the world's terminal greenhouse gas (GHG) pollution budget. We are doomed unless we can stop this Arctic CH_4 release [28].

Years Left To Zero Emissions (see World's Worst for GHG Pollution). Recent estimates of years left before the World exceeds the Terminal Carbon Pollution Budget range from 0-15 years [20-23]. If one considers Australia's Domestic plus Exported GHG pollution, Australia used up its "fair share" of the World's Terminal Carbon Budget in 2011 (3 years ago) [26].

Zero Emissions (see Years Left To Zero Emissions). The world must finally reach zero emissions in about 2050 but many highly polluting countries will have already exceeded their "fair share" of the World's Terminal Carbon Budget in the first decade of the 21 st century [26]. If one considers Australia's Domestic plus Exported GHG pollution, Australia used up its "fair share" of the World's Terminal Carbon Budget in 2011 (9 years ago) [1] and is now stealing the entitlement of all other countries on Earth – climate criminality and climate injustice. The World used up this budget in 2020.

2020 POSTSCRIPT

This chapter sets out the state of play 5-10 years ago and has involved some editing and updating. However in recent years things have got much worse in the perception of ordinary citizens, especially in relation to catastrophic biodiversity loss, loss of ice from glaciers, loss of sea ice, drought, forest fires, tropical storms, sea level rise, storm surges and elevated urban temperatures (circa 50 degrees C in some Indian cities). Scientists are typically conservative and qualified in their assessments but one now senses a new sense of urgency in their public statements in the face of a worsening Climate Emergency and Climate Crisis. Thus, for example, some very eminent climate scientists in a 2019 paper on irreversible "tipping points" concluded: "Act now. In our view, the evidence from tipping points alone suggests that we are in a state of planetary emergency: both the risk and urgency of the situation are acute… We argue that the intervention time left to prevent tipping [T] could already have shrunk towards zero, whereas the reaction time to achieve net zero emissions [τ] is 30 years at best. Hence we might already have lost control of whether tipping happens. A saving grace is that the rate at which damage accumulates from tipping — and hence the risk posed — could still be under our control to some extent. The stability and resilience of our planet is in peril. International action — not just words — must reflect this" (Timothy Lenton, Johan Rockstrom, Owen Gaffney, Stefan Rahmsdorf, Katherine Richardson, Will Steffen and Hans Joachim Schellnhuber, "Climate tipping points – too risky to bet against", Nature 575, 592-595, 27 November 2019: https://www.nature.com/articles/d41586-019-03595-0).

While outright climate change denial seems to be on the retreat, it is being replaced by effective climate change denial through resolute political inaction. Thus while President Donald Trump (who has uttered over 20,000 falsehoods in the last 4 years) appears to have retreated from his assertion that climate science is a "hoax", he remains a resolute and powerful supporter of expanded fossil fuel exploitation. Ditto the highly influential Murdoch media empire. Indeed while former president Barack Obama was rational, science-informed and a strong advocate for climate change action, he nevertheless presided over a massive coal-to-gas transition contra-indicated by science – at a systemic natural gas leakage of about 3% (as in the US) burning gas can be seen as dirtier greenhouse gas (GHG)-wise than burning coal either on a per mass basis or a per energy yield basis ("Gas is not clean energy": https://sites.google.com/site/gasisnotcleanenergy/.)

References.

[1]. Gideon Polya, "2011 Climate Change Course": https://sites.google.com/site/300orgsite/2011-climate-change-course.
[2]. Gideon Polya, "Body Count. Global avoidable mortality since 1950", that includes an avoidable mortality-related history of every country since Neolithic times and is now available for free perusal on the web: http://globalbodycount.blogspot.com.au/.
[3]. Gideon Polya, "Forest biomass-derived Biochar can profitably reduce global warming and bushfire risk", Yarra Valley Climate Action Group: http://sites.google.com/site/yarravalleyclimateactiongroup/forest-biomass-derived-biochar-can-profitably-reduce-global-warming-and-bushfire-risk.
[4]. Gideon Polya, editor, 300.org: https://sites.google.com/site/300orgsite/300-org.
[5]. Gideon Polya, editor, "300.org – return atmosphere CO_2 to 300 ppm": https://sites.google.com/site/300orgsite/300-org---return-atmosphere-co2-to-300-ppm.
[6]. 350.org: http://350.org/.
[7]. Gideon Polya, editor, "Boycott Murdoch Media": https://sites.google.com/site/boycottmurdochmedia/.
[8]. Gideon Polya, editor, "Boycott Apartheid Israel": https://sites.google.com/site/boycottapartheidisrael/.
[9]. Gideon Polya, "Australian carbon burning-related deaths and carbon burning subsidies => minimum Carbon Price of A\$555 per tonne carbon (C) or A\$150 per tonne CO_2-e ", Yarra Valley Climate Action Group: http://sites.google.com/site/yarravalleyclimateactiongroup/2011-carbon-burning.
[10]. DARA, "Climate Vulnerability Monitor. A guide to the cold calculus of a hot planet", 2012, Executive Summary pp2-3: http://daraint.org/climate-vulnerability-monitor/climate-vulnerability-monitor-2012/.
[11]. DARA report quoted by Reuters,"100 mln to die by 2030 if world fails to act on climate", 28 September 2012: http://in.reuters.com/article/2012/09/26/climate-inaction-idINDEE88P05P20120926.
[12]. Gideon Polya, "Expert Witness Testimony To Stop Gas-Fired Power Plant Installation", Countercurrents, 14 June, 2013: http://www.countercurrents.org/polya140613.htm.
[13]. Natural gas.org, "Natural gas and the environment": http://www.naturalgas.org/environment/naturalgas.asp.

[14]. WHO, "Air quality and health":
http://www.who.int/mediacentre/factsheets/fs313/en/.
[15]. Gideon Polya, editor, "Carbon Debt Carbon Credit":
https://sites.google.com/site/carbondebtcarboncredit/.
[16]. James Hansen,"Letter to PM Kevin Rudd by Dr James Hansen",
2008:
http://www.aussmc.org.au/documents/Hansen2008LetterToKevinRud
d_000.pdf.
[17]. Dr Chris Hope, "How high should climate change taxes be?",
Working Paper Series, Judge Business School, University of
Cambridge, 9.2011:
http://www.jbs.cam.ac.uk/media/assets/wp1109.pdf.
[18]. Gideon Polya, "Australian Coalition Government Sacrifices Key
Industries While Committing Hundreds Of Billions Of Dollars To
Carbon Pollution & War", Countercurrents, 10 February 2014:
http://www.countercurrents.org/polya100214.htm.
[19]. James Hansen, "Global warming twenty years later: tipping
points near", Address to the US Congress, 2008:
http://www.columbia.edu/~jeh1/2008/TwentyYearsLater_20080623.p
df.
[20]. 2009 Report of the German Advisory Council on Global Change
(WBGU, Wissenshaftlicher Beirat der Bundesregierung Globale
Umweltveränderungen) was entitled "Solving the climate dilemma:
the budget approach":
http://www.wbgu.de/fileadmin/templates/dateien/veroeffentlichungen/
sondergutachten/sn2009/wbgu_sn2009_en.pdf.
[21]. Australian Climate Commission, "The critical decade 2013: a
summary of climate change science, risks and responses", 2013, p7:
http://climatecommission.gov.au/wp-content/uploads/The-Critical-
Decade-2013-Summary_lowres.pdf.
[22]. Gideon Polya, "Doha climate change inaction. Only 5 years left
to act", MWC News, 9 December 2012:
http://mwcnews.net/focus/analysis/23373-gideonpolya-climate-
change.html.
[23]. "Cut carbon emissions 80% by 2020":
https://sites.google.com/site/cutcarbonemissions80by2020/.
[24]. Bill McKibben, "Global warming's terrifying new math",
Rolling Stone, 19 July 2012:
http://www.rollingstone.com/politics/news/global-warmings-
terrifying-new-math-20120719.
[25]. Gideon Polya, "Australia 's Huge Coal, Gas & Iron Ore Exports
Threaten Planet", Countercurrents, 15 May 2012:
http://www.countercurrents.org/polya150512.htm.

[26]. Gideon Polya, "Country By Country Analysis Of Years Left Until Science-demanded Zero Greenhouse Gas Emissions", Countercurrents, 11 June 2011: http://www.countercurrents.org/polya110611.htm.
[27]. Robert Goodland and Jeff Anfang. "Livestock and climate change. What if the key actors in climate change are … cows, pigs and chickens?", World Watch, November/December 2009: http://www.worldwatch.org/files/pdf/Livestock%20and%20Climate%20Change.pdf.
[28]. "Are we doomed? Too late to save earth?", 300.org: https://sites.google.com/site/300orgsite/are-we-doomed.
[29]. Gideon Polya, "Expert Witness Testimony To Stop Gas-Fired Power Plant Installation", Countercurrents, 14 June 2013: http://www.countercurrents.org/polya140613.htm.
[30]. John Kerry, "Remarks on climate change", US Department of State, 16 February 2014: http://www.state.gov/secretary/remarks/2014/02/221704.htm.
[31]. Synthesis Report from the March 2009 Copenhagen Climate Change Conference, Climate Change, Global risks, challenges & decisions", Copenhagen 10-12 March, 2009, University of Copenhagen, Denmark: http://lyceum.anu.edu.au/wp-content/blogs/3/uploads//Synthesis%20Report%20Web.pdf.
[32]. 2010 Open Letter by 255 members of the US National Academy of Sciences, "Open Letter: climate change and the integrity of science", Guardian, 6 May 2010: http://www.guardian.co.uk/environment/2010/may/06/climate-science-open-letter.
[33]. Phillip Levin, Donald Levin, "The real biodiversity crisis", American Scientist, January-February 2002: http://www.americanscientist.org/issues/pub/the-real-biodiversity-crisis.
[34]. Selina Ward, Ove Hoegh-Guldberg [preface by Anna Rose], "Lights out for the reef", prepared for Earth Hour, 8.30 pm Australian EST, Saturday 22 March 2014 [Saturday 29 March], WWF: http://earthhour.org.au/LIGHTS_OUT_FOR_THE_REEF_Earth_Hour_2014.pdf.
[35]. Gideon Polya, editor, "Climate Genocide": https://sites.google.com/site/climategenocide/.
[36]. US EIA, "Energy-related carbon dioxide emissions", Table A10: http://www.eia.gov/forecasts/ieo/emissions.cfm.
[37]. Intergovernmental Panel of Climate Change (IPCC) Fourth Assessment report: Climate Change 2007 (AR4): http://www.ipcc.ch/publications_and_data/publications_and_data_reports.shtml#.T-ZXUZHiK9K.

[38]. IPCC Fourth Assessment report: Climate Change 2007: Synthesis report for Policy Makers (2007): http://www.ipcc.ch/publications_and_data/ar4/syr/en/spm.html.

[39]. Professor Peter Wadhams quoted in Julia Horton, "Arctic sea ice will vanish within three years, says expert ", Scotsman, 29 August 2012: http://www.scotsman.com/news/environment/arctic-sea-ice-will-vanish-within-three-years-says-expert-1-2493681

[40]. Gail Whiteman, Chris Hope and Peter Wadhams, "Vast costs of Arctic change", Nature, 499, 25 July 2013: http://www.nature.com/nature/journal/v499/n7459/pdf/499401a.pdf and http://www.nature.com/nature/journal/v499/n7459/full/499401a.html.

[41]. "Trends in Atmospheric Carbon Dioxide": http://www.esrl.noaa.gov/gmd/ccgg/trends/.

[42]. Jorgen Randers, "Systematic short-termism: Climate, capitalism and democracy", Climate Code Red, 2012: http://www.climatecodered.org/2012/11/systematic-short-termism-climate.html?utm_source=feedburner&utm_medium=email&utm_campaign=Feed%3A+ClimateCodeRed+%28climate+code+red%29.

[43]. David Spratt and Phillip Sutton. "Climate Code Red. The case for emergency action": http://scribepublications.com.au/books-authors/title/climate-code-red/.

[44]. UN Genocide Convention: http://www.edwebproject.org/sideshow/genocide/convention.html.

[45]. Javier Sethness-Castro, "Imperiled life: revolution against climate catastrophe", AK Press, 2012: http://www.amazon.com/Imperiled-Life-Revolution-Catastrophe-Interventions/dp/1849351058.

[46]. Gideon Polya, Formal complaint to the Chief Prosecutor of the International Criminal Court re Australian Government involvement in Aboriginal Genocide, Iraqi Genocide, Afghan Genocide and Climate Genocide: http://climateemergency.blogspot.com/2008_02_01_archive.html.

[47]. James Hansen, Pushker Kharecha, Makiko Sato, Valerie Masson-Delmotte, Frank Ackerman, David J. Beerling, Paul J. Hearty, Ove Hoegh-Guldberg, Shi-Ling Hsu, Camille Parmesan, Johan Rockstrom, Eelco J. Rohling, Jeffrey Sachs, Pete Smith, Konrad Steffen, Lise Van Susteren, Karina von Schuckmann, James C. Zachos, "Assessing "dangerous climate change": required reduction of carbon emissions to protect young people, future generations and Nature", PLOS One, 8 (12), 3 December 2013: http://www.plosone.org/article/info%3Adoi%2F10.1371%2Fjournal.pone.0081648.

[48]. Gideon Polya, editor, "Biofuel Genocide":
https://sites.google.com/site/biofuelgenocide/
[49]. Timothy Searchinger et al., "Use of U.S. Croplands for Biofuels Increases Greenhouse Gases Through Emissions from Land-Use Change", Science 29 February 2008, Vol. 319. no. 5867, pp. 1238 – 1240: http://www.sciencemag.org/cgi/content/abstract/1151861.
[50]. Joseph Fargione and colleagues ("Land Clearing and the Biofuel Carbon Debt", Science 29 February 2008, Vol. 319. no. 5867, pp. 1235 – 1238:
http://www.sciencemag.org/cgi/content/abstract/1152747.
[51]. John Holdren, "The Science of climatic disruption":
http://www.usclimateaction.org/userfiles/JohnHoldren.pdf.
[52]. Brendan Mackey, Heather Keith, Sandra Berry, David Lindenmeyer (ANU), "Green Carbon. The role of natural forests in carbon storage":
http://epress.anu.edu.au/green_carbon/pdf/whole_book.pdf.
[53]. Carbon Tracker, "Unburnable carbon 2013: Wasted capital and stranded asserts": http://www.carbontracker.org/wastedcapital.
[54]. Gideon Polya, editor, "Divest from fossil fuels":
https://sites.google.com/site/300orgsite/divest-from-fossil-fuels.
[55]. Drew T. Shindell, Greg Faluvegi, Dorothy M. Koch, Gavin A. Schmidt, Nadine Unger and Susanne E. Bauer, "Improved Attribution of Climate Forcing to Emissions", Science, 30 October 2009: Vol. 326 no. 5953 pp. 716-718:
http://www.sciencemag.org/content/326/5953/716.
[56]. Shindell et al. (2009), Fig.2:
http://www.sciencemag.org/content/326/5953/716.figures-only.
[57]. Gideon Polya, editor, "Science and economics experts: Carbon tax needed Not Carbon Trading", 300.org:
https://sites.google.com/site/300orgsite/sciennce-economics-experts-carbon-tax-needed-not-carbon-trading.
[58]. James Hansen, "It's Possible To Avert The Climate Crisis", Counterurrents, 29 November, 2009:
http://www.countercurrents.org/hansen291109.htm.
[59]. "List of countries by greenhouse gas emissions per capita":
http://en.wikipedia.org/wiki/List_of_countries_by_greenhouse_gas_emissions_per_capita.
[60]. Gideon Polya, "Why PM Julia Gillard Must Go: 66,000 Preventable Australian Deaths Annually", Countercurrents:
http://www.countercurrents.org/polya210212.htm.
[61]. Gideon Polya, "Ongoing Aboriginal Genocide And Aboriginal Ethnocide By Politically Correct Racist Apartheid Australia ", Countercurrents, 16 February 2014:
http://www.countercurrents.org/polya160214.htm.

[62]. Gideon Polya, "One million Americans die preventably annually in USA", Countercurrents:
http://www.countercurrents.org/polya180212.htm.
[63]. Gideon Polya, "American Holocaust, millions of untimely American deaths and the $40 trillion cost of Israel to Americans", Countercurrents, 27 August 2013:
http://www.countercurrents.org/polya270813.htm.
[64]. Gideon Polya, editor, "100% renewable energy by 2020":
https://sites.google.com/site/100renewableenergyby2020/.
[65]. Gideon Polya, editor, "Cut carbon emissions 80% by 2020":
https://sites.google.com/site/cutcarbonemissions80by2020/.
[66]. James Hansen, Pushker Kharecha, Makiko Sato, Valerie Masson-Delmotte, Frank Ackerman, David J. Beerling, Paul J. Hearty, Ove Hoegh-Guldberg, Shi-Ling Hsu, Camille Parmesan, Johan Rockstrom, Eelco J. Rohling, Jeffrey Sachs, Pete Smith, Konrad Steffen, Lise Van Susteren, Karina von Schuckmann, James C. Zachos, "Assessing "dangerous climate change": required reduction of carbon emissions to protect young people, future generations and Nature", PLOS One, 8 (12), 3 December 2013:
http://www.plosone.org/article/info%3Adoi%2F10.1371%2Fjournal.pone.0081648.
[67]. J.E.N. Veron, O. Hoegh-Guldberg, T.M. Lenton, J.M. Lough, D.O. Obura, P. Pearce-Kelly, C.R.C. Sheppard, M. Spalding, M.G. Stafford-Smith and A.D. Rogers, "The coral reef crisis: the critical importance of <350 ppm CO_2", Marine Pollution Bulletin, vol. 58, (10), October 2009, 1428-1436:
http://www.sciencedirect.com/science?_ob=ArticleURL&_udi=B6V6N-4X9NKG7-3&_user=10&_rdoc=1&_fmt=&_orig=search&_sort=d&_docanchor=&view=c&_searchStrId=1072337698&_rerunOrigin=google&_acct=C000050221&_version=1&_urlVersion=0&_userid=10&md5=6858c5ff7172f9355068393496a5b35d.
[68]. Andrew Balmford et al., "Economic Reasons for Conserving Wild Nature", Science 9 August 2002: 950-953:
http://www.sciencemag.org/content/297/5583/950.abstract.
[69]. "Biofuel famine, biofuel genocide, meat & global food price crisis":
http://globalavoidablemortality.blogspot.com.au/2008/05/biofuel-famine-biofuel-genocide-meat.html.
[70]. Gideon Polya, editor, "Biofuel famine":
https://sites.google.com/site/biofuelgenocide/.
[71]. Jan Burck, Franziska Marten, Christoph Bals, "The Climate Change Performance Index – Results 2014":
https://germanwatch.org/en/download/8599.pdf.

[72]. Jan Burck, Ursula Hagen, Niklas Höhne, Leonardo Nascimento, Christoph Bals, "The Climate Change Performance Index – Results 2020: https://www.climate-change-performance-index.org/the-climate-change-performance-index-2020.

SECTION A: CLIMATE CRISIS, CLIMATE EMERGENCY & BIODIVERSITY CATASTROPHE.

"The continuing use of the atmosphere as an open sewer for industrial pollution has already added some 305 GtC to the atmosphere together with land clearing and animal-emitted methane. This raised CO_2 levels to 387 ppm CO_2 to date, leading toward conditions which existed on Earth about 3 million years (Ma) ago (mid-Pliocene), when CO_2 levels rose to about 400 ppm, temperatures to about 2–3 degrees C and sea levels by about 25 +/- 12 metres." Dr Andrew Glikson (paleo-climate and earth scientist, Australian National University, Canberra), 2008.

"We face a climate emergency." Dr James Hansen (top US climate scientist, former head, NASA's Goddard Institute for Space Studies), 2009.

"We are in real danger." Professor Peter Doherty (Nobel Prize-winning immunologist), 2009.

"The coral reef crisis: the critical importance of <350 ppm CO_2", Marine Pollution Bulletin, October 2009: "Temperature-induced mass coral bleaching causing mortality on a wide geographic scale started when atmospheric CO_2 levels exceeded 320 ppm. When CO_2 levels reached 340 ppm, sporadic but highly destructive mass bleaching occurred in most reefs world-wide, often associated with El Niño events. Recovery was dependent on the vulnerability of individual reef areas and on the reef's previous history and resilience. At today's level of 387 ppm, allowing a lag-time of 10 years for sea temperatures to respond, most reefs world-wide are committed to an irreversible decline." J.E.N. Veron and other top coral scientists, 2009.

"The risks of climate-induced impacts are projected to be higher at 2°C than those at global warming of 1.5°C (high confidence). Coral reefs, for example, are projected to decline by a further 70–90% at 1.5°C (high confidence) with larger losses (>99%) at 2°C (very high confidence). The risk of irreversible loss of many marine and coastal ecosystems increases with global warming, especially at 2°C or more (high confidence)." IPCC, "Global warming of 1.5 °C. Summary for Policymakers", 2018.

"How dare you!" Greta Thunberg at the UN, 2019.

Back in 2008 I concluded from dire predictions about loss of Arctic summer sea ice that "The World urgently needs a Declaration of Climate Emergency to meet the huge threat from accelerating and catastrophic polar ice melting. Climate scientists have recently discovered that the rate of polar ice loss is accelerating unexpectedly and that the current atmospheric carbon dioxide (CO_2) has reached a tipping point for complete loss of Arctic sea ice [1]. In 2009 I wrote: "The World is acutely threatened by man-made climate change and it may already be too late to stop catastrophe – however there is Good News as well as Bad News. The Bad News is that top UK climate scientists, Professor Kevin Anderson and Dr Alice Bows of the Tyndall Centre for Climate Change Research, University of Manchester, made the following shocking conclusions in a 2008 paper in the prestigious Philosophical Transactions of the Royal Society: "Unless economic growth can be reconciled with unprecedented rates of decarbonization (in excess of 6% per year), it is difficult to envisage anything other than a planned economic recession being compatible with stabilization at or below 650 ppmv CO_2 e"… The good news is that the best non-carbon and renewable energy options (solar, wind, geothermal, wave, tide and hydro-electric) currently yield power at roughly the same market price as coal burning-based power" [2].

In mid-February 2009 just after the tragic Black Saturday bushfires in Victoria I wrote about this catastrophe: "Australia's

State of Victoria (capital Melbourne), has just suffered record-breaking heat wave temperatures and a tragic bushfire disaster (over 180 people dead, over 1,000 homes destroyed, over 300,000 hectares burnt). This tragedy has occurred on top of a background of sustained drought, man-made global warming and global government inaction" [3]. 10 years later, in 2019/2020 South Eastern Australia was again devastated by catastrophic bushfires, with increasing public acceptance that man-made global warming is contributing to such disasters around the world, notably in Southern Europe, Siberia, the Amazon rainforest, California, Indonesia, and Australia.

By mid-2012 I had concluded that zero emissions was not enough and that massive draw-down of atmospheric CO_2 was required: "We all know that we have to reduce greenhouse gas (GHG) pollution and eventually reach zero emissions (by about 2050 according to European climate scientists). Indeed top climate scientists and biologists are telling us that reaching zero emissions is not enough – we then have to reduce atmospheric carbon dioxide (CO_2) concentration from the current 394 parts per million (ppm) to 350 ppm (according to 350.org) and thence to 300 ppm (according to the latest science-informed 300.org: https://sites.google.com/site/300orgsite/300-org---return-atmosphere-co2-to-300-ppm)" [4]. However the Anglosphere countries of the US, Australia and Canada continued (and still continue) to irresponsibly exploit fossil fuel resources of the Planet. I wrote (2012): "Australian, Canadian and US oil and gas developments, as well as existing huge coal exploitation, are threatening Humanity, the Great Barrier Reef of Australia and the Biosphere. Not content with massive existing fossil fuel exploitation, these countries are developing huge coal seam gas (CSG) reserves at immediate environmental cost and the US and Canada are developing oil sands that according to Dr James Hansen of NASA simply mean "game over" for the Planet" [5]. Instead of the atmospheric CO_2 decreasing to the pre-Industrial Revolution level of circa 300 ppm CO_2 required for a safe and sustainable environment for all peoples and all species, it has steadily increased to a present monthly mean of 415 ppm CO_2 (US NOAA, Mauna Loa:

<u>https://www.esrl.noaa.gov/gmd/ccgg/trends/</u>) i.e. we are heading in the wrong direction.

In the face of this resolute climate criminality it was reasonable for me in 2012 to ask the question "Are We Doomed?" and to state: "The World is facing a climate emergency due to global warming from man-derived greenhouse gas (GHG) pollution. The atmospheric carbon dioxide (CO_2) concentration is now 394 parts per million (ppm) and increasing at 2.4 ppm per year. Until recently the atmospheric CO_2 concentration was in the range of 180-300 ppm for the last 800,000 years, fluctuations in this range giving rise to successive glacial and inter-glacial periods that imposed selection pressures upon evolving humanity. The average surface temperature is now +0.8C above that in 1900 and this has already been associated with major climate, weather and biological disruptions. Indeed the species extinction rate is now 100-1,000 times greater than normal" [6]. My 2012 article "Are We Doomed? Too Late To Save Earth?" quotes the opinions of a selection of eminent experts on the seriousness of our predicament and is reproduced as Chapter 1 of this book.

The Intergovernmental Panel on Climate Change (IPCC) has published a series of detailed reports on man-made climate change over the years since it was founded in 1988. However the IPCC reports are conservative because of the need for inter-governmental consensus, and their estimations are qualified by loose probabilistic caveats. I wrote (2014): "The international consensus basis of the latest IPCC Summary for Policymakers (2014) has resulted in a report that softens the present acute seriousness of unaddressed man-made climate change. Thus the [latest] IPCC Summary argues for a limitation of temperature rise to $2^{o}C$ through limiting greenhouse gas (GHG) pollution of the atmosphere to 450 ppm CO_2-equivalent but hard evidence says that we have already reached 478 ppm CO_2-equivalent, that $2^{o}C$ is dangerous and essentially inevitable, and that the world will use up its Carbon Budget for avoiding $2^{o}C$ within about 4 years" [7].

Over the last 5 years I have continued to write carefully researched, science-informed articles about man-made climate change and in particular about the global Climate Crisis, Climate Emergency and the ever-worsening Biodiversity catastrophe [8-17]. Four of these articles are particularly information-rich and are reproduced in full in the book as Chapters 2-5, namely Chapter 2, "Revised Annual Per Capita Greenhouse Gas Pollution For All Countries – What Is Your Country Doing?" [9], Chapter 3, "Extrapolating 11,000 scientists Climate Emergency warning to 2030 catastrophe" [15], Chapter 4, "Climate scientists: Planetary Emergency, Planet in peril, act now" [16], and Chapter 5, "Wrong way go back – global sectoral greenhouse gas emissions are all in the wrong direction" [17].

A recurring and continuing problem in the Climate Crisis (and other serious matters) has been political inaction assisted by neoliberal Big Money-driven Mainstream mendacity and obfuscation (see "Mainstream media censorship": https://sites.google.com/site/mainstreammediacensorship/home and "Mainstream media lying": https://sites.google.com/site/mainstreammedialying/home). Writing after the 2015 Paris Climate Change Conference, I wrote (2015): "The Paris Climate Change Conference aims for international consensus on actions to keep global warming to less than 2 degrees C, but has betrayed humanity before it even began. Thus science informs that the internationally-agreed 2 degree C temperature target is both catastrophic and inevitable. Pre-conference national greenhouse gas (GHG) pollution reduction commitments to Paris imply a plus 2.7 degrees C temperature rise. Indeed the present circa plus 0.9 degree C temperature rise is already catastrophic for populations in low-lying island and mega-delta locations. This massive public deception has come about because of Mainstream journalist, politician and academic lying by commission and lying by omission" [8]. Man-made greenhouse gas (GHG) pollution derives substantially from land use as well as from transport and industry but is often officially underestimated by minimizing land use and the actual GHG effect of methane (CH_4). I attempted to determine revised annual per capita GHG pollution estimates for all countries taking these factors into account [9] (Chapter 2). No doubt a billion dollar institution could do much better but I am still waiting.

2015 marked the 200th anniversary of the Congress of Vienna that re-organized post-Napoleonic Europe. However a "Congress Of Vienna 2015 For Global Stability" remarkably failed to address the existential threats of climate change and nuclear weapons. My comment in a detailed analysis (2016): "200 years after the 1815 Congress of Vienna re-organized post-Napoleonic Europe, the Canadian Sheldon Chumir Foundation for Ethics in Leadership in collaboration with a number of leading policy organisations and universities hosted a lavish Congress of Vienna 2015 to explore policies for a stable world order that unfortunately was opened by mass murderer Henry Kissinger, and generated a Draft Report that - as set out in my tough critique, reality check and Alternative Report - is soft, wishy-washy, myopic and predicated on the Big Power- and Big Money-dominated status quo that kills 17 million people annually through deprivation and threatens to wipe out most of Humanity and the Biosphere. My science-informed position is that the world is acutely threatened by nuclear weapons (that could wipe out most of Humanity and the Biosphere at any time), poverty (that kills 17 million people each year) and by man-made climate change from greenhouse gas pollution (pollutants from carbon burning kill 7 million people each year but unaddressed man-made climate change may kill 10 billion people this century in an already worsening climate genocide)" [10].

The tightening of the climate noose was evident from a 2018 IPCC report on 1.5C versus 2.0C. In my analysis of this dire report I stated (2018): "The IPCC has just issued a report that details the numerous bad outcomes of a global +1.5 degree Centigrade (+1.5C) of warming versus the catastrophic outcomes from a +2C e.g. a further 70-90% decline of coral reefs at +1.5C versus more than 99% loss at +2C. Crucially, the IPCC report is a consensus and hence conservative document, and its propositions are probabilistic. Dauntingly, the IPCC Report says that for less than +1.5C coal burning must cease by

2050 but also declares that the terminal CO_2 pollution budget for a 66% chance of avoiding +1.5C will be used up in 10 years. Humanity and particularly young people must revolt with zero tolerance for coal-burning climate criminals epitomized by pro-coal Australia and Trump America" [11]. Pro-coal Australia and Trump America both rejected this dire IPCC Report [12].

In a detailed 2018 article (now reproduced as Chapter 3 of the book) I quantitatively addressed the key resource limitations summarized thus by Professor Dabo Guan (School of International Development, University of East Anglia, UK) (2016): "For everyone in the world to have an American lifestyle, we would need seven planets, and three to live as Europeans". Using our dying coral reefs as a "canary in the mine' I concluded that there must be negative carbon pollution (atmospheric CO_2 draw-down to about 300 ppm CO_2 from the present 415 ppm CO_2), negative population growth (population decline by about 50%), and negative economic growth (de-growth) of about 50% [13].

In 2017 over 15,000 scientists signed up to a detailed analysis of worsening global trends in numerous areas (from industrial pollution to the Biosphere), and in 2019 an expanded analysis was co-signed by 11,000 scientists. I carefully analysed and summarized the findings of both papers [14, 15]. My 2019 analysis of both papers concludes [with correcting editing]: "Over 11,000 scientists have signed up to a World scientists' warning of a Climate Emergency that sets out trends in 24 climate-related areas over the last 40 years. Scientists became aware of the climate change threat from greenhouse gas (GHG) pollution in the 1980s, but in 21 of these 24 areas the trends are (a) huge, (b) in the wrong direction, and (c) linear or quasi-linear functions of time, with this allowing extrapolation from the present climate emergency to a climate catastrophe in 2030" [15]. This detailed analysis is Chapter 4 of the book.

In 2019 a number of eminent climate scientists provided an incisive analysis of the time frame for exceeding key, irreversible tipping points and concluded: "Act now. In our view, the evidence from tipping points alone suggests that we are in a state of planetary emergency: both the risk and urgency of the situation are acute… We argue that the intervention time left to prevent tipping [*T*] could already have shrunk towards zero, whereas the reaction time to achieve net zero emissions [*τ*] is 30 years at best. Hence we might already have lost control of whether tipping happens. A saving grace is that the rate at which damage accumulates from tipping — and hence the risk posed — could still be under our control to some extent. The stability and resilience of our planet is in peril. International action — not just words — must reflect this". My summary of this important analysis [16] is Chapter 5 of the book.

Finally, in 2020 I published a detailed analysis of pollution in major economic areas that arrived at the following dire conclusion: "Vehicle drivers are familiar with crucial highway signs saying "WRONG WAY GO BACK". A detailed analysis shows that global greenhouse gas (GHG) emissions in all key economic sectors of energy production and non-energy production are increasing whereas the worsening Climate Emergency demands urgently decreasing GHG emissions and ultimately net zero GHG emissions ASAP. The world is driving headlong in the wrong direction to certain disaster" [17]. This detailed analysis is Chapter 6 of the book.

The following articles are reproduced as Chapters 1-6 of Section A:

Chapter 1: Gideon Polya, "Are We Doomed? Too Late To Save Earth?", Countercurrents, 4 September, 2012: https://www.countercurrents.org/polya040912.htm.
Chapter 2: Gideon Polya, "Revised Annual Per Capita Greenhouse Gas Pollution For All Countries – What Is Your Country Doing?", Countercurrents, 6 January 2016: https://countercurrents.org/polya060116.htm.
Chapter 3: Gideon Polya, "How much negative emissions, negative population growth & negative economic growth is needed to save Planet?", Countercurrents, 28 November 2018: https://countercurrents.org/2018/11/28/how-much-negative-carbon-emissions-negative-population-growth-negative-economic-growth-is-needed-to-save-planet/.

Chapter 4: Gideon Polya, "Extrapolating 11,000 scientists Climate Emergency warning to 2030 catastrophe", Countercurrents, 14 November 2019: https://countercurrents.org/2019/11/extrapolating-11000-scientists-climate-emergency-warning-to-2030-catastrophe.
Chapter 5: Gideon Polya, "Climate scientists: Planetary Emergency, Planet in peril, act now", Countercurrents, 3 December 2019: https://countercurrents.org/2019/12/climate-scientists-planetary-emergency-planet-in-peril-act-now.
Chapter 6: Gideon Polya, "Wrong way go back – global sectoral greenhouse gas emissions are all in the wrong direction", Countercurrents, 20 March 2020: https://countercurrents.org/2020/03/wrong-way-go-back-global-sectoral-greenhouse-gas-emissions-are-all-in-the-wrong-direction/.

References (those asterisked are reproduced as chapters in the book).

[1]. Gideon Polya, "Time To Declare Global Climate Emergency", Countercurrents, 11 February 2008.
[2]. Gideon Polya, "Good and bad climate news", Countercurrents, 14 January 2009: https://www.countercurrents.org/polya140109.htm.
[3]. Gideon Polya, "Global Warming Impacting Humanity", Countercurrents, 14 February, 2009: https://www.countercurrents.org/polya140209.htm.
[4]. Gideon Polya, "World Rapidly Running Out Of Time To Cease Greenhouse Gas Pollution", Countercurrents, 1 August 2011: https://www.countercurrents.org/polya010811.htm.
[5]. Gideon Polya, "Australian, Canadian & US Oil & Gas & Coal Threaten Great Barrier Reef, Humanity & Biosphere", Countercurrents, 4 March 2012: https://www.countercurrents.org/polya040312.htm.
*[6]. Gideon Polya, "Are We Doomed? Too Late To Save Earth?", Countercurrents, 4 September, 2012: https://www.countercurrents.org/polya040912.htm.
[7]. Gideon Polya, "International Consensus-Based IPCC Summary For Policymakers (2014) Downplays Acute Seriousness Of Climate Crisis", Countercurrents, 12 November 2014: https://www.countercurrents.org/polya121114.htm.
[8]. Gideon Polya, "Mainstream Media Lying Permits Betrayal Of Humanity By Paris Climate Change Conference Before It Begins", Countercurrents, 29 November 2015: https://www.countercurrents.org/polya291115.htm.
*[9]. Gideon Polya, "Revised Annual Per Capita Greenhouse Gas Pollution For All Countries – What Is Your Country Doing?", Countercurrents, 6 January 2016: https://countercurrents.org/polya060116.htm.
[10]. Gideon Polya, "Congress Of Vienna 2015 For Global Stability Fails To Address Key, Existential Nuclear, Poverty And Climate Change Threats", Countercurrents, 12 January 2016: https://www.countercurrents.org/polya120116.htm.
[11]. Gideon Polya, "IPCC +1.5C avoidance report – effectively too late, but stop coal burning for "less bad" catastrophes", Countercurrents, 12 October 2018: https://countercurrents.org/2018/10/12/ipcc-1-5c-avoidance-report-effectively-too-late-but-stop-coal-burning-for-less-bad-catastrophes/.
[12]. Gideon Polya, "Pro-coal Australia & Trump America reject dire IPCC Report & declare war on Terra", Countercurrents, 17 October

2018: https://countercurrents.org/2018/10/17/pro-coal-australia-trump-america-reject-dire-ipcc-report-declare-war-on-terra/.
*[13]. Gideon Polya, "How much negative emissions, negative population growth & negative economic growth is needed to save Planet?", Countercurrents, 28 November 2018: https://countercurrents.org/2018/11/28/how-much-negative-carbon-emissions-negative-population-growth-negative-economic-growth-is-needed-to-save-planet/.
[14]. Gideon Polya, "Over 15,000 scientists issue dire warning to humanity on catastrophic climate change and biodiversity loss", Countercurrents, 20 November 2017: https://countercurrents.org/2017/11/20/over-15000-scientists-issue-dire-warning-to-humanity-on-catastrophic-climate-change-and-biodiversity-loss/.
*[15]. Gideon Polya, "Extrapolating 11,000 scientists Climate Emergency warning to 2030 catastrophe", Countercurrents, 14 November 2019: https://countercurrents.org/2019/11/extrapolating-11000-scientists-climate-emergency-warning-to-2030-catastrophe.
*[16]. Gideon Polya, "Climate scientists: Planetary Emergency, Planet in peril, act now", Countercurrents, 3 December 2019: https://countercurrents.org/2019/12/climate-scientists-planetary-emergency-planet-in-peril-act-now.
*[17]. Gideon Polya, "Wrong way go back – global sectoral greenhouse gas emissions are all in the wrong direction", Countercurrents, 20 March 2020: https://countercurrents.org/2020/03/wrong-way-go-back-global-sectoral-greenhouse-gas-emissions-are-all-in-the-wrong-direction/.

CHAPTER 1: ARE WE DOOMED? TOO LATE TO SAVE EARTH?

[First published as Gideon Polya, "Are We Doomed? Too Late To Save Earth?", Countercurrents, 4 September, 2012: https://www.countercurrents.org/polya040912.htm.]

The World is facing a climate emergency due to global warming from man-derived greenhouse gas (GHG) pollution. The atmospheric carbon dioxide (CO_2) concentration is now 394 parts per million (ppm) and increasing at 2.4 ppm per year. Until recently the atmospheric CO_2 concentration was in the range of 180-300 ppm for the last 800,000 years, fluctuations in this range giving rise to successive glacial and inter-glacial periods that imposed selection pressures upon evolving humanity. The average surface temperature is now +0.8C above that in 1900 and this has already been associated with major climate, weather and biological disruptions. Indeed the species extinction rate is now 100-1,000 times greater than normal [1].

Both Dr James Lovelock FRS (Gaia hypothesis) and Professor Kevin Anderson (Director, Tyndall Centre for Climate Change Research, University of Manchester, UK) have recently estimated that only about 0.5 billion people will survive this century due to unaddressed, man-made global warming. Noting that the world population is expected to reach 9.5 billion by 2050 (UN Population Division), these estimates translate to a Climate Genocide involving deaths of 10 billion people this century, this including roughly twice the present population of particular mainly non-European groups, specifically 6 billion under-5 year old infants, 3 billion Muslims in a terminal Muslim Holocaust, 2 billion Indians, 1.3 billion non-Arab Africans, 0.5 billion Bengalis, 0.3 billion Pakistanis and 0.3 billion Bangladeshis [2].

Collective, national responsibility for this already commenced Climate Genocide is in direct proportion to per capita national pollution of the atmosphere with greenhouse gases (GHGs). Indeed, fundamental to any international agreement on national rights to pollute our common atmosphere and oceans should be the belief that "all men are created equal". However reality is otherwise: "annual per capita greenhouse gas (GHG) pollution" in units of "tonnes CO_2-equivalent per person per year" (2005-2008 data) is 0.9 (Bangladesh), 0.9 (Pakistan), 2.2 (India), less than 3 (many African and Island countries), 3.2 (the Developing World), 5.5 (China), 6.7 (the World), 11 (Europe), 16 (the Developed World), 27 (the US) and 30 (Australia; or 54 if Australia's huge Exported CO2 pollution is included) [2].

Basically we know what the problem is (man-made GHG pollution) and how to solve it. Fundamentally, as enunciated by 300.org, we need to reduce atmospheric CO_2 concentration to about 300 ppm for a safe planet for all peoples and all species [3].

To achieve 300 ppm CO_2 we must achieve the following [4]:

1. Change of societal philosophy to one of scientific risk management and biological sustainability with complete cessation of species extinctions and zero tolerance for lying.

2. Urgent reduction of atmospheric CO_2 to a safe level of about 300 ppm as recommended by leading climate and biological scientists.

3. Rapid switch to the best non-carbon and renewable energy (solar, wind, geothermal, wave, tide and hydro options that are currently roughly the same market price as coal burning-based power) and to energy efficiency, public transport, needs-based production, re-afforestation and return of carbon as biochar to soils coupled with correspondingly rapid cessation of fossil fuel burning, deforestation, methanogenic livestock production and population growth.

We understand the problem and have the technological solutions – the impending catastrophe simply does not have to happen. Are we going to be able to overcome the current political stasis and act before it is too late for Humanity and the Biosphere? Is it too late? Are we all doomed?

This article records the opinions of leading scientists and writers on this terminal question for Humanity.

[1]. Dr Gideon Polya,"Climate change course", 2011 via Yarra Valley Climate Action Group: http://yvcag.blogspot.com/2011_08_01_archive.html and https://sites.google.com/site/yarravalleyclimateactiongroup/2011-climate-change-\ course) and 300.org (see: http://300org.blogspot.com/2011_08_01_archive.html and https://sites.google.com/site/300orgsite/2011-climate-change-course.
[2]. "Climate Genocide": https://sites.google.com/site/climategenocide/.
[3]. 300.org: https://sites.google.com/site/300orgsite/.
[4]. "Climate crisis facts and required actions", Yarra Valley Climate Action Group: https://sites.google.com/site/yarravalleyclimateactiongroup/climate-crisis-facts-required-actions.

ANALYSIS BY COUNTRY: Analysis by country of years left [relative to 2012] to zero emissions for 75% chance of avoiding 2C temperature rise

The 2009 Report of the German Advisory Council on Climate Change (WBGU, Wissenshaftlicher Beirat der Bundesregierung Globale Umweltveränderungen) was entitled "Solving the climate dilemma: the budget approach" and crucially stated: "The budget of CO_2 emissions still available worldwide could be derived from the 2 degree C guard rail. By the middle of the 21st century a maximum of approximately 750 Gt CO_2 (billion metric tons) may be released into the Earth's atmosphere if the guard rail is to be adhered to with a probability of 67%. If we raise the probability to 75%, the cumulative emissions within this period would even have to remain below 600 Gt CO_2. In any case, only a small amount of CO_2 may be emitted worldwide after 2050. Thus, the era of an economy driven by fossil fuels will definitely have to come to an end within the first half of this century" (see: WBGU (German Advisory Council of Climate Change), "Solving the climate dilemma: the budget approach": http://www.wbgu.de/fileadmin/templates/dateien/veroeffentlichungen /sondergutachten/sn2009/wbgu_sn2009_en.pdf).

Using country by country data for greenhouse gas emissions per capita (tonnes CO_2-e per person per year) (see "List of countries by greenhouse gas emissions per capita", Wikipedia: http://en.wikipedia.org/wiki/List_of_countries_by_greenhouse_ gas_emissions_per_capita) one can readily determine for each country in the world how many years left to zero pollution.

The average world population in the period 2010 and 2050 will be 8.321 billion (see UN Population Division, 2010 Revision). Accordingly, the per capita share of this terminal CO_2 pollution budget is less than 600 billion tonnes CO_2/8.321 billion people = less than 72.1 tonnes CO_2 per person.

Years [relative to 2012] to the required "fair shares" total cessation of GHG pollution at current rates of pollution = 72.1 tonnes CO_2-e per person/ (tonnes CO_2-e per person per year).

Countries that must cease GHG pollution within 5 years.

Belize (0.8 years), Qatar (1.3), Guyana (1.4), Malaysia (1.9), United Arab Emirates (2.0), Kuwait (2.4), Papua New Guinea (2.5), Brunei (2.8), Australia (2.8; 1.1 if including its huge GHG Exports), Antigua & Barbuda (2.8), Zambia (2.9), Canada (3.0), Bahrain (3.0), United States (3.1), Trinidad & Tobago (3.3), Luxembourg (3.4), Panama (3.7), New Zealand (3.7), Estonia (4.0), Botswana (4.1), Ireland (4.3), Saudi Arabia (4.4), Venezuela (4.6), Indonesia (4.8), Equatorial Guinea (5.0), Belgium (5.0).

Countries that must cease GHG pollution within 5-10 years.

Turkmenistan (5.1 years), Singapore (5.1), Czech Republic
(5.2), Liberia (5.2), Netherlands (5.3), Russia (5.3), Nicaragua
(5.4), Finland (5.5), Oman (5.6), Palau (5.6), Brazil (5.6),
Uruguay (5.7), Denmark (5.8). Germany (5.9), Mongolia (6.1),
Israel (6.1), Nauru (6.2), Norway (6.3), South Korea (6.5),
Kazakhstan (6.6), United Kingdom of Great Britain and
Northern Ireland (6.6), Libya (6.7), Greece (6.7), Japan (6.7),
Myanmar (6.7), Taiwan (6.8), Cyprus (7.0), Slovenia (7.1),
Cambodia (7.1), Austria (7.2), Iceland (7.2), Peru (7.3),
Paraguay (7.3), Ukraine (7.4), Poland (7.5), South Africa (7.6),
Argentina (7.8), Slovakia (7.8), Spain (7.8), Italy (7.8), Central
African Republic (8.0), France (8.3), Suriname (8.4), Belarus
(8.4), Gabon (8.6), Ecuador (8.8), Bolivia (8.9), Cameroon
(9.5), Iran (9.5), Côte d'Ivoire (9.6), Sweden (9.6), Seychelles
(9.7), Guatemala (9.7), Bulgaria (9.7), Serbia & Montenegro
(9.7), Hungary (9.7), Congo, Democratic Republic (formerly
Zaire) (9.7), Uzbekistan (9.9), Portugal (10.0).

**Countries that must cease GHG pollution within 10-20
years.**

Switzerland (10.2 years), Azerbaijan (10.6), Angola (10.8),
Bahamas (10.9), Benin (11.1), Zimbabwe (11.1), Laos (11.3),
Mexico (11.3), Nepal (11.4), Colombia (11.4), Namibia (11.4),
Chile (11.4), Malta (11.8), Congo, Republic (12.0), Madagascar
(12.0), Croatia (12.2), Jamaica (12.2), Macedonia (12.4),
Barbados (12.4), Latvia (12.6), Mauritania (12.9), Turkey
(12.9), Romania (13.1), Lithuania (13.4), Costa Rica (13.4),
Lebanon (13.6), North Korea (13.9), Thailand (14.1), Jordan
(14.7), Honduras (15.3), Sudan (15.7), Bosnia & Herzegovina
(16.0), Algeria (17.2), Iraq (17.2), Sierra Leone (17.2), Syria
(18.0), China (18.5), Tunisia (19.5), Dominican Republic (20.6
years).

Countries that must cease GHG pollution within 20-30 years.

St Kitts & Nevis (21.8), Nigeria (21.8), Fiji (21.8), Guinea (22.5), Mauritius (22.5), Cuba (23.3), Togo (23.3), Vanuatu (24.0), Philippines (24.0), Malawi (24.0), Mali (24.9), Chad (24.9), Sri Lanka (25.8), Uganda (26.7), Dominica (26.7), St Lucia (26.7), Egypt (27.7), Niue (27.7), Ghana (27.7), Moldova (28.8), Grenada (28.8), El Salvador (30.0), Guinea-Bissau (30.0), Tanzania (30.0), Djibouti (30.0).

Countries that must cease GHG pollution within 30-50 years.

Pakistan (31.3 years), Samoa (31.3), Tonga (31.3), Morocco (32.8), Senegal (32.8), Albania (32.8), Georgia (32.8), Armenia (34.3), St Vincent & Grenadines (36.1), Kenya (36.1), Maldives (37.9), Kyrgyzstan (37.9), Burkina Faso (37.9), India (40.1), Cook Islands (40.1), Bhutan (42.4), Yemen (45.1), Tajikistan (45.1), Mozambique (45.1), Rwanda (45.1), Burundi (45.1), Lesotho (48.1), Swaziland (48.1).

Countries that must cease GHG pollution within about 50-120 years.

Eritrea (51.5), Haiti (51.5), Solomon Islands (65.5), Vietnam (65.5), Cape Verde (65.5), Niger (65.5), Ethiopia (65.5), São Tomé and Príncipe (72.1), Afghanistan (80.1), The Gambia (80.1), Bangladesh (80.1), Comoros (103.0), Kiribati (120.2).

<u>ANDERSON: Professor Kevin Anderson, Deputy Director of the Tyndall Centre for Climate Change Research</u>

Professor Kevin Anderson is the Deputy Director of the Tyndall Centre for Climate Change Research, holds a joint chair in Energy and Climate Change at the School of Mechanical, Aerospace and Civil Engineering at the University of Manchester and School of Environmental Sciences at University of East Anglia, is an honorary lecturer in Environmental Management at the Manchester Business School, and is an

adviser to the British Government on climate change (see:
http://en.wikipedia.org/wiki/Kevin_Anderson_%28scientist%29
)

Professor Anderson, together with Dr Alice Bows wrote an extremely important paper describing 6-8% annual GHG emissions reductions needed for 450 ppm CO_2-equivalent (CO_2-e): "According to the analysis conducted in this paper, stabilizing at 450 ppmv [carbon dioxide equivalent = CO_2-e, atmospheric concentration measured in parts per million by volume] requires, at least, global energy related emissions to peak by 2015, rapidly decline at 6-8% per year between 2020 and 2040, and for full decarbonization sometime soon after 2050 …Unless economic growth can be reconciled with unprecedented rates of decarbonization (in excess of 6% per year), it is difficult to envisage anything other than a planned economic recession being compatible with stabilization at or below 650 ppmv CO_2-e ... Ultimately, the latest scientific understanding of climate change allied with current emissions trends and a commitment to "limiting average global temperature increases to below $4^{o}C$ above pre-industrial levels", demands a radical reframing of both the climate change agenda, and the economic characterization of contemporary society" (see: Kevin Anderson & Alice Bows, "Reframing the climate change challenge in light of post-2000 emission trends", Proc. Trans. Roy. Soc, A, 2008: http://www.tyndall.ac.uk/publications/journal_papers/fulltext.pdf ; Gideon Polya, "Good and bad climate news", Green Blog, 2009: http://www.green-blog.org/2009/01/13/good-and-bad-climate-news/ ; and George Monbiot, "One shot left", Monbiot.com (also published in the UK Guardian, 2008): http://www.monbiot.com/archives/2008/11/25/one-shot-left/).

Professor Kevin Anderson on how many will survive the century in a "terrifying prospect" (November 2009): "For humanity it's a matter of life or death. We will not make all human beings extinct as a few people with the right sort of resources may put themselves in the right parts of the world and survive. But I think it's extremely unlikely that we wouldn't have mass death at 4C. If you have got a population of nine billion by 2050 and you hit 4C, 5C or 6C, you might have half a billion people surviving… The worst possible result at Copenhagen is a bad deal where the world leaders have to come home and say it's a good deal when its rubbish. That's the real danger – that they will feel under pressure to sign up to anything. That could lock us into something bad for the next ten years." [1].

[1]. Professor Kevin Anderson quoted by Jenny Fyall, "Warming "will wipe out billions"", The Scotsman, 29 November 2009:
http://news.scotsman.com/latestnews/Warming-will-39wipe-out-billions39.5867379.jp.

BIALEK: Professor Janusz W. Bialek, Professor of Electrical Engineering, University of Durham, UK

Professor Janusz W. Bialek is Professor of Electrical Engineering, Chair of Electrical Power and Control, University of Durham, UK (see:
http://www.newstatesman.com/perspectives-on-energy/2011/10/carbon-government-professor).

Professor Janusz Bialek in interview about tackling climate change and answering the question "Are we all doomed?" (2011): "The human race has survived all kinds of crises. Necessity is the mother of invention, and people will come through with new solutions. We are an infinitely inventive race. Engineering solutions and technical solutions are easy. Getting political consensus is the big challenge." [1].

[1]. Professor Janusz Bialek quoted in Samira Shackle, "Q&A with Professor Jausz Bialek", New Statesman, 17 October 2011:
http://www.newstatesman.com/perspectives-on-energy/2011/10/carbon-government-professor.

<u>BIROL: Dr Fatih Birol, Chief Economist of the International Energy Agency (IEA)</u>

Dr Fatih Birol (1958 Ankara, Turkey) is the Chief Economist of the International Energy Agency (IEA) and Director, Office of The Chief Economist, with overall responsibility for the organisation's economic analysis of energy and climate change policy. He oversees the annual World Energy Outlook series which is the flagship publication of the IEA and is recognized as an authoritative source for energy analysis and projections (see: http://en.wikipedia.org/wiki/Fatih_Birol).

Dr Fatih Birol, chief economist at the International Energy Agency commenting on the IAE Report warning that if fossil fuel infrastructure is not rapidly changed, the world will 'lose for ever' the chance to avoid dangerous climate change (2011): "The door is closing. I am very worried – if we don't change direction now on how we use energy, we will end up beyond what scientists tell us is the minimum [for safety]. The door will be closed forever… The shift away from nuclear worsens the situation." [1].

[1]. Dr Fatih Birol quoted by Fiona Harvey, "World headed for irreversible climate change in five years, IEA warns", The Guardian, 9 November 2011: http://www.guardian.co.uk/environment/2011/nov/09/fossil-fuel-infrastructure-climate-change.

<u>COX: Dr Brian Cox and Dr Jeff Forshaw, UK physicists on Science & "Are we all doomed?"</u>

Dr Brian Cox and Dr Jeff Forshaw are UK physicists who co-authored the book "The Quantum Universe" (see: http://www.newstatesman.com/scitech/2011/12/interview-science-cox-physics).

Dr Brian Cox and Dr Jeff Forshaw interviewed by Helen Lewis-Hasteley for the UK New Statesman (8 December 2011):

HL-H: How much of a responsibility do you feel to be an advocate for science generally?

BC: A lot. I think the peer-review process is the best way we have of giving our best view on how nature works. There are no absolute truths in science. Take a so-called controversy such as climate change: the correct thing to say is that we make measurements of the climate, we look at the data, we model it and here are a range of predictions. While it's easy to point out the flaws, in general it's unarguable that science works ... because we're not in fucking caves!

HL-H: What motivates climate sceptics and the rest?
BC: Carl Sagan pointed out that "Science challenges". And the natural human response from people who are educated, who have a title or position, is to assume their opinion is worth something. And science tells you that your opinion is worthless when confronted with the evidence. That's a difficult thing to learn. When you look back at the Greeks or Romans and think, "Why didn't they get science?", maybe it was that.
JF: As a theoretical physicist, most of my time is spent doing calculations that are wrong. It's a humbling exercise, a massive dose of humility.

...

HL-H: Are we all doomed?

JF: The only thing that will save us is fundamental physics, because we have to escape to a distant part of the universe.

BC: On the human timescale, the adoption of the scientific method - making rational decisions based on evidence - that's the important thing. Look at public policy, health policy, economics: there's a reluctance to be humble." [1].

[1]. Dr Brian Cox and Dr Jeff Forshaw interviewed by Helen Lewis-Hasteley for the UK New Statesman, "Science tells you that your opinion is worthless. That's difficult", UK New Statesman, 8 December 2011: http://www.newstatesman.com/scitech/2011/12/interview-science-cox-physics.

DIXON: Dr Richard Dixon, Head of World Wildlife Fund (WWF) Scotland

Dr Richard Dixon, Head of World Wildlife Fund (WWF) Scotland, has a BSc and PhD in Astrophysics, and an MSc in Energy Systems and Environment Management from Glasgow Caledonian University. Before joining WWF Scotland, he worked for Community Service Volunteers and Strathclyde Regional Council, and was Head of Research at Friends of the Earth Scotland. During his eight years at the latter, he worked on issues as diverse as climate change and fish farming. He helped to treble the number of Scottish beaches officially recognised by the government and wrote the first comprehensive report on air pollution in Scotland (see: http://www.wwf.org.uk/what_we_do/press_centre/spokespeople/head_of_wwf_scotland.cfm).

Dr Richard Dixon, director of WWF Scotland, re the huge melting of Arctic sea ice and the need for action over man-made climate change (2012): "This is not unexpected but it is extremely bad news. These satellite images show what we are doing to the planet and while climate change has gone off the political agenda recently, things have got worse and governments and big business must do more to fix that… Polar ice shapes the weather systems in Scotland and the north-west of Europe, and sea ice loss is one reason why we're getting colder summers and have had very cold winters in the last few years."

Dr Richard Dixon quoted in Julia Horton, "Arctic sea ice will vanish within three years, says expert", Scotsman, 29 August 2012: http://www.scotsman.com/news/environment/arctic-sea-ice-will-vanish-within-three-years-says-expert-1-2493681.

**HANSEN: Dr James Hansen, Head, NASA Goddard
Institute for Space Studies, New York, USA**

James Hansen is a leading US climate scientist and heads the
NASA Goddard Institute for Space Studies in New York City, a
part of the Goddard Space Flight Center in Greenbelt,
Maryland. He has held this position since 1981. He is also an
adjunct professor in the Department of Earth and Environmental
Sciences at Columbia University. He first warned the US
Congress about the danger from man-made climate change over
20 years ago. He published "Storms of My Grandchildren" in
2009 (see: http://en.wikipedia.org/wiki/James_Hansen).

Dr James Hansen on being asked before the political December
2009 Copenhagen Climate Change Conference "Is there any
real chance of averting the climate crisis? (2009): "Absolutely.
It is possible – if we give politicians a cold, hard slap in the
face. The fraudulence of the Copenhagen approach – "goals" for
emission reductions, "offsets" that render ironclad goals almost
meaningless, the ineffectual "cap-and-trade" mechanism – must
be exposed. We must rebel against such politics as usual.
Science reveals that climate is close to tipping points. It is a
dead certainty that continued high emissions will create a
chaotic dynamic situation for young people, with deteriorating
climate conditions out of their control. Science also reveals
what is needed to stabilise atmospheric composition and
climate. Geophysical data on the carbon amounts in oil, gas and
coal show that the problem is solvable, if we phase out global
coal emissions within 20 years and prohibit emissions from
unconventional fossil fuels such as tar sands and oil shale. Such
constraints on fossil fuels would cause carbon dioxide emissions
to decline 60% by mid-century or even more if policies make it
uneconomic to go after every last drop of oil. Improved forestry
and agricultural practices could then bring atmospheric carbon
dioxide back to 350 ppm (parts per million) or less, as required
for a stable climate. Governments going to Copenhagen claim to
have such goals for 2050, which they will achieve with the
"cap-and-trade" mechanism. They are lying through their teeth.
Unless they order Russia to leave its gas in the ground and
Saudi Arabia to leave its oil in the ground (which nobody has

proposed), they must phase out coal and prohibit unconventional fossil fuels." [1].

[1]. James Hansen, "Copenhagen summit: Is there any real chance of averting the climate crisis?" Observer/UK Guardian, Sunday 29 November 2009, http://www.guardian.co.uk/commentisfree/2009/nov/29/copenhagen-summit-climate-change.

IEA: The International Energy Agency (IEA) 2011 World Energy Outlook (WEO) report

The International Energy Agency (IEA) in its own words: "The International Energy Agency (IEA) is an autonomous organisation which works to ensure reliable, affordable and clean energy for its 28 member countries and beyond. Founded in response to the 1973/4 oil crisis, the IEA's initial role was to help countries co-ordinate a collective response to major disruptions in oil supply through the release of emergency oil stocks to the markets. While this continues to be a key aspect of its work, the IEA has evolved and expanded. It is at the heart of global dialogue on energy, providing authoritative and unbiased research, statistics, analysis and recommendations". The IEA publishes an annual World Energy Outlook series (see: https://www.iea.org/reports/world-energy-outlook-2019).

Press release by the IEA on the 2011 World Energy Outlook (WEO) report (2011): "Without a bold change of policy direction, the world will lock itself into an insecure, inefficient and high-carbon energy system, the International Energy Agency warned as it launched the 2011 edition of the World Energy Outlook (WEO). The agency's flagship publication, released today in London, said there is still time to act, but the window of opportunity is closing… The WEO presents a 450 Scenario, which traces an energy path consistent with meeting the globally agreed goal of limiting the temperature rise to 2°C. Four-fifths of the total energy-related CO_2 emissions permitted to 2035 in the 450 Scenario are already locked-in by existing capital stock, including power stations, buildings and factories. Without further action by 2017, the energy-related infrastructure then in place would generate all the CO_2 emissions allowed in the 450 Scenario up to 2035. Delaying action is a false economy: for every \$1 of investment in cleaner technology that is avoided in the power sector before 2020, an additional \$4.30 would need to be spent after 2020 to compensate for the increased emissions." [1].

[1]. IEA, "The world is locking itself into an unsustainable energy future which would have far-reaching consequences, IEA warns in its latest World Energy Outlook", IEA Press release, 9 November 2011: http://www.iea.org/pres/pressdetail.asp?PRESS_REL_ID=426.

LIVELY: Penelope Lively: "Tampering with the physical world is what we do supremely well…. Finis"

Dame Penelope Margaret Lively, DBE, FRSL (born 17 March 1933) is a major UK author of fiction for both children and adults. Her novel "Moon Tiger" won the Booker Prize (see: http://en.wikipedia.org/wiki/Penelope_Lively).

Penelope Lively's heroine Claudia Hampton on the end of the World in the novel "Moon Tiger" (1987): "Tampering with the physical world is what we do supremely well. – in the end, perhaps, we shall achieve it definitively. Finis. And history will indeed come to an end". [1].

[1]. Penelope Lively, "Moon Tiger" (Penguin), Chapter 1, page 13.

<u>LOVELOCK: Dr James Lovelock FRS, Gaia Hypothesis & atmospheric gas analysis</u>

Dr James Lovelock, Fellow of the Royal Society (FRS) is one of the UK's and the World's most eminent climate scientists. He is famous for his Gaia Hypothesis (mutually interacting physical world and biological feedbacks affecting atmosphere and climate) and for his invention of the electron capture detector which ultimately assisted in discoveries about the persistence of

CFCs and their role in stratospheric ozone depletion (see: http://en.wikipedia.org/wiki/James_Lovelock).

Dr James Lovelock re very few surviving this century (June 2009): "If we can keep civilization alive through this century perhaps there is a chance that our descendants will one day serve Gaia and assist her in the fine-tuned self-regulation of the climate and composition of our planet. We have enjoyed 12,000 years of climate peace since the last shift from a glacial age to an interglacial one. Before long, we may face planet-wide devastation worse even than unrestricted nuclear war between superpowers. The climate war could kill nearly all of us and leave the few survivors living a Stone Age existence. But in several places in the world, including the U.K., we have a chance of surviving and even of living well. For that to be possible, we have to make our lifeboats seaworthy now." [1].

Dr James Lovelock, quoted by interviewer (November 2007): "By 2100, Lovelock believes, the Earth's population will be culled from today's 6.6 billion to as few as 500 million, with most of the survivors living in the far latitudes -- Canada, Iceland, Scandinavia, the Arctic Basin." [2].

Dr James Lovelock re fewer than 1 billion surviving this century and in answer to the New Scientist interviewer question "Do you think that we will survive"(January 2009): "I'm an optimistic pessimist. I think it's wrong to assume we'll survive 2°C of warming: there are already too many people on Earth. At 4°C we could not survive with even one-tenth of our current population. The reason is we would not find enough food, unless we synthesised it. Because of this, the cull during this century is going to be huge, up to 90 per cent. The number of people remaining at the end of the century will probably be a billion or less. It has happened before: between the ice ages there were bottlenecks when there were only 2000 people left. It's happening again. I don't think humans react fast enough or are clever enough to handle what's coming up. Kyoto was 11 years ago. Virtually nothing's been done except endless talk and meetings." [3].

[1]. James Lovelock, "Climate war could kill nearly all of us, leaving survivors on the Stone Age", Guardian, 29 June 2009: http://www.guardian.co.uk/environment/2009/jun/29/climate-war-lovelock.
[2]. Jeff Goodell, "The Prophet of climate change", Rolling Stone, 1 November 2007: http://www.rollingstone.com/politics/story/16956300/the_prophet_of_climate_change_james_lovelock.
[3]. Gaia Vince, "One last chance to save mankind", New Scientist, 23 January 2009: http://www.newscientist.com/article/mg20126921.500-one-last-chance-to-save-mankind.html?full=true

MORRIGAN: Tariel Mórrígan: back to 300 ppm CO_2 & "current trends ... will significantly limit the human carrying capacity of the Earth"

Tariel Mórrígan is the Principal Research Associate, Global Climate Change, Human Security and Democracy, Global & International Studies, University of California, Santa Barbara, California, USA: http://globalchangewatch.weebly.com/.

Tariel Mórrígan on need to get back to 300 ppm CO_2 (2010): "Limiting the atmospheric CO_2 concentration to no greater than 350 ppm might prevent committed global warming to no more than 2.4°C in the long term, after the temporary delay by climate and ocean thermal inertia reach their peak potential climate forcing (i.e. warming). Stabilization at or below 350 ppm CO_2-eq provides a 93% probability of staying below 2°C above pre-industrial values (IPCC, 2007c ; Meinshausen, 2006). Therefore a CO_2 target as low as 300 ppm may be necessary to stabilize to prevent a dangerous warming of 2°C. Global average temperatures may stabilize within a likely range of 0.6-1.4°C above pre-industrial values at or below 350 ppm CO_2-eq (300 ppm CO_2) (IPCC, 2007c; Meinshausen, 2006)." [1].

Synopsis of "Peak energy, climate change, and the collapse of global civilization. The current oil crisis" by Tariel Morrigan (2010): "This report is a synthesis of the current state of knowledge on energy resources and global climate and environmental change. The findings clearly indicate that the convergence of peak energy resources and dangerous anthropogenic climate and environmental change will likely have a disastrous impact in the near- and long-term on the quantity and quality of human life on the planet… The findings suggest:

>> Global conventional oil production likely peaked around 2005 – 2011.

>> Peak global production of coal, natural gas, and uranium resources will likely occur by 2020 – 2030, if not sooner.

>> Energy resource shortages post-peak oil will likely cause a systemic collapse of global industrialized civilization in the near-term as the abundant fossil fuel energy resources used to develop and support industrialized economies become increasingly scarce.

>> Current trends in land, soil, water, and biodiversity loss and degradation, combined with potential climate change impacts, ocean acidification, a mass extinction event, and energy scarcity will significantly limit the human carrying capacity of the Earth." [2].

[1]. Tariel Mórrígan, "Target atmospheric greenhouse gas concentrations. Why humanity should aim for 350 ppm CO_2-e": http://www.global.ucsb.edu/climateproject/papers/pdf/Morrigan_2010_Target%20Atmospheric%20GHG%20Concentrations.pdf.

[2]. Synopsis of Tariel Mórrígan, "Peak energy, climate change, and the collapse of global civilization. The current oil crisis", University of California, Santa Barbara (2010): http://www.global.ucsb.edu/climateproject/papers/index.html#.

<u>**REES: Professor Martin Rees FRS, astronomer,
astrophysicist, cosmologist**</u>

Martin Rees, Baron Rees of Ludlow, OM, FRS (born 23 June 1942 in York) is a British cosmologist, astronomer and astrophysicist. He has been Astronomer Royal since 1995 and Master of Trinity College, Cambridge since 2004. He was President of the Royal Society between 2005 and 2010 (see: http://en.wikipedia.org/wiki/Martin_Rees,_Baron_Rees_of_Lud low).

Martin Rees on the question "Are we all doomed? (2011): "The threat of global nuclear annihilation involving tens of thousands of bombs has been in abeyance since the end of the cold war. But, in future decades, a global political realignment could lead to a stand-off between new superpowers which could be handled less well or less luckily than the Cuban crisis was. In the meantime, there is more risk than ever that smaller nuclear arsenals will be used in a regional context, or even by terrorists. But other threats loom larger. Devastation could arise insidiously rather than suddenly, through unsustainable pressure on energy supplies, food, water and other natural resources. The world's population is projected to reach nine billion by 2050. The bigger the population becomes, the greater these pressures will be - especially if the developing world narrows the gap between itself and the developed world in its per-capita consumption… Humankind's collective footprint is growing. It may irreversibly degrade our environment as our numbers grow and we each consume more. Advanced technology threatens us with other vulnerabilities… There seems to be no scientific impediment to achieving a sustainable world beyond 2050, in which the developing countries have narrowed the gap with the developed and all people benefit from further advances that could have as great and benign an impact on society as information technology. But the intractable politics and sociology - the gap between what could be and what really happens - engender pessimism." [1].

[1]. Martin Rees, "Are we all doomed?", New Statesman, 9 June 2011:

http://www.newstatesman.com/society/2011/06/world-planet-rees-future.

<u>VAN DER HOEVEN: Maria van der Hoeven, Executive Director of the International Energy Agency</u>

Maria Josephina Arnoldina van der Hoeven (born 13 September 1949) is a Dutch politician of the Christian Democratic Appeal (CDA) party and has been the Executive Director of the International Energy Agency since 1 September 2011 (see: http://en.wikipedia.org/wiki/Maria_van_der_Hoeven).

IEA Executive Director Maria van der Hoeven commenting on the IEA 2011 World Energy Outlook (WEO) report (2011): "Growth, prosperity and rising population will inevitably push up energy needs over the coming decades. But we cannot continue to rely on insecure and environmentally unsustainable uses of energy. Governments need to introduce stronger measures to drive investment in efficient and low-carbon technologies. The Fukushima nuclear accident, the turmoil in parts of the Middle East and North Africa and a sharp rebound in energy demand in 2010 which pushed CO_2 emissions to a record high, highlight the urgency and the scale of the challenge" [1].

[1]. Maria van der Hoeven quoted in IEA, "The world is locking itself into an unsustainable energy future which would have far-reaching consequences, IEA warns in its latest World Energy Outlook", IEA Press release, 9 November 2011: http://www.iea.org/press/pressdetail.asp?PRESS_REL_ID=426.

<u>WADHAMS: Peter Wadhams, professor of Ocean Physics, and Head of the Polar Ocean Physics Group in the Department of Applied Mathematics and Theoretical Physics, University of Cambridge, UK.</u>

Peter Wadhams is professor of Ocean Physics, and Head of the Polar Ocean Physics Group in the Department of Applied Mathematics and Theoretical Physics, University of Cambridge, UK. He is best known for his work on sea ice and for his early

warnings in 1990 about the thinning of Arctic summer sea ice.
Professor Wadhams is the president of the International
Association for the Physical Sciences of the Ocean Commission
on Sea Ice and Co-ordinator for the International Programme for
Antarctic Buoys (see:
http://en.wikipedia.org/wiki/Peter_Wadhams).

Professor Peter Wadhams on the record loss of Arctic sea ice
and the impending loss of all Arctic summer sea ice by 2015
(2012): "The entire ice cover is now on the point of collapse.
The extra open water already created by the retreating ice
allows bigger waves to be generated by storms, which are
sweeping away the surviving ice. It is truly the case that it will
be all gone by 2015. The consequences are enormous and
represent a huge boost to global warming" [1].

[1]. Professor Peter Wadhams quoted in Julia Horton, "Arctic
sea ice will vanish within three years, says expert", Scotsman,
29 August 2012:
http://www.scotsman.com/news/environment/arctic-sea-ice-
will-vanish-within-three-years-says-expert-1-2493681.

**WBGU: WBGU (Germany): no more than 600 Gt CO_2 more
to avoid 2C rise**

The 2009 Report of the German Advisory Council on Climate
Change (WBGU, Wissenshaftlicher Beirat der Bundesregierung
Globale Umweltveränderungen) was entitled "Solving the
climate dilemma: the budget approach" and crucially stated:
"The budget of CO_2 emissions still available worldwide could
be derived from the 2 degree C guard rail. By the middle of the
21st century a maximum of approximately 750 Gt CO_2 (billion
metric tons) may be released into the Earth's atmosphere if the
guard rail is to be adhered to with a probability of 67%. If we
raise the probability to 75%, the cumulative emissions within
this period would even have to remain below 600 Gt CO_2. In
any case, only a small amount of CO_2 may be emitted
worldwide after 2050. Thus, the era of an economy driven by
fossil fuels will definitely have to come to an end within the

first half of this century" (see WBGU, <u>"Solving the climate dilemma: the budget approach"</u>).

The consequences of this declaration of less than 600 Gt CO_2 in emissions for a 75% chance of avoiding 2 degree C temperature rise are profound. Thus, would you board a plane if it had a 25% chance of crashing? Further, the average world population in the period 2010 and 2050 will be 8.321 billion (see <u>UN Population Division, 2010 Revision</u>). Accordingly the per capita share of this terminal CO_2 pollution budget is less than 600 billion tonnes CO_2/8.321 billion people = less than 72.1 tonnes CO_2 per person.

Using data for the annual per capita greenhouse gas (GHG) (including land use change) for every country in the world in 2000 (see <u>"List of countries by greenhouse gas emissions per capita"</u>, Wikipedia) one can determine how many years left at current rates of GHG pollution (in units of CO_2-e or CO_2-equivalent i.e. taking other GHGs into account) before a given country uses up its "share". Thus for Australia 72.1 tonnes CO_2-e per person / 25.9 tonnes CO_2-e per person per year in 2000 = 2.8 years left, based on the 2000 data. Note that this analysis does not take into account historical pollution of the atmosphere.

In 2009 Australia's population was 22.0 million, Australia's GHG pollution was 600 Mt CO_2-e (CO_2 equivalent i.e. taking into account other greenhouse gases such as methane, CH_4, and nitrous oxide, N_2O). 600 Mt CO_2-e per year/ 22.0 million people = 27.3 t CO_2-e per person per year and at that rate of GHG pollution Australia would use up its 2010-2050 "share" in 72.1 t CO_2-e per person/ 27.3 t CO_2-e per person per year = 2.6 years.

However in 2009 Australia's Domestic plus Exported GHG pollution (in Mt CO_2-e) was 600 (Domestic) + 784 (coal exports) + 31 (LNG exports) = 1,415 Mt CO_2-e, this giving Australia an annual per capita Domestic plus Exported GHG pollution in 2009 of 1,415 Mt CO_2-e per year/ 22.0 million people = 64.3 tonnes CO_2-e per person per year, this being 64.3/0.9 = 71.4 times greater than the annual per capita of

Bangladesh (0.9 tonnes CO_2-e per person per year). Based on its 2009 Domestic plus Exported GHG pollution rate, Australia will take 72.1 Mt CO_2-e per person/ 64.3 t CO_2-e per person per year = 1.1 years in the period 2010-2050 to use up its "fair share" of the terminal 600 Gt CO_2-e carbon pollution budget i.e. Australia has already used up its "share" of the terminal greenhouse gas (GHG) pollution budget.

Of course there is no way that Australia will meet its "all men are created equal" global obligations and cease polluting after having already in July 2011 achieved its "fair share" of the terminal 600 Gt CO_2 global GHG pollution "budget". Australia is fundamentally committed to coal and gas use and exports. Thus about 92% of Australia's electricity derives from fossil fuel combustion, Australia is the world's biggest coal exporter, and Australia is a major liquid natural gas (LNG) exporter. The only major change adumbrated by the Gillard Labor Government is a coal to gas transition for electric power generation, this ignoring the reality that this will mean a doubling of greenhouse gas generation from the electricity sector because methane (CH_4) is 85% of natural gas, leaks at about 3.3% and is 105 times worse than CO_2 as a greenhouse gas on a 20 year timeframe and taking aerosol impacts into account.

The Australian Bureau of Agricultural and Resource Economics (ABARE) has projected that Australia's black coal exports will increase at an average rate of 2.6% per year over the next 20 years and that liquid natural gas (LNG) exports will increase at 9% per year over the same period (see <u>"Invest in Australia"</u>). Further, it is estimated that Australian exports of dried brown coal will reach 20 Mt by 2020, this corresponding to about 59 Mt CO_2-e after combustion.

Accordingly, by 2020 and based on Liberal-National Party Coalition Opposition and Labor Government (aka Lib-Lab) promises of "5% off Domestic GHG pollution by 2020" and ABARE projections (see ABARE, <u>"Australian energy: national and state projections to 2029-30"</u>), Australian Domestic plus Exported GHG pollution will be 621 Mt CO_2-e (Domestic)

(Australian Government, Treasury, "Strong Growth, Low Pollution. Modelling a Carbon Price", 2011) + 1.326 x 784 =1,039 Mt CO_2-e (coal exports) + 2.580 x 31 = 80 Mt CO_2-e (LNG exports) + 59 Mt CO_2-e (brown coal exports) = 1,799 Mt CO_2-e i.e. 127% of that in 2009 (see "Analysis: Australian Labor Government Carbon Price-ETS scheme fails & entrenches climate change inaction", Bellaciao, 16 July 2010).

Thus Australian policy flies in the face of science and "all men are created equal" which show that Australia has already used up it share of the 2010-2050 terminal GHG pollution budget. Instead Australia officially projects to increase its annual pollution by 2020 by about 27% over that in 2009. How does Australia's refusal to decrease its disproportionate GHG pollution compare with the conduct of other countries? The time (at 2000 pollution rates) for every country in the World to use up its "fair share" of the World's 600 Gt CO_2 terminal GHG pollution budget is set out above [by 2020 the World had used up this terminal budget].

Years to the required "fair shares" total cessation of GHG pollution at current rates of pollution = 72.1 tonnes CO_2-e per person/ (tonnes CO_2-e per person per year). The annual per capita GHG pollution for each country in 2000 with the land use contribution included (tonnes CO_2-e per person per year) was used (the available data for Uruguay was the 2005 per capita data without the land use contribution included). It should be noted that fossil fuel use, livestock production and deforestation variously contribute to annual per capita GHG pollution. Of course if you can access more up-to-date data (e.g. the example of Australia) and then you can use it to determine an updated time for zero emissions. Note that this analysis does not take into account historical industrial pollution of the atmosphere (73% due to European countries) (see 2008 Letter of Dr James Hansen, NASA GISS, to PM Kevin Rudd of Australia).

I must reiterate that there is no way that Australia will meet its global "fair shares" obligations because it is fundamentally committed to oil use and to coal and gas use and exports. Thus about 92% of Australia's electricity derives from fossil fuel combustion, Australia is the world's biggest coal exporter and a major liquid natural gas (LNG) exporter. Both the major parties, the Liberal –National Party Coalition Opposition (the Libs) and the Labor Government (the Labs) (collectively known as the Lib-Labs) are committed to a derisory policy of 5% off 2000 Domestic GHG pollution by 2020 but with greed-driven growth of coal and LNG Exports (at 2.6% pa and 9% pa, respectively). Australia is committed to a greedy and inhumane course of climate exceptionalism, climate racism and climate injustice. Having already used up its share of the terminal 600 Gt CO_2-e budget, climate criminal Australia is now greedily and disproportionately using up the quotas of other countries (climate racism), with serious global implications as set out below.

Both Dr James Lovelock FRS (Gaia hypothesis) and Professor Kevin Anderson (Director, Tyndall Centre for Climate Change Research, University of Manchester, UK) have recently estimated that only about 0.5 billion people will survive this century due to unaddressed, man-made global warming. Noting that the world population is expected to reach 9.5 billion by 2050, these estimates translate to a Climate Genocide involving deaths of about 10 billion people this century, mostly non-Europeans, this including about 6 billion under-5 year old infants, 3 billion Muslims in a terminal Muslim Holocaust, 2 billion Indians, 1.3 billion non-Arab Africans, 0.5 billion Bengalis, 0.3 billion Pakistanis and 0.3 billion Bangladeshis. Already 18 million people die avoidably every year in Developing countries (minus China) due to deprivation and deprivation-exacerbated disease and man-made global warming is already clearly worsening this global avoidable mortality holocaust. However 10 billion avoidable deaths due to global warming this century will yield an average global annual avoidable death rate of 100 million per year (see "Climate Genocide").

Where does your country come in this "years left until zero emissions" analysis? The World is badly running out of time. The World will have to take action against the more notorious climate criminal and climate racist countries such as Australia through Sanctions, Boycotts, Sporting Boycotts (as were successfully applied to Apartheid South Africa through exclusion from the Olympic Games and other events), Green Tariffs, International Court of Justice litigations and International Criminal Court prosecutions.

CHAPTER 2: REVISED ANNUAL PER CAPITA GREENHOUSE GAS POLLUTION FOR ALL COUNTRIES – WHAT IS YOUR COUNTRY DOING?

[First published as Gideon Polya, "Revised Annual Per Capita Greenhouse Gas Pollution For All Countries – What Is Your Country Doing?", Countercurrents, 6 January 2016: https://countercurrents.org/polya060116.htm.]

The Paris Climate Change Conference failed Humanity and has locked in a catastrophic temperature rise of about plus 2.7 degrees C. All ordinary folk can do is to boycott the worst polluters. World Bank analysts have revised annual greenhouse gas (GHG) pollution upwards by 50% to 64 billion tonnes CO_2-e by properly accounting for land use for animal husbandry and the same approach has been used here to properly re-calculate annual per capita GHG pollution for all countries and hence the best targets for global Boycotts, Divestment and Sanctions (BDS) to save the planet.

World Bank analysts carefully re-evaluated the contribution of livestock production to world annual GHG pollution and found that the world's annual total rose from 41.76 billion tonnes CO_2-equivalent (CO_2-e) as estimated by the Food and Agricultural Organisation (FAO) to 63.80 billion tonnes CO_2-e, with livestock production contributing over 51% of the higher figure [1]. A key element of their analysis was to use a Global Warming Potential (GWP) of methane (CH_4) relative to that of carbon dioxide (CO_2) of 72 in a 20-year time frame rather than the 25 on a 100 year time frame used by the FAO [1]. Indeed the World Bank analysis evidently still understates the GHG pollution because NASA scientists have re-evaluated the GWP of CH_4 as 105 in a 20 year time frame with aerosol impacts considered [2].

Relatively up to date and accurate data on global CO_2 emissions from cement manufacture and burning fossil fuels is available from the US Government [3] and from the EU Emission Database for Global Atmospheric Research (EDGAR) [4]. Global 2000 data on annual per capita greenhouse gas (GHG) pollution with and without land use is available on Wikipedia [5] and was used, as outlined below, to determine for every country a current corrected estimate of annual per capita greenhouse gas (GHG) emissions – data that is otherwise astonishingly unavailable for ready access by concerned citizens of the planet.

World non-land use CO_2 pollution in 2013 was 4.932 tonnes CO_2 per person per year [4], and with a world population of 7.137 billion in 2013 this corresponds to 35.23 billion tonnes CO_2-e. The World Bank estimate of 63.80 billion tonnes CO_2-e involves $63.80 - 35.23 = 28.57$ billion tonnes CO_2-e due to land use and livestock [1]. The previous FAO estimate of 7.52 billion tonnes CO_2-e due to land use and livestock is 3.8 times less than the revised World Bank estimate of 28.57 billion tonnes CO_2-e due to livestock; the FAO estimate of an annual total of 41.76 billion tonnes CO_2-equivalent (CO_2-e) minus 35.23 billion tonnes CO_2 (due to fossil fuel burning and cement) is 6.53 billion tonnes CO_2-e due to land use which is 4.4 times less than revised World Bank estimate of 28.57 billion tonnes CO_2-e due to livestock; accordingly, as described below, a correction factor of 4 was conveniently applied to make upwardly corrected estimates of annual per capita GHG pollution due to land use.

For each country, 2013 annual per capita CO_2 pollution from fossil fuel burning and cement (A) [4] was subtracted from 2000 annual per capita GHG pollution with land use included (B) [5] to get an estimate of annual per capita pollution due to land use only (C) which was corrected upwards by multiplying by a factor of 4 to get a corrected estimate of annual per capita GHG pollution due to land use (D). The sum of A plus D gave a corrected estimate (E) of current annual per capita GHG pollution for each country. In some cases, A was greater than B, in which case A was taken as the final current corrected estimate (E).

Below are listed revised annual per capita greenhouse gas (GHG) pollution for all countries (tonnes CO_2-e per person per year), the world average being 63.80 billion tonnes CO_2-e / 7.137 billion people in 2013 = 8.9 tonnes CO_2-e per person per year. The countries are grouped as the Very Good (below world average annual per capita GHG emissions), the Good (1-2 times world average), the Bad (2-4 times world average) and the Ugly (4-41 times greater than the word average).

The Ugly - about 4 to 41 times the world average:

Belize (366.9), Guyana (203.1), Malaysia (126.0), Papua New Guinea (114.7), Qatar (101.8), Zambia (97.5), Antigua & Barbuda (85.6), United Arab Emirates (82.4), Panama (68.0), Botswana (64.9), Liberia (55.0), Indonesia (53.6), New Zealand (53.2), Australia (52.9; 116 if including its huge GHG-generating exports), Nicaragua (51.2), Canada (50.1), Equatorial Guinea (47.5), Venezuela (45.2), Brazil (43.4), Myanmar (41.9), Ireland (41.4), United States (41.0), Cambodia (40.5), Kuwait (37.3), Paraguay (37.2), Central African Republic (35.7).

The Bad – between 2 and 4 times the world average:

Peru (34.8), Mongolia (32.2), Singapore (31.2), Bahrain (30.5), Trinidad & Tobago (29.8), Cameroon (29.5), Congo, Democratic Republic (formerly Zaire) (29.3), Côte d'Ivoire (29.1), Denmark (27.8), Brunei (27.4), Bolivia (27.3), Guatemala (26.9), Belgium (26.3), Ecuador (26.2), Estonia (25.4), Laos (25.3), Suriname (25.1), Netherlands (24.9), Libya (24.9), Nepal (24.6), Benin (24.5), Angola (23.8), Madagascar (23.7), Argentina (23.7), Uruguay (23.7)*, Luxembourg (23.6), Turkmenistan (23.5), Czech Republic (23.5), Zimbabwe (23.3), Gabon (23.1), Greece (21.9), United Kingdom of Great Britain and Northern Ireland (21.5), Cyprus (21.4), Congo, Republic (21.0), Spain (20.9), Finland (20.6), Israel (20.2), Norway

(20.1), Colombia (19.8), Namibia (19.8), Mauritania (19.7), South Africa (19.4), Ukraine (19.1), Germany (18.6).

The Good - between 1 and 2 times the world average:

France (17.7), Italy (17.6), Uzbekistan (17.5), Costa Rica (17.1), Sudan (16.8), Saudi Arabia (16.6), Slovenia (16.5), Azerbaijan (16.4), Russia (16.2), Sierra Leone (16.2), Slovakia (15.9), Honduras (15.8), Hungary (15.5), Kazakhstan (15.4), Portugal (15.0), Sweden (15.0), Iran (14.5), Iceland (14.2), Mexico (13.9), Oman (13.8), Malta (13.3), Austria (13.0), Poland (12.9), Jamaica (12.8), Palau (12.8), South Korea (12.7), Guinea (12.5), North Korea (12.1), Bahamas (12.1), Nigeria (11.7), Nauru (11.7), Malawi (11.7), Mali (11.6), Chad (11.6), Taiwan (11.6), Latvia (11.4), Vanuatu (11.1), Switzerland (11.0), Romania (10.9), Togo (10.9), Japan (10.7), Serbia & Montenegro (10.4), Seychelles (10.2), Bulgaria (10.1), Lebanon (9.8), Syria (9.4), Tanzania (9.3), Turkey (9.2), Barbados (9.1), Jordan (9.1), Occupied State of Palestine (9.1)*, Philippines (9.0), Guinea-Bissau (9.0).

The Very Good - at or below world average (8.9):

Ghana (8.9), Thailand (8.7), Chile (8.7), Fiji (8.7), Belarus (8.6), Sri Lanka (8.5), Macedonia (8.5), Tonga (7.4), Croatia (7.4), China (7.4), Burkina Faso (7.3), Bosnia & Herzegovina (7.2), Kenya (7.1), Dominican Republic (7.1), Senegal (7.0), Tunisia (7.0), Algeria (6.6), Grenada (6.4), Samoa (6.2), Rwanda (6.1), El Salvador (6.0), Lithuania (5.9), Mozambique (5.8), Lesotho (5.7), Burundi (5.5), Iraq (5.5), Eritrea (5.3), St Kitts & Nevis (5.1), Uganda (5.1), Haiti (5.0), Mauritius (5.0), Albania (4.3), Dominica (4.2), Bhutan (4.1), Niger (4.1), Ethiopia (4.1), Moldova (4.0), Georgia (4.0), Yemen (3.7), Tajikistan (3.7), Afghanistan (3.6), Swaziland (3.6), Cuba (3.5), Cape Verde (3.5), Kyrgyzstan (3.4), The Gambia (3.0), St Lucia (2.9), Bangladesh (2.7), Egypt (2.6), Niue (2.6), Pakistan (2.5), Morocco (2.5), Djibouti (2.4), St Vincent & Grenadines (2.4), Armenia (2.3), Maldives (2.1), India (2.1), Cook Islands (2.1), Vietnam (1.9), São Tomé and Príncipe (1.9), Comoros (1.6), Solomon Islands (1.4), Kiribati (1.2), Tuvalu (1.2)* (* indicates

an estimate based on that for an immediately contiguous,
ethnically-related country).

Conclusions.

The above analysis lists countries in order of corrected annual
per capita greenhouse gas (GHG) pollution and readily
distinguishes the Very Good (annual per capita GHG emissions
below the world average) from the Good (1-2 times greater
than the world average), the Bad (2-4 times the world average)
and the Ugly (4-41 times greater than the word average). The
countries that are merely Good GHG-wise deserve our
encouragement but the Bad and Ugly countries merit global
blowback via Boycotts, Divestment and Sanctions (BDS).

The weak, non-binding and dishonest Paris Climate Agreement
has betrayed our children, grandchildren, future generations, the
Developing World, Humanity and the Biosphere – the target of
1.5 to 2 degrees C is both unavoidable and catastrophic and key
matters are non-binding. The Paris betrayal demands a peaceful,
world-wide Climate Revolution involving peaceful actions via
the ballot box and via Boycotts, Divestment and Sanctions
(BDS) against all politicians, corporations and countries
disproportionately involved in ecocidal, speciescidal and
terracidal GHG pollution that amounts to state terrorism, state-
sanctioned corporate terrorism and state-sanctioned climate
terrorism [6, 7, 8].

The 2009 Report of the German Scientific Advisory Council on
Global Change (WBGU, Wissenshaftlicher Beirat der
Bundesregierung Globale Umweltveränderungen) entitled
"Solving the climate dilemma: the budget approach" crucially
stated: "The budget of CO_2 emissions still available worldwide
could be derived from the 2 degree C guard rail. By the middle
of the 21st century a maximum of approximately 750 Gt CO_2
(billion metric tons) may be released into the Earth's atmosphere
if the guard rail is to be adhered to with a probability of 67%. If
we raise the probability to 75%, the cumulative emissions
within this period would even have to remain below 600 Gt
CO_2. In any case, only a small amount of CO_2 may be emitted

worldwide after 2050. Thus, the era of an economy driven by fossil fuels will definitely have to come to an end within the first half of this century" [9].

The consequences of this declaration of less than 600 Gt CO_2 (600 billion tonnes CO_2) in emissions for a 75% chance of avoiding a catastrophic 2 degree C temperature rise are profound. Thus, would you board a plane if it had a 25% chance of crashing? Further, the average world population in the period 2010 and 2050 will be 8.321 billion. Accordingly the per capita share of this Terminal Carbon Pollution Budget is less than 600 billion tonnes CO_2/8.321 billion people = less than 72.1 tonnes CO_2 per person.

Using the above corrected data for the annual per capita greenhouse gas (GHG) emissions that properly accounts for land use and livestock impacts, one can determine how many years left at current rates of GHG pollution (in units of CO_2-e or CO_2-equivalent i.e. taking other GHGs into account) before a given country uses up its "fair share" of this Terminal Carbon Pollution Budget. Thus, for example, for Australia 72.1 tonnes CO_2-e per person / 52.9 tonnes CO_2-e per person per year = 1.4 years left relative to 2010, noting that this analysis does not take into account historical pollution of the atmosphere. Thus Australia used up its "fair share" of the world's Terminal Carbon Pollution in 2011, and since then has been stealing the entitlement of the other countries which have not yet used up their entitlement.

Indeed the whole world is very close to using up its Terminal Carbon Pollution Budget. Thus years left before the world uses up this budget = 72.1 tonnes CO_2 per person/ 8.9 tonnes CO_2-e per person per year = 8.1 years relative to 2010 and thus only about 2 years relative to 2016. We are badly running out of time to deal with man-made climate change, and sensible, humane, science-informed people can (a) inform everyone they can about the corrected annual per capita GHG pollution data presented here, and (b) urge and apply Boycotts, Divestment and Sanctions (BDS) against all people, politicians, parties, companies, corporations and countries disproportionately

involved in the greenhouse gas pollution that so acutely
threatens Humanity and the Biosphere [the World finally used
up its terminal budget in 2020].

References.

[1]. Robert Goodland and Jeff Anfang. "Livestock and climate change. What if the key actors in climate change are … cows, pigs and chickens?", World Watch, November/December 2009: http://www.worldwatch.org/files/pdf/Livestock%20and%20Climate%20Change.pdf.

[2]. Drew T. Shindell, Greg Faluvegi, Dorothy M. Koch, Gavin A. Schmidt, Nadine Unger and Susanne E. Bauer, "Improved Attribution of Climate Forcing to Emissions", Science, 30 October 2009: Vol. 326 no. 5953 pp. 716-718: http://www.sciencemag.org/content/326/5953/716.

[3]. "List of countries by carbon dioxide emissions per capita", Wikipedia: https://en.wikipedia.org/wiki/List_of_countries_by_carbon_dioxide_emissions_per_capita.

[4]. European Commission, "CO_2 time series 1990-2013 per capita for world countries", Emission Database for Global Atmospheric Research (EDGAR): http://edgar.jrc.ec.europa.eu/overview.php?v=CO2ts_pc1990-2013.

[5]. "List of countries by greenhouse gas emissions per capita", Wikipedia: https://en.wikipedia.org/wiki/List_of_countries_by_greenhouse_gas_emissions_per_capita.

[6]. Gideon Polya, "Paris Climate Agreement Betrays Humanity Which Must Apply Boycotts, Divestment And Sanctions (BDS) Against Climate Criminal People, Corporations & Countries", 14 December, 2015: http://www.countercurrents.org/polya141215.htm.

[7]. "Stop state terrorism": https://sites.google.com/site/stopstateterrorism/.

[8]. "State crime and non-state terrorism": https://sites.google.com/site/statecrimeandnonstateterrorism/

[9]. WBGU, "Solving the climate dilemma: the budget approach", 2009: http://www.wbgu.de/fileadmin/templates/dateien/veroeffentlichungen/sondergutachten/sn2009/wbgu_sn2009_en.pdf.

CHAPTER 3: HOW MUCH NEGATIVE EMISSIONS, NEGATIVE POPULATION GROWTH & NEGATIVE ECONOMIC GROWTH IS NEEDED TO SAVE PLANET?

[First published as Gideon Polya, "How much negative emissions, negative population growth & negative economic growth is needed to save Planet?", Countercurrents, 28 November 2018: https://countercurrents.org/2018/11/28/how-much-negative-carbon-emissions-negative-population-growth-negative-economic-growth-is-needed-to-save-planet/.]

Several recent reports collectively endorsed by thousands of expert scientists have warned the world that time is running out to save Humanity and the Biosphere from further catastrophic climate change and further massive biodiversity loss. Massive harm has already occurred due to continuing carbon pollution, population growth and economic growth and it is clear that zero growth in these areas is insufficient – there must be negative carbon pollution (atmospheric CO_2 draw-down), negative population growth (population decline) and negative economic growth (de-growth) to halt and reverse this worsening disaster. However the key question is how much reversal is needed for a safe and sustainable planet?

Of course the whole proposition of negative growth in relation to atmospheric greenhouse gases (GHGs) and economic activity is anathema to the present growth-based world order as neatly demonstrated by the results from Google Searches for the following terms: "zero emissions" (2.24 million) versus "CO_2 draw-down" (2,810), and "economic growth" (76.4 million) versus "negative "economic growth" (279,000). However a quantitatively similar degree of concern over "zero population growth" (213,000) versus "negative population growth" (132,000) is explicable because population decline is already a reality in many advanced economies to the consternation of the politically dominant, growth-committed neoliberals.

1. Atmospheric CO_2 population and economic growth existentially threaten Humanity and the Biosphere.

Before addressing the question of how much de-growth is required in relation to GHG pollution, population and economic activity, it is useful to summarize the worsening climate crisis, resource depletion crisis and biodiversity crisis. In 2017 a letter signed by over 15,000 scientists documented the following disastrous changes in 9 key areas in the last 24 years: (1) – 26.1% in freshwater resources per capita, (2) –6.4% in marine catch, (3) +75.3% in marine dead zones, (4) –2.8% in forest area, (5) –28.9% in vertebrate species abundance, (6) +62.1% in annual CO_2 emissions, (7) +167.6% in average temperature change, (8) +35.5% in human population, and (9) +20.5% in ruminant population [1, 2]. If these changes observed over a quarter century continue at the same rate then, for example, there will be a near 100% decline in non-human vertebrate species abundance in the coming century. We must not destroy what we cannot replace but the species extinction rate is presently 100-1,000 times greater than normal [3].

Similarly the World Wildlife Fund (WWF) "Living Planet Report 2018" [4] describes the unfolding human consumption disaster since 1750 as the Great Acceleration that is characterized by huge increases in 12 socio-economic trends (world population, real GDP, foreign direct investment, urban population, direct energy use, fertilizer consumption, large dams, global water use, paper production, transportation, telecommunications, and international tourism) and in 12 Earth system trends (carbon dioxide (CO_2), nitrous oxide (N_2O), methane (CH_4), stratospheric ozone (O_3) loss [a good result], surface temperature, ocean acidification, marine fish capture, shrimp aquaculture, nitrogen to coastal zone, tropical forest loss, domesticated land, and terrestrial biosphere degradation) (Figure 2, [4]). The WWF states: "Average abundance of

16,704 populations representing 4,005 species monitored across the globe have declined by 60%" [since 1970] (Figure 7 [4]). This huge human impact on the Biosphere has led to the present age being described as the Anthropocene Era.

In October 2018 the IPCC issued a dire report that details the numerous bad outcomes of a global +1.5 degree Centigrade (+1.5C) of warming versus the catastrophic outcomes from a +2C e.g. a further 70-90% decline of coral reefs at +1.5C versus more than 99% loss at +2C. The IPCC Report says that for less than +1.5C coal burning must cease by 2050 but also declares that the terminal CO_2 pollution budget for a 66% chance of avoiding +1.5C will be used up in 10 years [5-7]. However the climate change denialist American Trump Administration and the effectively climate change denialist, pro-coal and pro-gas Australian Government have effectively rejected the IPCC Report [7]. The atmospheric levels of the GHGs CO_2, methane (CH_4) and nitrous oxide (N_2O) continue to rise [8-12]. Indeed the world is acutely threatened by a methane bomb in the warming Arctic as methane is released from huge stores of methane-water clathrates in the tundra and the Arctic Ocean sea bed [13, 14]. In defiance of the science, electricity sector gas, oil and coal use has steadily increased over the period 1965-2017 [15]. Indeed the global national commitments to the Paris Agreement amounted to a horrendous 3.4C temperature rise.

The Red Cross World Disasters Report 2018 states that in the last 10 years there have been 3,751 reported major natural hazards (84% weather-related), and these have been associated with 2 billion people affected (95% weather-related), and a $1,658 billion cost (76.2% weather-related) (Figure 9 [16]). It has been estimated that 0.4 million people die annually from climate change [17] but this is a considerable under-estimate because already 15 million people die avoidably from deprivation each year in the Third World that is disproportionately impacted by man-made climate change [18]. 7 million people perish each year due to the long-term effects of pollutants from carbon fuel burning [19]. Several leading climate scientists have predicted that only 0.5 billion people will survive this century unless requisite climate action is taken, this translating to a worsening climate genocide in which 10 billion people will perish this century [20-24].

The 3 key existential threats against Humanity are nuclear weapons (a nuclear winter from nuclear war – that could happen any time – would wipe out most of Humanity and the Biosphere), poverty (already 15 million people die avoidably from deprivation in the Third World each year (this being compounded by climate change, population growth, resource depletion and competition for resources), and climate change (that unaddressed could kill 10 billion people this century) [18, 20, 24]. Rational world governance aside, 3 key processes to tackle these existential threats are negative carbon pollution (atmospheric CO_2 draw-down), negative population growth (population decline) and negative economic growth (de-growth). The following sections address how much reversal is needed for a safe and sustainable planet.

2. Negative carbon pollution – draw down of atmospheric carbon dioxide (CO_2) to circa 300 ppm CO_2.

The excellent climate activist organization 350.org argues that the atmospheric CO_2 should be drawn down from the present dangerous and damaging circa 410 ppm CO_2 to no more than 350 ppm CO_2. However numerous climate scientists, Arctic ice scientists, paleoclimatologists, biologists and coral scientists argue that the target should be 300 ppm CO_2, or roughly the pre-Industrial Revolution atmospheric CO_2 level [24, 25]. A pertinent sample of numerous such opinions is given below.

Thus Professor Hans Joachim Schellnhuber (director of the Potsdam Institute for Climate Impact Research, Germany): "It is a compromise between ambition and feasibility. A rise of $2^{o}C$ could avoid some of the big environmental disasters, but it is still only a compromise…It is a very sweeping argument, but nobody can say for sure that 330ppm is safe. Perhaps it will not matter whether we have 270ppm or 320ppm, but operating well outside the [historic] realm of carbon dioxide concentrations is risky as long as we have not fully understood the relevant feedback mechanisms" [26].

Professor James Hansen (NASA and Columbia University): "CO_2 ~300-325 ppm, may be needed to restore sea ice to its area of 25 years ago" [27, 28].

Dr Andrew Glikson (an Earth and paleo-climate research scientist at Australian National University, Canberra, Australia): "The amount of carbon stored in Arctic sediments and permafrost is estimated as 500–2500 Gigaton Carbon (GtC), as compared with the world's total fossil fuel reserves estimated as 5000 GtC. Compare with the 700 GtC of the atmosphere, which regulate CO_2 levels in the range of 180–300 parts per million and land temperatures in a range of about − 50 to + 50 degrees C, which allowed the evolution of warm blooded mammals. The continuing use of the atmosphere as an open sewer for industrial pollution has already added some 305 GtC to the atmosphere together with land clearing and animal-emitted methane. This raised CO_2 levels to 387 ppm CO_2 to date [415 ppm CO_2 in 2020], leading toward conditions which existed on Earth about 3 million years (Ma) ago (mid-Pliocene), when CO_2 levels rose to about 400 ppm, temperatures to about 2–3 degrees C and sea levels by about 25 +/- 12 metres" [28].

David Bielo (a contributing editor at Scientific American): "The last time CO_2 levels at Mauna Loa were this high [400 ppm CO_2], *Homo sapiens* did not live there. In fact, the last time CO_2 levels are thought to have been this high was more than 2.5 million years ago, an era known as the Pliocene, when the Canadian Arctic boasted forests instead of icy wastes. The land bridge connecting North America and South America had recently formed. The globe's temperature averaged about 3 degrees C warmer, and sea level lapped coasts 5 meters or more higher" [29].

Dr J.E.N. Veron and leading coral scientist colleagues: "The coral reef crisis: the critical importance of <350 ppm CO_2… Temperature-induced mass coral bleaching causing mortality on a wide geographic scale started when atmospheric CO_2 levels exceeded 320 ppm. When CO_2 levels reached 340 ppm, sporadic but highly destructive mass bleaching occurred in most reefs world-wide, often associated with El Niño events. Recovery was dependent on the vulnerability of individual reef areas and on the reef's previous history and resilience. At today's level of 387 ppm [415 ppm CO_2 in 2020], allowing a lag-time of 10 years for sea temperatures to respond, most reefs world-wide are committed to an irreversible decline. Mass bleaching will in future become annual, departing from the 4 to 7 years return-time of El Niño events. Bleaching will be exacerbated by the effects of degraded water-quality and increased severe weather events. In addition, the progressive onset of ocean acidification will cause reduction of coral growth and retardation of the growth of high magnesium calcite-secreting coralline algae" [30].

There are many ways by which atmospheric CO_2 can be drawn down [31-34]. The various Carbon Dioxide Removal (CDR) technologies suggested include Bio-energy with Carbon Capture and Storage (BECCS, involving CO_2 sequestration into cellulosic biomass by photosynthesis and safe storage of the biomass), Biochar (from high temperature anaerobic pyrolysis of biomass cellulose to generate carbon for safe storage), re-afforestation, Accelerating Weathering of Limestone (AWL, generating alkaline bicarbonate ions for dissolution in the ocean), Direct Air Capture (DAC, via removal of CO_2 from power plant exhaust through dissolution in alkaline sodium hydroxide solution), ocean fertilization (to promote CO_2 sequestration as algal cellulosic biomass), and Carbon Capture and Storage (CCS, that is very expensive and yet to be applied on a big scale). Because CO_2 removal is so expensive the simplest solution is immediate cessation of carbon fuel burning [31-34], noting that future generations will have to pay for this CO_2 draw-down [35-39]. Of course it is not just the atmospheric levels of the GHG CO_2 that must be reduced. Methane (CH_4) and nitrous oxide (N_2O) are variously generated from agriculture and land use and are much more potent as GHGs than CO_2. The climate crisis demands cessation of meat eating and methanogenic animal husbandry [10, 40-43].

3. **Negative population growth in the impoverished South must follow the present example of the rich North.**

Even if per capita GHG pollution remained the same, increasing population would see GHG pollution increase. Further, while low birth-rate, the rich North could live more modestly, the impoverished and high birth-rate South needs an increase in per capita income to enable circumstances in which birth rate can be reduced (high female literacy, infant survival and economic security [18]). UN Population Division projections show that this century the population of rich Europe will fall from 742 million (2018) to 653 million (2100) whereas the population of impoverished Africa will explode from 1,274 million (2018) to 4,468 million (2100) [44]. However in terms of relative carbon footprint the per capita GDP (nominal) is $25,596 for Europe versus $1,753 for Africa (2016), this translating to a total GDP of $19.3 trillion for Europe and $2.2 trillion for Africa [45]. For carbon footprint equality the European population would need to decrease 10-fold or the African per capita income would have to increase 10-fold. However Professor Dabo Guan (School of International Development, University of East Anglia, UK) (2016) has commented thus on inescapable limits to growth: "For everyone in the world to have an American lifestyle, we would need seven planets, and three to live as Europeans" [46]. Clearly the European population is not declining fast enough and the African population is expanding impossibly.

Drastic decrease in population is required for cessation of the present appalling level of ecosystem destruction (ecocide) that is associated with catastrophic species loss (speciescide) and leading to omnicide and terracide. If we take world coral as a "canary in the coal mine" then the 320 ppm CO_2 at which coral reefs started to decline [30] was reached in 1965 [8], at which time the world's population was 3.340 billion as compared to the present 7.5 billion [44]. One can therefore plausibly suggest that given the present carbon economy the world's population needs to roughly halve for a safe and sustainable environment for all peoples and all species.

4. **Negative economic growth in the North and initial limited economic growth in the South.**

As outlined above the impact of man on the Biosphere needs to be reduced by about 50% but most of the negative economic growth must happen in the North with initial limited economic growth in the South to enable a decent life and the possibility of negative population growth that has prerequisites of modest economic security, female literacy, good primary health care and low infant mortality. The world's GDP (nominal) is estimated at \$88 trillion (2018) [47] and with an eventual halving of world population there would be a corresponding halving of world GDP (nominal) to a steady state of about \$44 trillion.

Professor Kevin Anderson (Deputy Director of the Tyndall Centre for Climate Change Research, University of Manchester), together with Dr Alice Bows wrote an extremely important paper in 2008 describing 6-8% annual GHG emissions reductions needed for a 450 ppm CO_2-equivalent (CO_2-e) target (2008): "According to the analysis conducted in this paper, stabilizing at 450 ppmv [carbon dioxide equivalent = CO_2-e, atmospheric concentration measured in parts per million by volume] requires, at least, global energy related emissions to peak by 2015, rapidly decline at 6-8% per year between 2020 and 2040, and for full decarbonization sometime soon after 2050 …Unless economic growth can be reconciled with unprecedented rates of decarbonization (in excess of 6% per year), it is difficult to envisage anything other than a planned economic recession being compatible with stabilization at or below 650 ppmv CO_2-e … Ultimately, the latest scientific understanding of climate change allied with current emissions trends and a commitment to "limiting average global temperature increases to below $4^{o}C$ above pre-industrial levels", demands a radical reframing of both the climate change agenda, and the economic characterization of contemporary society" [48].

Unfortunately 10 years on in 2018 the atmospheric CO_2 is increasing [8] as is energy sector oil use, gas use and (arguably) coal use [15]. Further, the "stabilizing" 450 ppm CO_2-e proposed by Anderson and Bows in 2008 [48] has already been exceeded. Thus the 2014 IPCC Report stated that "About 450 ppm CO_2-e [is] likely to limit warming to 2°C above pre-industrial levels" [49] but according to Professor Ron Prinn (Professor of Atmospheric Science, Massachusetts Institute of Technology) an atmospheric 478 ppm CO2-e (i.e. also taking GHGs other than CO_2 into account) was already attained by 2013 [50]. Atmospheric CO_2 continues to rise, having already reached 410 ppm CO_2 [8], and atmospheric methane (CH_4, from methanogenic livestock, land use, systemic natural gas leakage and Arctic tundra melting) and nitrous oxide (N_2O, from oxidation of nitrogen fertilizers) are also increasing [14, 51]. Atmospheric CO_2, CH_4 and N_2O levels are presently about 1.4, 2.5 and 1.2 times, respectively, of those in 1800 [51], noting that the relative Global Warming Potential per unit mass for these GHGs on a 20-year time scale are 1, 86 and 268, respectively [52] and the present levels in 2018 are 410 ppm CO_2, 1.85 ppm CH_4 and 0.33 ppm N_2O [51].

In view of the continuing GHG pollution one concludes that exceedance of +2C is now effectively unavoidable but we are obliged to do everything we can to make the future "less bad" for future generations. As outlined below, a substantial effective de-growth in the North can be achieved by a North to South wealth transfer through application of a.substantial annual wealth tax.

5. **Effective North de-growth through an annual wealth tax to assist the South.**

Avoidable mortality (avoidable death, excess mortality, excess death, deaths that do not have to happen) can be defined as the difference between actual deaths in a country and deaths expected for a peaceful, decently-run country with the same demographics (i.e. similar birth rate and age distribution) [18]. For relatively high birth rate Developing World countries the baseline death rate is about 0.4% or 4 persons per 1,000 of population each year [18]. However for the Developing World (minus China) (2017-2018 population 4,886 million) the death rate is 7.1 deaths per 1,000 of population per year (2017-2018) [44], this yielding an avoidable death rate of 7.1 – 4.0 = 3.1 avoidable deaths per 1,000 of population per year, and 3.1 avoidable deaths per 1,000 of population per year x 4,885,541 thousand persons = 15.1 million avoidable deaths annually.

The annual avoidable deaths as a percentage of population is about 0.0% for Overseas European countries (the US, Canada, Australia, New Zealand, and Apartheid Israel), 0.01% (East Asia), 0.03% (Latin America and the Caribbean), 0.05% (Western Europe), 0.25% (Arab North Africa and Middle East), 0.26% (South East Asia), 0.26% (Turkey, Iran and Central Asia), 0.31% (Eastern Europe), 0.38% (South Asia), 0.39% (the Pacific), and 0.97% (non-Arab Africa) (2003 data, [18]).

Notably, annual avoidable deaths as a percentage of population is 0.0% for the US, Cuba and China [18] but these countries presently have annual per capita incomes (annual GDP per capita) of $57,808, $7,815 and $8,126, respectively (UN data, 2016, [45]), and under-5 infant mortality (under-5 infant deaths per 1,000 live births) of 6, 6 and 12, respectively (2017-2018, [44]).

It is glaringly obvious that zero avoidable mortality (as defined above) and extremely low infant mortality is possible in countries such as Cuba and China with annual per capita incomes about 7 times lower than that of the US provided the countries have good governance, high female literacy, good primary health care and peace [18]. Accordingly, it is useful to calculate how much it would cost in wealth transfer from the rich to the poor in order to bring all countries up to a Cuba level of annual per capita income and hence with rational governance abolish the Global Avoidable Mortality Holocaust that kills 15 million people each year through deadly deprivation [18].

I published such a calculation in 2014 using data made available on Wikipedia, specifically "List of countries by GDP (nominal) per capita" (US dollars, UN data, 2012) [45] and "List of countries by population" (2013-2014) [53]. For each country I simply subtracted the country's per capita from the Cuban per capita ($6,301) and multiplied this difference by the country's population. The result was a wealth tax in 2014 of $16, 016.5 billion or about $16 trillion [54] and about 4% of the total global wealth of $367 trillion as estimated by Thomas Piketty (page 438 [55]).

Since the Cuban GDP had increased from $6,301 (2012) to $7,815 (2016) we could estimate that the cost of lifting all the world to a Cuban per capita GDP would have proportionally increased to about $16,016 x $7,815 / $6,301 = $19,864 billion or about $20 trillion in 2016 and $25 trillion in 2020. Thus a $25 trillion annual donation from the North to the South to abolish Third World avoidable mortality would in effect be an annual North economic de-growth in terms of wealth transfer. Money means power and it is not going to happen under the present world order but the present calculation at least establishes a boundary condition for future action.

6. **Universal Basic Income and economic de-growth in rich countries through abolition of damaging and deadly economic pursuits.**

Annual avoidable mortality from deprivation is effectively zero in rich countries such as those of the Anglosphere. However such rich countries suffer a major burden of preventable deaths through "life-style choices" (e.g. smoking, drinking, obesity, air pollution, cars) and inadequate government action (e.g. adverse hospital events, inadequate mental health services, inadequate preventive medicine, and lack of medical insurance). Thus in the US each year about 1.7 million Americans die preventably each year from "life-style" or "governance" causes, the breakdown (with some overlaps) including (1) 443,000 from smoking-related causes, (2) 440,000 from adverse events in hospitals, (3) 300,000 from obesity-related causes, (4) 200,000 from air pollution (e.g. from coal burning, vehicle exhaust, carbon burning in general), (5) 75,000 from alcohol-related causes, (6) 45,000 from lack of medical insurance, (7) 38,000 US drug-related deaths annually, this including 21,000 US opiate drug-related deaths annually from US restoration and protection of the Taliban-destroyed Afghan opium industry), (8)

33,000 from motor vehicle accidents, (9) 31,000 gun-related deaths, (10) 30,000 suicides with 7,000 being US veterans, (11) 21,000 preventable under-5 year old infant deaths, and (12) 15,000 violent homicides with an average of only 4 due to jihadi psychopaths [56, 57]. Now the risk avoidance-based Value of a Statistical Life is about $7 million for Americans [58] and hence the cost of these 1.7 million preventable American deaths is $11.9 trillion or 58% of the US GDP (nominal, 2018) of $20.4 trillion [47].

Annual preventable deaths from "life-style" or "governance" causes in Australia, Canada, the UK and the US total, 85,000, 101,000, 150,000 and 1.7 million, respectively [56-66]. Such preventable deaths in Australia, Canada, UK and the US have thus totalled 1.4 million, 1.7 million, 2.6 million, and 29 million, respectively, since the US Government's false flag 9/11 atrocity in 2001 that killed about 3,000 people [67]. Pollutants from carbon fuel burning kill 7 million people annually (WHO) and have killed 119 million people thus since 9-11 [19]. There have been 3.4 million US air pollution deaths versus about 70 American deaths in the US from jihadi terrorism deaths since 9-11 [19, 68]. Clearly these preventable deaths in these rich countries could have been avoided by prohibition (e.g. of alcohol, tobacco, illicit drugs, guns, air pollution) and better social organization in relation to medical care access, mental health services, preventive medicine, and public education. Accordingly there is considerable scope for economic de-growth by abolishing damaging and deadly industries ranging from polluting enterprises (fossil fuels, cars, methanogenic livestock production, deforestation), to guns, licit drugs and illicit drugs.

These ignored and hidden costs of economic activity well illustrate Polya's 3 Laws of Economics that mirror the 3 Laws of Thermodynamics and are (1) Price minus COP (Cost of Production) equals profit (p); (2) Deception about COP strives to a maximum; and (3) No work, price or profit on a dead planet. These fundamental laws help expose the failure of neoliberal capitalism in relation to wealth inequality, massive tax evasion by multinational corporations, and horrendous avoidable deaths from poverty and pollution culminating in general ecocide, speciescide, climate genocide, omnicide and terracide [69]. Ignoring the huge human premature mortality associated with much economic activity means grossly under-estimating the actual Cost of Production (COP) and the ultimate profit (p) contributing to GDP.

In the Anglosphere countries the neoliberal politicians woo the voters with the mantra of "jobs and growth" that ignores the deadly downsides of some jobs and the environmental impact of "growth". Eliminating deadly and damaging "jobs" provides considerable scope for a circa 50% economic de-growth to minimize preventable deaths as well as supporting environmental sustainability. Conversely there are many extremely valuable jobs [e.g. child care] that are presently poorly paid or unpaid and not considered part of the "economy" as cogently analysed by Marilyn Waring [70, 71]. Further, negative economic growth is required to save what is left of wild nature that has a substantial economic value estimated at about $35 trillion annually (2002) or about half of the world's annual GDP [72]. Andrew Balmford and colleagues (2002): "Loss and degradation of remaining natural habitats has continued largely unabated. However, evidence has been accumulating that such systems generate marked economic benefits, which the available data suggest exceed those obtained from continued habitat conversion. We estimate that the overall benefit:cost ratio of an effective global program for the conservation of remaining wild nature is at least 100:1" [72].

Abolition of a huge number of damaging or unnecessary jobs (perhaps about 50% of the present economy in the North) would make a massive contribution to requisite North de-growth. This adumbrated contraction leads to consideration of Artificial Intelligence (AI), the future of work, work participation, government job guarantee, and Universal Basic Income. Fundamental to such considerations is how society is viewed. Neoliberalism demands maximal freedom for smart and advantaged individuals to exploit natural and human resources for private profit (and hence power), with an asserted "trickle down" benefit to the rest of society. In contrast, social humanism (socialism, democratic socialism, eco-socialism, the welfare state) seeks to sustainably maximize human happiness, dignity and opportunity for everyone through evolving intra-national and international social contracts. The worsening climate crisis and the existential threat to Humanity and the Biosphere means that neoliberalism has failed and must be replaced by sustainable social humanist models that provide a basic decent life for everyone in cooperative societies while also allowing for ambition and excellence [73, 74].

Rich, negative population growth, neoliberal Western societies are associated with an increasingly aged population, an ultimately declining participation rate (presently 65.7% in Australia) [75], and increasing part-time work for the young who are disempowered relative to the politically dominant One Percenters (who globally possess 50% of the wealth). Cooperative and humane societies with the social security of basic housing, universal health care, free education, government job guarantee and a Universal Basic Income [76-78] are conceivable in the context of substantial (circa 50%) economic de-growth. Happiness, dignity and opportunity would flow from constructive social involvement in activities from caring, community, sport, arts and altruism to scholarly, professional and technical excellence for the betterment of society. Substantial economic de-growth in cooperative societies based on "need not greed" can and must happen for environmental sustainability and avoidance of terracide [79].

Final comments.

In summary, variously as a result of overpopulation, excessive resource exploitation and huge greenhouse gas (GHG) pollution the world is already experiencing massive biodiversity loss and worsening calamities impacting Humanity (drought, forest fires, sea level rise, warming-exacerbated storms and storm surges). Unfortunately a catastrophic plus 2C temperature rise is now effectively unavoidable but we are obliged to do everything we can to make the future "less bad" for future generations. Crucial to this salvaging enterprise will be negative carbon pollution (atmospheric CO_2 draw-down to about 300 ppm CO_2), negative population growth by about 50% (population decline that is already beginning in the rich North but crucial for the impoverished but high birth-rate South), and negative economic growth (circa 50% de-growth) based on "need not greed" eco-socialism to halt and reverse the presently worsening climate and Biosphere disaster. Young people in particular can save their future by (a) informing everyone they can, and (b) re-democratizing and saving their world through demanding and achieving ecological sustainability with application of substantial wealth taxes on the presently politically dominant neoliberal One Percenters.

References.

[1]. William J. Ripple et al., 15,364 signatories from 184 countries, "World scientists' warning to Humanity: a second notice", Bioscience, 13 November 2017: https://academic.oup.com/bioscience/advance-article/doi/10.1093/biosci/bix125/4605229.

[2]. Gideon Polya, "Over 15,000 scientists issue dire warning to Humanity on catastrophic climate change and biodiversity loss", Countercurrents, 20 November 2017: https://countercurrents.org/2017/11/20/over-15000-scientists-issue-dire-warning-to-humanity-on-catastrophic-climate-change-and-biodiversity-loss/.

[3]. Phillip Levin, Donald Levin, "The real biodiversity crisis", American Scientist, January-February 2002: http://www.americanscientist.org/issues/pub/the-real-biodiversity-crisis.

[4]. World Wildlife Fund (WWF) "Living Planet Report 2018 aiming higher – summary": https://s3.amazonaws.com/wwfassets/downloads/lpr2018_summary_report_spreads.pdf.

[5]. IPCC, "Global warming of 1.5 °C. Summary for Policymakers", 8 October 2018: http://report.ipcc.ch/sr15/pdf/sr15_spm_final.pdf.

[6]. Gideon Polya, "IPCC +1.5C avoidance report – effectively too late but stop coal burning for "less bad" catastrophes", Countercurrents, 12 October 2018: https://countercurrents.org/2018/10/12/ipcc-1-5c-avoidance-report-effectively-too-late-but-stop-coal-burning-for-less-bad-catastrophes/.

[7]. Gideon Polya, "Pro-coal Australia and Trump America reject dire IPCC report & declare War on Terra", Countercurrents, 17 October 2018: https://countercurrents.org/2018/10/17/pro-coal-australia-trump-america-reject-dire-ipcc-report-declare-war-on-terra/

[8]. US NOAA, "Full Mauna Loa CO_2 record": https://www.esrl.noaa.gov/gmd/ccgg/trends/full.html.

[9]. "2011 climate change course": https://sites.google.com/site/300orgsite/2011-climate-change-course.

[10]. Robert Goodland and Jeff Anfang, "Livestock and climate change. What if the key actors in climate change are … cows, pigs and chickens?", World Watch, November/December 2009: http://www.worldwatch.org/files/pdf/Livestock%20and%20Climate%20Change.pdf.

[11]. Drew T. Shindell, Greg Faluvegi, Dorothy M. Koch, Gavin A. Schmidt, Nadine Unger and Susanne E. Bauer, "Improved Attribution

of Climate Forcing to Emissions", Science 30 October 2009: Vol. 326 no. 5953 pp. 716-718: http://www.sciencemag.org/content/326/5953/716.

[12]. "Gas is not clean energy": https://sites.google.com/site/gasisnotcleanenergy/.

[13]. Gail Whiteman, Chris Hope and Peter Wadhams, "Vast costs of Arctic change", Nature, 499, 25 July 2013: http://www.nature.com/nature/journal/v499/n7459/pdf/499401a.pdf.

[14]. "Methane Bomb Threat": https://sites.google.com/site/methanebombthreat/.

[15]. David Spratt, "Our energy challenge in 6 eye-popping charts", Climate Code Red, 17 June 2018: http://www.climatecodered.org/2018/06/our-energy-challenge-in-6-eye-popping.html.

[16]. Red Cross, "World Disasters Report 2018": https://media.ifrc.org/ifrc/wp-content/uploads/sites/5/2018/10/B-WDR-2018-EXECSUM-EN.pdf.

[17]. DARA, "Climate Vulnerability Monitor. A guide to the cold calculus of a hot planet", 2012, Executive Summary pp2-3: http://daraint.org/climate-vulnerability-monitor/climate-vulnerability-monitor-2012/.

[18]. Gideon Polya, "Body Count. Global avoidable mortality since 1950", that includes a succinct history of every country and is now available for free perusal on the web: http://globalbodycount.blogspot.com/.

[19]. "Stop air pollution deaths": https://sites.google.com/site/300orgsite/stop-air-pollution-deaths.

[20]. "Climate Genocide": https://sites.google.com/site/climategenocide/.

[21]. "Are we doomed?": https://sites.google.com/site/300orgsite/are-we-doomed.

[22]. "Too late to avoid global warming catastrophe": https://sites.google.com/site/300orgsite/too-late-to-avoid-global-warming.

[24]. "Nuclear weapons ban, end poverty & reverse climate change": https://sites.google.com/site/300orgsite/nuclear-weapons-ban.

[24]. 300.org: https://sites.google.com/site/300orgsite/300-org.

[25]. "300.org – return atmosphere CO_2 to 300 ppm CO_2": https://sites.google.com/site/300orgsite/300-org—return-atmosphere-co2-to-300-ppm.

[26]. Hans Joachim Schellnhuber quoted by David Adam, "Roll back time t safeguard climate, expert warns", Guardian, 15 September 2008:

http://www.guardian.co.uk/environment/2008/sep/15/climatechange.carbonemissions.

[27]. James Hansen, M. Sato, P. Kharecha, D. Beerling, R. Berner, V. Masson-Delmotte, M. Pagani, M. Raymo, D.L. Royer, and J.C. Zachos, "Target atmospheric CO2: Where should humanity aim?", Open Atmos. Sci. J., 2, 217-231 (abstract): http://pubs.giss.nasa.gov/abstracts/2008/Hansen_etal.html.

[28]. James Hansen et al., "Target atmospheric CO_2: Where should humanity aim?", Open Atmos. Sci. J., 2, 217-231: http://pubs.giss.nasa.gov/docs/2008/2008_Hansen_etal.pdf.

[29]. Andrew Glikson, "The Methane Time Bomb and the Triple Melt-down", Countercurrents, 2009 : http://www.countercurrents.org/glikson101008.htm.

[30]. J.E.N. Veron, O. Hoegh-Guldberg, T.M. Lenton, J.M. Lough, D.O. Obura, P. Pearce-Kelly, C.R.C. Sheppard, M. Spalding, M.G. Stafford-Smith and A.D. Rogers, "The coral reef crisis: the critical importance of <350 ppm CO_2", Marine Pollution Bulletin, vol. 58, (10), October 2009, 1428-1436: https://www.ncbi.nlm.nih.gov/pubmed/19782832.

[31]. Gideon Polya, "Intergenerational Theft – For Every $1 For Coal Today Future Generations Will Pay $1-$14 To Sequester CO2", Countercurrents, 8 April, 2015: https://countercurrents.org/polya080415.htm.

[32]. "Carbon dioxide removal", Wikipedia: https://en.wikipedia.org/wiki/Carbon_dioxide_removal#Economic_issues.

[33]. Clive Hamilton, "Earth Masters. Playing God with the climate", Allen & Unwin, 2013.

[34]. Gideon Polya, "Forest biomass-derived Biochar can profitably reduce global warming and bushfire risk": https://sites.google.com/site/yarravalleyclimateactiongroup/forest-biomass-derived-biochar-can-profitably-reduce-global-warming-and-bushfire-risk.

[35]. "Climate Revolution Now": https://sites.google.com/site/300orgsite/climate-revolution.

[36]. "Climate Justice & Intergenerational Equity": https://sites.google.com/site/300orgsite/climate-justice.

[37]. "Stop climate crime": https://sites.google.com/site/300orgsite/stop-climate-crime.

[38]. "Punish climate criminals": https://sites.google.com/site/punishclimatecriminals/.

[39]. "Carbon Debt Carbon Credit": https://sites.google.com/site/carbondebtcarboncredit/.

[40]. Gideon Polya, "Biofuel famine, biofuel genocide, meat & global food price crisis", Global avoidable mortality: http://globalavoidablemortality.blogspot.com.au/2008/05/biofuel-famine-biofuel-genocide-meat.html.

[41]. P.W. Gerbens-Leenes, M.M. Mekonnen and A.Y. Hoekstra, "The water footprint of poultry, pork and beef: a comparative study in different countries and production systems", Water Resources and Industry, Volumes 1–2, March–June 2013, Pages 25-36: https://www.sciencedirect.com/science/article/pii/S2212371713000024.

[42]. Bibi van der Zee, "What is the true cost of eating meat?", The Guardian, 7 May 2018: https://www.theguardian.com/news/2018/may/07/true-cost-of-eating-meat-environment-health-animal-welfare.

[43]. Gideon Polya, "Worsening Climate Emergency And Record CO2 Emissions Demand Vegetarian Diet For All To Help Save Planet", Countercurrents, 20 June, 2016: https://www.countercurrents.org/polya200616.htm.

[44]. UN Population Division, "World Population Prospects 2017" : https://population.un.org/wpp/.

[45]. "List of countries by GDP (nominal) per capita", Wikipedia: https://en.wikipedia.org/wiki/List_of_countries_by_GDP_(nominal)_per_capita.

[46]. Irene Banos Ruiz, "China's new love affair with dogs – as pets, not food – presents environmental problems", DW, 21 June 2016: https://www.dw.com/en/chinas-new-love-affair-with-dogs-as-pets-not-food-presents-environmental-problems/a-19197523.

[47]. "List of countries by projected GDP", Statistics Times, 6 May 2018: http://statisticstimes.com/economy/countries-by-projected-gdp.php.

[48]. Kevin Anderson & Alice Bows, "Reframing the climate change challenge in light of post-2000 emission trends", Proc. Trans. Roy. Soc, A, 2008: http://www.tyndall.ac.uk/publications/journal_papers/fulltext.pdf ;

[49]. The IPCC, "Climate Change 2014 Synthesis Report, Approved Summary for Policy Makers", 1 November 2014: http://www.ipcc.ch/pdf/assessment-report/ar5/syr/SYR_AR5_SPM.pdf.

[50]. Ron Prinn, "400 ppm CO_2? Add other GHGs and its equivalent to 478 ppm", Oceans at MIT, 6 June 2013: http://oceans.mit.edu/featured-stories/5-questions-mits-ron-prinn-400-ppm-threshold.

[51]. "Trends in atmospheric concentrations of CO_2, CH_4 and N_2O", European Environment Agency, 31 January 2018:

https://www.eea.europa.eu/data-and-maps/daviz/atmospheric-concentration-of-carbon-dioxide-3#tab-chart_5_filters=%7B%22rowFilters%22%3A%7B%7D%3B%22colu mnFilters%22%3A%7B%22pre_config_polutant%22%3A%5B%22C O2%20(ppm)%20%22%5D%7D%7D.

[52]. "Global Warming Potential", Wikipedia:
https://en.wikipedia.org/wiki/Global_warming_potential.

[53]. "List of countries by population (United Nations)", Wikipedia:
https://en.wikipedia.org/wiki/List_of_countries_by_population_(Unite d_Nations).

[54]. Gideon Polya, "4 % Annual Global Wealth Tax To Stop The 17 Million Deaths Annually", Countercurrents, 27 June, 2014:
https://www.countercurrents.org/polya270614.htm

[55]. Thomas Piketty, "Capital in the Twenty-First Century", Harvard University Press, 2014.

[56]. Gideon Polya, "West Ignores 11 Million Muslim War Deaths & 23 Million Preventable American Deaths Since US Government's False-flag 9-11 Atrocity", Countercurrents, 9 September, 2015:
https://countercurrents.org/polya090915.htm.

[57]. "Stop state terrorism":
https://sites.google.com/site/stopstateterrorism/.

[58]. Ryan Bosworth et al., "The Value of a Statistical Life: economics and politics", Strata, March 2017:
https://strata.org/pdf/2017/vsl-full-report.pdf

[59]. Gideon Polya, "Australian State Terrorism – Zero Australian Terrorism Deaths, 1 Million Preventable Australian Deaths & 10 Million Muslims Killed By US Alliance Since 9-11", Countercurrents, 23 September, 2014:
http://www.countercurrents.org/polya230914.htm.

[60]. Gideon Polya, "Pro-Zionist, Pro-war, Pro-Opium, War Criminal Canadian Government Defames Iran & Cuts Diplomatic Links", Countercurrents, 10 September, 2012:
http://www.countercurrents.org/polya100912.htm.

[61]. Gideon Polya, "UK Terror Hysteria exposed – Empirical Annual Probability of UK Terrorism Death 1 in 16 million", Countercurrents, 16 September, 2014:
http://www.countercurrents.org/polya160914.htm.

[62]. Gideon Polya, "West Ignores 11 Million Muslim War Deaths & 23 Million Preventable American Deaths Since US Government's False-flag 9-11 Atrocity", Countercurrents, 9 September, 2015:
http://www.countercurrents.org/polya090915.htm

[63]. Gideon Polya, "Corporate terrorism is state sanctioned, kills over 30 million people annually and dooms humanity by lying", State crime and non-state terrorism:

https://sites.google.com/site/statecrimeandnonstateterrorism/corporate
-terrorism.

[64]. Gideon Polya, "Save Humanity And The Biosphere Through Zero Tolerance For Deadly Neoliberalism And Remorseless Neoliberals", Countercurrents, 13 May, 2016: https://www.countercurrents.org/polya130516.htm.

[65]. Gideon Polya, "Pro-Apartheid Australia's New White Australia Policy & compulsory Australian values statement", Countercurrents, 12 May 2017: https://countercurrents.org/2017/05/12/pro-apartheid-australias-new-white-australia-policy-compulsory-australian-values-statement/

[66]. Gideon Polya, "Jingoistic, US Lackey Australia's Deadly Betrayal Of Its Traumatized Veterans", Countercurrents, 18 May, 2018: https://countercurrents.org/2018/05/18/26768/).

[67]. "Experts; US did 9-11": https://sites.google.com/site/expertsusdid911/.

[68]. "Carbon terrorism: 3 million US air pollution deaths versus 53 US political terrorism deaths since 9-11 (2001-2015)", State crime and non-state terrorism: https://sites.google.com/site/statecrimeandnonstateterrorism/carbon-terrorism.

[69]. Gideon Polya, "Polya's 3 Laws Of Economics Expose Deadly, Dishonest And Terminal Neoliberal Capitalism", Countercurrents, 17 October, 2015: https://countercurrents.org/polya171015.htm.

[70]. Marilyn Waring, "If Women Counted" (introduction by Gloria Steinem), 1988.

[71]. "Marilyn Waring", Wikipedia: https://en.wikipedia.org/wiki/Marilyn_Waring.

[72]. Andrew Balmford, A. Bruner, P. Cooper, R. Costanza, S. Farber, R. E. Green, M. Jenkins, P. Jefferiss, V. Jessamy, J. Madden, K. Munro, N. Myers, S. Naeem, J. Paavola, M. Rayment, S. Trumper and R. K. Turner, "Economic reasons for conserving wild nature", Science 297, 2002, 950-953: http://www.sciencemag.org/cgi/content/abstract/297/5583/950.

[73]. Brian Ellis,"Social Humanism. A New Metaphysics", Routledge, UK, 2012).

[74]. Gideon Polya, "Book Review: "Social Humanism. A New Metaphysics" By Brian Ellis – Last Chance To Save Planet?", Countercurrents, 19 August, 2012: https://countercurrents.org/polya190812.htm.

[75]. Australian Bureau of Statistics, "Labour force participation rate at all-time high", 19 April 2018: http://www.abs.gov.au/ausstats/abs@.nsf/lookup/6202.0Media%20Release1Mar%202018.

[76]. William Gerhardie, "God's Fifth Column. A biography of the Age 1890-1940" (Hodder & Stoughton, 1981; The Hogarth Press, 1990, with editing and an Introduction by Michael Holroyd and Robert Skidelsky).
[77]. Gideon Polya, "Book Review: "God's Fifth Column" By William Gerhardie", Countercurrents, 6 May, 2014: https://countercurrents.org/polya060514.htm.
[78]. "Basic income", Wikipedia: https://en.wikipedia.org/wiki/Basic_income.
[79]. "Economic de-growth – negative economic growth to save Humanity & Biosphere", Carbon Debt Carbon Credit: https://sites.google.com/site/carbondebtcarboncredit/economic-degrowth.

CHAPTER 4: EXTRAPOLATING 11,000 SCIENTISTS CLIMATE EMERGENCY WARNING TO 2030 CATASTROPHE

[First published as Gideon Polya, "Extrapolating 11,000 scientists Climate Emergency warning to 2030 catastrophe", Countercurrents, 14 November 2019: https://countercurrents.org/2019/11/extrapolating-11000-scientists-climate-emergency-warning-to-2030-catastrophe.]

Over 11,000 scientists have signed up to a World scientists' warning of a Climate Emergency that sets out trends in 24 climate-related areas over the last 40 years. Scientists became aware of the climate change threat from greenhouse gas (GHG) pollution in the 1980s, but in 21 of these 24 areas the trends are (a) huge, (b) in the wrong direction, and (c) linear or quasi-linear functions of time, with this allowing extrapolation from the present climate emergency to a climate catastrophe in 2030.

The 2019 paper by William Ripple et al., and co-signed by over 11,000 scientists from around the world [1], is a successor to a similar warning issued by William Ripple et al. in 2017 and co-signed by over 15,000 scientists [2, 3]. The 2017 warning presented data for the 56 year period of 1960-2016 showing a quasi-linear change as a function of time of increasing CO_2 emissions, increasing temperature, the number of ocean dead zone regions, increasing population increase, and increasing ruminant population. Paralleling these changes were quasi-linear decreases over recent decades of freshwater resources per capita, reconstructed annual marine catch, total forest area, and vertebrate species abundance (as a % of the 1970 abundance). In stark contrast to this worsening disaster in 8 areas was a decrease since 1992 of emissions of ozone depletors (in Mt CFC-11 equivalents per year), this demonstrating clearly that global action on serious environmental pollution was actually possible [2, 3].

In 2017 Ripple et al. stated: "Twenty-five years ago [in 1992], the Union of Concerned Scientists and more than 1,700 independent scientists, including the majority of living Nobel laureates in the sciences, penned the 1992 "World Scientists' Warning to Humanity"… These concerned professionals called on humankind to curtail environmental destruction and cautioned that "a great change in our stewardship of the Earth and the life on it is required, if vast human misery is to be avoided." In their manifesto, they showed that humans were on a collision course with the natural world … Since 1992, with the exception of stabilizing the stratospheric ozone layer, humanity has failed to make sufficient progress in generally solving these foreseen environmental challenges, and alarmingly, most of them are getting far worse… Especially troubling is the current trajectory of potentially catastrophic climate change due to rising GHGs from burning fossil fuels… deforestation… and agricultural production – particularly from farming ruminants for meat consumption … Moreover, we have unleashed a mass extinction event, the sixth in roughly 540 million years, wherein many current life forms could be annihilated or at least committed to extinction by the end of this century" [2].

2017 WARNING DATA.

Figure 1 of the 2017 Ripple et al. warning paper [2] showed trends over time for environmental issues identified in the 1992 scientists' warning to humanity. Percentage [quasi-linear] change since 1992 for the variables in each panel are as follows, noting that because these changes were roughly linear with time one can extrapolate to roughly estimate the situation in 2030. All trends except for the decrease in ozone depletors (*) were in the wrong direction from the perspective of Humanity and the Biosphere.

***(a) Ozone depletors (Mt CFC-11-equivalent per year):** 1.05 (1992) to 0.345 (2016), -62.3% overall, -26.0% per decade, -0.029 per year; all ozone depletors gone before 2030.

(b) Freshwater resources per capita (1,000 metre$^{3)}$: 8.0 (1992) to 5.9 (2014), -25.4% overall, − 12.0 % per decade, - 0.095 per year; 4,400 metre^3 per capita in 2030.

(c) Reconstructed marine catch (Mt per year): 121 (1992) to 113 (2011), -6.6%, overall, − 3.5% per decade, -0.42 per year; 105 million tonnes marine catch per year in 2030.

(d) Dead zones (number of affected regions): 360 (1992) to 660 (2008), +83.3% overall, +52% per decade, +11.5 per year; 913 ocean dead zones by 2030.

(e) Total forest (million hectares): 4.110 (1992) to 3.990 (2016), −2.9% overall, -1.2% per decade, -0.005 per year; 3.92 million hectares by 2030.

(f) Vertebrate species abundance (% of 1970): 59.5 (1992) to 43.5 (2010), − 26.9% overall, -14.9% per decade, -0.89 per year; 27.5% of 1970 level by 2030.

(g) CO_2 emissions (Gt CO_2 per year): 22.5 (1992) to 36.0 (2014), + 60.0% overall, +27.3% per decade, + 0.84 per year; 49.4 Gt CO_2 emitted in 2030.

(h) Temperature change (^{o}C): 0.30 (1992) to 0.90 (2016), +200% overall, +83.3% per decade, 0.0033 per year; + 0.95 ^{o}C by 2030.

(i) (a). Human population (billion individuals): 5.50 (1992) to 7.45 (2016), + 35.5% overall, +14.8% per decade, 0.081 per year; 8.58 billion human individuals in 2030.

(i) (b). Ruminant population (billion individuals): 3.25 (1992) to 3.90 (2014), +20.0% overall, +9.1% per decade, +.03 per year; 4.38 billion ruminant individuals in 2030 [2, 3].

2019 WARNING FIGURE 1 DATA.

Figures 1 and 2 of the 2019 Ripple et al. warning paper [1] sets out trends in 24 areas within a 40 year period from 1979 to 2019. Because the changes were quasi-linear as a function of time one could reasonably extrapolate the data to 2030 to (a) roughly estimate the seriousness of the situation in a mere decade's time, (b) compare the data with that of the prior 2017 study, and (c) provide absolute estimates for the various parameters. Most trends were in the wrong direction except for those asterisked (*).

Figure 1 (a). Human population (billions): 4.35 (1979) to 7.65 (2018), +75.9% overall, +19.5% per decade, + 0.085 billion per year, and an estimated 9.07 billion human population in 2030 (assuming an average increase of +84.6 million per year, noting that the increase in population was linear over the period 1979-2018).

Comments. This result is similar to that of the 2017 paper (8.58 billion) [2].The UN Population Division World Population Prospects 2019 estimates a human population of 8.55 billion in 2030 [4]. Drastic decrease in population is required for cessation of the present appalling level of ecosystem destruction (ecocide) that is associated with catastrophic species loss (speciescide) and leading to omnicide and terracide. If we take world coral as a "canary in the coal mine" then the 320 ppm CO_2 at which coral reefs started to decline [5] was reached in 1965 [6], at which time the world's population was 3.340 billion as compared to the present 7.5 billion [4]. One can therefore plausibly suggest that given the present carbon economy the world's population needs to roughly halve for a safe and sustainable environment for all peoples and all species [7]. However a number of climate scientists and demographers have variously estimated a 0.5-1.0 billion sustainable human population in 2100 with this implying 10 billion premature deaths in a Climate Genocide this century [8-11]. UN Population Division projections show that this century the population of rich Europe will fall from 742 million (2018) to 653 million (2100) whereas the population of impoverished

Africa will explode from 1,274 million (2018) to 4,468 million (2100) [4]. However in terms of relative carbon footprint, the per capita GDP (nominal) is $25,596 for Europe versus $1,753 for Africa (2016), this translating to a total GDP of $19.3 trillion for Europe and $2.2 trillion for Africa [12]. For carbon footprint equality the European population would need to decrease 10-fold or the African per capita income would have to increase 10-fold. However Professor Dabo Guan (School of International Development, University of East Anglia, UK) (2016) has commented thus on inescapable limits to growth: "For everyone in the world to have an American lifestyle, we would need seven planets, and three to live as Europeans" [13]. Clearly the rich European population is not declining fast enough and the impoverished African population is expanding impossibly.

***Figure 1 (b). Total fertility (births per woman):** 3.78 (1979) to 2.70 (2008) to 2.44 (2017). Births per woman per decade were 0.37 (1979-2008) to 0.24 (2008-2017) and an average of 0.35 (1979-2017) (about −10.9% per decade, -1.1% per year and 0.035 per year). Extrapolation from the 2008-2017 trend of − 0.024 per year yields 2.13 average births per woman in 2030.

Comments. Births per woman per decade fell into 2 quasi linear periods of 1979-2008 and 2008-2017, with the rate in the latter period being 64.9% of that in the earlier period. This is a useful trend but it disguises a continuing resource-utilization disaster in which (a) the GDP per capita is about 10 times greater in low birth rate, rich and mainly European countries than in impoverished, Developing World countries (minus China) [7, 12], and (b) while many prosperous countries have attained or are approaching zero population growth (ZPG), some areas (notably Africa) are out of control as discussed above for Figure 1 (b). The replacement fertility rate depends upon the death rate and thus is 2.1 in rich Developed World (zero avoidable mortality annually from deprivation) and about 3.0 in Developing countries (minus China) in which avoidable deaths from deprivation total 15 million each year [14, 15].

Figure 1 (c). Ruminant livestock (billion individuals): 2.86 billion (1979) to 3.93 billion (2017) (+ 8.72% per decade,

+0.87% per year and +0.028 per year), noting an apparent, short, low-growth period in the middle 1990s. Assuming the 1979-2017 average trend yields a value from extrapolation of 4.29 billion individual ruminants in 2030.

Comments. The ratio of livestock to human individuals has decreased from 0.66 (1979) to 0.52 (2017) but is predicted to flatten to 0.48 (2030) (presumably as more Chinese join the meat-eating middle class). However ruminant livestock are methanogenic and methane (CH_4) has a Global Warming Potential (GWP) relative to the same mass of CO_2 of 105 on a 20 year time frame taking aerosols into account [16], this being 5 times higher than the 21 with a 100 year timeframe [17, 18]. When World Bank analysts took land use and a 20 year-based GWP for CH_4 into account, the global annual greenhouse gas (GHG) pollution rose from 41.8 gigatonnes CO_2-equivalent (Food and Agricultural Organization, FAO) to 63.8 gigatonnes CO_2-equivalent [19]. One notes that Figure 1 (k) confusingly cites the 2018 global CO_2 emissions as 34 gigatonnes CO_2-equivalent per year [1]. The German WBGU (that advises the German Government) (2009) and the Australian Climate Commission (2013) have estimated that no more than 600 billion tonnes of CO_2 can be emitted between 2010 and zero emissions in 2050 if the world is to have a 75% chance of avoiding a catastrophic 2C temperature rise [20, 21]. However the upwardly revised estimates of greenhouse gas (GHG) pollution properly taking land use and methane (CH_4) into account means an annual GHG pollution of 63.8 Gt CO_2-equivalent and hence 638 Gt CO_2-equivalent over the next decade i.e. the world will have exceeded this Terminal Carbon Pollution Budget by 2020. Of course, one can well ask whether you would board a plane with a 75% chance of it crashing. A major imperative (as indeed perceived by Ripple et al. [1]) is for the world to adopt a vegetarian diet [22]. Australia is among world leaders in 15 areas of climate criminality, including export of methanogenically-derived meat [23, 24]. Climate criminal Australians (and other rich, meat-eating Westerners) would be well-advised to adopt vegetarianism or, failing that, urgently return to the up to 65,000 years old Indigenous

Australian cultures that supplemented their sustainably-derived diets with non-methanogenic kangaroo meat [25].

Figure 1 (d). Per capita meat production (kg per year): 30 (1979) to 44 (2018) (+11% per decade, +1.1% per year, +0.36 per year). This yields an estimated 62.3 kg per capita in 2030, noting that the remarkably linear progression from 1979 onwards gives greater confidence to this extrapolation.

Comments. This remarkable rise in per capita meat production contrasts with the roughly static ratio of livestock to human individuals (Figure 1 (c) [1] data considered above). This data reflects (a) the meat eating of the growing middle classes of the Developing World (including China) and (b) the rise in chicken production (especially in Asia). Chicken meat production in 2014- 2023 will increase at 2.3% per year to 134.5 million tonnes in 2023. By way of comparison, the growth rates are 1.2% per year for beef and pig meat, 2.3% per year for sheep meat and 1.6% per year for meat overall [26]. The conversion efficiency (kg grain to produce 1 kg gain in live weight) is as follows: herbivorous farmed fish (e.g. carp, tilapia, catfish; less than 2), chicken (2), pork (4), and beef (7). In 2003, 37% of the world grain harvest, or nearly 700 million tons, was used to produce animal protein [27]. A vegetarian diet is a must for an increasingly resource-challenged Humanity [22].

Figure 1 (e). World GDP (trillion current US$/year). The data appear to fall into to 2 successive upward linear phases with an upward inflection in about 2000: 10.0 (1979) to 32.5 (1998) and 37.5 (2001) to 86.3 (2018), with growth rates of $1.18 trillion per year and $2.87 trillion per year, respectively, and an overall growth of + 80.5% /10 years (+8.1% per year and +$1.96 trillion per year). Assuming that the latter faster rate applies over the next dozen years (+ $2.87 trillion per year), one can predict a World GDP of $120.7 trillion in 2030.

Comments. As discussed in relation to Figure 1 (a) [1] above, there needs to be a halving of population, and a consequent halving of the global economy through economic de-growth (negative economic growth). The major burden of such negative economic growth would have to be borne by rich Western countries to allow the global South to achieve a modest increase to permit a modestly decent life and cessation of the horrendous Global Avoidable Mortality Holocaust in which 15 million people die avoidably each year from imposed deprivation [15].

Figure 1 (f). Global tree cover loss (million hectares per year): The data fall into 2 successive upward linear phases with an apparent short low-rate period in the late 2000s: 14 (2001) to 18 (2005) and 18 (2010) to 27.5 (2018) with corresponding rates of loss of 1.0 and 1.2 million hectares per year, respectively. Extrapolation using the latter faster latter rate yields an estimated loss rate of 41.9 million hectares of tree cover loss per year by 2030.

Comments. This asserted tree loss would be happening in biodiverse tropical forests. However Xiao-Peng Song and colleagues in a 2018 paper in the prestigious journal Nature: "We analyse 35 years' worth of satellite data and provide a comprehensive record of global land-change dynamics during the period 1982–2016. We show that—contrary to the prevailing view that forest area has declined globally' tree cover has increased by 2.24 million km^2 [224 million hectares] (+7.1% relative to the 1982 level [3.155 million hectares]). This overall net gain is the result of a net loss in the tropics being outweighed by a net gain in the extratropics. Global bare ground cover has decreased by 1.16 million km^2 [116 million hectares] (−3.1%), most notably 2009-2018 in agricultural regions in Asia. Of all land changes, 60% are associated with direct human activities and 40% with indirect drivers such as climate change" [28]. One Tree Planet states: "Scientists have always speculated about how trees can help end the current climate crisis, but the question has always been how many. A new study from the Crowther Lab out of Swiss University ETH Zurich, has finally come up with an answer: 1 trillion more trees on Earth could sequester more than 60% of anthropogenic carbon emissions… Other technology like direct-air capture or CO_2 separation have proven to be comparatively ineffective and far exceed the costs of just planting more trees. In their study titled *The Global Tree Restoration Potential*, Crowther Lab determined our planet

could support 4.4 billion hectares of continuous tree coverage.
With current global tree coverage sitting around 2.8 billion
hectares, that means Earth could sustain 1.6 billion more
hectares of forest. Of those 1.6 billion hectares, Crowther Lab
discovered just under a billion hectares, excluding land
occupied by agriculture and cities (about 1 trillion trees worth),
inherently suitable for supporting reforestation efforts – that's a
land area almost the size of China" [29].

**Figure 1 (g). Brazilian Amazon forest loss (millions hectares
per year)** was in the range 1-3 million hectares per year in the
period 1988-2008 (very scattered data) and then fell to about
0.7- 0.9 million hectares per year in the period 2009-2018 (-
24.3% /10 years, -2.4% per year).

Comments. One could tentatively assume that Brazil might
keep to this rate of Amazon forest destruction of about 1 million
hectares per year by 2030, increased by the current burning of
the Amazon forest under neoliberal and neofascist President Jair
Bolsonaro and decreased by the consequent international outcry.

Figure 1 (h). Energy consumption (Gt oil equivalent/year).
Annual oil, coal and gas exploitation increased in a quasi-linear
fashion from 1979-2018 with overall increases of
+11.9%/decade, + 22.5%/decade and +30.1% /decade,
respectively. Oil usage rose from 3.0 Gt oe (1979) to 4.6 Gt oe
(2018), coal rose from 1.8 Gt oe (1979) to 3.9 Gt oe (2018) and
gas rose from 1.1 Gt oe (1979) to 3.2 Gt oe (2018).
Extrapolation based on these numbers would indicate usage in
2030 totalling 6.2 Gt oe for oil, 5.73 Gt oe for coal and 5.0 Gt
oe (gas).

Oil $(CH_2)_n$) is oxidized thus on combustion: $(CH_2)_n + 1.5n\ O_2 ->$
$nCO_2 + nH_2O$ i.e. 14 Gt oil yields 44 Gt CO_2 and 1 Gt oil yields
3.14 Gt CO_2. Accordingly, CO_2 from burning oil, coal and gas
in 2030 is projected to be 19.5 Gt CO_2 (oil), 18.0 Gt CO_2 (coal),
15.7 Gt CO_2 (gas) and 53.2 Gt CO_2 in total, as compared to the
figures for 2018 of 14.4 Gt CO_2 (oil), 12.2 Gt CO_2 (coal) and
10.0 Gt CO_2 (gas) with a total of 36.6 Gt CO_2 (similar to the
34.0 Gt CO_2-equivalent of emissions in 2018 according to

Figure 1(k)). Solar and wind use started to increase substantially in about 2000, and by 2018 had reached 0.45 Gt oe/year (Gt 1.4 Gt CO_2 emissions replaced in 2018) (+373%/decade) i.e. solar and wind renewable energy was only 3.7% of the total energy mix in 2018.

Comments. Public comments about the serious threat of global warming due to greenhouse gases (GHGs) like CO_2 have been made by scientists since the mid-1980s. It is accordingly quite shocking that oil, coal and gas usage has nevertheless remorselessly increased at a constant rate over this 40 year period. Indeed in 2017 the pro-coal Treasurer (and now present PM) of climate criminal Australia, Scott Morrison, brought a lump of coal into Parliament, idiotically declaring "This is coal. Don't be afraid. Don't be scared" [30]. A slightly more optimistic assessment of this bleak data is that annual coal use remained constant at 3.85 Gt oe (12.1 Gt CO_2) per year in the period 2011-2018, suggesting that global coal use, if not decreasing, may actually be finally flattening out. Further on the positive side, Frontier Group reports (2019): "America produces almost five times as much renewable electricity from the sun and the wind as in 2009, and currently wind and solar energy provide nearly 10 percent of our nation's electricity" [31]. Renewable Energy reports re China (2019):"Power generation from renewable energy sources reached 1,870 TWh in 2018, an increase of 170 TWh and making up 26.7 percent of the country's total. Hydro contributed 1,200 TWh (up 3.2 percent), wind – 366 TWh (up 20 percent), PV – 177.5 TWh (up 50 percent) and biomass – 90.6 TWh (up 14 percent)" [32]. Of course many companies, towns, cities and countries have already achieved 100% renewable energy [33]. The Germany-based Energy Watch Group: "The new study by the Energy Watch Group and [Finnish] LUT University is the first of its kind to outline a 1.5°C scenario with a cost-effective, cross-sectoral, technology-rich global 100% renewable energy system that does not build on negative CO_2 emission technologies. The scientific modelling study simulates a total global energy transition in the electricity, heat, transport and desalination sectors by 2050. It is based on four and a half years of research and analysis of data collection, as well as technical and financial modelling by 14 scientists. This proves that the transition to 100% renewable energy is economically competitive with the current fossil and nuclear-based system, and could reduce greenhouse gas emissions in the energy system to zero even before 2050" [34].

Figure 1 (i). Air transport (billion passengers carried per year) was biphasic linear, rising from 0.6 (1979) -> 1.60 (2000) (0.048 per year) to 2.3 (2007) -> 4.0 (2017) (0.170 per year) with a transition in the 2000s (overall + 63.3% per decade). Extrapolating from the most recent, faster rate of change in 2007-2017 yields an estimated 6.2 billion passengers carried per year in 2030.

Comments. This peripatesis is a scandalous profligacy. Thus, for example, I am domiciled in remote Melbourne Downunder in the Southern Hemisphere en route to Antarctica but am interviewed by a progressive North American radio station by Skype for free. Indeed Australia's Nobel Prize-winning writer Patrick White commented thus on public events to celebrate the bicentenary of the British invasion of Australia in 1788: "They are flogging camels from East to West and from West to East, and running from City to Surf and from Surf to City. Why don't they just stay home, cook themselves a nice meal and curl up with a good book?" One notes that our climate heroine, 16 year old Greta Thunberg, recently travelled from Sweden to New York to attend the 2019 UN Climate Summit on a high-technology yacht. Energy use in megajoules per person per kilometre (MJ/passenger-km) is about 0.1 (bicycle), 0.2 (walking person), 1 (train), 2 (plane), and 5 (car; a 2 tonne car transporting an 80kg person is highly inefficient) [35].

***Figure 1(j). Total institutional assets divested (trillion US dollars)** linearly increased from 0.2 (2014) to 7.5 (2018), an average of $1.8 billion /year (+9125% per decade for the period 2013 to 2018). Extrapolating from this linear increase yields an estimate of $28.9 trillion in assets divested in 2030.

Comments. To put this excellent outcome in context, world GDP was $86.3 trillion in 2018, extrapolates to $120.7 trillion in 2030 (see discussion about Figure 1 (e) [1] above), and thus total institutional assets divested as a per cent of GDP are estimated to rise from 8.7% in 2014 to 23.9% in 2030. These divestment decisions are not "political' or "ideological" as the climate change denialist Trumpists assert (notably in climate criminal Coalition-ruled Australia) but are made not just for saving the planet and "corporate image" reasons but also for critical, hard-headed, market-based reasons connected with avoidance of "stranded assets" e.g. coal reserves that will be unusable if a proper Carbon Price is applied [36-41]. Thus the fundamental 3 Laws of Thermodynamics state that (1) the energy of a closed system remains constant, (2) the entropy (disorder, chaos, lack of information content) increases to a maximum, and (3) zero entropy for a perfect crystal at absolute zero. Polya's 3 Laws of Economics correspondingly declare that (1) Profit equals Price minus Cost of Production, (2) deception about Cost of Production rises to a maximum, and (3) no jobs, production, price or profit on a dead planet [42]. The extraordinary deception about the Cost of Production in the World's present Carbon Economy is illustrated by the determination by the International Monetary fund (IMF) that the average global Carbon Price is presently a paltry circa $2 per tonne CO_2-equivalent [43, 44] whereas Dr Chris Hope of 90-Nobel Laureate Cambridge University has estimated a damage-related Carbon Price of $200 per tonne CO_2-equivalent [45]. The resultant global Carbon Debt (to be inescapably met by future generations) is about $200 trillion and increasing at about $10 trillion per year [37].

Figure 1(k), CO_2 emissions (gigatonnes CO_2 equivalent per year), showed a triphasic linear increase from 18.0 (1979) to 22.5 (1997) (0.25 per year), through 22.5 (1997) to 30.0 (2009) (0.63 per year) and thence 30.0 (2009) to 34.0 (2018) (0.44 per year). The overall increase was +17.9% per decade, 1.7% per year). Extrapolating using the most recent rate of 0.44 gigatonnes CO_2-equivalent per year yields emissions of 39.3 gigatonnes CO_2 equivalent per year in 2030.

Comments. As discussed above (Figure 1 (c)), when World Bank analysts took land use and a 20 year-based GWP for CH_4

into account, the global annual greenhouse gas (GHG) pollution rose from 41.8 gigatonnes CO_2-equivalent (Food and Agricultural Organization, FAO) to 63.8 gigatonnes CO_2-equivalent [19]. Applying a correction factor of $63.8/41.8 = 1.5$ to the data of Ripple et al. (2019) gives a present rate of increase of GHG pollution of $0.44 \times 1.5 = 0.66$ gigatonnes CO_2-equivalent per year. Applying this revised rate of increase to an estimated annual GHG pollution of 63.8 gigatonnes CO_2-equivalent in 2009 provides a revised estimate of 69.7 Gt CO_2-equivalent per year in 2018 and 77.7 Gt CO_2-equivalent per year in 2030. On this basis we can see that the 2010 onwards Terminal Carbon Pollution Budget for a 75% chance of avoiding +2C was exceeded in 2018 (602 Gt CO_2-equivalent emitted in the period 2010-2018).

Figure 1(l). Per capita emissions (tonnes CO_2 equivalent per person per year) fell from 4.23 (1979) to 3.83 (1997) (– 0.022 per year), then rose from 3.83 (1997) to 4.47 (2010) (+ 0.049 per year) and have remained roughly constant at 4.50 since 2006) (the overall average being 2.15% per decade, 0.2% per year). One could guess that if the post-2006 trend continues then per capita emissions would remain at about 4.5 tonnes CO_2 equivalent per person per year in 2030 i.e. with increases in GHG pollution (Figure 1 (k)) matching an increase in population (Figure 1(a)).

Comments. As discussed above in relation to Figure 1 (k), these per capita emissions estimates understate actual GHG pollution with land use and CH_4 properly considered. Thus taking such considerations into account increases annual GHG pollution from 41.8 gigatonnes CO_2-equivalent per year to 63.8 gigatonnes CO_2-equivalent per year in 2009 [19] and a world average GHG pollution in 2009 of 63.8 gigatonnes CO_2-equivalent per year /6.9 billion = 9.2 tonnes CO_2-equivalent per person [46, 47]. Annual per capita greenhouse gas pollution per se is a flawed measure of national culpability for this because it ignores rich countries outsourcing industrial pollution to China, and impoverished countries compelled to pollute to barely survive. A better measure of culpability is weighted annual per capita greenhouse gas pollution taking relative per capita income into account, this revealing that the worst polluters on this basis include the rich Anglosphere countries of the US, UK, Australia, New Zealand, Canada and Ireland [47].

***Figure 1(m). GHG emissions covered by carbon pricing (%)** rose linearly from 3.0 (2007) to 15.0 (2017) (+ 1.2% per year) (with an overall increase of 256% per decade or 25.6% per year). Extrapolating this linear increase yields 30.6% of GHG emissions covered by carbon pricing in 2030.

Comments. Of course the crucial matter is the magnitude of the carbon pricing. Thus Dr Chris Hope of 90-Nobel-Laureate University of Cambridge has estimated a damage-related Carbon Price of $200 per tonne CO_2-equivalent [44] whereas the IMF recently estimated that the global average Carbon Price is a paltry circa $2 per tonne CO_2-equivalent [43, 44].

Figure 1(n). Carbon Price ($ per tonne CO_2 emissions) was assertedly about 53 (1990-2005) and then decreased to a steady circa about 15 (2012-2018) (overall -33% per decade, -3.3% per year).

Comments. I have no idea how they arrived at these estimates. Thus as discussed above in relation to Figure 1(j), the IMF in 2019 noted an average global Carbon Price of about $2 per tonne CO_2-equivalent [43, 44].

Figure 1(o). Fossil fuel subsidies (billion US$ per year) increased from 440 (2010) to 550 (2012) (+55/year), thence decreased from 550 (2012) to 260 (- 73 /year) and then increased from 260 (2016) to 430 (+85/year). The overall change was -1.08% per decade (-0.1% per year).

Comments. Again it is not clear how exactly these numbers were arrived at but one assumes that they are an amalgam of such things as transport subsidies and tax relief components. However the IMF recently stated that the global average Carbon Price is a mere $2 per ton of CO_2 [43, 44] as compared to the estimated damage-related Carbon Price of $200 per tonne CO_2 [45]. The difference – $198 per tonne CO_2 – represents a huge direct subsidy of fossil fuel. In 2018 the carbon emissions were 69.7 gigatonnes per year (properly taking land use and CH_4 into account; see discussion re Figure 1(k)) and thus the subsidy in 2018 (ignoring transport subsidies and tax relief) was $198 per tonne CO_2 x 69.7 gigatonnes CO_2 per year = $13,800 billion per year or about 35 times greater than the mean fossil fuel subsidy of about $400 billion per year claimed in Figure 1(o).

2019 WARNING FIGURE 2 DATA.

Figure 2(a). Carbon dioxide (CO_2 parts per million) rose from 339 (1980) to 407 (2018) for an average of 1.79 ppm CO_2 per year (+ 4.98% per decade, +0.5% per year) and an extrapolation-derived 427 ppm CO_2 in 2030.

Comments. Numerous climate scientists, coral scientists, biologists and science-informed commentators argue that we need to draw down atmospheric CO_2 from the present deadly and dangerous 410 parts per million to the pre-Industrial Revolution level of about 300 ppm CO_2 for a safe and sustainable planet for all peoples and all species [48, 49]. Thus, for example, coral scientists inform us that world coral started dying when the atmospheric CO_2 reached 320 ppm CO_2 and that a return to about 300 ppm CO_2 is needed for sustainability of world coral reefs [48-51]. However while it is technically possible in all kinds of ways to draw down atmospheric CO_2 to 300 ppm CO_2, the cost of doing so is enormous [41, 52-54]. Professor James Hansen (of 96 Nobel Laureate Columbia University): "One ppm of CO_2 is 2.12 billion tons of carbon or about 7.77 billion tons of CO_2. Recently Keith et al. (2018) achieved a cost breakthrough in carbon capture, demonstrated with a pilot plant in Canada. Cost of carbon capture, not including the cost of transportation and storage of the CO_2, is $113-232 per ton of CO_2. Thus the cost of extracting 1 ppm of CO_2 from the atmosphere is $878-1803 billion. In other words, the cost, in a single year, of closing the gap between reality and the IPCC scenario that limits climate change to +1.5°C is already about $1 trillion. And that is without the cost of transporting and storing the CO_2, or consideration of whether there will be citizen objection to that transportation and storage.

This annual cost will rise rapidly, unless there is a rapid slowdown in carbon emissions… cost of CO_2 storage… has been estimated as $10-20/t$CO_2$" [52].Taking Professor Hansen's data, and including his estimates of the cost of transport and storage of CO_2, indicates that this "best so far" cost of atmospheric CO_2 draw-down is $123-252/t$CO_2$. Accordingly, the upper estimate of the cost of reducing the atmospheric CO_2 by 1 ppm = 1 ppm x 7.77 billion ton CO_2/ppm x $252/ton CO_2 = $1,958 billion. Thus the upper estimate of the cost of the present 1.79 ppm /year increase in atmospheric CO_2 is about $3,500 billion. Of course the GHGs include not just CO_2 but also methane (CH_4) and nitrous oxide (N_2O) [17]. While the atmospheric CO_2 is presently about 412 ppm CO_2, the CO_2-equivalent (taking other GHGs into account) is about 500 ppm CO_2-equivalent [55].

Figure 2(b). Methane (CH_4 parts per billion) increased in a quasi-linear fashion from 1,648 (1984) to 1,854 (2017) for an average of 6.2 ppb per year (+3.65% per decade, 0.37% per year). Extrapolating from this average yields 1,935 ppb CH_4 in 2030.

Comments. There are huge amounts of CH_4 present as frozen H_2O-CH_4 clathrates in the Arctic tundra and Arctic Ocean sea bed. It is estimated that 50 Gt of CH_4 will be released in the Arctic in coming decades [56]. The Global Warming Potential (GWP) of CH_4 on a 20 year time frame and with aerosol impacts considered is 105 times that of CO_2 [16]. The German WBGU (2009) and the Australian Climate Commission (2013) have estimated that no more than 600 billion tonnes of CO_2 can be emitted between 2010 and zero emissions in 2050 if the world is to have a 75% chance of avoiding a catastrophic 2C temperature rise [20, 21]. The 50 Gt (billion tonnes) CH_4 in the East Siberian Arctic Shelf is thus equivalent to 50 billion tonnes CH_4 x 105 tonnes CO_2-equivalent/tonne CH_4 = 5,250 tonnes CO_2-e or about nine (9) times more than the World's Terminal Carbon (GHG) Pollution Budget. We are doomed unless we can stop this Arctic CH_4 release [57]. CO_2 and CH_4 are the major GHG contributors to the atmospheric 500 ppm CO_2-equivalent [59].

Figure 2 (c). Nitrous oxide (N$_2$O parts per billion) increased in a quasi-linear fashion from 301 (1979) to 331 (2018) for an average of 0.77 ppb per year (+2.46% per decade, +0.28% per year). Using this average, extrapolation yields a 2030 value of 340 ppb N$_2$O.

Comments. N$_2$O is derived from oxidation of nitrogenous agricultural fertilizers and has a Global Warming Potential (GWP) relative to the same mass of CO$_2$ of 289 on a 20-year time frame (298 on a 100 year time frame) [17]. Nitrous oxide (N$_2$O, "laughing gas"), like other GHGs, is not a laughing matter.

Figure 2(d). Surface temperature change (^{o}C) increased in a quasi-linear fashion from 0.21 (1979) to 0.91 (2018) for an average of 0.18 ^{o}C per decade (0.018 ^{o}C per year). Conservatively extrapolating this 4-decade trend yields + 1.13 ^{o}C in 2030.

Comments. The global average surface temperature increase relative to 1880 is about +1.1 ^{o}C but according to the IPCC it will reach 1.5 ^{o}C in 10 years' time on the present pollution trajectory [95-97]. Indeed David Spratt and Ian Dunlop: "Another example [of scientific reticence] is the recent IPCC 1.5°C report, which projected that warming would continue at the current rate of ~0.2°C per decade and reach the 1.5°C mark around 2040. However the 1.5°C boundary is likely to be passed in half that time, around 2030, and the [catastrophic] 2°C boundary around 2045, due to accelerating anthropogenic emissions, decreased aerosol loading and changing ocean circulation conditions" [58]. Burning sulphur-containing coal generates sulphur dioxide (SO$_2$) which forms global dimming sulphate aerosols in the atmosphere – on cessation of coal burning, the washing out of sulphate aerosols from the atmosphere will result in a rapid increase in temperature. Paleoclimatologist and Earth scientist Dr Andrew Glikson (2019: "The planetary consequences of injecting > 910 billion tons CO$_2$ into the atmosphere are playing in real time. The Arctic Circle is suffering from an unprecedented number of wildfires in the latest sign of a climate crisis… As the globe warms, to date by a mean of near ~1.5 °C or ~2.0°C when the masking effects of sulphur dioxide and other aerosols are considered, and by a mean of ~2.3°C in the Polar Regions, the expansion of warm tropical latitudes and the polar-ward migration of climate zones ensue in large scale droughts in subtropical latitudes such as in inland Australia and southern Africa" [59].

Figure 2 (e). Minimum Arctic sea ice (million km^2) decreased from 7.4 (1979) to 4.5 (2018) for an average of –0.074 km^2 per year (- 11.7% per decade, -1.2% per year). Extrapolation using this 4 decade average yields 3.612 million km^2 by 2030. If this rate of loss was maintained then there would be zero (0) minimum (Summer) sea ice this century.

Comments. The melting of the high albedo (reflectivity) Arctic sea ice generates black, low albedo sea which absorbs more solar radiation, warms and thence melts more sea ice in a positive feed-back loop. In addition, broken ice enables further sea ice break-up in storms. I painted a huge painting called "Terra" that included the Arctic and Antarctic polar ice caps [60], but unfortunately the Arctic ice cap will be gone in coming decades. The Arctic summer sea ice will be gone before 2050 with Professor James Anderson of Harvard University predicting that the summer sea ice will disappear in the 2020s [61]. This sea ice loss has a big ecological impact (notably on polar bears) and the warming of the Arctic Ocean impacts ocean currents.

Figure 2(f). Greenland ice mass decrease (gigatonnes, Gt) was quasi-linear and went from – 200 Gt (2003) to -3,700 Gt (2016) for an average of -269 Gt per year (-2,610 Gt per decade as estimated by the paper). Extrapolating yields an ice loss of 7,466 Gt by 2030.

Comments. The Guardian has summarized consonant recent research findings (2019): "Greenland lost about 280bn tons of ice per year between 2002 and 2016, enough to raise the worldwide sea level by 0.03 inches annually. If all of Greenland's vast ice sheet, 3km thick in places, was [sic] to melt, global sea levels would rise by seven meters, or more than 20ft, drowning most coastal settlements. The rate of loss hasn't been even, however, with the ice melting four times faster in 2013 compared with 2003" [62].

Figure 2(g). Antarctic ice mass change (gigatons) decreased quasi-linearly from -150 (2002) to -1,750 (2018) for an average loss of -100 Gt per year (-1,230 Gt per decade estimated by the paper). Assuming an average loss of -100 Gt ice per year, extrapolation yields -2,950 Gt ice by 2030.

Comments. Madeleine Stone writing in National Geographic (2019): "The frozen mountains and icy plains of Antarctica hold enough water to raise global sea levels nearly 200 feet. Thankfully, over three-quarters of the continent is girded by ice shelves, the floating extensions of glaciers that protect the land-bound ice behind them… in rapidly-melting parts of West Antarctica, "upside down rivers" of warm water are gnawing away at the ice shelves' weak spots from below… [blue] lakes are bad news for ice. Because of their dark color, they absorb more of the sun's energy, triggering further warming. And under certain conditions, clusters of lakes can drain rapidly into the ice below them, causing it to break apart in a process known as "lake-induced hydrofracturing"… many of them [East Antarctica blue lakes] seemed to be clustered in regions of ice shelves that could be vulnerable to collapse via hydrofracturing" [63]. Georgia Rose Grant and Timothy Naish writing in Phys.Org (2019): "We know that our planet has experienced warmer periods in the past, during the Pliocene geological epoch around three million years ago. Our research, published today, shows that up to one third of Antarctica's ice sheet melted during this period, causing sea levels to rise by as much as 20 meters above present levels in coming centuries… The Pliocene was the last time atmospheric carbon dioxide concentrations were above 400 parts per million and Earth's temperature was 2°C warmer than pre-industrial times." [64]. The melting of Antarctic sea ice has the successive ecological impacts of a decrease in ice-associated phytoplankton, a decrease in krill and thence a decrease in krill-eating whales and penguins [65, 66]. Most of the Earth's ice resides in the East and West Antarctic ice sheets that comprise about 90 percent of Earth's fresh water in 26.5 million cubic kilometers of ice, and that if melted would raise sea level by 60 meters [67, 68]. Some scientists fear that the tipping point for long-term and irreversible West Antarctic ice sheet melting may have been reached [69].

Figure 2(h). Glacier thickness change (meter of water equivalent) was -1 (1979) to -21 (2018) for an overall average change of -0.15 /year. The quasi-linear change in the 2100s was -14 (2010) to -21 (2018), this yielding a change of -0.88 meters per year. Using the latter estimate, extrapolation yields a decrease by 2030 of 31.6 meters.

Comments. Glaciers are melting around the world [70]. The tipping point for long-term, centuries time-frame and irreversible West Antarctic ice sheet melting may have already been reached [69].

Figure 2(i). Ocean heat content change (10^{22} joules) monotonically increased from 0 (1979) to 21.75 (2016) for a change of 0.59 per year (+ 6.18 x 10^{22} joules per decade). However the change was quasi-linear from 7.75 (2000) to 21.75 (2016) for a change of 0.875 per year. Extrapolating this latter rate yields an increase in ocean heat content by 2030 of 34.0 x 10^{22} joules.

Comments. 90% of the extra heat has ended up in the ocean. We have great trouble grasping large and small numbers. Thus 1 joule is the heat needed to raise the temperature of 1 g of water by 0.24 °C [71] (e.g. roughly the warming of the dregs of a glass of cold water in 10 seconds at room temperature). 34.0 x 10^{22} joules is the energy of 10 million 1 megaton thermonuclear bombs [72]. The Guardian has estimated that the total heat taken up by the oceans in the last 150 years is equivalent to about 5 Hiroshima size (10 kiloton TNT equivalent) atomic bombs exploding per second [73].

Figure 2(j). Ocean acidity measured as pH ($-\log_{10}$ of H^+ concentration in moles per litre) declined in a linear fashion from pH 8.114 (1979) to pH 8.064 (2017) i.e. by 0.0013 pH per year (+4.12% acidity increase per decade, +0.4% acidity increase per year). Extrapolation yields a pH of 8.047 in 2013.

Comments. Ocean acidification impacts all ocean organisms with calcareous exoskeletons. About one-third of the carbon dioxide (CO_2) released into the atmosphere as a result of human activity has been absorbed by the oceans, where it partitions into the ionized forms of carbonic acid, this leading to ocean acidification, one of the major threats to marine ecosystems and particularly to calcifying organisms such as corals (animals with photosynthetic zooanthellae symbionts), foraminifera (single-celled eukaryote amoeboid protists with a calcareous external shell), coccolithophores (photosynthetic algae that generate calcareous plates) and lobster, crab, shrimp and krill crustaceans with food chain implications) [17].

Figure 2 (k). Sea level change relative to the 20-year mean (mm) increased linearly from − 34 (1993) to + 46 (2017) to give a rise of 3.33 mm per year (+31.4 mm per decade or +3.14 per year according to the paper). Extrapolation yields a sea level rise in 2030 of 89 mm (relative to that in 2005) and 123 mm (relative to that in 1993).

Comments. US NOAA Climate.org: "Since the start of the satellite sea level record in 1993, the average rate of sea level has been about one-eighth of an inch (3.1 mm) per year. The rising water level is mostly due to a combination of meltwater from glaciers and ice sheets and thermal expansion of seawater as it warms… Scientists are very confident that global mean sea level will rise at least 8 inches (0.2 meter), but no more than 6.6 feet (2.0 meters) by 2100, depending on the energy decisions we make" [74]. At the present about +1.1C low-lying Island Nations and low-lying parts of mega-delta nations (e.g. Bangladesh and India) are already being devastated by a combination of higher sea level, and warming-exacerbated and energized severe storms and storm surges [75, 76]. US Climate Central has estimated that 190 million people will be living in areas that are projected to be below the high-tide mark in 2100, and that 110 million people are presently living in these lands, protected by sea walls [77, 78].

Dr Andrew Glikson (an Earth and paleo-climate research scientist at Australian National University, Canberra, Australia) (2009): "For some time now, climate scientists warned that melting of subpolar permafrost and warming of the Arctic Sea (up to 4 degrees C during 2005–2008 relative to the 1951–1980) are likely to result in the dissociation of methane hydrates and the release of this powerful greenhouse gas into the atmosphere (methane: 62 times the infrared warming effect of CO_2 over 20 years and 21 times over 100 years) … The amount of carbon stored in Arctic sediments and permafrost is estimated as 500–2500 Gigaton Carbon (GtC), as compared with the world's total fossil fuel reserves estimated as 5000 GtC. Compare with the 700 GtC of the atmosphere, which regulate CO_2 levels in the range of 180–300 parts per million and land temperatures in a range of about – 50 to + 50 degrees C, which allowed the evolution of warm blooded mammals. The continuing use of the atmosphere as an open sewer for industrial pollution has already added some 305 GtC to the atmosphere together with land clearing and animal-emitted methane. This raised CO_2 levels to 387 ppm CO_2 to date [415 ppm CO_2 in 2020], leading toward conditions which existed on Earth about 3 million years (Ma) ago (mid-Pliocene), when CO_2 levels rose to about 400 ppm, temperatures to about 2–3 degrees C and sea levels by about 25 +/- 12 metres" [79]. The atmospheric CO_2 has now reached about 410 ppm CO_2 [5, 6].

Figure 2 (l). Area burned in the United States (million hectares per year) involves very scattered data. However, assuming the authors' quasi-linear best fit involving an increase from 0.9 (1983) to 3.2 (2018), one can estimate an increase of 0.066 million hectares per year (+ 44.1% per decade, +4.4% per year). Extrapolating from this average yields an estimate of 4.0 million hectares burned in 2030.

Comments. In 2019 alone there have been massive forest fires in California, the Amazon and continent-wide in Australia. The

Guardian Australia: "The link between rising greenhouse gas emissions and increased bushfire risk is complex but, according to major science agencies, clear. Climate change does not create bushfires but it can and does make them worse. A number of factors contribute to bushfire risk, including [elevated] temperature, fuel load, dryness, [and] wind speed and [decreased] humidity… The year coming into the 2019-20 summer has been unusually warm and dry for large parts of Australia. Above-average temperatures now occur most years and 2019 had the fifth-driest start to the year on record, and the driest since 1970. Australia recorded its hottest month in January 2019, its third-hottest July and its hottest October day in some areas, among other temperature records" [80]. Phys.org has listed 10 ways in which climate change can variously make wildfires worse: (1) hot, dry, windy weather [very low humidity], (2) more plant fuel, (3) change of plant species to more flammable, dryness-compatible species, (4) last-drop transpiration by thirsty plants, (5) more lightning, (6) differential warming of the Arctic and a weakened jet stream (this sinking drier and hotter air), (7) the periodic El Nino events lead to decreased rainfall, increased temperatures and increased fire risk in Indonesia and eastern Australia, (8) increase in the intensity of bushfires to unstoppable, (9) increased winter survival of pine beetles with increased flammable dead wood in Canadian boreal forests, (10) positive feedback from increased CO_2 -> increased warming -> increased forest fires -> increased CO_2 – > increased warming - > … [81]. The link between global warming and more forest fires has been documented by expert scientists for many years (e.g. see [82]).

In Australia the risk of bushfires has been decreased by fuel reduction burning of forest undergrowth, but with global warming the "fire seasons" have been extending into winter, and fuel reduction burning has been increasingly limited because of conditions being too impossibly wet or too dangerously dry. The rational alternative is large-scale mechanical harvesting of undesirable undergrowth, scrub and grasslands with subsequent profitable conversion of this cellulosic biomass into biochar (carbon, C) through renewable energy-driven anaerobic pyrolysis at 400-700C. The biochar can be either simply safely sequestered underground or used to add carbon to improve arable soils. It is estimated that potential biochar from existing agricultural and forestry waste is equivalent to annual carbon pollution from industry (12 Gt C per year) [83]. Such harvesting for biochar production would be profitable (circa $200 per tonne biochar) and would avoid the disastrous positive feedback loop from uselessly burning biomass (indeed such biomass could be harvested and used to generate renewable energy). In Australia unharvested grasslands are presently burning (and burning homes) while desperate farmers in drought-ravaged Eastern Australia are going broke paying for feed for their otherwise starving [methanogenic] livestock.

Figure 2(m). Extreme weather/climate/hydrological events (numbers per year) have been monotonically increasing in a tri-phasic quasi-linear fashion from 200 (1980) to 400 (1994) (an increase of 14.3 events per year), a constant of circa 400 per year in the period 1994-2004, and then a linear increase from 440 (2004) to 800 (2019) (an increase of 32.7 events per year) (+ 43.8% per decade and +4.4% per year overall). Assuming an increase of 32.7 events per year, extrapolation yields 1,160 events in 2030 or more than 3 such events per day.

Comments. What will it take before the murderously neoliberal One Percenters running the World – and obscenely owning half its wealth [84, 85] – bend before massively deadly inequality? Presently each year 15 million people die avoidably from deprivation in the Developing World (minus China) [15], 8 million people die from air pollution, this including 10,000 rich Australians and 75,000 people dying world-wide from the effects of pollutants from the burning of Australia's world-leading coal exports [88]. An International Organization for Migration (IOM) report (2008) quoted estimates of 25 million refugees by the mid-1990s (IPCC, 1990), 50 million environmental refugees by 2010 (Red Cross, 2001) and 200 million climate refugees by 2050 (Professor Norman Myers, University of Oxford). The IOM report (2008) stated that 192 million people or 3% of the world population now lived outside their place of birth [89]. Dina Ionesco (Head of the Migration, Environment and Climate Change (MECC) Division at the UN Migration Agency (IOM)) stated: "In 2018 alone, 17.2 million new displacements associated with disasters in 148 countries and territories were recorded (IDMC) and 764,000 people in Somalia, Afghanistan and several other countries were displaced following drought" [90]. Environmental migrants evidently now total about 200 million people.

Figure 2 (n). Annual losses due to weather/climate/hydrological events ($billions) and excluding 2 exceptional outlier events, increased in a quasi-linear fashion from 20 (1980) to 196 (2018) for a rate of 4.63 per year (+83.7% per decade, 8.4% per year). Extrapolation yields an annual cost of $2,452 billion in 2030.

Comments. These estimates of annual costs of pollution are in reality huge under-estimates. Thus we should be decreasing atmospheric CO_2 pollution rather than increasing it. In the discussion about Figure 2(a) the upper estimate of the cost of reversing a 1 ppm increase in atmospheric CO_2 is about $1,958 billion. Accordingly the cost of reversing the present annual increase of 1.79 ppm CO_2 is $3,500 billion. The upper estimate of draw-down of atmospheric CO_2 from the present observed monthly mean of 412 ppm CO_2 to a safe and sustainable 300 ppm CO_2 is 412 ppm CO_2 x $1,958 per ppm CO_2 = $807 trillion or about the 2 times the accumulated wealth of the world [86, 87]. Approaching the problem from the estimate of a damage-related carbon price of $200 per tonne CO_2 [45], one can estimate that the World has a Carbon Debt of about $200 trillion that is increasing at about $10 trillion per year. Australia has a Carbon Debt of $5 trillion that is increasing at $400 billion per year and at $40,000 per head per year for under-30 year old Australians [37]. This is an inescapable Carbon Debt that will have to be repaid by future generations. Unlike conventional debt (that can be variously evaded by default, bankruptcy, or printing money), this enormous Carbon Debt is inescapable. Thus unless sea walls are built, cities, towns and arable land will be inundated, and unless forests are protected by instant dousing of ignition points, global forests will disappear in unstoppable fire storms. In general, unless this Carbon Debt is met by requisite protective measures, human assets from homes, buildings and infrastructure to farms and forests will be uninsurable.

What needs to be urgently done.

This Second Warning to the World [1] concludes with suggested actions under 6 headings that are summarized below [with my numerically-based amplification comments appended].

(1). Energy. There should be a rapid transition away from dirty energy (coal, oil, gas and other carbon fuels) to 100% renewable energy, with this associated with assistance to developing economies, cessation of fossil fuel exploitation, and application of proper Carbon Prices.

Comment. There must be a rapid transition within about 10 years to 100% renewable energy, coupled with cessation of fossil fuel exploitation, carbon fuel-based public and private transport, and of deforestation and land clearing. The environmental and social cost of carbon pollution should be "fully borne" by the polluters with a damage-related Carbon Price of $200 per tonne CO_2-equivalent.

(2). Short-lived pollutants. Short-lived pollutants such as methane (CH_4), deadly carbon particles (e.g. $PM_{2.5}$) and hydrofluorocarbons (HFCs) need to be eliminated.

Comment. Major sources of CH_4 (notably anaerobic swamps from deforestation and methanogenic livestock), deadly carbon particulates (from carbon fuel burning including fossil fuel burning and forest fires) and man-made HFCs (as well as other man-generated GHGs such as SF_6 and NF_3) must be eliminated. 8 million annual pollution deaths from the long-term impacts of air pollution will be progressively eliminated in coming decades through elimination of all carbon fuel burning. The relatively long-term GHG pollutant CO_2 must be urgently drawn down from the present damaging and dangerous 412 ppm CO_2 in the atmosphere to yield a circa 300 ppm CO_2 that is safe and sustainable for all peoples and all species [48, 49].

(3). Nature. "We must protect and restore Earth's ecosystems. Phytoplankton, coral reefs, forests, savannas, grasslands, wetlands, peatlands, soils, mangroves, and sea grasses contribute greatly to sequestration of atmospheric CO_2" [1].

Comment. The presently ongoing and massive speciescide and ecocide of the Anthropocene Era is associated with an extinction rate 100-1,000 times greater than normal [91] and heads the world towards man-made omnicide and terracide. Any species is irreplaceable and accordingly priceless. There must be cessation of deforestation and land clearing with massive re-afforestation and ecosystem restoration. Dr Andre Balmford and colleagues have stated that "We estimate that the overall benefit:cost ratio of an effective global program for the conservation of remaining wild nature is at least 100:1"[92].

(4). Food. "Eating mostly plant-based foods while reducing the global consumption of animal products" [1].

Comment. There must be global vegetarian diet and replacement of cow-derived milk with plant substitutes (soybean and almond milk). The deadly food-for-fuel Biofuel Genocide obscenity must be stopped. Food adequacy involves cessation of global inequity, population growth, economic growth, GHG pollution, sea level rise, soil salinization, deforestation and damaging land use.

(5). Economy. "Our goals need to shift from GDP growth and the pursuit of affluence toward sustaining ecosystems and improving human well-being by prioritizing basic needs and reducing inequality" [1].

Comment. There must be rapid halving of economic output (economic de-growth) with the burden overwhelmingly borne by the rich global North to permit the global South to attain a decent state [7]. "For everyone in the world to have an

American lifestyle, we would need seven planets, and three to live as Europeans" [13].

(6). Population. "Still increasing by roughly 80 million people per year, or more than 200,000 per day (figure 1a–b), the world population must be stabilized—and, ideally, gradually reduced—within a framework that ensures social integrity." [1].

Comment. The rich European population is declining, but not fast enough, and the impoverished African population is expanding impossibly. The world population needs to halve as soon as possible [7]. Present projections are for a sustainable human population of only 0.5-1.0 million by 2100 [8].

Final comments.

The 2019 "World scientists' warning of a climate emergency" [1] sets out disastrous, quasi-linear trends in 21 out of 24 "parameter versus time plots" variously covering the 1979-2019 time period. The only hopeful trends were decline in human fertility (Figure 1(b)), an increase in institutional divestment from fossil fuels (Figure 1(j) and an increase in GHG emissions covered by carbon pricing (Figure 1(m)). Because these trends were quasi-linear it was possible to plausibly extrapolate to estimate levels of all these parameters in 2030.

Despite scientists being aware of the greenhouse effect for 150 years, and having warned politicians of the increasing seriousness of global warming since the mid-1980s, the World is now in a dire situation in which catastrophic plus 2C temperature rise is now effectively unavoidable [93-99]. Thus paleoclimatologist and earth scientist Dr Andrew Glikson: "As the globe warms, to date by a mean of near ~1.5 $^{\circ}$C, or ~2.0°C when the masking effects of sulphur dioxide and other aerosols are considered, and by a mean of ~2.3°C in the Polar Regions" [59].

Nevertheless decent people are obliged to do everything they can to make the future "less bad' for future generations. Eminent physicist Professor Stephen Hawking has succinctly warned [my emphasis added]: "We see great peril if governments and societies do not take action NOW to render nuclear weapons obsolete and to prevent further climate change" [93, 94].

What can people do? Informed by the science [1], decent people around the world must (a) inform everyone they can about the worsening Climate Emergency, Climate Genocide and Intergenerational Inequity, (b) urge a climate revolution (peaceful and non-violent of course) with hundreds of millions out in the streets inspired by the likes of teenage activist Greta Thunberg, and (c) urge and apply Boycotts, Divestment and Sanctions against all people, politicians, parties, collectives, corporations and countries disproportionately involved in the worsening Climate Genocide that is presently set to kill 10 billion people this century en route to a sustainable human population of merely 0.5-1.0 billion in 2100 [8].

References.

[1]. William Ripple et al., "World scientists' warning of a climate emergency", BioScience, 5 November 2019: https://academic.oup.com/bioscience/advance-article/doi/10.1093/biosci/biz088/5610806.

[2]. William J. Ripple et al., 15,364 signatories from 184 countries, "World scientists' warning to Humanity: a second notice", Bioscience, 13 November 2017: https://academic.oup.com/bioscience/advance-article/doi/10.1093/biosci/bix125/4605229.

[3]. Gideon Polya, "Over 15,000 scientists issue dire warning to humanity on catastrophic climate change and biodiversity loss", Countercurrents, 20 November 2017: https://countercurrents.org/2017/11/20/over-15000-scientists-issue-dire-warning-to-humanity-on-catastrophic-climate-change-and-biodiversity-loss/.

[4]. UN Population Division: https://population.un.org/wpp/DataQuery/.

[5]. US NOAA, "Monthly average Maua Loa CO_2": https://www.esrl.noaa.gov/gmd/ccgg/trends/.

[6]. US NOAA, "Full Mauna Loa CO_2 record": https://www.esrl.noaa.gov/gmd/ccgg/trends/full.html.

[7]. Gideon Polya, "How much negative carbon emissions, negative population growth & negative economic growth is needed to save Planet?" Countercurrents, 28 November 2018: https://countercurrents.org/2018/11/how-much-negative-carbon-emissions-negative-population-growth-negative-economic-growth-is-needed-to-save-planet.

[8]. "Climate Genocide": https://sites.google.com/site/climategenocide/.

[9]. "Are we doomed?": https://sites.google.com/site/300orgsite/are-we-doomed.

[10]. "Too late to avoid global warming catastrophe": https://sites.google.com/site/300orgsite/too-late-to-avoid-global-warming.

[11]. "Nuclear weapons ban, end poverty & reverse climate change": https://sites.google.com/site/300orgsite/nuclear-weapons-ban.

[12]. "List of countries by GDP (nominal) per capita", Wikipedia: https://en.wikipedia.org/wiki/List_of_countries_by_GDP_(nominal)_per_capita.

[13]. Irene Banos Ruiz, "China's new love affair with dogs – as pets, not food – presents environmental problems", DW, 21 June 2016:

https://www.dw.com/en/chinas-new-love-affair-with-dogs-as-pets-not-food-presents-environmental-problems/a-19197523.
[14]. "Zero population growth", Wikipedia:
https://en.wikipedia.org/wiki/Zero_population_growth.
[15]. Gideon Polya, "Body Count. Global avoidable mortality since 1950", this including an avoidable mortality-related history of every country since Neolithic times and now available for free perusal on the web: http://globalbodycount.blogspot.com.au/2012/01/body-count-global-avoidable-mortality_05.html.
[16]. Drew T. Shindell, Greg Faluvegi, Dorothy M. Koch, Gavin A. Schmidt, Nadine Unger and Susanne E. Bauer, "Improved Attribution of Climate Forcing to Emissions", Science, 30 October 2009: Vol. 326 no. 5953 pp. 716-718:
http://www.sciencemag.org/content/326/5953/716 and Fig.2: http://www.sciencemag.org/content/326/5953/716.figures-only.
[17]. "2011 climate change course":
https://sites.google.com/site/300orgsite/2011-climate-change-course.
[18]. "Gas is not clean energy":
https://sites.google.com/site/gasisnotcleanenergy/.
[19]. Robert Goodland and Jeff Anfang. "Livestock and climate change. What if the key actors in climate change are … cows, pigs and chickens?", World Watch, November/December 2009:
https://pdfs.semanticscholar.org/6704/c7a0777c82357704d82b9ae800 7c1197cb07.pdf?_ga=2.187734888.1597394103.1556059730-1006954717.1556059730.
[20]. WBGU, "Solving the climate dilemma: the budget approach": http://www.ecoequity.org/2009/10/solving-the-climate-dilemma-the-budget-approach/.
[21]. Australian Climate Commission, "The critical decade 2013: a summary of climate change science, risks and responses", 2013, p7: http://climatecommission.gov.au/wp-content/uploads/The-Critical-Decade-2013-Summary_lowres.pdf.
[22]. Gideon Polya, "Vegetarianism can help save Planet", MWC News, 24 January 2011: http://mwcnews.com/focus/analysis/8188-vegetarianism-can-help-save-planet.html.
[23]. Gideon Polya, "Millions join Global School Climate Strike – we are running out of time", Countercurrents, 22 September 2019: https://countercurrents.org/2019/09/millions-join-global-school-climate-strike-we-are-running-out-of-time.
[24]. Gideon Polya, "Australia rejects IMF Carbon Tax & preventing 4 million pollution deaths by 2030", Countercurrents, 15 October 2019: https://countercurrents.org/2019/10/australia-rejects-imf-carbon-tax-preventing-4-million-pollution-deaths-by-2030.
[25]. Bruce Pascoe, "Dark Emu", Magabala, 2014.

[26]. Jackie Linden, "Global poultry trends 2014: poultry set to become no. 1 meat in Asia", The Poultry Site, 2 September 2014: https://thepoultrysite.com/articles/global-poultry-trends-2014-poultry-set-to-become-no1-meat-in-asia.
[27]. "Biofuel famine, biofuel genocide, meat & global food price crisis", Global avoidable mortality: http://globalavoidablemortality.blogspot.com/2008/05/biofuel-famine-biofuel-genocide-meat.html.
[28]. Xiao-Peng Song et al., "Global land change from 1982 to 2016", Nature volume 560, pages 639–643, 2018: https://www.nature.com/articles/s41586-018-0411-9.
[29]. One Tree Planted, "New research: 1 billion hectares of forest could save the Planet ", One Tree Planted: https://onetreeplanted.org/blogs/stories/research-trees-climate.
[30]. Katharine Murphy, "Scott Morrison brings coal to question time: what fresh idiocy is this?", Guardian, 9 February 2017: https://www.theguardian.com/australia-news/2017/feb/09/scott-morrison-brings-coal-to-question-time-what-fresh-idiocy-is-this.
[31]. Jonathan Sundby, Gideon Weissman, and Rob Sargent, "Renewables on the rise 2019: a decade of progress towards a clean energy future", Frontier Energy, 22 August 2019: https://frontiergroup.org/reports/fg/renewables-rise-2019.
[32]. Liu Yuanyuan, "China's renewable energy installed capacity grew 12 percent across all sources in 2018", Renewable Energy World, 6 March 2019: https://www.renewableenergyworld.com/2019/03/06/chinas-renewable-energy-installed-capacity-grew-12-percent-across-all-sources-in-2018/.
[33]. "100% renewable energy by 2020": https://sites.google.com/site/100renewableenergyby2020/.
[34]. "New study: global energy system based on 100% renewable energy", Energy Watch Group, 12 April 2019: http://energywatchgroup.org/new-study-global-energy-system-based-100-renewable-energy.
[35]. "Energy efficiency in transport", Wikipedia: https://en.wikipedia.org/wiki/Energy_efficiency_in_transport.
[36]. "Divest from fossil fuels": https://sites.google.com/site/300orgsite/divest-from-fossil-fuels.
[37]. "Carbon Debt Carbon Credit": https://sites.google.com/site/carbondebtcarboncredit/.
[38]. "Climate Justice & Intergenerational Equity": https://sites.google.com/site/300orgsite/climate-justice.

[39]. "Science & economics experts: Carbon Tax needed NOT Carbon Trading": https://sites.google.com/site/300orgsite/sciennce-economics-experts-carbon-tax-needed-not-carbon-trading/.

[40]. "Stop climate crime": https://sites.google.com/site/300orgsite/stop-climate-crime.

[41]. Gideon Polya, "Inescapable $200-250 trillion global carbon debt increasing by $16 trillion annually", Countercurrents, 27 April 2019: https://countercurrents.org/2019/04/inescapable-200-250-trillion-global-carbon-debt-increasing-by-16-trillion-annually-gideon-polya

[42]. Gideon Polya, "Polya's 3 Laws Of Economics Expose Deadly, Dishonest And Terminal Neoliberal Capitalism", Countercurrents, 17 October, 2015: https://countercurrents.org/polya171015.htm.

[43]. International Monetary Fund (IMF), "Fiscal Monitor: how to mitigate climate change". Executive Summary", September 2019: file:///C:/Users/Gideon/AppData/Local/Temp/execsum-6.pdf.

[44]. Gideon Polya, "Australia rejects IMF Carbon Tax & preventing 4 million pollution deaths by 2030", Countercurrents, 15 October 2019: https://countercurrents.org/2019/10/australia-rejects-imf-carbon-tax-preventing-4-million-pollution-deaths-by-2030.

[45]. Chris Hope, "How high should climate change taxes be?", Working Paper Series, Judge Business School, University of Cambridge, 9.2011: http://www.jbs.cam.ac.uk/fileadmin/user_upload/research/workingpapers/wp1109.pdf.

[46]. Gideon Polya, "Revised Annual Per Capita Greenhouse Gas Pollution For All Countries – What Is Your Country Doing?", Countercurrents, 6 January 2016: https://countercurrents.org/polya060116.htm.

[47]. Gideon Polya, "Exposing And Thence Punishing Worst Polluter Nations Via Weighted Annual Per Capita Greenhouse Gas Pollution Scores", Countercurrents, 19 March, 2016: https://countercurrents.org/polya190316.htm.

[48]. 300.org:. https://sites.google.com/site/300orgsite/300-org.

[49]. "300.org – return atmosphere CO_2 to 300 ppm CO_2": https://sites.google.com/site/300orgsite/300-org—return-atmosphere-co2-to-300-ppm.

[50]. Output of the technical working group meeting, The Royal Society, London, 6th July, 2009, "The Coral Reef Crisis: scientific justification for critical CO_2 threshold levels of less than 350ppm": http://static.zsl.org/files/statement-of-the-coral-reef-crisis-working-group-890.pdf.

[51]. J.E.N. Veron, O. Hoegh-Guldberg, T.M. Lenton, J.M. Lough, D.O. Obura, P. Pearce-Kelly, C.R.C. Sheppard, M. Spalding, M.G. Stafford-Smith and A.D. Rogers, "The coral reef crisis: the critical

importance of <350 ppm CO_2", Marine Pollution Bulletin, vol. 58, (10), October 2009, 1428-1436: https://www.ncbi.nlm.nih.gov/pubmed/19782832.
[52]. James Hansen, "Climate change in a nutshell: the gathering storm", Columbia University, 18 December 2018: http://www.columbia.edu/~jeh1/mailings/2018/20181206_Nutshell.pdf.
[53]. Gideon Polya, "Intergenerational Theft – For Every $1 For Coal Today Future Generations Will Pay $1-$14 To Sequester CO_2", Countercurrents, 8 April, 2015: http://www.countercurrents.org/polya080415.htm
[54]. Gideon Polya, "Huge Carbon Debt and Intergenerational Injustice: CO_2 draw-down necessity. 300.org and 300 ppm CO_2 target", Global Research, 7 June 2018: https://www.globalresearch.ca/huge-carbon-debt-and-intergenerational-injustice-co2-drawdown-necessity/5643365.
[55]. Andrew Glikson, "The fatal nexus – atmospheric CO_2 and the mass extinction of species", Countercurrents, 6 November 2019: https://countercurrents.org/2019/11/the-fatal-nexus-atmospheric-co2-and-the-mass-extinction-of-species.
[56]. Gail Whiteman, Chris Hope and Peter Wadhams, "Vast costs of Arctic change", Nature, 499, 25 July 2013: http://www.nature.com/nature/journal/v499/n7459/pdf/499401a.pdf and http://www.nature.com/nature/journal/v499/n7459/full/499401a.html.
[57]. "Methane Bomb threat": https://sites.google.com/site/methanebombthreat/.
[58]. David Spratt and Ian Dunlop, "Existential climate-related security risk: a scenario approach", Breakthrough – National Centre for Climate Restoration, May 2019: https://docs.wixstatic.com/ugd/148cb0_90dc2a2637f348edae45943a88da04d4.pdf.
[59]. Andrew Glikson, "Inferno: from climate denial to planetary arson", Countercurrents, 8 September 2019: https://countercurrents.org/2019/09/inferno-from-climate-denial-to-planetary-arson.
[60]. Gideon Polya, "Terra" (painting): https://www.flickr.com/photos/gideonpolya/4289095912/
[61]. "Arctic sea ice decline", Wikipedia: https://en.wikipedia.org/wiki/Arctic_sea_ice_decline.
[62]. Oliver Milman, "Greenland's ice melting faster than scientists previously thought – study", Guardian, 22 January 2019: https://www.theguardian.com/world/2019/jan/21/greenland-ice-melting-faster-than-scientists-previously-thought-study.

[63]. Madeleine Stone, "How Antarctica is melting from above and below", National Geographic, 9 October 2019: https://www.nationalgeographic.com/science/2019/10/how-antarctic-melting-above-below-ice-sheet/.

[64]. Georgia Rose Grant and Timothy Naish, "If warming exceeds 2C, Antarctica's melting ice sheets could rise seas 20 meters in coming centuries", Phys. Org, 3 October 2019: https://phys.org/news/2019-10-antarctica-ice-sheets-seas-meters.html.

[65]. Stacey L. Deppeler and Andrew T. Davidson, "Southern Ocean phytoplankton in a changing environment", Frontiers of Marine Science, 16 February 2017: https://www.frontiersin.org/articles/10.3389/fmars.2017.00040/full.

[66]. Matthew Taylor, "Decline in krill threatens Antarctic wildlife, from whales to penguins", Guardian, 15 February 2018: https://www.theguardian.com/environment/2018/feb/14/decline-in-krill-threatens-antarctic-wildlife-from-whales-to-penguins.

[67]. Richard Aster and Valentina Roberta Barletta, "The West Antarctic ice sheet is in trouble – but the ground beneath it may buy it some time", The Conversation, 25 June 2018: https://theconversation.com/the-west-antarctic-ice-sheet-is-in-trouble-but-the-ground-beneath-it-may-buy-some-time-98368.

[68]. "West Antarctic Ice Sheet' Wikipedia: https://en.wikipedia.org/wiki/West_Antarctic_Ice_Sheet.

[69]. Adam Morton, "Glacial melting in Antarctica may become irreversible", Guardian, 9 July 2019: https://www.theguardian.com/world/2019/jul/09/glacial-melting-in-antarctica-may-become-irreversible.

[70]. "Retreat of glaciers around the world", Wikipedia: https://en.wikipedia.org/wiki/Retreat_of_glaciers_since_1850.

[71]. "Joule", Wikipedia: https://en.wikipedia.org/wiki/Joule.

[72]. "TNT equivalent", Wikipedia: https://en.wikipedia.org/wiki/TNT_equivalent.

[73]. Damian Carrington, "Global warming of oceans equivalent to an atomic bomb per second", Guardian, 8 January 2019: https://www.theguardian.com/environment/2019/jan/07/global-warming-of-oceans-equivalent-to-an-atomic-bomb-per-second.

[74]. US NOAA Climate.org, "climate change: global sea level": https://www.climate.gov/news-features/understanding-climate/climate-change-global-sea-level.

[75]. Gideon Polya, "Climate criminal Australia sabotages Pacific Islands Forum & threatens all Island Nations", Countercurrents, 24 August 2019: https://countercurrents.org/2019/08/climate-criminal-australia-sabotages-pacific-islands-forum-threatens-all-island-nations.

[76]. Pacific Islands Development Forum 4 September 2015 "Suva Declaration on Climate Change": http://pacificidf.org/wp-content/uploads/2013/06/PACIFIC-ISLAND-DEVELOPMENT-FORUM-SUVA-DECLARATION-ON-CLIMATE-CHANGE.v2.pdf.
[77]. Scott A. Kulp and Benjamin H. Strauss, "New elevation data triple estimates of global vulnerability to sea level rise and coastal vulnerability", Nature Communications, volume 10,
Article number: 4844, 2019: https://www.nature.com/articles/s41467-019-12808-z.
[78]. Jonathan Amos, "Climate change: sea level rise to affect "three times more people"", BBC, 30 October 2019:
https://www.bbc.com/news/science-environment-50236882.
[79]. Andrew Glikson, "The Methane Time Bomb and the Triple Melt-down", Countercurrents, 2009 :
http://www.countercurrents.org/glikson101008.htm.
[80]. "What are the links between climate change and bushfires? – explainer", Guardian, 11 November 2019:
https://www.theguardian.com/australia-news/2019/nov/11/what-are-the-links-between-climate-change-and-bushfires-explainer.
[81]. "Ten ways climate change can make wildfires worse", Phys.org, 12 Novemebr 2019: https://phys.org/news/2019-11-ten-ways-climate-wildfires-worse.html.
[82]. A.L. Westerling, H. G. Hidalgo, D. R. Cayan, T. W. Swetnam, Warming and Earlier Spring Increase Western U.S. Forest Wildfire Activity, *Science* 18 August 2006: Vol. 313. no. 5789, pp. 940 – 943 (DOI: 10.1126/science.1128834; see:
http://www.sciencemag.org/cgi/content/full/313/5789/940.
[83]. "Forest biomass-derived Biochar can profitably reduce global warming and bushfire risk", Yarra Valley Climate Action Group:
https://sites.google.com/site/yarravalleyclimateactiongroup/forest-biomass-derived-biochar-can-profitably-reduce-global-warming-and-bushfire-risk.
[84]. Oxfam, "Rapidly growing inequality is worsening poverty around the world", 20 January 2014:
https://www.oxfam.org.au/2014/01/rapidly-growing-inequality-is-worsening-poverty-around-the-world/.
[85]. Oxfam, "Working for the Few", 20 January 2014:
https://www.oxfam.org.au/wp-content/uploads/2014/01/bp-working-for-few-political-capture-economic-inequality-200114-embargo-en.pdf.
[86]. Thomas Piketty, "Capital in the Twenty-First Century", Harvard University Press, 2014.

[87]. Gideon Polya, "Key Book Review: "Capital In The Twenty-First Century" By Thomas Piketty", Countercurrents, 1 July, 2014: https://countercurrents.org/polya010714.htm.
[88]. "Stop air pollution deaths": https://sites.google.com/site/300orgsite/stop-air-pollution-deaths.
[89]. Oli Brown, "Migration and climate change", International Organization for Migration, 2008: file:///C:/Users/Gideon/AppData/Local/Temp/5866-4.pdf.
[90]. Dina Ionescu, "Let's talk about climate migrants, not climate refugees – think piece", Environmental Migration Portal, 6 June 2019: https://environmentalmigration.iom.int/blogs/let%E2%80%99s-talk-about-climate-migrants-not-climate-refugees-think-piece.
[91]. Phillip S. Levin and Donald A. Levin, "The real biodiversity crisis", Macroscope, January-February 2002: http://www.soc.duke.edu/~pmorgan/levin&levin.2002.the_real_biodiversity_crisis.html.
[92]. A. Balmford, A. Bruner, P. Cooper, R. Costanza, S. Farber, R. E. Green, M. Jenkins, P. Jefferiss, V. Jessamy, J. Madden, K. Munro, N. Myers, S. Naeem, J. Paavola, M. Rayment, S. Trumper and R. K. Turner "Economic reasons for conserving wild nature", Science 297: 950-953, 2002: http://www.sciencemag.org/cgi/content/abstract/297/5583/950.
[93]. Professor Stephen Hawking quoted in Will Dunham, "Nuclear, climate perils push Doomsday Clock ahead", Reuters, 22 January 2007: https://www.reuters.com/article/idUSN17314370.
[94]. Stephen Hawking, "Brief Answers to the Big Questions", John Murray, 2018, Chapter 7.
[95]. IPCC, "Global warming of 1.5 °C", 8 October 2018: http://www.ipcc.ch/report/sr15/.
[96]. IPCC, "Global warming of 1.5 °C. Summary for Policymakers", 8 October 2018: http://report.ipcc.ch/sr15/pdf/sr15_spm_final.pdf.
[97]. Gideon Polya, "IPCC +1.5C avoidance report – effectively too late but stop coal burning for "less bad" catastrophes", Countercurrents, 12 October 2018: https://countercurrents.org/2018/10/12/ipcc-1-5c-avoidance-report-effectively-too-late-but-stop-coal-burning-for-less-bad-catastrophes/.
[98]. Gideon Polya, "Millions join global school strike – we are running out of time" Coutercurrentds, 22 September 2019: https://countercurrents.org/2019/09/millions-join-global-school-climate-strike-we-are-running-out-of-time.
[99]. Andrew Glikson, "The IPCC's final warnings of extreme global warming", Countercurrents, 10 October 2018: https://countercurrents.org/2018/10/10/the-ipccs-final-warnings-of-extreme-global-warming/.

CHAPTER 5: CLIMATE SCIENTISTS: PLANETARY EMERGENCY, PLANET IN PERIL, ACT NOW

[First published as Gideon Polya, "Climate scientists: Planetary Emergency, Planet in peril, act now", Countercurrents, 3 December 2019: https://countercurrents.org/2019/12/climate-scientists-planetary-emergency-planet-in-peril-act-now.]

A paper co-authored by some eminent climate scientists and just published in the prestigious scientific journal Nature, analyses critical tipping points impacted by man-made climate change, and concludes: "Act now… the evidence from tipping points alone suggests that we are in a state of planetary emergency: both the risk and urgency of the situation are acute… The stability and resilience of our planet is in peril. International action — not just words — must reflect this".

A climate change-related "tipping point" for a particular phenomenon (e.g. loss of Arctic summer sea ice) is the point at which the change is irreversible. Successive IPCC reports over past decades warned of the likelihood of tipping points being reached at various degrees of global warming. However the authors provide a Figure showing that IPCC-estimated "High" to "Very High" risk of rapid and irreversible changes in the climate system has progressively occurred at lower average warming over the last 20 years. Thus an approximate simplification of this Figure is that IPCC-perceived "High" to "Very High" risk occurred at +5.5C (2001), +5C (2007), +4C (2013) and +2C (2018) [1].

The authors conclude that "If current national pledges to reduce greenhouse-gas emissions are implemented — and that's a big 'if' — they are likely to result in at least 3 °C of global warming. This is despite the goal of the 2015 Paris agreement to limit warming to well below 2 °C… if tipping points are looking more likely, then the 'optimal policy' recommendation of simple cost–benefit climate-economy models aligns with those of the recent IPCC report [2]. In other words, warming must be limited to 1.5 °C. This requires an emergency response" [1].

Noting that a "Moderate" risk of tipping points exceedances in the range of + 1-2C and that the global warming is now about +1C above the pre-industrial, one can well ask whether a sane individual would board a plane that had a "Moderate" risk of crashing.

The authors then considered some key tipping points and their assessments are summarized below (with global consequences and amplifications in brackets).

(1). West Antarctica (Amundsen Sea embayment): the tipping point may have been exceeded (this could destabilize the rest of the West Antarctic ice sheet leading to circa 3 metres of sea-level rise on a timescale of 100s to 1000s of years, noting that this has happened in the past).

(2). East Antarctic ice sheet (Wilkes Basin): approaching a tipping point (3-4 metres in sea level rise on a timescale of 100s of years).

(3). Greenland ice sheet: melting at an increasing rate and the tipping point may be the circa +1.5C expected in about 2030 (7 metres sea level rise over 1000s of years).

(4). Arctic sea ice: massive sea ice loss already (at +2C there is a 10–35% probability of near-total summer sea ice loss).

(5). Coral reefs: mass coral bleaching worldwide and loss of 50% of the coral of Australia's iconic Great Barrier Reef due to warming, ocean acidification, pollution and Crown of Thorn predator population explosion (over 99% of tropical corals are predicted to be lost at +2C with massive loss of marine biodiversity and fisheries).

(6). Amazon rain rainforest: estimated tipping point at 20% – 40% deforestation with about 17% lost since 1970 (the world's largest rainforest, it contains 1 in 10 of known species, has a continent-scale climate impact, and is presently burning on a huge scale).

(7). North American boreal forest: warming in the sub-Arctic has led to insect pest population explosion, boreal (high latitude) tree die-off and fires (some boreal forest regions are converting from being carbon sinks to a carbon sources).

(8). Arctic permafrost: irreversible thawing and release of carbon dioxide (CO_2) and methane (CH_4) with Arctic warming about 2 times bigger than the global average (CH_4 has an estimated Global Warming Potential (GWP) as low as 21 relative to the same mass of CO_2 on a 100 year time-scale, but on a 20 year time scale and with aerosol impacts included it is 105).

(9). Atlantic Meridional Overturning Circulation (AMOC): a key salt- and heat-conveying Atlantic Ocean circulation system subject to a 15% slow-down since the mid-20th century (Arctic sea-ice loss is increasing regional warming; Arctic warming and Greenland melting are putting fresh water into the North Atlantic; slowdown in the AMOC is destabilizing the West African monsoon, impacting drought in the Sahel region, drying the Amazon, disrupting the East Asian monsoon and warming the Southern Ocean with increased Antarctic ice loss).

The authors provide a further Figure that dramatically summarizes how the key changes summarized above can variously impact on each other, with AMOC having a key centrality. The authors comment: "We argue that cascading effects might be common. Research last year analysed 30 types of regime shift spanning physical climate and ecological systems, from collapse of the West Antarctic ice sheet to a switch from rainforest to savanna. This indicated that exceeding tipping points in one system can increase the risk of crossing them in others. Such links were found for 45% of possible interactions" [1].

The authors adduce paleo-climatological evidence that: "Atmospheric CO_2 is already at levels last seen around four million years ago, in the Pliocene epoch. It is rapidly heading towards levels last seen some 50 million years ago — in the Eocene — when temperatures were up to 14 °C higher than they were in pre-industrial times. It is challenging for climate models to simulate such past 'hothouse' Earth states. One possible explanation is that the models have been missing a key tipping point: a cloud-resolving model published this year suggests that the abrupt break-up of stratocumulus cloud above about 1,200 parts per million of CO_2 could have resulted in roughly 8 °C of global warming" [1].

Finally, the paper provides a compelling mathematical analysis: "We define emergency (E) as the product of risk and urgency. Risk (R) is defined by insurers as probability (p) multiplied by damage (D). Urgency (U) is defined in emergency situations as reaction time to an alert (τ) divided by the intervention time left to avoid a bad outcome (T). Thus: $E = R \times U = p \times D \times \tau / T$. The situation is an emergency if both risk and urgency are high. If reaction time is longer than the intervention time left ($\tau / T > 1$), we have lost control" [1].

The authors conclude: "**Act now.** In our view, the evidence from tipping points alone suggests that we are in a state of planetary emergency: both the risk and urgency of the situation are acute… We argue that the intervention time left to prevent tipping [T] could already have shrunk towards zero, whereas the reaction time to achieve net zero emissions [τ] is 30 years at best. Hence we might already have lost control of whether tipping happens. A saving grace is that the rate at which damage accumulates from tipping — and hence the risk posed — could still be under our control to some extent. The stability and resilience of our planet is in peril. International action — not just words — must reflect this".

Final comments.

The authors considered the Terminal Carbon Pollution Budget for a 50% chance of not exceeding +1.5C: "The world's remaining emissions budget for a 50:50 chance of staying within 1.5 °C of warming is only about 500 gigatonnes (Gt) of CO_2. Permafrost emissions could take an estimated 20% (100 Gt CO_2) off this budget, and that's without including methane from deep permafrost or undersea [CH_4-H_2O] hydrates. If forests are close to tipping points, Amazon dieback could release another 90 Gt CO_2 and boreal forests a further 110 Gt CO_2. With global total CO_2 emissions still at more than 40 Gt per year, the remaining budget could be all but erased already" [1].

In 2018 the IPCC issued a Report that detailed the numerous bad outcomes of a global +1.5 degree Centigrade (+1.5C) of warming versus the catastrophic outcomes from a +2C e.g. a further 70-90% decline of coral reefs at +1.5C versus more than 99% loss at +2C. Crucially, the IPCC Report says that for less than +1.5C coal burning must cease by 2050 but also declares that the Terminal Carbon Pollution Budget for a 66% chance of avoiding +1.5C (420 Gt CO_2) will be used up in 10 years at a rate of 42 Gt CO_2 per year [2-4].

Indeed the IPCC in its Fifth Assessment Report (AR5) stated (2014): "Emissions ranges for baseline scenarios and mitigation scenarios that limit greenhouse gas concentrations to low levels (about 450 ppm CO_2-eq, likely to limit warming to 2°C above pre-industrial levels) are shown for different sectors and gases in Figure SPM.14" [5]. However according to Professor Ron Prinn (from 100-Nobel-Laureate MIT) we may have already reached 478 ppm CO_2-equivalent in 2013 [6-10] (it is presently at nearly 500 ppm CO_2-equivalent).

The German WBGU (2009) and the Australian Climate Commission (2013) have estimated that no more than 600 billion tonnes of CO_2 can be emitted between 2010 and zero emissions in 2050 if the world is to have a 75% chance of avoiding a catastrophic +2C temperature rise [11, 12]. However a revised global annual GHG pollution of 64 Gt CO_2-equivalent (properly taking land use and CH_4 in to account) [13] means that this Terminal Carbon Pollution Budget was exceeded in 2019.

Australia (0.3% of world population, but with Domestic GHG pollution 2.1% of global GHG pollution and 4.5% with its huge Exported GHG pollution included) has a climate criminal policy (supported by both the ruling Coalition and the Labor Opposition) of unlimited coal, gas and iron ore exports, and it can be estimated that complete exploitation of Australia's huge resources in these areas would mean exceeding the whole world's 2009 Terminal Carbon Pollution Budget by a factor of 3 [14]. Similarly, the Global Warming Potential (GWP) of CH_4 on a 20 year time frame and with aerosol impacts considered is 105 times that of CO_2 [15] and the 50 Gt (billion tonnes) CH_4 in the East Siberian Arctic Shelf that is predicted to be released in coming decades [16] is thus equivalent to 50 billion tonnes CH_4 x 105 tonnes CO_2-equivalent/tonne CH_4 = 5,250 tonnes CO_2-equivalent or about nine (9) times more than the world's Terminal Carbon Pollution Budget. We are doomed unless we can stop this massive Arctic CH_4 release [4, 8-10].

Humanity and the Biosphere are existentially threatened by nuclear weapons and climate change [17]. Eminent physicist Professor Stephen Hawking has stated the problem and solution very succinctly: "We see great peril if governments and societies do not take action now to render nuclear weapons obsolete and to prevent further climate change" [18, 19]. A paper co-signed by over 11,000 scientists has detailed trends in 24 climate-related areas over the last 40 years. Scientists became aware of the climate change threat from greenhouse gas (GHG) pollution in the 1980s, but in 21 of these 24 areas the trends are (a) huge, (b) in the wrong direction, and (c) linear or quasi-linear functions of time, with this allowing extrapolation from the present climate emergency to a climate catastrophe in 2030 [20-23].

What can decent people do? It is effectively too late to avoid a catastrophic +2C temperature rise but decent people are obliged to do everything they can to make the future "less bad" for future generations. Decent people must act individually or better still act collectively (e.g. I am the secretary of the climate action group Banyule Climate Action Now (BCAN) that is based in Melbourne in the City of Banyule that has recently declared a Climate Emergency) [24].

Decent folk must (a) inform everyone they can about the worsening Climate Emergency, Climate Genocide and Intergenerational Inequity, (b) urge a climate revolution (peaceful and non-violent of course) with hundreds of millions out in the streets inspired by the likes of teenage activist Greta Thunberg, and (c) urge and apply Boycotts, Divestment and Sanctions (BDS) against all people, politicians, parties, collectives, corporations and countries disproportionately involved in the worsening Climate Genocide that is presently set to kill 10 billion people this century en route to a sustainable human population of merely 0.5-1.0 billion in 2100 [25]. There is no Planet B.

References.

[1]. Timothy Lenton, Johan Rockstrom, Owen Gaffney, Stefan Rahmsdorf, Katherine Richardson, Will Steffen and Hans Joachim Schellnhuber, "Climate tipping points – too risky to bet against", Nature 575, 592-595, 27 November 2019: https://www.nature.com/articles/d41586-019-03595-0.
[2]. IPCC, "Global warming of 1.5 °C", 8 October 2018: http://www.ipcc.ch/report/sr15/.
[3]. IPCC, "Global warming of 1.5 °C. Summary for Policymakers", 8 October 2018: http://report.ipcc.ch/sr15/pdf/sr15_spm_final.pdf.
[4]. Gideon Polya, "IPCC +1.5C avoidance report – effectively too late but stop coal burning for "less bad" catastrophes", Countercurrents, 12 October 2018: https://countercurrents.org/2018/10/12/ipcc-1-5c-avoidance-report-effectively-too-late-but-stop-coal-burning-for-less-bad-catastrophes/.
[5]. IPCC, "Climate Change 2014 Synthesis Report, Approved Summary for Policy Makers", 1 November 2014: http://www.ipcc.ch/pdf/assessment-report/ar5/syr/SYR_AR5_SPM.pdf).
[6]. Ron Prinn, "400 ppm CO_2? Add other GHGs and its equivalent to 478 ppm", Oceans at MIT, 6 June 2013: http://oceans.mit.edu/featured-stories/5-questions-mits-ron-prinn-400-ppm-threshold.
[7]. Gideon Polya, "International consensus-based IPCC Summary For Policymakers (2014) downplays acute seriousness of Climate Crisis", Countercurrents, 12 November, 2014: http://www.countercurrents.org/polya121114.htm.
[8]. "Are we doomed?": https://sites.google.com/site/300orgsite/are-we-doomed.
[9]. "Methane Bomb Threat": https://sites.google.com/site/methanebombthreat/.
[10]. "Too late to avoid global warming catastrophe": https://sites.google.com/site/300orgsite/too-late-to-avoid-global-warming.
[11]. WBGU, "Solving the climate dilemma: the budget approach": http://www.ecoequity.org/2009/10/solving-the-climate-dilemma-the-budget-approach/.
[12]. Australian Climate Commission, "The critical decade 2013: a summary of climate change science, risks and responses", 2013, p7: http://climatecommission.gov.au/wp-content/uploads/The-Critical-Decade-2013-Summary_lowres.pdf.

[13]. Robert Goodland and Jeff Anfang, "Livestock and climate change. What if the key actors in climate change are … cows, pigs and chickens?", World Watch, November/December 2009: http://www.worldwatch.org/files/pdf/Livestock%20and%20Climate%20Change.pdf.

[14]. Gideon Polya, "Australia's Huge Coal, Gas & Iron Ore Exports Threaten Planet", Countercurrents, 15 May 2012: http://www.countercurrents.org/polya150512.htm.

[15]. Drew T. Shindell, Greg Faluvegi, Dorothy M. Koch, Gavin A. Schmidt, Nadine Unger and Susanne E. Bauer, "Improved Attribution of Climate Forcing to Emissions", Science, 30 October 2009: Vol. 326 no. 5953 pp. 716-718: http://www.sciencemag.org/content/326/5953/716.

[16]. Gail Whiteman, Chris Hope and Peter Wadhams, "Vast costs of Arctic change", Nature, 499, 25 July 2013: http://www.nature.com/nature/journal/v499/n7459/pdf/499401a.pdf.

[17]. "Nuclear weapons ban, end poverty & reverse climate change": https://sites.google.com/site/300orgsite/nuclear-weapons-ban.

[18]. Professor Stephen Hawking quoted in Will Dunham, "Nuclear, climate perils push Doomsday Clock ahead", Reuters, 22 January 2007: https://www.reuters.com/article/idUSN17314370.

[19]. Stephen Hawking, "Brief Answers to the Big Questions", John Murray, 2018, Chapter 7.

[20]. William Ripple et al, "World scientists' warning of a climate emergency", BioScience, 5 November 2019: https://academic.oup.com/bioscience/advance-article/doi/10.1093/biosci/biz088/5610806.

[21]. Gideon Polya, "Extrapolating 11,000 scientists' climate emergency warning to 2030 climate catastrophe", Countercurrents, 14 November 2019: https://countercurrents.org/2019/11/extrapolating-11000-scientists-climate-emergency-warning-to-2030-catastrophe.)

[22]. William J. Ripple et al., 15,364 signatories from 184 countries, "World scientists' warning to Humanity: a second notice", Bioscience, 13 November 2017: https://academic.oup.com/bioscience/advance-article/doi/10.1093/biosci/bix125/4605229.

[23]. Gideon Polya, "Over 15,000 scientists issue dire warning to humanity on catastrophic climate change and biodiversity loss", Countercurrents, 20 November 2017: https://countercurrents.org/2017/11/20/over-15000-scientists-issue-dire-warning-to-humanity-on-catastrophic-climate-change-and-biodiversity-loss/.

[24]. "Banyule Climate Action Now":
https://sites.google.com/site/banyuleclimateactionnow/.
[25]. "Climate Genocide":
https://sites.google.com/site/climategenocide/.

CHAPTER 6: WRONG WAY GO BACK – GLOBAL SECTORAL GREENHOUSE GAS EMISSIONS ARE ALL IN THE WRONG DIRECTION

[First published as Gideon Polya, "Wrong way go back – global sectoral greenhouse gas emissions are all in the wrong direction", Countercurrents, 20 March 2020: https://countercurrents.org/2020/03/wrong-way-go-back-global-sectoral-greenhouse-gas-emissions-are-all-in-the-wrong-direction/.]

Vehicle drivers are familiar with crucial highway signs saying "WRONG WAY GO BACK". A detailed analysis shows that global greenhouse gas (GHG) emissions in all key economic sectors of energy production and non-energy production are increasing whereas the worsening Climate Emergency demands urgently decreasing GHG emissions and ultimately net zero GHG emissions ASAP. The world is driving headlong in the wrong direction to certain disaster.

The main sources of greenhouse gas (GHG) emissions (2013) are electricity and heat (31%), agriculture (11%), transportation (15%), forestry (6%) and manufacturing (12%). GHG emissions sources can be put into 2 categories (A) Non-energy production (28%) and (B) Energy production (72%).

(A). Non-energy production (28%) includes the following: Bunker fuels (Mt CO_2) 2.2%, Land-use Change & Forestry (Mt CO_2) 6.0%, Waste (Mt CO_2-e) 3.0%, Agriculture (Mt CO_2-e) 11.0%, Industrial processes (Mt CO_2-e) 6.0%.

(B). Energy production (72%) includes the following: Electricity & Heat (Mt CO_2) 31.0%, Transportation (Mt CO_2) 15.0%, Manufacturing & Construction (Mt CO_2) 12.4%, Other

Fuel Combustion (Mt CO_2-e) 8.4%, Fugitive Emissions (Mt CO_2-e) 5.2% [1].

Net global GHG emissions in 2013 totalled 48.3 Gt CO_2-e, steadily increasing in a quasi-linear fashion from 39.8 Gt CO_2-e in 2003 i.e. increasing by 8.5Gt CO_2-e or by +21.4% over 10 years, or by an average of 0.85Gt CO_2-e per year or an average of 1.9% per year [2]. From this we can estimate that net global GHG emissions totalled about 54.7 Gt CO_2-e in 2019.

Fundamental to the worsening Climate Emergency is the refusal of the dominant neoliberal One Percenters to allow the cost of past and present carbon pollution to be met, preferring to pass these immense, rising and inescapable costs (Carbon Debt) on to future generations. Eminent economist Lord Nicholas Stern has described this refusal as "the greatest market failure in history" [3]. Pope Francis has demanded that the environmental and social cost of pollution be "fully borne" by the polluters [4, 5]. Climate economist Dr Chris Hope (120 Nobel Laureate University of Cambridge) estimated a damage-related Carbon Price of $200 per tonne CO_2-e [6], with this equating to the upper estimate by Professor James Hansen (NASA and 96 Nobel Laureate Columbia University) from the cost of drawing down CO_2 from the atmosphere [7].

(A1). Bunker Fuel use (2.2% of total GHG emissions or 1.06 Gt CO_2-e in 2013) has increased in recent years at 8.3% per year in terms of value.

Current wrong way. Bunker Fuel is the fuel oil for all shipping. Allied Market Research: "The global bunker fuel market was valued at $137,215.5 million [$137.2 billion] in 2017 and is expected to reach $273,050.4 million [$273.1 billion] by 2025, registering a CAGR [Compound Annual Growth Rate] of 9.4% from 2018 to 2025" [8]. The increase in value in 2008-2025 is $135.9 billion (99.1% over 8 years) or an average of $17.0 billion pa or 8.3% per year. At a Carbon Price of $200 per tonne CO_2-e, removal of the 1.06 Gt CO_2 introduced into the atmosphere by burning Bunker Fuel in 2013 = 1.06 Gt CO_2 x $200/ tonne CO_2 = $212 billion, this exceeding the expected market value of Bunker Fuel in 2020 of $182 billion [8].

Required actions. Bunker Fuel use is increasing. Net zero GHG emissions for Bunker Fuel use could be achieved by compensatory Re-afforestation or better still by preserving existing old growth forests. While bulk movement of grains is crucial for survival of many, the movement of perishable vegetable matter (e.g. grapes from California to grape producing Australia) is an unwarranted and wasteful indulgence. Even within countries people should ideally use seasonal and local food if possible.

(A2). Waste CO_2-e from anaerobic bacterial decomposition of biological material may be 7.0 Gt CO_2-e per year and is increasing at 135 Mt CO_2-e per year.

Current wrong way. Landfill gas (LFG) is about 50% CH_4 and 50% CO_2 [9]. Waste CO_2-e is mainly CH_4 from anaerobic bacterial decomposition of biological material (e.g. in metropolitan waste landfill sites) and was estimated at 3.0% of 48.3 Gt CO_2-e (2013) or 1.4 Gt CO_2-e [1]. However the Global Warming Potential of CH_4 relative to the same mass of CO_2 (formerly taken as 21 on a 100 year time scale) has been re-assessed as 105 on a 20 year time scale and with aerosol impacts considered [10]. On this basis the Waste CO_2-e from anaerobic bacterial decomposition of biological material may be 5 x 1.4 Gt CO_2-e = 7.0 Gt CO_2-e per year (on a 20 year time frame). CH_4 emissions are further considered below under Land-use & Forestry and under Fugitive Emissions. The Global Methane Initiative (2010): "While methane is in the atmosphere for a shorter period of time and is emitted in smaller quantities than CO_2, its ability to trap heat in the atmosphere, which is called its "global warming potential," is 21 times greater than that of CO_2 [on a 100 year time frame] … Globally, landfills are the third largest anthropogenic source of methane, accounting for approximately 11 percent of estimated global methane emissions or nearly 799 Mt CO_2-e in 2010" [11]. Waste Management Research (2013): "Methane emissions from landfills can be estimated by using the Intergovernmental Panel on Climate Change (IPCC) methodology. Less than 5% of global GHG comes from the waste sector, of which landfill methane is the largest source. The IPCC has estimated that landfill methane will reach 795 Mt CO_2-e and 910 Mt CO_2-e in 2015 and 2020, respectively, out of [which] approximately 300 Mt CO_2-e emitted by landfills and dumps in developing countries in south and east Asia. In developed countries, LFG emissions have been stabilized in response to stringent regulations, yet methane emissions are expected to continue increasing in developing countries as more waste goes to landfills that lack sufficient (and in some cases, any) emission control systems" [12]. From this latter data one can estimate that landfill waste gas is increasing at 27 Mt CO_2-e per year i.e. 5 x 27 = 135 Mt CO_2-e per year on a 20 year time frame. At a Carbon Price of $200 per tonne CO_2-e, the estimated Carbon Debt from Waste is $200 per tonne CO_2-e x 7.0 Gt CO_2-e = $1,400 billion annually.

Required actions. The solutions are (a) to collect and usefully burn CH_4 from waste landfill, (b) to ensure that all waste biological material ends up in aerobic composting, and (c) to drastically cut down on food wastage (thus it is estimated that 40% of food is wasted in the US) [13].

(A3). Agricultural GHG emissions (CO_2-e) revised upwards to 13.8 Gt CO_2-e in 2013 and are increasing at 0.7% per year.

Current wrong way. 2013 GHG emissions from agriculture ostensibly totalled 0.11 x 48.3 Gt CO_2-e = 5.31 Gt CO_2-e, with 40% (2.12 Gt CO_2-e) due to CH_4 from livestock. The World Resources Institute (2019): "The majority of agricultural production emissions come from raising livestock. More than 70 billion animals are raised annually for human consumption. The biggest single source is methane from cow burps and manure. Enteric fermentation—a natural digestive process that occurs in ruminant animals such as cattle, sheep and goats—accounts for about 40% of agricultural production emissions in the past 20 years... Emissions from agricultural production currently account for

11% of global greenhouse gas emissions and have risen 14% since 2000" [14]. From this data we can estimate an annual percentage increase of 14%/19.25 years = 0.73% per year. However the estimate of 5.31 Gt CO_2-e of Agricultural GHG emissions in 2013 [14] assumes a Global Warming Potential (GWP) for methane (CH_4) of 21 relative to the same mass of CO_2 (on a 100 year time frame), whereas on a 20 year time frame and with aerosol impacts considered it is 105 [10]. Accordingly the ostensible 0.4 x 5.31 Gt CO_2-e = 2.12 Gt CO_2-e (40% of the total) due to livestock (2013) should actually be 5 times larger (i.e. 5 x 2.12Gt CO_2-e = 10.60 Gt CO_2-e), and the estimated Agricultural GHG emissions should be increased on this basis alone to 10.60 + (0.6 x 5.31) = 13.79 Gt CO_2-e per year (2013).

The following Agriculture-related parameters are going in the wrong direction in a fashion that is quasi-linear as a function of time and Humanity must urgently address this disastrous trend [15-18]:

Freshwater resources per capita (1,000 metre$^{3)}$: 8.0 (1992) to 5.9 (2014), -25.4% overall, – 12.0 % per decade, -0.095 per year; 4,400 metre^3 per capita in 2030.

Reconstructed marine catch (Mt per year): 121 (1992) to 113 (2011), -6.6%, overall, – 3.5% per decade, -0.42 per year; 105 million tonnes marine catch per year in 2030.

Ocean dead zones (number of affected regions): 360 (1992) to 660 (2008), +83.3% overall, +52% per decade, +11.5 per year; 913 ocean dead zones by 2030.

Vertebrate species abundance (% of 1970): 59.5 (1992) to 43.5 (2010), – 26.9% overall, -14.9% per decade, -0.89 per year; 27.5% of 1970 level by 2030.

Human population (billion individuals): 5.50 (1992) to 7.45 (2016), + 35.5% overall, +14.8% per decade, 0.081 per year; 8.58 billion human individuals in 2030.

Ruminant population (billion individuals): 3.25 (1992) to 3.90 (2014), +20.0% overall, +9.1% per decade, +.03 per year; 4.38 billion ruminant individuals in 2030

Ruminant livestock (billion individuals): 2.86 billion (1979) to 3.93 billion (2017) (+ 8.72% per decade, +0.87% per year and +0.028 per year).

Per capita meat production (kg per year): 30 (1979) to 44 (2018) (+11% per decade, +1.1% per year, +0.36 per year).

World GDP (trillion current US\$/year). The data appear to fall into to 2 successive upward linear phases with an upward inflection in about 2000: 10.0 (1979) to 32.5 (1998) and 37.5 (2001) to 86.3 (2018), with growth rates of \$1.18 trillion per year and \$2.87 trillion per year, respectively, and an overall growth of + 80.5% /10 years (+8.1% per year and +\$1.96 trillion per year).

Global tree cover loss (million hectares per year): The data fall into 2 successive upward linear phases with an apparent short low-rate period in the late 2000s: 14 (2001) to 18 (2005) and 18 (2010) to 27.5 (2018) with corresponding rates of loss of 1.0 and 1.2 million hectares per year, respectively.

Brazilian Amazon forest loss (millions hectares per year) was in the range 1-3 million hectares per year in the period 1988-2008 (very scattered data) and then fell to about 0.7- 0.9 million hectares per year in the period 2009-2018 (– 24.3% /10 years, -2.4% per year).

Nitrous oxide (N_2O parts per billion) increased in a quasi-linear fashion from 301 (1979) to 331 (2018) for an average of 0.77 ppb per year (+2.46% per decade, +0.28% per year). Using this average, extrapolation yields a 2030 value of 340 ppb N_2O, noting that N_2O is derived from oxidation of nitrogenous agricultural fertilizers and has a Global Warming Potential (GWP) relative to the same mass of CO_2 of 289 on a 20-year time frame (298 on a 100 year time frame).

Sea level change relative to the 20-year mean (mm) increased linearly from – 34 (1993) to + 46 (2017) to give a rise of 3.33 mm per year (+31.4 mm per decade or +3.14 per year according to the paper). Extrapolation yields a sea level rise in 2030 of 89 mm (relative to that in 2005) and 123 mm (relative to that in 1993).

Extreme weather/climate/hydrological events (numbers per year) have been monotonically increasing in a tri-phasic quasi-linear fashion from 200 (1980) to 400 (1994) (an increase of 14.3 events per year), a constant of circa 400 per year in the period 1994-2004, and then a linear increase from 440 (2004) to 800 (2019) (an increase of 32.7 events per year) (+ 43.8% per decade and +4.4% per year overall). Assuming an increase of 32.7 events per year, extrapolation yields 1,160 events in 2030 or more than 3 such events per day.

Annual losses due to weather/climate/hydrological events ($billions) and excluding 2 exceptional outlier events, increased in a quasi-linear fashion from 20 (1980) to 196 (2018) for a rate of 4.63 per year (+83.7% per decade, 8.4% per year). Extrapolation yields an annual cost of $2,452 billion in 2030.

Required actions. Conversion efficiency (kg grain to produce 1 kg gain in live weight) is as follows: herbivorous farmed fish (e.g. carp, tilapia, catfish; less than 2), chicken (2), pork (4), and beef (7) [19, 20]. Methanogenic beef production is contraindicated because of the huge inefficiency in grain to meat conversion as well as the huge contribution to GHG emissions. Indeed World Bank analysts have re-assessed the contribution of animal husbandry and associated land clearing to annual global GHG emissions, considering the GWP for CH_4 on a 20 year time frame – on this basis the total annual GHG emissions in 2009 rose from 41.8 Gt CO_2-e to 63.8 Gt CO_2-e [21]. Ripple et al. recommend re food: "Eating mostly plant-based foods while reducing the global consumption of animal products" [17]. There must be a global vegetarian diet and replacement of cow-derived milk with plant substitutes (soybean and almond milk). The deadly food-for-fuel Biofuel Genocide obscenity must be stopped. Food adequacy involves cessation of global inequity, population growth, economic growth, GHG pollution, sea level rise, soil salinization, deforestation and damaging land use [18, 19]. We are all one Humanity and all decent people accept the proposition of the American Declaration of Independence that "all men are created equal and have an inalienable right to life, liberty and the pursuit of happiness". However 2009 per capita meat consumption in kilograms per person was 120.2 for the US as compared to 4.4 for India [22], noting that most Agriculture sector emissions come from raising livestock [14].

(A4). Land-use Change and Forestry emissions revised upwards to 5.6 Gt CO_2-e and loss of 1.2 million hectares of forest photosynthetic CO_2 sequestration per year.

Current wrong way. Global Land-use and Forestry emissions in 2013 were estimated at 6.0% of the total i.e. at 0.06 x 48.3 Gt CO_2-e = 2.9 Gt CO_2. However World Bank analysts have re-assessed the contribution of animal husbandry and associated land clearing to annual global GHG emissions, considering the GWP for CH_4 on a 20 year time frame. On this basis the total annual GHG emissions in 2009 rose from 41.8 Gt CO_2-e to 63.8 Gt CO_2-e with "overlooked" land use GHG emissions amounting to over 4.2% of the revised total i.e. 0.042 x 63.8 Gt CO_2-e = 2.7 Gt CO_2-e [21]. Thus the revised global Land-use and Forestry emissions in 2013 total 2.9 + 2.7 = 5.6 Gt CO_2-e. There is presently loss of 1.2 million hectares of forest photosynthetic CO2 sequestration per year.

The following Land-use- and Forestry-related parameters are going in the wrong direction in a fashion that is quasi-linear as a function of time and Humanity must urgently address this disastrous trend [15-18]:

Total forest (million hectares): 4.110 (1992) to 3.990 (2016), −2.9% overall, -1.2% per decade, -0.12% per year; 3.92 million hectares left by 2030.

Vertebrate species abundance (% of 1970): 59.5 (1992) to 43.5 (2010), – 26.9% overall, -14.9% per decade, -0.89 per year; 27.5% of 1970 level by 2030.

Human population (billions): 4.35 (1979) to 7.65 (2018), +75.9% overall, +19.5% per decade, + 0.085 billion per year, and an estimated 9.07 billion human population in 2030 (assuming an average increase of +84.6 million per year, noting that the increase in population was linear over the period 1979-2018).

Global tree cover loss (million hectares per year): The data fall into 2 successive upward linear phases with an apparent short low-rate period in the late 2000s: 14 (2001) to 18 (2005) and 18 (2010) to 27.5 (2018) with corresponding rates of loss of 1.0 and 1.2 million hectares per year, respectively. Extrapolation using the latter faster latter rate yields an estimated loss rate of 41.9 million hectares of tree cover loss per year by 2030.

Brazilian Amazon forest loss (millions hectares per year) was in the range 1-3 million hectares per year in the period 1988-2008 (very scattered data) and then fell to about 0.7- 0.9 million hectares per year in the period 2009-2018 (- 24.3% /10 years, -2.4% per year).

Area burned in the United States (million hectares per year) involves very scattered data. However, assuming the authors' quasi-linear best fit involving an increase from 0.9 (1983) to 3.2 (2018), one can estimate an increase of 0.066 million hectares per year (+ 44.1% per decade, +4.4% per year). Extrapolating from this average yields an estimate of 4.0 million hectares burned in 2030.

Required actions. Ripple et al. (2019): "We must protect and restore Earth's ecosystems. Phytoplankton, coral reefs, forests, savannahs, grasslands, wetlands, peatlands, soils, mangroves, and sea grasses contribute greatly to sequestration of atmospheric CO_2" [17]. There must be cessation of land clearing and deforestation coupled with large-scale ecosystem restoration and re-afforestation. It is possible to sequester 12 Gt biochar (charcoal, carbon C) annually (from the anaerobic pyrolysis of forestry and agricultural cellulosic waste at 400-700C, and this can more than compensate for the present 9 Gt carbon (33 Gt CO_2-e) pollution annually from industry [23]. It should be noted that the south eastern Australian *Eucalyptus* forests are the best forest carbon sinks in the world and must be preserved [24]. Tropical forests have now reached or are close to reaching a tipping point of generating more CO_2 than they sequester [25-27]. Crucially it is necessary to halt and reverse the human population growth that underlies increased economic production, increased land use, increased deforestation and associated increased biodiversity loss. The species extinction rate is now 100-1,000 times greater than normal [28]. The 28.9% decline in vertebrate species abundance over the last quarter century means that if these changes continue at the same rate then there will be a near 100% decline in non-human vertebrate species abundance in the coming century [15-18, 29]. From a fundamental, qualitative perspective we must not destroy what we cannot replace, and all species are priceless.

(A5). Industrial processes (steel and cement manufacture) generated 6.9 Gt CO_2-e in 2019, with GHG emissions from steel and cement production presently increasing by 3.4% and 1.2%, respectively.

Current wrong way. In 2013 Industrial processes generated 6.0% of global GHG pollution or 0.06 x 48.3 Gt CO2-e = 2.9 Gt CO2-e [1, 2]. Steel and cement production are major sources of greenhouse gas (GHG) pollution. Thus global crude steel production reached 1.87 Gt in 2019, up by 3.4% compared to 2018 [30]. In 2008 steel production accounted for 4-5 % of global GHG emissions, with 1.9 tons of CO_2 produced per ton of steel produced [31]. Thus production of 1.87 Gt steel in 2019 was associated with emission of 3.55 Gt CO_2-e. Cement production from limestone ($CaCO_3$) is the source of 5-8% of global GHG pollution. Producing 1 tonne of cement from heating limestone ($CaCO_3$) yields 0.82 tonne CO_2. Global CO_2 emissions from cement manufacture was recently estimated to currently total 1.5 Gt CO_2 per year or 4.5% of global CO_2 production of 33 Gt CO_2. [32, 33]. Cement production increased from 3.31 Gt (2010) to 4.05 Gt (2018) and thence to 4.10 in 2019. While cement production increased by 23.9% over the last decade, it increased by only 1.2% in 2019 [34]. In 2019 CO_2 from cement and steel manufacture totalled 3.55 Gt (steel) + 3.36 Gt (cement) = 6.91 Gt CO_2 or 12.6 % of net global CO_2 emissions of 54.7 Gt CO_2-e in 2019. In 2019 GHG emissions from steel and cement production increased by 3.4% and 1.2% respectively.

Required actions. Steel can be manufactured based on renewable energy using hydrogen (H_2) as a fuel and reductant. Indeed Swedish steel giant Svenskt Stål AB (Swedish Steel or SSAB) is to use renewable hydrogen to produce fossil-free steel by 2026 [35]. Replacing cement (a source of 5-8% of global GHG pollution) will be more difficult. However a water-resistant magnesium oxychloride cement (MOC) is being developed as a green alternative to Portland cement [33]. Of course steel and cement use can be substantially decreased by reducing economic output through negative population growth and associated economic de-growth that are demanded by the worsening Climate Emergency [29].

Energy sector CO_2 emissions overall total 33.2 Gt CO_2 (2018) and are increasing at 1.7% per year (2018).

Current wrong way. The IEA has estimated energy sector GHG emissions over the period 1990-2018: "Global energy-related CO_2 emissions grew 1.7% in 2018 to reach a historic high of 33.1 Gt CO_2. It was the highest rate of growth since 2013, and 70% higher than the average increase since 2010" [36]. The energy sector GHG emissions rose in a quasi-linear fashion from 20.5 Gt CO_2 (1990) to 33.2 Gt CO_2 (2018) for an overall rate of 0.45 Gt CO_2 per year or an average increase of 1.7% per year.

Required actions. Ripple et al. (2019): "There should be a rapid transition away from dirty energy (coal, oil, gas and other carbon fuels) to 100% renewable energy, with this associated with assistance to developing economies, cessation of fossil fuel exploitation, and application of proper Carbon Prices" [17, 18].

(B1). Electricity & Heat generated 15.0 Gt CO_2-e or 31.0% of annual global GHG emissions (2013) and annual coal, gas, and oil use are presently increasing by 3%, 4.6% and 3.5%, respectively.

Current wrong way. In 2013 GHG emissions from the Electricity & Heat sector constituted 31.0%, of annual global GHG emissions i.e. 0.31 x 48.3 Gt CO_2-e = 15.0 Gt CO_2-e.

Coal. According to the International Energy Agency (IEA) (2019): "Coal power generation increased 3% in 2018 (similar to the 2017 increase), and for the first time crossed the 10 000 TWh [energy] mark. Coal remains firmly in place as the largest source of power at 38% [38.5%] of overall generation. Growth was mainly in Asia, particularly in China and India. That said, investment in coal-fired power declined by nearly 3% to the lowest level since 2004, and final investment decisions for new plants continue to decline. Coal-fired generation without CCUS [Carbon Capture Utilization and Storage] needs to decrease 5.8% per year to 2030 to be in line with the SDS [Sustainable Development Scenario 2000-2040]" [37, 38]. Total global energy production in 2018 was accordingly 10,000TWh/0.385 = 26,000 TWh [37]. GHG emissions from coal use went from 8.3 Gt CO_2 (1990) to 14.6 Gt CO_2 (2018) with an average increase of 0.23 Gt CO_2 per year or 2.0%. GHG emissions from coal use went from 14.2 Gt CO_2 (2016) to 14.6 Gt CO_2 (2018) with an average increase of 0.2 Gt CO_2 per year or 1.4% [36].

Gas. The IEA: "Natural gas had a remarkable year in 2018, with a 4.6% increase in consumption accounting for nearly half of the increase in global energy demand. Since 2010, 80% of growth has been concentrated in three key regions: the United States, where the shale gas revolution is in full swing; China, where economic expansion and air quality concerns have underpinned rapid growth; and the Middle East, where gas is a gateway to economic diversification from oil… Gas-fired power generation increased 4% in 2018, led by strong generation growth in the United States and China. At around 6100 TWh, gas accounts for 23% of overall power generation" [39]. Gas use has increased in a quasi-linear fashion in the period 1990-2017 from 43.892 x 10^{18} J in 1990 to 69.888 x 10^{18} J in 2017 at an average rate of 0.96 x 10^{18} joules (J) per year (844.1 Mt gas in 1990 to 1,344.0 Mt gas in 2017 at an average increase of 18.5 Mt gas per year or 1.7% pa) [38]. Extraordinarily, the IAE predicts very little change in gas usage between 2018 and 2040 [39].

Oil. Imported oil use increased from 1,532.6 ktoe (kilotonne of oil equivalent) in 1990 to 2,319.0 ktoe in 2017 with an average rate of increase of 29.1 ktoe per year. In the period 2014-2017 imported oil increased by 3.5% per year. The kilotonne of oil equivalent (ktoe) is a unit of energy defined as the amount of energy released by burning one kilotonne of crude oil and is about 42,000 gigajoules. The IEA predicts an increase in oil usage in the period 2018 to 2040: "Under the Stated Policies Scenario, oil use in passenger cars peaks in the late 2020s… However there is no definitive peak in oil use overall, as there are continued increases in petrochemicals, trucks and the shipping and aviation sector" [40].

Fugitive emissions of CH_4 from oil and gas use. The IEA states that "Methane emissions from the oil and gas sector reached close to 80 Mt (or 2.4 billion tonnes of CO_2 equivalent) in 2017. This is equal to 6% of global energy sector GHG emissions [i.e. 6% of 40 Gt CO_2–e per year = 2.4 Gt CO_2–e]. Emissions remain high despite initial industry-led initiatives and government policies announced recently" [40]. However this assumes a Global Warming Potential (GWP) for methane (CH_4) of 2,400 Mt CO_2-e/80 Mt CH_4 = 30 (on a 100 year time frame) whereas the GWP of CH_4 relative to the same mass of CO_2 is 105 with a 20 year time frame and including aerosol impacts [10]. Assuming a GWP for CH_4 of 105, at 2.6% CH_4 leakage the greenhouse effect of the leaked CH_4 is the same as the greenhouse effect from burning the remaining 97.4% of the gas [41, 42].

In 2018 global natural gas production was 3,868 bcm (billion cubic metres) or 3,326 Mt oil-equivalent and the growth rate was +5.2% per year [43]. Combustion of 1 tonne of CH_4 (85% of natural gas) yields 2.75 tonne CO_2 as compared to combustion of 1 tonne of carbon (about 90% of coal) yielding 3.67 tonne CO_2 and 1 tonne of saturated hydrocarbon $((CH_2)_n)$ yielding 3.1 t CO_2. Thus in 2018 CO_2 emissions from burning annual gas production of 3,326 Mt oil-equivalent = 3,326 Mt oil-equivalent x 3.1 t CO_2/t oil = 10,331 Mt CO_2. Assuming a GWP for CH_4 of 105, at 2.6% CH_4 leakage the greenhouse effect of the leaked CH4 is the same as the greenhouse effect from burning the remaining 97.4% of the gas [41, 42]. Accordingly, the GHG effect from a 2.6% gas leakage is 0.974 x 10,331 Mt CO_2 = 10.1 Gt CO_2-e. Applying the same approach, the estimate for Fugitive Emissions in 2013 was 8,732 Mt CO_2-e, which is 8.7 Gt/2.4 Gt = 3.6 times greater than the IEA estimate given above for oil and gas fugitive emissions (indeed this is still underestimated because gas leakage in the US is estimated at 5.4% [44-46]).

Required actions. Things that need to be urgently implemented in the Electricity & Heat sector include 100% renewable energy (with appropriate energy storage back-up), cessation of use of coal, gas, oil and other carbon fuels, and increased energy efficiency. Crucially, there must be cessation of public lying by fossil fuel lobbyists. Thus patently false lying by commission about asserted "clean coal", "clean gas", "cleaner gas" and a "coal to gas transition" must be called out, as indeed must be lying by omission about science-based accounting of GHG emissions (e.g. about fugitive emissions that are much greater than admitted).

(B2). Transportation generated 7.2 Gt CO_2 or 15.0% of annual global GHG emissions (2013) with annual growth of about 3.1% for the global truck market.

Current wrong way. Transport accounts for 7.2 Gt CO_2 or 15.0% of annual global emissions of 48.3 Gt CO_2 (2013). According to Deloitte: "From now to 2024, annual growth of > 3% is expected in the global truck market, mainly driven by global GDP growth, estimated at 3.3% per year… Between 2014 and 2024 the global truck market is expected to continue growing at a fairly low rate of 3.1% per year. The Indian market will be the main driver" [47]. According to IBIS World: "Over the five years to 2019, rising per capita income is expected to drive revenue growth for the Global Car and Automobile Sales industry. The industry's dependency on employment rates, global consumer spending, financing rates and the world price of crude oil make it vulnerable to economic shifts. During the period, the United States and many other developed nations have experienced stable economic growth, leading to increased consumer demand for big-ticket purchases, such as vehicles. As a result, industry revenue is expected to grow at an annualized rate of 2.7% to $4.0 trillion over the five years to 2019, including revenue growing 2.3% in 2019 alone" [48].

The IEA on rail: "Global demand for transport is growing fast. Given present trends, passenger and freight activity will more than double by 2050. Such growth is a token of social and economic progress, but it carries with it greater energy demand and increased CO_2 emissions and atmospheric pollutants… The transport sector is responsible for more than half of global oil demand and around one-quarter of global CO_2 emissions from fuel combustion. Therefore changes in transportation are fundamental to achieving energy transitions globally. Yet while rail is among the most energy efficient modes of transport for freight and passengers, it is often neglected in public debate… Today, three-quarters of passenger rail transport activity takes place on electric trains, which is an increase from 60% in 2000 – the rail sector is the only mode of transport that is widely electrified today. This reliance on electricity means that the rail sector is the most energy diverse mode of transport" [49]. In terms of trillion passenger-km the rail sector has steadily grown from 2.15 (1995) to 4.05 (2016), a growth rate of 0.1 trillion passenger-km per year. However the diesel train sector has also grown, albeit modestly, from 0.9 (1995) to 1.23 (2016) for an average of +0.016 trillion passenger-km per year. Of course a substantial proportion of the electrified global train systems is powered by burning fossil fuels (see B1).

Required actions. Clearly we should be moving towards 100% renewable energy sources, electric vehicles, rail not road, transport energy efficiency, and public transport not private transport.

(B3). Manufacturing & Construction emitted 6.0 Gt CO_2-e in 2013 and are increasing at 6.7% and 3.9% annually, respectively.

Current wrong way. McKinsey Global Group (2012): "A decade into the 21st century, the role of manufacturing in the global economy continues to evolve. We see a promising future. Over the next 15 years, another 1.8 billion people will enter the global consuming class and worldwide consumption will nearly double to $64 trillion. Developing economies will continue to drive global growth in demand for manufactured goods, becoming just as important as markets as they have been as contributors to the supply chain. And a strong pipeline of innovations in materials, information technology, production processes, and manufacturing operations will give manufacturers the opportunity to design and build new kinds of products, reinvent existing ones, and bring renewed dynamism to the sector" [50]. This estimate corresponds to an average annual growth for Manufacturing of 100/15 = 6.7% per year. Similarly PwC states re Construction : "PwC has sponsored a new report – Global Construction 2030 – which forecasts that the volume of construction output will grow by 85% to $15.5 trillion worldwide by 2030, with three countries, China, US and India, leading the way and accounting for 57% of all global growth. The benchmark global study, the fourth in a series from Global Construction Perspectives and Oxford Economics, shows average global construction growth of 3.9% pa to 2030, outpacing that of global GDP by over one percentage point, driven by developed countries recovering from economic instability and emerging countries continuing to industrialize" [51]. Manufacturing & Construction accounted for 0.124 x 48.3 Gt CO_2-e = 6.0 Gt CO_2 in 2013.

Required actions. Massive harm to the environment and biodiversity has occurred due to continuing carbon pollution, population growth and economic growth. It is clear that zero growth in these areas is insufficient and that there must be negative carbon pollution (atmospheric CO_2 draw-down to a safe and sustainable 300 ppm CO_2 from the present dangerous and damaging 412 ppm CO_2), negative population growth (circa 50% population decline) and negative economic growth (circa 50% de-growth) to halt and reverse this worsening disaster [29]. Taking world coral as a "canary in the coal mine", the 320 ppm CO_2 at which coral reefs started to decline was reached in 1965, at which time the world's population was 3.340 billion as compared to the present 7.6 billion [29]. Even if per capita GHG pollution remained the same, increasing population would see that GHG pollution from attendant economic activities would increase. Further, while the low birth-rate rich North could live more modestly, the impoverished and high birth-rate South needs an increase in per capita income to enable circumstances in which birth rate can be reduced (high female literacy, higher infant survival and economic security) [29]. Professor Dabo Guan (School of International Development, University of East Anglia, UK) (2016) has commented thus on inescapable limits to growth: "For everyone in the world to have an American lifestyle, we would need seven planets, and three to live as Europeans" [52]. The European population is declining but the African population is expanding impossibly.

(B4). Other Fuel Combustion was 4.1 Gt CO_2 in 2013 and "food for fuel" biofuel is increasing at 9.7% pa.

Current wrong way. In 2013 Other Fuel Combustion was 0.084 x 48.3 Gt CO_2-e = 4.1 Gt CO_2-e. Other Fuel Combustion includes burning of biofuel (e.g. ethanol and biodiesel), wood, other biomass and of cattle dung (the latter notably in Developing countries in Africa and South Asia). The Trump-subverted US Environment Protection Authority (EPA) has falsely declared that burning wood is carbon neutral – it is not because felling and burning trees not only releases sequestered carbon but also exposes huge stores of soil carbon to oxidation and stops further centuries of ongoing photosynthesis by the trees thus destroyed and subsequent carbon storage in wood and soil [53]. The World Bioenergy Association (2016): "Biomass for energy originates from a variety of sources classified into forestry, agriculture and waste streams. Some of the potential sources include: crops for biofuels, energy grass, short rotation forests, woody biomass and residues, herbaceous by-products and municipal solid waste. Globally, in 2012, the biggest share of biomass for energy came from forests – almost 49 EJ out of a total supply of 56.2 EJ. The current global energy supply is about 560 EJ [560 x 10^{18} J] … In 2012, biomass from forests and other areas with trees contributed 49EJ to the global energy supply. The lion share of this biomass is used in Asia and Africa… The contribution of biomass to primary energy can be increased from 56 EJ in 2012 to about 150 EJ by 2035 [7.3% per year]" [54].

The total global land area (2012) is 13,019 million hectares (13,019 Mha), this including 4,922 Mha (agriculture), 4,022 Mha (forest) and 4,075 Mha (other land). Of the agricultural land 1,396 Mha is arable, 164 Mha are for permanent crops, and 3,362 Mha are for pastures and meadows. In the period 2000 – 2005 the average annual gain of forest area was 5.7 Mha, this corresponding to an annual increase of 5.7 Mha x 100/ 4,000 Mha = 0.14% pa [54]. The World Bioenergy Association: "The 1.5% increase in total primary energy supply during 2016-17 has been matched by coal, oil and natural gas while renewables are lagging behind (0.7%). This trend appears to continue to 2018 and 2019 as well… Among [assertedly] renewable energy sources, bioenergy (energy from bio-based sources) is the largest. In 2017, bioenergy accounted for 70% of the renewable energy consumption… In 2017, 596 TWh of biopower [electricity from biomass] was produced globally which is an increase of 25TWh (+ 4%) over the previous year" [55].

Global biofuel production grew at an average rate of 9.2% per year over the period 2000-2018 and grew by 9.7% pa in 2017-2108 [56]. Biofuel is bad because it is food for fuel in a hungry world that is set to become hungrier as population increases, the Developing World middle class increases, arable land decreases, and water for agriculture decreases [57]. Biofuel, including bio-ethanol and bio-diesel production, generates a huge Carbon

Debt from destruction of the balanced Carbon Cycle in swathes of wild nature [58-60].

Ignored in this analysis is inadvertent fuel combustion in ever-worsening, climate change-driven forest fires in California, Southern Europe, the Amazon, South East Asia, and Australia. Australia is among world leaders in 16 areas of climate criminality, and with a mere 0.33% of the world's population contributes 5.4% of global annual GHG emissions (exported GHG emissions included) [61, 62]. Taking land use and CH_4 emissions into account, Australia's annual Domestic GHG emissions total 1,423 Mt CO_2-e (ignoring the 2019-20 bushfire contribution of 750 Mt CO_2-e in 1 year [23]) – but the endlessly mendacious Australian Coalition Government claims annual Domestic emissions of 540 Mt CO_2-e by failing to properly account for CH_4 leakage (underestimating likely fugitive emissions by a factor of 8-54), land use, the CH_4 Global Warming Potential on a 20 year time frame of 105, and ignoring the huge contribution by bushfires [61, 62].

Required actions. Cellulosic biomass, sugar and plant-derived oil should simply not be used as biofuel in a hungry world. Human energy requirements can be met by genuine renewable energy (hydroelectric, solar PV, solar thermal, wind, tide, wave and geothermal) with hydrological, molten salts, battery and other energy storage. Nuclear waste and security issues aside, conventional nuclear power is not an option because the uranium and thorium ore resources are non-renewable, and because in a Carbon Economy the overall nuclear cycle (excluding the actual nuclear fission) generates large amounts of CO_2.

(B5). Fugitive Emissions from energy-related activities are re-assessed as a huge 10.1 Gt CO_2-e in 2018 and are increasing at 5.2% pa.

Current wrong way. There is massive under-estimation of fugitive emissions of methane (CH_4) from coal mines, gas extraction sites, and the gas distribution systems. Thus energy-related Fugitive Emissions in 2013 were estimated by the IEA as 0.052 x 48.3 Gt CO_2-e = 2.51 Gt CO_2-e [1] whereas GHG emissions from fugitive emission in the energy sector were calculated in B1 above as 8.75 Gt CO_2-e in 2013 (4.2 times greater than the IEA estimate) and 10.1 Gt CO_2-e (2018). Fugitive Emissions are directly proportional to gas production and hence the 5.2% pa increase in natural gas production is associated with a 5.2% pa increase in Fugitive Emissions.

As a check on this result, consider a separate set of data from DNV-GL on Liquid Natural Gas (LNG) alone (2018): "DNV GL forecasts global LNG production will increase from 250 million tonnes per year in 2016 to around 630 million tonnes per year by 2050 [+11.2% pa]" [63]. Combustion of 1 tonne of CH_4 (85% of natural gas) yields 2.75 tonne CO_2. Thus CO_2 emissions from burning annual gas production of 250 Mt gas = 250 Mt gas x 2.75 t CO_2/t CH_4 = 687.5 Mt CO_2. Assuming a GWP for CH_4 of 105, at 2.6% CH_4 leakage the greenhouse effect of the leaked CH_4 is the same as the greenhouse effect from burning the remaining 97.4% of the gas [41, 42]. Accordingly, the overall GHG effect with a 2.6% gas leakage from LNG production is 0.974 x 687.5 Mt CO_2 = 670 Mt CO_2-e = 0.67 Gt CO_2-e pa (indeed this may be under-estimated because gas leakage in the US is estimated at 5.4% [44-46]).

Required actions. There must be rapid cessation of gas exploitation. Because gas leaks and the GWP for CH_4 is 105 relative to that of the same mass of CO_2 on a 20 year time frame with aerosol impacts considered, at 2.6% leakage the Fugitive Emissions from natural gas exploitation have roughly the same greenhouse gas effect as the CO_2 from burning the residual gas. Indeed in typical circumstances using gas is worse GHG-wise that using coal and hence a "coal-to-gas transition" is a dangerous absurdity.

Final comments.

Eminent physicist and cosmologist Professor Stephen Hawking (of 120 Nobel laureate University of Cambridge) has commented succinctly on the existential threats to Humanity posed by nuclear weapons and global warming: "We see great peril if governments and societies do not take action now to render nuclear weapons obsolete and to prevent further climate change" [64]. The operative word here is "now".

Scientists warn that "net zero emissions by 2050" is required to keep under a catastrophic +2C of global warming. However, for example, the anti-science, anti-environment, climate criminal and Trumpist Australian Coalition Government has attacked the Labor Opposition for adopting this goal [65]. The science-informed and humanitarian Australian Greens state that: "A safe climate will require a return to an atmospheric concentration of 350 parts per million or lower of greenhouse gases (and CO_2 equivalents)... Net zero or net negative Australian greenhouse gas emissions by no later than 2040" [66].

As set out in this detailed analysis, global greenhouse gas (GHG) emissions in all key economic sectors of energy production and non-energy production are increasing whereas the worsening Climate Emergency demands urgently decreasing GHG emissions and ultimately net zero GHG emissions ASAP. All that sane, science-informed people can do is to endlessly re-state "Wrong Way Go Back".

Refernces.

[1]. Center for Climate and Energy Solutions, "Global emissions", 2017: https://www.c2es.org/content/international-emissions/.
[2]. New Zealand Government, "Global greenhouse gas emissions", http://archive.stats.govt.nz/browse_for_stats/environment/environmental-reporting-series/environmental-indicators/Home/Atmosphere-and-climate/global-greenhouse-gases.aspx.
[3]. Alison Benjamin, "Stern: climate change a "market failure"", Guardian, 29 November 2007: https://www.theguardian.com/environment/2007/nov/29/climatechange.carbonemissions.
[4]. Pope Francis, Encyclical Letter "Laudato si'", 2015: http://w2.vatican.va/content/francesco/en/encyclicals/documents/papa-francesco_20150524_enciclica-laudato-si.html.
[5]. Gideon Polya, "Green Left Pope Francis Demands Climate Action "Without Delay" To Prevent Climate "Catastrophe"", Countercurrents, 10 August, 2015: https://countercurrents.org/polya100815.htm.
[6]. Chris Hope, "How high should climate change taxes be?", Working Paper Series, Judge Business School, University of Cambridge, 9, 2011: http://www.jbs.cam.ac.uk/media/assets/wp1109.pdf.
[7]. James Hansen, "Climate change in a nutshell: the gathering storm", Columbia University, 18 December 2018: http://www.columbia.edu/~jeh1/mailings/2018/20181206_Nutshell.pdf.
[8]. Allied Market Research, "Bunker Fuel Market", January 2019: https://www.alliedmarketresearch.com/bunker-fuel-market.
[9]. "Landfill gas", Wikipedia: https://en.wikipedia.org/wiki/Landfill_gas.
[10]. Drew T. Shindell, Greg Faluvegi, Dorothy M. Koch, Gavin A. Schmidt, Nadine Unger and Susanne E. Bauer, "Improved Attribution of Climate Forcing to Emissions", Science, 30 October 2009: Vol. 326 no. 5953 pp. 716-718: http://www.sciencemag.org/content/326/5953/716.
[11]. Global Methane Initiative, "Landfill methane: reducing emissions, advancing recovery and use opportunities", September 2011: https://www.globalmethane.org/documents/landfill_fs_eng.pdf.
[12]. Waste Management Research editorial, "Landfilling in developing countries", Waste Management Research, 31(1) 1–2, 2013: https://journals.sagepub.com/doi/pdf/10.1177/0734242X12469169.

[13]. Erica Gies, "Landfills have a huge greenhouse gas problem. Here's what we can do about it", ENSIA, 25 October 2016: https://ensia.com/features/methane-landfills/.

[14]. Aleksandra Arcipowska, Emily Mangan, You Lyu and Richard Waite, "5 questions about agricultural emissions, answered ", World Resources Institute, 29 July 2019: https://www.wri.org/blog/2019/07/5-questions-about-agricultural-emissions-answered.

[15]. William J. Ripple et al., 15,364 signatories from 184 countries, "World scientists' warning to Humanity: a second notice", Bioscience, 13 November 2017: https://academic.oup.com/bioscience/advance-article/doi/10.1093/biosci/bix125/4605229.

[16]. Gideon Polya, "Over 15,000 scientists issue dire warning to humanity on catastrophic climate change and biodiversity loss", Countercurrents, 20 November 2017: https://countercurrents.org/2017/11/20/over-15000-scientists-issue-dire-warning-to-humanity-on-catastrophic-climate-change-and-biodiversity-loss/.

[17]. William Ripple et al., "World scientists' warning of a climate emergency", BioScience, 5 November 2019: https://academic.oup.com/bioscience/advance-article/doi/10.1093/biosci/biz088/5610806.

[18]. Gideon Polya, "Extrapolating 11,000 scientists' climate emergency warning to 2030 catastrophe", Countercurrents, 14 November 2019: https://countercurrents.org/2019/11/extrapolating-11000-scientists-climate-emergency-warning-to-2030-catastrophe.

[19]. Gideon Polya, "Worsening Climate Emergency And Record CO_2 Emissions Demand Vegetarian Diet For All To Help Save Planet", Countercurrents, 20 June, 2016: https://countercurrents.org/polya200616.htm.

[20]. Gideon Polya, "Biofuel famine, biofuel genocide, meat & global food price crisis", Global avoidable mortality: http://globalavoidablemortality.blogspot.com.au/2008/05/biofuel-famine-biofuel-genocide-meat.html.

[21]. Robert Goodland and Jeff Anfang. "Livestock and climate change. What if the key actors in climate change are … cows, pigs and chickens?", World Watch, November/December 2009: https://awellfedworld.org/wp-content/uploads/Livestock-Climate-Change-Anhang-Goodland.pdf.

[22]. "List of countries by meat consumption", Wikipedia: https://en.wikipedia.org/wiki/List_of_countries_by_meat_consumption.

[23]. Gideon Polya, "Trumpist climate change denials Australian bushfires, fuel reduction, biochar & Carbon Debt", Countercurrents, 10 January 2020: https://countercurrents.org/2020/01/trumpist-climate-change-denial-australian-bushfires-fuel-reduction-biochar-carbon-debt.

[24]. H. Keith, B.G. Mackey and D. B. Lindenmayer., "Re-evaluation of forest biomass carbon stocks and lessons from the world's most carbon-dense forests", PNAS, 14 July 2009, 106 (28) 11635-11640: https://www.pnas.org/content/106/28/11635.

[25]. Hubau, W., Lewis, S.L., Phillips, O.L. et al. "Asynchronous carbon sink saturation in African and Amazonian tropical forests",

[26]. Baccini et al., 2017, "Tropical forests are a net carbon source based on aboveground measurements of gain and loss", Science, http://science.sciencemag.org/content/early/2017/09/27/science.aam5962.

[27]. "Tropical forests are now carbon source, not carbon sinks", Countercurrents, 5 March 2020: https://countercurrents.org/2020/03/tropical-forests-are-now-carbon-source-not-carbon-sinks.

[28]. Phillip Levin and Donald Levin, "The real biodiversity crisis", American Scientist, January-February 2002: http://www.americanscientist.org/issues/pub/the-real-biodiversity-crisis.

[29]. Gideon Polya, "How much negative carbon emissions, negative population growth & negative economic growth is needed to save Planet?" Countercurrents, 28 November 2019: https://countercurrents.org/2018/11/how-much-negative-carbon-emissions-negative-population-growth-negative-economic-growth-is-needed-to-save-planet.

[30]. World Steel Association, "Global steel output increases by 3.4% in 2019", 27 January 2020: https://www.worldsteel.org/media-centre/press-releases/2020/Global-crude-steel-output-increases-by-3.4–in-2019.html.

[31]. M. Kundak, L. Lazic, and J. Crnko, "CO_2 emissions in the steel industry", Metabk, 48 (3), 193-197, 2009: file:///C:/Users/Gideon/AppData/Local/Temp/MET_48_3_193_197_Kundak-2.pdf.

[32]. "Portland cement", Wikipedia: https://en.wikipedia.org/wiki/Portland_cement.

[33]. Yyxia (Sarah) Zhang, Khin Soe, and Yingying Guo, "Green cement a step closer to being a game-changer for construction emissions", The Conversation, 19 November 2019: https://theconversation.com/green-cement-a-step-closer-to-being-a-game-changer-for-construction-emissions-126033.

[34]. Statista, "Cement production globally and in the U.S. from 2010 to 2019", 2019: https://www.statista.com/statistics/219343/cement-production-worldwide/.

[35]. Michael Mazengarb, "Nordic steel giant to use renewable hydrogen to produce fossil- free steel by 2026", Renew Economy, 30 January 2020: https://reneweconomy.com.au/nordic-steel-giant-to-use-renewable-hydrogen-to-produce-fossil-free-steel-by-2026-2026/.

[36]. IAE, "Global energy and CO_2 status report 2019": https://www.iea.org/reports/global-energy-co2-status-report-2019/emissions.

[37]. IEA, "Tracking power: coal power", May 2019: https://www.iea.org/reports/tracking-power-2019/coal-fired-power.

[38]. IEA, "Coal", 2019: https://www.iea.org/fuels-and-technologies/coal.

[39]. IEA, "Gas", 2019: https://www.iea.org/fuels-and-technologies/gas.

[40]. IEA, "Oil", 2019: https://www.iea.org/fuels-and-technologies/oil.

[41]. "Gas is dirty energy": https://sites.google.com/site/gasisnotcleanenergy/gas-is-dirty-energy.

[42]. "Gas is not clean energy": https://sites.google.com/site/gasisnotcleanenergy/home.

[43]. "BP statistical review of world energy 2019": https://www.bp.com/content/dam/bp/business-sites/en/global/corporate/pdfs/energy-economics/statistical-review/bp-stats-review-2019-full-report.pdf

[44]. Robert Howarth, "Methane emissions and climatic warming risk from hydraulic fracturing and shale gas development: implications for policy", Energy & Emission Control Technologies, 8 October 2015: https://www.eeb.cornell.edu/howarth/publications/f_EECT-61539-perspectives-on-air-emissions-of-methane-and-climatic-warmin_100815_27470.pdf.

[45]. "Gas leakage – systemic gas leakage in the US is about 5.8%", Gas is not clean energy: https://sites.google.com/site/gasisnotcleanenergy/gas-leakage.

[46]. Gideon Polya, "Rampant Orwellian falsehood in neoliberal Australia – and in your country too?", Countercurrents, 1 March 2020: https://countercurrents.org/2020/03/rampant-orwellian-falsehood-in-neoliberal-australia-and-in-your-country-too.

[47]. Deloitte, "The global truck market is optimistic", Truck Market 2024: https://www2.deloitte.com/content/dam/Deloitte/tr/Documents/process-and-operations/truck-studie-2014-s.pdf.

[48]. IBISWorld, "Global car & automobile industry – market research report", November 2019: https://www.ibisworld.com/global/market-research-reports/global-car-automobile-sales-industry/.

[49]. IEA, "The future of rail", 2019: https://www.iea.org/reports/the-future-of-rail.

[50]. "Manufacturing the future: The next era of global growth and innovation", Executive Summary, November 2012: https://www.mckinsey.com/~/media/McKinsey/Business%20Functions/Operations/Our%20Insights/The%20future%20of%20manufacturing/MGI_Manufacturing%20the%20future_Executive%20summary_Nov%202012.ashx

[51]. PwC, "Global construction 2030: A global forecast for the construction industry to 2030": https://www.pwc.com/vn/en/industries/engineering-and-construction/pwc-global-construction-2030.html.

[52]. Irene Banos Ruiz, "China's new love affair with dogs – as pets, not food – presents environmental problems", DW, 21 June 2016: https://www.dw.com/en/chinas-new-love-affair-with-dogs-as-pets-not-food-presents-environmental-problems/a-19197523.

[53]. Jason Daley, "The EPA declared that burning wood is carbon neutral. It's actually a lot more complicated" Smithsonian Magazine Smart News, 24 April, 2018: https://www.smithsonianmag.com/smart-news/epa-declares-burning-wood-carbon-neutral-180968880/.

[54]. World Bioenergy Association, "Global biomass potential towards 2035", March 2016: https://worldbioenergy.org/uploads/Factsheet_Biomass%20potential.pdf.

[55]. World Bioenergy Association, "Global bioenergy statistics 2019": https://worldbioenergy.org/uploads/191129%20WBA%20GBS%202020 19_LQ.pdf.

[56]. Statista, "Global biofuel production from 2000 to 2018": https://www.statista.com/statistics/274163/global-biofuel-production-in-oil-equivalent/.

[57]. "Biofuel Genocide": https://sites.google.com/site/biofuelgenocide/home.

[58]. Timothy Searchinger et al., "Use of U.S. Croplands for Biofuels Increases Greenhouse Gases Through Emissions from Land-Use Change", Science 29 February 2008, Vol. 319. no. 5867, pp. 1238 – 1240: http://www.sciencemag.org/cgi/content/abstract/1151861.

[59]. Joseph Fargione et al., "Land Clearing and the Biofuel Carbon Debt", Science 29 February 2008, Vol. 319. no. 5867, pp. 1235 – 1238: http://www.sciencemag.org/cgi/content/abstract/1152747.
[60]. Gideon Polya, "Exposing The Greenwash Falsehood Of "Renewable Ethanol"", Countercurrents, 19 April, 2014: https://www.countercurrents.org/polya190414.htm.
[61]. Gideon Polya, "Rampant Orwellian falsehood in neoliberal Australia – and in your country too?", Countercurrents, 1 March 2020: https://countercurrents.org/2020/03/rampant-orwellian-falsehood-in-neoliberal-australia-and-in-your-country-too.
[62]. Gideon Polya, "Methane leakage makes Australia a world leading per capita greenhouse gas polluter", Countercurrents, 18 February 2020: https://countercurrents.org/2020/02/methane-leakage-makes-australia-a-world-leading-per-capita-greenhouse-gas-polluter.
[63]. Steel Guru, "DNV-GL forecasts global LNG production to increase to around 630 million tonnes by 2050", Gasoil News, 5 April 2019: https://steelguru.com/gas-oil/dnv-gl-forecasts-global-lng-production-to-increase-to-around-630-million-tonne-by-2050/537692.
[64]. Stephen Hawking, "Brief Answers to the Big Questions", John Murray, 2018, Chapter 7.
[65]. Mike Foley, "Australia must hit net zero by 2050 to meet Paris: scientist", Sydney Morning Herald, 24 February 2020: https://www.smh.com.au/politics/federal/australia-must-hit-net-zero-by-2050-to-meet-paris-scientists-20200224-p543ss.html.
[66]. The Greens, "Climate change and energy", November 2018: https://greens.org.au/policies/climate-change-and-energy.

SECTION B: DROUGHT, FLOOD, FIRE, STORM, SEA SURGE, HEAT & WATER CRISES

"We are now at 50 percent natural ecosystem loss globally. I conclude that Earth needs to maintain some two-thirds of its land area as natural and seminatural ecosystems to meet local needs and to maintain local and global ecological sustainability. Along with a number of other planetary boundaries, including climate change and biodiversity loss, Earth is already in ecological overshoot and will collapse unless we pull back from the brink." Dr Glen Barry (US ecologist), 2013.

"Very crudely, we are now committed to global sea level rise equivalent to a permanent [2012] Hurricane Sandy storm surge… The possibility that we have already committed to 3 or more meters of sea level rise from West Antarctica will be disquieting to many people, even if the rise waits centuries before arriving." Dr Richard Alley (glaciologist, Pennsylvania State University), Science Now, 2014.

"We, the Leaders of the Pacific Islands Development Forum… declare that we… Are gravely distressed that climate change poses irreversible loss and damage to our people, societies, livelihoods, and natural environments creating existential threats to our very survival… but appropriate responses are lacking…See and suffer from the adverse impacts of climate change, including but not limited to increased intensity of tropical cyclones, sea level rise, severe storm surges, more frequent and more extreme weather events, coral bleaching, saltwater intrusions, higher king tides, coastal erosion, changing precipitation patterns, submersion of islands, and ocean acidification, with scientific evidence clearly informing us these impacts will further intensify over time." Suva Declaration by Leaders of the Pacific Islands Development Forum, 2015.

"We have certainly committed to some dangerous changes in climate already and it's possible we have already warmed the oceans enough that they're going to melt the ice shelves around Antarctica enough to destabilize enough of that Antarctic ice to give us ten feet or eleven feet of sea level rise by the end of the century. That is a very real possibility." Professor Michael Mann (Professor of Meteorology, Pennsylvania State University), 2015.

"Ice melt, sea level rise and superstorms: evidence from paleoclimate data, climate modelling, and modern observation that 2 degree C global warming could be dangerous." Dr James Hansen et al., Atmospheric Chemistry and Physics, 2016.

As indicated in the Preface, for 30 years scientists have been aware of the serious threats to Humanity from man-made climate change as manifested in global warming-impacted drought, floods, tropical storms, sea surges, damaging heat, and decreased agricultural output. Writing about the 6-7 million victim but "forgotten" WW2 Bengali Holocaust (WW2 Bengal Famine) in a detailed and documented essay, I concluded (1995): "History may yet repeat itself with respect to the Bengalis. Australia is per capita one of the most profligate contributors to "greenhouse" gas emission and hence global warming. There is unfortunately a current bipartisan political consensus in Australia opposed to effective short-term reduction of such emission. The recent Berlin Climate Change Conference, heavily influenced by the U.S. and similarly-inclined Australia and Canada, concluded with no agreed targets on greenhouse gas emission and produced merely a motherhood commitment to continuing assessment. While global warming may have advantages for some, there is a widespread concern that global warming and consequent sea level rises will have a major impact on low-lying regions such as densely populated and impoverished West Bengal and Bangladesh" [1, 2]. I sent a summary of that humanitarian essay to Australia MPs, journalists and academics but the Silence was Deafening.

However Anglo-Indian Greens Senator Christabel Chamarette who had worked in Bangladesh made a speech in the Australian Senate about my submission, which she tabled in the Australian Senate (page 192 [3]).

By 1998 my concerns expressed in this academic paper (published in International Network on Holocaust and Genocide [2]) had translated into a huge book, "Jane Austen and the Black Hole of British History. Colonial rapacity, holocaust denial and the crisis in biological sustainability" [3]. A 2008 second edition with updating postscripts for each chapter is now available for free perusal on the Web [4]. In Chapter 16 of this book I stated (1998, 2008): "The coming holocaust…These coming events will have a much more energy-rich environment to feed upon and their victims will now regularly experience extraordinarily destructive events that they would previously have expected to encounter only once in their lifetimes. Thus Bangladesh experienced a major cyclonic disaster in 1970 that killed about 300,000 people in coastal regions. In 1991 a similar event claimed 140,000 lives in Bangladesh. The hurricanes of a much hotter world will be much more likely to be extraordinarily destructive super-hurricanes of this kind, and the destructiveness of these events will be seriously compounded by increases in population density and by the effects of sea level rises brought about through global warming"(page 169 [5]).

My country Australia is particularly susceptible to heat, drought, fire and flood, and has been famously described thus by poet Dorothy Mackellar in her poem "My Country": "I love a sunburnt country/A land of sweeping plains,/Of ragged mountain ranges,/Of droughts and flooding rains/… Core of my heart, my country! / Land of the Rainbow Gold,/ For flood and fire and famine,/ She pays us back three fold - /". In 2009 I wrote about the Black Saturday bushfires in Victoria in which 173 people died, 450,000 ha were burnt, and over 2,000 houses were destroyed: "We can now realistically expect 450 ppm atmospheric CO_2 by about 2030 (i.e. in about 20 years' time assuming [plus] 3 ppm CO_2/y, or earlier due to positive feedbacks) with a change in temperature (ΔT) 2°C above that in 1900. Above 450 ppm CO_2 there is intensification of existing conditions ([nearly] all Arctic summer sea ice will have gone by 2015, massive hurricanes, storm surges, droughts, mega bushfires, coastal and inland flooding, food shortages, huge mass mortality) plus increasingly major damage to coral reefs – including Australia's Great Barrier Reef - which will be dying due to ocean warming and acidification above 450 ppm CO_2 - with increasing damage to already stressed fisheries and agriculture with consequent mass starvation"[6].

In 2010 I wrote: "The world is facing a worsening water crisis that is compounded by population expansion, aquifer depletion and the effects of mainly First World-imposed global warming (increasing temperature, drought and melting of the Himalayan and other glaciers that are major water sources for the Developing World)" [7]. A crucial point is that while variability of weather means that any particular catastrophe cannot be attributed solely to climate change, the underlying warming trend contributes to the intensity of heat, drought, fire, storm and flood events. Thus in relation to climate change and devastating Queensland floods I wrote (2011): "Because of weather variability [one] cannot attribute any specific weather event to climate change. However the successive events of burning fossil fuels, increased atmospheric greenhouse gases, increased sea temperature, increased evaporation, increased precipitation and floods simply mean stop burning fossil fuels ASAP as urged by the Climate Emergency Network (CEN)" [8].

In several articles published in recent years I criticized use of the term "natural disaster" when both insufficient threat mitigation and man-made global warming contribute to drought, flood and fire disasters [9-12]. Thus in 2018 I wrote: "Despite a global warming of plus 1C, the unavoidability of a catastrophic plus 2C and now regular, high energy hurricane catastrophes in tropical Island Nations, the atmospheric carbon dioxide (CO_2) is relentlessly increasing as is carbon fuel burning for energy. This climate criminal greenhouse gas (GHG) pollution and climate inaction is enabled by the remorseless mendacity of One Percenter-dominated Mainstream media, politician, commentariat and academic presstitutes who ignore horrendous current Hazard Response Debt, huge, inescapable and worsening Carbon Debt, and the palpable reality that Disasters Are Not Natural. While personal, corporate or national debt can be expunged or evaded in various ways, Hazard Response Debt and Carbon Debt are inescapable – unless that sea wall is built, the population will drown" [9].

My 2019 article [13] makes a fundamental point that deaths in a particular catastrophic event are best accounted by excess deaths associated with the event, in this instance heat stress deaths: "More than 500 Australians presently die from heat stress each year with this disaster disproportionately impacting the elderly, the impoverished and Indigenous First Nations peoples. Australia is currently beset continent-wide with climate change-exacerbated heat waves with sustained temperatures above 40 degrees Centigrade, extraordinary, nation-wide bush fire emergencies commencing in Winter, and a devastating and widespread drought in Eastern Australia that has lasted for years… It was estimated from public health records that 432 people died from heat stress in the heat wave in Victoria and South Australia leading up to and including the Black Saturday bushfire catastrophe in which 173 people perished in the Victorian bushfire holocaust" [13]. It is estimated that "In August 2003, a heatwave in Europe caused 44,878 excess deaths", with "excess deaths" ("excess mortality") being defined as "a comparison of expected deaths with ones that actually happened" (Giuliana Viglione, "How many people has the coronavirus killed?", Nature, 1 September 2020: https://www.nature.com/articles/d41586-020-02497-w).

For a country or region excess death (excess mortality, avoidable death, avoidable mortality, premature death, untimely death, death that should not have happened) can be measured as the difference between actual deaths and deaths expected for a peaceful, decently-run country or region with the same demographics (birth rate, proportion of children). Thus annual avoidable deaths from deprivation in the Developing World (minus China) total 15 million annually [14]. However if the worsening Climate Crisis and Climate Genocide are not urgently addressed then 10 billion people may die this century en route to a sustainable human population of only 0.5-1.0 billion in 2100 i.e. for an average annual mortality from deprivation this century of 100 million per annum (see "Climate Genocide": https://sites.google.com/site/climategenocide/). Presently 9 million people die annually due to air pollution from the burning of carbon fuels (see "Stop air pollution deaths": https://sites.google.com/site/300orgsite/stop-air-pollution-deaths.)

Four articles are reproduced in full as Chapters in this section:

Chapter 7: "Kerala flood disaster is not natural but man-made – warning for India & World" (about Indian floods) [10].
Chapter 8: "Water crisis, Global Avoidable Mortality Holocaust, Water Apartheid, Global Warming & Mina Guli" (about the worsening global water and drought crisis) [11].
Chapter 9: "Trumpist climate change denial, Australian bushfires, fuel reduction, biochar & Carbon Debt" (about the catastrophic 2019-2020 South Eastern Australian bushfires coming on top of years of horrendous drought) [12].
Chapter 10. "Australian climate criminality, heat stress deaths & Australian Aboriginal Ethnocide" (about heat stress deaths) [13].

References (those asterisked are reproduced in full in Section B).

[1]. Gideon Polya "The Forgotten Holocaust - The 1943/44 Bengal Famine", 1995: http://globalavoidablemortality.blogspot.com.akesau/2005/07/forgotten-holocaust-194344-bengal.html.
[2]. Gideon Polya, "The famine of history: Bengal 1943", International Network on Holocaust and Genocide vol.10, 10-15 (an edited version of "The Forgotten Holocaust" [1]).
[3].Gideon Polya, "Jane Austen and the Black Hole of British History. Colonial rapacity, holocaust denial and the crisis in biological sustainability", G.M. Polya, Melbourne, 1998.
[4]. Gideon Polya, "Jane Austen and the Black Hole of British History. Colonial rapacity, holocaust denial and the crisis in biological sustainability", G.M. Polya, Melbourne, 2008, second edition now available for free perusal on the Web: http://janeaustenand.blogspot.com/
[5]. Gideon Polya, "Global warming and the unthinkable world of 2050", Chapter 16, "Jane Austen and the Black Hole of British History. Colonial rapacity, holocaust denial and the crisis in biological sustainability", 2008: http://janeaustenand.blogspot.com/2012/03/jane-austen-and-black-hole-chapter-16.html.
[6]. Gideon Polya, "Global Warming Impacting Humanity", Countercurrents, 14 February, 2009: https://www.countercurrents.org/polya140209.htm.
[7]. Gideon Polya, "Water Crisis, Palestinian Genocide And Climate Genocide", Countercurrents, 27 April 2010: https://www.countercurrents.org/polya270410.htm.
[8]. Gideon Polya, "CEN, Climate Change, Queensland Floods: Stop Burning Fossil Fuels ASAP", Countercurrents, 15 January, 2011: https://www.countercurrents.org/polya150111.htm.
[9]. Gideon Polya, "Mendacious Mainstream presstitutes ignore huge carbon debt & horrendous hazard response debt: disasters are not natural", Countercurrents, 13 August 2018: https://countercurrents.org/2018/08/13/mendacious-mainstream-presstitutes-ignore-huge-carbon-debt-horrendous-hazard-response-debt-disasters-are-not-natural/.
*[10]. Gideon Polya, "Kerala flood disaster is not natural but man-made – warning for India & World", Countercurrents, 21 August 2018: https://countercurrents.org/2018/08/21/kerala-flood-disaster-is-not-natural-but-man-made-warning-for-india-world/.

*[11]. Gideon Polya, "Water crisis, Global Avoidable Mortality Holocaust, Water Apartheid, Global Warming & Mina Guli", Countercurrents, 17 May 2019: https://countercurrents.org/2019/05/water-crisis-global-avoidable-mortality-holocaust-water-apartheid-global-warming-mina-guli.

*[12]. Gideon Polya, "Trumpist climate change denial, Australian bushfires, fuel reduction, biochar & Carbon Debt", Countercurrents, 10 January 2020: https://countercurrents.org/2020/01/trumpist-climate-change-denial-australian-bushfires-fuel-reduction-biochar-carbon-debt.

*[13]. Gideon Polya, "Australian climate criminality, heat stress deaths & Australian Aboriginal Ethnocide", Countercurrents, 19 December 2019: https://countercurrents.org/2019/12/australian-climate-criminality-heat-stress-deaths-australian-aboriginal-ethnocide/.

[14]. Gideon Polya, "Body Count. Global avoidable mortality since 1950", including an avoidable mortality-related history of every country from Neolithic times and now available for free perusal on the web: http://globalbodycount.blogspot.com.au/.

CHAPTER 7: KERALA FLOOD DISASTER IS NOT NATURAL BUT MAN-MADE – WARNING FOR INDIA & WORLD

[First published as Gideon Polya, "Kerala flood disaster is not natural but man-made – warning for India & World", Countercurrents, 21 August 2018: https://countercurrents.org/2018/08/21/kerala-flood-disaster-is-not-natural-but-man-made-warning-for-india-world/.]

The dreadful Kerala flood disaster (350 deaths and 1 million people displaced) is not a natural disaster – it is a man-made disaster involving local Indian contributions (inadequate water storage hazard response and landslides from deforestation) and a major international contribution from global warming due to ever-increasing greenhouse gas (GHG) pollution, with the prosperous First World and Anglosphere countries being disproportionately high contributors on a per capita basis.

Below is my understanding as a biological chemist of this dreadful flood disaster in otherwise water-blessed Kerala that presently masks a worsening man-made disaster of worsening rainfall deficit (drought) in South Asia and elsewhere, including presently drought-ravaged, fire-ravaged, high-polluting and rich Australia. The hazard response failure in Kerala should be a cogent warning to India and the World.

1. **The Kerala flood disaster is not natural but is man-made.**

Despite a global warming of plus 1C, the unavoidability of a catastrophic plus 2C and now regular, high energy hurricane catastrophes in tropical Island Nations and low-lying nations, the atmospheric carbon dioxide (CO_2) is relentlessly increasing as is carbon fuel burning for energy. This relentless climate criminal greenhouse gas (GHG) pollution, climate inaction and consequential floods and droughts are enabled by political short-termism that ignores horrendous current Hazard Response Debt, huge, inescapable and worsening Carbon Debt, and the palpable reality that Disasters Are Not Natural. While personal, corporate or national debt can be expunged or evaded in various ways, Hazard Response Debt and Carbon Debt are inescapable – human populations will suffer grievously unless there is adequate risk management, drought-proofing and flood-proofing [1].

However there are stark national differences in per capita responsibility for climate hazards facing the world and in per capita ability to respond to such hazards. Thus:

(a) annual per capita GHG pollution in tonnes CO_2-equivalent per person is 2.1 for India versus 41.0 for the United States and 52.9 for Australia (116 if including its huge GHG-generating exports) [2] ;

(b) a weighted annual pollution score taking income into account is 0.3 for India versus 207.1 for the United States and 306.8 for Australia (or 672.8 if including its huge GHG-generating exports) [3];

(c) the historical, damage-related Carbon Debt (1751-2006) in billion tonnes CO_2 [Gt CO_2] is 9.01 for India versus 99.12 for the United States and 5.59 for Australia [4, 5] – at a damage-related Carbon Price of $200 per tonne CO_2 [6] this translates to $1.8 trillion for India versus $19.8 trillion for the United States and $1.1 trillion for Australia;

(d) the per capita, historical, damage-related Carbon Debt (1751-2006) in tonnes CO_2 per head of 2018 population is 6.6 for India (population 1,356 million), 303.0 for the United States (population 327.1 million) and 225.4 for Australia (population 24.8 million) – at a damage-related Carbon Price of $200 per tonne CO_2 [6] this translates [USD per person] to $1,320 for

India versus \$60,600 for the United States and \$45,080 for Australia.

(e) The per capita GDP is \$1,983 for India versus \$59,501 for the United States and \$55,707 for Australia [7].

However climate change does not recognize borders, and in a globalized and steadily robotizing world economy the cost of materials for Hazard Response are much the same in the Third World as in the Developed World. Increasingly flood-, drought- and elevated temperature-impacted India is hugely impoverished and thus is hugely response-limited relative to the Developed World and Anglosphere countries which, further, have a disproportionately high responsibility for these horrendous impacts. Indeed the people of Kerala are hugely response-limited relative to the disproportionately hugely wealthy Indian ruling class who, in a carbon economy context, are disproportionately hugely complicit in the floods, droughts and elevated temperatures impacting India.

Dr V. S. Vijayan (Environmental scientist and a member of the expert panel on the Western Ghats) on a man-made flood disaster in Kerala: "This was waiting to happen. Insensible use of land, soil and rocks led to this deluge. Landslips and flash floods happened in areas that witnessed widespread human incursions. I hope everyone will learn a lesson from this. Due to climate change, such tragedies are bound to increase. Nobody can stop rains or control floods. But we can take measures to lower the intensity of such impacts" [8].

Dr V. S. Vijayan is absolutely correct about the need for requisite hazard perception and hazard response [8]. Rational risk management is crucial for societal safety and successively involves (a) accurate information, (b) science-based analysis, this involving the critical testing of potentially falsifiable hypotheses, and (c) science-informed systemic change to minimize risk. Unfortunately, greed and short-termism frequently intervene to replace science-based risk management with (a) censorship, intimidation, lying and spin, (b) anti-science, spin-based analysis involving the selective use of asserted facts to support a partisan position, and (c) propaganda and "blame and shame" that inhibit the primary need for accurate reportage. Thus in a global context, 16 million people die avoidably from deprivation each year (4 million in India) on Spaceship Earth with the flight deck run by One Percenters who own 50% of the world's wealth [9-11].

2. Probabilistic attribution of extreme floods and droughts to man-made climate change.

Man-made global warming is associated with increased Sea Surface Temperature (SST), increased evaporation, increased atmospheric moisture and thence increased precipitation. Evaporation is associated with heat absorption (why we perspire to cool down) and, conversely, precipitation is associated with an exothermic release of latent heat which increases the energy of storms. Thus man-made global warming through GHG pollution is increasing the energy of tropical storms (hurricanes, cyclones) and increasing sea level and hence the damaging effects of storm surges. Increased temperature over land is likewise associated with soil drying and severe drought in the absence of precipitation.

However there is always considerable variability in the day-to-day weather, and there is a probabilistic element in attributing significant weather events to man-made global warming. Thus Quirin Schiermeier writing in the Scientific American (2018): "The idea behind attribution science is simple enough. Disasters such as record-breaking heatwaves and extreme rainfall are likely to become more common because the build-up of greenhouse gases is altering the atmosphere. Warmer air contains more water vapour and stores more energy; the increasing temperatures can also change large-scale atmospheric circulation patterns. But extreme weather can also arise from natural cycles, such as the El Niño phenomenon that periodically warms sea surface temperatures in the tropical Pacific Ocean" [12].

Eastern Australia is presently suffering a devastating, multi-year drought that is variously impacted by man-made global warming, a shift southwards of a major westerly "Roaring Forties" system (that swept European sailors across the Indian Ocean and aided European invasion and colonization of India, Indonesia and Australia), and by the El Niño phenomenon in the Pacific (ocean warming in the Western Pacific leading to drought in Eastern Australia and floods in the Western parts of the Americas, the converse system being called La Niña). Dr Andrew King and Professor David Karoly (School of Earth Sciences, University of Melbourne): "Whenever we experience heatwaves or cold snaps, droughts or floods, people want to know: was this due to climate change? And until about a decade ago the response was: we don't know. We can't link specific extreme weather events to climate change. Now things have changed. While we can't say climate change caused an extreme event, we can estimate how much more or less likely the event has become due to human influences on the climate. This relatively new and rapidly developing area of science is called "Event Attribution"… Australian climate scientists have been leading the way in event attribution and are at the forefront of new developments in the field. In part, this is because Australia is already experiencing a rapidly changing climate with many extreme events occurring in the last few years. We have assessed the role of human-forced climate change and natural variability (related to El Niño and La Niña) in a range of recent extreme events including the heavy rains of 2011/12, the "Angry Summer" of 2013, and the heatwaves and droughts of the last three years… [e.g.] human activities have exacerbated the impacts of tropical cyclones through sea level rise causing larger storm surges" [13].

The US National Academies of Sciences have similarly stated (2016): "As climate has warmed over recent years, a new pattern of more frequent and more intense weather events has unfolded across the globe. Climate models simulate such changes in extreme events, and some of the reasons for the changes are well understood. Warming increases the likelihood of extremely hot days and nights, favors increased atmospheric moisture that may result in more frequent heavy rainfall and snowfall, and leads to evaporation that can exacerbate droughts. Even with evidence of these broad trends, scientists cautioned in the past that individual weather events couldn't be attributed to climate change. Now, with advances in understanding the climate science behind extreme events and the science of extreme event attribution, such blanket statements may not be accurate. The relatively young science of extreme event attribution seeks to tease out the influence of human-cause climate change from other factors, such as natural sources of variability like El Niño, as contributors to individual extreme events" [14].

Giuling Wang et al. have recently provided a powerful new insight into the basis of increasing rainfall rates with increasing temperature (2017): "Theoretical models predict that, in the absence of moisture limitation, extreme precipitation intensity could exponentially increase with temperatures at a rate determined by the [thermodynamics] Clausius–Clapeyron (C–C) relationship. Climate models project a continuous increase of precipitation extremes for the twenty-first century over most of the globe. However, some station observations suggest a negative scaling of extreme precipitation with very high temperatures, raising doubts about future increase of precipitation extremes. Here we show for the present-day climate over most of the globe, the curve relating daily precipitation extremes with local temperatures has a peak structure, increasing as expected at the low–medium range of temperature variations but decreasing at high temperatures. However, this peak-shaped relationship does not imply a potential upper limit for future precipitation extremes. Climate models project both the peak of extreme precipitation and the temperature at which it peaks (T_{peak}) will increase with warming; the two increases generally conform to the C–C scaling rate in mid- and high-latitudes, and to a super C–C scaling in most of the tropics. Because projected increases of local mean temperature (T_{mean}) far exceed projected increases of T_{peak} over land, the conventional approach of relating extreme precipitation to T_{mean} produces a misleading sub-C–C scaling rate" [15].

John P. Abraham (engineer, climate scientist and professor of thermal and fluid sciences at the University of St. Thomas

School of Engineering, Minnesota) has explained this very clearly in lay terms: "It's expected that in general, air will get moister as the Earth warms – provided there is a moisture source. This may cause more intense rainfalls and snow events, which lead to increased risk of flooding. But warmer air can also more quickly evaporate water from surfaces. This means that areas where it's not precipitating dry out more quickly. In fact, it's likely that some regions will experience both more drought *and* more flooding in the future (just not at the same time!). The dry spells are longer and with faster evaporation causing dryness in soils. But, when the rains fall, they come in heavy downpours potentially leading to more floods… Traditionally, we have related precipitation events to the local average temperature (T_{mean}). However, it's clear that there's a strong relationship between the peak temperature [for precipitation, T_{peak}] and the precipitation rates. In fact, relations reveal that precipitation rates are increasing between 5 and 10% for every degree C increase [in T_{peak}]. The expected rate of increase, just based on thermodynamics is 7%… in some parts of the globe, the relationship is even stronger. For instance, in the tropics, there's more than a 10% increase in precipitation for a degree Celsius increase in temperature. This is not unexpected because precipitation releases latent heat, which can in turn invigorate storms. From a practical standpoint, this helps us plan for climate change (it is already occurring) including planning resiliency… storing water from times when there is too much for the inevitable times when we have too little (drought), results in better water management and multiple benefits. This shows why climate science is so important. The [Trump] US government is in the process of decimating our climate science infrastructure" [16].

3. **Global warming, Indian Ocean warming, Kerala floods and worsening rain deficits in much of India.**

The Times of India has reported findings by the Indian Institute of Tropical Meteorology that warming of the Indian Ocean has been associated with a weakening land-sea thermal gradient over South Asia, a weakening monsoon and worsening rain deficits in much of India (2015): "The rapid warming in the

Indian Ocean has been weakening the monsoon in Central India consistently which has resulted in 10% to 20% deficient rain over Uttar Pradesh, Jhakhand, Madhya Pradesh, Uttarakhand, and Chhattisgarh during the past century… rapid warming of the India Ocean is playing an important role in weakening the monsoon circulation and the rainfall… rainfall has been decreasing over central South Asia – from the south of Pakistan through central India to Bangladesh. The drop is highly significant over central India where agriculture is mostly rain-fed with a reduction of up to 10%-20% in the mean rainfall over the last 112 years… In an ideal global warming scenario, the monsoon drivers should get stronger, causing increased rainfall. One of the main monsoon drivers is the land-sea temperature difference in summer, which drives the monsoon circulation towards the subcontinent… the land-sea temperature difference over South Asia domain has in fact reduced in the past decades… primarily contributed by a strong warming in the Indian Ocean" [17].

Mathew Koll Roxy (Centre for Climate Change Research, Indian Institute of Tropical Meteorology, Pune) on the notable warming of the western tropical Indian Ocean (2014): "Recent studies have pointed out an increased warming over the Indian Ocean warm pool (the central-eastern Indian Ocean characterized by sea surface temperatures greater than $28.0°C$) during the past half-century, although the reasons behind this monotonous warming are still debated. The results here reveal a larger picture—namely, that the western tropical Indian Ocean has been warming for more than a century, at a rate faster than any other region of the tropical oceans, and turns out to be the largest contributor to the overall trend in the global mean sea surface temperature (SST). During 1901–2012, while the Indian Ocean warm pool went through an increase of $0.7°C$, the western Indian Ocean experienced anomalous warming of $1.2°C$ in summer SSTs. The warming of the generally cool western Indian Ocean against the rest of the tropical warm pool region alters the zonal SST gradients, and has the potential to change the Asian monsoon circulation and rainfall, as well as alter the marine food webs in this biologically productive region. The current study using observations and global coupled ocean–

atmosphere model simulations gives compelling evidence that, besides direct contribution from greenhouse warming, the long-term warming trend over the western Indian Ocean during summer is highly dependent on the asymmetry in the El Niño–Southern Oscillation (ENSO) teleconnection, and the positive SST skewness associated with ENSO during recent decades… the land-sea thermal gradient over South Asia has been decreasing, due to rapid warming in the Indian Ocean and a relatively subdued warming over the subcontinent" [18].

Mathew Koll Roxy and colleagues from the Indian Institute of Tropical Meteorology have warned of the prospect of large-scale drought in South Asia linked to Indian Ocean warming (2015): "There are large uncertainties looming over the status and fate of the South Asian summer monsoon, with several studies debating whether the monsoon is weakening or strengthening in a changing climate. Our analysis using multiple observed datasets demonstrates a significant weakening trend in summer rainfall during 1901–2012 over the central-east and northern regions of India, along the Ganges-Brahmaputra-Meghna basins and the Himalayan foothills, where agriculture is still largely rain-fed. Earlier studies have suggested an increase in moisture availability and land-sea thermal gradient in the tropics due to anthropogenic warming, favouring an increase in tropical rainfall. Here we show that the land-sea thermal gradient over South Asia has been decreasing, due to rapid warming in the Indian Ocean and a relatively subdued warming over the subcontinent. Using long-term observations and coupled model experiments, we provide compelling evidence that the enhanced Indian Ocean warming potentially weakens the land-sea thermal contrast, dampens the summer monsoon Hadley circulation, and thereby reduces the rainfall over parts of South Asia… overall weakening trend of monsoon rainfall over South Asia is a matter of grave concern since the socio-economic livelihood in this region, including agriculture, water resources and power generation are irrevocably dependent on it. The role of Indian Ocean warming might not be solely limited to weakening the Asian monsoon. Research based on climate proxy records and models point out that the climate signals forced by warm SST anomalies over the tropical Indian and western Pacific oceans synergistically contribute to widespread drying over South Asia and mid-latitudes, an ideal scenario for a large-scale drought" [19].

Kerala is a very high rainfall area of India but in the last 6 weeks rainfall in Kerala increased 37% above normal whereas rainfall for the rest of India ranged from plus 16% (Sikkim) and plus 15% (Jammu and Kashmir) to minus 18% (Bihar), minus 19% (West Bengal, Delhi and Haryana), minus 24% (Tripura), minus 25% (Assam), minus 27% (Gujarat and Jharkhand), minus 29% (Nagaland), minus 38% (Arunachal Pradesh), minus 43% (Meghalaya) and minus 66% (Manipur), with a third of India by area receiving below-average rainfall [8]. Crucially, one notes that evapotranspiration from vegetation in the Western Ghats (along the western side of India from Tamil Nadu through Kerala, Karnataka and Goa to Maharashtra) accounts for about 25% of rainfall in the Indian Peninsular and, for example, supplies up to half the rainfall in Tamil Nadu [20]. Professor Raghu Murtugudde (Earth System Science Interdisciplinary Center, University of Maryland, US*)*: "Any reduction in rainfall due to deforestation of the Western Ghats would lead to a warming of peninsular India as well. This can be expected since the monsoon rains typically bring a significant cooling over India by dragging down the dry cool air from the upper atmosphere. This cooling is a lifesaving relief from the scorching pre-monsoon temperatures and heatwaves. Continued deforestation is of great concern for the ecological community due to the loss of precious flora and fauna. This study puts a finer point on the value of the Western Ghats biodiversity as a significant source of moisture for rainfall over parts of India that are constantly struggling for water for agriculture as well as domestic and industrial use. The battle cry for protecting biodiversity hotspots and the overall forest cover over the Western Ghats just got louder" [20].

Final comments.

Man-made global warming is evidently impacting Humanity around the world with severe flooding, severe droughts, sea level rise, more intense tropical storms and catastrophic forest fires. My country, Australia, is experiencing a drying trend in the south as a key rain-bearing weather system shifts southwards, with increased precipitation in the tropical north, increased sea level and drought. Presently a large part of Eastern Australia (New South Wales and Queensland) is in the grip of a circa 5-year drought and New South Wales is experiencing severe forest fires (bushfires) in winter. The Australian Climate Council: "Climate change is likely making drought conditions in southwest and southeast Australia worse. Climate change has contributed to a southward shift in weather systems that typically bring cool season rainfall to southern Australia. Since the 1970s late autumn and early winter rainfall has decreased by 15 percent in southeast Australia, and Western Australia's southwest region has experienced a 15 percent decline in cool season rainfall. Climate change is also driving an increase in the intensity and frequency of hot days and heatwaves in Australia, exacerbating drought conditions" [21].

In India global warming has meant dangerous and deadly temperatures of circa 50C in some urban centres [22, 23], sea level rise, more intense tropical storms, and a weakening of the summer monsoon that is associated with a decrease in the land-sea thermal gradient arising from a warming Indian Ocean and a relatively decreased land warming (with this in turn linked to anthropogenic land cooling from industrial, transport and agricultural air pollution that is also associated with about 1 million Indian air pollution deaths annually [24, 25]). Over 2 decades ago an international agronomy study concluded that cereal production in tropical and sub-tropical areas of the Americas, Africa and Asia would be badly impacted by increasing atmospheric CO_2 and global warming [26, 27]. Thus Ian Chambers and John Humble (2007): "Climate Change potentially reduces the ability of the human race to produce the three main crops which provide the backbone of food production – wheat, rice and corn. For example, a 1 degree centigrade rise has been shown to reduce crop yields from between 5 and 7 per cent in major crops, such as rice. A forecast increase of 2 degrees [C] in global temperature, as a result of global warming, could significantly reduce the ability of crops

to be pollinated to produce the grain that are harvested for food"
[28].

Mathew Koll Roxy and colleagues from the Indian Institute of
Tropical Meteorology have warned of the prospect of large-
scale drought in South India linked to Indian Ocean warming
(2015): "Overall weakening trend of monsoon rainfall over
South Asia is a matter of grave concern since the socio-
economic livelihood in this region, including agriculture, water
resources and power generation are irrevocably dependent on it"
[19]. Presently 16 million people (4 million of them Indians) die
preventably from deprivation each year. However several
leading climate scientists have warned that failure to address
man-made global warming may kill all but 0.5 billion humans
this century in a worsening Climate Genocide in which about 10
billion people (including 2 billion Indians) will die prematurely
this century at an average of 100 million deaths per year [29].

The dreadful Kerala floods (350 people dead, and about 1
million people displaced) have been impacted by (a) global
warming-enhanced precipitation and (b) inadequate hazard
response. Thus the BBC reports: "Officials and experts have
said the floods in Kerala would not have been so severe if
authorities had gradually released water from at least 30 dams.
The state has 44 rivers flowing through it" [30]. The Kerala
disaster was not natural – it was man-made. However the
present Kerala flood disaster masks a worsening man-made
disaster for populous India involving a weakening summer
monsoon, a worsening rainfall deficit and drought in South
Asia. The fundamental problem is inadequacy of hazard
response in India and the world as a whole.

Professor Hans Joachim Schellnhuber (founding director of the
Potsdam Institute for Climate Impact Research, and a senior
advisor to Pope Francis, German Chancellor Angela Merkel and
the European Union) has bluntly stated: "Climate change is now
reaching the end-game, where very soon humanity must choose
between taking unprecedented action, or accepting that it has
been left too late and bear the consequences" [31]. It is now
effectively too late to avoid a catastrophic plus 2C temperature

rise [32-35] but we are all obliged to do everything we can to make the future "less bad" for ourselves, our children and future generations.

References.

[1]. Gideon Polya, "Mendacious Mainstream Presstitutes Ignore Huge Carbon Debt & Horrendous Hazard Response Debt: Disasters Are Not Natural", Countercurrents, 13 August 2018: https://countercurrents.org/2018/08/13/mendacious-mainstream-presstitutes-ignore-huge-carbon-debt-horrendous-hazard-response-debt-disasters-are-not-natural/.
[2]. Gideon Polya, "Revised Annual Per Capita Greenhouse Gas Pollution For All Countries – What Is Your Country Doing?", Countercurrents, 6 January, 2016: https://countercurrents.org/polya060116.htm.
[3]. Gideon Polya, "Exposing And Thence Punishing Worst Polluter Nations Via Weighted Annual Per Capita Greenhouse Gas Pollution Scores", Countercurrents, 19 March, 2016: https://countercurrents.org/polya190316.htm.
[4]. "Carbon Debt Carbon Credit": https://sites.google.com/site/carbondebtcarboncredit/.
[5]. James Hansen, "Letter to PM Kevin Rudd by Dr James Hansen", 2008: http://www.aussmc.org.au/documents/Hansen2008LetterToKevinRudd_000.pdf.
[6]. Chris Hope, "How high should climate change taxes be?", Working Paper Series, Judge Business School, University of Cambridge, 9.2011: http://www.jbs.cam.ac.uk/fileadmin/user_upload/research/workingpapers/wp1109.pdf.
[7]. "List of countries by GDP (nominal) per country", Wikipedia: https://en.wikipedia.org/wiki/List_of_countries_by_GDP_(nominal)_per_capita.
[8]. Chetan Chauhan and Ramesh Babu, "For Kerala's flood disaster, we have ourselves to blame", Hibndustan Times, 17 August 2018: https://www.hindustantimes.com/india-news/what-is-behind-the-kerala-monsoon-fury/story-2NxvHfTDAmS10k9hHofiiO.html.
[9]. Gideon Polya, "Body Count. Global avoidable mortality since 1950", that includes a succinct history of every country and is now available for free perusal on the web: http://globalbodycount.blogspot.com/.
[10]. Gideon Polya, "4 % Annual Global Wealth Tax To Stop The 17 Million Deaths Annually", Countercurrents, 27 June, 2014: http://www.countercurrents.org/polya270614.htm.
[11]. "1% ON 1%: one percent annual wealth tax on One Percenters": https://sites.google.com/site/300orgsite/1-on-1.

[12]. Quirin Schiermeier, "Droughts, heat waves and floods: how to tell when climate change is to blame" ", Scientific American, 30 July 2018: https://www.scientificamerican.com/article/droughts-heat-waves-and-floods-how-to-tell-when-climate-change-is-to-blame/.

[13]. Andrew King and Professor David Karoly, "How we can link some extreme weather to climate change', Pursuit, 18 March 2016: https://pursuit.unimelb.edu.au/articles/how-we-can-link-some-extreme-weather-to-climate-change.

[14]. US National Academies of Sciences, Engineering, and Medicine, "Attribution of extreme weather events in the context of climate change:, Washington, DC: The National Academies Press, 2016: http://www.nap.edu/catalog/21852/attribution-of-extreme-weather-events-in-the-context-of-climate-change?utm_source=Division+on+Earth+and+Life+Studies&utm_campaign=7c39de3f77-BASC_PRB_OSB_Extreme_Events3_11_2016&utm_medium=email&utm_term=0_3c0b1ad5c8-7c39de3f77-233946077&mc_cid=7c39de3f77&mc_eid=a55d8dfb66.

[15]. Giuling Wang et al., "The peak structure and future changes of the relationships between extreme precipitation and temperature", Nature Climate Change volume 7, pages 268–274, 2017: https://www.nature.com/articles/nclimate3239.

[16]. John Abraham, "Global warming is increasing rainfall rates", The Guardian, 22 March 2017: https://www.theguardian.com/environment/climate-consensus-97-per-cent/2017/mar/22/global-warming-is-increasing-rainfall-rates.

[17]. Neha Madaan, "Warming Indian Ocean impedes rains: Indian Institute of Tropical Meteorology", Times of India, 17 June 2015: https://timesofindia.indiatimes.com/india/Warming-Indian-Ocean-impedes-rains-Indian-Institute-of-Tropical-Meteorology/articleshow/47698910.cms.

[18]. Mathew Koll Roxy, "The curious case of Indian Ocean warming", Journal of Climate, 4 November 2014: https://journals.ametsoc.org/doi/10.1175/JCLI-D-14-00471.1.

[19]. Mathew Koll Roxy, Kapoor Ritika, Pascal Terray, Raghu Murtugudde, Karumuri Ashok, & B. N. Goswami, "Drying of Indian sub-continent by rapid Indian Ocean warming and a weakening land-sea thermal gradient", Nature Communications, volume 6, Article number: 7423, 2015: https://www.nature.com/articles/ncomms8423.

[20]. Raghu Murtugudde, "Western Ghat's biodiversity is a major source of moisture for monsoon", The Hindu Business Line, 4 May 2018: https://www.thehindubusinessline.com/news/science/western-

ghats-biodiversity-is-a-significant-source-of-moisture-for-monsoon/article23772839.ece.

[21]. Climate Council, "Fact sheet: climate change and drought", June 2018: https://www.climatecouncil.org.au/wp-content/uploads/2018/06/CC_MVSA0146-Fact-Sheet-Drought_V2-FA_High-Res_Single-Pages.pdf.

[22]. Agence France Presse, "India records its hottest day ever as temperature hits 51C (that's 123.8F)", The Guardian, 20 May 2016: https://www.theguardian.com/world/2016/may/20/india-records-its-hottest-day-ever-as-temperature-hits-51c-thats-1238f

[23]. BBC News, "Anger over India's "poor response" to heatwave crisis", 27 May 2015: https://www.bbc.com/news/world-asia-india-32895602

[24]. "Stop air pollution deaths": https://sites.google.com/site/300orgsite/stop-air-pollution-deaths.

[25]. Gideon Polya, "Latest lancet data imply Adani Australian coal project will eventually kill 1.4 million Indians", Countercurrents, 21 April 2017: https://countercurrents.org/2017/04/21/latest-lancet-data-imply-adani-australian-coal-project-will-kill-1-4-million-indians/.

[26]. Cynthia Rosenzweig & Martin L. Parry, "potential impact of climate change on world's food supply", Nature volume 367, pages 133–138, 13 January 1994: https://www.nature.com/articles/367133a0.

[27]. Gideon Polya, "Jane Austen and the Black Hole of British History. Colonial rapacity, holocaust denial and the crisis in biological sustainability", G.M. Polya, Melbourne, 1998 and 2008, that is now available for free perusal on the web: http://janeaustenand.blogspot.com/.

[28]. Ian Chambers and John Humble, "Plan for the Planet: A Business Plan for a Sustainable World", Gower Green Economics and Sustainable Growth Series, 2007.

[29]. "Climate genocide": https://sites.google.com/site/climategenocide/.

[30]. Navin Singh Khadka, "Why the Kerala floods proved so deadly", BBC News, 21 August 2018: https://www.bbc.com/news/world-asia-india-45243868.

[31]. David Spratt and Ian Dunlop, "Take unprecedented action or bear the consequences, says eminent scientist and advisor", Renew Economy, 20 August 2018: https://reneweconomy.com.au/take-unprecedented-action-or-bear-the-consequences-says-eminent-scientist-and-advisor-45081/.

[32]. "Too late to avoid global warming catastrophe": https://sites.google.com/site/300orgsite/too-late-to-avoid-global-warming.

[33]. "Are we doomed?": https://sites.google.com/site/300orgsite/are-we-doomed.

[34]. "Methane Bomb Threat": https://sites.google.com/site/methanebombthreat/.

[35]. "Nuclear weapons ban, end poverty and reverse climate change": https://sites.google.com/site/drgideonpolya/nuclear-weapons-ban.

CHAPTER 8: WATER CRISIS, GLOBAL AVOIDABLE MORTALITY HOLOCAUST, WATER APARTHEID, GLOBAL WARMING & MINA GULI

[First published as Gideon Polya, "Water crisis, Global Avoidable Mortality Holocaust, Water Apartheid, Global Warming & Mina Guli", Countercurrents, 17 May 2019: https://countercurrents.org/2019/05/water-crisis-global-avoidable-mortality-holocaust-water-apartheid-global-warming-mina-guli.]

The world faces a worsening global water crisis that is compounded by remorselessly increasing carbon pollution, population, and economic output. For the 5.0 billion people of the impoverished global South (the Developing World minus China) the worsening water crisis is a present reality. However endless growth in a carbon economy means that all 7.6 billion of Humanity are threatened. Dire warning is now being given by the need for bottled potable water in some towns in the rich global North. Unfortunately rational and humane solutions are being trumped by ruthless and deadly neoliberalism.

The UN World Water Assessment Programme (WWAP) has summarized Humanity's dilemma thus (2018): "The stakes are high. Current trends suggest that around two thirds of forests and wetlands have been lost or degraded since the beginning of the 20th century. Soil is eroding and deteriorating in quality. Since the 1990s, water pollution has worsened in almost all rivers in Africa, Asia and Latin America. These trends pose broader challenges from the increased risk of floods and droughts, which, in turn, has an impact on our ability to adapt to climate change. We know also that water scarcity can lead to civil unrest, mass migration, and even to conflict within and between countries" [1].

The worsening global water crisis has many major elements that are systematically and succinctly dealt with in this well-documented essay under 30 headings as set out below.

(1) Global fresh water resources. 2.00-2.75% % of the Earth's water is fresh water (less than 500 ppm NaCl) including about 1.75-2.00% as glaciers, ice or snow, 0.5–0.75% fresh groundwater and soil moisture, less than 0.01% surface water in rivers, swamps and fresh water lakes (lakes constituting 87% of liquid fresh surface water), and less than 0.04% in the atmosphere [2]. Humanity derives nearly all of its fresh water (about 4,600 km^3 pa [1]) from rain, lakes, rivers and groundwater [2, 3]. There is a hydrological cycle of solar energy-driven evaporation with subsequent precipitation.

(2). Desalination by reverse osmosis. Reverse osmosis, the most widely used desalination technique, involves pumping water through a succession of membranes that allow passage of water (H_2O) but not of larger molecules [3]. The world's largest reverse osmosis desalination plant is at Sorek, Israel, has an output of 624,000 cubic metre per day that is sold for about $0.5 per cubic metre ($500/ML i.e. $500 per million Litres), and provides potable (drinkable) water for over 1.5 million people, this constituting 20% of the municipal water demand in Israel [4, 5]. In 2015 there were 18,426 desalination plants globally, producing 86.8 million cubic meters per day of potable water for 300 million people [6]. Reclamation of used water for sanitation and agricultural use is now widespread [7].

(3) Water use. Global water demand (2018) is 4,600 km^3 [cubic kilometers] per year and projected to increase by 20%–30% to 5,500-6,000 km^3 per year by 2050. Agriculture uses about 70% of global water withdrawals (i.e. about 3,220 km^3 pa), with most being used for used for irrigation. Industrial water use is about 20% of global withdrawals (i.e. about 920 km^3), 75% for energy production (690 km^3 pa) and 25% for manufacturing (230 km^3). Water demand for manufacturing will increase by 400% over the period 2000–2050. Global water withdrawals for energy production have been projected to rise by 20% in the period 2010–2035. Domestic water use is 10% of total global water withdrawals (i.e. about 460 km^3), and is expected to increase greatly over the 2010–2050 period (3-fold in Africa, 2-fold in Latin America, but with little change in zero population growth Western Europe) [1].

The UN World Water Assessment Programme (WWAP) (2018): "Throughout the early-mid 2010s, about 1.9 billion people (27% of the global population) lived in potential severely water-scarce areas and in 2050 this could increase to some 2.7–3.2 billion. However, if monthly variability is taken into account, 3.6 billion people worldwide (nearly half the global population) are already living in potential water-scarce areas at least one month per year and this could increase to some 4.8–5.7 billion in 2050. About 73% of the affected people live in Asia (69% by 2050). Factoring in adaptive capacity, 3.6–4.6 billion people (43–47%) will be under water stress in the 2050s with 91–96% living in Asia, mainly Southern and Eastern, and 4–9% in Africa, mainly in the north... Groundwater use globally, mainly for agriculture, amounts to 800 km^3 per year in the 2010s, with India, the United States of America (USA), China, Iran and Pakistan (in descending order) accounting for 67% of total abstractions worldwide … Water withdrawals for irrigation have been identified as the primary driver of groundwater depletion worldwide... A large surge in groundwater abstractions amounting to 1,100 km^3 per year has been predicted to occur by the 2050s, corresponding to a 39% increase over current levels... at about 4,600 km^3 per year, current global withdrawals are already near maximum sustainable levels... A third of the world biggest groundwater systems are already in distress... An estimated 80% of all industrial and municipal wastewater are released to the environment without any prior treatment, resulting in a growing deterioration of overall water quality with detrimental impacts on human health and ecosystems. Globally, the most prevalent water quality challenge is nutrient loading, which depending on the region is often associated with pathogen loading" [1].

The enormous, quasi-linear increase in world water use in the period 1900-2000 has been graphically presented by the World Water Council, with usage dominated by agricultural use, thence much lesser industrial use and then relatively minor municipal use [8]. As outlined later below, water use is already severely stressing reservoirs, aquifers and rivers. Thus by 2025 1.8 billion people will experience absolute water scarcity, 60% of European cities with populations greater than 0.1 million are using groundwater faster than the replenishment rate, and in the last 20 years 55% of China's rivers have run dry due to industrial over-exploitation [9].

(4) Clean drinking water. In 2015 89% of people globally had access to potable water, 40-80% of sub-Saharan Africans had access to potable water, nearly 4.2 billion people worldwide had access to tap water, and 2.4 billion had access to wells or public taps. About 1- 2 billion people presently lack safe drinking water, this being associated with avoidable deaths of 1.5 million people each year [10]. By 2025 1.8 billion people will experience absolute water scarcity [9].

(5) Sanitation. According to the World Health Organization (WHO) (2018): "In 2015, 39% of the global population (2.9 billion people) used a safely managed sanitation service – defined as use of a toilet or improved latrine, not shared with other households, with a system in place to ensure that excreta are treated or disposed of safely; 27% of the global population (1.9 billion people) used private sanitation facilities connected to sewers from which wastewater was treated; 13% of the global population (0.9 billion people) used toilets or latrines where excreta were disposed of in situ; 68% of the world's population (5.0 billion people) used at least a basic sanitation service; 2.3 billion people still do not have basic sanitation facilities such as toilets or latrines; of these, 892 million still defecate in the open, for example in street gutters, behind bushes or into open bodies of water; at least 10% of the world's population is thought to consume food irrigated by wastewater; poor sanitation is linked to transmission of diseases such as cholera, diarrhoea, dysentery, hepatitis A, typhoid and polio; inadequate sanitation is estimated to cause 280,000 diarrhoeal deaths annually and is a major factor in several neglected tropical diseases, including intestinal worms, schistosomiasis, and trachoma. Poor sanitation also contributes to malnutrition. Hygienic sanitation facilities are crucial for public health. Since 1990, the number of people gaining access to improved sanitation has risen from 54% to 68% but some 2.3 billion people still do not have toilets or improved latrines" [11]. India has embarked on a massive increase in sanitation in which 0.5 million villages been declared Open Defecation Free (ODF) in the past 5 years. PM Modi has stated (2019): "Sanitation coverage has crossed 98 percent, over 100 million toilets have been built for our people" [12].

(6) Water deficits, the Global Avoidable Mortality Holocaust and crippling Carbon Debt. An estimated 15 million people die avoidably from deprivation each year in a Global Avoidable Mortality Holocaust [13] that is set to worsen as the climate emergency [14-16] and climate genocide [17] both worsen. That deadly deprivation includes lack of water for basic food production, lack of clean drinking water, and lack of sanitation. About 1 million people die from climate change each year with too little water (droughts) or too much water (floods and storm surges) being key agents. There is a worsening climate genocide in which the direst estimates are for 10 billion deaths this century in the absence of requisite action on climate change [17]. The damage-related cost of carbon pollution has been estimated at about $200 per tonne CO_2-equivalent (CO_2-e) as determined by climate economist Dr Chris Hope from 118 Nobel Laureate Cambridge University [18, 19] and independently by leading climate scientist Professor James Hansen from 96 Nobel Laureate Columbia University [18, 20]. This huge damage-related carbon pollution cost means that the world has an inescapable Carbon Debt of $200- $250 trillion that is increasing at $19 trillion per year [18], with this inevitably impacting the capacity of the world to realistically tackle climate change.

(7) Aquifer depletion and toxic contamination. Groundwater aquifers are major sources of usable fresh water [21] but major problems are aquifer depletion, salinization of coastal aquifers due to lowering of the water table, high saline and other mineral content, and contamination of groundwater (notably by toxic free arsenic arising from anaerobic bacterial action, this being a major problem in South Asia, South East Asia, East Asia and the US) [22-25]. Aquifer depletion has been cited by Professor Lester Brown as one of the causes of food price rises (2011): "Meanwhile aquifer depletion is fast shrinking the amount of irrigated area in many parts of the world; this relatively recent phenomenon is driven by the large-scale use of mechanical pumps to exploit underground water. Today, half the world's people live in countries where water tables are falling as overpumping depletes aquifers. Once an aquifer is depleted, pumping is necessarily reduced to the rate of recharge unless it is a fossil (nonreplenishable) aquifer, in which case pumping ends altogether. But sooner or later, falling water tables translate into rising food prices… The Arab Middle East is the first geographic region where spreading water shortages are shrinking the grain harvest. But the really big water deficits are in India, where the World Bank numbers indicate that 175 million people are being fed with grain that is produced by overpumping. In China, overpumping provides food for some 130 million people. In the United States, the world's other leading grain producer, irrigated area is shrinking in key agricultural states such as California and Texas… It is no longer conflict between heavily armed superpowers, but rather spreading food shortages and rising food prices — and the political turmoil this would lead to — that threatens our global future. Unless governments quickly redefine security and shift expenditures from military uses to investing in climate change mitigation, water efficiency, soil conservation, and population stabilization, the world will in all likelihood be facing a future with both more climate instability and food price volatility. If business as usual continues, food prices will only trend upward" [26].

60% of European cities with populations greater than 0.1 million are using groundwater faster than the replenishment rate [9]. India (population 1.34 billion) illustrates the problem for the global South. Thus Caleb Gorton (Global Food and Water Crises Research Programme): "India is the world's highest user

of groundwater. It consumes over a quarter of the global total –
equivalent to 230 cubic kilometres per year. Groundwater from
over 30 million access points supplies 85 per cent of drinking
water in rural areas and 48 per cent of water requirements in
urban areas. Most groundwater is used for irrigation, which
accounts for 88 per cent of total groundwater usage.
Groundwater is required for the daily needs of around 700
million Indians living in the country's villages. An assessment
of 6,607 groundwater units in 2011 found that 1,017 were
"overexploited", indicating the rate of groundwater extraction
exceeded replenishment. Around one-third of all units in India
were under stress. The World Bank predicts that by 2032,
around 60 per cent of aquifers in the country will be in a critical
state" [27].

In Palestine, Apartheid Israel controls water supply, notably that
from the largely West Bank-based Mountain Aquifer with 8.9
million Israelis getting 87% of Mountain Aquifer water whereas
5 million Occupied Palestinians get a mere 13%. While the
WHO minimum daily per capita water allocation is 100 litres,
Israelis get 240-300 litres and West Bank Palestinians get 73
litres. Israelis have deliberately demolished 50 water extraction
facilities in the West Bank and by bombing have destroyed
water and sewerage infrastructure in the Gaza Concentration
Camp to the point that it is approaching unliveable conditions
[28- 30]. In relation to this massive and deliberately imposed
domestic water deficit, a serial war criminal Apartheid Israel is
grossly violating Articles 55 and 56 of the Geneva Convention
Relative to the Protection of Civilians in Time of War that state
that an Occupier must provide the Conquered Subjects with life
sustaining food, medical and other conditions "to the fullest
extent of the means available to it" [31]. Apartheid Israel is
stealing Mountain Aquifer water that derives from rain falling
on the West Bank and in doing so is violating the Hague
Regulations of 1907, which prohibit an occupying power from
expropriating the resources of occupied territory [30].

**(8). Crop agriculture via drip irrigation and closed
greenhouses.** Agriculture consumes 70% of global water
withdrawals [1] but there is massive water loss associated with

evaporation. Fertilizer and water loss can be greatly decreased through plastic-based drip irrigation systems developed initially in the US and Israel. In California there was 5% drip irrigation in 1988 and 40% by 2010 [32]. Of massive potential are closed system glass houses for hydroponic horticulture [33].

(9). Evaporation. Massive water loss occurs through evaporation. In 1995 evaporative losses from reservoirs totalled 200 kilometer^3 and of the 10% of water from renewable resources withdrawn for human use, about 50% was lost through run off and evaporation. Depending on the technology, consumption can range from 30–40% for flood irrigation to 90% for drip irrigation. In 1995 11% of water withdrawn for agricultural use was lost through evaporation [8]. In hot countries like Australia over 33% of stored water can be lost by evaporation from surface storages, with this leading to proposals for underground storage of surplus water in aquifers [34]. Evaporation from surface dams can be reduced by 80% through use of six-sided floating disks that spread out to cover water surfaces and also cut down on algal growth [35].

(10) Salinization. The World Water Council (2001): "Unless carefully managed, irrigated areas risk becoming waterlogged and building up salt concentrations that could eventually make the soil infertile. This process probably caused the downfall of ancient irrigation-based societies and threatens the enormous areas brought under irrigation in recent decades. By the late 1980s an estimated 50 million hectares of the world's irrigated areas, or more than 20%, had suffered a buildup of salts in the soil" [8].

10. Cost of water. The cheapest water in the world is that from the Mountain Aquifer in Palestine that is simply expropriated (stolen) by the international law-violating, war criminal Occupier, Apartheid Israel, for $0.0 per megalitre (ML, million litres) [30]. The most expensive water is bottled water which costs about $1 per litre or $1,000,000/ML in a Western supermarket or $200,000/ML in India. However there are numerous other rates in between these extremes.

Thus, for example, the Murray-Darling Basin in south eastern Australia is a major irrigation area and "Australia's food bowl". However overuse of water has resulted in decreased flows in the Murray River and the Darling River, with the latter now reduced to a succession of dead pools. A politically contentious Federal Murray-Darling Plan has involved "water buy-back" for the environment from users whose forebears (like the Israelis today and their ongoing Palestinian Genocide) had stolen the land and the water rights from Indigenous people in a 2-century Australian Aboriginal Genocide [36]. Reviewers of the scheme stated (2018): "Overall, the average cost of water recovery by RTB [Restoring The Balance] since 2007–2008 was A\$2,026/ML [US\$1,418/ML], while infrastructure cost was A\$4,970/ML [US\$3,479/ML]. But the costs of water recovery from subsidies from infrastructure could, in fact, be many times more costly per megaliter of water recovered than these figures suggest because of the decline in return flows associated with upgrades in infrastructure" [37]. In a worsening "Watergate scandal", a corporate entity with links to the National Party (the ecocidal farmer's party that is part of the Australian Coalition Government) and with corporate structures involving the Cayman Islands tax haven, sold 28,700 ML of "overflow" surface water to the Australian government for A\$78.9 million, this yielding a water price of A\$2,749 /ML (US\$1.924/ML) (this "overflow" water from rare excessive precipitation would normally have flowed to downstream creeks, rivers and users, but was retained by artificial levees for private profit [38], and one notes that the cost of expensively desalinated water is about \$500/ML [4, 5]).

11. Water Apartheid in Apartheid Israel and North versus South. While the state-of-the-art Sorek desalination plant in Apartheid Israel supplies high quality drinking water at \$500/ML, the Occupied Palestinians have to pay huge prices for dangerously insufficient and dangerously low quality water as revealed by the World Bank (2009): "[West Bank] Water prices that before the Intifada were generally in the range 5-10 NIS/m^3 were now typically in the range 10-20 NIS/m^3 [\$2,800 – \$5,600/ML; 1 New Israeli Shekel = 0.28 USD; 1 m^3 = 1000 L]… There is a high incidence of water related diseases. Water-borne disease is a major problem for Palestinians, creating substantial costs and losses ... The health impacts can be gauged by the high incidence of diarrhoea amongst infants, and the health costs of poor water and sanitation services have been estimated at 0.4% of GDP ... In Jenin, farmers are paying up to NIS $12/m^3$ [$3,360/ML] for water to irrigate plastic houses... These prices are considerably higher than those paid by (competing) Israeli farmers, who pay less than 1 NIS/m^3 (0.818 NIS/m^3) [$229/ML]... [In Gaza] between 5% and 10% of water supplied through the network meets potable standards" [39].

This deadly, Zionist-imposed Water Apartheid is a direct consequence of the Zionist-imposed political Apartheid in Palestine. Of about 14 million Palestinians (50% children, 75% women and children), 7 million are forbidden to even step foot in their own country, 5 million are highly abusively imprisoned in the blockaded and bombed Gaza Concentration Camp (2.0 million) or in ever-dwindling West Bank ghettoes (3.0 million), and 1.9 million live as Third Class citizens as Israeli Palestinians under over 60 Nazi-style Apartheid Israeli race laws. GDP per capita is $3,000 for Occupied Palestinians as compared to $40,000 for Israelis. Through imposed deprivation, each year Apartheid Israel passively murders about 2,700 under-5 year old Palestinian infants and passively murders 4,200 Occupied Palestinians in general who die avoidably from deprivation each year under Israeli Apartheid (in the 21st century Israelis have also violently killed an average of 550 Occupied Palestinians each year). There is a circa 10 year life expectancy gap between Occupied Palestinians and Israelis, this grossly violating Articles 55 and 56 of the Geneva Convention Relative to the Protection of Civilian Persons in Time of War that demand that an Occupier must provide life-sustaining food and medical services to the Occupied "to the fullest extent of the means available to it". Nuclear terrorist, serial war criminal, genocidally racist, democracy-by-genocide Apartheid Israel determines that 72% of its now 50% Indigenous Palestinian subjects who are Occupied Palestinians cannot vote for the government ruling them i.e. egregious Apartheid [40-44].

While the US lackey and Zionist -subverted West
largely ignores these horrible Zionist crimes, numerous anti-
racist Jewish and non-Jewish humanitarians endlessly protest
the gross violation of Palestinian human rights by Apartheid
Israel in its ongoing Palestinian Genocide [45-53]. Of course the
same moral indignation in decent people is also elicited by the
desperate lot of 1 million Rohingya refugees [54], of 70 million
refugees worldwide [55], and of 5.0 billion people in the
Developing World (minus China) who are suffering deprivation
in general and variously suffering deadly deprivation in relation
to water quantity and quality [13]. Thus 15 million people die
avoidably from deprivation each year on Spaceship Earth with
One Percenters in charge of the flight deck – deadly Apartheid
indeed, with Water Apartheid being a major factor and directly
causing about 10% of these deaths [10].

12. Per capita water use by country. There is wide variation in
per capita water use for domestic, industrial and agricultural
purposes [56]. However a stand-out is Australia at 5,104 m^3 per
person per year as compared to 1,600 (the US), and 305
(Apartheid Israel) [56]. Under Apartheid Israeli occupation the
Occupied Palestinians score a deadly 91.5 m^3 per person per
year [57] (an Olympic swimming pool contains 2,500,000 L
or 2,500 m^3 of water, and 1 m^3 is equivalent to 3.2 forty-four
gallon drums).

13. Aqua nullius and terra nullius. Just as the genocidally
racist Zionists stole all of Palestine and all of its surface and
underground water, so the genocidally racist British invaders
did the same to Indigenous Australians. Thus R. Quentin
Grafton and Sarah Jane Wheeler (2018): "[Re] Australia's First
Peoples, water is a part of everything in their culture, such as
their dreaming stories, art, songs, and dance, and there is
evidence that they have been managing water in the MDB
[Murray-Darling Basin] for perhaps as long 40,000 years.
Indeed, the Brewarrina fish traps (Baiame's Ngunnhu) on the
Barwon River are possibly the oldest continuously used human
water construction in the world… By comparison, European
settlers typically viewed water as a resource to be used,
manipulated, and harnessed for economic benefits. These two

vastly different worlds have often clashed. Along with the legal fiction of terra nullius, there was also the myth of "aqua nullius" that rendered existing Indigenous water relationships invisible" [37]. Capitalist greed by the descendants of the genocidal British invaders has meant that the Darling River has now been reduced to a succession of putrid water holes [58]. For the dispossessed and impoverished survivors of the Aboriginal Genocide along the Darling River and its tributaries, the Barwon and Namoi Rivers (that meet at the town of Walgett that means "the meeting of the two waters"), there is no water to drink, fish or swim in [58-60].

14. Water footprint for meat. The water footprint in litres per kilogram (L/kg) is as follows for milk (1,020), eggs (3,265), chicken meat (4,325), butter (5,553), pig meat (5,988), sheep or goat meat (8,763) and beef meat (15,415) [61, 62]. In contrast, the water footprint for vegetables and potatoes is 300-400 L/kg [62]. Animal-based food is also very expensive in terms of conversion of plant material to animal meat. Thus consider the following conversion efficiencies (kg grain to produce 1 kg gain in live weight): herbivorous farmed fish (e.g. carp, tilapia, catfish; less than 2), chicken (2), pork (4), and beef (7) [63]. Annual per capita meat consumption (in kg per person per year) [64] very roughly correlates with annual per capita income (in USD per person per year) (UN, 2014) [65, 66] (while noting the exceptions of livestock-based but impoverished Mongolia and the fishing-rich but impoverished Island States of Saint Lucia, Saint Vincent and the Grenadines and Samoa).

Annual global greenhouse gas (GHG) emissions presently total 64 billion tonnes CO_2-equivalent with methanogenic livestock production and associated land use contributing over 50%. With annual emissions of CO_2 per se from industry at a record high, grossly insufficient global climate change action, and with Humanity and the Biosphere existentially threatened by the Methane Bomb of the warming Arctic in coming decades [67], Humanity must urgently adopt a vegetarian or vegan diet to help save the Planet [66]. Several recent reports collectively endorsed by thousands of expert scientists have warned the world that time is running out to save Humanity and the

Biosphere from further catastrophic climate change and further massive biodiversity loss [68-75]. Massive harm has already occurred due to continuing carbon pollution, population growth and economic growth, and it is clear that zero growth in these areas is insufficient – there must be negative carbon pollution (atmospheric CO_2 draw-down to about 300 ppm CO_2) [76-80], negative population growth (50% population decline) [81], and negative economic growth (50% de-growth with most of the burden in Developed countries) [71] to halt and reverse this worsening disaster.

Much of eastern Australia has recently been in the grip of years of severe drought, and sheep and cattle production have suffered in huge swathes of country without a blade of grass. Yet neoliberal governments have moved to support continued methanogenic livestock production that is immensely environmentally destructive. The Indigenous Australians had a 65,000 years old farming civilization based on use of fire to encourage pasture for kangaroos and other soft-footed macropod marsupials, with areas of strict conservation as emergency resources in times of drought. The British invaders actively and passively exterminated most of the Indigenous people in an ongoing Aboriginal Genocide and Aboriginal Ethnocide, and engaged in massive ecosystem destruction (ecocide) and speciescide. The hard hooves of the ungulate sheep and cattle compacted soil and introduced feral animals ([horses, buffalo,] rabbits, rats, cats, dogs, foxes, camels, cane toads and pigs) wiped out much of the Indigenous fauna. Professor Jared Diamond in his seminal book "Collapse" described this carnage as "mining the land" and predicted that Australia would eventually become a food importer [82]. Australia is among world leaders in 14 areas of environmental destruction and climate criminal activities or parameters, to whit: (1) annual per capita greenhouse gas pollution, (2) live methanogenic livestock exports, (3) natural gas exports, (4) recoverable shale gas reserves that can be accessed by hydraulic fracturing (fracking), (5) coal exports, (6) land clearing and deforestation, (7) speciescide – species extinction, (8) coral reef destruction, (9) whale killing and extinction threat through global warming, (10) terminal carbon pollution budget exceedance, (11) per capita carbon debt, (12) GHG generating iron ore exports, (13) climate change inaction, and (14) climate genocide [83, 84].

15. Crop agriculture water footprint. M.M. Mekonnen and A.Y. Hoekstra (2010): "The water footprint of a product is defined as the total volume of freshwater that is used to produce the product… The bluewater footprint refers to the volume of surface and groundwater consumed (evaporated) as a result of the production of a good; the greenwater footprint refers to the rainwater consumed. The grey water footprint of a product refers to the volume of freshwater that is required to assimilate the load of pollutants based on existing ambient water quality standards" [85]. The water footprint in litres per kilogram (L/kg) is as follows for various food crops: sugar (197), vegetables (322), starchy roots (387), fruits (962), cereals (1,644), oil crops (2,364), pulses (4,055) and nuts (9,063) [62, 85].

The water footprint is 9,113 L/kg for cotton lint and 9,982 L/kg for cotton textile [85]. Australia is heavily involved in arid zone cotton production and in cotton export that amounts to an absurd export of water from the world's most arid inhabited continent. Upstream irrigation for cotton has massively contributed to the near- death of the Darling River that is at the heart of the Murray-Darling Basin that is "Australia's food bowl". Neoliberal greed and corruption are at the core of this terracidal disaster. Neoliberalism demands maximal freedom for the rich and advantaged to exploit natural and human resources for maximum private profit. Conversely, social humanism (socialism, democratic socialism, eco-socialism, the welfare state) aims to sustainably maximize happiness, opportunity and dignity for everyone through evolving intra-national and international social contracts [86, 87].

16. Manufacturing industry water footprint. Water Footprint: "The average Briton drinks about 2 to 5 litres of water each day, and then uses another 145 or so litres for cooking, cleaning, washing, and flushing. Does that sound like much to you? Now multiply that figure by 23, and you have just uncovered a hidden truth: the average Briton really consumes about 3,400 litres of water every day – well over a million litres a year!… Embedded water refers to the amount of water necessary to produce a product. It takes about 140 litres to grow one cup of coffee; about 11,000 litres to produce a pair of jeans; and about 400,000 litres to build a car. When we add to the amount of water running from our taps the amount hidden in everything we consume, our true water consumption is exposed – about 3,400 litres per Briton per day" [88]. Between 2000 and 2050 manufacturing industry use of water will increase by 400% [9].

17. Deforestation. Forests have major roles in the hydrological cycle (involving transpiration, water evaporation and subsequent water precipitation) [89] and the carbon cycle (involving photosynthetic CO_2 fixation in cellulosic and non-cellulosic carbohydrates and subsequent CO_2 release through oxidation of carbohydrates) [90-93]. Trees consume water through photolysis, release water through respiration, and move water from roots to escape via leaf stomata through transpiration. Trees also retain water in living cells and in the soil through generation of water-retaining soil carbon. Water from transpiration in the Amazon rainforest is so large as to affect continent-scale weather and precipitation. Deforestation and land clearing (most notably in countries like Brazil, Australia and Indonesia [94-96]) has a huge impact on the water cycle, the carbon cycle, biodiversity, greenhouse gas (GHG) pollution, global warming, drought, precipitation and soil water retention.

Climate change economist Sir Nicholas Stern critically stated (2006):"The science tells us that GHG emissions are an externality; in other words, our emissions affect the lives of others. When people do not pay for the consequences of their actions we have market failure. This is the greatest market failure the world has seen" [96, 97], and also argued for an international programme to combat deforestation, which contributes 15-20% of greenhouse gas emissions (2006): "For $10-15bn (£4.8-7.2bn) per year, a programme could be constructed that could stop up to half the deforestation" [97].

18. Global warming. Since the beginning of the Industrial Revolution in circa 1750, the atmospheric level of CO_2 has increased from 280 ppm CO_2 to the present circa 410 ppm CO_2. Increases in other GHGs such as methane (CH_4) and nitrous oxide (N_2O) mean that the CO_2-equivalent in the atmosphere is presently about 490 ppm CO_2-e, and the average global surface temperature has increased by about 1 degree Centigrade (1C). This has been associated with the world's Sixth Mass Extinction event in which species extinction is 100-1,000 greater than normal in what is now described as the Anthropocene Era [66-81]. The world already has an inescapable Carbon Debt of $200-250 trillion that – instead of being urgently reduced by massive atmospheric CO_2 draw-down – is relentlessly increasing at about $16 trillion per year [18]. In a domestic household budget setting this would be utter lunacy – in a global setting it is mass suicide of Humanity. This utterly disastrous present scenario means that a catastrophic plus 2C temperature rise is now evidently unavoidable – we have only 10 years at the current rate of GHG pollution before a 50% probability of a catastrophic plus 2C exceedance [72-74]. However we are obliged to do everything we can to make the future "less bad" for future generations. 90% of the extra heat has gone into the oceans [72-74].

As further detailed below, global warming is variously associated with increased drought and forest fires in some regions, and increased evaporation, increased atmospheric H_2O and thence increased precipitation in others. Global warming-driven changes in ocean currents and stratospheric jet streams have paradoxically been associated with colder winters in Europe. In Australia global warming has been associated with increased precipitation in the tropical north and decreased precipitation and drought in southern Australia. Global warming and increased sea temperature in the tropics has been associated with more intense storms (cyclones, hurricanes). Global warming-driven sea level rise is associated with more damaging storm surges impacting coastal parts of Island Nations and of low-lying, mega-delta countries (notably north eastern India and Bangladesh most recently devastated by Cyclone Fani [98]).

19. Floods. Increased ocean temperature leads to increased evaporation and thence increased precipitation, floods and flash floods that are compounded by the effects of deforestation [99-101]. Lin et al. (2018): "Weather extremes have widespread harmful impacts on ecosystems and human communities with more deaths and economic losses from flash floods than any other severe weather-related hazards. Flash floods attributed to storm runoff extremes are projected to become more frequent and damaging globally due to a warming climate and anthropogenic changes" [99]. The recent huge Kerala flood disaster (350 deaths and 1 million people displaced) was not a natural disaster – it was a man-made disaster involving local Indian contributions (inadequate water storage hazard response and landslides from deforestation) and a major international contribution from global warming due to ever-increasing greenhouse gas (GHG) pollution, with the prosperous First World and Anglosphere countries being disproportionately high contributors on a per capita basis [101].

20. Drought and soil loss. Global warming variously impacts drought and precipitation. The overall global drought risk as variously determined from tree rings and meteorological data shows an increase in the period 1900-1950, a paradoxical decrease in the period 1950-1975, and a steady state since 1975. Nevertheless increasing global drought risk is implicit in the data [102, 103]. Thus in specific areas (the Sahel, the Middle East and Southern Australia) drought is having a big impact. The African Sahel has experienced a 50% increase in record dry months in recent decades [104]. Famine from drought is impacting 30 million people from Niger to Yemen [105] and is compounded by the obscene war on starving Somalis and starving Yemenis by the obscenely wealthy US Alliance [106]. Drought in the Middle East has contributed to the Syrian War catastrophe [107] that was also driven by US and Apartheid Israeli desire for fossil fuel resources and hegemony. Loss of vegetation means that topsoil in drought-ravaged South East Australia has been blown in huge dust storms into the Tasman Sea [108]. The Australian Climate Council states that climate change has exacerbated the present disastrous drought, and that Southern Australia is predicted to become drier. The Australian

Murray-Darling Basin, which produces more than 33% of Australia's food, has experienced a 41% decline in streamflow over the past 2 decades [109, 110].

21. Ice sheet melting, glacier melting and sea level rise. Global warming has resulted in a circa 25 cm rise in sea level since 1880 through the effects of thermal expansion and ice melting, with the current rate being about 30 cm per century and a predicted rise to 1 metre by 2100 [111]. There is massive loss of Arctic summer sea ice that is predicted to disappear completely over the next several decades [112]. There is also massive melting of the Antarctic ice sheet with about 159 Gt being lost per year [114]. Greenland contains enough ice that when melted would raise sea levels by 7 metres, and has already lost about 10,000 Gt [115]. Complete melting of the Antarctic ice sheet would increase sea level by 57 metres. Glaciers around the world are melting and western North American glaciers are losing 12.3 Gt annually [115]. 360 billion tonnes of ice melting would increase sea level by 1 mm [115]. In the Arctic there are huge stores of methane (CH_4) that exists with water (H_2O) in clathrate crystal arrays. CH_4 has 105 times the Global Warming Potential (GWP) of CO_2 (on a 20 year time frame with aerosol impacts considered). The predicted released of 50 Gt CH_4 (5,250 Gt CO_2-equivalent) from the Arctic in coming decades means game over for Humanity and the Biosphere [15]. Compounding this nightmare is a positive feedback loop in which warming -> decreased sea ice -> decreased albedo (reflectivity) -> increased light absorbance – > increased warming.

22. Paleoclimatology and sea level. Dr Andrew Glikson (an Earth and paleo-climate research scientist at Australian National University, Canberra, Australia) (2009) has commented thus on the long-term implications of the present CO_2 level rise : "The continuing use of the atmosphere as an open sewer for industrial pollution has already added some 305 GtC to the atmosphere together with land clearing and animal-emitted methane. This raised CO_2 levels to 387 ppm CO_2 to date, leading toward conditions which existed on Earth about 3 million years (Ma) ago (mid-Pliocene), when CO_2 levels rose to about 400 ppm,

temperatures to about 2–3 degrees C and sea levels by about 25
+/- 12 metres" [79, 80, 116].

23. Storm surges, inundation and salinization. 90% of the
extra heat from the greenhouse effect has gone into the oceans.
Ocean surface warming of circa 2C means increased water
evaporation and increased energy of tropical storms associated
with precipitation. Global warming has meant increased energy
of tropical hurricanes (cyclones) [117] that with increased sea
level means inundation of coastal areas with destructive and
deadly sea surges. Higher sea level and storm surges mean
increasing salinization of coastal areas, and consequent decrease
in crucial agricultural productivity. Low-lying tropical Island
Nations and mega-delta regions (e.g. Bengal) are existentially
threatened by sea level rise, high energy hurricanes and storm
surges [17, 118].

**24. Forest fires including tropical and temperate rainforest
combustion.** Global warming and dry conditions have been
associated with increased forest fires around the world, notably
in western North America, southern Europe, Indonesia and
Australia. Climate change is a major contributor to this worldwide conflagration [119-121]. Warmer and drier conditions lead
to a greater probability of undergrowth ignition due to loss of
moisture and humidity. There is a positive feedback loop
associated with this massive ecosystem destruction: greenhouse
gases (GHGs) -> global warming -> dryness -> increased
wildfires -> more CO_2 release -> more warming. Alarmingly,
wet, tropical forests are now burning due to deforestation and
drought. Thus forest fires in the Amazon emit twice the GHGs
due to deforestation [122]. Edges of the extraordinary temperate
rainforests of southwest Tasmania may now be susceptible to
burning due to drought [123].

25. Death of rivers and lakes. In the last 20 years 55% of
China's rivers have run dry due to industrial over-exploitation
[9]. In Australia the Darling, Barwon and Namoi Rivers of the
key food-producing Murray-Darling Basin have become
successions of putrid puddles due to drought and over-
exploitation of scarce riverine water for irrigation agriculture in

a very dry region (notably for cotton production) [58-60]. In Central Asia, the Aral Sea was once the fourth largest lake in the world but due to water use for agriculture (notably for cotton production) has almost disappeared [124, 125]. Salinization of rivers is a major problem across America because of the use of salt (sodium choride, NaCl) to clear roads of snow and ice. In Flint, Michigan, the authorities stopped sourcing water from Detroit and to save money took it from the Flint River. Consequent corrosion of lead piping lead to poisoning of 27,000 children [126, 127], this being reminiscent of the waste methylmercury neurotoxic poisoning of the people of Minamata in Japan [128]. The Jordan River, immortalized in Christianity and in African American spiritual songs, has become an agricultural run-off-, sewerage-, and saline-polluted stream [129], and the Jordan Valley has been largely ethnically cleansed of Indigenous Palestinians by the genocidally racist Zionists [50].

26. Desertification. Drought and over-exploitation of water resources is contributing to the growing cancer of desertification around the world. The regions of high to highest vulnerability to desertification are those contiguous with existing desert regions in the Western USA, Central Asia, the Sahara (North Africa and the Sahel), southern Africa, the Middle East, Australia and parts of South America and South Asia. Desertification is increasing while population and the need for arable land is also increasing. Drylands represent about 40% of the Earth's land area, and are inhabited by about 2 billion people. About 1 billion people are presently threatened by desertification [130].

27. Ecocide, speciescide, omnicide and terracide. The Earth is presently undergoing a 6th mass extinction event in what has been described as the Anthropocene Era [68-71, 131-133]. The Intergovernmental Science-Policy Platform on Biodiversity and Ecosystem Services (IPBES) re Summary of the 2019 Global Assessment Report on Biodiversity and Ecosystem Services (2019): "75%: terrestrial environment "severely altered" to date by human actions (marine environments 66%)… >85%: of wetlands present in 1700 had been lost by 2000 – loss of wetlands is currently three times faster, in percentage terms, than forest loss… 8 million: total estimated number of animal and plant species on Earth (including 5.5 million insect species)… Tens to hundreds of times: the extent to which the current rate of global species extinction is higher compared to average over the last 10 million years, and the rate is accelerating… Up to 1 million: species threatened with extinction, many within decades… >500,000 (+/-9%): share of the world's estimated 5.9 million terrestrial species with insufficient habitat for long term survival without habitat restoration … >40%: amphibian species threatened with extinction… Almost 33%: reef forming corals, sharks and shark relatives, and >33% marine mammals threatened with extinction… 25%: average proportion of species threatened with extinction across terrestrial, freshwater and marine vertebrate, invertebrate and plant groups that have been studied in sufficient detail… 33%: marine fish stocks in 2015 being harvested at unsustainable levels; 60% are maximally sustainably fished; 7% are underfished… >55%: ocean area covered by industrial fishing… 70%: proportion of cancer drugs that are natural or synthetic products inspired by nature" [131].

The summary for policy makers of the global assessment report on the biodiversity and ecosystem services of the Intergovernmental Science-Policy Platform on Biodiversity and Ecosystem Services (IPBES) (2019): "Scenarios project mostly adverse climate change effects on biodiversity and ecosystem functioning, which worsen, in some cases exponentially, with incremental global warming. Even for global warming of 1.5°C to 2°C, the majority of terrestrial species ranges are projected to shrink profoundly. Changes in ranges can adversely affect the capacity of terrestrial protected areas to conserve species, greatly increase local species turnover and substantially increase the risk of global extinctions. For example, a synthesis of many studies estimates that the fraction of species at risk of climate-related extinction is 5 per cent at 2°C warming, rising to 16 per cent at 4.3°C warming. Coral reefs are particularly vulnerable to climate change and are projected to decline to 10-30 per cent of former cover at 1.5°C warming and to less than 1 per cent at 2°C warming. Therefore, scenarios show that limiting global warming to well below 2°C plays a critical role in reducing adverse impacts on nature and its contribution to people" [132].

Warming, acidification and pollution of the oceans is killing the world's coral reefs, complex ecosystems associated with many species. Fertilizer run-off and warming are creating huge dead zones in the oceans. Over-fishing is driving species to extinction. More complex pathways are threatening ocean species. Thus, for example, man-made global warming is decreasing Antarctic sea ice and hence the substratum for photosynthetic algae that are consumed by krill, with the consequent decline in krill impacting krill-eating species such as whales. Massive land clearing, forest fires, desertification, wetland "reclamation" and urbanization are destroying habitats, ecosystems and species. Violation of water systems is variously associated with ecocide and speciescide, with this putting the planet on a path towards omnicide and terracide.

28. Water theft and water wars. The Palestinian Genocide (commonly described by the Mainstream media as the "Israeli-Palestinian Conflict") has been about land, fossil fuel and water resources as well as race-based colonization and ethnic cleansing [39-55]. Drought, displacement and disaffection had a major role in the related US Alliance- and Apartheid Israel-backed Syrian Civil War (aka Syrian Holocaust and Syrian Genocide) [54, 55]. Major potential conflict arises from multi-national upstream exploitation and damming of major rivers such as the Nile, the Indus, the Ganges and the Mekong.

29. Australian water and environmental lunacy. Australia is the driest inhabited continent but is profligate in the use of water and in the effective export of water through export of goods with a high water footprint (e.g. cotton and cattle). The worsening water crisis and the worsening climate emergency are intimately connected. Rich Australia is among world leaders for the following 14 climate criminal activities or parameters: (1) annual per capita greenhouse gas pollution [134-136], (2) live methanogenic livestock exports [82, 137-144], (3) natural gas exports [145-147], (4) recoverable shale gas reserves that can be accessed by hydraulic fracturing (fracking) [148], (5) coal exports [149-153], (6) land clearing, deforestation and ecocide [154, 155], (7) speciescide – species extinction [156], (8) coral reef destruction [78-80, 156-163], (9) whale killing and extinction threat through global warming [164], (10) terminal carbon pollution budget exceedance [165-168], (11) per capita carbon debt [18-20, 169-172], (12) GHG generating iron ore exports [173-175], (13) climate change inaction [176] and (14) climate genocide and approach towards omnicide and terracide [17]. Thanks to the homicidal greed of climate criminal countries such as Australia, the present plus 1C is already devastating Island Nations and a catastrophic plus 2C warming is now unavoidable [14-17, 177].

30. Mina Guli, activism and Mainstream media unresponsiveness. Mina Guli is an Australian science and law graduate with an impressive business background in carbon trading and energy policy. She founded the water use advocacy and education organization Thirst, and courageously attempted to run an extraordinary 100 marathons over 100 days to bring global attention to the water crisis, but was stopped after 60 by a bone fracture [178, 179]. It is a sad testament to entrenched denialism in public life in neoliberal Western democracies that an informed, skilled and articulate professional such as Mina Guli found it necessary to embark on such an extraordinary physical endeavour in order to achieve effective free speech and be able educate the public on the worsening water crisis. Indeed I was inspired by Mina Guli's courageous and ethical example to research and write this wide-ranging, science-based and documented sketch of the water crisis in the interests of Humanity.

Final comments.

The world faces a worsening water crisis that is inextricably linked to a worsening climate emergency and climate genocide. Presently 15 million people die avoidably from deprivation and deprivation-exacerbated disease in the Developing World minus China [13]. This Global Avoidable Mortality Holocaust is significantly linked to water-related impacts including drought, storms, fires, inundation, biosphere degradation, lack of water for agriculture, and lack of potable water. However a worsening, water-related climate genocide may see up to 10 billion people dying this century if man-made global warming is not requisitely addressed [17]. Technological solutions exist for water deficiency problems in rich countries e.g. urban water recycling, hydroponic agriculture, desalination, and rational industrial and domestic water use. However the linked water crisis, biodiversity crisis and climate crisis require drastic global action, specifically negative greenhouse gas pollution (CO_2 draw-down to 300 ppm CO_2 from the present dangerous and damaging 410 ppm CO_2), negative population growth (a halving of population), and negative economic growth (a 50% decrease with the burden to be largely borne by the rich North) [81]. Fundamentally these water-related crises require replacement of neoliberal facilitation of untrammelled individual greed with balanced and sustainable resource utilization for the benefit of all peoples and all species. Please inform everyone you can.

References.

[1]. WWAP (United Nations World Water Assessment Programme)/UN-Water. 2018. The United Nations World Water Development Report 2018: Nature-Based Solutions for Water. Paris, UNESCO: https://reliefweb.int/sites/reliefweb.int/files/resources/261424e.pdf.
[2]. "Fresh water", Wikipedia: https://en.wikipedia.org/wiki/Fresh_water.
[3]. "Water distribution on earth", Wikipedia: https://en.wikipedia.org/wiki/Water_distribution_on_Earth.
[4]. "Reverse osmosis", Wikipedia: https://en.wikipedia.org/wiki/Reverse_osmosis.
[5]. "Sorek desalination plant": https://www.ide-tech.com/en/our-projects/sorek-desalination-plant/?data=item_1.
[6]. "Desalination", Wikipedia: https://en.wikipedia.org/wiki/Desalination.
[7]. "Reclaimed water", Wikipedia: https://en.wikipedia.org/wiki/Reclaimed_water.
[8]. World Water Council, "The use of water today", 2001: http://www.worldwatercouncil.org/fileadmin/wwc/Library/WWVision/Chapter2.pdf
[9]. "Thirst": http://www.thirstforwater.org/water-crisis.html.
[10]. "Drinking water", Wikipedia: https://en.wikipedia.org/wiki/Drinking_water.
[11]. World Health Organization (WHO), "Sanitation", 2018: https://www.who.int/news-room/fact-sheets/detail/sanitation.
[12]. "India's sanitation market to reach US$60 billion by 2021: official", Jakarta Post, 25 April 2019: https://www.thejakartapost.com/news/2019/04/25/indias-sanitation-market-to-reach-60-billion-usd-by-2021-official.html.
[13]. Gideon Polya, "Body Count. Global avoidable mortality since 1950" that includes an avoidable mortality-related history of every country since Neolithic times and is now available for free perusal on the web: http://globalbodycount.blogspot.com/.
[14]. Are we doomed?": https://sites.google.com/site/300orgsite/are-we-doomed.
[15]. "Methane Bomb Threat": https://sites.google.com/site/methanebombthreat/.
[16]. "Nuclear weapons ban, end poverty and reverse climate change": https://sites.google.com/site/drgideonpolya/nuclear-weapons-ban.

[17]. "Climate Genocide":
https://sites.google.com/site/climategenocide/.
[18]. Gideon Polya, "Inescapable $200-$250 trillion global Carbon Debt increasing by $16 trillion annually", Countercurrents, 27 April 2019: https://countercurrents.org/2019/04/27/inescapable-200-250-trillion-global-carbon-debt-increasing-by-16-trillion-annually-gideon-polya/.
[19]. Chris Hope, "How high should climate change taxes be?", Working Paper Series, Judge Business School, University of Cambridge, 9.2011:
http://www.jbs.cam.ac.uk/fileadmin/user_upload/research/workingpapers/wp1109.pdf.
[20]. James Hansen, "Climate change in a nutshell: the gathering storm", Columbia University, 18 December 2018:
http://www.columbia.edu/~jeh1/mailings/2018/20181206_Nutshell.pdf.
[21]. "Aquifer", Wikipedia: https://en.wikipedia.org/wiki/Aquifer.
[22]. "Arsenic contamination of groundwater", Wikipedia:
https://en.wikipedia.org/wiki/Arsenic_contamination_of_groundwater
.
[23]. Matt McGrath, ""Alarmingly high" levels of arsenic in Pakistan's groundwater", BBC, 23 August 2017:
https://www.bbc.com/news/science-environment-41002005.
[24]. Prosun Bhattacharya, David Polya, Dragana Jovanovic, "Best Practice Guide on the Control of Arsenic in Drinking Water", IWA, 2017.
[25]. Farhana S. Islam, Andrew G. Gault, Christopher Boothman, David A. Polya, John M. Charnock, Debashis Chatterjee & Jonathan R. Lloyd, "Role of metal-reducing bacteria in arsenic release from Bengal delta sediments", Nature, volume 430, pages 68-71, 2004:
https://www.researchgate.net/profile/Jon_Lloyd2/publication/8478243_Role_of_metal-reducing_bacteria_in_arsenic_release_from_Bengal_delta_sediments/links/00b49515adf73e2e30000000.pdf.
[26]. Lester Brown, "The great food crisis of 2011", Foreign Policy, 10 January 2011: https://foreignpolicy.com/2011/01/10/the-great-food-crisis-of-2011/.
[27]. Caleb Gorton, "India's groundwater crisis: the consequences of unsustainable pumping", Future Directions, 17 July 2017:
http://www.futuredirections.org.au/wp-content/uploads/2017/07/Indias-Groundwater-Crisis-The-Consequences-of-Unsustainable-Pumping.pdf.
[28]. Camilla Corradin, "Israel: water as a tool to dominate Palestinians", 23 June 2016:

https://www.aljazeera.com/news/2016/06/israel-water-tool-dominate-palestinians-160619062531348.html.
[29]. "Gaza Concentration Camp":
https://sites.google.com/site/palestiniangenocide/gaza-concentration.
[30]. "Water supply and sanitation in the State of Palestine",
Wikipedia:
https://en.wikipedia.org/wiki/Water_supply_and_sanitation_in_the_St
ate_of_Palestine
[31]. Geneva Convention Relative to the Protection of Civilians in
Time of War: http://www.unhchr.ch/html/menu3/b/92.htm.
[32]. "Drip agriculture", Wikipedia:
https://en.wikipedia.org/wiki/Drip_irrigation.
[33]. "Hydroponics", Wikipedia:
https://en.wikipedia.org/wiki/Hydroponics.
[34]. Andrew Ross, "Banking water underground for our future", The
Conversation, 7 January 2017: https://theconversation.com/banking-
water-underground-for-our-future-11031.
[35]. Daron Jacks, "Hexa-Cover discs help to reduce evaporation,
algae in water storages", Weekly Times, 20 February 2019:
https://www.weeklytimesnow.com.au/machine/crop-gear/heaxacover-
discs-help-to-reduce-evaporation-algae-in-water-storages/news-
story/dc886edad7d87083f1348e640f9d0d67.
[36]. "Aboriginal Genocide":
https://sites.google.com/site/aboriginalgenocide/.
[37]. R. Quentin Grafton and Sarah Jane Wheeler, "Economics of
water recovery in the Murray-Darling Basin, Australia", Annual
Review of Resource Economics, 2018:
https://www.mdbrc.sa.gov.au/sites/g/files/net3846/f/mdbrc-exhibit-
61-grafton-and-wheeler-economics-of-water-recovery-in-the-murray-
darling-basin-5-march-2018.pdf?v=1530682071.
[38]. "Joyce "doesn't care" about inquiry into $80 million water deal,
as auditor general to investigate", SBS News, 2 May 2019:
https://www.sbs.com.au/news/joyce-doesn-t-care-about-inquiry-into-
80-million-water-deal-as-auditor-general-to-investigate.
[39]. World Bank Report No. 47657-Gz, "West Bank and Gaza.
Assessment of restrictions on Palestinian water sector development",
April 2009:
http://documents.worldbank.org/curated/en/775491468139782240/pdf
/476570SR0P11511nsReport18Apr2009111.pdf.
[40]. Gideon Polya, "Gideon Polya: pro-Apartheid Israel Australian
Labor Party scraps outstanding anti-Apartheid candidate, Melissa
Parke", Countercurrents, 17 April 2019:
https://countercurrents.org/2019/04/17/pro-apartheid-israel-australian-

labor-party-scraps-outstanding-anti-apartheid-candidate-melissa-parke/.
[41]. Gideon Polya, "Israeli Jewish Nation-State Law enshrines Apartheid and genocidal racism", Countercurrents, 24 July 2018: https://countercurrents.org/2018/07/24/israeli-jewish-nation-state-law-enshrines-apartheid-and-genocidal-racism/.
[42]. Gideon Polya, "Democratic one-state solution (unitary state, bi-national state) for post-Apartheid Palestine", Countercurrents, 22 December 2018: https://countercurrents.org/2018/12/22/democratic-one-state-solution-unitary-state-bi-national-state-for-post-apartheid-palestine/.
[43]. Gideon Polya, "70th anniversary of Apartheid Israel & commencement of large-scale Palestinian Genocide", Countercurrents, 11 May 2018: https://countercurrents.org/2018/05/11/70th-anniversary-of-apartheid-israel-commencement-of-large-scale-palestinian-genocide/.
[44]. Gideon Polya, "Israeli-Palestinian & Middle East conflict – from oil to climate genocide", Countercurrents, 21 August 2017: https://countercurrents.org/2017/08/21/israeli-palestinian-middle-east-conflict-from-oil-to-climate-genocide/.
[45]. "Boycott Apartheid Israel": https://sites.google.com/site/boycottapartheidisrael/.
[46]. "Gaza Concentration Camp": https://sites.google.com/site/palestiniangenocide/gaza-concentration.
[47]. "Jews Against Racist Zionism": https://sites.google.com/site/jewsagainstracistzionism/.
[48]. "Non-Jews Against Racist Zionism": https://sites.google.com/site/nonjewsagainstracistzionism/.
[49]. "Nuclear weapons and Israel", Wikipedia: https://en.wikipedia.org/wiki/Nuclear_weapons_and_Israel.
[50]. "Palestinian Genocide": https://sites.google.com/site/palestiniangenocide/.
[51]. Apartheid Israeli state terrorism: (A) individuals exposing Apartheid Israeli state terrorism, and (B) countries subject to Apartheid Israeli state terrorism.", Palestinian Genocide: https://sites.google.com/site/palestiniangenocide/apartheid-israeli-state-terrorism.
[52]. "Stop state terrorism": https://sites.google.com/site/stopstateterrorism/.
[53]. "State crime and non-state terrorism": https://sites.google.com/site/statecrimeandnonstateterrorism/.
[54]. "Muslim Holocaust Muslim Genocide": https://sites.google.com/site/muslimholocaustmuslimgenocide/.

[55]. "Refugee", Wikipedia:
https://sites.google.com/site/muslimholocaustmuslimgenocide/.
[56]. "List of countries by freshwater withdrawal", Wikipedia:
https://en.wikipedia.org/wiki/List_of_countries_by_freshwater_withdr
awal.
[57]. FAO Aquastat, "Occupied Palestinian Territory":
http://www.fao.org/nr/water/aquastat/countries_regions/PSE/PSE-
CP_eng.pdf.
[58]. "Cry me a river", ABC News, 2 May 2018:
https://www.abc.net.au/news/2019-05-02/cry-me-a-river/11074384.
[59]. Michael Vincent, "Walgett has 2 rivers but no water left to
drink", ABC News, 19 December 2018:
https://www.abc.net.au/news/2018-12-19/walgett-has-two-rivers-but-
no-water-left-to-drink/10558428.
[60]. Danielle Bonica, "The Barwon River at Walgett is just a series
of stagnant pools at the moment", ABC News, 11 December 2018:
https://www.abc.net.au/news/2018-12-11/walgett-water-victor-
murray/10599352.
[61]. M.M. Mekonnen and A.Y. Hoekstra, "The green, blue and grey
water footprint of farm animals and animal products. Volume 2:
Appendices", UNESCO Institute for Water Education, December
2010: https://waterfootprint.org/media/downloads/Report-48-
WaterFootprint-AnimalProducts-Vol2_1.pdf.
[62]. Water Footprint Network, "Water footprint of crop and animal
products: a comparison": https://waterfootprint.org/en/water-
footprint/product-water-footprint/water-footprint-crop-and-animal-
products/.
[63]. Gideon Polya, "Biofuel famine, biofuel genocide, meat & global
food price crisis", Global avoidable
mortality:http://globalavoidablemortality.blogspot.com.au/2008/05/bi
ofuel-famine-biofuel-genocide-meat.html.
[64]. Current worldwide annual meat consumption per
capita", ChartsBin: http://chartsbin.com/view/12730.
[65]. "List of countries by GDP (nominal) per capita", Wikipedia:
https://en.wikipedia.org/wiki/List_of_countries_by_GDP_(nominal)_
per_capita.
[66]. Gideon Polya, "Worsening climate emergency and record CO2
emissions demand vegetarian diet for all to help save planet",
Countercurrents, 20 June 2016:
https://countercurrents.org/2016/06/20/worsening-climate-emergency-
and-record-co2-emissions-demand-vegetarian-diet-for-all-to-help-
save-planet/.
[67]. "Methane Bomb
Threat": https://sites.google.com/site/methanebombthreat/.

[68]. William J. Ripple et al., 15,364 signatories from 184 countries, "World scientists' warning to Humanity: a second notice", Bioscience, 13 November 2017: https://academic.oup.com/bioscience/advance-article/doi/10.1093/biosci/bix125/4605229.

[69]. Gideon Polya, "Over 15,000 scientists issue dire warning to Humanity on catastrophic climate change and biodiversity loss", Countercurrents, 20 November 2017: https://countercurrents.org/2017/11/20/over-15000-scientists-issue-dire-warning-to-humanity-on-catastrophic-climate-change-and-biodiversity-loss/.

[70]. Phillip Levin, Donald Levin, "The real biodiversity crisis", American Scientist, January-February 2002: http://www.americanscientist.org/issues/pub/the-real-biodiversity-crisis.

[71]. World Wildlife Fund (WWF) "Living Planet Report 2018 aiming higher – summary": https://s3.amazonaws.com/wwfassets/downloads/lpr2018_summary_report_spreads.pdf.

[72]. IPCC, "Global warming of 1.5 °C. Summary for Policymakers", 8 October 2018: http://report.ipcc.ch/sr15/pdf/sr15_spm_final.pdf.

[73]. Gideon Polya, "IPCC +1.5C avoidance report – effectively too late but stop coal burning for "less bad" catastrophes", Countercurrents, 12 October 2018: https://countercurrents.org/2018/10/12/ipcc-1-5c-avoidance-report-effectively-too-late-but-stop-coal-burning-for-less-bad-catastrophes/.

[74]. Gideon Polya, "Pro-coal Australia and Trump America reject dire IPCC report & declare War on Terra", Countercurrents, 17 October 2018: https://countercurrents.org/2018/10/17/pro-coal-australia-trump-america-reject-dire-ipcc-report-declare-war-on-terra/

[75]. US NOAA, "Full Mauna Loa CO_2 record": https://www.esrl.noaa.gov/gmd/ccgg/trends/full.html.

[76]. "Are we doomed?": https://sites.google.com/site/300orgsite/are-we-doomed.

[77]. "Too late to avoid global warming catastrophe": https://sites.google.com/site/300orgsite/too-late-to-avoid-global-warming.

[78]. Output of the technical working group meeting, The Royal Society, London, 6th July, 2009, "The Coral Reef Crisis: scientific justification for critical CO_2 threshold levels of less than 350ppm": http://static.zsl.org/files/statement-of-the-coral-reef-crisis-working-group-890.pdf.

[79]. 300.org: https://sites.google.com/site/300orgsite/300-org.

[80]. "300.org – return atmosphere CO_2 to 300 ppm CO_2":
https://sites.google.com/site/300orgsite/300-org—return-atmosphere-co2-to-300-ppm.

[81]. Gideon Polya, "How much negative carbon emissions, negative population growth & negative economic growth is needed to save planet?", Countercurrents, 28 November 2018: https://countercurrents.org/2018/11/28/how-much-negative-carbon-emissions-negative-population-growth-negative-economic-growth-is-needed-to-save-planet/.

[82]. Jared Diamond, "Collapse", Penguin, 2011.

[83]. Gideon Polya, "Pro-coal Australia & Trump America Reject Dire IPCC Report & Declare War on Terra", Countercurrents, 17 October 2018: https://countercurrents.org/2018/10/17/pro-coal-australia-trump-america-reject-dire-ipcc-report-declare-war-on-terra/.

[84]. Gideon Polya, "Offences of Pentecostal Christian Scott Morrison, PM after Australia's fourth PM-removing coup in 8 years", Countercurrents, 18 September 2019: https://countercurrents.org/2018/09/18/offences-of-pentecostal-christian-scott-morrison-pm-after-australias-fourth-pm-removing-coup-in-8-years/.

[85]. M.M. Mekonnen and A.Y. Hoekstra, "The green, blue and grey water footprint of crops and crop-derived products. Volume 1: main report", December 20101: https://waterfootprint.org/media/downloads/Report47-WaterFootprintCrops-Vol1.pdf

[86]. Brian Ellis, "Social Humanism. A New Metaphysics", Routledge, UK, 2012.

[87]. Gideon Polya, "Book Review: "Social Humanism. A New Metaphysics" by Brian Ellis – last chance to save Planet?", Countercurrents, 19 August, 2012: https://www.countercurrents.org/polya190812.htm.

[88]. Water Footprint, "Hidden waters", February 2007: https://waterfootprint.org/media/downloads/Zygmunt_2007_1.pdf.

[89]. "Water cycle", Wikipedia: https://en.wikipedia.org/wiki/Water_cycle.

[90]. "Carbon cycle", Wikipedia: https://en.wikipedia.org/wiki/Carbon_cycle.

[91]. "2011 climate change course": https://sites.google.com/site/300orgsite/2011-climate-change-course.

[92]. David L. Nelson and Michael M. Cox, "Lehninger Principles of Biochemistry", Worth, 2000.

[93]. Gideon Polya, "Biochemical Targets of Plant Bioactive Compounds", Taylor & Francis/CRC Press, 2003.

[94]. "Fact check: is Queensland clearing land as fast as Brazil?", Fact Check, 16 July 2018: http://www.abc.net.au/news/2017-12-01/fact-check-queensland-land-clearing-brazilian-rainforest/9183596.

[95]. Michael Slezak, "Global deforestation hotspot": 3m hectares of Australian forest to be lost in 15 years", Guardian, 5 March 2018: https://www.theguardian.com/environment/2018/mar/05/global-deforestation-hotspot-3m-hectares-of-australian-forest-to-be-lost-in-15-years.

[96]. Sir Nicholas Stern, quoted in "Climate change: "the greatest market failure the word has seen"", New Economist, 30 October 2006: http://neweconomist.blogs.com/new_economist/2006/10/stern_review_2.html.

[97]. Alison Benjamin, "Stern: climate change a "market failure"", Guardian, 29 November 2009: https://www.theguardian.com/environment/2007/nov/29/climatechange.carbonemissions.

[98]. "Cyclone Fani", Wikipedia: https://en.wikipedia.org/wiki/Cyclone_Fani.

[99]. "Rising temperatures and human activity are increasing storm run off and flash floods", Science Daily, 22 October 2018: https://www.sciencedaily.com/releases/2018/10/181022085820.htm.

[100]. Jiabo Yin, Pierre Gentine, Sha Zhou, Sylvia C. Sullivan, Ren Wang, Yao Zhang, and Shenglian Guo, "Large increase in global storm runoff extremes driven by climate and anthropogenic changes", Nature Communications, 9, article no. 4389, 2018: https://www.nature.com/articles/s41467-018-06765-2.

[101]. Gideon Polya, "Kerala flood disaster is not natural but man-made – warning for India & World", Countercurrents, 21 August 2018: https://countercurrents.org/2018/08/21/kerala-flood-disaster-is-not-natural-but-man-made-warning-for-india-world/.

[102]. Daisy Dunne, "Climate change has influenced global drought risk for "more than a century"", World Economic Forum, 3 May 2019: https://www.weforum.org/agenda/2019/05/climate-change-has-influenced-global-drought-risk-for-more-than-a-century/.

[103]. Kate Marvel, Benjamin I. Cook, Céline J. W. Bonfils, Paul J. Durack, Jason E. Smerdon & A. Park Williams, "Twentieth-century hydroclimate changes consistent with human influence", Nature volume 569, pages 59–65, 2019: https://www.nature.com/articles/s41586-019-1149-8.

[104}. Laurie Goering, "Hike in record-dry months for Africa's Sahel worries scientists", Reuters, 14 December 2018: https://www.reuters.com/article/us-climatechange-africa-

drought/hike-in-record-dry-months-for-africas-sahel-worries-scientists-idUSKBN1OC1PT.
[105]. "Famine and hunger crisis", Oxfam, 2019:
https://www.oxfam.org/en/emergencies/famine-and-hunger-crisis.
[106]. Gideon Polya, "Paris Atrocity Context: 27 Million Muslim Avoidable Deaths From Imposed Deprivation In 20 Countries Violated By US Alliance Since 9-11", Countercurrents, 22 November, 2015: https://countercurrents.org/polya221115.htm.
[107]. Aled Jones, "Food security: how drought and rising prices led to conflict in Syria", The Conversation, 26 January 2017:
https://theconversation.com/food-security-how-drought-and-rising-prices-led-to-conflict-in-syria-71539.
[108]. "Dust storms blanket NSW and ACT as air quality reaches hazardous levels":, Guardian, 13 February 2019:
https://www.theguardian.com/australia-news/2019/feb/13/dust-storms-blanket-nsw-and-act-as-air-quality-reaches-hazardous-levels.
[109]. "New report: current drought exacerbated by climate change", Climate Council, 2019:
https://www.climatecouncil.org.au/resources/new-report-current-drought-exacerbated-by-climate-change/.
[110]. Climate Council, "Deluge and drought: Australia's water security in a changing climate", 2019:
https://www.climatecouncil.org.au/wp-content/uploads/2018/11/Climate-Council-Water-Security-Report.pdf
[111]. "Sea level rise", Wikipedia:
https://en.wikipedia.org/wiki/Sea_level_rise.
[112]. "Arctic sea ice decline", Wikipedia:
https://en.wikipedia.org/wiki/Arctic_sea_ice_decline.
[113]. Bethan Davies, "Mass balance of the Antarctic ice sheet from 1992 to 2017", Antarctic Glaciers, 13 June 2018:
http://www.antarcticglaciers.org/2018/06/mass-balance-antarctic-ice-sheet-1992-2017/.
[114]. "Decline of West Antarctic glaciers appears irreversible", Earth Observatory, 2014:
https://earthobservatory.nasa.gov/images/83672/decline-of-west-antarctic-glaciers-appears-irreversible.
[115]. Stephen Leahy, "Greenland's ice is melting four times faster than thought – what it means", National Geographic, 21 January 2019:
https://www.nationalgeographic.com/environment/2019/01/greeland-ice-melting-four-times-faster-than-thought-raising-sea-level/.
[116]. Andrew Glikson, "The Methane Time Bomb and the Triple Melt-down", Countercurrents, 2009 :
http://www.countercurrents.org/glikson101008.htm).

[117]. "Storm intensity increase", Climate signals, 25 April 2019: https://www.climatesignals.org/climate-signals/storm-intensity-increase.

[118]. Pacific Islands Development Forum 4 September 2015 "Suva Declaration on Climate Change": http://pacificidf.org/wp-content/uploads/2013/06/PACIFIC-ISLAND-DEVELOPMENT-FORUM-SUVA-DECLARATION-ON-CLIMATE-CHANGE.v2.pdf.

[119]. Kendra Pierre-Louis and Nadja Popovitch, "Climate change is fuelling wildfires nationwide, new report warns", New York Times, 27 November 2018: https://www.nytimes.com/interactive/2018/11/27/climate/wildfire-global-warming.html.

[120]. "How climate change is increasing forest fires around the world", DW, https://www.dw.com/en/how-climate-change-is-increasing-forest-fires-around-the-world/a-19465490.

[121]. "Wildfires will become more frequent due to rising temperaturesm but study finds changes will be far from uniform", PhysOrg, 5 April 2018: https://phys.org/news/2018-04-wildfires-frequent-due-temperatures-uniform.html.

[122]. Luiz Aragao, Jos Barlow, and Liana Anderson, "Amazon forests that were once fire-proof have become flammable", The Conversation, 14 February 2018: https://theconversation.com/amazon-rainforests-that-were-once-fire-proof-have-become-flammable-91775.

[123]. Sam Wood, "The 2019 Tasmanian fires so far: what has burned and where/", FireCentre, 2019: https://firecentre.org.au/the-2019-tasmanian-fires-so-far-what-has-burned-and-where/.

[124]. "World of change: shrinking Aral Sea", NASA Earth Observatory, 2018: https://earthobservatory.nasa.gov/world-of-change/aral_sea.php.

[125]. "Aral Sea", Wikipedia: https://en.wikipedia.org/wiki/Aral_Sea.

[126]. Tafline Laylin, "How Michigan's Flint River came to poison a city", Guardian, 19 January 2016: https://www.theguardian.com/environment/2016/jan/18/michigan-flint-river-epa-lead-contamination-mdeq-pollutants-water-safety-health.

[127]. Michael Moore, "Fahrenheit 11/9", 2018 movie, Wikipedia: https://en.wikipedia.org/wiki/Fahrenheit_11/9.

[128]. "Minamata disease", Wikipedia: https://en.wikipedia.org/wiki/Minamata_disease.

[129]. Kieran Cooke, "The dead river? How the waters of the Jordan run foul", Middle East Eye, 28 September 2017: https://www.middleeasteye.net/opinion/dead-river-how-waters-jordan-run-foul.

[130]. "Desertification", Wikipedia:
https://en.wikipedia.org/wiki/Desertification.
[131]. Intergovernmental Science-Policy Platform on Biodiversity and Ecosystem Services (IPBES) re summary of the 2019 Global Assessment Report on Biodiversity and Ecosystem Services, "Media Release: nature's dangerous decline "unprecedented"; species extinction rates "accelerating", May 2019:
https://www.ipbes.net/news/Media-Release-Global-Assessment.
[132]. "Summary for policy makers of the global assessment report on the biodiversity and ecosystem services of the Intergovernmental Science-Policy Platform on Biodiversity and Ecosystem Services (IPBES)", 6 May 2019:
https://www.ipbes.net/sites/default/files/downloads/spm_unedited_advance_for_posting_htn.pdf
[133]. Global Assessment report on Biodiversity and Ecosystem Services", Wikipedia:
https://en.wikipedia.org/wiki/Global_Assessment_Report_on_Biodiversity_and_Ecosystem_Services.
[134]. Gideon Polya, "2011 climate change course":
https://sites.google.com/site/300orgsite/2011-climate-change-course.
[135]. Gideon Polya, "Revised Annual Per Capita Greenhouse Gas Pollution For All Countries – What Is Your Country Doing?", Countercurrents, 6 January, 2016:
http://www.countercurrents.org/polya060116.htm.
[136]. Gideon Polya, "Exposing And Thence Punishing Worst Polluter Nations Via Weighted Annual Per Capita Greenhouse Gas Pollution Scores", Countercurrents, 19 March, 2016:
https://countercurrents.org/polya190316.htm.
[137]. Alison Penfold, "Australia leads the world in livestock export", Issues, June 2013: http://www.issuesmagazine.com.au/article/issue-june-2013/australia-leads-world-livestock-export.html.
[138]. Khushboo Sheth, "Top 20 exporters of live cattle", World Atlas, 25 April 2017: https://www.worldatlas.com/articles/top-20-exporters-of-live-cattle.html.
[139]. Mike Keogh and Adam Tomlinson, "Australia risks missing a big livestock export and animal welfare opportunity", Australian Farm Institute, August 2013:
http://www.farminstitute.org.au/newsletter/2013/August_2013/August_2013_featurearticle.html.
[140]. Robert Goodland and Jeff Anfang. "Livestock and climate change. What if the key actors in climate change are … cows, pigs and chickens?", World Watch, November/December 2009:
http://www.worldwatch.org/files/pdf/Livestock%20and%20Climate%20Change.pdf.

[141]. "List of countries by meat consumption", Wikipedia: https://en.wikipedia.org/wiki/List_of_countries_by_meat_consumptio n.

[142]. Gideon Polya, "Biofuel famine, biofuel genocide, meat & global food price crisis", Global Avoidable Mortality, 9 May 2008: http://globalavoidablemortality.blogspot.com/2008/05/biofuel-famine-biofuel-genocide-meat.html.

[143]. Gideon Polya, "Worsening Climate Emergency And Record CO_2 Emissions Demand Vegetarian Diet For All To Help Save Planet", Countercurrents, 20 June, 2016: https://www.countercurrents.org/polya200616.htm.

[144]. "Live animal exports", RSPCA: https://www.rspca.org.au/campaigns/live-animal-export.

[145]. "List of countries by natural gas exports", Wikipedoa: https://en.wikipedia.org/wiki/List_of_countries_by_natural_gas_expo rts.

[146]. "Liquefied natural gas export market share worldwide in 2017, by country", Statista, 2017: https://www.statista.com/statistics/722846/lng-export-market-share-worldwide-by-country/.

[147]. "Gas is not clean energy": https://sites.google.com/site/gasisnotcleanenergy/.

[148]. "List of countries by recoverable shale gas", Wikipedia: https://en.wikipedia.org/wiki/List_of_countries_by_recoverable_shale _gas.

[149]. Daniel Workman, "Coal exports by country", World's top exports, 24 August 2018: http://www.worldstopexports.com/coal-exports-country/.

[150]. Katharine Murphy, "Scott Morrison brings coal to question time: what fresh idiocy is this?", Guardian, 9 February 2017: https://www.theguardian.com/australia-news/2017/feb/09/scott-morrison-brings-coal-to-question-time-what-fresh-idiocy-is-this.

[151]. Climate Council, "Too risky: stop Adani's mine": https://www.climatecouncil.org.au/actions/too-risky-stop-adanis-mine/?c=1#action-form-wrapper.

[152]. "Stop air pollution deaths": https://sites.google.com/site/300orgsite/stop-air-pollution-deaths.

[153]. Gideon Polya, "Latest Lancet data imply Adani Australian coal project will kill 1.4 million Indians", Countercurrents, 21 April 2017: https://countercurrents.org/2017/04/21/latest-lancet-data-imply-adani-australian-coal-project-will-kill-1-4-million-indians/.

[154]. "Fact check: is Queensland clearing land as fast as Brazil?", Fact Check, 16 July 2018: http://www.abc.net.au/news/2017-12-01/fact-check-queensland-land-clearing-brazilian-rainforest/9183596.

[155]. Michael Slezak, "'Global deforestation hotspot': 3m hectares of Australian forest to be list in 15 years", Guardian, 5 March 2018: https://www.theguardian.com/environment/2018/mar/05/global-deforestation-hotspot-3m-hectares-of-australian-forest-to-be-lost-in-15-years.

[156]. Ben Spraggon, "Chart of the day: Australia named as fourth-worst country for species extinctions", ABC News, 20 July 2018: http://www.abc.net.au/news/2018-07-20/australia-fourth-on-animal-extinction-list/10002380.

[157]. Great Barrier Reef Marine Park Authority, "Climate change impacts on coral reefs": http://www.gbrmpa.gov.au/managing-the-reef/threats-to-the-reef/climate-change/what-does-this-mean-for-habitats/coral-reefs.

[158]. "Great Barrier Reef", Wikipedia: https://en.wikipedia.org/wiki/Great_Barrier_Reef.

[159]. Louisa Rebgetz and Laura Gartry, "Great Barrier Reef to get $500m to tackle pollution and breed resilient coal", ABC News, 29 April 2018: http://www.abc.net.au/news/2018-04-29/great-barrier-reef-$500m-package-to-preserve-area/9708230.

[160]. "Great Barrier Reef Foundation": https://www.barrierreef.org/.

[161]. Output of the technical working group meeting, The Royal Society, London, 6th July, 2009, "The Coral Reef Crisis: scientific justification for critical CO_2 threshold levels of less than 350ppm": http://static.zsl.org/files/statement-of-the-coral-reef-crisis-working-group-890.pdf.

[162}. J.E.N. Veron, O. Hoegh-Guldberg, T.M. Lenton, J.M. Lough, D.O. Obura, P. Pearce-Kelly, C.R.C. Sheppard, M. Spalding, M.G. Stafford-Smith and A.D. Rogers, "The coral reef crisis: the critical importance of <350 ppm CO2", Marine Pollution Bulletin, vol. 58, (10), October 2009, 1428-1436: http://www.sciencedirect.com/science?_ob=ArticleURL&_udi=B6V6N-4X9NKG7-3&_user=10&_rdoc=1&_fmt=&_orig=search&_sort=d&_docanchor=&view=c&_searchStrId=1072337698&_rerunOrigin=google&_acct=C000050221&_version=1&_urlVersion=0&_userid=10&md5=6858c5ff7172f9355068393496a5b35d.

[163]. Professor Ove Hoegh-Guldberg quoted in "Scientists call for urgent "global cooling" to save coral reefs", University of Queensland News, 8 November 2009: https://www.uq.edu.au/news/article/2009/11/scientists-call-urgent-global-cooling-save-coral-reefs.

[164]. Matthew Taylor, "Decline in krill threatens Antarctic wildlife, from whales to penguins", Guardian, 15 February 2018:

https://www.theguardian.com/environment/2018/feb/14/decline-in-krill-threatens-antarctic-wildlife-from-whales-to-penguins.
[165]. WBGU, "Solving the climate dilemma: the budget approach": http://www.ecoequity.org/2009/10/solving-the-climate-dilemma-the-budget-approach/.
[166]. Gideon Polya, "Current Carbon Debt Or Carbon Credit For All Countries: Australia, Canada And US Default On Carbon Debt", Countercurrents, 4 October, 2013: https://www.countercurrents.org/polya041013.htm
[167]. "Carbon Debt Carbon Credit": https://sites.google.com/site/carbondebtcarboncredit/.
[168]. James Hansen, "Letter to PM Kevin Rudd by Dr James Hansen", 2008: http://www.aussmc.org.au/documents/Hansen2008LetterToKevinRudd_000.pdf.
[169]. "Climate Justice & Intergenerational Equity": https://sites.google.com/site/300orgsite/climate-justice.
[170]. "Stop climate crime": https://sites.google.com/site/300orgsite/stop-climate-crime.
[171]. "Climate Revolution Now": https://sites.google.com/site/300orgsite/climate-revolution.
[172]. Gideon Polya, "Resolutely promised prosecutions of climate criminals may force urgent climate action", Global Research, 8 February 2019: https://www.globalresearch.ca/resolutely-promised-prosecutions-of-climate-criminals-may-force-urgent-climate-action/5667949.
[173]. "Iron ore", Wikipedia: http://en.wikipedia.org/wiki/Iron_ore#Production_and_consumption.
[174]. "Top iron ore producing countries in the world", World Atlas: https://www.worldatlas.com/articles/top-iron-ore-producing-countries-in-the-world.html.
[175]. Gideon Polya, "Australia's Huge Coal, Gas & Iron Ore Exports Threaten Planet", Countercurrents, 15 May 2012: http://www.countercurrents.org/polya150512.htm.
[176]. Climate Watch Performance Index, "Results 2018": https://www.germanwatch.org/sites/germanwatch.org/files/publication/20504.pdf.
[177]. "Too late to avoid global warming catastrophe": https://sites.google.com/site/300orgsite/too-late-to-avoid-global-warming.
[178]. "Mina Guli", Wikipedia: https://en.wikipedia.org/wiki/Mina_Guli.
[179]. "Thirst": http://www.thirstforwater.org/.

CHAPTER 9: TRUMPIST CLIMATE CHANGE DENIAL, AUSTRALIAN BUSHFIRES, FUEL REDUCTION, BIOCHAR & CARBON DEBT

[First published as Gideon Polya, "Trumpist climate change denial, Australian bushfires, fuel reduction, biochar & Carbon Debt", Countercurrents, 10 January 2020: https://countercurrents.org/2020/01/trumpist-climate-change-denial-australian-bushfires-fuel-reduction-biochar-carbon-debt.]

Australia has been devastated by catastrophic bushfires in all states across the continent. The Australian fire season used to encompass the 6 months of Spring and Summer but now encompasses 9 months from late Winter to early Autumn. The effective climate change denialist and anti-science Coalition Government has a major responsibility. Mechanical fuel reduction coupled with biochar (charcoal, carbon, C) generation is a possible mitigation avenue in a worsening, climate change-, drought- and warming-driven disaster prospect.

(1). The catastrophic 2019-2020 Australian bushfires by the numbers.

As of 10 January 2020 the numbers of hectares burned are as follows: Australian Capital Territory (0), New South Wales (4,900,000), Northern Territory (not available), Queensland (2,500,000), South Australia (274,000), Tasmania (32,000), Victoria (1,200,000), and Western Australia (1,700,000) for a total of 10,700,000 hectares.

As of 10 January 2020 the numbers of people tragically killed are as follows: Australian Capital Territory (1), New South Wales (20), Northern Territory (0), Queensland (0), South

Australia (4), Tasmania (0), Victoria (3), and Western Australia (0) for a total of 28 bushfire-related deaths.

As of 10 January 2020 the numbers of homes destroyed are as follows: Australian Capital Territory (0), New South Wales (1,687), Northern Territory (5), Queensland (48), South Australia (161), Tasmania (2), Victoria (300), and Western Australia (1) for a total of 2,204 homes destroyed [1].

These terrible numbers are set to increase with the global warming-impacted "new normal" of an Australian bushfire season likely to extend for a further 3-4 months i.e. through the remainder of Summer and early Autumn [2]. Large areas of eastern New South Wales and eastern Victoria have been subject to declarations of emergency, the Australian Defence Force (ADF) has been involved on land, air and sea, and about 100,000 people have been evacuated to safety. These Australian bushfires have variously commenced in July (Winter) 2019 and are likely to extend into March and April (Autumn) 2020 [2]. The extension of the "new normal" bushfire season into cooler and wetter Autumn and Winter months has reduced the ability of various fire services to safely conduct fuel hazard reduction burning to minimize the forest and grassland fuel load.

There are 100 hectares per square kilometer so the area of Australia burned so far (10,700,000 hectares, 107,000 square kilometers) is 1.4% of the area of Australia (7.7 million square kilometres), 44.1% of the area of England (242,495 square kilometres), and 93.4% of the combined area of Belgium, the Netherlands and Denmark (114,600 square kilometres).

The burning continues.

(2) The 2019-2020 Australian bushfires have so far emitted 750 Mt CO_2-e.

According to the official data of the Australian Government Department of Energy and the Environment, Australia's Domestic greenhouse gas (GHG) emissions had been steadily decreasing under successive Labor Governments (2007-2013) but have been steadily rising since the pro-coal, pro-gas, pro-fossil fuels and variously climate change denialist or effective climate change denialist Liberal Party-National Party Coalition came to power on the basis of a dishonest, anti-science and environmentally disastrous slogan of "No Carbon Tax" [3].

Dr Niels Andela, NASA Goddard Space Flight Center, and a collaborator in the Global Fire Emissions Database (GFED), has estimated from satellite data that in the period 1 August 2019 – 13 December 2019 the NSW bushfires have emitted 195 Mt CO_2-equivalent (CO_2-e) and the Queensland fires another 55 Mt CO_2-e, this corresponding to 47% of Australia's annual Domestic carbon pollution of 532 Mt CO_2-e [4].

However the Sydney Morning Herald, based on updated estimates from Dr Pep Canadell, a senior research scientist for CSIRO and the executive director of the Global Carbon Project (2 January 2020): "Australia's annual industrial emissions budget in 2018-19 was 532 million tonnes of carbon dioxide equivalent. This season's bushfires, which have burnt through more than 5 million hectares across the country, are estimated to have released two-thirds of this amount – or about 350 million tonnes – of carbon dioxide into the atmosphere so far. Until recently, Australia's forests were thought to reabsorb all the carbon released in bushfires, meaning they achieved net zero emissions. But scientists say climate change, drought and escalating fires combined are reducing forests' ability to reabsorb carbon… The 350 million tonne figure was extrapolated from data on NASA's Global Fire Emissions Database issued two weeks ago for NSW's bushfires, which at that stage had burnt 2.7 million hectares and emitted 190 million tonnes of carbon dioxide" [5].

However Wikipedia reports that as of 8 January 2020, in the Australian 2019–20 bushfire season 10.7 million hectares have burned since June 2019 (a bushfire season of 10 months rather than the previous Spring-Summer norm of 6 months)[1].

Accordingly total bushfire emissions = 190 Mt CO_2-e x 10.7 Mha/ 2.7 Mha = 753 Mt CO_2-e or 1.4 times Australia's annual

Domestic CO_2-e emissions (ditto, 350 Mt CO_2-e x 10.7 Mha/ 5 Mha = 749 Mt CO_2-e) i.e. Australia's total emissions jumped this 2019/20 financial year from 532 Mt CO_2-e to 532 Mt CO_2-e + 753 Mt CO_2-e = 1,285 Mt CO_2-e with 3-4 months to go.

(3). Huge loss of forest photosynthetic CO_2 capture capacity.

Missing in this accounting is the immense and prolonged loss of photosynthetic CO_2 capture capacity by South East Australian eucalyptus forests that can be the best forest carbon sinks in the world as reported by Keith et al. (2009): "From analysis of published global site biomass data ($n = 136$) from primary forests, we discovered (i) the world's highest known total biomass carbon density (living plus dead) of 1,867 tonnes carbon per ha (average value from 13 sites) occurs in Australian temperate moist *Eucalyptus regnans* forest" [6]. The bushfires are not only generating GHG emission greater than the national annual emissions, they are also causing a major [temporary] loss of the ability of 107,000 square kilometres of horrendously burned bush to sequester CO_2 through photosynthesis.

(4). Serious long-term health impacts from toxic pollutants from bushfires.

Far more serious for humans is the long-term impact on health of prolonged exposure to elevated CO (carbon monoxide), N_2O (nitrous oxide), NO_2 (nitrogen dioxide), SO_2 (sulphur dioxide) and $PM_{2.5}$ (deadly, fine carbon particulates), especially in smoke-blanketed, population-dense Sydney, Canberra and Melbourne. 8 million people die world-wide from air pollution each year (WHO) [7], with this including about 3.8 million from indoor, household pollution, 4.2 million from ambient (outdoor) pollution, 10,000 Londoners, 10,000 Australians and 75,000 people from the long-term effects of pollutants from the burning of Australia's world-leading coal exports [8]. These presently catastrophic Australian bushfires will have a deadly harvest in the long-term.

(5). Mechanical fuel load reduction (MFLR) may help reduce the likelihood and severity of bushfires.

F. Ximines et al. (2017): "Australia is the most fire-prone of all continents, with large areas of the country affected by bushfires each year. Bushfires can have profound impacts on communities and on the environment. Traditionally, prescribed burns have been used as a fuel reduction treatment. However, the proportion of land that is subjected to prescribed burns in Australia has decreased since 1990. This is partly due to the increasingly smaller windows of opportunity available to conduct burns safely, concerns about operational costs, health impacts from smoke, and the social acceptability of prescribed burning. These issues suggest a need for considering alternative fuel reduction approaches in select urban and rural areas, such as the potential for mechanical fuel load reduction (MFLR) treatments. The aim of MFLR is to reduce size, likelihood and severity of bushfires" [9].

(6). The US market price of biochar (charcoal, carbon, C) from forestry and agricultural waste is \$2,580 per ton.

Once cellulosic biomass is mechanically harvested it can be transported and converted to biochar (charcoal, carbon, C) through anaerobic pyrolysis at 400-700 C. The biochar can then be added back to soils, safely sequestered underground (e.g. in old coal mines) or used for other purposes (e.g. as an activated carbon substitute for filtration of polluted air or water).

Farm Energy (2019): "Biochar is charred biomass [charcoal, carbon, C], which is added to soil to improve soil health and soil quality. The solid, charcoal-like substance is created using a process known as [anaerobic] pyrolysis, which is basically the heating of biomass to a high temperature ($>400°$ degrees C) in the absence of oxygen. The addition of biochar to soils is actually an ancient technique that was used by early civilizations to enhance soil productivity. The potential importance of biochar was recognized when anthropologists discovered dark, highly fertile soils in the middle of the Amazon rain forest. It is speculated that thousands of years ago, farmers living in the Amazon engineered the terra preta soils (called "terra preta," or "black earth" in Portuguese) by adding biochar along with fish bones and other organic residues to the otherwise infertile soils. Similar soil enhancement practices have also been discovered by early farming cultures in Japan, South America, and Africa. The biochar-enriched soils have been shown to have significantly higher organic carbon content, higher water-holding capacity, higher nutrient holding capacity, and lower nutrient leaching rates compared to adjacent soils that have not been amended with biochar…The report found that a majority of the global market share comes from the US (65%), followed by Europe (25%), Asia (7%), and Africa (3%). Ninety percent (90%) of commercial activity is focused on small specialty retail markets—mostly nursery and garden centers. A small percentage (10%) is focused on larger scale markets, such as agriculture or land remediation projects. It has been reported that Japan has a growing commercial biochar industry, with approximately 15,000 tons traded annually (Okimori et al., 2003). In 2011, 94,159 tons of compost, a common soil amendment, was sold at a price of $30-50 per ton (USDA ERS, 2011). The price per ton for biochar is far higher and the volumes far lower. Companies responding to the IBI survey reported sales of only 911 tons in 2013, sixty-five percent (597 tons) of which was produced in North America. The average price for biochar in the US was $1.29 per pound or $2,580 per ton. In the near term, due to price point, biochar will likely only be used in high-end specialty markets, but there is growing interest in developing biochar into new products that take advantage of its unique chemical properties. For example, today wastewater treatment facilities purchase thousands of tons of activated carbon annually to help absorb potential contaminants and reduce odors. Biochar has been shown to be a cost-effective alternative" [10].

(7). A possible 9-12 Gt biochar carbon from forestry and agricultural waste annually versus 9 Gt carbon pollution annually from industry.

I wrote the following a decade ago on the amount of biochar that could be potentially generated annually from waste forestry and agricultural cellulosic biomass (2009): "Thus p224, Progress in Thermochemical Biomass Conversion, volume 1, IAE Bioenergy, ed. A.V. Bridgewater (Blackwell Science) [11] informs us that we could obtain 1.7 GtC/yr (straw from agriculture) + 4.2 GtC/yr (total grass upgrowth from grasslands upgrowth) + 6 GtC/yr (possible sustainable woodharvest) = 11.9 GtC/yr. From this one can see why biochar expert Professor Johannes Lehmann of Cornell University is correct in calculating that it is realistically possible to fix 9.5bn tonnes of carbon per year using biochar, noting that global annual production of carbon from fossil fuels is 8.5bn tonnes. According to a UK Guardian report: "Johannes Lehmann of Cornell University has calculated that it is realistically possible to fix 9.5bn tonnes of carbon per year using biochar. The global production of carbon from fossil fuels stands at 8.5bn tonnes" [12, 13].

One can see also why Professor James Lovelock FRS is also correct in his assessment: "There is one way we could save ourselves and that is through the massive burial of charcoal. It would mean farmers turning all their agricultural waste – which contains carbon that the plants have spent the summer sequestering – into non-biodegradable charcoal, and burying it in the soil. Then you can start shifting really hefty quantities of carbon out of the system and pull the CO_2 down quite fast … The biosphere pumps out 550 gigatonnes [550 billion tonnes] of carbon [carbon dioxide, CO_2] yearly; we put in only 30 gigatonnes [CO_2]. Ninety-nine per cent of the carbon that is fixed by plants is released back into the atmosphere within a year or so by consumers like bacteria, nematodes and worms. What we can do is cheat those consumers by getting farmers to burn their crop waste at very low oxygen levels to turn it into charcoal, which the farmer then ploughs into the field. A little CO_2 is released but the bulk of it gets converted to carbon. You get a few per cent of biofuel as a by-product of the combustion process, which the farmer can sell. This scheme would need no subsidy: the farmer would make a profit. This is the one thing we can do that will make a difference, but I bet they won't do it" [14].

Professor Lovelock's estimates are consonant with those of Dr J.A. Harrison, specifically a terrestrial carbon fixation of 121.3 GtC/y (449 Gt CO_2 = 449 billion tonnes of CO_2) of which about half returns annually to the atmosphere through animal and plant respiration, and most of the remaining half returns to the air through the [oxidative] action of soil fungi and bacteria [15, 16].

(8). Massive and fraudulent ignoring of carbon pollution "externality" costs is leading to ecocide, speciescide, climate genocide, and ultimately to omnicide and terracide.

Fundamental to chemistry and physics are the 3 Laws of Thermodynamics, specifically (1) the energy of a closed system is constant, (2) the entropy (disorder, chaos, lack of information content) strives to a maximum, and (3) zero motion in a perfect crystal at absolute zero degrees Kelvin (-273.15 degrees Centigrade). Polya's 3 Laws of Economics mirror the 3 Laws of Thermodynamics and are (1) Price minus COP (Cost of Production) equals profit; (2) Deception about COP strives to a maximum; and (3) No work, price or profit on a dead planet. These fundamental laws help expose the failure of neoliberal capitalism in relation to wealth inequality, massive tax evasion by multinational corporations, and horrendous avoidable deaths from poverty and pollution culminating in general ecocide, speciescide, climate genocide, omnicide and terracide [17].

It is now effectively too late to avoid a catastrophic plus 2 degrees Centigrade temperature rise but we are nevertheless obliged to do everything we can to make the future "less bad" for future generations. Critical to intergenerational justice and intergenerational equity is the notion of Carbon Debt (Carbon Price) that is simply the damage-related cost of greenhouse gas (GHG) pollution that if not addressed now will inescapably have to be paid by future generations. Thus, for example, if sea walls are not built then coastal populations will drown or be displaced. However GHG emissions continue to rise inexorably [18-20], and there is presently no global program to draw down CO_2 from the atmosphere to a safe and sustainable pre-Industrial Revolution level of about 300 ppm CO_2 as advocated

by many scientists and science-informed activists [21, 22].
Nevertheless GHG pollution and negative consequences in
numerous areas (notably ecocide and speciescide) are
relentlessly increasing to the point of exceedance or near-
exceedance of critical tipping points for massive irreversible
changes to hugely impact Humanity and the Biosphere (18-20,
23-38]. Eminent physicist Professor Steven Hawking has
succinctly summarized our dire circumstance thus: "We see
great peril if governments and societies do not take action now
to render nuclear weapons obsolete and to prevent further
climate change" [37, 38]. Facing this worsening existential
threat [39-43], young people, as exemplified by teenage activist
Greta Thunberg, are now vociferously demanding massive
climate action [44- 47].

Mendacious neoliberal capitalism falsely boasts of its efficiency
and efficacy but, in keeping with Polya's Second law of
Economics, utterly ignores the huge economic externality of the
social and environmental cost of pollution that should be added
to the Cost of Production. This constitutes the greatest fraud in
Australian and indeed human history. Climate change
economist Sir Nicholas Stern critically stated (2006): "The
science tells us that GHG emissions are an externality; in other
words, our emissions affect the lives of others. When people do
not pay for the consequences of their actions we have market
failure. This is the greatest market failure the world has seen"
[48, 49], and also argued for an international programme to
combat deforestation, which contributes 15-20% of greenhouse
gas emissions (2006): "For $10-15bn (£4.8-7.2bn) per year, a
programme could be constructed that could stop up to half the
deforestation" [49]. Social conservative but science-trained
Pope Francis has demanded that the environmental and social
cost of pollution be "fully borne" by the polluters [50, 51].
Further, noting the massive biodiversity loss, and the 100-1,000-
fold greater species extinction rate in the present Anthropocene
era [52, 53], we must not destroy what we cannot replace, and
any species is priceless [54].

(9). Damage-related cost of carbon pollution $200 per tonne CO_2 – saving 12 Gt C as biochar annually saves $8.8 trillion per annum.

Dr Chris Hope (Cambridge) has estimated a damage-related Carbon Price of $200 per tonne CO_2 [55] but the IMF has determined that average global Carbon Price is presently a mere $2 per tonne CO_2 [56, 57]. Dr Chris Hope (Judge Business School, 90 Nobel Laureate University of Cambridge, UK) has estimated the damage-related cost of carbon pollution (2011): "If the best current scientific and economic evidence is to be believed, and climate change could be a real and serious problem, the appropriate response is to institute today a climate change tax equal to the mean estimate of the damage caused by a tonne of CO_2 emissions. The raw calculations from the default PAGE09 model suggest the tax should be about $100 per tonne of CO_2 in the EU. But correcting for the limited time horizon of the model, and bringing the calculations forward to 2102 [2012], in year 2012 dollars, brings the suggested tax up to about $150 per tonne of CO_2. There are good arguments for setting the initial tax at about $250 per tonne of CO_2 in the US, while starting off at a much lower level, maybe $15 per tonne of CO_2, in the poorest regions of the world, all in year 2012 dollars. That such policy advice would not pass the laugh test, particularly in the US [and Australia], shows that the rhetoric about getting to grips with climate change has not been seriously thought through to its logical conclusion. As a result, rather than falling, greenhouse gas emissions are continuing to rise (Le Quere et al., 2009). A fiscally neutral significant climate change tax is the best chance we have of bringing the climate change problem under control" [55].

An International Monetary Fund (IMF) report on climate change mitigation advocates a Carbon Tax of $75 per ton of CO_2 by 2030 that, if progressively implemented in the G20 countries alone, would prevent an estimated 4 million air pollution deaths by 2030 [56, 57]. However the effective climate change denialist Coalition Government of Australia, a G20 country that is among world leaders in 15 areas of climate criminality (for detailed documentation see [58]), has flatly rejected a global Carbon Tax [57, 59, 60]. If all G20 countries followed Australia's rejection of a global Carbon Tax then they would be complicit in an unconscionable Climate Genocide and Climate Holocaust killing 4 million people over the next decade [56, 57].

Professor James Hansen et al. (of 96 Nobel Laureate Columbia University) have also arrived at a Carbon Price estimate of about $200 per tonne CO_2 (2018): "One ppm of CO_2 is 2.12 billion tons of carbon or about 7.77 billion tons of CO_2. Recently Keith et al. (2018) achieved a cost breakthrough in carbon capture, demonstrated with a pilot plant in Canada. Cost of carbon capture, not including the cost of transportation and storage of the CO_2, is $113-232 per ton of CO_2. Thus the cost of extracting 1 ppm of CO_2 from the atmosphere is $878-1803 billion. In other words, the cost, in a single year, of closing the gap between reality and the IPCC scenario that limits climate change to +1.5°C is already about $1 trillion. And that is without the cost of transporting and storing the CO_2, or consideration of whether there will be citizen objection to that transportation and storage. This annual cost will rise rapidly, unless there is a rapid slowdown in carbon emissions… cost of CO_2 storage… has been estimated as $10-20/t$CO_2$" [30]. Based on a Carbon Price of $200 per tonne CO_2, requisite atmospheric CO_2 draw-down from the present 407 ppm CO_2 over the next decade will cost up to $12.7 to $24.9 trillion annually for various 342 to 280 ppm CO_2 goals [54].

(10). At $200 per tonne CO_2 the real cost of the 2019-2020 Australian bushfires (so far, and lives lost aside) is $150 billion.

At $200 per tonne CO_2 the real cost of the Australian bushfires (so far) is 750 Mt CO_2-e x $200 per tonne CO_2 = $150 billion (this ignoring lives lost). By way of comparison, the climate criminal Australia Government has committed about $2 billion over 2 years for a National Bushfire Recovery Agency headed by Andrew Colvin, APM, OA, a former Commissioner of the Australian Federal Police (AFP) [61].

However the $150 billion cost of the 2019-20 Australian bushfires is just the tip of the iceberg of the cost of the climate criminality of the Coalition Australian Government. Thus Australia's annual Domestic carbon pollution is asserted by the Australian Government to be 532 Mt CO_2-e [4, 5] and at $200 per tonne CO_2 this amounts to a cost of $106 billion per year.

But it gets worse. Taking land use into account and considering the Global Warming Potential (GWP) of methane (CH_4) on a 20 year time frame (circa 80 relative to the same mass of CO_2) rather than on a 100 year time frame (25), Australia's annual per capita Domestic GHG pollution in tonnes CO_2-equivalent per person per year is 52.9, or 116 if including its huge GHG-generating exports [62-65]. Assuming a present Australian population of 25 million this translates to 1,323 Mt CO_2-e (or 2,900 Mt CO_2-e if one includes Australia's Exported GHG pollution). Assuming a damage-related Carbon Price of $200 per tonne CO_2, this in turn translates to a real cost of $265 billion (or $580 billion if one includes Australia's Exported GHG pollution).

To put this into a wider Australian national context, the Australian GDP was $1.9 trillion (2019) [66] and the gross Australian government debt was $552 billion (2017). Thus Australia's annual additional Carbon Debt ($580 billion per year) is similar to the accumulated gross Australian government debt ($552 billion). Of course Humanity shares one ocean and one atmosphere with the Australians, and thus Australia's annual additional Carbon Debt of $580 billion per year is imposed on all of Humanity, and represents an inescapable burden on all future generations. At some point Humanity (and especially the young) will tell climate criminal Australia to stop it [44-47].

(11). 85,000 preventable Australian deaths annually from "lifestyle" or "political choice" reasons.

Air pollution kills 8 million people each year [7], with this including 75,000 annual deaths from the long-term impact of pollutants from the burning of Australia's world-leading coal exports [8]. About 10,000 Australians die each year from air pollution, with 200,000 having died thus this century [8]. However this is just the tip of an iceberg of Australian preventable deaths. Thus each year about 85,000 Australians die preventably from "life style" and "political choice" reasons, the breakdown (including some overlaps) being as follows: (1) 26,000 annual Australian deaths from adverse hospital events, (2) 17,000 obesity-related Australian deaths, (3) 15,500 smoking-related Australian deaths, (4) 10,000 carbon burning pollution-derived Australian deaths, (5) 4,000 avoidable Indigenous Australian deaths, (6) 5,600 Australian alcohol-related deaths, (7) 2,900 Australian suicides (circa 100 being veterans), (8) 1,400 Australian road deaths, (9) 630 Australian opiate drug-related deaths with 570 linked to US restoration of the Taliban-destroyed Afghan opium industry, (10) 550 heat stress deaths, and (11) 300 Australian homicides (80 being of women killed domestically) [67-71].

(12). Livestock deaths.

Of Australia's cattle (24 million beef cattle and 2.6 million dairy cows), 9% are in the present bushfire area. Of Australia's 71 million sheep and lambs, 12% are in the present bushfire zone [72, 73]. It is estimated that 100,000 animals may have perished so far in the bushfires [72]. There have been massive losses of wildlife, notably of the iconic koala in Queensland and New South Wales. Harvesting of grasslands in bushfire threatened areas could have provided desperately needed fodder for livestock in drought-impacted areas. Considerations of the confinement, feeding, watering and humane protection of livestock in bushfire-threatened areas raise major humane and ethical questions.

(13). Experts: climate change is exacerbating drought and forest fires.

Climate change is exacerbating drought and forest fires world-wide (as well as causing ice melting, sea level rise and exacerbation of the intensity of tropical storms). Thus the expert Australian Climate Council (formerly the Australian Climate

Commission that was scrapped by the effective climate change denialist Coalition Government in 2013) on climate change impacting worsening drought in Australia (2018): "Climate change is likely making drought conditions in southwest and southeast Australia worse. Climate change has contributed to a southward shift in weather systems that typically bring cool season rainfall to southern Australia. Since the 1970s late autumn and early winter rainfall has decreased by 15 percent in southeast Australia, and Western Australia's southwest region has experienced a 15 percent decline in cool season rainfall. Climate change is also driving an increase in the intensity and frequency of hot days and heatwaves in Australia, exacerbating drought conditions. Queensland and New South Wales are currently in the grip of severe drought, with drought declared for 16.4 percent of New South Wales and 57.6 percent of Queensland. Current drought conditions come after a 2016/2017 summer characterised by record-breaking temperatures, followed by a record dry winter. Rainfall over southern Australia during autumn 2018 was the second lowest on record. Time spent in drought is projected to increase in the future across southern Australia. Future drying trends in Australia will be most pronounced over southwest Western Australia, with total reductions in autumn and winter precipitation potentially as high as 50 percent by the late 21st century" [74].

Similarly the Australian Climate Council has issued a report on the exacerbation of bushfire risk due to climate change (2019): "Key Findings. 1. The catastrophic, unprecedented fire conditions currently affecting NSW and Queensland have been aggravated by climate change. Bushfire risk was exacerbated by record breaking drought, very dry fuels and soils, and record-breaking heat. 2. Bushfire conditions are now more dangerous than in the past. The risks to people and property have increased and fire seasons have lengthened. It is becoming more dangerous to fight fires in Australia. 3. The fire season has lengthened so substantially that it has already reduced opportunities for fuel reduction burning. This means it is harder to prepare for worsening conditions. 4. The costs of fighting fires are increasing. Australia relies on resource sharing arrangements between countries and states and territories within

Australia. As seasons overlap and fires become more destructive, governments will be increasingly constrained in their ability to share resources and the costs of tackling fires will increase. 5. The government must develop an urgent plan to (1) prepare Australian communities, health and emergency services for escalating fire danger; and (2) rapidly phase out the burning of coal, oil and gas which is driving more dangerous fires" [75].

In 2019 over 11,000 scientists signed up to a World scientists' warning of a Climate Emergency that sets out trends in 24 climate-related areas over the last 40 years [26]. Reproduced here is my analysis of the bushfire-related section of [26]: "Figure 2 (l). Area burned in the United States (million hectares per year)" involves very scattered data. However, assuming the authors' quasi-linear best fit involving an increase from 0.9 (1983) to 3.2 (2018), one can estimate an increase of 0.066 million hectares per year (+ 44.1% per decade, +4.4% per year). Extrapolating from this average yields an estimate of 4.0 million hectares burned in 2030. Comments. In 2019 alone there have been massive forest fires in California, the Amazon and continent-wide in Australia. The Guardian Australia: "The link between rising greenhouse gas emissions and increased bushfire risk is complex but, according to major science agencies, clear. Climate change does not create bushfires but it can and does make them worse. A number of factors contribute to bushfire risk, including [elevated] temperature, fuel load, dryness, [and] wind speed and [decreased] humidity… The year coming into the 2019-20 summer has been unusually warm and dry for large parts of Australia. Above-average temperatures now occur most years and 2019 had the fifth-driest start to the year on record, and the driest since 1970. Australia recorded its hottest month in January 2019, its third-hottest July and its hottest October day in some areas, among other temperature records" [76]. Phys.org has listed 10 ways in which climate change can variously make wildfires worse: (1) hot, dry, windy weather, (2) more plant fuel, (3) change of plant species to more flammable, dryness-compatible species, (4) last-drop transpiration by thirsty plants, (5) more [dry] lightning, (6) differential warming of the Arctic and a weakened jet stream (this sinking drier and hotter air), (7) the periodic El Nino events lead to decreased rainfall, increased temperatures and increased fire risk in Indonesia and eastern Australia, (8) increase in the intensity of bushfires to unstoppable, (9) increased winter survival of pine beetles with increased flammable dead wood in Canadian boreal forests, (10) positive feedback from increased CO_2 -> increased warming -> increased forest fires -> increased CO_2 – > increased warming - > … [77]. The link between global warming and more forest fires has been documented by expert scientists for many years (e.g. see [78])" [27].

(14). Coalition-ruled Australia is among world leaders in 15 areas of effective climate change denialism and climate criminality.

A climate criminal Australia is among world leaders in 15 areas relating to GHG pollution and hence climate criminality. As confirmed by the results of the recent Australian Federal election (that returned an anti-science, climate criminal Coalition Government), a rich, greedy and anti-science Australia is mindlessly committed to coal and gas exploitation. Thanks to the homicidal greed of climate criminal countries such as Australia, the present plus 1.1C temperature rise is already devastating Island Nations, and a catastrophic plus 2C warming is now effectively unavoidable on present trends. Climate criminal Australia is among world leaders for the following 15 climate criminal activities or parameters: (1) annual per capita greenhouse gas pollution, (2) live methanogenic livestock exports, (3) natural gas exports, (4) recoverable shale gas reserves that can be accessed by hydraulic fracturing (fracking), (5) coal exports, (6) land clearing, deforestation and ecocide, (7) speciescide or species extinction], (8) coral reef destruction, (9) whale killing and extinction threat through global warming impacting on krill stocks, (10) terminal carbon pollution budget exceedance, (11) per capita Carbon Debt], (12) ultimately GHG generating iron ore exports, (13) climate change inaction, (14) climate genocide and approach towards omnicide and terracide, and (15) increasing Domestic GHG pollution despite Paris commitments to lower GHG pollution (for detailed documentation see [58]).

PM Scott "Scomo" Morrison heads a pro-coal, pro-gas, pro-oil, pro-fossil fuels, anti-science, anti-environment, and right-wing Coalition Government. (one third climate change deniers, and two thirds effective climate change deniers through resolute climate change inaction). Morrison notoriously held up a lump of coal in the Australian Parliament, idiotically declaring "This is coal. Don't be afraid, don't be scared" [79]. PM Morrison was on holiday overseas as the bushfire tragedy unfolded, as was the Defence Minister and the New South Wales Emergency Minister. On his return after public outcry, Morrison attended a bushfires press conference at which the fire and police chiefs and the New South Wales premier were grim faced but Scomo grinned throughout. His first major announcement on his return was that of 2 gas-fired power stations to be built in Gatton Queensland and Dandenong, Melbourne, Victoria, respectively, with the possibility of some coal-fired power stations also being built – utter irresponsibility and ignoring the reality of fossil fuel burning driving global warming, drought and bushfires (see section 13 above). Pro-coal Coalition MPs got in the act and declared that it was "not the time" to discuss climate change during the bushfire emergency. However Greens Leader, Dr Di Natale was not intimidated, declaring: "Yet with Queensland and New South Wales burning, the Coalition government refuses to acknowledge this scientific reality [of climate change driving drought and bushfires] and instead wants to use taxpayer dollars to fund new coal-fired power stations. Every politician, lobbyist, pundit and journalist who has fought to block serious action on climate change bears responsibility for the increasing risk from a heating planet that is producing these deadly bushfires" [79].

The Coalition then politically exploited the bushfires by drafting 3,000 army reservists (a much more expensive option than using the regular defence forces), and announced financial measures for the bushfires, including $2 billion over 2 years for a National Bushfire Recovery Agency. However the Coalition then politicized these actions by boasting about them on a slick social media presentation. A major fire chief was evidently upset over not having been consulted. Retired fire chiefs who had collectively (but unsuccessfully) sought timely meetings with government to warn of the danger were effectively ignored. Some people involved on the ground in the bushfire tragedy refused to shake hands with Morrison. And then a Coalition MP and critic of climate change action outraged UK media by ignorantly denying any connection between climate change and the Australian bushfires [81].

Final comments.

The Australian bushfires are a nation-wide tragedy involving (as of 10 January 2020). 10,700,000 hectares burned, 28 deaths, 2,204 homes destroyed and massive killing of native animals and livestock. Yet the burning may continue for several more months of Summer and thence into the beginning of Autumn. While scientists, including the Australian Climate Council, had been warning of worsening drought and bushfire risk for years, the effective climate change denialist Coalition Government backed by right-wing oligopoly media, ignored the science. Under the climate criminal Coalition Australia is among world leaders in 15 areas of climate criminality [58]. Australia with 0.3% of the world's population is responsible for 4.5% of GHG emissions (Exports included) [62]. Climate criminal Australia's anti-science profligacy also impacts the whole world through climate change and air pollution deaths. Thus, for example, the Australian Coalition Government approved a modified Adani coal mining, coal-for-India scheme in Queensland, but it was estimated that toxic pollutants from burning the coal from the full-blown initial proposal would eventually kill 1.4 million Indians over the life-time of the huge mine [8].

To reiterate the words of Australian Greens Leader, Dr Di Natale: "Every politician, lobbyist, pundit and journalist who has fought to block serious action on climate change bears responsibility for the increasing risk from a heating planet that is producing these deadly bushfires" [79]. What will threatened Humanity do around the world? The world will eventually lose patience with a climate criminal Australia and take action via Green Tariffs, International Criminal Court prosecutions, International Court of Justice litigations, and Boycotts, Divestment and Sanctions (BDS). Decent Australians will utterly reject the anti-science, climate criminal and Australian-killing Coalition, vote 1 Green and put the Coalition last.

References.

[1]. "2019-20 Australian bushfire season", Wikipedia:
https://en.m.wikipedia.org/wiki/2019%E2%80%9320_Australian_bus
hfire_season.
[2]. "2019-20 Australian bushfire season", Global Fire Data:
https://globalfiredata.org/pages/category/australia/.
[3]. Australian Government Department of Energy and the
Environment, "Australia's emissions projections 2019" (Figure 5,
page 9):
https://www.environment.gov.au/system/files/resources/4aa038fc-
b9ee-4694-99d0-c5346afb5bfb/files/australias-emissions-projections-
2019-report.pdf.
[4]. Graham Redfearn, "Australia's bushfires have emitted 250 m
tonnes of CO_2, almost half of country's annual emissions", Guardian,
13 December 2019:
https://www.theguardian.com/environment/2019/dec/13/australias-
bushfires-have-emitted-250m-tonnes-of-co2-almost-half-of-countrys-
annual-emissions.
[5]. Mike Foley, "Bushfires spew two-thirds of national carbon
emissions in one season", Sydney Morning Herald, 2 January 2020:
https://www.smh.com.au/politics/federal/bushfires-spew-two-thirds-
of-national-carbon-emissions-in-one-season-20200102-p53oez.html.
[6]. H. Keith, B.G. Mackey and D. B. Lindenmayer, "Re-evaluation
of forest biomass carbon stocks and lessons from the world's most
carbon-dense forests", PNAS, 14 July 2009, 106 (28) 11635-11640:
https://www.pnas.org/content/106/28/11635.
[7]. World Health Organization (WHO), "Air pollution":
https://www.who.int/airpollution/household/en/.
[8]. "Stop air pollution deaths":
https://sites.google.com/site/300orgsite/stop-air-pollution-deaths.
[9]. F. Ximines, M. Stephens, M. Brown, B. Law, M. Mylek, J.
Schirmer, "Mechanical fuel load reduction in Australia: a potential
tool for bushfire mitigation", Australian Forestry, Volume 80, Issue 2,
2017:
https://www.tandfonline.com/doi/abs/10.1080/00049158.2017.131120
0?src=recsys&journalCode=tfor20.
[10]. Farm Energy, "Biochar: prospects of commercialization", Farm
Energy Extension, 3 April 2019: https://farm-
energy.extension.org/biochar-prospects-of-commercialization/.
[11]. Progress in Thermochemical Biomass Conversion, volume 1,
IAE Bioenergy, ed. A.V. Bridgewater (Blackwell Science), page 224:
http://books.google.com.au/books?id=rdqGX0LEg7sC&pg=PA224&l

pg=PA224&dq=Gt++biomass+%22arable+land%22&source=bl&ots
=KfEmoUUg6T&sig=EuLvPTf4uJHK6Wq7jbpQ3WLHcnM&hl=en
&ei=UdzISZXlDpmMsQPH3cyMAQ&sa=X&oi=book_result&resnu
m=1&ct=result#PPP1,M1.

[12]. Alok Jha, ""Biochar" goes industrial with giant microwaves to lock carbon in charcoal", Guardian (13 March 2009): http://www.guardian.co.uk/environment/2009/mar/13/charcoal-carbon.

[13]. Johannes Lehmann, Biochar for mitigating climate change: carbon sequestration in the black": http://www.geooekologie.de/download_forum/forum_2007_2_spfo07 2b.pdf.

[14]. Gaia Vince, "One last chance to save mankind", New Scientist, 23 January 2009: http://www.newscientist.com/article/mg20126921.500-one-last-chance-to-save-mankind.html?full=true and http://biocharfund.com/images/hansen%2C%20target%20atmospheric %20c02.pdf.

[15]. J.A. Harrison, "The carbon cycle ", Vision Learning: http://www.visionlearning.com/library/module_viewer2.php?mid=95 &l=&let1=Ear.

[16]. Gideon Polya, "Forest biomass-derived Biochar can profitably reduce global warming and bushfire risk", Yarrra Valley Climate Action group, 2009: https://sites.google.com/site/yarravalleyclimateactiongroup/forest-biomass-derived-biochar-can-profitably-reduce-global-warming-and-bushfire-risk.

[17]. Gideon Polya, "Polya's 3 Laws Of Economics Expose Deadly, Dishonest And Terminal Neoliberal Capitalism", Countercurrents, 17 October 2015: https://countercurrents.org/polya171015.htm.

[18]. US National Oceanic and Atmospheric Administration, "Trends in Atmospheric Carbon Dioxide": https://www.esrl.noaa.gov/gmd/ccgg/trends/.

[19]. William Ripple et al., "World scientists' warning of a climate emergency", BioScience, 5 November 2019: https://academic.oup.com/bioscience/advance-article/doi/10.1093/biosci/biz088/5610806.

[20]. Gideon Polya, "Extrapolating 11,000 scientists' climate emergency warning to 2030 climate catastrophe", Countercurrents, 14 November 2019: https://countercurrents.org/2019/11/extrapolating-11000-scientists-climate-emergency-warning-to-2030-catastrophe.

[21]. 300.org:. https://sites.google.com/site/300orgsite/300-org.

[22]. "300.org – return atmosphere CO_2 to 300 ppm CO_2":
https://sites.google.com/site/300orgsite/300-org—return-atmosphere-co2-to-300-ppm.

[23]. Timothy Lenton, Johan Rockstrom, Owen Gaffney, Stefan Rahmsdorf, Katherine Richardson, Will Steffen and Hans Joachim Schellnhuber, "Climate tipping points – too risky to bet against", Nature 575, 592-595, 27 November 2019:
https://www.nature.com/articles/d41586-019-03595-0.

[24]. IPCC, "Global warming of 1.5 °C", 8 October 2018: http://www.ipcc.ch/report/sr15/.

[25]. IPCC, "Global warming of 1.5 °C. Summary for Policymakers", 8 October 2018: http://report.ipcc.ch/sr15/pdf/sr15_spm_final.pdf.

[26]. William Ripple et al., "World scientists' warning of a climate emergency", BioScience, 5 November 2019:
https://academic.oup.com/bioscience/advance-article/doi/10.1093/biosci/biz088/5610806.

[27]. Gideon Polya, "Extrapolating 11,000 scientists' climate emergency warning to 2030 climate catastrophe", Countercurrents, 14 November 2019: https://countercurrents.org/2019/11/extrapolating-11000-scientists-climate-emergency-warning-to-2030-catastrophe.

[28]. William J. Ripple et al., 15,364 signatories from 184 countries, "World scientists' warning to Humanity: a second notice", Bioscience, 13 November 2017: https://academic.oup.com/bioscience/advance-article/doi/10.1093/biosci/bix125/4605229.

[29]. Gideon Polya, "Over 15,000 scientists issue dire warning to humanity on catastrophic climate change and biodiversity loss", Countercurrents, 20 November 2017: https://countercurrents.org/2017/11/20/over-15000-scientists-issue-dire-warning-to-humanity-on-catastrophic-climate-change-and-biodiversity-loss/.

[30]. James Hansen, "Climate change in a nutshell: the gathering storm", Columbia University, 18 December 2018: http://www.columbia.edu/~jeh1/mailings/2018/20181206_Nutshell.pdf.

[31]. Gideon Polya, "IPCC +1.5C avoidance report – effectively too late, but stop coal burning for "less bad" catastrophes', Countercurrents, 12 October 2018: https://countercurrents.org/2018/10/ipcc-1-5c-avoidance-report-effectively-too-late-but-stop-coal-burning-for-less-bad-catastrophes.

[32]. Andrew Glikson, "Inferno: from climate denial to planetary arson", Countercurrents, 8 September 2019: https://countercurrents.org/2019/09/inferno-from-climate-denial-to-planetary-arson.

[33]. Clive Hamilton, "Earth Masters. Playing god With the Climate", 2013.

[34]. WBGU, "Solving the climate dilemma: the budget approach": http://www.ecoequity.org/2009/10/solving-the-climate-dilemma-the-budget-approach/.

[35]. Australian Climate Commission, "The critical decade 2013: a summary of climate change science, risks and responses", 2013, p7: http://climatecommission.gov.au/wp-content/uploads/The-Critical-Decade-2013-Summary_lowres.pdf.

[36]. Robert Goodland and Jeff Anfang. "Livestock and climate change. What if the key actors in climate change are ... cows, pigs and chickens?", World Watch, November/December 2009: https://pdfs.semanticscholar.org/6704/c7a0777c82357704d82b9ae8007c1197cb07.pdf?_ga=2.187734888.1597394103.1556059730-1006954717.1556059730

[37]. Professor Stephen Hawking quoted in Will Dunham, "Nuclear, climate perils push Doomsday Clock ahead", Reuters, 22 January 2007: https://www.reuters.com/article/idUSN17314370.

[38]. Stephen Hawking, "Brief Answers to the Big Questions", John Murray, 2018, Chapter 7.

[39]. "Are we doomed?": https://sites.google.com/site/300orgsite/are-we-doomed.

[40]. "Methane Bomb Threat": https://sites.google.com/site/methanebombthreat/.

[41]. "Too late to avoid global warming catastrophe": https://sites.google.com/site/300orgsite/too-late-to-avoid-global-warming.

[42]. "Climate Genocide": https://sites.google.com/site/climategenocide/.

[43]. "Nuclear weapons ban, end poverty and reverse climate change": https://sites.google.com/site/drgideonpolya/nuclear-weapons-ban.

[44]. Greta Thunberg, "No one is too small to make a difference", Penguin, 2019.

[45]. "Climate Justice & Intergenerational Equity": https://sites.google.com/site/300orgsite/climate-justice.

[46]. "Stop climate crime": https://sites.google.com/site/300orgsite/stop-climate-crime.

[47]. "Climate Revolution Now": https://sites.google.com/site/300orgsite/climate-revolution.

[48]. Sir Nicholas Stern, quoted in "Climate change: "the greatest market failure the word has seen"", New Economist, 30 October 2006:

http://neweconomist.blogs.com/new_economist/2006/10/stern_review_2.html.

[49]. Alison Benjamin, "Stern: climate change a "market failure"", Guardian, 29 November 2009: https://www.theguardian.com/environment/2007/nov/29/climatechange.carbonemissions.

[50]. Pope Francis, Encyclical Letter "Laudato si", 2015: http://w2.vatican.va/content/francesco/en/encyclicals/documents/papa-francesco_20150524_enciclica-laudato-si.html.

[51]. Gideon Polya, "Green Left Pope Francis Demands Climate Action "Without Delay" To Prevent Climate "Catastrophe"", Countercurrents, 10 August, 2015: https://countercurrents.org/polya100815.htm.

[52]. Phillip Levin, Donald Levin, "The real biodiversity crisis", American Scientist, January-February 2002: http://www.americanscientist.org/issues/pub/the-real-biodiversity-crisis.

[53]. World Wildlife Fund (WWF) "Living Planet Report 2018 aiming higher – summary": https://s3.amazonaws.com/wwfassets/downloads/lpr2018_summary_report_spreads.pdf.

[54]. Gideon Polya, "Inescapable $200-$250 trillion global Carbon Debt increasing by $16 trillion annually", Countercurrents, 27 April 2019: https://countercurrents.org/2019/04/inescapable-200-250-trillion-global-carbon-debt-increasing-by-16-trillion-annually-gideon-polya.

[55]. Chris Hope, "How high should climate change taxes be?", Working Paper Series, Judge Business School, University of Cambridge, 9.2011: http://www.jbs.cam.ac.uk/fileadmin/user_upload/research/workingpapers/wp1109.pdf.

[56]. International Monetary Fund (IMF), "Fiscal Monitor: how to mitigate climate change". Executive Summary", September 2019: file:///C:/Users/Gideon/AppData/Local/Temp/execsum-6.pdf.

[57]. Gideon Polya, "Australia rejects IMF Carbon Tax & preventing 4 million pollution deaths by 2030", Countercurrents, 15 October 2019: https://countercurrents.org/2019/10/australia-rejects-imf-carbon-tax-preventing-4-million-pollution-deaths-by-2030.

[58]. Gideon Polya, "War criminal & climate criminal Australian deception at UN General Assembly", Countercurrents, 29 September 2019: https://countercurrents.org/2019/09/war-criminal-climate-criminal-australian-deception-at-un-general-assembly.

[59]. Paul Karp, "Josh Frydenberg rejects IMF report that Australia will fail to meet Paris target", Guardian, 11 October 2019:

https://www.theguardian.com/environment/2019/oct/11/australia-will-fail-to-meet-paris-target-even-with-carbon-price-of-us75-a-ton-imf-says.
[60]. Josh Frydenberg," Interview with Sabra Lane, AM, ABC Melbourne", Josh Frydenberg, 11 October 2019: http://ministers.treasury.gov.au/ministers/josh-frydenberg-2018/transcripts/interview-sabra-lane-am-abc-melbourne.
[61]. "National Bushfire Recovery Agency", Wikipedia: https://en.wikipedia.org/wiki/National_Bushfire_Recovery_Agency.
[62]. Gideon Polya, "Revised Annual Per Capita Greenhouse Gas Pollution For All Countries – What Is Your Country Doing?", Countercurrents, 6 January, 2016: http://www.countercurrents.org/polya060116.htm.
[63]. Gideon Polya, "Exposing And Thence Punishing Worst Polluter Nations Via Weighted Annual Per Capita Greenhouse Gas Pollution Scores", Countercurrents, 19 March, 2016: https://countercurrents.org/polya190316.htm.
[64]. Gideon Polya, "Australia shocked by cricket ball tampering but ignores horrendous Australian crimes from child abuse to genocide", Countercurrents, 24 April 2018: https://countercurrents.org/2018/04/australia-shocked-by-cricket-ball-tampering-but-ignores-horrendous-australian-crimes-from-child-abuse-to-genocide.
[65]. Gideon Polya, "Annual per capita greenhouse gas emissions by country (2013)", Gideon Polya: https://sites.google.com/site/drgideonpolya/annual-per-capita-greenhouse.
[66]. "Economy of Australia", Wikipedia: https://en.wikipedia.org/wiki/Economy_of_Australia.
[67]. Gideon Polya, "Horrendous Cost For Australia Of US War On Terror", Countercurrents, 14 October, 2012: https://countercurrents.org/polya141012.htm.
[68]. Gideon Polya, "Australian state terrorism (4). Jingoistic, US Lackey Australia's Deadly Betrayal Of Its Traumatized Veterans", Stop state terrorism, 2018: https://sites.google.com/site/stopstateterrorism/australian-state-terrorism-4.
[69]. Gideon Polya, "Advance Australia Fair" Hides Australian Racism, Theft, Genocide, Ecocide, Speciescide & Terracide", Countercurrents, 1 July 2019: https://countercurrents.org/2019/07/advance-australia-fair-hides-australian-racism-theft-genocide-ecocide-speciescide-terracide.
[70]. "Stop state terrorism": https://sites.google.com/site/stopstateterrorism/.

[71]. "Exposing Australia":
https://sites.google.com/site/exposingaustralia/home.
[72]. Sarah Jane Bell, "Farmers impacted by bushfires count "heartbreaking" cost as livestock losses climb", ABC News, 7 January 2020: https://www.abc.net.au/news/2020-01-07/farmers-recount-heartbreaking-toll-of-bushfire-livestock-losses/11844696.
[73]. Australian Bureau of Statistics, "Livestock":
https://www.abs.gov.au/ausstats/abs@.nsf/Latestproducts/7121.0Main%20Features612017-18?opendocument&tabname=Summary&prodno=7121.0&issue=2017-18&num=&view=.
[74]. Climate Council, ""Fact Sheet. Climate change and drought June 2018": https://www.climatecouncil.org.au/wp-content/uploads/2018/06/CC_MVSA0146-Fact-Sheet-Drought_V2-FA_High-Res_Single-Pages.pdf.
[75]. Climate Council of Australia, ""This is Not Normal": Climate change and escalating bushfire risk", November 2019: https://www.climatecouncil.org.au/wp-content/uploads/2019/11/bushfire-briefing-paper_18-november.pdf.
[76]. "What are the links between climate change and bushfires? – explainer", Guardian, 11 November 2019: https://www.theguardian.com/australia-news/2019/nov/11/what-are-the-links-between-climate-change-and-bushfires-explainer.
[77]. "Ten ways climate change can make wildfires worse", Phys.org, 12 November 2019: https://phys.org/news/2019-11-ten-ways-climate-wildfires-worse.html.
[78]. A.L. Westerling, H. G. Hidalgo, D. R. Cayan, T. W. Swetnam, Warming and Earlier Spring Increase Western U.S. Forest Wildfire Activity, Science 18 August 2006: Vol. 313. no. 5789, pp. 940 – 943: http://www.sciencemag.org/cgi/content/full/313/5789/940.
[79]. Katharine Murphy, "Scott Morrison brings coal to question time: what fresh idiocy is this?", Guardian, 9 February 2017: https://www.theguardian.com/australia-news/2017/feb/09/scott-morrison-brings-coal-to-question-time-what-fresh-idiocy-is-this.
[80]. Colin Brinsden, "Greens ramp up climate war as fires burn", Canberra Times", 10 November 2019: https://www.canberratimes.com.au/story/6484064/greens-ramp-up-climate-war-as-fires-burn/?cs=14231.
[81]. Sarah Martin, Craig Kelly interview: senior government MPs distance themselves after Piers Morgan lashing "", Guardian, 7 January 2020: https://www.theguardian.com/australia-news/2020/jan/07/craig-kelly-interview-piers-morgan-calls-mp-disgraceful-for-denying-climate-link-to-bushfires.

CHAPTER 10: AUSTRALIAN CLIMATE CRIMINALITY, HEAT STRESS DEATHS & AUSTRALIAN ABORIGINAL ETHNOCIDE

[First published as Gideon Polya, "Australian climate criminality, heat stress deaths & Australian Aboriginal Ethnocide", Countercurrents, 19 December 2019: https://countercurrents.org/2019/12/australian-climate-criminality-heat-stress-deaths-australian-aboriginal-ethnocide/.]

More than 500 Australians presently die from heat stress each year with this disaster disproportionately impacting the elderly, the impoverished and Indigenous First Nations peoples. Australia is currently beset continent-wide with climate change-exacerbated heat waves with sustained temperatures above 40 degrees Centigrade, extraordinary, nation-wide bush fire emergencies commencing in Winter, and a devastating and widespread drought in Eastern Australia that has lasted for years.

Heat stress deaths.

I am very conscious of heat stress deaths because I am 75, my home city of Melbourne (with rest of Australia) is suffering a heat wave (43 degrees Centigrade predicted for Melbourne tomorrow on Friday 20 December), and I am well aware of the need to keep out of the sun, keep cool and drink plenty of water.

According to Dr Richard Kidd, chairperson of the Australia Medical Association of Queensland Council of General Practice, more than 500 people die of heat stress in Australia each year, more than total annual deaths from floods, fires and cyclones and with about 22% occurring on 27 January in the middle of the Australian Summer, the day after the national

"

holiday of Australia Day (Invasion Day to Indigenous Australians and which marks the anniversary of the invasion of Australia by the genocidal British on 26 January 1788) [1, 2].

Australian researchers have collated data on 5,300 heat stress deaths in Australia over the period 1840-2009, with decadal heat stress deaths steadily rising to a maximum of about 400-800 per decade in the 6-decade period of 1890- 1949, thence falling and then peaking again to over 500 in the decade 2000-2009 [3, 4].

The worst events (state location and heat stress deaths in brackets) were in October 1895- January 1896 (Western Australia, South Australia, Victoria, Queensland, New South Wales; 435), January 1908 (Victoria, South Australia, New South Wales; 213), January 1939 (New South Wales, Victoria, South Australia; 420), January-February 1959 (Melbourne, Victoria; 145), and January-February 2009 (Victoria, South Australia; 432) [4].

It was estimated from public health records that 432 people died from heat stress in the heat wave in Victoria and South Australia leading up to and including the 2009 Black Saturday bushfire catastrophe in which 173 people perished in the Victorian bushfire holocaust [4-6].

Jacobs et al. (Monash University, Melbourne) have expertly summarized the heat stress circumstances of the Black Saturday holocaust (2015): "The Black Saturday bushfire event of February 7, 2009, devastated the state of Victoria, Australia, resulting in 173 deaths. On this day, the maximum temperature in Melbourne (state capital of Victoria, population 4 million people) exceeded 46 °C, there were wind gusts of over 80 km h(-1) and the relative humidity dropped below 5 %. We investigated the severe meteorological conditions of Black Saturday and the risk of heat stress and dehydration for the residents of Melbourne. This was through the analysis of weather station data, air pollution data, the apparent temperature (AT) and the COMfort FormulA human energy budget model. A very strong pressure gradient caused hot and dry air to be advected to Melbourne from the desert interior of Australia creating the extreme weather conditions. The AT showed that on Black Saturday, heat stress conditions were present, though underrepresented due to assumptions in the AT formula. Further investigation into the human energy budget revealed that the conditions required a sweating rate of 1.4 kg h(-1) [1.4 kg per hour] to prevent heat accumulation into the body. If sweating stopped, hyperthermia could occur in 15 min. Sensitivity tests indicated that the dry air and strong winds on Black Saturday helped to release latent heat, but the required sweating rate was virtually unattainable for an average person and would result in intense dehydration. Air particulates were at dangerous concentrations in Melbourne on Black Saturday, further intensifying the stresses to the human body. In the future, we recommend that the AT is not used as a thermal comfort measure as it underestimates the physical stress people experience" [5].

Dr Karl Kruszelnicki (medical doctor and celebrated Australian science educator)(2016): "In 2009, the terrible Black Saturday bushfires killed 173 people. What most Australians don't realise is that the crippling heat around the horrendous bushfires killed 374 people. In the European heatwave of 2003, 50,000-70,000 people died between June and August. The Russian heatwave of 2010 killed about 55,000 people… In Paris alone in 2003, some 15,000 people died from heat. They were overwhelmingly elderly women, living alone, and in the upper levels of apartments. "Elderly" often implies smaller reserves of strength. "Living alone" meant that another person wasn't available to help. "Upper levels" meant that their apartments were getting the full brunt of the heat. Another factor in Europe is that houses are designed to keep the heat in, not out. Excessive heat is especially harmful to the very young and the very old – and to those with chronic diseases and mental illnesses. Other risk factors are being obese, very malnourished, or very unfit. Drugs – both legal and illegal – can worsen your risk. Dehydration from alcohol can contribute. If the electrical grid crashes, and you lose air-conditioning, the heat in poorly designed houses can be fatal" [6].

Peter Gardner (Australian climate change activist) (2019): "The number of people killed by political terrorism in Australia over the last one hundred years can be counted on your fingers. As a comparison Melbourne climate commentator Dr Gideon Polya noted "Presently about 80,000 Australians die [preventably] … each year in Australia from "life-style" and "political choice" causes (e.g. 15,500 pa from smoking, 10,000 pa from air pollution, 500 pa from heat stress) as compared to 0.2 pa from jihadi terrorism [in Australia] this century." You have more chance of dying falling out of bed. Also of note is the 500 fatalities per annum due to heat stress – an extreme weather event heavily influenced by global warming. Polya continued "However globally about 7.5 million people die avoidably (prematurely) each year due to the effects [of] carbon burning pollutants (7.0 million) (WHO) or to climate change (0.5 million). This latter estimate of presently about 0.5 million climate change-related deaths may be an under-estimate… [with] impoverished, tropical or sub-tropical countries already being severely impacted by global warming" [7].

Gideon Polya (me) writing about heat stress deaths 10 years ago (2009): "Older people are peculiarly threatened by the worsening climate emergency in three key areas that can be summarized by the "three Ds" of Devaluation (GDP growth to prevent devaluation of pensions, investments, superannuation and family support is only sustainable in a non-carbon, renewable energy-based economy), Death (older people are frailer and more susceptible to heat stress death because of a weakened brain signalling system) and Descendants (their descendants will hate them for what they have done to the planet)… The World must urgently take the advice of top climate scientists and analysts who are advocating "100% renewable energy by 2020", "cut carbon emissions 80% by 2020" and a rapid return of atmospheric CO_2 to about 300 ppm for a planet that is safe and sustainable for all peoples and species… If Humanity really wants to save itself and the Biosphere – including Australia's Great Barrier Reef, the Kakadu Wetlands, temperate and tropical Rainforests and the Murray-Darling River System – then it must urgently apply Sanctions, Boycotts, Green Tariffs, Reparation Demands and

International Criminal Court (ICC) prosecutions against racist, climate genocidal Apartheid Australia and other climate criminal countries" [8].

Gideon Polya (2009): "Australian electricity consumers pay for electricity @ $1.6 million per fellow Australian killed by coal [burning][4,859 per year, ignoring 500 heat stress deaths pa]"[9, 10].

Gideon Polya in a Submission to a Senate Inquiry on Men's Health) (2009): "3. Australian (especially older Australian) deaths from global warming-related excessive heat (about 50% male deaths). The week before the Black Saturday February 7 2009 bushfire disaster saw a sustained heat wave in SE Australia (population about 6 million) with temperatures exceeding 43°C for 3 successive days in Melbourne (Wednesday January 28, Thursday January 29 and Friday January 30). In the late January 2009 heatwave over 100 people died in Melbourne and over 200 died in South East Australia (Victoria, South Australia and Northern Tasmania) as determined by Professor Neville Nicholls, Monash University, by comparing before and after Monday and Tuesday Death Notices…The consequent devastating bushfire tragedy in Victoria on Saturday February 7 killed more than 209 people, with 500 injured, 100 in hospital with burns, over 1,834 homes destroyed, thousands of homes damaged, and over 450,000 hectares burned … By way of comparison, the 2003 European heat wave occurred in one of the hottest summers on record in Europe (Western European population 392 million in 2003). The heat wave led to health crises in several countries and combined with drought to create a crop production shortfall in Southern Europe. More than 52,000 Europeans died as a result of the heat wave with ~15,000 dying in France… Heat waves will differentially kill elderly people (e.g. the heat wave in Europe that killed 35,000-52,000 people in Europe and nearly 15,000 in France). Older people are differentially affected, the problems being that older people are frailer, more prone to heat stress and have diminished brain signalling of dehydration stress… Public education is urgently needed about these risks" [11].

I am a half-century career scientist who has always been
prepared to speak out in the public interest but was suddenly
rendered "invisible" in Australia about a dozen years ago with
Australian Intelligence "nutters" (described thus by Labor PM
Paul Keating), mendacious, US- and Zionist-subverted,
oligopoly Mainstream media gate-keepers, and the Apartheid
Israel Lobby being the obvious culprits from an evidence-based
perspective. Yet if Australians had been able to listen to me
(and no doubt to numerous other like-minded and science-
informed people) and taken appropriate action, then conceivably
about 12 years x 500 heat stress deaths per year = 6,000 such
Australian heat stress deaths could have been saved. One notes
that before 2014 zero (0) Australians had been killed in
Australia by jihadi non-state terrorists, and beginning in 2014
only 4 Australians have been killed thus.

Howard Florey Institute (University of Melbourne)
neuroscience researchers Dr Michael Farrell, Associate
Professor Gary Egan and Professor Derek Denton discovered
that a region in the brain called the mid cingulate cortex predicts
how much water a person needs, but this region malfunctions in
older people. Dr Farrell described how their study involved
drinking water by older and younger people: "Although all
participants had the same level of thirst, the older people only
drank half as much water as the younger subjects. Using PET
[Positron-Emission Tomography] imaging we found in the older
people, the mid cingulate cortex was 'turned off' much earlier
by drinking small volumes. This discovery helps explain why
the elderly can become easily dehydrated" [12, 13].

If only "They" had listened to these medical experts… I am
reminded that back in 1990 Professor Fred Mendelsohn
(medical doctor, eminent neuroscience researcher and later head
of the Howard Florey Institute) courageously wrote a letter
published by The Age newspaper (Melbourne) opposing the
coming US Alliance war on Iraq because it would kill huge
numbers of children. This fine medical scientist and
humanitarian was correct – the subsequent Iraqi Holocaust and
Iraqi Genocide took the lives of 4.6 million Iraqis, half of them
children, in the period 1990-2011 ([14-16; see also [17]). Indeed

in the only US admission of the immensity of this genocidal crime, in 1996 anti-racist Jewish American journalist Lesley Stahl publicly asked Jewish American UN Ambassador Madeleine Albright "We have heard that half a million [Iraqi] children have died. I mean, that's more children than died in Hiroshima. And, you know, is the price worth it?", to which Albright replied "We think the price is worth it" ([14, 18]. If only "They" had listened…

Heat stress deaths and the ongoing Australian Aboriginal Genocide and Aboriginal Ethnocide.

Heat stress disproportionately impacts the elderly and poor people who cannot afford to have or run air conditioning. The latter category includes Australia's impoverished Indigenous Aboriginal and Torres Strait Islander people who largely live in remote areas and rural towns as well as in the concrete heat sinks of Australia's capital cities. Since the British invasion in 1788 about 2 million Indigenous Australians died through dispossession, deprivation, introduced disease and violence (the latter killing about 0.1 million Aborigines). This Aboriginal Genocide continues with about 4,200 Indigenous Australians dying avoidably from deprivation each year (on a global comparative scale) out of an Indigenous population of 0.7 million, as compared to about 4,200 Indigenous Palestinians dying avoidably from deprivation out of an Occupied Palestinian population of 5 million in the ongoing Palestinian Genocide. Indeed the avoidable deaths as a percentage of population is 0.6% for Indigenous Australians as compared to 0.1% for Occupied Indigenous Palestinians, 1.0% for non-Arab Africans, 0.4% for Indians and zero % (0%) for non-Indigenous Australians [2, 19-21].

However there is also an ongoing Aboriginal Ethnocide conducted by a genocidally racist White Australia. Thus of 350-750 Aboriginal languages and dialects in 1788, only 150 survive and of these all but 20 are endangered [2]. This ongoing Aboriginal Ethnocide is promoted by English-only education [26, 27], Coalition hostility to remote Aboriginal communities living "on Country"(i.e. on their sacred tribal homeland in continental Australia), and the forced removal of children from their mothers (with this presently running at a record rate notwithstanding former Labor PM Kevin Rudd's famous "Sorry" for the so-called Stolen Generations that involved about 0.1 million Indigenous children being forcibly removed from their mothers, forbidden to speak their language, and trained to be effective quasi-slave labour in many instances) [27-30].

There is now a well-founded fear that Indigenous Australians will be forced out of remote communities and become Australia's first internal climate refugees because of unsustainably hot living conditions, constraints on potable water supply, lack of effective cooling of substandard and crowded housing, and poverty-related cut-off of electricity supply [31]. Lorena Allam: "Across central Australia, people are bracing themselves for another scorching summer of drought. At least nine remote communities and outstations are running out of water. A further 12 have reported poor quality drinking water as aquifers run low and the remaining supply is saline. Temperature records have already been broken. In the year to July 2019, Alice Springs had 129 days over 35C, and 55 days over 40C... Houses that don't cool down overnight create big health and social problems... Predictions by the Central Australian Aboriginal Congress for the health impacts of heat are dire. In its submission to the NT [Northern Territory] government's climate change policy discussion paper, it outlined some of them: "Increased sickness and mortality due to heat stress, increased food insecurity and malnutrition, increased risk from infectious disease, poorer mental health and an increased potential for social conflict"" [31].

Melissa Sweet writing in The Guardian cites evidence from northwest Queensland Indigenous Health worker Renee Blackman: "Blackman describes elderly Aboriginal people with multiple health problems stuck in inadequate housing without air-conditioning during increasingly frequent extreme heatwaves. Sometimes it is so hot, she says, the bitumen melts, making it difficult for her health teams to reach communities in times of high need. As well, patients are presenting to Gidgee Healing clinics with conditions such as dehydration that might be preventable if they could afford their power bills and had appropriate housing" [32].

Of course it is not just Indigenous Australians who are under threat from the present callous, neoliberal, Liberal Party-National Party Coalition Government. About 26% of Australians do not have air conditioning [33]. Public housing for poor people in Melbourne provides air conditioning to people with serious medical conditions but not to others [34].

Further, one cannot avoid consideration of neoliberal political responsibility for the man-made climate change that is presently impacting Australia with prolonged, widespread and catastrophic drought in Eastern Australia, with rivers running out of water (the Darling, the Barwon and the Namoi Rivers) [35], towns running out of drinking water, sustained, deadly and continent-wide high temperatures above 40 degrees Centigrade, and horrendous bushfires in Western Australia, South Australia, New South Wales and Queensland, with other states and territories facing similar catastrophes this Summer. The Aboriginal Genocide was premised on racism and the obscene and genocidal concept of "terra nullius" or an "empty land" for British settlers. And having been robbed of their family members, family, land, language, and culture under "terra nullius", Indigenous Australians in the New South Wales outback can no longer drink from or wash, swim or fish in the Darling, Barwon and Namoi Rivers that have now run dry under the deadly dichotomy of neoliberal greed and the obscene racist concept of "aqua nullius" (no non-agricultural human or environmental right to water) [35].

Thus under the climate criminal, anti-science, corrupt and anti-environment Coalition Government, while Australia has only 0.3% of the world's population it is responsible for 4.5% of global greenhouse gas (GHG) pollution (with its Exported GHG pollution included)[36, 37]. Further, Australia is among world leaders in 15 areas of climate criminality, specifically (1) annual per capita greenhouse gas pollution, (2) live methanogenic livestock exports, (3) natural gas exports, (4) recoverable shale gas reserves that can be accessed by hydraulic fracturing (fracking), (5) coal exports, (6) land clearing, deforestation and ecocide, (7) speciescide or species extinction, (8) coral reef destruction, (9) whale killing and extinction threat through global warming impacting on krill stocks, (10) terminal carbon pollution budget exceedance, (11) per capita Carbon Debt, (12) ultimately GHG generating iron ore exports, (13) climate change inaction, (14) climate genocide and approach towards omnicide and terracide, and (15) increasing GHG pollution despite Paris commitments to lower GHG pollution [36-44].

Australia's PM Scott "Scomo" Morrison is Australia's version of the populist buffoonery of the UK's Boris Johnson and America's Donald Trump. His only claim to fame outside of politics was an advertising campaign for Tourism Australia that starred a scantily clad young woman who crassly asked: "Where the bloody hell are we?" With bushfires raging across Australia, rivers running dry, catastrophic drought, shocking heat waves, a debt-burdened and faltering economy, and hundreds likely to die from heat stress in coming months, Pentecostal Christian but fervently neoliberal Scott "Scomo" Morrison (aka Scum-o, Skim-o, Scam-o and Scheme-o) had chosen to go overseas to a secret location for a holiday, this prompting Australian media to ask "Where the bloody hell are you?" One could well also ask what happened to the "Love thy neighbour as thyself" proposition of that wonderful Palestinian humanitarian, Jesus.

PM Scott "Scomo" Morrison's Coalition Government is extremely secretive and frequently invokes "national security" to justify not answering questions from the Opposition or journalists. However science-based, rational risk management, that is crucial for public safety and security, successively involves (a) accurate information, (b) scientific analysis (this involving the critical testing of potentially falsifiable hypotheses), and (c) systemic change to minimize harm when bad circumstances inevitably occur. Now "Scomo" is perfectly entitled to believe in anything he likes, but in the interests of Australian national security PM Scott "Scomo" Morrison, as leader of an anti-science and effective climate change denialist Coalition Government, should be asked in Parliament whether he believes in (a) Biblical Literalism, (b) Creationism, and (c) Miracles.

Of course heat stress-related deaths (presently over 500 such deaths per year in Australia) are but one part of a continuing disaster of 85,000 preventable Australian deaths each year due to "lifestyle" or "political choice" reasons, the breakdown (including some overlaps) being as follows: (1) 26,000 annual Australian deaths from adverse hospital events, (2) 17,000 obesity-related Australian deaths, (3) 15,500 smoking-related Australian deaths, (4) 10,000 carbon burning pollution-derived Australian deaths, (5) 4,000 avoidable Indigenous Australian deaths, (6) 5,600 Australian alcohol-related deaths, (7) 2,900 Australian suicides (circa 100 being veterans), (8) 1,400 Australian road deaths, (9) 630 Australian opiate drug-related deaths with 570 linked to US restoration of the Taliban-destroyed Afghan opium industry, (10) 500 heat stress- deaths, and (11) 300 Australian homicides (80 being of women killed domestically) [45-49].

Final comments.

About 7 million people die from air pollution each year with this including 10,000 Australians and 75,000 people dying each year from the long-term effects of pollutants from the burning of Australia's world-leading coal exports [50]. About 1 million people die from climate change each year although this may be an under-estimate because 15 million people die annually from deprivation in the Developing World (minus China), countries that are disproportionately impacted by man-made global warming [19]. The world is facing a worsening climate genocide in which 10 billion people may die this century en route to a sustainable human population of only about 0.5-1.0 billion by 2100 [51]. While Australia has only 0.3% of the world's population it is responsible for 4.5% of global greenhouse gas (GHG) pollution (with its Exported GHG pollution included)[36, 37] i.e. it is disproportionately contributing to this existential disaster to a 15 times greater extent than the rest of the world on a per capita basis.

Nevertheless the climate criminal Coalition Australian Government, that is dominated by an anti-science, climate change denialist, extremist minority, resolutely rejects a global warming basis for the current heat wave, drought and bush fire disaster that has been impacting Australia. The climate criminal Coalition, backed by the dominant US Murdoch media Empire, has been able to fool half the population. However the heat wave conditions in addition to exacerbating drought and bushfires is actually killing Australians with impoverished Indigenous Australians being disproportionately impacted. Indeed a racist and climate criminal Australia that has disproportionately contributed to generating over 20 million climate change-displaced persons per each year [52], is now on the verge of making large numbers of impoverished Indigenous Australian climate refugees within Australia in a foreshadowed jump in the Australian Aboriginal Genocide and Aboriginal Ethnocide [2]. Surely Indigenous Australians have suffered enough in over 230 years of genocidal racism [2].

Australia under the racist and climate criminal Coalition faces the prospect of international action via Green Tariffs, Boycotts, Divestment and Sanctions (BDS), International Criminal Court prosecutions and International Court of Justice litigations. Decent Australians will utterly reject the racist, climate criminal, anti-science and Australian-killing Coalition, vote 1 Green and put the Coalition last.

References.

[1]. Meecham Philpott and Sophie Kesteven, "Heat stress deaths rise following Australia Day, warns Queensland doctor", ABC News, 25 January 2016: https://www.abc.net.au/news/2016-01-25/heat-stress-deaths-rise-following-australia-day/7113030.
[2]. "Aboriginal Genocide":
https://sites.google.com/site/aboriginalgenocide/.
[3]. Andrew Gissing and Lucinda Coates, "Heatwaves are Australia's deadliest natural hazard and many of us are unprepared", ABC News, 18 January 2018: https://www.abc.net.au/news/2018-01-18/heatwaves-australias-deadliest-hazard-why-you-need-plan/9338918.
[4]. Lucinda Coates et al. "Exploring 167 years of vulnerability: An examination of extreme heat events in Australia 1844-2010", Environmental Science & Policy, Volume 42, October 2014, Pages 33-44:
https://www.sciencedirect.com/science/article/pii/S1462901114000999.
[5]. S. J. Jacobs, T. Vihma, and A.B. Pezza, "Heat stress during the Black Saturday event in Melbourne, Australia", Int. J. Biometeorol. 2015 Jun, 59(6):759-70:
https://www.ncbi.nlm.nih.gov/pubmed/25172086.
[6]. Karl Kruszelnicki, "Heat waves are more deadly than bushfires and they are going to get worse", Sydney Morning Herald, 13 December 2016: https://www.smh.com.au/opinion/heatwaves-are-more-deadly-than-bushfires-and-theyre-going-to-get-worse-20161212-gt9fyl.html.
[7]. Peter Gardner, "Scare campaigns and the Federal election", Peter Gardner, 13 March 2019: http://petergardner.info/2019/03/scare-campaigns-and-the-federal-election/.
[8]. Gideon Polya, "300.org Message To World – Act Urgently Over Climate Emergency And Climate Genocide", Countercurrents, 23 September, 2009: https://countercurrents.org/polya230909.htm.
[9]. Gideon Polya, "Coal is King: Australia CO_2 pollution fact sheet": http://sites.google.com/site/yarravalleyclimateactiongroup/%E2%80%9Ccoal-isking%E2%80%9D-australia-co2-pollution-fact-sheet.
[10]. Gideon Polya, "Senate Inquiry Submission by Dr Gideon Polya, biological chemist, to the Senate Select Committee on Climate Policy", 2009:
file:///C:/Users/Gideon/AppData/Local/Temp/sub273_pdf-3.pdf.

[11]. Gideon Polya, "Submission by Dr Gideon Polya to the Senate Select Committee on Men's Health" (2009): file:///C:/Users/Gideon/AppData/Local/Temp/sub40-2.pdf.
[12]. "Brain malfunction explains dehydration in elderly", Medindia, 25 December 2005: https://www.medindia.net/news/brain-malfunction-explains-dehydration-in-elderly-31069-1.htm.
[13]. Gideon Polya, "Global warming dangers and solutions for older people", Yarra Valley Climate Action Group, 2009: https://sites.google.com/site/yarravalleyclimateactiongroup/global-warming-dangers–solutions-for-older-people.
[14]. "Iraqi Holocaust, Iraqi Genocide": https://sites.google.com/site/iraqiholocaustiraqigenocide/.
[15]. "Muslim Holocaust Muslim Genocide": https://sites.google.com/site/muslimholocaustmuslimgenocide/.
[16]. "Iraqi Genocide Iraqi Holocaust articles by Gideon Polya", 2004- : https://sites.google.com/site/drgideonpolya/iraqi-genocide-iraqi-holocaust.
[17]. Ross Caputi, Richard Hil, and Donna Mulhearn, "The Sacking of Fallujah. A People's History", University of Massachusetts Press, 2019.
[18]. Lesley Stahl and Madeleine Albright quoted in "Madeleine Albright", Wikipedia: http://en.wikipedia.org/wiki/Madeleine_Albright.
[19]. Gideon Polya, "Body Count. Global avoidable mortality since 1950", that includes a succinct history of every country and is now available for free perusal on the web: http://globalbodycount.blogspot.com/.
[20]. "Palestinian Genocide": https://sites.google.com/site/palestiniangenocide/.
[21]. Gideon Polya, "Apartheid Israel's Palestinian Genocide & Australia's Aboriginal Genocide compared", Countercurrents, 20 February 2018: https://countercurrents.org/2018/02/20/apartheid-israels-palestinian-genocide-australias-aboriginal-genocide-compared/.
[22]. Colin Tatz, "With Intent to Destroy. Reflecting on Genocide", Verso, London, 2003,
[23]. Colin Tatz, "Genocide in Australia", AIATSIS Discussion Paper, Number 8, 1999: http://www.aiatsis.gov.au/research/docs/dp/DP08.pdf.
[24]. Gideon Polya, "Review: "The Cambridge History Of Australia" Ignores Australian Involvement In 30 Genocides", Countercurrents, 14 October, 2013: https://www.countercurrents.org/polya141013.htm.
[25]. Gideon Polya, "Australian Day is Invasion Day: will Australia join a Trump US War on China?", Countercurrents, 27 January 2017:

https://countercurrents.org/2017/01/27/australia-day-is-invasion-day-will-australia-join-a-trump-us-war-on-china/

[26]. Jo Caffery, Patrick McConvell, and Jane Simpson, "Gaps in Australia's Indigenous language policy", AIATSIS, December 2009: https://aiatsis.gov.au/publications/products/gaps-australias-indigenous-language-policy-dismantling-bilingual-education-northern-territory.

[27]. Gideon Polya, "Film Review: "Utopia" By John Pilger Exposes Genocidal Maltreatment Of Indigenous Australians By Apartheid Australia", Countercurrents, 14 March, 2014: https://countercurrents.org/polya140314.htm.

[28]. Paddy Gibson, "Stolen futures", Overland, Spring 2013: http://overland.org.au/previous-issues/issue-212/feature-paddy-gibson/.

[29]. Katharine Murphy, "Indigenous child removal rate risks "second stolen generation", Kevin Rudd warns", Guardian, 13 February 2017: https://www.theguardian.com/australia-news/2017/feb/13/indigenous-child-removal-rate-risks-second-stolen-generation-kevin-rudd-warns.

[30]. David Shoebridge, "Stolen Generation continues – time to break the silence", 13 February 2014: http://davidshoebridge.org.au/2014/02/13/stolen-generation-continues-time-to-break-the-silence/."

[31]. Lorena Allam, Nick Evershed and Mike Bowers, "Too hot for humans? First Nations people fear becoming Australia's first climate refugees", Guardian, 18 December 2019: https://www.theguardian.com/australia-news/2019/dec/18/too-hot-for-humans-first-nations-people-fear-becoming-australias-first-climate-refugees.

[32]. Melissa Sweet, "Inequality and climate change –the perfect storm threatening the health of Australia's poorest, The Guardian, 14 May 2-10: https://www.theguardian.com/australia-news/2019/may/14/inequality-and-climate-change-the-perfect-storm-threatening-the-health-of-australias-poorest.

[33]. "Air conditioning: amazing facts", Blue NRG, 15 November 2018: https://www.bluenrg.com.au/energy-insights/air-conditioning-16-amazing-facts.

[34]. Miki Perkins, ""Desperately hot" summer drives push to keep public housing cooler", The Age, 18 April 2019: https://www.theage.com.au/national/victoria/desperately-hot-summer-drives-push-to-keep-public-housing-cooler-20190406-p51bjq.html.

[35]. Gideon Polya, "Water Crisis, Global Avoidable Mortality Holocaust, Water Apartheid, Global Warming & Mina Guli", Countercurrents, 17 May 2019:

https://countercurrents.org/2019/05/water-crisis-global-avoidable-mortality-holocaust-water-apartheid-global-warming-mina-guli.
[36]. Gideon Polya, "Revised Annual Per Capita Greenhouse Gas Pollution For All Countries – What Is Your Country Doing?", Countercurrents, 6 January 2016: https://countercurrents.org/polya060116.htm.
[37]. Gideon Polya, "Exposing And Thence Punishing Worst Polluter Nations Via Weighted Annual Per Capita Greenhouse Gas Pollution Scores", Countercurrents, 19 March, 2016: https://countercurrents.org/polya190316.htm.
[38]. "Renewable energy in China", Wikipedia: https://en.wikipedia.org/wiki/Renewable_energy_in_China.
[39]. Gideon Polya, "Millions join Global School Climate Strike – we are running out of time", Countercurrents, 22 September 2019: https://countercurrents.org/2019/09/millions-join-global-school-climate-strike-we-are-running-out-of-time.
[40]. Gideon Polya, "Australia rejects IMF Carbon Tax & preventing 4 million pollution deaths by 2030", Countercurrents, 15 October 2019: https://countercurrents.org/2019/10/australia-rejects-imf-carbon-tax-preventing-4-million-pollution-deaths-by-2030.
[41]. Gideon Polya, "Redaction: Mainstream media censorship & self-censorship in pre-police-state Australia", Countercurrents, 24 October 2019: https://countercurrents.org/2019/10/redaction-mainstream-media-censorship-self-censorship-in-pre-police-state-australia.
[42]. Climate Change Performance Index 2020: https://germanwatch.org/sites/germanwatch.org/files/CCPI-2020-Results_0.pdf.
[43]. Australian Government, Department of the Environment and Energy, "Australia's emissions projections 2018": http://www.environment.gov.au/system/files/resources/128ae060-ac07-4874-857e-dced2ca22347/files/australias-emissions-projections-2018.pdf.
[44]. Petra Stock, Australian Climate Council, "Let's get something straight – Australia is not on track to meet its Paris Climate target", Australian Climate Council, 21 December 2018: https://www.climatecouncil.org.au/australia-not-on-track-to-meet-climate-targets/.
[45]. Gideon Polya, "Horrendous Cost For Australia Of US War On Terror", Countercurrents, 14 October, 2012: https://countercurrents.org/polya141012.htm.
[46]. Gideon Polya, "Australian state terrorism (4). Jingoistic, US Lackey Australia's Deadly Betrayal Of Its Traumatized Veterans", Stop state terrorism, 2018:

https://sites.google.com/site/stopstateterrorism/australian-state-terrorism-4.
[47]. Gideon Polya, "Advance Australia Fair" Hides Australian Racism, Theft, Genocide, Ecocide, Speciescide & Terracide", Countercurrents, 1 July 2019: https://countercurrents.org/2019/07/advance-australia-fair-hides-australian-racism-theft-genocide-ecocide-speciescide-terracide.
[48]. "Exposing Australia": https://sites.google.com/site/exposingaustralia/home.
[49]. Gideon Polya, "Australia rejects IMF Carbon Tax & preventing 4 million pollution deaths by 2030", Countercurrents, 15 October 2019: https://countercurrents.org/2019/10/australia-rejects-imf-carbon-tax-preventing-4-million-pollution-deaths-by-2030.
[50]. "Stop air pollution deaths": https://sites.google.com/site/300orgsite/stop-air-pollution-deaths.
[51]. "Climate Genocide": https://sites.google.com/site/climategenocide/home.
[52]. Steve Trent, "For climate refugees, tomorrow will be too late", Environmental Justice Foundation, 3 August 2018: https://ejfoundation.org/news-media/for-climate-refugees-tomorrow-will-be-too-late.

SECTION C: EXISTENTIAL THREAT, 2100 CARRYING CAPACITY, CLIMATE HOLOCAUST & CLIMATE GENOCIDE

"In the present Convention, genocide means any of the following acts committed with intent to destroy, in whole or in part, a national, ethnic, racial or religious group, as such: a) Killing members of the group; b) Causing serious bodily or mental harm to members of the group; c) Deliberately inflicting on the group conditions of life calculated to bring about its physical destruction in whole or in part; d) Imposing measures intended to prevent births within the group; e) Forcibly transferring children of the group to another group." Article 2 of the UN Genocide Convention, 1948.

"With an increase of 2 degrees, there will be widespread crop failures, drought, desertification, and flooding throughout Asia, Africa, Europe, North and South America, and Australia. Of plant and animal species, 15-40 percent are likely to go extinct. But more importantly, a 2-degree warming will constitute "a dangerous anthropogenic interference" as it will initiate a series of climate feedbacks that are likely to take the earth beyond a set of "tipping points". Beyond these tipping points, global warming will become a self-sustaining process out of human control, leading to massive catastrophes that could wipe out most of the species on earth." (Minqi Li, "The rise of China and the demise of the capitalist world economy", 2008.

"For humanity it's a matter of life or death. We will not make all human beings extinct as a few people with the right sort of resources may put themselves in the right parts of the world and survive. But I think it's extremely unlikely that we wouldn't have mass death at 4C. If you have got a population of nine billion by 2050 and you hit 4C, 5C or 6C, you might have half a billion people surviving." Professor Kevin Anderson (Tyndall Centre for Climate Change, University of Manchester), 2009.

"The first law of humanity is not to kill your children" and "It was a scientific conference, preceding the infamous Copenhagen conference in 2009, and actually I talked about the carrying capacity of the Earth, which is an interesting issue. What I said is, if global warming is not in any way mitigated, and we go into a six or eight degrees warmer world, then our planet will probably only be able to support a billion people." Hans Joachim "John" Schellnhuber (founding director of the Potsdam Institute for Climate Impact Research, and EU chief climate change adviser), 2015.

"According to Hans Joachim Schellnhuber, chief climate advisor to the European Union, "We're simply talking about the very life support system of this planet". As fascism and the horror of murderous hate crimes are on the rise, governments are presiding over runaway climate change leading toward mass extinctions of species, costing the lives of billions and the demise of much of nature, while children are protesting against the betrayal of their future." Dr Andrew Glikson, 2019.

As set out in Sections A and B above, the World is facing a worsening Climate Crisis and Climate Emergency that is already associated with mass mortality. Over the last 25 years I have published scores of articles and generated numerous websites variously dealing with the existential threat posed by man-made climate change to most of Humanity and the Biosphere [1-40]. Put simply, unless requisite urgent action is taken there is the awful prospect of 10 billion people dying prematurely due to climate change en route to a sustainable human population by 2100 of merely 0.5-1 billion [21-23]. Presently about 9 million people die each year due to the long-term effects of pollutants from the burning of carbon fuels [28]. Annual deaths from climate change per se are of the order of 1 million, but this may be a considerable under-estimate because 15 million people die avoidably from deprivation each year in tropical and sub-tropical Developing Countries (minus China) that are disproportionately impacted by man-made climate change [37]. The worst case scenario of 10 billion climate change-related deaths this century would mean an annual average of 100 million such deaths – a veritable Climate Holocaust and Climate Genocide as set out below.

Holocaust from man-made climate change simply means death of a huge number of people as exampled by the WW2 Jewish Holocaust (5-6 million people killed), the WW2 European Holocaust in general (30 million Slavs, Jews and Gypsies killed), the WW2 Bengali Holocaust (6-7 million Indians deliberately starved to death by the British with Australian complicity), and the 35-40 million Chinese killed associated with the Japanese occupation of China in the period 1937-1945 (WW2 Chinese Holocaust) [21-23, 41]. However the definition of "genocide" by Article 2 of the UN Genocide Convention involves Assessment of "intent' by those responsible and states: "In the present Convention, genocide means any of the following acts committed with intent to destroy, in whole or in part, a national, ethnic, racial or religious group, as such: a) Killing members of the group; b) Causing serious bodily or mental harm to members of the group; c) Deliberately inflicting on the group conditions of life calculated to bring about its physical destruction in whole or in part; d) Imposing measures intended to prevent births within the group; e) Forcibly transferring children of the group to another group" (UN Genocide Convention: http://www.edwebproject.org /sideshow/genocide/convention.html). The resolute climate change inaction by governments for 3 decades, despite increasingly desperate warnings from the world's scientists, surely constitutes "intent".

In a 1995 paper published in a scholarly genocide journal I predicted deadly mass mortality from climate change in Bengal that would be far greater than that from starvation in the WW2 Bengali Holocaust (6-7 million deaths): "History may yet repeat itself with respect to the Bengalis. Australia is per capita one of the most profligate contributors to "greenhouse" gas emission and hence global warming. There is unfortunately a current bipartisan political consensus in Australia opposed to effective short-term reduction of such emission. The recent Berlin Climate Change Conference, heavily influenced by the U.S. and similarly-inclined Australia and Canada, concluded with no agreed targets on greenhouse gas emission and produced merely a motherhood commitment to continuing assessment. While global warming may have advantages for some, there is a widespread concern that global warming and consequent sea level rises will have a major impact on low-lying regions such as densely populated and impoverished West Bengal and Bangladesh. It is quite conceivable that continuing global greed and irresponsibility in relation to greenhouse gas emissions, desertification and forest destruction will have an even more devastating impact on Bengal in the 21st century [31-35] than that of ruthless imperialism in past centuries. We must learn from the appalling consequences of the blindness, complacency and insensitivity with which the world responded at the time to the Jewish Holocaust and to the now-forgotten Bengali Holocaust" [1, 2]. I subsequently reiterated this plea for international action to avoid a Climate Holocaust and Climate Genocide in successive editions of a huge book entitled "Jane Austen and the Black Hole of British History. Colonial rapacity, holocaust denial and the crisis in biological sustainability" [3-5], in numerous articles [6-22], and in key climate-related websites [23-40].

Space does not permit even a succinct review of this huge body of climate-related writing, but the titles convey the essential concerns [1-40]. A key point must be made that in dealing with mass mortality it is important to document reliable sources and not to under-estimate deaths. Indeed while scepticism is at the heart of science, scientists cannot endlessly "re-invent the wheel" and hence science is built on an edifice of peer-reviewed prior published research, established phenomenology and methodology. While science is about the critical testing of potentially falsifiable hypotheses, scientists routinely adopt well-established hypotheses as a matter of operational convenience, and have a good sense of who are eminent scientists, top institutions and top scientific journals. I have generated numerous numerically- and alphabetically-organized compendia of expert opinion relating to climate change in a number of websites [23-40]. In relation to Climate Holocaust, Climate Genocide and predicted mass mortality associated with worsening climate change, of particular note are the compendia "Are we doomed?" [29], "Too late to avoid climate catastrophe?" [32] and "Climate Genocide" [23] that record opinions of numerous scientists and science-informed activists.

An edited version of the website "Climate Genocide" [24] is reproduced as Chapter 11 of this book. Chapter 11 includes a numerically- and alphabetically-organized compendium of expert opinions about the 2100 human carrying capacity of the planet, predicted mass mortality in the absence of requisite action, the presently acute threat to Developing World countries, and the existential threat to tropical Island Nations and to mega-deltaic coastal nations in particular. Climate change aside, with finite available exploitable resources, the world can only have more people at the price of less per head. Professor Dabo Guan (School of International Development, University of East Anglia, UK) (2016): "For everyone in the world to have an American lifestyle, we would need seven planets, and three to live as Europeans" (Irene Banos Ruiz, "China's new love affair with dogs – as pets, not food – presents environmental problems", DW, 21 June 2016: https://www.dw.com/en/chinas-new-love-affair-with-dogs-as-pets-not-food-presents-environmental-problems/a-19197523.)

References.

[1]. Gideon Polya,"The Forgotten Holocaust - The 1943/44 Bengal Famine", 1995:
http://globalavoidablemortality.blogspot.com.au/2005/07/forgotten-holocaust-194344-bengal.html.
[2]. Gideon Polya, "The famine of history: Bengal 1943", International Network on Holocaust and Genocide vol.10, 10-15 (an edited version of "The Forgotten Holocaust" [1]).
[3].Gideon Polya, "Jane Austen and the Black Hole of British History. Colonial rapacity, holocaust denial and the crisis in biological sustainability", G.M. Polya, Melbourne, 1998.
[4]. Gideon Polya, "Jane Austen and the Black Hole of British History. Colonial rapacity, holocaust denial and the crisis in biological sustainability", G.M. Polya, Melbourne, 2008, second edition now available for free perusal on the Web:
http://janeaustenand.blogspot.com/.
[5]. Gideon Polya, "Global warming and the unthinkable world of 2050", Chapter 16, "Jane Austen and the Black Hole of British History. Colonial rapacity, holocaust denial and the crisis in biological sustainability", 2008:
http://janeaustenand.blogspot.com/2012/03/jane-austen-and-black-hole-chapter-16.html.
[6]. Gideon Polya, "War On Terra: US Empire Versus Gaia And Infants", Countercurrents, 26 February 2006:
https://www.countercurrents.org/polya260206.htm.
[7]. Gideon Polya, "Climate holocaust, climate genocide & Australia's Secret Genocide History", Climate Genocide:
https://sites.google.com/site/climategenocide/climate-holocaust-climate-genocide ; republished from
Gideon Polya, "Australia's Secret Genocide History. La Trobe, "Bundoora Eucalyptus" & Black Crimes of White Australia", MWC News (Media With Conscience News), 30 April 2008 (for an image of the huge "Bundoora Eucalyptus" painting see:
http://www.flickr.com/photos/gideonpolya/4318154968/).
[8]. Gideon Polya, "G8 Failure Means Climate Genocide For Developing World", Countercurrents, 11 July 2009:
https://www.countercurrents.org/polya110709.htm.
[9]. Gideon Polya, "We Are Running Out Of Time To Save Humanity And The Biosphere", Countercurrents, 20 November 2009:
https://www.countercurrents.org/polya201109.htm.

[10]. Gideon Polya, "Water Crisis, Palestinian Genocide And Climate Genocide", Countercurrents, 27 April 2010: https://www.countercurrents.org/polya270410.htm.
[11]. Gideon Polya, "US-Complicit 9-11, War On Terror And Terminal War On Terra", Countercurrents, 10 September 2011: https://www.countercurrents.org/polya100911.htm.
[12]. Gideon Polya, "GetUp Time Capsule Letter To My Great Great Grandchildren About Australia And Climate Genocide", Countercurrents, 2 November 2011: https://www.countercurrents.org/polya021111.htm.
[13]. Gideon Polya, "Australia 's Huge Coal, Gas & Iron Ore Exports Threaten Planet", Countercurrents, 15 May 2012: https://countercurrents.org/polya150512.htm.
[14]. Gideon Polya, "To Save Planet Stop Canada-US Keystone XL Pipeline", Countercurrents, 23 August 2011: https://www.countercurrents.org/polya230811.htm.
[15]. Gideon Polya,"Terra" painting, Flickr: https://www.flickr.com/photos/gideonpolya/4289095912/.
[16]. Gideon Polya, "25 Ways World-Leading Climate Criminal Australia Threatens Planet And Invites Boycotts, Divestment & Sanctions (BDS)", Countercurrents, 6 June, 2015: http://www.countercurrents.org/polya060615.htm.
[17]. Gideon Polya, "Australian Commitment To Unlimited Natural Gas Exploitation Threatens Planet & Invites Global Blowback", Countercurrents, 12 April 2017: http://www.countercurrents.org/2017/04/08/australian-commitment-to-unlimited-natural-gas-exploitation-threatens-planet-invites-global-blowback/.
[18]. Gideon Polya, "Advance Australia Fair" hides Australian racism, theft, genocide, ecocide, speciescide & terracide", Countercurrents, 1 July 2019: https://countercurrents.org/2019/07/advance-australia-fair-hides-australian-racism-theft-genocide-ecocide-speciescide-terracide.
[19]. Gideon Polya, "Climate criminal Australia sabotages Pacific Islands Forum & threatens all Island Nations", Countercurrents, 24 August 2019: https://countercurrents.org/2019/08/climate-criminal-australia-sabotages-pacific-islands-forum-threatens-all-island-nations.
[20]. Gideon Polya, "Climate criminal Australia sabotages Pacific Islands Forum & threatens all Island Nations", Countercurrents, 24 August 2019: https://countercurrents.org/2019/08/climate-criminal-australia-sabotages-pacific-islands-forum-threatens-all-island-nations.
[21]. Gideon Polya, "US-Imposed Post-9/11 Muslim Holocaust & Muslim Genocide", 400 pages, Korsgaard Publishing, Germany, 4

June 2020: https://www.amazon.com/US-Imposed-Post-9-Muslim-Holocaust-Genocide/dp/8793987056, especially Chapters 19 and 21.
[22]. Gideon Polya, "Racist Mainstream ignores "US-Imposed Post-9/11 Muslim Holocaust & Muslim Genocide"", Countercurrents, 17 July 2020: https://countercurrents.org/2020/07/racist-mainstream-ignores-us-imposed-post-9-11-muslim-holocaust-muslim-genocide/.
*[23]. Gideon Polya, editor, "Climate Genocide": https://sites.google.com/site/climategenocide/.
[24]. Gideon Polya, editor, "Biofuel Genocide": https://sites.google.com/site/biofuelgenocide/.
[25]. Gideon Polya, editor, "Carbon Debt Carbon Credit": https://sites.google.com/site/carbondebtcarboncredit/.
[26]. Gideon Polya, editor, "Climate Revolution Now": https://sites.google.com/site/300orgsite/climate-revolution.
[27]. Gideon Polya, editor, "Stop climate crime": https://sites.google.com/site/300orgsite/stop-climate-crime.
[28]. Gideon Polya, editor, "Stop air pollution deaths": https://sites.google.com/site/300orgsite/stop-air-pollution-deaths
[29]. Gideon Polya, editor, "Are we doomed?": https://sites.google.com/site/300orgsite/are-we-doomed.
[30]. Gideon Polya, editor, "Methane Bomb Threat": https://sites.google.com/site/methanebombthreat/.
[31]. Gideon Polya, editor, "Nuclear weapons ban, end poverty & reverse climate change": https://sites.google.com/site/300orgsite/nuclear-weapons-ban.
[32]. Gideon Polya, editor, "Too late to avoid global warming catastrophe": https://sites.google.com/site/300orgsite/too-late-to-avoid-global-warming.
[33]. Gideon Polya, editor, "Stop state terrorism" : https://sites.google.com/site/stopstateterrorism/ [state and corporate complicity in worsening climate genocide and 7 million annual air pollution deaths from carbon fuel burning].
[34]. Gideon Polya, editor, "State crime and non-state terrorism": https://sites.google.com/site/statecrimeandnonstateterrorism/.

[35]. Gideon Polya, "Climate terrorism: 400,000 climate change-related deaths globally annually versus an average of 4 US deaths from political terrorism annually since 9-11": https://sites.google.com/site/statecrimeandnonstateterrorism/climate-terrorism.

[36]. Gideon Polya, "Carbon terrorism: 3 million US air pollution deaths versus 53 US political terrorism deaths since 9-11 (2001-2015)": https://sites.google.com/site/statecrimeandnonstateterrorism/carbon-terrorism.

[37]. "Gideon Polya": https://sites.google.com/site/drgideonpolya/home.

[38]. "Climate change articles by Gideon Polya": https://sites.google.com/site/drgideonpolya/climate-change-articles.

[39]. "Climate change websites created by Gideon Polya": https://sites.google.com/site/drgideonpolya/climate-change-websites.

[40]. Gideon Polya, "2011 climate change course": https://sites.google.com/site/300orgsite/2011-climate-change-course.

[41]. Gideon Polya, "Body Count. Global avoidable mortality since 1950",G.M. Polya, Melbourne, 2007, that includes a succinct history of every country from Neolithic times and is now available for free perusal on the Web: http://globalbodycount.blogspot.com/.

CHAPTER 11: WAR ON TERRA, OMNICIDE, TERRACIDE, 2100 CARRYING CAPACITY, CLIMATE GENOCIDE & CLIMATE HOLOCAUST

[An edited version of Gideon Polya, editor, "Climate Genocide": https://sites.google.com/site/climategenocide/].

Climate Holocaust and Climate Genocide.

Summary. Presently about 9.5 million people die avoidably (prematurely) each year due to the effects carbon burning pollutants (9.0 million) (WHO) or to climate change (0.5 million). This latter estimate of presently about 0.5 million climate change-related deaths may be an under-estimate because UN Population Division data indicate that presently 15 million people die avoidably (prematurely) each year (half of them children) due to poverty in the Developing World (minus China), with these impoverished, tropical or sub-tropical countries already being severely impacted by global warming. Both Dr James Lovelock FRS (atmospheric composition and Gaia hypothesis) and Professor Kevin Anderson (Deputy Director, Tyndall Centre for Climate Change Research, University of Manchester, UK) have estimated that only about 0.5 billion people will survive this century due to unaddressed, man-made global warming. Noting that the world population is expected to reach 9.5 billion by 2050 (UN Population Division), these estimates translate to a Climate Genocide involving deaths of 10 billion people this century, this including roughly twice the present population of particular mainly non-European groups, specifically 6 billion under-5 year old infants, 3 billion Muslims in a terminal Muslim Holocaust, 2 billion Indians, 1.3

billion non-Arab Africans, 0.5 billion Bengalis, 0.3 billion Pakistanis and 0.3 billion Bangladeshis. Already 15 million people (about half of them children) die avoidably every year due to deprivation and deprivation-exacerbated disease, and man-made global warming is already impacting this global avoidable mortality holocaust. However 10 billion avoidable deaths due to global warming this century yields an average annual avoidable death rate this century of 100 million per year (see Gideon Polya, "Body Count. Global avoidable mortality since 1950", that includes a history of every country from Neolithic times and is now available for free perusal on the web: http://globalbodycount.blogspot.com/ and "Climate Genocide": https://sites.google.com/site/climategenocide/).

 Humanity demands that we speak out, bear witness, and tell the truth for a better world (see "Gideon Polya Writing": https://sites.google.com/site/gideonpolyawriting/ and "Gideon Polya": https://sites.google.com/site/drgideonpolya/home). Gideon Polya on near-terminal Climate Genocide (2020): "While the presently dominant and deadly neoliberalism demands maximal freedom to exploit human and natural resources for private profit, this rapacious course now existentially threatens Humanity and the Biosphere as the world careers towards a Climate Genocide in which [with inaction] 10 billion people will die avoidably this century en route to a sustainable population in 2100 of a mere 0.5- 1.0 billion (page 358, Gideon Polya, "US-Imposed Post-9/11 Muslim Holocaust & Muslim Genocide", 400 pages, Korsgaard Publishing, Germany, 4 June 2020: https://www.amazon.com/US-Imposed-Post-9-Muslim-Holocaust-Genocide/dp/8793987056 ; for a summary of the book see Gideon Polya, "Racist Mainstream ignores "US-Imposed Post-9/11 Muslim Holocaust & Muslim Genocide"", Countercurrents, 17 July 2020: https://countercurrents.org/2020/07/racist-mainstream-ignores-us-imposed-post-9-11-muslim-holocaust-muslim-genocide/).

This site documents authoritative opinions of leading climate scientists, climate economists, climate analysts and world figures who predict a massive death toll from unaddressed, man-made global warming. Climate holocaust (huge numbers of deaths already from carbon burning and climate change) and climate genocide (huge numbers of deaths already from intentional carbon burning and climate change despite knowledge of the consequences) should be key terms in the public discussion of the worsening climate emergency - but are not due to Mainstream media, politician and academic lying by omission.

Presently about 10 million people die avoidably (prematurely) each year from carbon fuel burning due to the deadly effects of carbon burning pollutants (9.0 million) or climate change (1 million). This latter estimate of presently about 1 million climate change-related deaths annually may be a huge under-estimate because UN Population Division data (UN Population Division: http://esa.un.org/wpp/unpp/p2k0data.asp) indicate that 15 million people die avoidably (prematurely) each year (half of them children) due to poverty in the Developing World (minus China), with these impoverished, tropical or sub-tropical countries already being severely impacted by global warming (see Gideon Polya, "Body Count. Global avoidable mortality since 1950", that includes a history of every country from Neolithic times and is now available for free perusal on the web: http://globalbodycount.blogspot.com/). The World Health Organization (WHO) estimates that 7 million people die from the effects of pollution each year (World Health Organization (WHO), "7 million premature deaths annually linked to air pollution":
http://www.who.int/mediacentre/news/releases/2014/air-pollution/en/).

The DARA 2012 Report commissioned by 20 countries states: "This report estimates that climate change causes 400,000 deaths on average each year today, mainly due to hunger and communicable diseases that affect above all children in developing countries. Our present carbon-intensive energy system and related activities cause an estimated 4.5 million deaths each year linked to air pollution, hazardous occupations and cancer… Continuing today's patterns of carbon-intensive energy use is estimated, together with climate change, to cause 5 million deaths per year by 2030, close to 700,000 of which would be due to climate change. This implies that a combined climate-carbon crisis is estimated to claim 100 million lives between now and the end of the next decade. A significant share of the global population would be directly affected by inaction on climate change" (DARA, "Climate Vulnerability Monitor. A guide to the cold calculus of a hot planet", 2012, Executive Summary pp2-3: http://daraint.org/climate-vulnerability-monitor/climate-vulnerability-monitor-2012/ and DARA report quoted by Reuters, "100 mln to die by 2030 if world fails to act on climate", 28 September 2012: http://in.reuters.com/article/2012/09/26/climate-inaction-idINDEE88P05P20120926)]. As outlined below, as many as 10 billion people will die avoidably (prematurely) this century if man-made climate change is not requisitely addressed.

 Both Dr James Lovelock FRS (Gaia hypothesis) and Professor Kevin Anderson (Deputy Director, Tyndall Centre for Climate Change Research, University of Manchester, UK) have estimated that only about 0.5 billion people will survive this century due to unaddressed, man-made global warming. Thus just before the 2009 Copenhagen summit, the Scotsman reported on Professor Anderson's estimates of "only half a billion surviving" (see: http://en.wikipedia.org/wiki/Kevin_Anderson_%28scientist%29): "Current Met Office projections reveal that the lack of action in the intervening 17 years – in which emissions of climate changing gases such as carbon dioxide have soared – has set the world on a path towards potential 4C rises as early as 2060, and 6C rises by the end of the century. Anderson, who advises the government on climate change, said the consequences were 'terrifying'. 'For humanity it's a matter of life or death,' he said. 'We will not make all human beings extinct as a few people with the right sort of resources may put themselves in the right parts of the world and survive. But I think it's extremely unlikely that we wouldn't have mass death at 4C. If you have got a population of nine billion by 2050 and you hit 4C, 5C or 6C, you might have half a billion people surviving'."

Dr James Lovelock FRS (top UK climate scientist, Fellow of the prestigious Royal Society, famous for his Gaia Hypothesis and atmospheric research): "If we can keep civilization alive through this century perhaps there is a chance that our descendants will one day serve Gaia and assist her in the fine-tuned self-regulation of the climate and composition of our planet. We have enjoyed 12,000 years of climate peace since the last shift from a glacial age to an interglacial one. Before long, we may face planet-wide devastation worse even than unrestricted nuclear war between superpowers. The climate war could kill nearly all of us and leave the few survivors living a Stone Age existence. But in several places in the world, including the U.K., we have a chance of surviving and even of living well. For that to be possible, we have to make our lifeboats seaworthy now" (see: James Lovelock, "Climate war could kill nearly all of us, leaving survivors on the Stone Age", Guardian, 29 June 2009: http://www.guardian.co.uk/environment/2009/jun/29/climate-war-lovelock).

Noting that the world population is expected to reach 9.5 billion by 2050 (UN Population Division), these estimates translate to a Climate Genocide involving deaths of 10 billion people this century, this including roughly twice the present population of particular mainly non-European groups, specifically 6 billion under-5 year old infants, 3 billion Muslims in a terminal Muslim Holocaust, 2 billion Indians, 1.3 billion non-Arab Africans, 0.5 billion Bengalis, 0.3 billion Pakistanis and 0.3 billion Bangladeshis. [1].

Already 15 million people (about half of them under-5 year old infants) die avoidably every year due to deprivation and deprivation-exacerbated disease – and man-made global warming is already clearly worsening this global avoidable mortality holocaust. However 10 billion avoidable deaths due to global warming this century yields an average annual avoidable death rate of 100 million per year [2].

Collective, national responsibility for this already commenced Climate Holocaust is in direct proportion to per capita national pollution of the atmosphere with greenhouse gases (GHGs). Indeed, fundamental to any international agreement on national rights to pollute our common atmosphere and oceans should be the belief that "all men are created equal". However reality is otherwise: "annual per capita greenhouse gas (GHG) pollution" in units of "tonnes CO_2-equivalent per person per year" (2005-2008 data without correction for land use and methane) is 0.9 (Bangladesh), 0.9 (Pakistan), 2.2 (India), less than 3 (many African and Island countries), 3.2 (the Developing World), 5.5 (China), 6.7 (the World), 11 (Europe), 16 (the Developed World), 27 (the US) and 25 (Australia; or 89 in 2015 if Australia's huge Exported CO_2 pollution is included) [3, 4].

 However there is also a past carbon pollution debt for carbon pollution from the start of the Industrial Revolution over 2 centuries ago. European First World countries are responsible for about 73% of historical carbon pollution of the atmosphere from 1751-2006, with India, Japan and China contributing 2.5%, 3.9% and 8.2%, respectively [5].

 A further measure of responsibility into the future comes from proposals to limit GHG pollution put to the [2009] Copenhagen Climate Conference (COP15) that was successfully sabotaged by 2 of the World's worst per capita GHG polluters, Australia and the US, acting in collaboration with other major polluters.

The fairest plan at the Copenhagen Climate Conference (COP15) has been the Alliance of Small Island States (AOSIS) proposal for "55% of 1990 levels by 2020 for Developed Countries" which means that Developed Countries' per capita would go from 16 to 6.5 tonnes CO_2-e per person per year while Developing Countries would "in aggregate aim to achieve significant deviations from baselines by 2020" i.e. less than a predicted 3.8 for the Developing World by 2020 [6].

The UN Draft Proposal is for a less just Developed 2020 per capita of 7.1- 8.9 from the present 16, but the US has proposed a 2020 per capita for itself of 12.4 from its current 27 [6].

The most unjust plan at Copenhagen comes from world's worst per capita GHG polluter, Australia, which has made a highly conditional best offer of "75% of 2000 value by 2020" which would mean that Australia's domestic per capita would go from 30 to 17.2 in 2020 and its domestic plus exported per capita would increase 15% from 54 to 62 tonnes CO_2-e per person per year in 2020 [4, 6, 7].

Holocaust simply means death of a huge number of people_as exampled by the WW2 Jewish Holocaust (5-6 million people killed), the WW2 Holocaust in general (30 million Slavs, Jews and Gypsies killed), the WW2 Bengali Holocaust (6-7 million Indians deliberately starved to death by the British with Australian complicity) and [the WW2 Chinese Holocaust] (the 35-40 million Chinese killed associated with the Japanese occupation of China in 1937-1945) [2]. However the definition of "genocide" by Article 2 of the UN Genocide Convention involves Assessment of "intent' by those responsible and states "In the present Convention, genocide means any of the following acts committed with intent to destroy, in whole or in part, a national, ethnic, racial or religious group, as such: a) Killing members of the group; b) Causing serious bodily or mental harm to members of the group; c) Deliberately inflicting on the group conditions of life calculated to bring about its physical destruction in whole or in part; d) Imposing measures intended to prevent births within the group; e) Forcibly transferring children of the group to another group" [8].

"Intent" to commit mass murder is only rarely explicitly expressed. Thus British writer, patriot and genocidal racist Charles Dickens expressed a desire to exterminate the Indian people [300 million] after the so-called Indian Mutiny of 1857: "I wish I were Commander in Chief over there [India]! I would address that Oriental character which must be powerfully spoken to, in something like the following placard, which should be vigorously translated into all native dialects, "I, The Inimitable, holding this office of mine, and firmly believing that I hold it by the permission of Heaven and not by the appointment of Satan, have the honor to inform you Hindoo gentry that it is my intention, with all possible avoidance of unnecessary cruelty and with all merciful swiftness of execution, to exterminate the Race from the face of the earth, which disfigured the earth with the late abominable atrocities

[2,000 British killed in the 1857 Indian War of Independence aka the 1857 Indian Mutiny]" [9]. In the event, the British killed an estimated 10 million Indians as reprisals for the so-called Indian Mutiny [10, 11].

"Intent" in genocide can also be established by sustained, remorseless complicity in destruction (including forced removal of populations from Homelands) and mass avoidable death – as exampled by Nazi German responsibility for the WW2 Genocide of Slavs, Jews and Gypsies (1939-1945), Japanese responsibility for the Chinese Genocide of the 1930s and 1940s, and British responsibility for the WW2 Bengali Genocide, 1943-1945 [12].

Sustained, remorseless "intent" to kill literally billions of non-Europeans through global warming is apparent in the sustained, remorseless pollution of the atmosphere by the First World, and by First World refusal at the disastrous 2009 Copenhagen Climate Change Conference to emplace legally-binding effective carbon pollution measures in the face of dire warnings from the world scientific community [13].

Just as violent mass murder, mass paeodocide and genocide invite genocide trials before the International Criminal Court, so does remorseless commitment by the First World to climate holocaust and climate genocide. Indeed a formal complaint has been made to the International Criminal Court over Australian complicity in the ongoing Aboriginal Genocide (9,000 avoidable deaths annually out of an Indigenous Australian population of 0.5 million), Iraqi Genocide (4.4 million violent and non-violent excess deaths, 1990-2009), Afghan Genocide (4.5 million violent and non-violent excess deaths, 2001-2009) and Climate Genocide (10 billion predicted to die this century) [14].

This site represents a valuable resource for decent, civilized humanity – it documents expert, authoritative opinions of leading climate scientists, climate economists, climate analysts and world figures who predict a massive, knowingly and deliberately caused death toll from unaddressed, man-made global warming this century.

A numerically- and alphabetically-ordered list of quotations re Climate Genocide from individuals and organizations (for documentation Google the quote or see Gideon Polya, editor, "Climate Genocide": https://sites.google.com/site/climategenocide/):

300.org (2009): "Developing Nations must urgently defend themselves from Australia-, US- and First World-imposed climate genocide."

350.org (2015): "According to the Intergovernmental Panel on Climate Change a two degree increase in the global mean temperature will mean a three or more degree increase for temperatures in Africa. Such an increase in temperature would lead to widespread devastation including predictions of a 50% reduction in crop yields in some areas, cutting food outputs in half, more than 600 million people left without adequate water supplies, and massive damage to coastlines, rural communities and cities. Marching through the Conference Center African groups chanted: "Two degrees is suicide" "One Africa, One Degree" and "No to Climate Colonialism, No to Climate Genocide" in response to the proposal."

9 January 2010 Formal Complaint by Dr Gideon Polya to the International Criminal Court (ICC) (2010): "Chief Prosecutor Luis Moreno-Ocampo… Dear Sir, Formal complaint to the Chief Prosecutor of the International Criminal Court re various US Alliance, NATO, EU, Australia, New Zealand and UK involvement in Palestinian Genocide, Iraqi Genocide, Afghan Genocide, Muslim Genocide, Aboriginal Genocide, Biofuel Genocide and Climate Genocide…"

AFRICA, African delegates to the UN Climate Change Conference in Copenhagen as reported by the UK BBC report (2009): "African delegates have made a desperate plea to the UN Climate Change Conference in Copenhagen, claiming that ignoring the continent's anticipated problems with climate change could be tantamount to genocide."

ANDERSON, Kevin. Professor Kevin Anderson (Tyndall Institute, University of Manchester) (2009): "If you have got a population of nine billion by 2050 and you hit 4°C, 5°C or 6°C, you might have half a billion people surviving."

ANDERSON, Will. Will Anderson (author of "This is Hope: Green Vegans and the New Human Ecology") (2013): "What is our sustainable human population from this biocentric perspective? Estimates vary… Anthropologist J. Kenneth Smail believes the Earth's long term sustainable population is 2 to 3 billion. Some feel the sustainable population is much higher, though there is little said about what Earth would lose in the process. Opinions go as low as 500 million."

ASLAM. Mohamed Aslam, (Minister of Housing and Environment, Maldives) (2011): "We wanted a legally binding deal to guarantee our survival. It's the only viable solutions. Current pledges are not concurrent with what parties agreed in Cancun. These pledges have us on a path to four degrees of warming. This will definitely wipe my country and many other islands off the map. It's time to act decisively… For my country it is extremely important to get action today."

AVERY. John Scales Avery (theoretical chemist in quantum chemistry, thermodynamics, evolution and history of science, and notable for his peace activism associated with the Nobel Prize-winning Pugwash Conferences on Science) (2017): "8.1 Climate change as genocide. Climate change does not affect all parts of the world equally. The harshest effects of the extreme weather that we are already experiencing are disproportionately felt by the poorest people of the world. In March, 2017 the Security Council was informed that 20 million people in four countries, Nigeria, Somalia, South Sudan and Yemen, were in danger of dying unless provided with immediate help. The cost of the necessary aid was estimated to be $4.4 billion. The developed world's response has been a shrug of indifference. If the more affluent parts of the world continue to produce greenhouse gases in a business-as-usual scenario, and if they continue to ignore calls for help from starving people, these actions will amount to genocide."

BAINIMARAMA. Frank Bainimarama (PM of Fiji) commenting on Australian politicians saying that Pacific Islanders could move to higher ground or migrate to Australia (2019): "Despite the enormous difficulties of these decisions, Fiji is lucky we even have the higher ground to allow for relocation at all. I'm keen to hear what the honourable member believes the people of Kiribati should do in the face of rising seas, when the highest point in their country sits at just 1.8 metres above sea level… In a time when we must be future-facing we can hardly tolerate such insensitive, neo-colonial prescriptions. I implore leaders of Australia to visit these communities and see them firsthand before they propose solutions that are so blatantly out of touch with the reality that Pacific Islanders live with on the ground, day in and day out… Let's be clear, the costs of climate change to our industries, our people, our potential, are crippling. This climate crisis kills. [It] destroys the lands we love."

BANGLADESH CIVIL SOCIEY ORGANIZATIONS. Bangladesh Civil Society Organizations (CSOs) Press Release at the Doha Climate Change Conference (2012): "Today Bangladeshi CSO climate networks namely BAPA, BIPNetCCBD, CCDF, CSRL, CFGN, EquityBD and NCCB conducted a press conference in QNCC, Doha where CoP 18 UNFCCC Climate Conference are happening, there are only one day remain to conclude the conference. The civil societies expressed frustration and condemned the attitude of developed countries especially of USA, they urged global leaders to make a deal based on science and consider due compensation as repaying of climate debt, they have also expressed worried as if there are no deals means somehow a collapse of multilateralism, meaning accepting climate genocide in poor countries."

BASSEY. Nnimmo Bassey (Nigerian environmentalist and chair of Friends of the Earth International) on Copenhagen COP15 (2009): "Justice has not been done. By delaying action, rich countries have condemned millions of the world's poorest people to hunger, suffering and loss of life as climate change accelerates. The blame for this disastrous outcome is squarely on the developed nations. We are disgusted by the failure of rich countries to commit to the emissions reductions they know are needed, especially the US, which is the world's largest historical emitter of greenhouse gases. In contrast African nations, China and others in the developing world deserve praise for their progressive positions and constructive approach. Major developing countries cannot be blamed for the failure of rich industrialised countries. Instead of committing to deep cuts in emissions and putting new, public money on the table to help solve the climate crisis, rich countries have bullied developing nations to accept far less. Those most responsible for putting the planet in this mess have not shown the guts required to fix it and have instead acted to protect short-term political interests."

BRADSHAW & BROOK. Corey Bradshaw and Barry Brook (Australian scientists) (2014): "The inexorable demographic momentum of the global human population is rapidly eroding Earth's life-support system. There are consequently more frequent calls to address environmental problems by advocating further reductions in human fertility. To examine how quickly this could lead to a smaller human population, we used scenario-based matrix modeling to project the global population to the year 2100. Assuming a continuation of current trends in mortality reduction, even a rapid transition to a worldwide one-child policy leads to a population similar to today's by 2100. Even a catastrophic mass mortality event of 2 billion deaths over a hypothetical 5-y window in the mid-21st century would still yield around 8.5 billion people by 2100. In the absence of catastrophe or large fertility reductions (to fewer than two children per female worldwide), the greatest threats to ecosystems—as measured by regional projections within the 35 global Biodiversity Hotspots—indicate that Africa and South Asia will experience the greatest human pressures on future ecosystems."

BRING CLIMATE CRIMINALS TO JUSTICE. Bring Climate Criminals to Justice (BCCJ) (UK) on the proposed

Climate Genocide Criminal Prosecutions Bill: "We seek the introduction of a Bill to Parliament that will facilitate the retrospective prosecution of key Climate Criminals. These are Government Ministers and others who, through their policies and activities, have made significant contribution to the mass loss of life. The Bill will permit the prosecution for the deaths that have occurred and those that will be caused in the future. The Bill will permit the introduction of unprecedented penalties which are proportionate to the scale of the crime (Penalties: Upon prosecution and conviction.) The current and next administrations will not support such legislation but we want Government Ministers and key business players to know that retrospective legislation will be introduced in the future. As public opinion hardens and campaigners demand justice, new politicians will be elected who will deliver legislation that will tackle those who are guilty of the most bestial of all crimes against humanity. We are hopeful that prompt action in Bangladesh to introduce such legislation will encourage other nations to follow suit."

BROWN, James. James Brown (ecologist, and a Distinguished Professor of Biology at the University of New Mexico) quoted by Professor Paul Ehrlich (2013): "Can humanity avoid a starvation-driven collapse? Yes, we can – though we currently put the odds at just 10 percent. As dismal as that sounds, we believe that, for the benefit of future generations, it is worth struggling to make it 11 percent. One of our most distinguished colleagues, biogeographer and energy expert James Brown of the University of New Mexico, disagrees. He puts the odds of sustaining human civilization at 1 percent, but thinks that it's worth trying to increase it to 1.1 percent."

BROWN, Lester. Lester Russel Brown (US environmentalist, founder of the Worldwatch Institute, founder and president of the Earth Policy Institute, author or co-author of over 50 books on global environmental issues, most recently "Plan B 4.0: Mobilizing to Save Civilization") (2011): "Averting Environmental Genocide - Plan B 4.0."

BUTTON. James Button et al. (Monash University Sustainable Development Institute) reviewing health impacts of climate change (2019): "For example [World Health Organization opines], between 2030 and 2050 climate change will cause an additional 250,000 deaths per year from heat stress, malaria, malnutrition and diarrhoea, according to some estimates. To summarise, "there is a widespread scientific consensus that the world's climate is changing [and] addressing these occurrences is a pressing challenge for public health"."

CASTRO. Fidel Castro (famed Cuban leader) on the Copenhagen COP15 Conference (2009): "The youth is more interested than anyone else in the future. Until very recently, the discussion revolved around the kind of society we would have. Today, the discussion centers on whether human society will survive. These are not dramatic phrases. We must get used to the true facts. Hope is the last thing human beings can relinquish. With truthful arguments, men and women of all ages, especially young people, have waged an exemplary battle at the Summit and taught the world a great lesson."

CHOMSKY. Professor Noam Chomsky (anti-racist Jewish American linguist, philosopher, cognitive scientist, historian, political activist, author of over 100 books, social critic, and Professor Emeritus at the 93-Nobel-Laureate Massachusetts Institute of Technology (MIT)) (2018): "The most striking features [of the 2018 US mid-term elections] are brutally clear. Humanity faces two imminent existential threats: environmental catastrophe and nuclear war. These were virtually ignored in the campaign rhetoric and general coverage. There was plenty of criticism of the Trump administration, but scarcely a word about by far the most ominous positions the administration has taken: increasing the already dire threat of nuclear war, and racing to destroy the physical environment that organized human society needs in order to survive. These are the most critical and urgent questions that have arisen in all of human history. The fact that they scarcely arose in the campaign is truly stunning — and carries some important, if unpleasant, lessons about our moral and intellectual culture."

CHAVEZ. Hugo Chávez (former socialist president of Venezuela and leader of the Bolivarian Revolution) at Copenhagen COP15 Conference (2009): "The total income of the 500 richest people in the world is greater than the 450 million poorest living on $2 a day. We have to change direction. How long are we going to tolerate the current international economic order, and allow the hungry not to have food? Let's eradicate poverty and bring in climate justice. If capitalism resists we have to do battle with it. If we do not, then mankind, the greatest creation in the universe, will disappear."

CLIMATE GENOCIDE ACT NOW. "Climate Genocide Act Now" (2016): "Over ten years ago the Trade Union movement fought for the introduction of legislation to tackle the number of employees being killed at work. Employer's groups worked hard to delay and undermine the legislation but the Corporate Manslaughter and Corporate Homicide Act 2007 was enacted. When they knew what was coming, directors decided they didn't want to end up in prison and suffer heavy fines so they transformed safety procedures at work, saving lives on building sites, down tunnels, etc. The legislation was a success. We are working to bring forward a private member's Bill in 2017, "The Climate Genocide Bill". Business leaders will fight this hard but we will win. We will enact legislation to criminalise those who kill others by expanding polluting activities and industries. The legislation will include provisions for the seizure of the proceeds of "climate crime" – those profits that are derived from industries that kill."

CONNOR. Paul Connor (Philosophy and Psychology student at the University of Melbourne who fasted for 44 days as part of the international Climate Justice Fast for Copenhagen (2009): "Today, my only prayer is that one day we will look back upon the current period of history and we will remember a time when the threat of climate change rendered our future uncertain. We will remember feeling fear as we watched the desperate warnings of scientists ignored by our leaders at COP15, and disbelief as our irreplaceable planet was sacrificed for meaningless profits. And we will remember our frustration as we worked to awaken a world that often seemed willfully ignorant of the enormous danger it faced. But this will not be all. I believe that one day, when we look back, we will also remember something incredible. We will remember how our generation made a collective decision to rise as one, all across the globe, and refused to let shortsightedness and greed destroy our future."

DARA. The DARA 2012 Report commissioned by 20 countries (2012): "This report estimates that climate change causes 400,000 deaths on average each year today, mainly due to hunger and communicable diseases that affect above all children in developing countries. Our present carbon-intensive energy system and related activities cause an estimated 4.5 million deaths each year linked to air pollution, hazardous occupations and cancer… Continuing today's patterns of carbon-intensive energy use is estimated, together with climate change, to cause 5 million deaths per year by 2030, close to 700,000 of which would be due to climate change. This implies that a combined climate-carbon crisis is estimated to claim 100 million lives between now and the end of the next decade. A significant share of the global population would be directly affected by inaction on climate change."

DE ABREU. Alcinda António de Abreu (Minister for Coordination of Environmental Affairs, Mozambique) statement at the Lima Climate Conference, Lima (2014): "Mozambique wants to express its full alignment with the statement made by the representatives of African group, G77 and China and LDCS Countries. Excellencies. The world is being witnessing the impacts of climate change almost every day with people dying or displaced and communities losing their livelihood and properties resulting in more people becoming vulnerable and poorest. The science has been telling us that if we continue to increase emissions, in the near future no single country will stand safe from the negative impact of climate change. Today we have the honor and privilege to revert this trend by adopting and implementing decisions that will stabilize and revert the current trend. None of us would like to be appointed by the present and the future generations as part of

those that did nothing to keep the planet safe, and as promoters
of climate genocide."

DERER. Patrícia Dérer (sustainability expert) (2018):
"Targets chosen in the well-known study of Daily et al. include
sufficient wealth, access to resources, universal human rights,
preservation of biodiversity and cultural diversity, and support
for intellectual, artistic and technological creativity. Estimating
the amount of energy to satisfy these human needs while
keeping ecosystems and resources intact, they calculated the
optimal population size in the vicinity of 1.5 – 2 billion
people… Pimentel et al. considered a comfortable consumption
based on European living standard and a sustainable use of
natural resources, suggesting only 2 billion people as
appropriate size… Lianos and Pseiridis attempt to estimate
optimal population size using an objective criterion designed to
assure that human resource use does not deplete Earth's natural
capital. This is the unitary value of the ecological footprint-
biocapacity ratio (L). The ecological footprint measures the
demand that human consumption places on the biosphere.
Biocapacity represents the biosphere's regenerative capacity;
i.e., it measures the productivity of various ecosystems.
Between 1961–2009, their ratio L increased dramatically [quasi-
linearly versus time from 0.7 to 1.5]. In the beginning of this
period, the world had a substantial ecological reserve. That
disappeared after about 10 years and since then we have been
operating in deficit mode. Today the demand for resources
exceeds the available supply by 50 % (L=1.5). (Fig.1)… The
authors calculate the maximum gross world product (GWP, the
combined gross national product of all the countries in the
world), the production of which would leave the natural capital
of the Earth and other species' populations intact (L=1). In order
not to exceed this maximum GWP, but keep a comfortable
European average per capita GWP level ($11,000), we should
reduce our population to 3.1 billion. If we wish to keep
population at 7 billion, the per capita product must be radically
reduced to $4,950, from the current $16,100."

**DI-APING. Lumumba Stanislaus Di-Aping (Sudanese lead
negotiator for the G77 (Developing World) countries at the**

2009 Copenhagen Climate Conference (COP15) on the US-inspired Copenhagen COP15 Accord (code-named L9) (2009): "L9 asks Africa to sign a suicide pact. It is a solution based on the values which, in our opinion, channelled six million people in Europe to the [Nazi] furnaces."

DUNLOP. Ian Dunlop (a member of the Club of Rome and formerly an international oil, gas and coal industry executive, chairman of the Australian Coal Association and chief executive of the Australian Institute of Company Directors (2017): "Nowhere in the debate is the critical issue even raised: the existential risk of climate change, which such development now implies. Existential means a risk posing large negative consequences to humanity that can never be undone. One where an adverse outcome would either annihilate life, or permanently and drastically curtail its potential. This is the risk to which we are now exposed unless we rapidly reduce global carbon emissions… Dangerous climate change, which the Paris agreement and its forerunners seek to avoid, is happening at the 1.2-degree increase already experienced as extreme weather events, and their economic costs, escalate. A 1.6-degree increase is already locked in as the full effect of our historic emissions unfolds. Our current path commits us to a 4 to 5-degree temperature increase. This would create a totally disorganised world with a substantial reduction in population, possibly to less than one billion people from 7.5 billion today."

EHRLICH. Paul Ehrlich (Bing Professor of Population Studies and President of Center for Conservation Biology, Stanford University. Fellow, National Academy of Sciences and Royal Society)(2017): "Our research group has concluded two decades ago that with foreseeable technologies an "optimum" population might be 1.5 to 2 billion people. The idea was that would be enough people to have big cities for those who like fine restaurants and opera, and few enough people to permit lots of wildlands for hunters and hermits. Overall, we thought that number might be long-term sustainable."

FLYVBJERG. Bent Flyvbjerg (Professor at Säid Business School, University of Oxford) on a 2-3 billion sustainable

population (2017): "From an anthropocentric perspective, to calculate the Earth's ideal population size, one would first need to establish an ideal benchmark for what we think is a good life, and calculate the resources it takes to sustain that lifestyle. As a first approximation, let's take the French lifestyle as a benchmark. According to the Global Footprint Network's calculator, if all of humanity were to live like the French, we would need about 2.5 Earths to sustain that lifestyle. Any lifestyle that cannot be universalized to the rest of humanity cannot be just. Every newborn should be able to enjoy their fair share of the world's resources. Thus, to ensure the ideal population size so that everyone could enjoy a comparable way of life, taking the French lifestyle as our benchmark, we would need to reduce the world population to about 3 billion people (4.6 billion less than today's population). If, instead, we chose the lifestyle of people in the USA as the benchmark, then the world population would need to be reduced to 1.9 billion. It is unclear, however, that the American lifestyle is any better than the French, in terms of well-being. On the contrary: it looks like Europeans are faring better than Americans."

FORENSIC ARCHITECTURE. Forensic Architecture (FA, a research agency based at Goldsmiths, University of London, that undertakes advanced architectural and media research on behalf of international prosecutors, human rights organisations, as well as political and environmental justice groups) on climate genocide (2017): "Climate crimes. Two accusations of genocide in the Sahel: The first issued by the International Criminal Court (ICC) in 2008 regarding war crimes in Sudan; the second issued 2009 by the Sudanese diplomat Lumumba Di-Aping directed at the world's developed nations. The first favors the West. The second deflects and returns the claim and thereby it raises the specter of a new form of violence. This work tests what it would take to support Di-Aping's claim and in doing so raises a number of questions about the violence wrought by climate change, especially the forums in which it is debated and eventually legitimized. What will be the role of forensic climatology in reconnecting the causes of environmental violence with their effects? And what will be the political consequences? Drawing on recent scientific research that shows a correlation between aerosol emission in the northern hemisphere and desertification in the Sahel, this project makes visible a new geopolitical cartography that ties together distant fates, linking industrialization in the North to deprivation in the South. In this way, it demonstrates that Di-Aping's claim is a legitimate one."

GERMAN CLIMATE CHANGE DEMONSTRATORS. German climate action demonstrators reported by AP thus (2017): "Germany (AP) — Thousands of demonstrators have marched through Bonn to protest the use of fossil fuels ahead of a global climate conference being held in the western German city next week. Participants in Saturday's event carried banners with slogans that included "Revolution Not Pollution," "Frack Off Our Land" and "Trump: Climate Genocide." Protest organizers say some 25,000 people took part in the demonstration."

GERRARD. Michael Gerrard (director, of the Sabin Center for Climate Change, Columbia Law School) (2018): "Toward the end of this century, if current trends are not reversed, large parts of Bangladesh, the Philippines, Indonesia, Pakistan, Egypt and Vietnam, among other countries, will be under water."

GLIKSON. Dr Andrew Glikson (Australian paleo-climate and earth scientist at the Australian National University) on climate sceptics and the likely death of billions and end of civilization (2009): "As indicated by Clive Hamilton in New Matilda there is little evidence the "climate change skeptics" worry their misunderstanding of climate science may lead to the death of billions and the likely demise of civilization. The legal status of disinformation campaigns aimed at the promotion of substances of proven fatal consequences, such as ozone-destroying CFCs, or the release of CO_2 to levels over 350 ppm, may yet prove to be the Achilles heel of global civilization" and **Dr Glikson (2020):** "Many governments are listening to medical science, implementing essential measures to combat the [Covid-19] plague, instigating social isolation and economic support systems in order to avoid a potential demise of hundreds of thousands or even millions of lives. Climate change is already causing deaths, according to a new report global warming would cause an additional 250,000 deaths per-year from heat and extreme weather events, yet most authorities continue to ignore the scientific evidence of climate disruption that threatens to exceed +2 degrees Celsius and toward 4 degrees Celsius. Potentially this is leading to a demise of billions of lives and many species through extreme weather events."

GOODMAN. Ellen Goodman (Pulitzer Prize-winning American journalist, columnist and author of many books) (1967): "By every measure, the UN 's Intergovernmental Panel on Climate Change raises the level of alarm. The fact of global warming is "unequivocal." The certainty of the human role is now somewhere over 90 percent. Which is about as certain as scientists ever get. I would like to say we're at a point where global warming is impossible to deny. Let's just say that global warming deniers are now on a par with Holocaust deniers, though one denies the past and the other denies the present and future."

GUAN. Professor Dabo Guan (School of International Development, University of East Anglia, UK) (2016): "For everyone in the world to have an American lifestyle, we would need seven planets, and three to live as Europeans" (Irene Banos Ruiz, "China's new love affair with dogs – as pets, not food – presents environmental problems").

GUARDIAN. The UK Guardian newspaper editorial on the 2018 IPCC Report "Global warming of 1.5 °C" (2018): "Climate change is an existential threat to the human race… The latest report of the Intergovernmental Panel on Climate Change tells us that there are only a dozen or so years in which to change our economies radically if we are to keep the effects of the warming already under way to manageable proportions. That would require the countries of the world to live up to the most ambitious of the goals of the Paris climate change agreement, and keep the rise in average global temperatures to 1.5C above preindustrial levels. A rise of even half a degree above that, to 2C, will have effects that are very much worse.

Already this seems much more likely. All corals will disappear, as will many insects and plants… All these risks make it quite credible that we will end with a warming of 3C, 4C or even worse – and the consequences will be globally terrible, and everywhere unavoidable. Hundreds of millions of people may die through droughts on land, and flooding at the coasts, through the loss of marine species due to acidification of the oceans, and probably through the disruption of long-term weather patterns around which the world's agriculture has been shaped."

HALLAM. Roger Hallam (co-founder of Extinction Rebellion) (2019): "Teenagers are shitting themselves about what's happening for the future, they've got another 50, 60, 70 years to live on this planet, by that time there could only be a billion people left. I mean that's six billion people that have died from starvation or slaughtered in war."

HAMILTON. Clive Hamilton (Australian climate economist and Professor of Public Ethics at the Centre for Applied Philosophy and Public Ethics (CAPPE) and the Vice-Chancellor's Chair in Public Ethics at Charles Sturt University) (2010): "One way or another, humans will have to adapt to life in a hotter world. Many plausible scenarios suggest a sharp decline in the number of people that will survive in the long term. Some suggest a billion or a few hundred million will remain in a century or two, but one guess is as good as another. One thing is certain: the transition to a new stage of stability will be long and brutal, especially for the poorest and most vulnerable whose survival will be threatened with food shortages, extreme weather events and disease" and **Dr Hamilton (2009):** "If the climate deniers were to succeed, and stopped the world responding to the mountain of evidence for human-induced global warming, then hundreds of millions of mostly impoverished people around the world would die from the effects of climate change. They will die from famine, flood and disease caused by our unwillingness to act. The Stern report provides some sobering estimates: an additional 30-200 million people at risk of hunger with warming of only 2-3°C; an additional 250-500 million at risk if temperatures rise above

3°C; some 70-80 million more Africans exposed to malaria; and an additional 1.5 billion exposed to dengue fever. Instead of dishonouring the deaths of six million in the past, climate deniers risk the lives of hundreds of millions in the future. Holocaust deniers are not responsible for the Holocaust, but climate deniers, if they were to succeed, would share responsibility for the enormous suffering caused by global warming. It is a ghastly calculus, yet it is worth making because the hundreds of millions of dead are not abstractions, mere chimera until they happen. We know with a high degree of certainty that if we do nothing they will die."

HANSEN. Dr James Hansen, (former head, NASA Goddard Institute for Space Studies in New York City and adjunct professor in the Department of Earth and Environmental Sciences at 100-Nobel-Laureate Columbia University) (2009): "After the ice has gone, would the Earth proceed to the Venus syndrome, a runaway greenhouse effect that would destroy all life on the planet, perhaps permanently? While that is difficult to say based on present information, I've come to conclude that if we burn all reserves of oil, gas, and coal, there is a substantial chance we will initiate the runaway greenhouse. If we also burn the tar sands and tar shale, I believe the Venus syndrome is a dead certainty."

HASAN. Mohammad Rakibul Hasan (professional photographer, independent journalist, short film maker and cinematographer based in Bangladesh who studied in Film & Video Production at UBS Film School, and maker of the video entitled "The Climate Genocide" with shocking images on storm-wracked coastal parts of Bangladesh) (2010): "Climate genocide is already begun in Bangladesh ..." **and further (2010):** "Climate genocide is already begun in Bangladesh. The populations in the South and South-East Asia coastline extending from the east coast of India to the coast of Myanmar have tasted the salt taste of annual cyclones from the Bay of Bengal with ever increasing tidal floods. Due to its existence in the middle of the coastline, Bangladesh is either the worst or the common victim irrespective of the locations where the cyclones make the landfall. Cyclones are not only resulting in human casualties and destruction of properties but also leaving behind perpetual tidal floods. Around one million people have been rendered homeless due to river erosion in the mainland river basins over the last three decades as the mighty River Brahmaputra-Jamuna continues to widen due to decrease in its depth for heavy rush of sediments from the upstream and poor erosion management in the downstream. Climate change is likely to lead to increasing rates of generation of climate refugees, and it is vital that evolving frameworks for climate change adaptation address issues for compliance by national and international communities to peacefully resettle those climate refugees. The whole World should work together on this issue to prevent climate genocide and ensure environmental sustainability."

HASHIAN. Adam Hashian (connected with Lucis Lumen that is dedicated to advancing innovative technology) (2015): "Climate genocide … Certainly, the current terms "climate change" and "global warming" do not accurately convey the seriousness of the issue we are facing. This is why I am calling this human induced global phenomena "climate genocide". I do not think this is harsh, exaggerated, or unrealistic. A quick google search for the definition of genocide reveals: the deliberate killing of a large group of people, especially those of a particular ethnic group or nation. And from Encarta World English Dictionary, genocide refers to the systematic killing of all the people from a national, ethnic, or religious group, <u>or an attempt to do this</u>. Note the underlined: or an attempt to do this. This is quite literally what is happening now. The nations and ethnic groups which are in the most serious danger are the "Small Island Developing States" or "SIDS"… Another quick google search reveals: "A report commissioned by the governments of more than 20 countries found that more than 100 million people will die as a result of climate change by 2030 if the world stays on its current path." It's very possible that 100 million is a low estimate. This would be higher than the total deaths of World War 2. Everyone in the world is affected by this. The world is so interconnected now that a major imbalance in one country can cause economic systems in another country across the globe to collapse. Essentially we are fighting World War 3 right now, and the common global enemy is humanity's bad habits and outdated systems."

HAWKING. Professor Stephen Hawking (eminent theoretical physicist and cosmologist) addressing the question "Will we survive on Earth?"(2018): "In January 2018, the Bulletin of the Atomic Scientists, a journal founded by some of the physicists who had worked on the Manhattan Project to produce the first atomic weapons, moved the Doomsday Clock, their measurement of the imminence of catastrophe – military or environmental – facing our planet, forward to two minutes to midnight… In 1947, the clock was set at seven minutes to midnight. It is now closer to Doomsday than at any time since then, save in the early 1950s at the start of the Cold War… We see great peril if governments and societies do not take action now to render nuclear weapons obsolete and to prevent further climate change."

HIL & POLYA. Dr Richard Hil (progressive author and adjunct sociology professor, Griffith University, Queensland, Australia) and Dr Gideon Polya (writer, author and former associate professor of biochemistry, La Trobe University, Melbourne, Australia) (2019): "The silencing of the climate emergency, and what we should do about it" (New Matilda, 5 March 2019: https://newmatilda.com/2019/03/05/silencing-climate-emergency/) critically asserts that: "We also need to reframe our language and acknowledge the potential for what many commentators and climate victims (for instance, in the Marshall Islands) refer to as "climate genocide". References also abound to "climate violence", "violent policy making", and increasingly, "climate criminals", with many legal experts linking climate policy failures to specific breaches of environmental and other laws. Increasingly, there are calls for politicians, corporate and media chiefs to be held to account so that there is no statute of limitations and no place to hide for those promoting policies and actions that lead to speciescide, ecocide, sociocide, omnicide and terracide – words that have begun to spill into the public sphere… The struggle over language, and its place in the domain of common sense, is integral to the struggle against violent power. Breaking the silence of the climate emergency is key to our future survival."

HOWES. Charlotte Howes (US writer) (2017): "Trump's climate genocide … When asked, "Do you believe in climate change?" Trump responded: "I think there'll be a little change here, it'll go up, it'll get a little cooler, it'll get a little warmer. Like it always has, for millions of years. It's called weather. I think it's a big scam for a lot of people to make a lot of money."… Trump plans to cancel the remaining sum to be paid to the Climate Change Fund and any future UN climate change programmes, as laid out in his plan for the first 100 days in office. The manifesto also proposes that the funds be used instead on 'fixing America's water and environmental infrastructure'. His sixth statement under 'actions to protect the American worker' declares his intention of reviving the pipeline projects [Keystone and Dakota] and lifting other 'Obama-Clinton roadblocks' to allow energy infrastructure projects to move forward on Americans' own finite fossil fuel reserves. This, in turn, will re-open the floodgates to up to $50 trillion dollars (£39.58 trillion) worth of shale, oil, natural gas and coal… It appears that Trump is set on waging war against the environment, at a time when if we don't start investing in green energy then we may miss our chance to preserve our only planet."

HUNZIKER. Robert Hunziker (a US writer) (2015): "Climate genocide… Not only that, temperature changes of only two (2C) degrees Celsius warmer than today can, and did, equate to 16 feet of water for NYC whereas only four (4C) degrees Celsius cooler than today can, and did, equate to a block of ice 1.24 miles thick surrounding NYC. Those two seemingly small numbers, 2C and 4C, are examples used in Sir David Attenborough's film, Are We Changing Planet Earth? BBC Natural History Unit, BBC One. Attenborough's film unequivocally answers the question its title poses: "Yes, we are"… Paleoclimatologists study ice core and sediment samples to determine with remarkable accuracy how the climate changed millions of years ago. In fact, as recently as tens-of-thousands of years ago, Greenland's temperature shot up by 5C-6C within a couple decades, not over hundreds of years and not over thousands of years…Furthermore, only recently, scientists from Rutgers University changed the course of scientific thought. Previously, scientists thought that 55 million years ago global temperatures took 10,000 years to increase by 5C. Now, a seminal sediment study by Rutgers University scientists has proven it only took 13 years for global temperatures to increase by 5C…Climate change is most prominent where people do not see it! Consequently, there are serious, informed scientists who believe the Arctic will be ice-free during the month of September, its annual minimal, within a few years. The corollary for the climate could be horrendous, devastating, and deathly. In other words, a climate holocaust may be lurking in the shadows. Such as, Arctic sea ice loss causing massive release of methane, smothering the Northern Hemisphere and life changing forever, as the planet heats up in a "geological instant."

HWANG. Andrew D. Hwang (Associate Professor of Mathematics, College of the Holy Cross, US) (2018): According to the Worldwatch Institute, an environmental think tank, the Earth has 1.9 hectares of land per person for growing food and textiles for clothing, supplying wood and absorbing waste. The average American uses about 9.7 hectares. These data alone suggest the Earth can support at most one-fifth of the present population, 1.5 billion people, at an American standard of living."

INTERNATIONAL MONETARY FUND. The International Monetary Fund (IMF) mitigation of climate change report (September 2019): "Limiting global warming to 2°C or less requires policy measures on an ambitious scale, such as an immediate global carbon tax that will rise rapidly to $75 a ton of CO_2 in 2030… The shift from fossil fuels will not only transform an economy but also profoundly change the lives of households, businesses, and communities. Importantly, the shift would generate additional and immediate domestic environmental benefits, such as lower mortality from air pollution (725,000 fewer premature deaths in 2030 for a $75 a ton tax for G20 countries alone)… The average price on global emissions is currently $2 a ton, a tiny fraction of what is needed for the 2°C target."

ISLAND STATES. Report from UN COP17 Climate Conference, Durban, December 2011, on dire warnings from Island States representatives about "climate genocide" from unaddressed, man-made climate change (2011): "Action must be taken immediately to avoid "climate genocide", according to several low-lying island states at COP17. Addressing ministers and heads of states for the first time this morning, parties representing at risk islands used the opportunity to issue dire warnings."

JOHNSTONE. A. Johnstone (UK socialist and activist) on system change to avert climate genocide (2011): "Yet it [terracide] needn't be. An Australian engineering team called Beyond Zero Emissions has released its 5 year study that shows how Australia can have 100% renewable energy by 2020 using renewable technologies of wind power and concentrated solar thermal with molten salts energy storage for 24/7, baseload operation. Professor Mark Jacobson of Stanford University, California, and Mark A. Delucchi of University of California Davis have produced a plan for 100% renewable energy plan for the whole world by 2020. James Hansen in answer to the question "Is there any real chance of averting the climate crisis?", has stated: "Absolutely. It is possible – if we give politicians a cold, hard slap in the face. The fraudulence of the Copenhagen approach – 'goals' for emission reductions, 'offsets' that render ironclad goals almost meaningless, the ineffectual 'cap-and-trade' mechanism – must be exposed. We must rebel against such politics as usual." Capitalism denies - or through resolute inaction effectively denies - the acute problem of man-made climate change. Capitalism – that brought us wars and holocausts – has been unable or unwilling to address man-made climate change and now threatens a climate genocide. It must be system change, not climate change."

KAROLY. Professor David Karoly (climate scientist, honorary professor at the University of Melbourne and CSIRO's climate change hub leader) on the sustainable

population of people on the Earth and in response to the question "Could the [present] mass extinction include humans?" (2019): "So, we're talking about break-up of the landscape corridors that would allow movement of animals, movement of plants, that would allow the connectivity to move southward [in the Southern Hemisphere] into cooler environments. That's a critical factor. For people, we can move. But, certainly, there's at least one – probably more – climate scientists in Europe that have said that the long-term sustainable population of people on the Earth is about 1 billion people in 2100 – not the foreshadowed United Nations population estimates of about 10 to 12 billion people. That's not good news."

KLARE. Michael Klare (Five College Professor of Peace and World Security Studies at Hampshire College in Amherst, Massachusetts and author of "The Race for What's Left: The Global Scramble for the World's Last Resources", "Rising Powers, Shrinking Planet: The New Geopolitics of Energy" and "Blood and Oil: The Dangers and Consequences of America's Growing Dependence on Imported Petroleum") (2017): "If the world's response to the current famine catastrophe [from Nigeria to Yemen] and the escalating fears of refugees in wealthy countries are any indication, people will die in vast numbers without hope of help. In other words, failing to halt the advance of climate change—to the extent that halting it, at this point, remains within our power—means complicity with mass human annihilation. We know, or at this point should know, that such scenarios are already on the horizon. We still retain the power, if not to stop them, then to radically ameliorate what they will look like, so our failure to do all we can means that we become complicit in what—not to mince words— is clearly going to be a process of climate genocide. How can those of us in countries responsible for the majority of greenhouse gas emissions escape such a verdict? And if such a conclusion is indeed inescapable, then each of us must do whatever we can to reduce our individual, community, and institutional contributions to global warming."

KORTEN. Dr David Korten (author of "Agenda for a New Economy" and "When Corporations Rule the World", board chair of YES! magazine, co-chair of the New Economy Working Group, a founding board member of the Business Alliance for Local Living Economies, president of the Living Economies Forum, and a member of the Club of Rome) (2018): "When I ask an audience, "Who believes we are on a path to self-extinction?" nearly every hand goes up. It's a sign of a growing awareness that humanity is on a path to self-imposed environmental and social collapse. For me, that awareness is a source of hope. I recently discovered an even deeper source of hope on a trip to South Korea. There I was involved in a remarkable series of international discussions on the transition to "ecological civilization"… The concept is gaining traction elsewhere as well. China has embedded its commitment to ecological civilization in its constitution."

LAHOUD. Adrian Lahoud writing for Forensic Architecture (FA) (a research agency, based at Goldsmiths, University of London, that undertakes advanced architectural and media research on behalf of international prosecutors, human rights organisations and political and environmental justice groups) (2009): "Climate crimes. Two accusations of genocide in the Sahel: The first issued by the International Criminal Court (ICC) in 2008 regarding war crimes in Sudan; the second [Climate Genocide] issued 2009 by the Sudanese diplomat Lumumba Di-Aping directed at the world's developed nations. The first favors the West. The second deflects and returns the claim and thereby it raises the specter of a new form of violence. This work tests what it would take to support Di-Aping's claim and in doing so raises a number of questions about the violence wrought by climate change, especially the forums in which it is debated and eventually legitimized. What will be the role of forensic climatology in reconnecting the causes of environmental violence with their effects? And what will be the political consequences? Drawing on recent scientific research that shows a correlation between aerosol emission in the northern hemisphere and desertification in the Sahel, this project makes visible a new geopolitical cartography that ties together distant fates, linking industrialization in the North to deprivation in the South…. In this context, can we begin to think about forums like the COP as crime scenes?"

LDC & MVC CIVIL RIGHTS GROUPS. Bangladesh Report on LDC (Least Developed Countries) and MVC (Most Vulnerable Countries) Civil Rights Groups' dismay over the Marrakech Climate Conference failure (2016): "Today, just one day before of the conclusion of CoP 22 Marrakech climate conference, civil society rights group representing LDCs and MVCs have organized a press conference in CoP 22 Marrakech climate conference at press conference room Dakhla, they express dismay on possible conclusion of CoP 22 Marrakech climate conference. In their group position paper, they said that US president elect Donald Trump's possible threats on possible pull out from Paris Agreement (PA) the climate multilateral deal, indecisions of developed countries in respect of enhancing mitigation ambitions, very little or no financing commitment in respect of financing especially for adaptation and loss and damages, finally inaction in respect of climate induced displacement from this conference will lead and aggravate climate genocide in LDCs and MVCs. They cite examples from a recent (14th Nov 2016) World Bank and Global Facility of Disaster Reduction and Recovery (GFDRR)'s released report that, there will be 26 million new poor and loss of $520 billion worth asset and services annually due to natural disasters" ("Civil Society Right Groups from LDC (Least Developed Countries) and Most Vulnerable Countries (MVC) express dismay on CoP 22 Marrakech Climate Conference. Developed countries indecisions in Marrakech will lead Climate Genocide in LDCs and MVCs.")

LEIFER. Dr Ira Leifer (Atmospheric Science at the Marine Sciences Institute, University of California, Santa Barbara): "Some scientists are indicating we should make plans to adapt to a 4C hotter world. While prudent, one wonders what portion of the population could adapt to such a world. My view is that it's just a few thousand people seeking refuge in the Arctic or Antarctica."

LI. Minqi Li (a Chinese political economist, world-systems analyst, and historical social scientist, currently an associate professor of Economics at the University of Utah, and known as an advocate of the Chinese New Left and a Marxian economist) (2008): "According to models adopted by the Intergovernmental Panel on Climate Change (IPCC), if atmospheric CO_2-equivalent rises to 445-90 ppm [500 ppm CO_2-e in 2020], then the global average temperature is likely to rise to 2-2.4 degrees above the pre-industrial level (IPCC, 2007c). With an increase of 2 degrees, there will be widespread crop failures, drought, desertification, and flooding throughout Asia, Africa, Europe, North and South America, and Australia. Of plant and animal species, 15-40 percent are likely to go extinct. But more importantly, a 2-degree warming will constitute "a dangerous anthropogenic interference" as it will initiate a series of climate feedbacks that are likely to take the earth beyond a set of "tipping points". Beyond these tipping points, global warming will become a self-sustaining process out of human control, leading to massive catastrophes that could wipe out most of the species on earth."

LONDON PRE-COPENHAGEN CONFERENCE DEMONSTRATORS. Breitbart London report on huge London demonstration just before the Copenhagen Climate Conference (2015): "Thousands took to the streets of London today, led by Labour leader Jeremy Corbyn, to protest climate change. The event was part of a worldwide day of protest with almost 2,500 similar events taking place globally ahead of a United Nations (UN) climate summit in Paris. Organisers claimed that about 50,000 people took part in the London march this afternoon… other UK events took place in cities including Newcastle and Belfast. Campaigners from groups such as Greenpeace, Oxfam, Friends of the Earth, ActionAid and the World Wildlife Fund marched alongside feminist groups and campaigners from local campaigns such as a group protesting against fracking in Lancashire. They held signs that read "No climate genocide" and "I am a climate migrant". Some chanted "we are the wretched of the earth" and "save our planet"."

LOVELOCK. James Lovelock (FRS, atmospheric gas analysis scientist and proposer of the Gaia Hypothesis) (2006): "We are not all doomed. An awful lot of people will die, but I don't see the species dying out ... A hot Earth couldn't support much over 500 million."

LYONS. Kate Lyons (Guardian reporter) on climate deaths (2019): "Climate change is "absolutely" already causing deaths, according to a new report on the health impacts of the climate crisis, which also predicts climate-related stunting, malnutrition and lower IQ in children within the coming decades. The report, From Townsville to Tuvalu, produced by Monash University in Melbourne, pulled together scientific research from roughly 120 peer-reviewed journal articles to paint a picture of the health-related impacts of the climate emergency in Australia and the Pacific region. It pointed to a 2018 report from the World Health Organisation, which predicted that between 2030 and 2050, global warming would cause an additional 250,000 deaths per year from heat stress, malnutrition, malaria and diarrhoea."

MAIN. Douglas Main (US journalist) commenting in Newsweek on study, by Corey Bradshaw and Barry Brook, published in the Proceedings of the National Academy of Sciences and re 2 billion people by 2153 if one child per woman on average by 2100 (2014): "The researchers wrote that reducing fertility rates (defined as the number of offspring the average woman has over her lifetime) to 2 from the current rate of 2.37 by 2020 would lead to 777 million fewer "people to feed planet-wide by 2050," for example. But they emphasized that worldwide population would take a long time to stabilize. Several other studies suggest that a world population of between 1 billion and 2 billion "might ensure that all individuals [live] prosperous lives, assuming limited change in per capita consumption and land/materials use." If humans reduced fertility rates to one child per woman on average by 2100, there could be as few as 2 billion people by 2153, they calculated."

MEHTA. Suketu Mehta (Gujarati Indian American writer, an associate professor of journalism at New York University, and author of "Maximum City: Bombay Lost

and Found" and "This Land is Our Land. An Immigrant's Manifesto") (2019): "But migration driven by climate change isn't something that's safely in the future; it's been dramatically increasing in the recent past. Since 1992, 4.2 billion people have been affected by droughts, floods and storms. Today, 1.8 billion people are suffering the effects of drought, land degradation, and desertification… Migrants come to work because they can't work at home. Heat waves took almost a million people out of the global workplace in 2016, half of them in India alone. They come to eat because they can't eat at home. For every degree Celsius increase in temperature, wheat yields have been falling by 6 percent and rice yields by 3 percent. 1.5 degrees, corn yields shrink by 10 percent."

MOHADJI. Fouad Mohadji, Vice President of the Comoros at the COP17 UN Climate Conference in Durban, December 2011, saying that millions of people were living with a death sentence over their heads: "History will condemn us for causing climate genocide."

MORALES. Evo Morales (former president of Bolivia) commenting on the day before the conclusion of the December 2010 Cancun Climate Conference (he has argued for cuts of up to 50 per cent in global carbon emissions by 2020) (2010): "We need limited industrial development, rational development, not industries to kill like we have now, but rather industries to save lives… It [failure to act] is leaving the world without ecology. I called it ecolocide, which will lead to genocide… It's important to create an international court of climate justice. Industries, social movements, governments and international organisations who do not meet the norms that are established must be sanctioned…The most important thing is that the proposals of the people of the world are being debated, the rights of mother earth are being debated… What's never been in a debate before is how to live in harmony with Mother Earth - that's the debate."

MORNINGSTAR. Cory Morningstar (Canadian environmentalist) on genocide of Brazilian Indians by the Bel Monte hydrolectric project, noting that "genocide"

according to Article 2 of the UN Genocide Convention is defined as "acts committed with intent to destroy, in whole or in part, a national, ethnic, racial or religious group" (2011): "Bel Monte/ Climate Genocide will not be Stopped with Petitions/Letters … The chief Raoni cries when he learns that Brazilian president Dilma released the beginning of construction of the hydroelectric plant of Belo Monte, even after tens of thousands of letters and emails addressed to her and which were ignored as [were] the more than 600 000 signatures. That is, the death sentence of the peoples of Great Bend of the Xingu River is enacted. Belo Monte will inundate at least 400,000 hectares of forest, an area bigger than the Panama Canal, thus expelling 40,000 indigenous and local populations and destroying habitat valuable for many species – all to produce electricity at a high social, economic and environmental cost, which could easily be generated with greater investments in energy efficiency."

PACIFIC ISLANDS FORUM. Pacific Islands Development Forum (2019): "We the People of the Pacific islands Development Forum, Striving to advance the sustainable and inclusive development of Pacific Island nations … Declare that we … 3. Underscore the serious concerns and stark warnings, documented by the IPCC Special Report on 1.5°C and the Special Report on Oceans and Cryosphere, GHG emissions must be reduced immediately. The science warns of the real possibility that coral atoll nations could become uninhabitable as early as 2030. By 2100, the coral atoll nations of the Republic of the Marshall Islands, Tuvalu, Kiribati, Tokelau and the Maldives and many SIDS could be submerged."

PATIENCE. Agodo Shabella Patience (a Ugandan Lawyer, founder and Executive Director of Green Teso Initiative a climate change Advocacy NGO based in Eastern Uganda) (2015): "For generations, Uganda has grappled with political instability, poverty and inequality. Today however, as she starts to witness a semblance of progress, she is rudely awakened to a climate change reality. A reality that threatens to plunge her into a much worse state than she has ever witnessed before. This unprecedented climate change reality promises the distraction [destruction] of systems essential for human livelihood such as water systems, ecosystems, health, food security and infrastructure. This unprecedented climate change reality promises the distraction of systems essential for human livelihood such as water systems, ecosystems, health, food security and infrastructure. And with over 70% of the populace dependent on climate fed agriculture, Uganda is headed for a climate genocide. To evidence these threats, the Ugandan weather has added twists and turns to its expected scope causing many parts of the country to suffer an avalanche of rains and droughts."

POLYA. Dr Gideon Polya on near-terminal Climate Genocide (2020): "While the presently dominant and deadly neoliberalism demands maximal freedom to exploit human and natural resources for private profit, this rapacious course now existentially threatens Humanity and the Biosphere as the world careers towards a Climate Genocide in which 10 billion people will die avoidably this century en route to a sustainable population in 2100 of a mere 0.5- 1.0 billion."

POPE FRANCIS. Pope Francis in his encyclical "Laudato si'" ("Praise be") has quite exceptionally gone further than any other major leader by alluding to the millions of "premature deaths" (2015): "[Section 20] Some forms of pollution are part of people's daily experience. Exposure to atmospheric pollutants produces a broad spectrum of health hazards, especially for the poor, and causes millions of premature deaths … [Section 48] The human environment and the natural environment deteriorate together; we cannot adequately combat environmental degradation unless we attend to causes related to human and social degradation... The impact of present imbalances is also seen in the premature death of many of the poor, in conflicts sparked by the shortage of resources, and in any number of other problems which are insufficiently represented on global agendas... [Section 195] The principle of the maximization of profits, frequently isolated from other considerations, reflects a misunderstanding of the very concept of the economy. As long as production is increased, little concern is given to whether it is at the cost of future resources or the health of the environment; as long as the clearing of a forest increases production, no one calculates the losses entailed in the desertification of the land, the harm done to biodiversity, or the increased pollution. In a word, businesses profit by calculating and paying only a fraction of the costs involved. 'Yet only when the economic and social costs of using up shared environmental resources are recognized with transparency and fully borne by those who incur them, not by other peoples or future generations,' [Benedict XVI] can those actions be considered ethical."

PRASHAD. Vijay Prashad (professor of international studies at Trinity College in Hartford, Connecticut and the author of 18 books, including "Arab Spring, Libyan Winter", "The Poorer Nations: A Possible History of the Global South" and "The Death of a Nation and the Future of the Arab Revolution") (2017): "It is not as if the Paris or Kyoto agreements would have been sufficient to stem the tide of adverse climate change. Even those were too mild, too friendly to corporations that make their money destroying the planet. But at least these agreements forced governments to accept that human activity—namely industrial capitalism—had hastened the destruction of nature. Now, Trump's Energy Secretary Rick Perry says openly that carbon dioxide emissions are not the main drivers of climate change. Perry pointed the finger of blame at 'ocean waters,' allowing industrial capitalism an exit from responsibility. Why bother with alternatives to carbon when there is no 'evidence' that such energy sources bring the planet closer to annihilation?…Will a 'small' nuclear exchange be contemplated for the Korean Peninsula and for Eastern Asia in general? We are between climate catastrophe and wars of extinction, with the final administration provoking both at hyper-speed. Trump plays the role of Judas in Gaudi's sculpture. Jesus speaks to him about betrayal. But he is looking over Judas' shoulder. He is asking the rest of us if we are participants in the betrayal. What are you doing today to prevent Trump's agenda from driving our planet closer to extinction?"

PROGRESSIVE RADO NETWORK. Progressive Radio Network (2016): "Climate genocide. The planet has warmed by 0.85C since the industrial revolution, or since 1880, according

to the Intergovernmental Panel on Climate Change (IPCC),
which is the baseline for measurement by the scientific
community. That doesn't seem like much; it's such a small
number, less than one (1). But, remarkably, the increase of
0.85C happened within 150 years, whereas historically it
normally took much, much longer for the planet's temperature
to increase by that amount. Ostensibly, it's speeding up by quite
a bit. Beware! The 0.85C temp increase is mere triviality when
compared to some prior events in the paleoclimatic record
books. An ominous event much more threatening than a 0.85C
increase may be lurking in the shadows. Paleoclimatic studies
confirm Earth's climate has turned nasty "on a dime of
geological time," within little more than a decade, more on this
later. Not only that, temperature changes of only two (2C)
degrees Celsius warmer than today can, and did, equate to 16
feet of water for NYC whereas only four (4C) degrees Celsius
cooler than today can, and did, equate to a block of ice 1.24
miles thick surrounding NYC. Those two seemingly small
numbers, 2C and 4C, are examples used in Sir David
Attenborough's film, Are We Changing Planet Earth? BBC
Natural History Unit, BBC One. Attenborough's film
unequivocally answers the question its title poses: "Yes, we
are." As explained in the film, seemingly small temperature
changes have huge planetary impact, for example: 160,000
years ago when temperatures were 4C cooler than today, NYC
would have been in a block of ice 1.25 miles thick. Then,
30,000 years later, when temperatures were 2C warmer than
today, NYC would have been in 16 feet of water."

**RAHMAN. Dr Atiq Rahman of the Bangladesh Centre for
Advanced Studies who has investigated how climate change
will affect Bangladesh (2004):** "It's a flat, flat, flat country. The
flow of water coming from the Himalayas - which is huge -
depends on the differential of height. When the sea level is
higher, the flow of that water will be restricted. So when you
hear now of Bangladesh being a flood-prone country - it will be
a much more flood-prone country in future ... No contribution,
highest impact [by rich, polluting countries] - that makes it a
huge case of moral inequality against which the global citizenry,

the global nation states, must take action. If not we'll be calling it climatic genocide. That's where we're heading."

RANAWAKA. Patali Champika Ranawaka on Developed World climate racism, climate terrorism and climate genocide (November 2009): "Climate racism … During the pre-industrial era when oil or coal was not used as a main energy source, carbon concentration in the atmosphere was 260ppm. Now it has risen to 390ppm. During the last century the mean temperature has risen by 0.73^{o}C. It is now predicted that the temperature would rise to a point between $1.5^{o} - 6^{o}$ Celsius, during the 21st century. Therefore, scientists introduced a carbon budget to avoid catastrophic environmental disaster which may end the humanity on the planet. According to their calculations, if we could be able to go back to the 350ppm and limit the temperature rise to 1.5^{o}C relative the pre-industrial era, there would be an over 90% probability to avert an environmental calamity that could devastate this planet… When the US administration talk about human right violation, global terrorism, genocide etc., they should realize that its mirror image is much worse environmental human right violation, global climate terrorism and climate genocide over our children etc. Joseph Stieglitz, the ex-president of the World Bank and Nobel prize winner for economics rightly pointed out that any agreement should be based on emissions per dollar of GDP and emissions per-capita… So in Copenhagen, developing countries will have a moral high ground and rational argument. On the other hand, developed countries only have brute force and concepts like Climate Nazism, Climate Racism and Climate Terrorism."

RANDERS. Jorgen Randers (professor at the Norwegian Business School BI and co-author of "The Limits to Growth" report to The Club of Rome (1972), and author of "2052- A Global Forecast for the Next Forty Years" (2018): "I am a climate pessimist. I believe (regrettably) that humanity will not meet the climate challenge with sufficient strength to save our grandchildren from living in a climate-damaged world. Humanity (regrettably) will not make what sacrifice is necessary today in order to ensure a better life for our ancestors forty years hence. The reason is that we are narrowly focused on maximum well-being in the short term. This short-termism is reflected in the systems of governance that we have chosen to dominate our lives: Both democracy and capitalism place more emphasis on costs today than on benefits forty years in the future… Otherwise, I predict, it will be the Chinese who solve the global climate challenge - singlehandedly. Through a sequence of 5-year plans established with a clear long term vision, and executed without asking regular support from the Chinese. They are already well on the way, for the benefit of our grandchildren."

RINCKER. Frits Rincker (Den Haag, Netherlands) commenting on a geo-engineering article by Professor Tim Lenton (an Earth system scientist based in the School of Environmental Sciences, University of East Anglia, Norwich, UK) (2009): "Black soil [biochar] and iron fertilizing are good ideas [for CO_2 removal by geo-engineering], but the 'could' mentality is killing us. Why not say we will have no environmentally lethal products or activities from NOW on. Every uncompensated molecule of CO_2 anywhere in the productive chain is one too much. Would life stop? I doubt it strongly. The choice is between having the blood of the coming climate genocide on our hands or not."

ROCKSTROM. Johan Rockström (director of the Potsdam Institute for Climate Impact Research, Germany) on a hotter world (2019): "We will have lost all the [coral] reefs decades before 2100 – at somewhere between 2C and 4C … A combination of climate change and deforestation could push it [the Amazon rain forest] into a savannah state … It's difficult to see how we could accommodate eight billion people or maybe even half of that. There will be a rich minority of people who survive with modern lifestyles, no doubt, but it will be a turbulent, conflict-ridden world… The reason is primarily making enough food, but also we would have lost the biodiversity we're dependent on and be facing a cocktail of negative shocks all the time, from fires to droughts."

ROMERO. Pablo Solón Romero, the Bolivian representative at the December 2010 Cancun Climate Conference (2010): "We're talking about a [combined] reduction in emissions of 13-16%, and what this means is an increase of more than 4C. Responsibly, we cannot go along with this - this would mean we went along with a situation that my president [Evo Morales] has termed "ecocide and genocide"."

SCHELLNHUBER. Hans Joachim Schellnhuber (founding director of the Potsdam Institute for Climate Impact Research, Germany and key adviser to Angela Merkel in the design of Germany's transition to a low-carbon economy) (2015): "In the world of Google you can raise any myth and you can raise any lie. If you go to the internet then I will be accused or applauded for asking to reduce the world population to one billion. Here is what I really said. It was a scientific conference, preceding the infamous Copenhagen conference in 2009, and actually I talked about the carrying capacity of the Earth, which is an interesting issue. What I said is, if global warming is not in any way mitigated, and we go into a six or eight degrees warmer world, then our planet will probably only be able to support a billion people."

SETHNESS. Javier Sethness on climate genocide in reviewing Richard A. Koenigsberg's book "Nations Have the Right to Kill: Hitler, the Holocaust and War" (October 2009): "As is well-known, climate change stands to threaten agricultural production across much of the globe, radically diminish the global supply of freshwater, inundate low-lying coastal settlements currently home to hundreds of millions of people, prompt widespread desertification, and literally eradicate some countries that today exist. The specter of such life-negating realities seems to find its genesis in capitalist society, a form of totalitarianism that essentially values profitability above all else. The response of nearly every advanced-capitalist country to the now well-established reality of climate change has been entirely inadequate toward the end of allowing much of humanity and life itself the chance to flourish or even survive the projected consequences of anthropogenic global warming; their lack of meaningful action on this question—a lack which results from the desire to hold existing society more or less unchanged—is systematic. It cannot merely then be stated that the mass murder—the rendering-impossible of human life—that follows from reformist inaction is a mistake, an unintended consequence, an 'externality.' Such horrifying consequences are today essentially inevitable in contemporary capitalism; as such, dominant Western treatment of these questions bears much in common with other genocidal episodes of human history…What is currently occurring, then, is the mass-murder of the global South by much of the global North. There has of course been a marked tendency toward this dynamic now for some time in human history, but it seems climate genocide constitutes the most final of these historical denials."

SKRBINA. David Skrbina (Senior Lecturer in Philosophy in the Department of Philosophy at the University of Michigan) (2020): "A world of, say, 5 billion people is more likely to be sustainable than one of 10 billion; and a world of 1 billion is likely more sustainable still. All things being equal, a world with fewer people will allow for a more robust planetary ecosystem, and a higher quality of life for humans, than a world with more people… First, we have been pumping carbon gases into the air for over 300 years, and neutrality only means no further additions; at some point, we need to start pulling carbon out of the air, and return to a stable condition. We need a "carbon negative" plan, not a carbon neutral one. Second, the implication is that population can continue its unhindered growth—as if a world of 10 billion "carbon neutral" people could be sustainable. It can't. But this raises an important question: How many people can the planet sustain? We can do a quick analysis, comparable to what I have done above. The total usable land area on Earth is around 11.2 billion ha. If we set aside half (5.6 bha) for nature and live on the other half, and we assume the current global average footprint of 2.8 ha/person, then the planet can support just (5.6 / 2.8 =) 2 billion people at current living standards. This is a decline of 74% from the current 7.7 billion."

SPRATT. David Spratt (leading Australian climate change activist and Ian Dunlop (leading Australian business man and climate change activist) (2017): "The first responsibility of a government is to safeguard the people and their future well-being. The ability to do this is threatened by climate change, whose accelerating impacts will also drive political instability and conflict, posing large negative consequences to human society which may never be undone. This report looks at climate change and conflict issues through the lens of sensible risk management to draw new conclusions about the challenge we now face. From tropical coral reefs to the polar ice sheets, global warming is already dangerous. The world is perilously close to, or passed, tipping points which will create major changes in global climate systems. The world now faces existential climate-change risks which may result in "outright chaos" and an end to human civilisation as we know it."

SUVA DECLARATION ON CLIMATE CHANGE BY THE PACIFIC ISLANDS DEVELOPMENT FORUM. Suva Declaration (2015): "We, the Leaders of the Pacific Islands Development Forum following open, transparent and inclusive discussions with stakeholders undertaken during the Pacific Islands Development Forum Third Annual Summit held in Suva, Fiji between 2-4 September 2015 declare that we: 1. Are gravely distressed that climate change poses irreversible loss and damage to our people, societies, livelihoods, and natural environments creating existential threats to our very survival and other violations of human rights to entire Pacific Small Island Developing States…"

UK ANTI-TRUMP PROTESTERS. Independent newspaper report: "UK anti-Trump protesters unfurl "A 100m banner on the embankment opposite the Houses of Parliament, next to St Thomas' Hospital, dropped by protesters during Donald Trump's UK visit" (July 2018) reading: "TRUMP: CLIMATE GENOCIDE"."

VINCE. Gaia Vince (UK climate journalist writing in the UK Guardian) (2019): "Many scientists think it's highly unlikely that we will stay below 2C (above pre-industrial levels) by the end of the century, let alone 1.5C. Most countries are not making anywhere near enough progress to meet these internationally agreed targets. Climate models predict we're currently on track for a heating of somewhere between 3C and 4C for 2100, although bear in mind that these are global average temperatures – at the poles and over land (where people live), the increase may be double that. Predictions are tricky, however, as temperatures depend on how sensitive the climate is to carbon dioxide (CO_2). Most models assume that it is not very sensitive – that's where the lower 3C comes from – but a whole new set of models to be published in 2021 finds much greater sensitivity. They put heating at around 5C by the end of the century, meaning people could be experiencing as much as 10C of heating over land… Higher sea levels will make today's low-lying islands and many coastal regions, where nearly half the global population live, uninhabitable, generating an estimated 2 billion refugees by 2100. Bangladesh alone will lose one-third of its land area, including its main breadbasket… It would mean abandoning huge tracts of the globe and moving Earth's human population to the high latitudes: Canada, Siberia, Scandinavia, parts of Greenland, Patagonia, Tasmania, New Zealand and perhaps newly ice-free parts of the western Antarctic coast."

WALKER. Patrick Walker (US writer)(2019): "Russiagate versus Climate Genocide: how Democrats cover up Trump's worst crime… Russiagate's Beauty for Dems: Dissing Trump without Discussing Issues. The more important an issue is to humanity, the likelier it is Democrats *don't* want to discuss it. Thus, it's hardly surprising that Noam Chomsky felt obliged to point out the "moral depravity"of *both* major parties for discussing neither climate change nor the increasing risk of nuclear war during the recent midterm election campaigns. It's likewise hardly surprising that the Republican Party, which Chomsky rightly terms "the most dangerous organization in human history," likes to stay mum on the real nature of its wantonly destructive policies. But what's truly amazing—until one habituates to the mammoth corruption of current U.S. politics—is how utterly unwilling today's Democratic Party is to discuss issues that would paint Republicans in not merely in an unflattering, but in a criminally insane light."

WALLACE-WELLS. David Wallace-Wells (US journalist who writes on the worsening climate emergency) on the 2018 IPCC Report "Global warming of 1.5 °C" (2018): "UN says climate genocide is coming. It's actually worse than that… Just two years ago, amid global fanfare, the Paris climate accords were signed — initiating what seemed, for a brief moment, like the beginning of a planet-saving movement. But almost immediately, the international goal it established of limiting global warming to two degrees Celsius began to seem, to many of the world's most vulnerable, dramatically inadequate; the Marshall Islands' representative gave it a blunter name, calling two degrees of warming "genocide". The alarming new report you may have read about this week from the UN's Intergovernmental Panel on Climate Change — which examines just how much better 1.5 degrees of warming would be than 2 — echoes the charge… Barring the arrival of dramatic new carbon-sucking technologies, which are so far from scalability at present that they are best described as fantasies of industrial absolution, it will not be possible to keep warming below two degrees Celsius — the level the new report describes as a climate catastrophe. As a planet, we are coursing along a trajectory that brings us north of four degrees by the end of the century. The IPCC is right that two degrees marks a world of climate catastrophe. Four degrees is twice as bad as that. And that is where we are headed, at present — a climate hell twice as hellish as the one the IPCC says, rightly, we must avoid at all costs".

WATSON. Paul Watson (founder of the Sea Shepherd Conservation Society) (2007): "We need to radically and intelligently reduce human population to fewer than one billion… Curing a body of cancer requires radical and invasive therapy, and therefore, curing the biosphere of the human virus will also require a radical and invasive approach."

WEIZMAN. Ines Weizman on Climate Genocide (2013): "Climate genocide… As the climate summit disbanded, with news of Di-Aping's claim of "climate genocide" spreading, hundreds of thousands of protestors gathered outside the Bella Centre in Copenhagen. The fracturing of a political accord mirrored the disaggregation of the climate model, both caught in the ambiguous scales of environmental politics. Di-Aping's responses to the colonial mentality of the G20 ("we had been asked to sign a suicide pact") and his invocation of the death camps of the Second World War ("climate genocide") were unexpected. Like a particle fired into view from an imperceptible background, Di-Aping's dissident utterance aimed and succeeded in breaking all protocol. The idea that a Sudanese person would speak out in such a way in the context of a major international summit elicited derisory comments that suggested a lack of responsibility, disingenuousness and unproductive hyperbole. Di-Aping's status as a speaking subject and chairman of 132 nations was called into further question because of the continuing criminal proceedings regarding war crimes in Darfur. What many commentators failed to realize is that it was precisely Di-Aping's proximity to events in Darfur that provided him with a visceral sense of what the change in temperature meant, since as recent climate research has demonstrated, much of the desertification of the Sahel is attributable to increased greenhouse gas concentrations in the atmosphere."

WESTRA. Professor Laura Westra (Professor Emerita (Philosophy), University of Windsor, Osgoode Hall Law School, Faculty of Law, University of Windsor, Toronto, Canada) in "Faces of state terrorism" states (2012): "As well, there are millions of avoidable deaths globally, because of deprivations and lack of health care: "Whether a child dies from neocon bombs or bullets or from neocon-imposed deprivation, the result is the same" (Polya 2011:2). Climate change exacerbates the condition of poor people everywhere. The increase in temperature permitted in the most recent COP meeting in Cancun – that is 4°C – encourages (and in fact guarantees) a situation that can be termed "ecocide" or "genocide", according to Bolivia's president. Hence the ways of supporting wholesale harms, in both war and peace, using the might of the state and that of its alliances, within and without the UN, are multiple and not simply limited to the support of some South American dictatorship."

WORLD BANK. World Bank Climate Report overview (2019): "Countries need sustainable economic growth and good development outcomes and climate change puts both at risk. Natural disasters cost about $18 billion a year in low- and middle-income countries through damage to power generation and transport infrastructure alone. They also trigger wider disruptions for households and firms costing at least $390 billion a year. The impact of extreme natural disasters is equivalent to a global $520 billion loss in annual consumption, and forces some 26 million people into poverty each year. Without urgent action, climate impacts could push an additional 100 million people into poverty by 2030. By 2050, it could mean that as many as 143 million people across three developing regions will become climate migrants, with individuals, families and even whole communities forced to seek more viable and less vulnerable places to live. Under the Paris Agreement, the world committed to limiting the rise in global temperatures to well below 2 degrees Celsius above pre-industrial levels by the end of the century."

WORLD HEALTH ORGANIZATION. The World Health Organization (WHO) reported (2014) : "25 March 2014 | Geneva - In new estimates released today, WHO reports that in 2012 around 7 million people died - one in eight of total global deaths – as a result of air pollution exposure. This finding more than doubles previous estimates and confirms that air pollution is now the world's largest single environmental health risk. Reducing air pollution could save millions of lives" [9 million such deaths annually estimated in 2020].

YATES. Oliver Yates (former Clean Energy Finance Corporation chief executive, former bank executive, former member of the Australian Liberal Party (conservative party) and independent candidate for the blue-ribbon Liberal seat of Kooyong in Melbourne) (2018): "If we don't address climate change and start to reduce our emissions, then it's likely that billions of families could be forced to move home unnecessarily… [I] cannot understand how Liberals would knowingly inflict damage on others when they have a perfectly workable economic cure in front of them [clean energy]."

References.

[1]. "Climate racism, climate injustice & climate genocide –
Australia, US and EU sabotage Copenhagen COP15":
http://bellaciao.org/en/spip.php?article19422.
[2]. Gideon Polya, "Body Count. Global avoidable mortality since
1950", G.M. Polya, Melbourne, 2007; see:
http://globalbodycount.blogspot.com/.
[3]. "List of countries by greenhouse gas emissions per capita":
http://en.wikipedia.org/wiki/List_of_countries_by_greenhouse_gas_e
missions_per_capita.
[4]. "Climate justice & climate injustice: Australia wants a 2020 per
capita GHG pollution 15 times greater than Developing World's":
http://sites.google.com/site/yarravalleyclimateactiongroup/climate-
justice.
[5]. Letter to PM Rudd by Dr James Hansen (Head, NASA's Goddard
Institute for Space Studies):
http://www.aussmc.org.au/documents/Hansen2008LetterToKevinRud
d_000.pdf.
[6]. "Copenhagen Greenhouse Gas reduction proposals - quantitative
comparisons":
http://sites.google.com/site/yarravalleyclimateactiongroup/copenhage
n-greenhouse-gas-reduction-proposals---quantitative-comparisons.
[7]. "Climate racism, climate injustice & climate genocide –
Australia, US and EU sabotage Copenhagen COP15":
http://bellaciao.org/en/spip.php?article19422.
[8]. UN Genocide Convention: http://www.edwebproject.org
/sideshow/genocide/convention.html.
[9]. Grace Moore (2004), Dicken and the Empire. Discourses of class,
race, and colonialism in the works of Charles Dickens" (Ashgate
Publishing, Aldershot, UK): http://www.amazon.com/Dickens-
Empire-Discourses-Colonialism-
Nineteenth/dp/glanceands=books/0754634124_ and
http://books.google.com.au/books?id=ZUxw0x84cGwC&pg=PA94&l
pg=PA94&dq=exterminate+dickens+%22
commander%22&source=bl&ots=HYRvwOuGJK&sig=gTCZ7lGSD
Kf20lcjsseDgicVm7k&hl=en&ei=xjkXStmlK6j8tgOC97HbCA&sa=
X&oi=book_result&ct=result&resnum=5#PPA94,M1.
[10]. Gideon Polya, "UK BBC holocaust denial", Countercurrents:
http://www.countercurrents.org/polya240509.htm.
[11]. Amaresh Misra, "War of Civilizations: India AD 1857", Volume
I, "The Road to Delhi" & Volume II, "The Long Revolution" (2008)
RUPA, India).

[12]. Dr Gideon Polya, Professor Amartya Sen, Dr Sanjoy Bhattacharya et al., BBC (2008), The things we forgot to remember. The Bengal Famine:: http://www.open2.net/thingsweforgot/bengalfamine_programme.html).

[13]. "Copenhagen: "Imperial" climate deal rejected by poor-country delegates", LINKS, 19 December 2009: http://links.org.au/node/1418.

[14]. Formal complaint to the Chief Prosecutor of the International Criminal Court re Australian Government involvement in Aboriginal Genocide, Iraqi Genocide, Afghan Genocide and Climate Genocide: http://climateemergency.blogspot.com/2008_02_01_archive.html.

SECTION D: GAS IS DIRTY ENERGY, GAS IS DIRTIER THAN COAL GHG-WISE & ARCTIC METHANE BOMB

"When uncounted tons [from livestock, land use and methane] are added to the global inventory of atmospheric GHGs, that inventory rises from 41,755 million tons to 63,803 million tons." Robert Goodland and Jeff Anfang, World Watch, 2009.

"The budget of CO_2 emissions still available worldwide could be derived from the 2 degree C guard rail. By the middle of the 21st century a maximum of approximately 750 Gt CO_2 (billion metric tons) may be released into the Earth's atmosphere if the guard rail is to be adhered to with a probability of 67%. If we raise the probability to 75%, the cumulative emissions within this period would even have to remain below 600 Gt CO_2." WBGU (German Advisory Council on Global Change), 2009.

"Economic time bomb. As the amount of Arctic sea ice declines at an unprecedented rate, the thawing of offshore permafrost releases methane. A 50-gigatonne (Gt) reservoir of methane [5,250 Gt CO_2-equivalent], stored in the form of hydrates, exists on the East Siberian Arctic Shelf. It is likely to be emitted as the seabed warms, either steadily over 50 years or suddenly." Professor Peter Wadhams and colleagues (University of Cambridge), 2013.

"With no gas leakage, burning 1 tonne of CH_4 yields 2.75 tonnes CO_2 as compared to burning 1 tonne of carbon (C) yielding 3.67 tonnes CO_2. However with a mere 2.6% of systemic gas leakage (it is about 3% in the US), the burning of gas results in 5.36 tonnes CO_2-equivalent i.e. with systemic gas leakage considered, burning natural gas is about 1.5 times as dirty GHG-wise than burning coal on a per mass basis." Dr Gideon Polya, Green Left, 2020.

"A safer climate will require a return to an atmospheric concentration of greenhouse gases equivalent to 350 parts per million of CO_2, or lower… A nationwide, systematic response is required to drastically reduce emissions from all sectors, draw down greenhouse gases, and be greenhouse gas neutral or negative by 2035." Australian Greens, 2020.

Methane (CH_4) is about 85% of natural gas and has a Global Warming Potential (GWP) that is 105 times greater than that of the same mass of CO_2 on a 20 year time frame and with aerosol impacts considered. Indeed depending upon the degree of systemic leakage (fugitive emissions), burning gas can actually be dirtier Greenhouse Gas (GHG)-wise than burning coal. Methane is making a huge contribution to the worsening Climate Crisis through systemic leakage in natural gas exploitation, and emissions from land use and the warming Arctic. I have set out my concerns in a number of detailed and documented articles and websites over the last dozen years [1-17] that are summarized below.

In 2009 I wrote: "Australia is the world's biggest coal exporter and a world leading annual per capita greenhouse gas polluter. It is now described by its Labor Government as a global energy superpower because of its major role in coal, liquid natural gas (LNG) and uranium exports. However, in view of the public perception of a worsening climate emergency, the Australian Government evidently believes that it has to appear to be doing something about climate change while not substantially affecting the wealth of its corporations or of its conservative citizenry" [1]. In 2009 the major political parties (the opposition Liberal-National Party Coalition and the governing Labor, aka the Lib-Labs) were united in support for unlimited coal and gas exports, with various politicians incorrectly asserting that gas is "clean energy" or "cleaner energy" [1]. In 2020 with the Coalition ruling and Labor in opposition, the situation remains essentially the same in Australia, with both government and opposition committed to unlimited exports of coal, gas and methanogenically-derived meat [10, 11]. There are some differences – about 1/3 of the Coalition MPs are anti-science climate change deniers and the Coalition government is effective climate change denialist through resolute inaction. Labor recognizes the reality of climate change and supports more renewable energy while supporting unlimited exports of coal, gas and methanogenically-derived meat. Cowardly, pragmatic and neoliberal-dominated Labor dishonestly hides behind a conveniently "far away" policy of "zero net emissions by 2050". The climate criminal Coalition government admits "zero net emissions" sometime in the latter half of the 21st century. The science-informed Australian Greens (10% of the vote) want massive climate action, negative GHG emissions ASAP, and understand the danger from methane [10, 11; Australian Greens, "Climate Change and Energy", 2020: https://greens.org.au/policies/climate-change-and-energy].

Another aspect of this false and disastrous coal-to-gas transition in Europe, East Asia, North America and Australia is the environmental devastation and aquifer contamination associated with gas exploitation extending from "conventional gas" to "unconventional gas" from the hydrological fracturing ("fracking") of gas-rich shale deposits in coal seam gas exploitation [2-7]. This obscenity has taken over America and spread to Canada, Australia and India [2-7]. In relation to India (2020 population 1.4 billion) I wrote (2011): "In the path to zero emissions (and beyond) India must not make the same mistakes as the European World which is responsible for 71% of post-Industrial Revolution carbon dioxide (CO_2) pollution in the atmosphere. India has already made huge mistakes in relation to permitting cigarette smoking and cars. However India has huge shale gas deposits (see "Shale gas": http://en.wikipedia.org/wiki/Shale_gas) and for the sake of India and the Planet Indians must ensure that this gas is not exploited – it represents a huge threat to India and to Humanity. India has huge reserves of natural gas trapped in shale fields. Such natural gas is being accessed on a huge scale in the US through horizontal drilling and "fracking", whereby shale deposits are fractured to release the gas through pumping water

and a cocktail of chemicals at huge pressure into shale rock
formations to push gas into pockets for easier recovery" [3].

Writing about the false and dangerous US coal-to-gas transition
I stated (2014): "The coal-to-gas transition touted by the pro-gas
Obama Administration and the US gas industry means replacing
one dirty energy source (coal) with another dirty energy source
(natural gas) that, as exampled above, may indeed be dirtier
than coal GHG-wise depending upon the extent of systemic
natural gas leakage. However non-scientist Obama is unfazed
by or ignorant of such considerations and was happy to falsely
declare that natural gas is "clean energy"(2013): "Today, we use
more clean energy – more renewables and natural gas – which is
supporting hundreds of thousands of good jobs". To reiterate:
natural gas is dirty energy and can indeed be dirtier GHG-wise
than coal. One can at least give some credit to the pro-gas,
climate criminal Obama Administration for "baby steps" in the
right direction and for good rhetoric about the need for climate
change action, in contrast to the climate change denialism and
effective climate change denialism of the terracidal fossil fuel
industry and the terracidal Republicans. Thus Obama (2013):
"Convince those in power to reduce our carbon pollution. Push
your own communities to adopt smarter practices.
Invest. Divest" [7].

Australia's eminent Chief Scientist has unfortunately argued for
a long-term role for gas in the so-called Finkel Review that I in
turn reviewed (2017): "Science is fundamentally sceptical in
involving the critical testing of potentially falsifiable
hypotheses. However scientists cannot endlessly re-invent the
wheel and science operationally involves qualified respect for
authoritative consensus opinion. In a science-based consensus
view the Synthesis Report of the 2,500-delegate March 2009
Copenhagen Scientific Climate Change Conference stated of
climate change inaction: "Inaction is inexcusable", and a 2010
Open Letter by 255 members of the US National Academy of
Sciences, including 11 Nobel Laureates, stated: "Delay is not an
option". This global scientific urgency is in stark contrast to the
substantively tardy or climate inaction policies of the politically
dominant Australian Lib-Labs (Coalition Government and

Labor Opposition) that is reflected in the Finkel Review, chaired by Australia's Chief Scientist, that in the words of Australian Greens leader Dr Richard Di Natale "Provides a long term lifeline to coal and gas" [9].

Unfortunately the Chief Scientist of Australia has again incorrectly supported gas, as I set out in a further detailed analysis (2017): "[Dr Finkel stated] "But, there is a limit to how much solar and wind we can use and still retain a reliable system. Ultimately, we will need to complement solar and wind with a range of technologies such as high levels of storage, long-distance transmission, and much better efficiency in the way we use energy. But, while these technologies are being scaled up, we need an energy companion today that can react rapidly to changes in solar and wind output. An energy companion that is itself relatively low in emissions, and that only operates when needed. In the short-term, as the Prime Minister and Minister Angus Taylor have previously stated, natural gas will play that critical role. In fact, natural gas is already making it possible for nations to transition to a reliable, and relatively low emissions, electricity supply." However as demonstrated in this essay, gas is not "relatively low in emissions" as asserted by Dr Finkel because (a) combustion of 1 tonne of CH_4 (85% of natural gas) yields 2.75 tonne CO_2 as compared to combustion of 1 tonne of carbon (about 90% of coal) yielding 3.67 tonne CO_2, and (b) depending upon the degree of gas systemic leakage, gas burning can actually be much dirtier GHG-wise than coal burning" [10].

Methane (CH_4) is a gas, leaks, is 85% of natural gas, and has a Global Warming Potential (GWP) of 105 relative to the same mass of CO_2 on a 20 year time frame and with aerosol impacts considered. On this basis and assuming that natural gas is 100% CH_4 or an equivalent hydrocarbon, a gas leakage of 2.6% means that the greenhouse gas (GHG) warming effect from the leaked gas is the same as that of the CO_2 from burning the remaining 97.4% of the gas.

With no gas leakage, burning 1 tonne of CH_4 yields 2.75 tonnes CO_2 as compared to burning 1 tonne of carbon (C) yielding 3.67 tonnes CO_2.

However suppose we have 100 tonnes (100 t) of methane (CH_4) and A tonnes leaks so that (100 – A) t CH_4 x 2.75 t CO_2-equivalent / t CH_4 = A t CH_4 x 105 t CO_2-equivalent/ t CH_4 -> 275 – 2.75A = 105 A - > 107.5A = 275 -> A = 275/107.5 -> A = 2.56 or about 2.6.

Thus with a mere 2.6% of systemic gas leakage (it is about 3% in the US), the burning 1 tonne of gas results in 2 x 0.974 x 2.75 = 5.36 tonnes CO_2-equivalent i.e. with 2.6% systemic gas leakage, burning natural gas is about 5.36/3.67 = 1.5 times dirtier GHG-wise than burning coal on a per mass basis [17].

Gas is dirty energy, gas is not clean energy, gas is not cleaner than coal GHG-wise, and indeed gas is dirtier than coal GHG-wise. The present absurd, dangerous and deadly coal-to-gas transition is testament to neoliberal One Percenter and Mainstream greed and mendacity.

While the falsehood of "gas is cleaner than coal GHG-wise" can be attributed to stupidity and ignorance on the part of scientifically illiterate pro-gas politicians, in the face of expert scientific advice (e.g. from smart, first year high school students) it thence becomes a lie, and utterly unforgivable intellectual child abuse when imposed on children by politicians - for which they must be removed from Parliament and all public office by an indignant electorate. Peace is the only way but silence kills and silence is complicity.

This picture changes a bit if we compare GHG emission per GJ (Gigajoule) generated from burning gas or coal, but on this basis too, gas is dirtier than coal GHG-wise as the gas leakage increases above 2.6%.

Combustion of methane produces 50.1kJ/g CH_4 (50.1 GJ/tCH_4) as compared to that from carbon of 32.6 GJ/t for pure carbon and the best anthracite (black coal), and about 17.0 GJ/t for brown coal or lignite (as used brown coal-based power stations in my state of Victoria, Australia; see https://personal.utdallas.edu/~metin/Merit/Folios/coal.pdf).

The CO_2-equivalent from burning CH_4 at 2.6% leakage is 5.36 t CO_2-e /t CH_4, from burning carbon is 3.67 t CO_2-e/t carbon, and 0.65 x 3.67 = 2.39 t CO_2-e/t brown coal (lignite that is about 65% carbon).

Accordingly, on this basis, the "CO_2-equivalent per GJ" is 5.36/ 50.1 = 0.107 for gas, 3.67/32.6 = 0.113 for pure carbon and the best anthracite, but 2.39/ 17.0 = 0.141 for brown coal or lignite.

Thus at 2.6% gas leakage, the "CO_2-equivalent per GJ" is much the same for methane and carbon, but is slightly (1.3 times) bigger for brown coal (as in Victoria, Australia).

However the "CO_2-equivalent per GJ" for burning gas increases as you increase the degree of systemic leakage above 2.6% (e.g. an estimated 4% lifetime" leakage" with fracking in the US; see Robert W. Howarth, "Ideas and perspectives: is shale gas a major driver of recent increase in global atmospheric methane?" Biogeosciences, 16(15):3033-3046, August 2019: https://www.researchgate.net/publication/335182626_Ideas_and _perspectives_is_shale_gas_a_major_driver_of_recent_increase _in_global_atmospheric_methane).

My conclusion is that with 2.6% systemic gas leakage, burning natural gas is about 5.36/3.67 = 1.5 times dirtier GHG-wise than burning coal on a mass basis. On a "CO_2-equivalent per GJ" basis and with 2.6% leakage, gas and top quality coal are about the same but brown coal is 1.3 times as bad as gas - but on a per GJ basis gas use becomes dirtier than black coal and brown coal GHG-wise as the degree of systemic gas leakage increases above about 3% and 4%, respectively.

Australia (2020 population 25.5 million) is now the world's leading exporter of both coal and Liquefied Natural Gas (LNG). In a 2017 article I argued that Australian commitment to unlimited natural gas exploitation threatens the Planet and invites global blowback [8]. From a detailed analysis taking methane into account I concluded (2020): "Methane leakage makes Australia a world-leading per capita greenhouse gas polluter… Australia with 0.33 % of the world population has revised annual Domestic greenhouse gas (GHG) emissions [1,573 Mt CO_2-e] that are 2.5% of the world's, and annual Domestic plus Exported GHG emissions [3,438 Mt CO_2-e] that are 5.4% of the world's annual GHG pollution" [10]. However the Australian Government conveniently ignores or insufficiently considers land use, methane emissions, the revised methane GWP, and Exports to estimate Domestic GHG emissions of a mere 535 Mt CO_2-e that is assertedly only 1% of global emissions. Of course for all the present-day political correctness, fascoid Australian exceptionalism and racism is still alive and well, and the neoliberal Coalition Australian Government sees nothing wrong with Australia having, on its own estimate, 3 times more than its fair share of global GHG pollution on its own flawed accounting.

Methane is emitted into the atmosphere from anaerobic bacterial metabolism in land fill and swamps, but is increasingly being released from the melting of methane-water clathrates in the tundra and Arctic Ocean seabed, and from anaerobic bacterial action in thawed tundra. There are 2 notable positive feedback loops present in the warming Arctic. The so-called "albedo flip" involves GHG-driven warming -> white, light-reflecting snow ice melts -> black sea -> greater light absorption -> more warming -> etc. In addition, GHG-driven warming -> the melting of methane-water clathrates -> increased atmospheric GHG -> more warming -> etc. The Global Warming Potential (GWP) of CH_4 on a 20 year time frame and with aerosol impacts considered is 105 times that of CO_2. The German WBGU (2009) and the Australian Climate Commission (2013) have estimated that no more than a Terminal Carbon Pollution Budget of 600 billion tonnes of CO_2 can be emitted between 2010 and zero emissions in 2050 if the world is to have a 75% chance of avoiding a catastrophic 2C temperature rise. The 50 Gt (50 billion tonnes) CH_4 in the East Siberian Arctic Shelf is thus equivalent to 50 billion tonnes CH_4 x 105 tonnes CO_2-equivalent/tonne CH_4 = 5,250 billion tonnes CO_2-e or about nine (9) times more than the world's Terminal Carbon Pollution Budget. We are doomed unless we can stop this Arctic CH_4 release [13]. However one notes that as of 2020 the Terminal Carbon Pollution Budget of the whole world for avoiding a catastrophic plus 2C has already been exceeded [13].

Continuing, remorselessly increasing GHG emissions mean that exceedance of a catastrophic plus 2C of warming is now essentially unavoidable. However we are obliged to do everything we can to make the future "less bad" for our children, grandchildren and future generations. As a public service in 2013 I published a useful analysis of expert witness testimony to stop a huge gas-fired power plant project, stating: "Indeed this analysis derives in part from my experience as a much-published, 5-decade career biological chemist acting as an Expert Witness (free of charge, of course) for concerned citizens in giving carefully researched evidence in court against the imposition on rural town residents of hugely polluting gas-fired power generation. While the emphasis and procedurally or legally allowable grounds for discussion may differ in the various jurisdictions (e.g. those in the various jurisdictions and specialist courts in America and the British Commonwealth), in general the major issues to be considered in the court would be negative impacts of a proposed gas-fired power station (GFPP) on (A) human health, (B) human amenity (e.g. finances, happiness, visual amenity, and rural versus industrial quality of life), and (C) the environment (flora, fauna, ecosystems, greenhouse gas pollution, global warming and climate change). Some key lines of evidence under these three headings for activists, legal counsel and Expert Witnesses are outlined below as a public service" [6].

The damage related price of GHG pollution has been estimated at about $200 per tonne CO_2-e [8, 15, 16]. In a key 2017 analysis I concluded: "Oil & Gas Australia states that "If current oil prices are maintained, Energy Quest estimates that the value of Australian LNG exports will double to around A\$36 billion [US\$27 billion] in 2017" … If there is a 2.6% leakage then in 2017 Australia will incur a gas exploitation Carbon Debt within Australia of 322 Mt CO_2-e x \$200/ tonne CO_2-e = \$64.4 billion, over twice the profit from exporting LNG…Trump America, Trumpist Australia and neo-Trumpist Canada are acutely threatening the world with unlimited greenhouse gas (GHG) pollution from unlimited fossil fuel exploitation. For the US, Canada and Australia, full exploitation of presently recoverable fossil fuel reserves would generate GHG pollution vastly exceeding (37-fold) the whole world's remaining Terminal Carbon Pollution Budget that must not be exceeded for a 75% chance of avoiding a catastrophic 2C temperature rise" [8]. Future generations will have to meet the cost of this massive and criminally permitted GHG pollution.

"Expert Witness Testimony To Stop Gas-Fired Power Plant Installation" [6], "Australian Commitment To Unlimited Natural Gas Exploitation Threatens Planet & Invites Global Blowback" [8], and "Methane leakage makes Australia a world-leading per capita greenhouse gas polluter" [10], are reproduced in Section D as Chapters 12, 13 and 14 respectively.

References (those marked with an asterisk * are reproduced as Chapters 12, 13 and 14 in Section D):

[1]. Gideon Polya, "Australia Absurdly Declares Methane Burning Clean And Renewable", Countercurrents, 26 August 2009: https://www.countercurrents.org/polya160809.htm.
[2]. Gideon Polya, "The Awful Truth Must Stop Global Gasland Disaster", Countercurrents, 27 December 2010: https://www.countercurrents.org/polya271210.htm.
[3]. Gideon Polya, "India And World Threatened By US-led Shale Gas Fracking", Countercurrents, 9 January 2011: https://www.countercurrents.org/polya090111.htm.
[4]. Gideon Polya, "Gas Dirtier Than Coal", Countercurrents, 20 March 2011: https://www.countercurrents.org/polya200311.htm.
[5]. Gideon Polya, "Science Says World Must Stop Coal Seam Gas Exploitation", Countercurrents, 23 October 2011: https://www.countercurrents.org/polya231011.htm.
*[6]. Gideon Polya, "Expert Witness Testimony To Stop Gas-Fired Power Plant Installation", Countercurrents, 14 June, 2013: https://countercurrents.org/polya140613.htm. [7]. Gideon Polya, "Pro-gas Obama's EPA-based Plan To Reduce Coal-based Pollution Amounts To Climate Change Inaction", Countercurrents, 7 June, 2014: https://www.countercurrents.org/polya070614.htm.
*[8]. Gideon Polya, "Australian Commitment To Unlimited Natural Gas Exploitation Threatens Planet & Invites Global Blowback", Countercurrents, 12 April 2017: http://www.countercurrents.org/2017/04/08/australian-commitment-to-unlimited-natural-gas-exploitation-threatens-planet-invites-global-blowback/.
[9]. Gideon Polya, "Australia's Finkel Review ignores 25 key Climate Emergency realities & advocates long term gas & coal use", Countercurrents, 17 June 2017: http://www.countercurrents.org/2017/06/17/australias-finkel-review-ignores-25-key-climate-emergency-realities-advocates-long-term-gas-coal-use/.
*[10]. Gideon Polya "Methane leakage makes Australia a world-leading per capita greenhouse gas polluter", Countercurrents, 18 February 2020: https://countercurrents.org/2020/02/methane-leakage-makes-australia-a-world-leading-per-capita-greenhouse-gas-polluter.
[11]. Gideon Polya, "Scores Of Huge Realities Resolutely Ignored By Mendacious, US Lackey Mainstream Australia", Countercurrents, 11 August 2020: https://countercurrents.org/2020/08/scores-of-huge-

realities-resolutely-ignored-by-mendacious-us-lackey-mainstream-australia/.
[12]. Gideon Polya, editor, "Gas is not clean energy":
https://sites.google.com/site/gasisnotcleanenergy/.
[13]. Gideon Polya, editor, "Methane Bomb Threat":
https://sites.google.com/site/methanebombthreat/.
[14]. Gideon Polya, editor, "Nuclear weapons ban, end poverty & reverse climate change":
https://sites.google.com/site/300orgsite/nuclear-weapons-ban.
[15].Gideon Polya, editor, "Carbon Debt Carbon Credit":
https://sites.google.com/site/carbondebtcarboncredit/.
[16]. Gideon Polya, editor, "2011 climate change course":
https://sites.google.com/site/300orgsite/2011-climate-change-course.
[17]. Gideon Polya, "The lie that "Gas is cleaner"", Green Left, 7 October 2020: https://www.greenleft.org.au/content/lie-gas-cleaner.

CHAPTER 12: EXPERT WITNESS TESTIMONY TO STOP GAS-FIRED POWER PLANT INSTALLATION

[First published as Gideon Polya, "Expert Witness Testimony To Stop Gas-Fired Power Plant Installation", Countercurrents, 14 June, 2013: https://countercurrents.org/polya140613.htm.]

Despite the worsening global warming crisis due to ever-increasing greenhouse gas (GHG) pollution, the world is paradoxically experiencing a gas boom and a gas a rush throughout the world. While the world urgently needs to implement 100% renewable energy and concomitantly decrease and eventually cease GHG pollution, massive corporate spin falsely asserts that "gas is clean" and that a first step to a clean energy future is to replace coal burning for power with gas burning. These are extraordinary falsehoods because methane (CH_4; a major constituent of natural gas) leaks (2-8% in the US) and has a Global Warning Potential (GWP) 105 times greater than that of carbon dioxide (CO_2) on a 20 year time frame and considering aerosol impacts. Thus, depending upon the degree of gas leakage, burning gas for power can be much dirtier GHG-wise than burning coal [1].

Nevertheless, the egregious falsehoods that "gas is clean", that "gas burning is cleaner greenhouse gas-wise than coal burning" and that "a coal to gas transition is a way to tackle climate change" have massive currency in the capitalist Murdochracies, Lobbyocracies and Corporatocracies in which Big Money buys politicians, parties, policies, public perception of reality and political power. As a consequence, gas burning for power is remorselessly increasing through the construction and operation of gas-fired power plants (GFPPs). However such plants have to be located somewhere, and in an increasingly crowded world that often means location near small suburbs or towns. While the populations of cities have the requisitely massive collective political muscle to stop such contiguous sources of massive and toxic pollution emission, small towns and urban fringe suburbs are much more vulnerable to the gas juggernaut.

Modern societies have Environment Protection Authorities (EPAs), Departments of the Environment, and other regulatory bodies and laws dedicated to protecting human health, human amenity and the environment from toxic pollution of the atmosphere. Citizens faced with proposals to introduce massive sources of pollution into their locality will have to enter into political lobbying at the local government, state government and national government levels, and finally enter the legal processes for opposing installation of gas-fired power plants. The legal processes can be very expensive and the Big Guns on the side of anti-pollution citizens' action groups will typically be like-minded town or shire local governments. Depending on the jurisdiction and the nature of the litigation involved, the legal processes of appeal and counter-appeal can be complex, very expensive and constrained in relation to allowable grounds for complaint and argument.

However small communities faced with the conflation of corporate greed and Corporatist Establishments can defend themselves by exposing the Truth – the Truth that, while it can be subjected to bald contradiction and spin, will simply never go away. Indeed while high polluting countries like the US, Australia and Canada still refuse to seriously tackle climate change, it is clear that they will eventually be forced to severely curtail and thence stop fossil fuel burning. Thus while fossil fuel burning corporations may continue to win in the immediate short term, in the longer term (e.g. within 10 years) their investment into abominations such as a \$1 billion, 40-year-lifetime, 1,000 MW, gas-fired power plant (GFPP) will prove to be very transient, uneconomic and expensive.

Communities can lobby, and while they don't have the power of the Corporatist oligopoly media they have the power of one-man-one-vote democracy applying to various levels of government. Indeed the Corporatist Establishment is not a united church, and it is likely that a corporate developer gaining approval to build a $1 billion, 40-year-lifetime, 1,000 MW, gas-fired power plant (GFPP) will have gained this licence to pollute at the expense of other major corporate greenhouse gas polluters, in particular coal- and oil-based energy generation, cement manufacture, fossil fuel-based transport, other industrial polluters, forestry, and agriculture (notably methanogenic livestock production). Indeed any such approval of fossil fuel-based power generation will also have finite impacts on societal fundamentals such as population growth, economic growth, and retention of the value of retirement investments.

Ultimately, communities will end up opposed to corporate developers in court in David and Goliath legal processes. While corporations can afford the very top legal counsel charging $10,000 per hearing day, small communities can acquire brilliant, young legal counsel charging $1,000 per hearing day but with everything to gain from the combat in terms of experience and professional exposure. Again, while corporations can afford the very best of Expert Witnesses to argue their case in court at a cost of $10,000 to $100,000 each, small communities can call upon dedicated Expert Witnesses who will be prepared to research and present the Court with alternative considerations for no charge at all, and who will be simply inspired by a sense of social responsibility and the aphorism attributed to Edmund Burke that "evil happens when good men do nothing".

Indeed this analysis derives in part from my experience as a much-published, 5-decade career biological chemist acting as an Expert Witness (free of charge of course) for concerned citizens in giving carefully researched evidence in court against the imposition on rural town residents of hugely polluting gas-fired power generation.

While the emphasis and procedurally or legally allowable grounds for discussion may differ in the various jurisdictions (e.g. those in the various jurisdictions and specialist courts in America and the British Commonwealth), in general the major issues to be considered in the court would be negative impacts of a proposed gas-fired power station (GFPP) on (A) human health, (B) human amenity (e.g. finances, happiness, visual amenity, and rural versus industrial quality of life), and (C) the environment (flora, fauna, ecosystems, greenhouse gas pollution, global warming and climate change). Some key lines of evidence under these three headings for activists, legal counsel and Expert Witnesses are outlined below as a public service.

(A). Detriment to human health.

1. Gas-fired power plants (GFPPs) are clean-er than coal-fired power plants (CFPPs) in the sense that per unit of electrical energy they emit lower amounts of pollutants such as the greenhouse gas carbon dioxide (CO_2), and the very toxic pollutants carbon monoxide (CO), nitrogen oxides (NOx), sulphur dioxide (SO_2), particulates (fine soot to ultrafine particles), and heavy metals such as mercury (Hg). Thus according to Naturalgas.org [2], the fossil fuel-derived pollutant emissions levels (in pounds per Btu of energy input) are as follows (noting that this is for average gas burning and not specifically for gas-turbine power stations) - for CO_2: 117,000 (gas), 164,000 (oil) and 208,000 (coal); for CO: 40 (gas), 33 (oil) and 208 (coal); for NOx: 92 (gas), 448 (oil) and 457 (coal); for SO_2: 1 (gas), 1,122 (oil), and 2,591 (coal); for particulates: 7 (gas), 84 (oil) and 2,744 (coal); and for Hg: 0.000 (gas), 0.007 (oil) and 0.016 (coal). In addition, radioactive material is a significant pollutant from coal burning but not from gas burning. However, while Hg and radioactivity are negligible pollutants for gas burning and SO_2 emission is very low, CO_2, NOx and particulate matter (PM)[e.g. $PM_{2.5}$] are significant pollutants from GFPPs. Qualitatively, in terms of greenhouse gas (GHG) pollution and emission of toxic pollutants, gas burning is certainly not clean – indeed it is dirty GHG-wise and dangerous to human health [1].

2. These days in the US gas-fired power stations on average produce about 0.475 t CO_2/MWh (versus the current 1.4 t CO_2/MWh for coal-fired power stations). Accordingly, a 1,000 MW gas-fired power station operating at a capacity factor of

100% (i.e. operating at 100% of full annual capacity as compared to the realistic 90% capacity factor) would produce 0.475 tCO_2 per MWh x 1,000 MW x 1.0 (capacity factor) x 365 days per year x 24 hours per day = 4,161,000 t CO_2/year.

3. From the relative data in 1 and the CO_2 emissions data in 2 above, one can estimate that a 1,000 MW GFPP at 100% capacity factor would each year produce 4,161,000 t CO_2 x 40 t CO/117,000 tCO_2 = 1,422.6 t CO; 4,161,000 t CO_2 x 92 t NOx/117,000 tCO_2 = 3,271.9 t NOx; 4,161,000 t CO_2 x 1 t SO_2/117,000 tCO_2 = 35.6 t SO_2; and 4,161,000 t CO_2 x 7 t particulate matter (PM)/117,000 tCO_2 = 248.9 t PM.

4. For a small, non-industrial rural town with a population of 10,000 people one can estimate that the annual CO_2 pollution from oil-based transport would be 71,857 t CO_2. Accordingly, the annual transport oil burning-derived pollution would be 71,857 t CO_2 x 33t CO/ 164,000 t CO_2 = 14.5 t CO; 71,857 t CO_2 x 448t NOx/ 164,000 t CO_2 = 196.3 t NOx; 71,857 t CO_2 x 1,122t SO_2/ 164,000 t CO_2 = 491.6 t SO_2; and 71,857 t CO_2 x 84 t particulate matter (PM)/ 164,000 t CO_2 = 36.8 t PM.

5. We can accordingly compare the annual pollutant output from oil burning for transport in our notional rural town with the huge pollutant output from the proposed 1,000 MW GFPP within the town limits (the latter in parenthesis): 14.5 t CO (1,422.6 t CO or 98 times more); 196.3 t NOx (3,271.9 t NOx or 17 times more); 491.6 t SO_2 (35.6 t SO2 or 0.07 times less); and 36.8 t PM (248.9 t PM or 6.8 times more).

6. Crucially, while vehicle-derived pollution is emitted at ground level, that from a GFPP is emitted at high velocity via a 30 metre high stack. Accordingly, the degree of atmospheric dispersion becomes critical. Dispersion is typically modelled by complex, theoretical equations having the general form C (pollutant concentration at coordinates x, y and z) = F (a complex mathematical function involving x, y and z) x O (pollutant output from the stack) / W (wind speed). Pollution at a given spot will be greater the higher the output and the lower the wind speed. Proponents of highly polluting plants will

model dispersion, and regulators will assess pollutant output at the stack and similarly model consequent ground level pollution. However, as discussed below, modelling alone is surely insufficient – local residents require hard, empirical data from similar GFPPs to provide assurance that safety limits are not being exceeded on an hourly, daily or annual basis. To quote an important Australian regulator: "The main outcomes of land-use planning that are of interest in managing air quality are the separation of heavy industry and residential or other areas that may be adversely affected by industrial activities" [3]. It must be further noted that any pollutants dispersed from the locality of a GFPP have to end up somewhere else with likely adverse outcomes.

7. Science involves the critical testing of potentially falsifiable hypotheses (models). Models are tested in the scientific process and if found to be incompatible with experimental data are consequently altered to be more consonant with reality. Sole reliance by polluters or indeed regulators on dispersion modelling is grossly insufficient. Thus manufacturers of processed foods for human ingestion have to measure and report the levels of additives. Regulators must perform analyses to check on compliance with serious action taken over non-compliance. Converse examples of non-compliance include the Toxic Oil Syndrome associated with the adulterated Spanish olive oil disaster in which 600 people died [4], and the New Zealand case in which 2 New Zealand schoolgirls determined that Ribena blackcurrant cordial contained almost no vitamin C contrary to the claims of the manufacturer [5]. In view of the huge amount of pollutants being produced by a 1,000 MW GFPP there is a clear need for hard empirical data from similar GFPPs to allay health fears. However there is a paucity of such publicly-accessible data world-wide [6].

8. The World Health Organization (WHO) safety limits for ground level CO, NOx and PM (e.g. PM_{10} are less than 10 micrometer in diameter) in micrograms per cubic meter are 10,000 for CO (8 hour mean), 40 for NOx (annual mean), 10 for $PM_{2.5}$ (annual mean) and 20 for PM_{10} (annual mean) [7]. However observed levels of CO, NOx and PM_{10} in a typical

urban area can be 20%, 35% and 75%, respectively, of these safety limits in the absence of a 1,000 MW GFPP. Given the enormously increased additional output pollution produced by such a GFPP (see above) there is a real fear than these safety limits will be exceeded, this reinforcing the crucial demand that ground level pollution in the vicinity of similar GFPPs should be determined. These standards are routinely exceeded in Beijing [8].

9. Further to the above concerns about safety limits being exceeded, one notes that human beings are different and some people (e.g. asthmatics and those suffering from chronic obstructive pulmonary disorder (COPD)) may be particularly susceptible to pollutants produced by a GFPP [9].

10. Carbon monoxide (CO) is a highly toxic gas that acts by binding to haemoglobin and thus preventing oxygen (O_2) transport from the lungs to tissues [10]. CO in urban areas can approach 10% of the safe level, and a 1000 MW GFPP will emit about 1,400 t CO annually, this emphasizing the need for measurement of ground level CO in the vicinity of existing GFPPs similar to that proposed.

11. Nitrogen oxides (NOx) are major toxic pollutants from GFPPs. Nitric oxide (NO) generated by GFPPs is rapidly converted to NO_2 which affects the respiratory system and which with sunlight and Volatile Organic Compounds (VOCs, from coal and oil burning pollution) is involved in photochemical generation of ozone (O_3) which is injurious to plant life as well as to human respiration. According to the WHO the safe limit of O_3 is 100 $\mu g/m^3$ (8-hour mean) [7] but, for example, the Maximum One-Hour Mean Concentration of Ozone in Beijing is 2 times this level [11]. NO at very low levels acts as a signalling compound in animals and plants [10]. The Department of the Environment and Climate Change in New South Wales has determined that "uncontrolled emissions from around 200 MW of [gas-fired] cogeneration would result in the health based nitrogen dioxide goal being exceeded across the [Sydney] CBD" [12, 13].

12. Particulate matter (PM) is generated by GFPPs and CFPPs but gas-fired power plants disproportionately produce fine PM that is of particular concern because it can penetrate deep into the lungs. Thus according to the World Health Organization (WHO) "no threshold for PM has been identified below which no damage to health is observed." [7].

13. Rural towns can be close to highways, gas pipelines, gas compressor stations, transmission lines, and electric substations, infrastructure that can be used to argue for the efficient co-location of GFPPs. However the precautionary principle in relation to human health and the potential fire hazard of a GFPP (e.g. associated with forest fires or explosion) demands that GFPPs should be located well away from urban areas.

14. Placement of a huge 1,000 MW GFPP within the limits of a bucolic rural town effectively turns the town into an industrial zone with the likelihood of further industrial development and pollution detrimental to human health.

(B). Detriment to human amenity.

1. A fundamental human amenity of living in a rural town is living close to nature away from the literal big smoke. Bringing a huge industrial polluter to the vicinity of a rural town helps demolish that amenity, and further inevitable industrialization driven by cheaper power will further promote the transformation.

2. A wonderful amenity of rural living is that on a moonless night one can see frequent shooting stars and millions of stars in the sky whereas in similar conditions city folk can see only several dozen stars. Light pollution and other pollution from a huge GFPP would largely eliminate this wonderful and fundamental amenity.

3. While a huge GFPP can be camouflaged from speeding drivers by painting it appropriately, there will be an inevitable loss of visual amenity to people on the ground due to a huge

industrial plant located in a rural and rural residential environment.

4. While proponents of a huge GFPP can argue for the local financial benefits of cheaper power and local employment in construction, plant operation, security and maintenance, these benefits would be outweighed by possible increases in morbidity (sickness) and avoidable mortality associated with possible toxic pollution from the plant.

5. Similarly, proponents of a huge GFPP will argue that it allows for beneficial increases in electricity to the community and the capacity of a GFPP for on-off function as a provider of peaking power to meet demand (e.g. the demand elicited by hot weather and increased use of air conditioners). However this "peak provider" argument is based on the incorrect premise that renewable energy (e.g. wind or solar) cannot provide 24/7 baseload power. Thus concentrated solar thermal power plants can store energy as heat using molten salts [14], and energy can be stored hydrologically by pumping water into elevated storages using solar or wind energy [15]. Indeed, noting that Australia currently has a 50,000 MW electricity capacity, the Beyond Zero Emissions (BZE) plan for 100% renewable stationary energy for Australia by 2020 (Zero Carbon Australia by 2020, ZCA 2020) involves 40% wind energy, 60% concentrated solar thermal (CST) with molten salts energy storage for 24/7 baseload power, biomass and hydroelectric backup (for days of no wind and low sunshine) and a HV DC and HV AC national power grid. The BZE scheme was costed at $370 billion over 10 years, with roughly half spent on CST, one quarter on wind, and one quarter on the national electricity grid [14]. A related scheme for 100% renewable energy for Australia has been set out by top electrical engineer Professor Peter Seligman (a major player in development of the bionic ear). Professor Seligman's scheme involves wind, solar thermal, other energy sources, hydrological energy storage (in dams on the Nullabor Plain in Southern Australia), an HV AC and HV DC electricity transmission grid, and a cost over 20 years of $253 billion [15]. Ignoring cost-increasing energy storage and transmission grid costs and cost-decreasing economies of scale

for a 2- to10-fold size increase, here are 2 similar cost estimates for installation of wind power for 80% of Australia's projected 325,000 GWh of annual electrical energy by 2020: (1) 90,000 MW capacity, 260,000 GWh/year, $200 billion/10 years (10-fold scale-up from GL Garrad Hassan), and (2) 96,000 MW, 260,000 GWh/year, $144 billion (2-fold scale up from BZE) [16].

6. Imposition of a huge GFPP on a rural community carries major financial implications for individuals in relation to reduced property values, and obviation of long-term investment and retirement plans, these burdens violating the principle that new developments should not be at the expense of existing residents by retrospectively damaging their carefully-planned long-term financial positions.

7. Imposition of a huge GFPP on a rural community carries major financial implications for the community as a whole in relation to short- and long-term town planning, and established zoning (e.g. for industry and for rural residential housing). Thus an attractive green, rural image of importance for future development is eliminated for starters.

8. The very process of community objection is expensive in terms of expert witness and legal representation costs. Thus top legal counsel in Australia could charge $10,000 per day and Expert Witnesses could charge $50,000 for their research, written reports and court testimony.

9. Hard to evaluate is the psychological impact on concerned residents of the imposition of a huge GFPP on their rural community.

10. Top climate scientists and biologists argue that we must urgently return the atmospheric CO_2 from the current damaging and dangerous 400 ppm CO_2 to the pre-Industrial Era (pre-Anthropocene) level of no more than 300 ppm CO_2 for a safe planet for all peoples and all species [17], noting that the current species extinction rate is 100-1,000 greater than normal [18]. Sequestration of CO_2 as cellulose (in wood) and thence

conversion of cellulosic material to biochar (carbon, charcoal, C) through anaerobic pyrolysis at 400-700C is a major means of achieving this with the current achievable rate of about 9 Gt biochar per year approximating the current industrial pollution rate [19]. Intergenerational equity is analysed below in relation to the enormous CO_2 debt for coal burning (similar to that for gas burning) as expressed in the ultimate cost of sequestering the generated CO_2 as biochar.

(a). The current price of Australian thermal coal is about $90 per tonne or $90 per t C x (12t C/ 44 t CO_2) = $24 per tonne of CO_2 released on eventual combustion.

(b). However top climate scientists and biologists say that the atmospheric CO_2 concentration must be rapidly returned from the present circa 400 ppm CO_2 [415 ppm CO_2 in 2020] to about 300 ppm CO_2 for a safe planet for all peoples and all species [17].

(c). Currently, apart from re-afforestation a major way of returning atmospheric CO_2 back to 300 ppm CO_2 is through producing biochar (carbon, charcoal, C) through anaerobic pyrolysis of cellulosic biomass waste in renewable energy-driven microwave furnaces at 400-700C (with existing agricultural and forestry waste this could achieve about 9 Gt per year, roughly the same as the annual industrial output) [19].

(d).The cost of conversion of cellulosic waste to biochar in the US mid-West is about US$49-US$74 per tonne CO_2 (US$210-US$303 per tonne CO_2 in the UK) [20].

(e). Ergo, for every $1 received for coal by the coal industry our children, grandchildren and further generations will have to spend $2 to $3 at present prices converting the consequent CO_2 back to biochar to save the Planet. One can accordingly similarly determine (for Australia) that the biochar production-related debt for future generations for every $1 received by the gas-based electricity industry will be about $0.6-$1.0 at present prices (these estimates must be multiplied by 4 for the cost of biochar production in the UK).

11. Leaving aside despoilation of the environment due to greenhouse gas pollution and resultant climate change impacting future generations, there are also other key intergenerational equity considerations. Thus eminent US climate scientist Dr James Hansen (former head, NASA's Goddard Institute of Space Studies and adjunct professor at 82-Nobel Laureate Columbia University) has commented (2009): "Is it feasible to phase out coal and avoid use of unconventional fossil fuels? Yes, but only if governments face up to the truth: as long as fossil fuels are the cheapest energy, their use will continue and even increase on a global basis. Fossil fuels are cheapest because they are not made to pay for their effects on human health, the environment and future climate… Are we going to stand up and give global politicians a hard slap in the face, to make them face the truth? It will take a lot of us – probably in the streets. Or are we going to let them continue to kid themselves and us and cheat our children and grandchildren? Intergenerational inequity is a moral issue. Just as when Abraham Lincoln faced slavery and when Winston Churchill faced Nazism, the time for compromises and half-measures is over. Can we find a leader who understands the core issue and will lead?" [21]. Children of a rural town subject to imposition of a huge GFPP will have to bear their share of the cost of remediation of pollution in which they had no part, an appalling example of retrospective moral and financial taxation.

12. A fundamental human amenity is that of good self-regard but the imposition of a huge GFPP on a small rural community will make its inhabitants the very worst annual per capita greenhouse gas polluters in the world. Thus a 1000 MW GFPP operating at a theoretical 100% capacity factor would emit 4.16 Mt CO_2/year and for a town of 10,000 people the additional annual per capita GHG pollution would be 416 t CO_2-e per person per year. Australia has an annual per capita GHG pollution of 25 t per person per year and accordingly the presence of a 1000 MW GFPP would increase this to 25 + 416 = 441 t per person per year, this being about 17 times greater than

the Australian average of 25, 62 times greater than the world average of 6.7, 4.4 times worse than that of the world 's worst GHG polluter Belize (93.9), and 7.6 times greater than that of Qatar (54.7), the world's second worst GHG polluter [22]. Thus against their fervently expressed will, the pro-environment people of this rural town will have by far the world's worst carbon footprint as measured by annual per capita greenhouse gas pollution in a world facing a worsening climate crisis. School children in such a town would grow up knowing that their town is the worst greenhouse gas polluter of the world.

(C). Environmental impact.

1. A fatuous argument could be advanced that the proposed GFPP operating at 90% capacity factor would emit only 3.7 Mt CO_2 per year, this being 3.7 Mt x 100/63,803 Mt CO_2-e = 0.006% of the world's annual GHG pollution. However if this defence were successfully applied to all other specific, localized sources of GHG pollution there would be zero action world-wide on climate change.

2. A GFPP cannot be considered in isolation and must be considered in the context of the overall process from gas extraction to ultimate reversal of atmospheric CO_2 pollution. Accordingly one must also consider the ultimate environmental impacts not just of global warming but also of gas extraction (either conventional or unconventional e.g. from coal seam gas or fracking) e.g. land salinization, landscape despoilation, aquifer depletion and pollution, and loss of forests and other ecosystems.

3. Through a cogent examination of livestock-related GHG pollution (in particular by assessing the GWP of methane on a 20 year time frame) World Bank experts have re-assessed annual global GHG pollution as 63,803 Mt CO_2-e, this being about 50% larger than the prior estimate of 41,755 Mt CO_2-e [23]. CO_2 pollution is remorselessly increasing and recently reached 400 ppm at the US observatory on Mauna Loa [24]. The US Energy Information Administration predicts unabated global GHG pollution and indeed the global GHG pollution is

so high that with projected increases the world will take only 5 years to exceed the terminal 2010-2050 GHG pollution budget of 600 billion t CO_2 that must not be exceeded if the world is to have a 75% chance of avoiding a disastrous 2C temperature rise [25]. The world must be decreasing and not increasing GHG pollution.

4. Climate change is already impacting ecosystems with many examples now of species moving polewards to lower average ambient temperatures.

5. NO_2 generated by a GFPP will give rise to ozone (O_3) which can cause leaf damage as well as respiratory irritation to fauna.

6. NO acts at very low concentrations as a signalling compound in both animal and plant systems [10], suggesting further biological consequences in the vicinity of GFPPs.

7. Additional noise, odour and the loss of being able to see stars due to GFPP light pollution may also negatively impact animal behaviour.

Conclusions.

1. The fundamental reason why gas-fired power plants (GFPPs) should not be built is that in view of the worsening climate crisis the world should be decreasing and not increasing fossil fuel burning and other greenhouse gas (GHG) pollution. Building a GFPP with a lifetime of 40 years simply locks society into more fossil fuel burning and correspondingly blocks the uptake of renewable energy.

2. A GFPP located near a rural town will impact (A) human health, (B) human amenity and (C) the environment in the many ways outlined above for the benefit of environmental activists, legal counsel and Expert Witnesses involved in GFPP-related legal proceedings.

3. It must be appreciated that the political and legal debate about fossil fuel burning is conducted with a background of

extraordinary, anti-science media misinformation in the Western Murdochracies, Lobbyocracies, [Plutocracies, Kleptocracies, Dollarocracies,] and Corporatocracies.

4. The bottom line of any analysis of public policy is avoidable death as described in detail in my book "Body Count. Global avoidable mortality since 1950" that is now available for free perusal on the Web [26]. It is estimated that already about 5 million people die annually worldwide from climate change (0.5 million) and from carbon burning pollutants (4.5 million) with this carnage set to total 100 million by 2030 in the absence of climate change action [27, 28], and that 10 billion people will perish this century if climate change is not addressed [22].

5. Ultimately for societal safety we must seriously consider the expert views of scientists. Thus a 2010 Open Letter from 255 eminent scientists of the prestigious US National Academy of Sciences (including 11 Nobel Prize winners) concluded: "Delay is not an option" [27, 29], and the Synthesis Report of the 2,500-delegate 2009 scientific Copenhagen Climate Change Conference concluded: "Inaction is inexcusable" [27, 30]. Investing huge amounts of scarce resources into burning more fossil fuels like gas flies in the face of expert scientific evidence that we should be urgently engaged in cessation of fossil fuel burning in the interests of Humanity and the Biosphere. Professor David Shearman (emeritus professor of Medicine at the University of Adelaide, Adelaide, South Australia) has succinctly stated: "The International Energy Agency has expressed concern about gas replacing renewable energy sources. This would delay any chance of early curtailment of greenhouse emissions. Climate change is accepted as a huge threat to health world wide" [27, 31].

References.

[1]. "Gas is dirty energy":
https://sites.google.com/site/gasisnotcleanenergy/gas-is-dirty-energy.
[2]. Natural gas.org, "Natural gas and the environment":
http://www.naturalgas.org/environment/naturalgas.asp.
[3]. Department of Environment and Resource Management, "Clean and healthy air for Gladstone", November 2011:
http://www.ehp.qld.gov.au/air/gladstone/pdf/chag-final-report-web.pdf.
[4]. "Toxic oil syndrome", Wikipedia:
http://en.wikipedia.org/wiki/Toxic_oil_syndrome.
[5]. Jeevan Vasagar, "Schoolgirls rumble Ribena vitamin C claims", The Guardian, 27 March 2007:
http://www.guardian.co.uk/world/2007/mar/27/schoolsworldwide.foodanddrink.
[6]. Dr Stephanie Shaw, Electric Power Research Institute (EPRI), "Air quality impacts from natural gas extraction and combustion", Power, 13 March 2013:
http://www.powermag.com/environmental/Air-Quality-Impacts-from-Natural-Gas-Extraction-and-Combustion_5417.html.
[7]. WHO, "Air quality and health":
http://www.who.int/mediacentre/factsheets/fs313/en/.
[8]. Greening Beijing:
http://www.unep.org/sport_env/documents/beijingreport07/chapter5.pdf.
[9]. The Lung Association, "Pollution & air quality":
http://www.lung.ca/protect-protegez/pollution-pollution/outdoor-exterior/index_e.php.
[10]. Gideon Polya, "Biochemical Targets of Pant Bioactive Compounds. A pharmacological reference guide to sites of action and biological effects" (Taylor & Francis, CRC Press, London & New York, 2003).
[11]. Xuxuan Xie, Shiqiu Zhang, Jianhua Zhu, Dan Wu, Tong Zhuc, "Cost effective control of ground-level ozone pollution in and around Beijing :
http://www.cerdi.org/uploads/sfCmsContent/html/323/Xie.pdf.
[12]. New South Wales Department of Energy and Climate Change (DECC), "Interim DECC nitrogen oxide policy for cogeneration in Sydney and the Illawarra":
http://www.environment.nsw.gov.au/resources/air/inp09124.pdf.
[13]. Drew Warne Smith, "Health fears over gas-fired generators", The Australian, 12 November 2010:

http://www.theaustralian.com.au/news/nation/health-fears-over-gas-fired-generators/story-e6frg6nf-1225952326880.

[14]. Beyond Zero Emissions Zero (BZE), Zero Carbon Australia by 2020 Report (BZE ZCA2020 Report), 2010: http://www.beyondzeroemissions.org/about/bze-brand.

[15]. Peter Seligman, "Australian sustainable energy – by the numbers", Melbourne Energy Institute, University of Melbourne, 2010: http://energy.unimelb.edu.au/ozsebtn/.

[16]. Gideon Polya, "2011 Climate Change Course": https://sites.google.com/site/300orgsite/2011-climate-change-course.

[17]. "300.org – return atmosphere CO2 to 300 ppm": https://sites.google.com/site/300orgsite/300-org---return-atmosphere-co2-to-300-ppm.

[18]. Phillip Levin, Donald Levin, "The real biodiversity crisis", American Scientist, January-February 2002: http://www.americanscientist.org/issues/pub/the-real-biodiversity-crisis.

[19]. Gideon Polya, "Forest biomass-derived Biochar can profitably reduce global warming and bushfire risk", Yarra Valley Climate Action Group: http://sites.google.com/site/yarravalleyclimateactiongroup/forest-biomass-derived-biochar-can-profitably-reduce-global-warming-and-bushfire-risk.

[20]. Simon Shackley, Jim Hammond, John Gaunt and Rodrigo Ibarrollo, "The feasibility and costs of biochar deployment in the UK", Carbon Management, 2(3), 335-356 (2011): http://www.geos.ed.ac.uk/homes/sshackle/CostsBiochar.pdf.

[21]. James Hansen, "It's Possible To Avert The Climate Crisis", Countercurrents, 29 November 2009: http://www.countercurrents.org/hansen291109.htm.

[22]. Climate Genocide: https://sites.google.com/site/climategenocide/.

[23]. Robert Goodland and Jeff Anfang. "Livestock and climate change. What if the key actors in climate change are … cows, pigs and chickens?", World Watch, November/December 2009: http://www.worldwatch.org/files/pdf/Livestock%20and%20Climate%20Change.pdf.

[24]. NOAA, "CO_2 at NOAA's Mauna Loa Observatory reaches new milestone: Tops 400 ppm", 10 May 2103: http://www.esrl.noaa.gov/gmd/news/7074.html.

[25]. Gideon Polya, "Doha climate change inaction. 5 years left to act", MWC News, 9 December 2012: http://mwcnews.net/focus/analysis/23373-gideonpolya-climate-change.html.

[26]. Gideon Polya, "Body Count. Global avoidable mortality since 1950" : http://globalbodycount.blogspot.com.au/2012/01/body-count-global-avoidable-mortality_05.html.

[27]. "Are we doomed?": https://sites.google.com/site/300orgsite/are-we-doomed.

[28]. DARA, "Climate Vulnerability Monitor. A guide to the cold calculus of a hot planet", 2012, Executive Summary pp2-3: http://daraint.org/climate-vulnerability-monitor/climate-vulnerability-monitor-2012/.

[29]. 2010 Open Letter by 255 members of the US National Academy of Sciences, "Open Letter: climate change and the integrity of science", Guardian, 6 May 2010: http://www.guardian.co.uk/environment/2010/may/06/climate-science-open-letter.

[30]. Synthesis Report of the March 2009 Copenhagen Scientific Climate Change Conference: http://climatecongress.ku.dk/pdf/synthesisreport.

[31]. David Shearman, "Dealing with the health risks of unconventional gas", The Conversation, 28 November 2012: http://theconversation.com/dealing-with-the-health-risks-of-unconventional-gas-10987.

CHAPTER 13: AUSTRALIAN COMMITMENT TO UNLIMITED NATURAL GAS EXPLOITATION THREATENS PLANET & INVITES GLOBAL BLOWBACK

[First published as Gideon Polya, "Australian Commitment To Unlimited Natural Gas Exploitation Threatens Planet & Invites Global Blowback", Countercurrents, 12 April 2017: http://www.countercurrents.org/2017/04/08/australian-commitment-to-unlimited-natural-gas-exploitation-threatens-planet-invites-global-blowback/.]

Australia is a major exporter of coal, gas and iron ore, and is set to become the world's largest exporter of Liquid Natural Gas (LNG). However, depending upon the degree of systemic gas leakage, burning gas for power may be worse greenhouse gas (GHG)-wise than burning coal. However a remorselessly neoliberal, anti-science and anti-environment Australia is committed to massive exploitation of conventional and non-conventional natural gas reserves for export and domestic use.

Australia is a world leader in annual per capita GHG pollution and both its Coalition Government and Labor Opposition are committed to unlimited coal and gas exploitation [1, 2]. Australia is a key player in a dangerous global coal to gas transition that is a deadly and dishonest neoliberal alternative to the complete cessation of fossil fuel exploitation demanded by the worsening climate emergency. The ideal target of no more than a 1.5C temperature rise agreed to at the 2015 Paris Climate Change Conference is now set to be exceeded within 4-10 years. A 2C temperature rise – that all governments (except for the idiotic, dangerous, climate change denialist US Trump Administration) agree would be catastrophic – is now unavoidable. While Humanity can still take action to make the now inevitable plus 2C future "less bad", there is a looming threat of global warming causing massive release of methane (CH_4) from the Arctic, a ticking "Methane Bomb" set to utterly devastate Humanity and indeed all life on earth in the coming century.

It is accordingly important to determine the greenhouse gas (GHG) pollution associated with this unlimited Australian exploitation of its gas reserves and associated with combustion or leakage of natural gas. Methane (CH_4) can be over 90% of natural gas but in the following calculations we will assume for simplicity that (a) natural gas is 100% CH_4 and (b) that all gas leakage from Australian natural gas exploitation is occurring within Australia. Of course the arguments developed here can be applied to other countries exploiting natural gas resources that should remain in the ground.

1. **2.6% CH_4 leakage is as polluting GHG-wise as burning the remaining CH_4.**

Methane (CH_4) has a molecular weight of about 16 and CO_2 has a molecular weight of about 44. Combustion of 16 tonnes CH_4 yields 44 tonnes CO_2. However CH_4 is a gas, leaks and has a Global Warming Potential (GWP) 105 times that of CO_2 on a 20 year time frame and with aerosol impacts considered [3]. One can readily calculate on this basis that a systemic gas leakage of 2.6% would contribute as much GHG pollution as generating the greenhouse gas CO_2 by burning the remaining 97.4% of the gas [4].

2. **Australia's annual Domestic GHG pollution is 1,270 million tonnes (Mt) CO_2-e.**

Australia's annual Domestic GHG pollution as reported by the Australian Government is about 600 Mt CO_2-equivalent [5]. CO_2-equivalent includes CO_2 and the contributions of other GHGs such as CH_4, expressed as CO_2-equivalent [CO_2-e].

However this estimate is based on a GWP for CH_4 of 21 relative to that of CO_2 and is accordingly deficient in assessing the GHG contribution of CH_4. A revised estimate of Australia's annual Domestic GHG pollution taking methanogenic livestock and associated land use into account is 52.9 tonnes per person x 24 million persons = 1,270 Mt CO_2-e [1].

3. **In 2017 Australia will export about 60 Mt gas and use 30 Mt gas domestically.**

Australian Mining reported (2017): "Australia's liquefied natural gas (LNG) exports surged in 2016 as a plethora of new projects in Western Australia, the Northern Territory and Queensland triggered a flood of new shipments to global markets, a report from EnergyQuest has revealed. Total exports in 2016 reached 36.8 million tonnes (Mt), a 37.7 per cent increase on the 26.7Mt recorded in the previous year, the report found. EnergyQuest expects 2017 to be even stronger for the export market. In its monthly LNG report, the consultancy outlined that as the Australia Pacific LNG and Gorgon operations continue to ramp up, and new projects like Wheatstone and Ichthys come into production, Australia's 2017 exports will be close to 60Mt, up 63 per cent on 2016" [6]. In 2014-2015 Australian gas consumption was 1,610 PJ = 29 Mt natural gas. Australian gas consumption in 2017 would accordingly be about 30 Mt [7, 8].

4. **At 2.6% gas leakage total GHG pollution within Australia from Australia's 90 Mt gas production in 2017 would be 322 Mt CO2-e.**

If there were no CH_4 leakage then combustion of 60 Mt of LNG exports (2017 expectation) outside of Australia would yield 60 Mt CH_4 x 44 t CO_2/ 16 t CH_4 = 165 Mt CO_2 (we can choose here to ignore this as part of Australia's annual GHG pollution, but one notes that "cause and effect" analysis shows that if Australia didn't extract and sell the gas then this GHG pollution would not have been generated).

If there were 2.6% CH_4 leakage, Exported GHG pollution from eventual CH_4 combustion would be 0.974 x 60 x 44/16 = 161 Mt CO_2. The Domestic GHG pollution from a 2.6% systemic CH_4 leakage from this 60 Mt of LNG would also be 161 Mt CO2-equivalent (see #1 above; this calculation assumes for simplicity that all the gas is CH_4 and ignores GHG pollution associated with extraction of Natural Gas (NG) and generation of Liquefied Natural Gas (LNG) from NG). Thus the total GHG pollution generated within Australia from this 60 Mt of exported LNG would be 161 million tonnes CO_2-equivalent.

With a 2.6% systemic CH_4 leakage from the 30 Mt gas used domestically, GHG pollution within Australia would be 80.5 (from CH_4 leakage) + 80.5 (from burning the residual 29.2 Mt CH_4) = 161 million tonnes CO_2-e.

Thus total GHG pollution within Australia from 2017 Australian gas production would be 161 + 161 = 322 Mt CO_2-e or about 25.4% of Australia's revised annual Domestic GHG pollution of 1,270 Mt CO_2-e.

5. **At 1.2% gas leakage, total GHG pollution within Australia from 2017 Australian gas production would be 195 Mt CO2-e.**

At 1.2% gas leakage total GHG pollution within Australia from 2017 Australian gas production would be (a) 0.012 x 60 x 105 = 75.6 Mt CO_2-e (CH_4 leaked from gas to be exported) plus (b) 0.988 x 30 x 44 / 16 = 81.5 (CO_2 from domestically burning gas) plus (c) 0.012 x 30 x 105 = 37.8 (CH_4 leaked from gas to be used domestically) for a total of 195 Mt CO_2-e or 15.4% of Australia's revised annual Domestic GHG pollution of 1,270 Mt CO_2-e.

6. **US gas leakage is 1.2% -2.4% overall and 3.6% - 7.9% from "fracking" for shale gas.**

The US EPA estimated 160.0 Mt CO_2-e (6.4 Mt CH_4 at a GWP of 25) gas leakage in 2015 [9] out of a total of 548 Mt gas extracted in 2015 [10], this corresponding to a 6.4 x 100/ 548.0

= 1.2% gas leakage. According to Professor Steven Cohen (executive director of 100-Nobel-laureate Columbia University's Earth Institute and a professor in the practice of public affairs at Columbia University's School of International and Public Affairs): "The EPA estimated 2.4 per cent leakage for total natural gas production in 2009 (Tollefson, 2013)" [11].

Professor Howarth and colleagues from 54-Nobel-laureate Cornell University have estimated that "3.6% to 7.9% of the methane from shale-gas production escapes to the atmosphere in venting and leaks over the life-time of a well" [12].

7. Unknown degree of gas leakage causing elevated atmospheric CH4 in Australia.

Two Australian scientists, Dr Isaac Santos and Dr Damien Maher from Southern Cross University's Centre for Coastal Biogeochemistry Research in the School of Environment, Science and Engineering, found 3.5-fold elevated atmospheric methane in a coal seam gas (CSG) extraction area. According to Dr Damien Maher: "Typically background concentrations of methane in the atmosphere are approximately 1.8 parts per million. In Tara [a coal seam gas extraction area in Queensland] the concentrations are consistently higher than two parts per million and approach seven parts per million in a few locations. This is about three and a half times higher than expected if there was no change in the atmosphere. These results are higher than values reported for conventional gas production fields in Siberia, one of the world's largest natural gas production areas." According to Dr Isaac Santos: "Despite commercial production starting in the mid-1990s, this is the first publicly available data on concentrations of methane in the atmosphere of an Australian CSG production areas. We used cutting edge technology to make the measurements. Our work highlights the need for further research to adequately quantify the emissions and their source" [13]. The degree of gas leakage in Australia has evidently not yet been accurately determined.

In 2015 the Department of the Environment of the Australian Government estimated total fugitive emissions (gas leakage) from fossil fuels of 46.5 Mt CO_2-e for 2017-2018 of which one third, 15.5 Mt CO_2-e, was from natural gas and oil [14]. However this estimate was based on a GWP of 21 for CH_4 – assuming a GWP of 105, as in the present analysis, would bring this up to 77.5 Mt CO_2-e, 2.5 times lower than the GHG pollution estimated from a 1.2% leakage from Australian gas (see #5 above) and 4.2 times lower than the GHG pollution estimated from a 2.6% leakage (see #4 above). As detailed below, hugely under-estimating gas leakage when Australia had a Carbon Tax amounted to huge Carbon Tax evasion.

8. **Gas leakage in Australia and huge Carbon Tax evasion.**

The major political parties in Australia, the presently ruling extreme Right Liberal Party and National Party Coalition and the Rightist-dominated Labor Party Opposition, both support massive conventional natural gas developments and non-conventional Coal Seam Gas (CSG) and shale gas developments variously involving "fracking" (hydrological fracturing of underground strata). To its credit, the more science-responsive former Labor Government introduced a Carbon Tax to penalize fugitive emissions (gas leakage). However the endlessly dishonest and climate criminal Labor Government understated gas leakage as 0.12% (rather than the science-based estimates of 3%-7.9% for gas produced non-conventionally) [12], and under-estimated the Global Warming Potential (GWP) of methane (CH_4, the bulk of natural gas) as 21 times that of [the same mass of] carbon dioxide (CO_2) (whereas it is actually 105 times worse than CO_2 as a greenhouse gas (GHG) on a 20 year time frame and considering aerosol impacts) [3]. This lead to an immense but unreported Carbon Tax Scandal in which a pro-gas Labor Government understated the Carbon Tax obligations of the corporate gas producers by a factor of 137- to 329-fold, collecting a mere $21 million annually rather than the science-indicated Carbon Tax obligation of about $3-7 billion each year from leakage of natural gas (fugitive emissions) [15]. Of course this huge, Elephant in the Room Carbon Tax Scandal was resolutely non-reported by Australian Mainstream media that are detestable purveyors of "Mainstream media fake news through lying by omission". Further, the Labor Government was ousted by the climate criminal Coalition who campaigned successfully on abolishing the Carbon Tax i.e. not even the $21 million would be collected.

9. Carbon Price, Carbon Debt, Australia's gas exploitation and intergenerational injustice.

Climate economist Dr Chris Hope of 90-Nobel-Laureate Cambridge University has estimated a damage-related Carbon Price in US dollars of $200 per tonne CO_2-equivalent [16]. Using estimates from Professor James Hansen (a leading climate scientist from 85-Nobel-Laureate Columbia University) of national contributions to Historical Carbon Debt, and assuming a damage-related Carbon Price of $200 per tonne CO_2-e, the World has a damage-related Carbon Debt of $370 trillion that is increasing at $13 trillion per year, and Australia has a Carbon Debt of $7.5 trillion that is increasing at $400 billion per year and at $40,000 per head per year for under-30 year old Australians [17]. Unlike conventional debt – that can be variously expunged by default, bankruptcy, printing money or the debtor running away to a secret new life – Carbon Debt via a damage-related Carbon Price is inescapable e.g. if future generations do not build sea walls, coastal cities will be submerged by increasing sea levels from global warming.

If there is no CH_4 leakage then in 2017 Australia's gas exploitation would incur a Carbon Debt within Australia from Domestic combustion of 30 Mt natural gas to yield 30 x 44/16 = 82.5 Mt CO_2-e and a consequent Carbon Debt of 82.5 Mt CO_2-e x $200 / tonne CO2-e = $16.5 billion – in today's money, and significant in the context of compounding interest on a "conventional" Australian gross national debt of $600 billion that is increasing at about $40 billion per year.

By way of comparison, Oil & Gas Australia states that "If current oil prices are maintained, EnergyQuest estimates that the value of Australian LNG exports will double to around A$36 billion [US$27 billion] in 2017" [18]. Thus, (assuming no gas leakage), gas exploitation would leave Australia with an annual increment in damage-related Carbon Debt of presently at least $16.5 billion, whereas one supposes that most of the present $27 billion in profits from Australian export of LNG will end up overseas in tax havens or capital investments.

However, if there is a 2.6% leakage then in 2017 Australia will incur a gas exploitation Carbon Debt within Australia of 322 Mt CO_2-e x $200/ tonne CO_2-e = $64.4 billion, over twice the profit from exporting LNG.

If there is a 1.2% leakage, then in 2017 Australia will incur a gas exploitation Carbon Debt within Australia of 195 Mt CO_2-e x $200/ tonne CO_2-e = $39.0 billion, this being 1.4 times the profit from exporting LNG.

Carbon Debt, including that arising from gas leakage, represents massive intergenerational inequity [20] to which the young must surely respond with Climate Revolution now (peaceful of course, but a revolution for all that) [21, 22].

10. **Dirty gas and Australia's disproportionate contribution to a worsening Climate Genocide.**

Trump America, Trumpist Australia and neo-Trumpist Canada are acutely threatening the world with unlimited greenhouse gas (GHG) pollution from unlimited fossil fuel exploitation. For the US, Canada and Australia, full exploitation of presently recoverable fossil fuel reserves would generate GHG pollution vastly exceeding (37-fold) the whole world's remaining Terminal Carbon Pollution Budget that must not be exceeded for a 75% chance of avoiding a catastrophic 2C temperature rise [23].

Australia's annual domestic plus exported greenhouse gas (GHG) pollution is so high that it exceeded its "fair share" of the whole world's Terminal Carbon Pollution Budget by mid-2011 and since then Australia has been stealing the entitlement of all other nations. Australia's commitment to unlimited gas, coal and iron ore exports (supported by both the extreme Right governing Coalition and the Right-dominated Labor

Opposition) means that it is committed to polluting the atmosphere with 20 times the whole world's remaining Terminal Carbon Pollution Budget of 88 billion tonnes CO_2-e as of mid-2017. Yet 24 million Australians represent a mere 0.3% of the world's population of 7.4 billion [23].

Recent revised estimates taking land use into account indicate that Australia's annual Domestic per capita greenhouse gas (GHG) pollution (52.9 tonnes CO_2-e per person per year) is about 20 times greater than the annual per capita GHG pollution of Bangladesh (2.7 tonnes CO_2-e per person per year) [1]. However, Australia 's annual Domestic plus Exported per capita greenhouse gas (GHG) pollution (116 tonnes CO_2-e per person per year) is about 43 times greater than the annual per capita GHG pollution of Bangladesh (2.7 tonnes CO_2-e per person per year) [1]. Of course, a look-the-other-way Australia remorselessly excludes from public discussion its huge Exported GHG pollution (1,511 Mt CO_2-e Exported GHG pollution in 2015 as compared to about 600 Mt CO_2-e Domestic GHG pollution) [5].). However if there is merely 1.2% of gas leakage then Australia's already huge annual Domestic GHG pollution increases by 30% – and if there is a 2.6% leakage it increases by about 50%.

Global CO_2 pollution is increasing at a record rate of 3 ppm CO_2 per year and atmospheric CO_2 is at a record level of about 406 ppm CO_2 [24]. The evident failure to tackle climate change is contributing to a worsening Climate Genocide in which 10 billion people may perish this century if global warming is not requisitely addressed [25]. Already 7.5 million people die each year from air pollution (7 million) or climate change (0.5 million) [26, 27]. However the latter figure may be an under-estimate because 17 million people presently die avoidably each year from deprivation in Developing Countries that are already disproportionately impacted by climate change [27]. Presently resolutely ignored by Mainstream media is the Arctic "Methane Bomb" threat – 50 billion tonnes CH_4 in the East Siberian Arctic Shelf is predicted to be released in coming decades and is equivalent to 50 billion tonnes CH_4 x 105 tonnes CO_2-equivalent/tonne CH_4 = 5,250 billion tonnes CO_2-e or about 60 times more than the world's remaining Terminal Carbon Pollution Budget of 88 billion tonnes CO_2 as of 2017 [19, 29]. We are doomed unless we can stop this Arctic CH_4 release.

Conclusions.

In 2015 a climate criminal Australia ranked second worst after Saudi Arabia for climate change inaction in the German Climate Change Performance Index [30]. Australia is one of the very worst climate criminal nations and its commitment to unlimited gas, coal and iron ore exports (supported by both the extreme Right governing Coalition and the Right-dominated Labor Opposition) means that it is committed to polluting the one common atmosphere of all countries with over 3 times the world's total Terminal Carbon Pollution Budget as of mid-2009 (600 billion tonnes CO_2-e) or 20 times the whole world's remaining Terminal Carbon Pollution Budget of 88 billion tonnes CO_2-e as of mid-2017. This Terminal Carbon Pollution Budget must not be exceeded if we are to have a 75% chance of avoiding a catastrophic 2C temperature rise [19, 23].

Australia is set to be the world's number 1 exporter of LNG and is presently exporting 60 Mt of LNG each year as well as using 30 Mt of gas each year domestically. Fanatical opposition to a Carbon Tax by anti-science, pro-fossil fuels and effective climate change denialist Coalition Governments under PM Tony Abbott and thence PM Malcolm Turnbull has led to great market uncertainty about investment in coal-fired electricity generation, with the slack due to the closure of aging coal-fired plants increasingly being taken up by investment in renewable energy (good) and gas-fired power (extremely bad). Rational, science-based schemes for rapid progression to 100% renewable energy for Australia include energy storage for "when the wind doesn't blow and the sun doesn't shine") variously involving molten salts, water storage, gas compression and batteries, together with a highly efficient national electricity grid [5, 31, 32]. In the absence of timely action by the anti-science, pro-fossil fuels and effective climate change denialist Federal Coalition Government, recent power blackout experiences prompted the Labor South Australian State Government to resort to dirty, gas-fired power as well as clean battery storage for base-load back-up in predicted future emergencies. The derelict and climate criminal Coalition Federal Government backs massive exploitation of gas for domestic purposes as well as for export.

Both the Australian Coalition Government and the Labor Opposition are committed to Australia remaining a world leading exporter of coal, gas and iron ore and to expansion of domestic use of dirty gas (e.g. conventional gas or non-conventional gas from fracking). Hydrological fracturing ("fracking") in particular has been opposed by environmental activists from urban Greens to farmers because it despoils farming land and threatens vital aquifers in drought-impacted Australia. Neoliberal politicians and lobbyists falsely argue that "gas is clean" and similarly falsely talk of "clean coal", yet the harsh reality ignored in neoliberal and look-the-other-way Mainstream Australia is that, depending upon the degree of gas leakage, gas burning for power can be dirtier GHG-wise than burning coal.

Australia's massive exploitation of dirty gas can be seen as a detonator for the Arctic Methane Bomb that threatens to destroy much of Humanity and the Biosphere in the coming century. Australia's ever-worsening, greed-driven climate criminality will inevitably invite global Climate Blowback through Boycotts, Divestment and Sanctions (BDS), International Court of Justice (ICJ) litigations, International Criminal Court (ICC) prosecutions, Eco-tariffs and Green Tariffs.

References.

[1]. Gideon Polya, "Revised Annual Per Capita Greenhouse Gas Pollution For All Countries – What Is Your Country Doing?", Countercurrents, 6 January, 2016: https://countercurrents.org/polya060116.htm.
[2]. Gideon Polya, "Exposing And Thence Punishing Worst Polluter Nations Via Weighted Annual Per Capita Greenhouse Gas Pollution Scores", Countercurrents, 19 March, 2016: https://countercurrents.org/polya190316.htm.
[3]. Drew T. Shindell, Greg Faluvegi, Dorothy M. Koch, Gavin A. Schmidt, Nadine Unger and Susanne E. Bauer, "Improved Attribution of Climate Forcing to Emissions", Science, 30 October 2009: Vol. 326 no. 5953 pp. 716-718: http://www.sciencemag.org/content/326/5953/716.
[4]. "Gas is not clean energy": https://sites.google.com/site/gasisnotcleanenergy/home.
[5]. "2011 climate change course": https://sites.google.com/site/300orgsite/2011-climate-change-course ; see section G.
[6]. Australian Mining, "Australian LNG developments trigger export growth: EnergyQuest", 17 January 2017: https://www.australianmining.com.au/news/australian-lng-developments-trigger-export-growth-energyquest/.
[7]. Australian Government, "Australian energy projections to 2049-50", November 2014: https://industry.gov.au/Office-of-the-Chief-Economist/Publications/Documents/aep/aep-2014-v2.pdf.
[8]. "Conversion factors relevant to petrochemical activities": http://www.chemlink.com.au/conversions.htm.
[9]. EPA, "Draft inventory of U.S. greenhouse gas emissions and sinks: 1990-2019", 2017: https://www.epa.gov/ghgemissions/draft-inventory-us-greenhouse-gas-emissions-and-sinks-1990-2015.
[10]. "Natural gas", BP statistical review of world energy, June 2016: http://www.bp.com/content/dam/bp/pdf/energy-economics/statistical-review-2016/bp-statistical-review-of-world-energy-2016-natural-gas.pdf.
[11]. Steven Cohen, "Understanding Environmental Policy", Columbia University Press, 2014, page 110.
[12]. Robert W. Howarth, Renee Santoro and Anthony Ingraffea, "Methane and the greenhouse-gas footprint of natural gas from shale formations", Climatic Change", 12 November 2010: http://www.eeb.cornell.edu/howarth/Howarth%20et%20al%20%2020 11.pdf.

[13]. "SCU releases first independent methane observations in Australian CSG [Coal Seam Gas] fields", Southern Cross University, 15 November 2012: http://www.scu.edu.au/news/media.php?item_id=6041&action=show _item&type=M.

[14]. Australian Government, Department of the Environment, "Fugitive emissions projections 2014-15, August 2015.

[15]. Gideon Polya, "Australia's Carbon Tax scandal", MWC News, 23 November 2012: http://mwcnews.net/focus/analysis/23026-gideonpolya-carbon-tax.html.

[16]. Chris Hope, "How high should climate change taxes be?", Working Paper Series, Judge Business School, University of Cambridge, 9, 2011: http://www.jbs.cam.ac.uk/media/assets/wp1109.pdf.

[17]. "Carbon Debt Carbon Credit": https://sites.google.com/site/carbondebtcarboncredit/.

[18]. Oil & Gas Australia, "Australian LNG exports to hit [A]$36 billion in 2017". 17 January 2017: http://www.oilandgasaustralia.com.au/australian-lng-exports-36-billion-2017/.

[19]. Gideon Polya, "Trump America, Trudeau Canada & Turnbull Australia Doom World With Unlimited Greenhouse Gas Pollution", Countercurrents, 5 April 2017: https://countercurrents.org/2017/04/05/trump-america-trudeau-canada-turnbull-australia-doom-world-with-unlimited-greenhouse-gas-pollution/.

[20]. "Climate Justice & Intergenerational Equity": https://sites.google.com/site/300orgsite/climate-justice.

[21]. "Divest from fossil fuels": https://sites.google.com/site/300orgsite/divest-from-fossil-fuels.

[22]. Climate Revolution Now": https://sites.google.com/site/300orgsite/climate-revolution.

[23]. Gideon Polya, "Australia's Huge Coal, Gas & Iron Ore Exports Threaten Planet", Countercurrents, 15 May 2012: https://countercurrents.org/polya150512.htm.

[24]. NOAA, "Trends in atmospheric CO_2": https://www.esrl.noaa.gov/gmd/ccgg/trends/

[25]. "Climate Genocide": https://sites.google.com/site/climategenocide/.

[26]. "Nuclear weapons ban, end poverty & reverse climate change": https://sites.google.com/site/300orgsite/nuclear-weapons-ban.

[27]. "Stop air pollution deaths": https://sites.google.com/site/300orgsite/stop-air-pollution-deaths

[28]. Gideon Polya, "Body Count. Global avoidable mortality since 1950", that includes a succinct history of every country and is now available for free perusal on the web: http://globalbodycount.blogspot.com/.
[29]. "Methane Bomb Threat": https://sites.google.com/site/methanebombthreat/.
[30]. German Climate Watch Index 2015: https://germanwatch.org/en/download/10407.pdf.
[31]. Beyond Zero Emissions Zero (BZE), Zero Carbon Australia by 2020 Report (BZE ZCA2020 Report), 2010: http://www.beyondzeroemissions.org/about/bze-brand.
[32]. Peter Seligman, "Australian sustainable energy – by the numbers", Melbourne Energy Institute, University of Melbourne, 2010: http://energy.unimelb.edu.au/ozsebtn/.

CHAPTER 14: METHANE LEAKAGE MAKES AUSTRALIA A WORLD-LEADING PER CAPITA GREENHOUSE GAS POLLUTER

[First published as Gideon Polya "Methane leakage makes Australia a world-leading per capita greenhouse gas polluter", Countercurrents, 18 February 2020: https://countercurrents.org/2020/02/methane-leakage-makes-australia-a-world-leading-per-capita-greenhouse-gas-polluter.]

Methane (CH_4) is 85% of natural gas, leaks, and has a Global Warming Potential (GWP) 105 times that of the same mass of carbon dioxide (CO_2) on a 20 year time frame with aerosol impacts included. Such considerations reveal that Australia with 0.33 % of the world population has revised annual Domestic greenhouse gas (GHG) emissions that are 2.5% of the world's, and annual Domestic plus Exported GHG emissions that are 5.4% of the world's annual GHG pollution.

Australia is among world leaders in annual per capita greenhouse gas (GHG) pollution [1, 2], is a major exporter of GHG pollution-implicit coal, gas and iron ore, and has become the world's largest exporter of Liquid Natural Gas (LNG) as well as of coal. However, depending upon the degree of systemic gas leakage, burning gas for power may be worse greenhouse gas (GHG)-wise than burning coal because methane (CH_4, about 85% of natural gas) has a global warming potential (GWP) that is 105 times that of the same mass of carbon dioxide (CO_2) on a 20 year time frame with aerosol impacts included [3-6]. However a remorselessly neoliberal, anti-science and anti-environment Australia is committed on a bipartisan political basis (i.e. with the support of the Right-Extreme Right

Liberal Party-National Party Coalition Government and the Right-Centrist Labor Party Opposition) to massive exploitation of conventional and non-conventional natural gas reserves for Export and Domestic use. Only the Greens oppose this Gadarene, ecocidal, speciescidal, and potentially omnicidal and terracidal profligacy that is driven by remorseless neoliberal greed and racism.

Australia is a world leader in annual per capita GHG pollution [1, 2] and both its Coalition Government and Labor Opposition are committed to unlimited coal and gas exploitation for Export [1, 2]. Australia is a key player in a dangerous global coal to gas transition that is a deadly and dishonest neoliberal alternative to the complete cessation of fossil fuel exploitation demanded by scientists in the face of the worsening climate emergency. The ideal target of no more than a 1.5C temperature rise agreed to at the 2015 Paris Climate Change Conference is now set to be exceeded on present trends within 10 years [3, 4]. A plus 2C temperature rise – that all governments (except for the idiotic, dangerous, anti-science and climate change denialist US Trump Administration) agree would be catastrophic – is now effectively unavoidable [5-7]. While Humanity can still take action to make the now inevitable plus 2C future "less bad", there is a looming threat of global warming causing massive release of methane (CH_4) from the Arctic, a ticking "Methane Bomb" set to utterly devastate Humanity and indeed all life on earth in the coming century because CH_4 has a Global Warming Potential 105 times greater than that of carbon dioxide (CO_2) on a 20 year time frame with aerosol effects included [8-11].

Australia continues to be devastated by high intensity, destructive and deadly 2019-2020 bushfires across the continent [6], conflagrations that recently threatened lives and homes in Australia's national capital Canberra in which an unprecedented emergency was declared. Scientists have been warning for decades that global warming and consequent increased temperature, dryness and drought will increase the probability of forest fires [6, 13-19]. However this is variously contested by the climate change denialist or effective climate change denialist Coalition politicians ruling Australia [20, 21]. Indeed at the height of the Australian bushfire catastrophe, pro-coal PM Scott Morrison (who notoriously flourished a lump of coal in Parliament, idiotically declaring "This is coal. Don't be afraid, don't be scared" [22]) announced Government underwriting of 2 new gas-fired power stations next to population centres, and raised the possibility of backing some new coal-fired power stations as well [23]. Utter stupidity.

Thanks to the homicidal greed of climate criminal countries such as Australia, the present plus 1.1C temperature rise is already devastating Island Nations, and a catastrophic plus 2C warming is now effectively unavoidable on present trends. Climate criminal Australia is among world leaders for the following 16 climate criminal activities or parameters: (1) annual per capita greenhouse gas pollution, (2) live methanogenic livestock exports, (3) natural gas exports, (4) recoverable shale gas reserves that can be accessed by hydraulic fracturing (fracking), (5) coal exports, (6) land clearing, deforestation and ecocide, (7) speciescide or species extinction, (8) coral reef destruction, (9) whale killing and extinction threat through global warming impacting on krill stocks, (10) terminal carbon pollution budget exceedance, (11) per capita Carbon Debt, (12) ultimately GHG generating iron ore exports, (13) climate change inaction, (14) climate genocide and approach towards omnicide and terracide, (15) increasing Domestic GHG pollution despite Paris commitments to lower GHG pollution, and (16) complicity in 8 million annual air pollution deaths from burning carbon fuels, Australia's share being 75,000 overseas and 10,000 Domestically [24-26](for detailed documentation see [27]). Australia with 0.3% of the world's population contributes about 4.5% of global GHG pollution (including that due to the burning of Australia's world leading gas and coal exports) [1].

Australian actions to "tackle climate change" would involve mitigatory action in all 16 areas but for the climate criminal Australian Coalition Government it is "business as usual" (BAU) – the climate criminal Australian dog-in-the-manger is simply BAU-wowing to a world facing a worsening Climate Emergency and a worsening Climate Genocide (already 1 million people die from climate change each year in a worsening Climate Genocide that will involve 10 billion deaths this century en route to a sustainable human population in 2100 of merely 0.5-1.0 billion) [28].

Now in his latest anti-science atrocity Australian PM Scott "Scomo" Morrison has announced a $2 billion [Australian dollars] "gas deal" with Premier Gladys Berejiklian of Australia's largest state, New South Wales (NSW). Phillip Coorey of the Australian Financial Review: "The federal government and NSW have reached a $2 billion energy deal which will require NSW to free up massive amounts of gas for domestic use in return for the construction of new interconnectors, the underwriting of new non-coal power generation, and funding for emissions reduction projects... Pivotal to the deal will be the NSW government having to find an extra 70 petajoules of gas per year [1.29 Mt gas per year] for the east coast domestic market. This could be done by either the government importing more gas through Port Kembla but it is far more likely to give the green light to extract gas from the Narrabri [NSW coal seam] gas fields" [29].

Prime Minister Scott Morrison utterly incorrectly stated: "There is no credible plan to lower emissions and keep electricity prices down that does not involve the greater use of gas as an important transition fuel" [30]. However his utterly false position has been slammed by science-informed critics. Thus Georgina Woods (from the anti-fracking, anti-coal seam gas, farmer's group, "Lock The Gate": "Rural communities should not be forced to sacrifice land, water and their economic security in the name of quick and dirty resource exploitation. Coal seam gas is a heavily polluting industry that leaks vast amounts of methane and won't do anything to bring down carbon emissions" [30]. NSW Greens senator Mehreen Faruqi: "It threatens the Great Artesian Basin, farmer's livelihoods, food security and the mighty biodiverse Pilliga Forest. It's clear that the federal and NSW governments have already made a political decision to allow this project to go ahead" [30]. Adam Bandt (Federal Greens MP): "NSW is doing a climate deal with the devil, locking in pollution that will blow Australia's emissions targets and put us on a path to climate catastrophe. As a global warming gas, methane is up to 86 times more powerful than carbon dioxide. The Prime Minister is trying to hoodwink people with his supposed climate action, but today's announcement amounts to little more than climate criminality" [30].

How does this latest bit of Australian Coalition climate criminality stack up with the science? Set out below is a detailed quantitative analysis showing (among many other surprising things) that the investment of a once-off A\$2 billion of taxpayer funds into the PM Morrison-Premier Berejiklian "gas deal" will result in an inescapable annual Carbon Debt of A\$3.1 billion for future generations, or A\$31 billion over the next decade (noting that the annual Australian defence budget is about A\$35 billion and a similar amount is spent annually on subsidies for organized religion).

(1). 2.6% CH_4 leakage is as polluting GHG-wise as burning the remaining CH4.

Methane (CH_4) is the major constituent of natural gas and has a molecular weight of about 16, CO_2 has a molecular weight of about 44, and carbon (C) has an atomic weight of 12. Combustion of CH_4 yields CO_2 and H_2O (CH_4 + $2O_2$ -> CO_2 + $2H_2O$) and thus 16 tonne CH_4 yields 44 tonnes of CO_2 and combustion of 1 tonne CH_4 yields 2.75 tonne CO_2. By way of comparison, combustion of coal (carbon, C) yields CO_2 (C + O_2 -> CO_2) and thus 12 tonnes C yields 44 tonnes CO_2 and combustion of 1 tonne C (coal) yields 3.67 tonnes CO_2. Thus per tonne combusted, coal yields 1.3 times more CO_2 than gas. Further, coal burning produces more toxic pollutants than gas burning, notably carbon monoxide (CO), sulphur dioxide (SO_2), nitrous oxide (N_2O), nitrogen dioxide (NO_2), radioactivity, heavy metals and fine carbon particulates (notably $PM_{2.5}$). Accordingly the fossil fuel industry and their Mainstream media, politician, academic and commentariat supporters advocate a transition from coal to an assertedly "cleaner" gas en route to an eventual zero fossil fuels future. However they are wrong – while gas burning produces less toxic pollutant than coal burning, massive systemic gas leakage (5.4% in the US) and a Global Warming Potential for CH_4 105 times that of CO_2 (on a 20 year time frame) means that gas burning can be dirtier than coal burning GHG-wise, as set out below.

(2) At 2.6% systemic gas leakage, burning gas yields 2 times more CO_2-equivalent as burning coal.

CH_4 is a gas, leaks and has a Global Warming Potential (GWP) 105 times that of the same mass of CO_2 on a 20 year time frame and with aerosol impacts considered [8]. One can readily calculate (assuming gas to be 100% CH_4 or CH_4 equivalent) that on this basis a systemic gas leakage of 2.6% would contribute as much GHG pollution as generating the greenhouse gas CO_2 by burning the remaining 97.4% of the gas [9]. Thus burning 1 tonne carbon yields 3.67 tonnes CO_2 and combustion of 1 tonne CH_4 with zero leakage yields 2.75 tonne CO_2. However combustion of 1 tonne CH_4 with 2.6% leakage yields 2.68 tonne CO_2 (from burning 97.4% of the CH_4) plus 2.68 tonne CO_2-equivalent (from the GHG effect of the leaked CH_4) = 5.36 tonnes CO_2-equivalent. One can crudely estimate that with a mere 2.6% of systemic leakage, burning 1 tonne of gas generates 1.5 times the CO_2-equivalent produced from burning 1 tonne of coal.

(3). Australian Government and business grossly under-estimate CH_4 leakage from unconventional production at 0.1% (54 times less than overall gas leakage in the US).

One notes that systemic gas leakage in the Boston urban region in the US is about 2.7% [31]. It is estimated that gas leakage in the US is about 2.3% of overall production [34]. Dr Robert Howarth (54 Nobel Laureate Cornell University) in an extensive review states (2015): "Over the past decade, shale gas production has increased from negligible to providing 40% of national gas and 14% of all fossil fuel energy in the USA in 2013... emissions from the natural gas industry, including both conventional gas and shale gas, could best be characterized as averaging 5.4% (±1.8%) for the full life cycle from well to consumer" [33, 34]. However according to a report from the

Melbourne Energy Institute authored by gas expert Tim Forcey, Australia claims gas leakage from unconventional gas production at a mere 0.1% [35], 54 times less than the 5.4% overall gas leakage in the US [33, 34]. Tim Forcey: "Looking specifically at methane emission rates from unconventional gasfields, measurements in the US are up to 10-25 times higher than rates reported by the Australian Government to the UNFCCC [United Nations Framework Convention on Climate Change]" [35].

(4). Damage-related Carbon Price US\$200 per tonne CO_2-equivalent (considering all major GHGs excepting H_2O).

Climate economist Dr Chris Hope of 90-Nobel-Laureate Cambridge University has estimated a damage-related Carbon Price (in US dollars) of \$200 per tonne CO_2-equivalent [16]. Professor James Hansen (of 96 Nobel Laureate Columbia University): "One ppm of CO_2 is 2.12 billion tons of carbon or about 7.77 billion tons of CO_2. Recently Keith et al. (2018) achieved a cost breakthrough in carbon capture, demonstrated with a pilot plant in Canada. Cost of carbon capture, not including the cost of transportation and storage of the CO_2, is \$113-232 per ton of CO_2. Thus the cost of extracting 1 ppm of CO_2 from the atmosphere is \$878-1,803 billion. In other words, the cost, in a single year, of closing the gap between reality and the IPCC scenario that limits climate change to +1.5°C is already about \$1 trillion. And that is without the cost of transporting and storing the CO_2, or consideration of whether there will be citizen objection to that transportation and storage. This annual cost will rise rapidly, unless there is a rapid slowdown in carbon emissions… cost of CO_2 storage… has been estimated as \$10-20/tCO_2" [37]. Taking Professor Hansen's data, and including his estimates of the cost of transport and storage of CO_2, indicates that this "best so far" cost of atmospheric CO_2 draw-down is \$123-252/tCO_2, similar to Dr Chris Hope's econometrics-based estimate of \$200 per tonne CO_2-equivalent [36].

(5). For a 300 ppm CO_2 draw-down target, the world has an upper estimate Carbon Debt of \$200 trillion that is increasing at \$13 trillion per year.

Many scientists and science-informed activists demand a reduction of atmospheric CO_2 to a safe and sustainable level for all peoples and all species of about 300 ppm CO_2 (roughly the pre-Industrial Revolution level and the maximum observed over the last 1 million years until recent decades) [38, 39]. The upper estimate of the Carbon Debt for a transition from the present monthly mean of 412 ppm CO_2 (and increasing at 2-3 ppm CO_2 per year) [40] to 300 ppm CO_2 is 112 ppm CO_2 x \$1,803 billion per ppm CO_2 = \$202 trillion.

This inescapable Carbon Debt for future generations is increasing at 2-3 ppm per year x \$1,803 billion per ppm CO_2 = \$3.6-5.4 trillion per year [40]. However this estimate does not take other GHGs, notably CH_4, into account. World Bank analysts have reconsidered annual GHG pollution taking land use into account and assuming a GWP for CH_4 of 86 on a 20 year time frame, this estimate increasing annual GHG pollution from 41.8 Gt CO_2-e per year to 63.8 Gt CO_2-e per year [41]. Thus on this basis the global Carbon Debt is increasing annually at 63.8 billion tonnes CO_2-e x \$200 per tonne CO_2-e = \$12.8 trillion per year.

(6). Australia's 2017-18 Domestic and Exported GHG emissions from natural gas exploitation alone totalled 471 Mt CO_2-e.

Australian liquefied natural gas (LNG) exports totalled 59.7 Mt in 2017–18 [42]. Assuming for computational and didactic simplicity that this is all CH_4 (or CH_4 equivalents), then on combustion it would yield 59.7 Mt CH_4 x 44 t CO_2/ 16 t CH_4 = 164.2 Mt CO_2. However assuming a gas leakage of only 2.6%, the warming effect of the leaked CH_4 equals that from burning the remaining CH_4 (see (1)). Thus the total warming effect of Australia's LNG Exports in 2017-18 is that of 2 x 0.974 x 164.2 Mt CO_2-e = 320 Mt CO_2-e.

However gas used Domestically in Australia in 2017-18 totaled 28.2 Mt CH_4 [43] that on combustion yielded 28.2 Mt CH_4 x 44 t CO_2/ 16 t CH4 = 77.6 Mt CO_2. The total warming effect of Australia's Domestic gas use in 2017-18 is that of 2 x 0.974 x 77.6 Mt CO_2-e = 151.2 Mt CO_2-e.

Accordingly the GHG emissions due to Australia's Domestic and Exported gas alone in 2017-18 = 320 Mt CO_2 + 151 Mt = 471 Mt CO_2 as compared to the total annual GHG emissions of about 535 Mt CO_2-e in 2017-18 as reported by the Australian Government (it has been steadily rising, contrary to Paris Agreement demands, since the Coalition Government was elected in 2013) [44-48]. One notes that the Australian Government conveniently ignores Australia's huge Exported GHG emissions, largely ignores huge fugitive CH_4 emissions [leakage], ignores huge GHG contributions from bushfires [49], and assumes a GWP for CH_4 on a 100 year time frame (initially 21, now 25 and 4-5 times lower than the [NASA asserted] 105 on a 20 year time frame with aerosol impacts considered).

(7). Australia's 2018-19 Domestic and Exported GHG emissions from natural gas exploitation alone totalled 502 Mt CO_2-e (similar to the government's asserted total Domestic emissions of 540 Mt CO_2-e in 2018-19).

In 2018-19 total Australian gas production was 93.6 Mt CH_4 (5,082 petajoules) and there was a record LNG output [for Export] of 75.0 Mt (million tonnes) (4,070 PJ). Domestic gas use in 2018-2019 was accordingly 18.6 Mt CH_4 (1,012 PJ) [50]. In 2019 Australia exported 77.5 Mt LNG worth A$49 billion and became the largest LNG exporter in the world [51].

The 75 Mt gas exported in 2018-19 would on combustion yield 75 Mt CH_4 x 44 t CO_2/ 16 t CH_4 = 206.3 Mt CO_2. Again, assuming a gas leakage of only 2.6%, the warming effect of the leaked CH_4 equals that from burning the remaining CH_4 (see (1)). Accordingly the total warming effect of Australia's LNG Exports in 2017-18 is that of 2 x 0.974 x 206.3 Mt CO_2-e = 401.9 Mt CO_2-e.

The gas used Domestically in 2018-19 = 18.6 Mt CH_4 x 44 t CO_2/ 16 t CH_4 = 51.2 Mt CO_2 on combustion. The total warming effect of Australia's Domestic gas use in 2018-19 is that of 2 x 0.974 x 51.2 Mt CO_2 = 99.7 Mt CO_2-e. Accordingly the GHG emissions due to Australia's Domestic and Exported gas alone in 2018-19 = 402 Mt CO_2 + 100 Mt CO_2-e = 502 Mt CO_2-e. By way of comparison, the Australian Government's asserted total annual GHG emissions totalled about 540 Mt CO_2-e in 2018-19 [52].

Several scholars have predicted that Australia's Domestic GHG emissions are set to fall to about 530 Mt CO_2-e by mid-2021 if renewables deployment continues at the present rate [52]. However Australian LNG export production may max out at about 88 Mt LNG per year [53], with this translating (if realized in the coming decade) to an annual 472 Mt CO_2-e Exported plus about 100 Mt CO_2-e from Domestic use for a total of 572 Mt CO_2-e in emissions from gas alone in the coming few years.

(8). Australian Coalition Government's one-off A\$2 billion investment for gas exploitation in New South Wales (NSW) will add an estimated Carbon Debt of A\$2 billion per year, A\$20 billion per decade…

Australian PM Scott Morrison is spending \$2 billion on a "gas deal" that will inject an extra 70 petajoules of gas per year (1.29 Mt gas per year) for Domestic use [28, 29]. This means CO_2 release on combustion of 1.29 Mt CH_4 x 44 t CO_2/ 16 t CH_4 = 3.5 Mt CO_2. Assuming a leakage of 2.6%, the GHG effect of this = 2 x 0.974 x 3.5 Mt CO_2 = 6.8 Mt CO_2-e. At a damage-related Carbon Price of US\$200 per tonne CO_2-e (A\$299) the cost of this climate criminal adventure to future generations will be A\$299 per tonne CO_2-e x 6.8 Mt CO_2-e per year = A\$2.0 billion per year. However while the Australian Government is making a once-off investment of A\$2 billion, the cost to young Australians of the future will be A\$2 billion per year, A\$20 billion for the next decade, and A\$100 billion over the 50 year life-time of the gas-exploiting infrastructure (coal seam gas extraction systems, pipelines and gas-fired power stations) [54-56].

(9). Revised annual GHG emissions (Gt CO_2-e): 1.57 (Australia Domestic), 3.15 (Australia Domestic plus Exported) and 63.8 (world).

Australia's annual per capita GHG pollution as reported by the Australian Government is presently (2018-19) 538.9 Mt CO_2-e / 25.2 million people = 21.4 tonnes CO_2-e per person per year [57]. The world population is presently 7.7 billion (2019) and the world's greenhouse gas emissions total 45.3 Gt CO_2 (2019) [58, 59]. Wikipedia reports that in 2017 Australia's GHG emissions totalled 580 Mt CO_2-e and represented 1.3% of the world's total of 45.3 Gt CO_2-e [60], noting that Australia's population is 25.2 million x 100/ 7,700 million = 0.33% of the world's population i.e. rich Australia disproportionately pollutes the world GHG-wise by a factor of 3.9 [more than its fair share]. However this disparity gets much worse if one considers the global warming impact of fugitive emissions (leakage) of CH_4 from natural gas exploitation as set out below.

The present Australian Government in estimating annual GHG emissions of 540 Mt CO_2-e conveniently ignores or underestimates GHG contributions from (a) land use (Australia is among world leaders in land clearing [61, 62]), (b) fugitive emissions of CH_4 (it formerly estimated this at 0.1%, and more recently revised this to 0.7% [57], whereas it is 5.4% in the US [33, 34]), (c) global warming potential of CH_4 (it assumed 21 and revised this recently to 25 relative to the same mass of CO_2 on a 100 year time frame, whereas it is 105 on a 20 year time frame with aerosol impacts included [8]), and (d) it ignores emissions from bushfires (that have, so far, added an estimated 750 Mt CO_2-e to Australia's annual GHG pollution in financial year 2019-2020 [63]).

World Bank analysts carefully re-evaluated the contribution of livestock production to world annual GHG pollution and found that the world's annual total rose from 41.76 billion tonnes CO_2-equivalent (CO_2-e) as estimated by the Food and Agricultural Organisation (FAO) to 63.80 billion tonnes CO_2-e, with livestock production contributing over 51% of the higher figure [41]. A key element of their analysis was to use a Global

Warming Potential (GWP) of methane (CH_4) relative to that of carbon dioxide (CO_2) of 72 on a 20-year time frame rather than the 25 on a 100 year time frame used by the FAO [41]. Indeed the World Bank analysis evidently still understates the GHG pollution because NASA scientists have re-evaluated the GWP of CH_4 as 105 on a 20 year time frame with aerosol impacts considered [8].

Accordingly, more properly taking land use into account Australia's revised annual per capita GHG pollution was estimated in 2015 (t CO_2-e per person) at 52.9 and 116 if including its huge GHG-generating exports [1, 2]. Assuming a population of 25 million this adjusts Australia's annual GHG pollution to 1,323 Mt (Domestic), 1,577 Mt CO_2-e (Exported), and 2,900 Mt (Domestic plus Exported).

However to this we could add a further 250 Mt CO_2-e due to the fugitive emissions of CH_4 from gas exploitation (assuming 2.6% leakage and thus contributing about 50% of Australia's 500 Mt CO_2-e of GHG emissions due to Australia's Domestic use and Export of gas as set out in (7) above). Assuming Australian responsibility for gas fugitive emissions both at home and on route to foreign consumers, then this would adjust Australia's annual GHG pollution to 1,573 Mt CO_2-e (Domestic) and 3,150 Mt CO_2-e (Domestic plus Exported). However …

(10). Australia (0.33% of world population) generates 2.2% of upwardly revised global GHG emissions (Australian Domestic use only) and 5.4% (Australian Domestic plus Exported GHG emissions).

Assuming the revised estimate of global GHG emissions of 63.8 Gt CO_2-e [1, 2, 41], and revised estimates of Australia's GHG pollution taking land use into account [1, 2], one can estimate that Australia (0.33% of world population) has
Domestic emissions that are 1.573 Gt x 100/63.8 Gt = 2.5% of the world total, and Domestic plus Exported emissions that are 3.15 Gt x 100/63.8 = 4.9% of global emissions. Thus Australia disproportionately pollutes GHG-wise 7.6 fold more (Domestic pollution) [than the World average] and 14.8-fold more

(considering Domestic plus Exported pollution). However it must be revised on closer inspection.

(a). The land use-accommodating, revised estimate of Australian annual Domestic GHG emissions (1,323 Mt CO_2-e; see section 9) must be revised upwards by adding the fugitive emissions from Domestic gas exploitation (99.7 Mt CO_2-e; see section 7) to yield a total of 1,423 Mt CO_2-e.

(b). The revised estimate of Exported GHG emissions (1,577 Mt CO_2-e; see section 9) must be updated as follows:

(i). Australian coal exports totalled 391.2 Mt (2016) and on combustion generated 996.8 Mt CO_2-e [64].

(ii). Australian oil crude exports totalled 10.3 Mt (2016), and on combustion generated 33.4 Mt CO_2-e [64].

(iii) Australian exported liquefied petroleum gas (LPG) in 2016 that on combustion generated 3.5 Mt CO_2-e [64].

(iv) Australia exported 75 Mt LNG in 2017-18 that corresponded on combustion to 401.9 Mt CO_2-e (this taking an assumed 2.6% leakage into account; see section 7).

(v) Australia exported 830 Mt of iron ore (Fe_2O_3) in 2018 [65], this corresponding to 579.2 Mt CO_2-e (based on an upper estimate of steel manufacture being responsible for an upper estimate of 5% of global CO_2 emissions) [65, 66].

The total Exported GHG emissions is 2,015 Mt CO_2-e. Domestic GHG plus Exported GHG = 1,423 + 2,015 = 3,438 Mt CO_2-e, this corresponding to 3,438 x100/ 63,800 = 5.4% of the global annual total of 63,800 Mt CO_2-e [41].

(11). Australia's Domestic plus Exported GHG pollution make it the third worst annual per capita GHG polluter in the world.

Australia's annual per capita GHG pollution (t CO_2-e per person per year) taking fugitive emissions into account is 1,423 Mt CO_2-e/25 million persons = 56.9 (considering Domestic pollution only) and 3,438 Mt CO_2-e/25 million persons = 137.5 (considering Domestic plus Exported GHG pollution). By way of comparison, 137.5 t CO_2-e per person per year puts Australia third in the world after Belize (366.9) and, Guyana (203.1). In t CO_2-e per person per year China is 7.4 and India 2.1 (2015 analysis) [1, 2].

This is set to get worse. Thus Australia's Domestic and Exported GHG pollution through gas exploitation is set to increase significantly in coming years [53], notwithstanding pleas from scientists that the world must rapidly stop fossil fuel exploitation [9, 11, 67-71]. Eminent physicist and cosmologist Professor Stephen Hawking (90-Nobel–Laureate University of Cambridge) has succinctly identified the 2 existential threats to Humanity and the solutions: "We see great peril if governments and societies do not take action now to render nuclear weapons obsolete and to prevent further climate change" [71].

 (12). Gas is dirty energy, gas burning can be dirtier GHG-wise than coal burning, a coal-to-gas transition is disastrous: stop burning all fossil fuels ASAP.

Gas is not clean energy [9-12, 64] and, as outlined above, gas burning can be dirtier GHG-wise than coal burning [on a per mass basis]. However pro-gas politicians and commentators arguing for a coal-to-gas transition are arguing for massive investment in 30-year-lifetime gas-fired power plants that may be worse GHG-wise than coal-fired power plants depending upon the degree of gas leakage [9]. Yet in climate criminal Australia the Coalition PM Scott Morrison responded to the horrific bushfire tragedy by promising government support for 2 new gas-fired power stations, and indeed did not rule out such support for new coal-fired power stations [23, 72]. Indeed President Barack Obama oversaw a massive shift from coal to gas in the US based on the false premise that gas was "clean-er" whereas it is not only dirty but can in fact be much dirtier than coal GHG-wise depending on the degree of gas leakage (see section 2 above) [73].

There is indeed a strictly limited interim role for gas [using existing plant] as an emergency back-up for solar and wind-based power until hydrological, battery, solar thermal, [gas compression,] and hydrogen-based storage systems are emplaced on a large scale. Australia's Chief Scientist Dr Alan Finkel: "But, there is a limit to how much solar and wind we can use and still retain a reliable system. Ultimately, we will need to complement solar and wind with a range of technologies such as high levels of storage, long-distance transmission, and much better efficiency in the way we use energy. But, while these technologies are being scaled up, we need an energy companion today that can react rapidly to changes in solar and wind output. An energy companion that is itself relatively low in emissions, and that only operates when needed. In the short-term, as the Prime Minister and Minister [for Energy and Emissions Reduction] Angus Taylor have previously stated, natural gas will play that critical role. In fact, natural gas is already making it possible for nations to transition to a reliable, and relatively low emissions, electricity supply" [74]. However as demonstrated in this essay, gas is not "relatively low in emissions" as asserted by Dr Finkel because (a) combustion of 1 tonne of CH_4 (85% of natural gas) yields 2.75 tonne CO_2 as compared to combustion of 1 tonne of carbon (about 90% of coal) yielding 3.67 tonne CO_2, and (b) depending upon the degree of gas systemic leakage, gas burning can actually be much dirtier GHG-wise than coal burning [on a per mass basis](see section 2).

Final comments on combatting falsehood, deceit and climate change inaction.

As perceived by the 2-day Australian National Climate Emergency Summit 2020 held on Friday 14 and Saturday 15 February 2020, Australia and the world are facing a Climate Emergency demanding urgent action (see [75, 76]). Unfortunately the fossil fuel Lobby supported by an army of Mainstream journalist, politician, academic, commentariat and lobbyist supporters has the political upper hand, most notably in climate criminal Trump America under climate change denialist Donald Trump and in its pro-coal, pro-gas lackey Australia under an effective climate change denialist pro-coal Coalition Government. Nevertheless the science is clear and indeed is obvious to any sensible, science-informed high school student, as exampled by the wonderfully articulate and straight-talking Greta Thunberg [77].

The success of the denialists and effective climate change denialists is a deadly and disastrous example of Polya's Second Law of Economics, to whit "Deceit about the Cost of Production strives to a maximum". The Second Law of Economics is based on the fundamental Second Law of Thermodynamics that states that entropy (disorder, chaos, lack of information content) strives to a maximum [78]. The International Monetary Fund (IMF) has exposed massive deceit in stating that while a damage-related Carbon Tax of $75 per ton CO_2 would be an effective way of addressing the climate threat, the present global average Carbon Price is only $2 per ton CO_2 : "The average price on global emissions is currently $2 a ton, a tiny fraction of what is needed for the 2°C target" [25, 54, 79]. Science-trained Pope Francis has stated: "Yet only when the economic and social costs of using up shared environmental resources are recognized with transparency and fully borne by those who incur them, not by other peoples or future generations, can those actions be considered ethical" [54, 80, 81]. Climate economist Dr Chris Hope (of 120-Nobel-Laureate Cambridge University) and climate scientist Professor James Hansen (of 96-Nobel-Laureate Columbia University) have independently estimated a damage-related Carbon Price of about $200 per tonne CO_2-e [36, 37, 54].

Eminent economist Lord Nicholas Stern has described this massive deceit thus: "The problem of climate change involves a fundamental failure of markets: those who damage others by emitting greenhouse gases generally do not pay. Climate change is a result of the greatest market failure the world has seen. The evidence on the seriousness of the risks from inaction or delayed action is now overwhelming. We risk damages on a scale larger than the two world wars of the last century. The problem is global and the response must be a collaboration on a global scale" [82]. This massive corporate and political deceit in ignoring the gigantic economic externality measured by a damage-related Carbon Price has created a huge, inescapable and assiduously ignored Carbon Debt for future generations of $200-250 trillion that is increasing each year by 63.8 Gt CO_2-e per year x $200 /t CO_2-e = $13 trillion annually [55, 64].

Young Australians will have to pay a gigantic Carbon Debt that has been estimated at $40,000 per head per year for under-30 year old Australians [54]. However this estimate needs correction taking fugitive emissions, land use and a 20 year-based Global Warming Potential (GWP) for CH_4 into account. Thus Australia's revised annual Domestic plus Exported GHG pollution is 3,438 Mt CO_2-e that corresponds to 3,438 Mt CO_2-e x $200 /t CO_2-e = $688 billion per year. The Carbon Debt for each Australian is thus increasing at [$688 billion/24 million =] $27,520 (A$41,000) per head per year for every Australian, at $70,000 (A$105,000) per head per year for 9.816 million under-30 year old Australians [83], and at $146,000 (A$218,000) per head per year for 4.7 million 0-14 year old Australian children [84]. The annual increase in Australia's Carbon Debt of $688 billion will ultimately be borne by these 0-14 year old children and is increasing at the rate of US$146,000 (A$218,000) per head per year.

Young Australians are increasingly aware of how badly they have been betrayed by their profligate elders but when they are cognizant of an inescapable Carbon Debt that is increasing at over A$100,000 per head per year for under-30s they will be out in the streets in their millions. Unlike Conventional Debt, which can be expunged by default, bankruptcy, [running away] or printing money, Carbon Debt is inescapable because, for example, unless sea walls are built at huge expense, arable land and cities will be inundated as the world heads towards a long-term equilibrium sea rise of 25 +/- 12 metres from present conditions of increased CO_2 and warming similar to those of the Pliocene era 4 million years ago [85]. Young Greta Thunberg's "How dare you!" just begins to express the indignation to come over this massive intergenerational injustice [55, 86] that is heading towards a Climate Revolution (peaceful and non-violent one hopes) [84]. For the world as a whole (population 7.6 billion) the inescapable Carbon Debt is increasing at about $12.8 trillion annually or at $1,684 per head per year, noting that the GDP (nominal) per capita for the World is merely $11,355 and that for India it is merely $2,171 [87]. Global warming is a commonly shared imposition and many countries are already failing to match the Carbon Debt imposed on them annually by rich, profligate countries like Australia.

A damaging plus 1.5C of warming will come in the coming decade, and a catastrophic plus 2C temperature rise is now effectively unavoidable [68-71], but we are obliged to do everything we can to make the future "less bad" for future generations. In Australia and other profligately climate criminal countries, decent people will utterly reject the climate criminal climate change deniers and effective climate change deniers at the ballot box. Decent people around the world will subject disproportionately climate criminal people, politicians, parties, collectives, corporations and countries to Boycotts, Divestment and Sanctions (BDS). Decent countries will subject climate criminal people, corporations and countries to legal actions via the International Criminal Court and the International Court of Justice. Time is running out.

References.

[1]. Gideon Polya, "Revised Annual Per Capita Greenhouse Gas Pollution For All Countries – What Is Your Country Doing?", Countercurrents, 6 January, 2016: https://countercurrents.org/polya060116.htm.

[2]. Gideon Polya, "Exposing And Thence Punishing Worst Polluter Nations Via Weighted Annual Per Capita Greenhouse Gas Pollution Scores", Countercurrents, 19 March, 2016: https://countercurrents.org/polya190316.htm.

[3]. IPCC, "Global warming of 1.5 °C. Summary for Policymakers", 8 October 2018: http://report.ipcc.ch/sr15/pdf/sr15_spm_final.pdf.

[4]. Gideon Polya, "IPCC +1.5C avoidance report – effectively too late, but stop coal burning for "less bad" catastrophes", Countercurrents, 12 October 2018: https://countercurrents.org/2018/10/ipcc-1-5c-avoidance-report-effectively-too-late-but-stop-coal-burning-for-less-bad-catastrophes.

[5]. Andrew Glikson, "Inferno: from climate denial to planetary arson", Countercurrents, 8 September 2019: https://countercurrents.org/2019/09/inferno-from-climate-denial-to-planetary-arson.

[6]. Gideon Polya, "Trumpist climate change denials Australian bushfires, fuel reduction, biochar & Carbon Debt", Countercurrents, 10 January 2020: https://countercurrents.org/2020/01/trumpist-climate-change-denial-australian-bushfires-fuel-reduction-biochar-carbon-debt.

[7]. 81 Australian Research Council Laureates, "Laureates Open Letter. An open latter on Australian bushfires and climate: urgent needs for deep cuts in carbon emissions", 2020: https://laureatebushfiresclimate.wordpress.com/.

[8]. Drew T. Shindell, Greg Faluvegi, Dorothy M. Koch, Gavin A. Schmidt, Nadine Unger and Susanne E. Bauer, "Improved Attribution of Climate Forcing to Emissions", Science, 30 October 2009: Vol. 326 no. 5953 pp. 716-718: http://www.sciencemag.org/content/326/5953/716.

[9]. "Gas is not clean energy": https://sites.google.com/site/gasisnotcleanenergy/home.

[10]. "2011 climate change course": https://sites.google.com/site/300orgsite/2011-climate-change-course.

[11]. "Methane Bomb Threat": https://sites.google.com/site/methanebombthreat/.

[12]. Gideon Polya, "Australian commitment to unlimited gas exploitation threatens planet & invites global blowback",

Countercurrents, 8 April 2017: https://countercurrents.org/2017/04/australian-commitment-to-unlimited-natural-gas-exploitation-threatens-planet-invites-global-blowback.

[13]. Climate Council, "Fact Sheet. Climate change and drought June 2018": https://www.climatecouncil.org.au/wp-content/uploads/2018/06/CC_MVSA0146-Fact-Sheet-Drought_V2-FA_High-Res_Single-Pages.pdf.

[14]. Climate Council of Australia, ""This is Not Normal": Climate change and escalating bushfire risk", November 2019: https://www.climatecouncil.org.au/wp-content/uploads/2019/11/bushfire-briefing-paper_18-november.pdf.

[15]. "What are the links between climate change and bushfires? – explainer", Guardian, 11 November 2019: https://www.theguardian.com/australia-news/2019/nov/11/what-are-the-links-between-climate-change-and-bushfires-explainer.

[16]. "Ten ways climate change can make wildfires worse", Phys.org, 12 November 2019: https://phys.org/news/2019-11-ten-ways-climate-wildfires-worse.html.

[17]. A.L. Westerling, H. G. Hidalgo, D. R. Cayan, T. W. Swetnam, Warming and Earlier Spring Increase Western U.S. Forest Wildfire Activity, Science 18 August 2006: Vol. 313. no. 5789, pp. 940 – 943: http://www.sciencemag.org/cgi/content/full/313/5789/940.

[18]. William Ripple et al., "World scientists' warning of a climate emergency", BioScience, 5 November 2019: https://academic.oup.com/bioscience/advance-article/doi/10.1093/biosci/biz088/5610806.

[19]. Gideon Polya, "Extrapolating 11,000 scientists' climate emergency warning to 2030 climate catastrophe", Countercurrents, 14 November 2019: https://countercurrents.org/2019/11/extrapolating-11000-scientists-climate-emergency-warning-to-2030-catastrophe.

[20]. Sarah Martin, Craig Kelly interview: senior government MPs distance themselves after Piers Morgan lashing", Guardian, 7 January 2020: https://www.theguardian.com/australia-news/2020/jan/07/craig-kelly-interview-piers-morgan-calls-mp-disgraceful-for-denying-climate-link-to-bushfires.

[21]. Colin Brinsden, "Greens ramp up climate war as fires burn", Canberra Times", 10 November 2019: https://www.canberratimes.com.au/story/6484064/greens-ramp-up-climate-war-as-fires-burn/?cs=14231.

[22]. Katharine Murphy, "Scott Morrison brings coal to question time: what fresh idiocy is this?", Guardian, 9 February 2017: https://www.theguardian.com/australia-news/2017/feb/09/scott-morrison-brings-coal-to-question-time-what-fresh-idiocy-is-this.

[23]. Amy Remeikis, "Morrison Government to underwrite two new gas power stations", Guardian, 23 December 2019: https://www.theguardian.com/australia-news/2019/dec/23/morrison-government-to-underwrite-two-new-gas-power-stations.
[24]. "Stop air pollution deaths": https://sites.google.com/site/300orgsite/stop-air-pollution-deaths.
[25]. Gideon Polya, "Australia rejects IMF Carbon Tax & preventing 4 million pollution deaths by 2030", Countercurrents, 15 October 2019: https://countercurrents.org/2019/10/australia-rejects-imf-carbon-tax-preventing-4-million-pollution-deaths-by-2030.
[26]. Gideon Polya, "Latest Lancet data imply Adani Australian coal project will kill 1.4 million Indians", Countercurrents, 21 April 2017: https://countercurrents.org/2017/04/latest-lancet-data-imply-adani-australian-coal-project-will-kill-1-4-million-indians.
[27]. Gideon Polya, "War criminal & climate criminal Australian deception at UN General Assembly", Countercurrents, 29 September 2019: https://countercurrents.org/2019/09/war-criminal-climate-criminal-australian-deception-at-un-general-assembly.
[28]. "Climate Genocide": https://sites.google.com/site/climategenocide/.
[29]. Phillip Coorey, "Scott Morrison strikes $2b gas deal with NSW", Australian Financial Review, 31 January 2020: https://www.afr.com/politics/federal/scott-morrison-strikes-2b-gas-deal-with-nsw-20200130-p53wa7."
[30]. Emma Elsworthy, "NSW strikes "landmark" energy deal with Federal Government, Greens MP calls it climate criminality ", ABC News, 31 January 2020: https://www.abc.net.au/news/2020-01-31/nsw-strikes-landmark-energy-deal-with-federal-government/11916314.
[31]. Kathryn McKain et al. "Methane emissions from natural gas infrastructure and use in the urban area of Boston, Massachusetts", PNAS, 112 (7) 1941-1946, February 17, 2015: https://www.pnas.org/content/112/7/1941.abstract).
[32]. Megan Geuss, "Study: US oil and gas methane emissions have been dramatically underestimated", Ars Technika, 23 June 2018: https://arstechnica.com/science/2018/06/study-us-oil-and-gas-methane-emissions-have-been-dramatically-underestimated/.
[33]. Robert Howarth, "Methane emissions and climatic warming risk from hydraulic fracturing and shale gas development: implications for policy", Energy & Emission Control Technologies, 8 October 2015: https://www.eeb.cornell.edu/howarth/publications/f_EECT-61539-perspectives-on-air-emissions-of-methane-and-climatic-warmin_100815_27470.pdf.

[34]. "Gas leakage – systemic gas leakage in the US is about 5.8%",
Gas is not clean energy:
https://sites.google.com/site/gasisnotcleanenergy/gas-leakage.
[35]. Sophie Vorrath, "Australia's new carbon bomb: uncounted coal
seam gas emissions", Renew Economy, 26 October 2016:
https://reneweconomy.com.au/australias-new-carbon-bomb-
uncounted-coal-seam-gas-emissions-14457/.
[36]. Chris Hope, "How high should climate change taxes be?",
Working Paper Series, Judge Business School, University of
Cambridge, 9,
2011: http://www.jbs.cam.ac.uk/media/assets/wp1109.pdf.
[37]. James Hansen, "Climate change in a nutshell: the gathering
storm", Columbia University, 18 December 2018:
http://www.columbia.edu/~jeh1/mailings/2018/20181206_Nutshell.pd
f.
[38]. 300.org: https://sites.google.com/site/300orgsite/300-org.
[39]. "300.org – return atmosphere CO_2 to 300 ppm CO_2":
https://sites.google.com/site/300orgsite/300-org—return-atmosphere-
co2-to-300-ppm.
[40]. US National Oceanic and Atmospheric Administration (NOAA),
"Trends in atmospheric carbon dioxide":
https://www.esrl.noaa.gov/gmd/ccgg/trends/.
[41]. Robert Goodland and Jeff Anfang. "Livestock and climate
change. What if the key actors in climate change are … cows, pigs
and chickens?", World Watch, November/December 2009:
https://pdfs.semanticscholar.org/6704/c7a0777c82357704d82b9ae800
7c1197cb07.pdf?_ga=2.187734888.1597394103.1556059730-
1006954717.1556059730.
[42]. Ewen Hosie, "Australian LNG exports surge to nearly 60 Mt in
2017-2018", Australian Mining, 17 July 2018:
https://www.australianmining.com.au/news/australian-lng-exports-
surge-nearly-60mt-2017-18/.
[43]. Australian Government, "Australian Energy Update 2018":
https://www.energy.gov.au/sites/default/files/australian_energy_updat
e_2018.pdf.
[44]. Lisa Cox, "Australia's emissions reach the highest on record",
Guardian 9 July 2019: https://www.theguardian.com/australia-
news/2019/jul/09/australias-emissions-reach-the-highest-on-record-
driven-by-electricity-sector.
[45]. Lisa Cox, "Australia's carbon emissions highest on record, data
shows", Guardian, 13 December 2018:
https://www.theguardian.com/australia-news/2018/dec/13/australias-
carbon-emissions-highest-on-record-data-shows.

[46]. Penny Timms and Michael Slezak, "Australia's greenhouse gas emissions rise again, according to delayed Federal Government data", ABC News, 6 June 2019: https://www.abc.net.au/news/2019-06-06/australian-emissions-rise-again-delayed-government-data-shows/11184906.

[47]. Michael Slezak, "Australia's greenhouse gas emissions soar in latest figures", Guardian, 4 August 2017: https://www.theguardian.com/australia-news/2017/aug/04/australias-greenhouse-gas-emissions-soar-in-latest-figures.

[48]. Australian Government, Department of the Environment and Energy, "Quarterly update of Australia's national greenhouse gas inventory: March 2019", March 2019: https://www.environment.gov.au/system/files/resources/6686d48f-3f9c-448d-a1b7-7e410fe4f376/files/nggi-quarterly-update-mar-2019.pdf.

[49]. Gideon Polya, "Trumpist climate change denial, Australian bushfires, fuel reduction, biochar & Carbon Debt", Countercurrents, 10 January 2020: https://countercurrents.org/2020/01/trumpist-climate-change-denial-australian-bushfires-fuel-reduction-biochar-carbon-debt.

[50]. Danica Cullinane, "Australia hits oil and gas production record, over a billion barrels", Small Caps, 10 September 2019: https://smallcaps.com.au/australia-oil-gas-production-record-over-billion-barrels/.

[51]. "Australia becomes the largest liquefied natural gas exporter in the world", Canberra Times, 7 January 2020: https://www.canberratimes.com.au/story/6568957/australia-now-the-worlds-largest-natural-gas-exporter/?cs=14231.

[52]. Andrew Blakers and Matthew Stocks,"Some good news for a change: Australia's greenhouse gas emissions are set to fall", The Conversation, 24 October 2019: https://theconversation.com/some-good-news-for-a-change-australias-greenhouse-gas-emissions-are-set-to-fall-125559.

[53]. Rachel Williamson, "Australia is the new queen of LNG exports, but can it last?", Stockhead, 6 January 2020: https://stockhead.com.au/energy/australia-is-the-new-queen-of-lng-exports-but-can-it-last/.

[54]. "Carbon Debt Carbon Credit": https://sites.google.com/site/carbondebtcarboncredit/.

[55]. "Climate Justice & Intergenerational Equity": https://sites.google.com/site/300orgsite/climate-justice.

[56]. "Stop climate crime": https://sites.google.com/site/300orgsite/stop-climate-crime.

[57]. Australian Government, "Quarterly update of Australia's national greenhouse gas inventory for March 2019": https://www.environment.gov.au/climate-change/climate-science-data/greenhouse-gas-measurement/publications/quarterly-update-australias-nggi-mar-2019.

[58]. Chelsea Harvey, "CO2 emissions will claim another record in 2019", Scientific American, 4 December 2019: https://www.scientificamerican.com/article/co2-emissions-will-break-another-record-in-2019/.

[59]. Global Carbon Project, "Global carbon project": https://www.globalcarbonproject.org/carbonbudget/19/highlights.htm.

[60]. "List of countries by greenhouse gas emissions", Wikipedia: https://en.wikipedia.org/wiki/List_of_countries_by_greenhouse_gas_emissions.

[61]. "Fact check: is Queensland clearing land as fast as Brazil?", Fact Check, 16 July 2018: http://www.abc.net.au/news/2017-12-01/fact-check-queensland-land-clearing-brazilian-rainforest/9183596.

[62]. Michael Slezak,"Global deforestation hotspot": 3m hectares of Australian forest to be lost in 15 years", Guardian, 5 March 2018: https://www.theguardian.com/environment/2018/mar/05/global-deforestation-hotspot-3m-hectares-of-australian-forest-to-be-lost-in-15-years.

[63]. Gideon Polya, "Trumpist climate change denial, Australian bushfires, fuel reduction, biochar & Carbon Debt", Countercurrents, 10 January 2020: https://countercurrents.org/2020/01/trumpist-climate-change-denial-australian-bushfires-fuel-reduction-biochar-carbon-debt.

[64]. Reserve Bank of Australia, "Box B. The recent increase in iron ore prices and implications for the Australian economy", August 2019: https://www.rba.gov.au/publications/smp/2019/aug/box-b-the-recent-increase-in-iron-ore-prices-and-implications-for-the-australian-economy.html

[65]. Gideon Polya, "Australia's Huge Coal, Gas & Iron Ore Exports Threaten Planet", Countercurrents, 15 May, 2012: https://www.countercurrents.org/polya150512.htm.

[66]. SSAB: http://www.ssab.com/en/Investor–Media/Sustainability/32/322/.

[67]. Gideon Polya, "Inescapable $200-250 trillion global Carbon Debt increasing at $16 trillion annually", Countercurrents, 27 April 2019: https://countercurrents.org/2019/04/inescapable-200-250-trillion-global-carbon-debt-increasing-by-16-trillion-annually-gideon-polya.

[68]. "Are we doomed?": https://sites.google.com/site/300orgsite/are-we-doomed.

[69]. "Nuclear weapons ban, end poverty and reverse climate change": https://sites.google.com/site/drgideonpolya/nuclear-weapons-ban.
[70]. "Too late to avoid global warming catastrophe": https://sites.google.com/site/300orgsite/too-late-to-avoid-global-warming.
[71]. Stephen Hawking, "Brief Answers to the Big Questions", John Murray, 2018, Chapter 7.
[72]. Simon Holmes à Court, "Scott Morrison is stuck in a time warp – more gas is not the answer", Guardian, 2 February 2020: https://www.theguardian.com/commentisfree/2020/feb/01/scott-morrison-is-stuck-in-a-time-warp-more-gas-is-not-the-answer.
[73]. Gideon Polya, "Pro-gas Obama's EPA-based Plan To Reduce Coal-based Pollution Amounts To Climate Change Inaction", Countercurrents, 7 June, 2014: https://www.countercurrents.org/polya070614.htm.
[74]. Alan Finkel, "National Press Club Address: The orderly transition to the electric planet", Australia's Chief Scientist, 12 February 2020: https://www.chiefscientist.gov.au/news-and-media/national-press-club-address-orderly-transition-electric-planet.
[75]. David Spratt, "A climate reality update at 2020 emergency summit", Climate Code Red, 17 February 2020: http://www.climatecodered.org/2020/02/a-climate-reality-update-at-2020.html?utm_source=feedburner&utm_medium=feed&utm_campaign=Feed:+ClimateCodeRed+(climate+code+red)&m=1.
[76]. National Climate Emergency Summit, "The Safe Climate Declaration", 15 February 2020: https://www.climateemergencysummit.org/declaration/.
[77]. Greta Thunberg, "No one is too small to make a difference", Penguin, 2019.
[78]. Gideon Polya, "Polya's 3 Laws Of Economics Expose Deadly, Dishonest And Terminal Neoliberal Capitalism", Countercurrents, 17 October, 2015: https://www.countercurrents.org/polya171015.htm.
[79]. International Monetary Fund (IMF), "Fiscal Monitor: how to mitigate climate change". Executive Summary", September 2019: file:///C:/Users/Gideon/AppData/Local/Temp/execsum-6.pdf.
[80]. Pope Francis, Encyclical Letter "Laudato si'", 2015: http://w2.vatican.va/content/francesco/en/encyclicals/documents/papa-francesco_20150524_enciclica-laudato-si.html.
[81]. Gideon Polya, "Green Left Pope Francis Demands Climate Action "Without Delay" To Prevent Climate "Catastrophe"", Countercurrents, 10 August, 2015: https://countercurrents.org/polya100815.htm.

[82]. Alison Benjamin, "Stern: climate change a "market failure"",
Guardian, 29 November 2007:
https://www.theguardian.com/environment/2007/nov/29/climatechang
e.carbonemissions.
[83]. "UN Population Division World Population Prospects":
https://population.un.org/wpp/.
[84]. Australian Bureau of Statistics, "Twenty years of population
change", 2019:
https://www.abs.gov.au/ausstats/abs@.nsf/0/1CD2B1952AFC5E7AC
A257298000F2E76.
[85]. Andrew Glikson, "The climate Titanic and the melting
icebergs", Countercurrents, 30 June 2016:
http://www.countercurrents.org/2016/06/30/the-climate-titanic-and-
the-melting-icebergs/.
[86]. "Climate Revolution, Now":
https://sites.google.com/site/300orgsite/climate-revolution.
[87]. "List of countries by GDP (nominal) per capita", Wikipedia:
https://en.wikipedia.org/wiki/List_of_countries_by_GDP_(nominal)_
per_capita.

SECTION E: CARBON PRICE, CARBON TAX, CARBON DEBT, TERMINAL CARBON BUDGET, ECO-SOCIALISM NOT NEOLIBERALISM

"The problem of climate change involves a fundamental failure of markets: those who damage others by emitting greenhouse gases generally do not pay. Climate change is a result of the greatest market failure the world has seen. The evidence on the seriousness of the risks from inaction or delayed action is now overwhelming. We risk damages on a scale larger than the two world wars of the last century. The problem is global and the response must be a collaboration on a global scale." Lord Nicholas Stern, 2007.

"Perhaps it will not matter whether we have 270 ppm or 320 ppm [atmospheric CO_2], but operating well outside the [historic] realm of carbon dioxide concentrations is risky as long as we have not fully understood the relevant feedback mechanisms." Professor Hans Joachim Schellnhuber (director of the Potsdam Institute for Climate Impact Research, Germany), 2008.

"Among the smartest and most promising - not to mention controversial - proposals is "climate debt," the idea that rich countries should pay reparations to poor countries for the climate crisis. In the world of climate-change activism, this marks a dramatic shift in both tone and content… Setting aside the morality of building high-tech fortresses to protect ourselves from a crisis we inflicted on the world, those enclaves and resource wars won't come cheap. And unless we pay our climate debt, and quickly, we may well find ourselves living in a world of climate rage." Naomi Klein, "Climate rage", Rolling Stone, 2009.

"Polya's 3 Laws of Economics are (1) Profit = Price – Cost of Production (COP), (2) deceit about COP strives to a maximum, and (3) zero jobs on a dead planet." Gideon Polya, "Polya's 3 Laws Of Economics Expose Deadly, Dishonest And Terminal Neoliberal Capitalism", Countercurrents, 2015.

""Yet only when the economic and social costs of using up shared environmental resources are recognized with transparency and fully borne by those who incur them, not by other peoples or future generations" [Benedict XVI] can those actions be considered ethical." Pope Francis in Encyclical Letter "Laudato si'", 2018).

The fundamental reason that Humanity and the Biosphere are on the edge of the Climate Crisis precipice is neoliberal greed. This key section of the book attempts in various ways to cost this speciescidal and, ecocidal drive towards omnicide and terracide i.e. to put a price on carbon pollution in US dollars [1-35]. The great paradox of the Enlightenment is that while it opened the door to reason, science and the manifold benefits to Humanity from science-based altruism, it also freed capitalism from the constraints of altruistic memes and unleashed deadly greed backed by high technology weaponry. The mindset that would offer food to a stranger was replaced with the unspeakable perversion enabling the instantaneous killing of 200,000 civilians in the "proof of principle" demonstration atomic bombing of Hiroshima and Nagasaki. Humanity and the Biosphere are existentially threatened by nuclear weapons and man-made climate change [33], this prompting famed scientist Stephen Hawking to issue the following blunt warning: "We foresee great peril if governments and societies do not take action now to render nuclear weapons obsolete and to prevent further climate change" (Stephen Hawking, "Brief Answers to the Big Questions", John Murray, 2018, Chapter 7).

As discussed in Section C, the planet has finite sustainably exploitable resources and thus a limited "human carrying capacity" that is only of the order of 1 billion by 2100 as compared to present demographic trend towards a population of over 10 billion this century. Whatever the asserted production efficiencies of neoliberalism and the "hidden hand of the market", there is a basic moral imperative of "sustainably sharing" limited resources (eco-socialism) in a climate change-damaged world [2, 3]. Put simply, neoliberalism demands maximal freedom for the smart and advantaged to exploit human and natural resources of the word for private profit. In contrast, social humanism (socialism, eco-socialism, human rights-observant communism, the welfare state, Universal Basic Income) seeks to maximize happiness, opportunity and dignity for everyone through evolving and culturally sensitive international and intra-national social contracts (Brian Ellis, "Social humanism. A new metaphysics", Routledge, 2012). Writing on the notable date of 1/1/11, I stated (2011):"The primary New Year's Resolution of all decent people around the World must be to (a) inform, everyone they can about the acute threat to Humanity and (b) urge and apply Sanctions and Boycotts against the climate criminal people, politicians, corporations and countries of the climate genocidal US Alliance. Peace is the only way but silence kills and silence is complicity. Acute concerns for Humanity, the Biosphere, our children, grandchildren, and intergenerational equity oblige us to take such action to help save the Planet before it is too late" [2]. As perceived 14 years ago, there is an ongoing War on Terra involving the US Empire versus Gaia (the living Earth) and infants who will inherit the mess [1].

In 2009 the German WBGU (that advises the German Government on climate and other matters) estimated that the Terminal Carbon Pollution Budget between 2010 and 2050 for a 75% probability of avoiding a catastrophic plus 2C of warning was about 600 billion tonnes of CO_2. I accordingly did a careful, country-by-country analysis of years left at current rates until zero greenhouse gas (GHG) emissions (2011): "An analysis of every country in the World reveals that for a high probability of avoiding a catastrophic 2 degree C temperature rise (EU policy), at current rates of greenhouse gas (GHG) pollution the World on average must achieve zero emissions in less than 20 years, Australia must cease GHG pollution within 5 years, and Bangladesh's low per capita pollution means that at current rates it has 139 years to use up its "fair share" allocation of atmospheric pollution" [5]. In another 2011 article I concluded that both the US and Australia must stop GHG pollution in 5 years [4]. However the notion of a Terminal Carbon Pollution Budget has now become academic because the whole world by 2020 has finally exceeded its 600 billion tonnes CO_2 carbon budget. Thus World Bank experts revised global GHG pollution to be 63.8 billion tonnes CO_2-equivalent per year (Robert Goodland and Jeff Anfang. "Livestock and climate change. What if the key actors in climate change are … cows, pigs and chickens?", World Watch, November/December 2009: https://pdfs.semanticscholar.org/6704/c7a0777c82357704d82b9ae8007c1197cb07.pdf?_ga=2.187734888.1597394103.1556059730-1006954717.1556059730) i.e. 10 x 63.8 = 638 billion tonnes CO_2-equivalent over the last decade [34].

The notion of a Terminal Carbon Pollution Budget raises the crucial issue of Carbon Price, the damage-related price of carbon pollution [5-9]. No less an eminent person than science-trained Pope Francis has asserted that the social and environmental cost of pollution should be "fully borne" by the polluters [10-12]. Eminent economist Lord Stern famously commented on the resolute failure of the polluters to include the "externality" of the social and environmental cost of pollution in the market price of goods: "The problem of climate change involves a fundamental failure of markets: those who damage others by emitting greenhouse gases generally do not pay. Climate change is a result of the greatest market failure the world has seen. The evidence on the seriousness of the risks from inaction or delayed action is now overwhelming. We risk damages on a scale larger than the two world wars of the last century. The problem is global and the response must be a collaboration on a global scale" (Alison Benjamin, "Stern: climate change a "market failure"", Guardian, 29 November 2007: https://www.theguardian.com/environment/2007/nov/29/climatechange.carbonemissions) [13-18].

Polya's 3 Laws of Economics are modelled on the 3 Laws of Thermodynamics of chemistry and physics, to whit (1) the energy of a closed system is constant, (2) entropy (S, disorder, chaos, lack of information content) strives to a maximum, and (3) zero molecular mobility at absolute zero (0 degrees Kelvin or minus 273.15 degrees Centigrade). Polya's analogous 3 Laws of Economics are (1) Profit = Price – Cost of Production (COP), (2) deceit about COP strives to a maximum, and (3) zero jobs on a dead planet [13]. The success of the greed-driven denialists and effective climate change denialists in ignoring the cost of carbon pollution is a deadly and disastrous example of Polya's Second Law of Economics, to whit deception about the Cost of Production strives to a maximum [13-18].

The International Monetary Fund (IMF) has exposed the massive deceit of the neoliberal denialists in stating that while a damage-related Carbon Tax of $75 per ton CO_2 would be an effective way of addressing the climate threat, the present global average Carbon Price is merely only $2 per ton CO_2, a tiny fraction of what is needed for keeping below the 2°C target [18]. Climate economist Dr Chris Hope (of 120-Nobel-Laureate Cambridge University) and climate scientist Professor James Hansen (of NASA and 96-Nobel-Laureate Columbia University) have independently estimated a damage-related Carbon Price of about $200 per tonne CO_2-equivalent (International Monetary Fund (IMF), "Fiscal Monitor: how to mitigate climate change". Executive Summary", September 2019: file:///C:/Users/Gideon/AppData/Local/Temp/execsum-6.pdf ; Chris Hope, "How high should climate change taxes be?", Working Paper Series, Judge Business School, University of Cambridge, 9, 2011: http://www.jbs.cam.ac.uk/media/assets/wp1109.pdf ; James Hansen, "Climate change in a nutshell: the gathering storm", Columbia University, 18 December 2018: http://www.columbia.edu/~jeh1/mailings/2018/20181206_Nuts hell.pdf)[15-17].

While profligate Western industrial countries have acquired massive Carbon Debt since the start of the Industrial Revolution in the mid-18th century, countries devastatingly de-industrialized under colonialism (notably those of South Asia) and countries non-industrialized under colonialism have acquired relatively little Carbon Debt and indeed could be seen to have Carbon Credit in having a disproportionately high share of the Terminal Carbon Pollution Budget. I made estimates of this differential Carbon Debt and Carbon Credit in 2013 [7, 8]. By using up their "fair share" of the Terminal Carbon Pollution Budget, the climate criminal countries of Australia, Canada and the US can be seen to have defaulted on their Carbon Debt and to be stealing from the rest of the World [4, 9]. Indeed in 2011 it was estimated that the US-Canada Keystone XL pipeline to carry Alberta tar sands oil to Texas would ultimately generate enough carbon pollution to mean "game over" for the Planet [6].

For obvious reasons the profligate European industrialized countries choose to ignore historical Carbon Debt and hence grossly under-estimate the Cost of Production. While regular debt can be evaded by default, bankruptcy or printing money, Carbon Debt is inescapable – future generations will have to pay the enormous cost of drawing down atmospheric CO_2 to safe and sustainable levels of about 300 ppm CO_2, and the horrendous cost of sea walls to protect cities, towns and arable land from rising sea levels [21, 34]. My related analysis concluded (2015): "The pro-One Percenter Eurozone and IMF demand timely repayment of Greece's Government Debt but ignore the huge Carbon Debt of major industrialized countries like Germany that measures huge and deadly global environmental damage. The per capita Government Debt plus Carbon Debt is much greater for Germany than for Greece which should insist with indebted Developing Countries that Carbon Debt is also put on the repayment and global action discussion table. The wonderful Palestinian humanitarian Jesus famously stated "Render unto Caesar the things that are Caesar's, and unto God the things that are God's" (Matthew 22, 21, The Holy Bible). For civilized pro-Humanity and pro-Biosphere environmentalists of the 21st century, from animists and theists to Buddhists and secular humanists, this should become "Render unto the Bank the things that are the Bank's, and unto the Planet the things that are the Planet's". The pro-One Percenter Eurozone and the IMF demand the timely repayment of Government Debt by Greece (and thence also Italy, Spain, Portugal, Ireland etc) but the very survival of Humanity and the Biosphere demands proportionate repayment and elimination of a gigantic and steadily increasing Carbon Debt that measures the impact of greenhouse gas (GHG) pollution on the one common atmosphere, ocean and land of all peoples" [14].

Similarly, honest consideration of the "dirty secret" of Carbon Price yields a radically different perception of reality in relation to "dumping" in international trade, as I set out in a 2020 article: "Despite China being Australia's biggest trading partner, US lackey Australia has joined Trump America in its anti-China trade war and military posturing policies. China has been offended by this increasing hostility from US lackey Australia and coincidentally for asserted anti-dumping reasons has applied tariffs to Australian beef and barley with threats of extending this to Australian wine. In actuality there is massive dumping by climate criminal Australia through non-application of a Carbon Price. In 2019 the Australian Gross Domestic Product (GDP) in USD was $1,393 billion for a population of 25.5 million and a per capita GDP $54,627 (albeit quite unevenly distributed). However rich Australia ignores the huge Carbon Debt associated with that productivity. Australia has a revised annual Greenhouses Gas (GHG) pollution of 1,423 Mt CO_2-e (Domestic) and 3,431 Mt CO_2-e (Domestic plus Exported). Assuming a damage-related Carbon Price of $200 per tonne of CO_2-equivalent means that Australia has an inescapable Carbon Debt of $5 trillion that is increasing at $686 billion annually and at $70,000 per head per year for under-30 year old Australians (USD). Thus Australia's annual GDP of $1,393 billion should be inflated to $1,393 billion + $686 billion Carbon Debt = $ 2,079 billion. The ratio of Carbon Debt/GDP = $686 billion/ $1,393 billion = 0.49 i.e. for every $1.0 billion of goods and services generated by Australia each year (or exported to China each year), there is a deliberately hidden but inescapable subsidy of about $0.5 billion to be paid by future generations" [17].

The World as a whole has an inescapable $200-$250 trillion global Carbon Debt that is increasing by $16 trillion annually [15], an horrendous imposition on future generations and a worsening scenario of climate criminality, corporate and state climate terrorism, intergenerational injustice and intergenerational inequity [19-33]. Of course while Carbon Debt has been variously estimated at $200 per tonne CO_2-e [15], any one species of the millions under acute threat in the worsening Biodiversity loss in the present Anthropocene Era is priceless. The present rate of species loss (speciescide) is 100-1,000 times greater than normal. Carbon Debt is a crucial concept in quantitatively assessing the obscene food-for-fuel obscenity of the Biofuel Genocide. Biofuel famine from global food price rises due to competition between cars and people is threatening biofuel genocide of billions of people in the Developing World, noting that the risk avoidance-based Value of a Statistical Life (VSL) is about $7 million per person for Americans. Further, expert biologists have determined that conversion of ecosystems to exploitation for biomass and biofuel is associated with a huge long-term Carbon Debt (Timothy Searchinger and colleagues, "Use of U.S. Croplands for Biofuels Increases Greenhouse Gases Through Emissions from Land-Use Change", Science 29 February 2008, Vol. 319. no. 5867, pp. 1238 – 1240: http://www.sciencemag.org/cgi/content/abstract/1151861, and Joseph Fargione and colleagues, "Land Clearing and the Biofuel Carbon Debt", Science 29 February 2008, Vol. 319. no. 5867, pp. 1235 – 1238: http://www.sciencemag.org/cgi/content/abstract/1152747) [27].

The "no-brainer" and fundamentally just proposition advanced by science-informed economists and ethicists that the social and environmental cost of pollution should be "fully borne" by the polluters is resolutely ignored or obfuscated by the neoliberal politicians running the world. Thus faced with growing public clamour for urgent climate action, some politicians have opted for elaborate Emissions Trading Schemes (ETSs) that add a modest Carbon Price to Business As Usual (BAU). However numerous economists, scientists and science-informed activists have concluded that an ETS approach is flawed, often dishonest and counterproductive, and that a Carbon Tax is needed and not Carbon Trading [25]. Indeed in Australia the Labor Opposition supports unlimited coal and gas exports while supporting renewable energy and having a time-remote policy of "zero net emissions by 2050". However in order to make itself more electorally attractive in mindlessly pro-coal and pro-gas Australia, the Labor leader recently announced that they were discarding any plans for a price on carbon. One despairs.

Fundamental to the general failure to put a proper price on carbon pollution is greed-driven rejection of reality disguised by a coat of mendacious, "we are tackling climate change" rhetoric. As set out in detail in Section F, Australia is among world leaders in 16 areas of climate criminality. However in a recent detailed analysis entitled "Scores Of Huge Realities Resolutely Ignored By Mendacious, US Lackey Mainstream Australia", climate change matters, including Carbon Debt and Carbon Price, were items 34-40 out of 40 topics [16].

The following articles are reproduced as chapters in the book as indicated (asterisked in References):

Chapter 15, "Current Carbon Debt Or Carbon Credit For All Countries: Australia, Canada And US Default On Carbon Debt" [9].
Chapter 16, "Polya's 3 Laws Of Economics Expose Deadly, Dishonest And Terminal Neoliberal Capitalism" [13].
Chapter 17, "Inescapable $200-$250 trillion global carbon debt increasing by $16 trillion annually" [15].
 Chapter 18, "Carbon Debt & Dumping – Climate Criminal Australia Hugely Subsidizes Meat, Grain & Wine Exports To China" [17].
Chapter 19, "Australia rejects IMF Carbon Tax & preventing 4 million pollution deaths by 2030" [18].

References.

[1]. Gideon Polya, "War on Terra: US Empire Versus Gaia And Infants", Countercurrents, 26 February 2006: https://www.countercurrents.org/polya260206.htm.

[2]. Gideon Polya, "New Year's Day 1/1/11 Message: Equal Shares on Our One Planet", Countercurrents, 2 January 2011: https://www.countercurrents.org/polya020111.htm.

[3]. Gideon Polya, "Science And Equality Dictate Eco-Socialism, Climate Socialism And Green Socialism To Save Humanity And Biosphere", Countercurrents, 17 February 2011: https://www.countercurrents.org/polya170211.htm.

[4]. Gideon Polya, "US And Australia Must Stop Greenhouse Gas Pollution In 5 Years", Countercurrents, 6 June 2011: https://www.countercurrents.org/polya060611.htm.

[5]. Gideon Polya, "Country By Country Analysis Of Years Left Until Science-demanded Zero Greenhouse Gas Emissions", Counterurrents, 11 June 2011: https://countercurrents.org/polya110611.htm.

[6]. Gideon Polya, "To Save Planet Stop Canada-US Keystone XL Pipeline", Countercurrents, 23 August 2011: https://www.countercurrents.org/polya230811.htm.

[7]. Gideon Polya, "Country By Country Analysis Of Fossil Fuel Burning-Based Carbon Debt And Carbon Credit", Countercurrents, 24 January 2012: https://sites.google.com/site/climatedebtclimatecredit/analysis-by-country.

[8]. Gideon Polya, "Global Climate Injustice: Massive European Carbon Debt Versus South Asian Carbon Credit", Countercurrents, 29 September, 2013: https://www.countercurrents.org/polya290913.htm.

*[9]. Gideon Polya, "Current Carbon Debt Or Carbon Credit For All Countries: Australia, Canada And US Default On Carbon Debt", Countercurrents, 4 October, 2013: https://www.countercurrents.org/polya041013.htm.

[10]. Gideon Polya, "Pope Francis Demands "Fully Borne" Cost of Pollution (Carbon Price) To Prevent "Millions Of Premature Deaths", Countercurrents, 29 July 2015: https://countercurrents.org/polya290715.htm.

[11]. Gideon Polya, "Western Mainstream Media Censor Green Left Pope Francis' "Laudato Si'" Message For Urgent Action On Climate Change", Countercurrents, 20 August 2015: https://countercurrents.org/polya200815.htm.

[12]. Gideon Polya, "Green Left Pope Francis Demands Climate Action "Without Delay" To Prevent Climate "Catastrophe"",

Countercurrents, 10 August 2015: https://countercurrents.org/polya100815.htm.
*[13]. Gideon Polya, "Polya's 3 Laws Of Economics Expose Deadly, Dishonest And Terminal Neoliberal Capitalism", Countercurrents, 17 October 2015: https://countercurrents.org/polya171015.htm.
[14]. Gideon Polya, "Germany's Per Capita Government Debt Plus Carbon Debt Greatly Exceeds Greece's", Countercurrents, 8 July 2015: https://countercurrents.org/polya080715.htm.
*[15]. Gideon Polya, "Inescapable $200-$250 trillion global carbon debt increasing by $16 trillion annually", Countercurrents, 27 April 2019: https://countercurrents.org/2019/04/inescapable-200-250-trillion-global-carbon-debt-increasing-by-16-trillion-annually-gideon-polya.
[16]. Gideon Polya, "Scores Of Huge Realities Resolutely Ignored By Mendacious, US Lackey Mainstream Australia", Countercurrents, 11 August 2020: https://countercurrents.org/2020/08/scores-of-huge-realities-resolutely-ignored-by-mendacious-us-lackey-mainstream-australia/.
*[17]. Gideon Polya, "Carbon Debt & Dumping – Climate Criminal Australia Hugely Subsidizes Meat, Grain & Wine Exports To China", Countercurrents, 31 August 2020: https://countercurrents.org/2020/08/carbon-debt-dumping-climate-criminal-australia-hugely-subsidizes-meat-grain-wine-exports-to-china/.
*[18]. Gideon Polya, "Australia rejects IMF Carbon Tax & preventing 4 million pollution deaths by 2030", Countercurrents, 15 October 2019: https://countercurrents.org/2019/10/australia-rejects-imf-carbon-tax-preventing-4-million-pollution-deaths-by-2030.
[19]. Gideon Polya (editor), 300.org: https://sites.google.com/site/300orgsite/300-org.
[20]. Gideon Polya (editor), "300.org – return atmosphere CO2 to 300 ppm CO2": https://sites.google.com/site/300orgsite/300-org---return-atmosphere-co2-to-300-ppm.
[21]. Gideon Polya, editor, "Carbon Debt Carbon Credit": https://sites.google.com/site/carbondebtcarboncredit/.
[22]. Gideon Polya, editor, "Climate Revolution Now": https://sites.google.com/site/300orgsite/climate-revolution.
[23]. Gideon Polya, editor, "Divest from fossil fuels": https://sites.google.com/site/300orgsite/divest-from-fossil-fuels.
[24]. Gideon Polya, editor, "Climate Justice & Intergenerational Equity": https://sites.google.com/site/300orgsite/climate-justice.
[25]. Gideon Polya, editor, "Science & economics experts: Carbon Tax needed NOT Carbon Trading":

https://sites.google.com/site/300orgsite/sciennce-economics-experts-carbon-tax-needed-not-carbon-trading/.
[26]. Gideon Polya, editor, "Stop climate crime":
https://sites.google.com/site/300orgsite/stop-climate-crime.
[27]. Gideon Polya, editor, "Biofuel Genocide":
https://sites.google.com/site/biofuelgenocide/.
[28]. Gideon Polya, editor, "Stop air pollution deaths":
https://sites.google.com/site/300orgsite/stop-air-pollution-deaths.
[29].Gideon Polya, editor, "Stop state terrorism":
https://sites.google.com/site/stopstateterrorism/ [state and corporate complicity in worsening climate genocide and 7 million annual air pollution deaths from carbon fuel burning].
[30]. Gideon Polya, editor, "State crime and non-state terrorism":
https://sites.google.com/site/statecrimeandnonstateterrorism/ [state and corporate complicity in worsening climate genocide and 7 million annual air pollution deaths from carbon fuel burning].
[31]. Gideon Polya, editor, "Climate terrorism: 400,000 climate change-related deaths globally annually versus an average of 4 US deaths from political terrorism annually since 9-11":
https://sites.google.com/site/statecrimeandnonstateterrorism/climate-terrorism.
[32]. Gideon Polya, editor, "Carbon terrorism: 3 million US air pollution deaths versus 53 US political terrorism deaths since 9-11 (2001-2015)":
https://sites.google.com/site/statecrimeandnonstateterrorism/carbon-terrorism.
[33]. Gideon Polya, editor, "Nuclear weapons ban, end poverty & reverse climate change":
https://sites.google.com/site/300orgsite/nuclear-weapons-ban.
[34]. Gideon Polya, editor, "Cut Carbon Emissions 80% by 2020":
https://sites.google.com/site/cutcarbonemissions80by2020/2017.
[35]. Gideon Polya, editor, "2011 climate change course":
https://sites.google.com/site/300orgsite/2011-climate-change-course.

CHAPTER 15: CURRENT CARBON DEBT OR CARBON CREDIT FOR ALL COUNTRIES: AUSTRALIA, CANADA AND US DEFAULT ON CARBON DEBT

[First published as Gideon Polya, "Current Carbon Debt Or Carbon Credit For All Countries: Australia, Canada And US Default On Carbon Debt", Countercurrents, 4 October, 2013: https://www.countercurrents.org/polya041013.htm. This article has been overtaken by events as the World has rapidly used up its Carbon Credit or the amount of CO_2 that can be released between 2010 and zero emissions in 2050 for a 75% chance of avoiding a catastrophic plus 2C temperature rise. Thus the Terminal Carbon Pollution Budget was 600 billion tonnes CO_2 in 2010 but has been decreasing at the rate of about 64 billion tonnes CO_2-e emitted per year (this including methane and land use), and thus was finally used in up 2020. This 2013 analysis ignored methane and land use and thus 200 billion tonnes CO_2 of the Terminal Budget was left by 2020 on this basis. However the article gives a useful view of relative historical carbon pollution profligacy]

The World is acutely threatened by a worsening climate crisis but the recent Durban COP17 climate change conference failed to achieve requisite international action. Fossil fuel burning yielding the greenhouse gas (GHG) carbon dioxide (CO_2) is a major component of man-made global warming. In relation to carbon pollution from the burning of fossil fuels, Net Carbon Debt is equal to the Historical Carbon Debt (from fossil fuel burning since the start of the Industrial Revolution in circa 1750) minus the Carbon Credit (the residual carbon pollution from fossil fuel burning permitted between now and zero emissions in 2050). As outlined below and based on fossil fuel burning, Net Carbon Debt (Net Climate Debt) has been estimated for all Carbon Debtor countries (notably the US and Australia) and Net Carbon Credit (Net Climate Credit) has been estimated for all Carbon Creditor countries (notably China and India). This information is crucial for climate justice as the World faces a worsening climate crisis born of GHG profligacy and climate change inaction.

The Historical Carbon Debt (aka Climate Debt) of the World (see "Climate Debt", Wikipedia: http://en.wikipedia.org/wiki/Climate_debt) has been estimated at 12 Gt CO_2 (12 billion tonnes CO_2) in 1751-1900 and 334 Gt CO_2-e for 1901-2008, for a total of 346 Gt CO_2 in the period 1751-2008 (see "Carbon dioxide in the atmosphere", Wikipedia: http://en.wikipedia.org/wiki/Carbon_dioxide_in_Earth%27s_atmosphere). Most of this greenhouse gas (GHG) pollution has occurred in the last half century.

In a 2008 letter to Australian PM Kevin Rudd, NASA's Dr James Hansen provided a breakdown of global responsibility for fossil fuel-derived CO_2 pollution between 1751 and 2006 (see "Letter to PM Kevin Rudd by Dr James Hansen", 2008: http://www.aussmc.org.au/documents/Hansen2008LetterToKevinRudd_000.pdf) that is summarized below as a percentage (%) of the Historical Climate Debt (1751-2006) of 346 Gt CO_2.

Ships/air (4%): 4% of 346 Gt CO_2 = 13.84 G CO_2. This has been allocated proportionately to the other groups as shown below.

India (2.5%) = (0.025 x 346 = 8.65) + (2.5 x 13.84/96 = 0.36) = 9.01 Gt CO_2.

Japan (3.9%) = 13.49 + 0.56 = 14.05 Gt CO_2.

UK (6.0%) = 20.76 + 0.87 = 21.63 Gt CO_2.

Germany (6.6%) = 22.84 + 0.95 = 23.79 Gt CO_2.

Russia (7.4%) = 25.60 + 1.07 = 26.67 Gt CO_2.

China (8.2%) = 28.37 + 1.18 = 29.55 Gt CO_2.

USA (27.5%) = 95.15 + 3.97 = 99.12 Gt CO_2.

Canada-Australia (3.1%) = 10.73 + 0.45 = 11.18 Gt CO_2 -> Canada 5.59 Gt CO_2 & Australia 5.59 Gt CO_2.

Rest of Europe (18.0%) (population 451.2 million) = 62.28 + 2.60 = 64.88 Gt CO_2.

Rest of World (12.8%) (population 3,197.1 million) = 44.29 + 1.85 = 46.14 Gt CO_2.

Post-2010 Carbon Credits (aka Climate Credits) relate to the last amount of GHG pollution the World can sustain before zero emissions in 2050 if it is to avoid a disastrous 2 degree Centigrade temperature rise. In 2009 the WBGU which advises the German Government on climate change estimated that for a 75% chance of avoiding a disastrous 2C temperature rise (EU policy), the World must emit no more than 600 billion tonnes of CO_2 between 2010 and zero emissions in 2050. From this information it was possible to use data for annual per capita GHG pollution (i.e. of CO_2-e; see "List of countries by greenhouse gas emissions per capita": http://en.wikipedia.org/wiki/List_of_countries_ by_greenhouse_gas_emissions_per_capita) to calculate years left to zero emissions for every country in the world (see Gideon Polya, "Shocking analysis by country of years left to zero emissions", Green Blog, 1 August 2011: http://www.green-blog.org/2011/08/01/shocking-analysis-by-country-of-years-left-to-zero-emissions/). This analysis was based on current per capita pollution of CO_2-e (CO_2-equivalent i.e. considering GHGs such as methane and nitrous oxide in addition to CO_2) and was used to estimate Carbon Debt (Climate Debt) in US dollars for most countries (see "Climate Debt, Climate Credit": https://sites.google.com/site/climatedebtclimatecredit/home and https://sites.google.com/site/climatedebtclimatecredit/net-climate-debt).

However a simpler and much more comprehensive analysis
of Carbon Debt (Climate Debt) for all countries of the
World is presented below that reports Carbon Debt in
millions of tonnes of CO_2 from fossil fuel burning alone (and
ignores GHG pollution deriving from land use (agriculture
and forestry), methane, nitrous oxide (N_2O) and other
GHGs).

Net Carbon Debt (aka Net Climate Debt) and Net Carbon Credit
(aka Net Climate Credit) can be estimated from the difference
between Historical Carbon Debt and post-2010 Carbon Credits.
Thus, by way of example, if one accepts that "all men are
created equal", the Carbon Credit for India (population 1,210.2
million out of a total global population of 6,983.2 million) is
600 billion tonnes CO_2 x 1,210.2 million/6,983.2 million =
103.981 billion tonnes CO_2. The Net Carbon Debt for India is
therefore 9.010 billion tonnes CO_2 (Historical Carbon Debt) –
103.981 billion tonnes CO_2 (post-2010 Carbon Credit) = -
94.971 billion tonnes Net Carbon Debt or a Net Carbon Credit
of + 94.971 billion tones CO_2.

Conversely, the Carbon Credit [fair share] for the US
(population 312.8 million out of a total global population of
6,983.2 million) is 600 billion tonnes CO_2 x 312.8
million/6,983.2 million = 26.876 billion tonnes. The Net Carbon
Debt for the US is therefore 99.120 billion tonnes CO_2
(Historical Carbon Debt) – 26.876 billion tonnes CO_2 (post-
2010 Carbon Credit) = 72,244 billion tonnes CO_2 Net Carbon
Debt.

For "Rest of Europe" countries the Net Carbon Debt is 64,880
million tonnes CO_2 /451.2 million people = 143.79 million
tonnes CO_2/person (Historical Carbon Debt) - 600,000 million
tonnes /6,983,2 persons = 85.92 tonnes per person (Carbon
Credit) = 57.49 tonnes CO_2 per person i.e. there is a positive Net
Carbon Debt which is in magnitude 57.87 x100/85.92 = 67.4%
of the 2010-2050 per capita Carbon Credit [and from which one

can calculate Net Carbon Debt for individual "Rest of Europe" countries].

For "Rest of World" countries the Net Carbon Debt is 46,140 million tonnes CO_2/3,197.1 million persons = 14.43 million tonnes CO_2/person (Historical Carbon Debt) – 85.92 tonnes per person (Climate Credit) = -71.49 tonnes per person i.e. there is a positive Net Carbon Credit which is in magnitude 71.49 x100/85.92 = 83.2% of the 2010-2050 per capita Carbon Credit [and from which one can calculate Net Carbon Debt for individual "Rest of World" countries].

Net Carbon Debt (millions of tonnes of CO_2) of Climate Debtor countries (descending order).

United States (72,244), Germany (16,765), United Kingdom (16,277), Russia (14,392), France (3,763), Australia (3,631), Japan (3,069), Italy (3,515), Spain (2,671), Ukraine (2,643), Canada (2,617), Poland (2,204), Romania (1,241), Netherlands (967), Belgium (627), Greece (624), Czech Republic (611), Portugal (611), Hungary (578), Belarus (548), Sweden (548), Austria (487), Switzerland (455), Bulgaria (426), Serbia (412), Denmark (323), Slovakia (315), Finland (313), Norway (289), Ireland (265), Croatia (248), Macedonia (241), Bosnia & Herzegovina (222), Moldova (206), Lithuania (186), Albania (164), Latvia (128), Macedonia (119), Slovenia (119), Estonia (78), Cyprus (46), Montenegro (36), Luxembourg (30), Malta (24), Iceland (18), Jersey (5.7), Andorra (4.9), Isle of Man (4.8), Guernsey (3.6), Greenland (3.3), Faroe Islands (2.8), Liechtenstein (2.1). Monaco (2.1), San Marino (1.9), Gibraltar (1.7), Saint Barthélemy (0.5), Saint Pierre et Miquelon (0.4), Falklands Islands (0.2), Vatican City (0.05).

Net Carbon Credit (millions of tonnes of CO_2) of Climate Creditor countries (ascending order).

Tokelau (0.07), Niue (0.07), Saint Helena Ascension and Trista da Cunha (0.3), Montserrat (0.4), Tuvalu (0.7), Nauru (0.7), Cook Islands (0.8), Wallis & Futuna (1.0), Anguilla (1.1), Palau (1.5), British Virgin Islands (2.0), Saint Martin (2.7), Turks and

Caicos Islands (3.0), Saint Kitts and Nevis (3.7), Northern Mariana Islands (3.9), Marshall Islands (3.9), Cayman Islands (3.9), American Samoa (4.0), Bermuda (4.5), Dominica (5.1), Antigua and Barbuda (6.4), Seychelles (6.5), Saint Vincent and the Grenadines (7.2), Kiribati (7.2), Aruba (7.3), Federated States of Micronesia (7.3), Tonga (7.5), United States Virgin Islands (7.6), Grenada (7.9), Curaçao (10), Guam (11), Saint Lucia (12), São Tomé and Principe (12), Samoa (13), Mayotte (15), French Guiana (16), Vanuatu (17), New Caledonia (18), French Polynesia (20), Barbados (20), Belize (22), Maldives (23), Bahamas (25), Martinique (28), Guadeloupe (29), Brunei (30), Cape Verde (35), Suriname (38), Western Sahara (39), Macau (40), Bhutan (51), Equatorial Guinea (51), Comoros (54), Guyana (56), Réunion (58), Fiji (62), Djibouti (65), Timor-Leste (76), Swaziland (86), Bahrain (88), Mauritius (92), Trinidad and Tobago (94), Guinea-Bissau (101), Gabon (110), Qatar (119), Gambia (127), Botswana (145), Lesotho (157), Namibia (166), Jamaica (193), Mongolia (196), Oman (198), Kuwait (201), Armenia (234), Mauritania (239), Uruguay (241), Panama (243), Liberia (249), Puerto Rico (266), Republic of the Congo (296), Occupied Palestinian Territories (298), Lebanon (304), Costa Rica (308), New Zealand (317), Georgia (319), Central African Republic (321), Turkmenistan (365), Singapore (371), Eritrea (387), Kyrgyzstan (389), Togo (411), Nicaragua (416), Sierra Leone (429), El Salvador (445), Jordan (447), Paraguay (453), Laos (454), Libya (459), Papua New Guinea (501), Hong Kong (508), Tajikistan (544), Israel (558), Honduras (587), United Arab Emirates (591), South Sudan (591), Burundi (613), Benin (651), Azerbaijan (651), Dominican Republic(670), Somalia (683), Haiti (721), Guinea (731), Bolivia (745), Tunisia (763), Rwanda (766), Cuba (804), Chad (806), Zimbabwe (912), Senegal (919), Zambia (933), Malawi (935), Cambodia (958), Ecuador (1,035), Mali (1,038), Guatemala (1,052), Niger (1,125), Burkina Faso (1,125), Kazakhstan (1,188), Chile (1,233), Madagascar (1,349), Cameroon (1,387), Angola (1,402), Sri Lanka (1,476), Syria (1,527), Côte d'Ivoire (1,530), Mozambique (1,648), Taiwan (1,660), Yemen, (1,704), North Korea (1,719), Ghana (1,732), Nepal (1,903), Saudi Arabia (1,940), Uzbekistan (2,002), Malaysia (2,026), Venezuela (2,108), Peru (2,130), Sudan

(2,209), Iraq (2,295), Afghanistan (2,313), Morocco (2,317), Uganda (2,355), Algeria (2,595), Kenya (2,760), Argentina (2,868), Tanzania (3008), Colombia (3,310), Myanmar (3,456), South Korea 3,473), South Africa (3,616), Congo, Democratic Republic (formerly Zaire) (4,844), Thailand (4,970), Turkey (5,270), Iran (5,429), Egypt (5,811), Ethiopia (5,865), Vietnam (6,137), Philippines (6,721), Mexico (8,028), Bangladesh (10,173), Nigeria (11,617), Pakistan (12,737), Brazil (13,753), Indonesia (16,989), China (85,558), India (94,971).

Some major observations arise from this data set

1. Some will argue that it is "unfair" to the major polluters of the European countries to saddle them with the Carbon Debt of previous generations. However these same countries have no problem with continuing to run up huge national debts, with demanding debt repayment by vulnerable countries (as in the current Eurozone crisis) or with crippling Third World countries with massive debt (for a damning account read John Perkins' "Confessions of an Economic Hit Man"). Indeed Germany finally paid its last reparations for World War 1 (1914-1918) in 2010, and 96.5% of the 1751-2008 Historical Carbon Debt considered in this analysis was generated between 1901 and 2008. It should be also noted that this analysis is actually rather unfair to India, China, the "Rest of World" and indeed much of the "Rest of Europe" because it ignores the reality that most of these countries were variously subject in this period of 1751-2006 to colonial subjugation or crippling hegemony by the major polluters, namely the UK, France, Germany, the USA, Russia and Japan.

2. This analysis is only concerned with available data on Carbon Debt arising from the burning of fossil fuels and ignores Carbon Debt from greenhouse gas (GHG) production from deforestation and methanogenic livestock production. Using the data that methane (CH_4) has 72 times the global warming potential (GWP) of carbon dioxide (CO_2) on a 20 year time frame (as compared to 25 times worse on a 100 year time frame) World Bank analysts have re-assessed annual global GHG pollution as 50% bigger than hitherto thought with methanogenic livestock

production contributing over 51% of the bigger figure (see Robert Goodland and Jeff Anfang, "Livestock and climate change. What if the key actors in climate change are … cows, pigs and chickens?", World Watch, November/December 2009: http://www.worldwatch.org/files/pdf/Livestock%20and%20Climate%20Change.pdf). However this re-assessment in turn needs further re-assessment because Dr Drew Shindell and colleagues at NASA have shown that CH_4 is actually 105 times worse than CO_2 as a GHG on a 20 year time frame when aerosol impacts are taken into account (see Drew T. Shindell, Greg Faluvegi, Dorothy M. Koch, Gavin A. Schmidt, Nadine Unger and Susanne E. Bauer, "Improved Attribution of Climate Forcing to Emissions", Science, 30 October 2009: Vol. 326 no. 5953 pp. 716-718: http://www.sciencemag.org/content/326/5953/716 and Shindell et al. (2009), Fig.2: http://www.sciencemag.org/content/326/5953/716.figures-only).

3. The set of all the Carbon Debtor (Climate Debtor) countries includes all the European countries and Japan. The set of all the Carbon Creditor (Climate Credit) countries includes all the non-European countries, excluding Japan, as well as the European colonies New Zealand and Israel (that could arguably be put in the "Rest of Europe" category).

4. One can convert the Carbon Debt or Carbon Credit from units of "million tonnes of CO_2" simply by multiplying by whatever carbon price you desire in, say, US dollars. Thus a genuine Carbon Price of US$100 per tonne of CO_2 would demand a transition from coal- and gas-burning for electric power. Using this value the Carbon Debt of the US would be 72, 244 million tonnes CO_2 x $100/ tonne CO_2 = $7,200, 244 million = $7.2 trillion. Likewise the Carbon Credit of China and India would be $8.6 trillion and $9.5 trillion, respectively.

5. The US is steadily increasing its current $15.3 trillion national debt (see: http://www.usdebtclock.org/) and is devaluing this debt by printing money. Conversely, the US has a 72,244 million tonne CO_2 ($7.2 trillion @ $100 per tonne CO_2)

Net Carbon Debt, but is steadily increasing this debt at the rate of 6,946 million tonnes CO_2-e per year (2008) i.e. the US Carbon Debt is increasing at about 10% per year. The US under Obama shows no indication of reducing its GHG pollution profligacy. Obama's declining to approve the current Keystone XL pipeline proposal to carry oil from Canadian tar sands to Texas may only be a temporary reprieve to keep pro-environmentalists on side in a Presidential election year. According to leading US climate scientist Dr James Hansen, exploitation of the Canadian tar sands will mean "game over" for the Planet.

6. Australia is the worst annual per capita GHG polluter of the Carbon Debtor countries but shows no indication of changing its disproportionate GHG pollution. Australia's Domestic plus exported GHG pollution was 1,077 million tonnes CO_2-e in 2000 but under the Australian Labor Government's dishonest "Carbon Tax-ETS Scheme" this is estimated to increase to 1,799 million tonnes by 2020 (a 1.7-fold increase) and to 4,490 million tonnes CO_2-e by 2050 (a 4.2-fold increase). In vain top US, UK, German and Australian climate scientists and biologists demand that global GHG pollution must be rapidly reduced to [net] zero emissions in about 2050 and that the atmospheric CO_2 concentration must return to about 300 parts per million (ppm) from the current damaging 394 ppm (increasing at 2.4 ppm per year) (see "300,org – return atmosphere CO_2 to 300 ppm": https://sites.google.com/site/300orgsite/300-org---return-atmosphere-co2-to-300-ppm). Australia's Net Carbon Debt (3,631 million tonnes CO_2) is currently increasing at about 1,415 million tonnes CO_2-e per year i.e. at 39% per year.

7. "Annual per capita greenhouse gas (GHG) pollution" in units of "tonnes CO_2 -equivalent per person per year" (2005-2008 data) is 0.9 (Bangladesh), 0.9 (Pakistan), 2.2 (India), less than 3 (many African and Island countries), 3.2 (the Developing World), 5.5 (China), 6.7 (the World), 11 (Europe), 16 (the Developed World), 27 (the US) and 30 (Australia; 54 if Australia's huge Exported CO_2 pollution is included, 64 being the 2010 figure). The major Climate Creditor countries are

vastly lower in per capita GHG pollution than Australia (see "Climate Genocide": https://sites.google.com/site/climategenocide/). Thus Australia's current annual per capita of 64 tonnes CO_2-e per person per year (with Exported GHG included) is 71 times that of Bangladesh.

8. The Carbon Debtors are stealing from the poor Carbon Creditors that are increasingly threatened by the worsening climate crisis. The Carbon Debtors (Climate Debtors) should be held to account through public advocacy, boycotts, sanctions, green tariffs, International Court of Justice (ICJ) litigations and International Criminal Court (ICC) prosecutions applied against Climate Debtor countries by Climate Creditor countries, notably the numerous Island States and major mega-delta countries such as Myanmar, Thailand, Cambodia, Vietnam, China, Egypt, Nigeria, India, Pakistan and Bangladesh. The climate criminals and Carbon Debtors (Climate Debtors) must be brought to account before it is too late.

Outstanding Canadian human rights activist and writer Naomi Klein has written thus about climate rage in response to non-recognition and non-payment of Climate Debt (2009): "Among the smartest and most promising - not to mention controversial - proposals is "climate debt", the idea that rich countries should pay reparations to poor countries for the climate crisis. In the world of climate-change activism, this marks a dramatic shift in both tone and content… Setting aside the morality of building high-tech fortresses to protect ourselves from a crisis we inflicted on the world, those enclaves and resource wars won't come cheap. And unless we pay our climate debt, and quickly, we may well find ourselves living in a world of climate rage" (see Naomi Klein, "Climate rage", Rolling Stone, November 2009: http://www.naomiklein.org/articles/2009/11/climate-rage). The fossil fuel-based Carbon Debt analysis given above provides a quantitative basis for such reparations payments, and should be used by Island States, mega-delta countries and other threatened Climate Creditor countries to force urgently needed climate change action.

CHAPTER 16: POLYA'S 3 LAWS OF ECONOMICS EXPOSE DEADLY, DISHONEST AND TERMINAL NEOLIBERAL CAPITALISM

[First published as Gideon Polya, "Polya's 3 Laws Of Economics Expose Deadly, Dishonest And Terminal Neoliberal Capitalism", Countercurrents, 17 October 2015: https://countercurrents.org/polya171015.htm.]

Polya's 3 Laws of Economics mirror the 3 Laws of Thermodynamics and are (1) Price minus COP (Cost of Production) equals profit; (2) Deception about COP strives to a maximum; and (3) No work, price or profit on a dead planet. These fundamental laws help expose the failure of neoliberal capitalism in relation to wealth inequality, massive tax evasion by multinational corporations, and horrendous avoidable deaths from poverty and pollution culminating in general ecocide, speciescide, climate genocide, omnicide and terracide.

Fundamental to the physical sciences and the explosion of the Industrial Revolution in the 19th century are the 3 Laws of Thermodynamics that can be simply stated as follows:

The First Law of Thermodynamics states that the energy of a closed system is constant i.e. we can have different amounts of energy in different forms in different parts of the system (a human body, the planet, the universe), but the total energy remains a constant.

The Second Law of Thermodynamics states that the entropy (disorder) of the world strives to a maximum i.e. the world inexorably tends to randomness, chaos, disorder and minimum information content. We are familiar with so many examples of this e.g. (a) the drops of ink in a glass of water disperse until the ink particles are uniformly distributed; (b) despite inbuilt cellular biochemical repair and replication systems, we inevitably age and die; and (c) the accelerating expansion of the universe.

The Third Law of Thermodynamics states that the entropy (disorder) of a pure crystal of a pure chemical at the absolute zero temperature of zero (0) degrees Kelvin (minus 273.15 degrees Centigrade) is zero i.e. in a universe full of motion and increasing disorder there is a boundary state of zero disorder.

The discovery of the 3 Laws of Thermodynamics enabled the massive industrialization of the last few centuries, and set the world on an increasingly high technology economic path culminating in the present sustainability crisis in which the species extinction rate is 100-1,000 times bigger than normal, and there is a worsening climate emergency in which a globally politically agreed plus 2 degrees C temperature rise is both catastrophic and inevitable.

Our presently destructive economic system can be described by laws akin to the 3 Laws of Thermodynamics as set out below:

Polya's First Law of Economics states that the Price (P) minus Cost of Production (COP) equals profit (p). Just as the First Law of Thermodynamics allows one to determine the magnitude in energy units (e.g. joules) of the different components of an energy inventory (e.g. light energy, heat energy, kinetic energy) in a particular physical system, so the First Law of Economics allows one to determine the magnitude in wealth units (e.g. USD, US dollars) of P, COP and p in a particular economic system (e.g. a slave, a worker, a company, a nation, or indeed the global economy) [p = P − COP].

Polya's Second Law of Economics states that deception about Cost of Production (COP) strives to a maximum. Just as the Second Law of Thermodynamics states that disorder and hence lack of information inevitably increase, so the Second Law of Economics declares that disinformation and deception strive to a maximum in an economy based on a market of

Buyers and Sellers, Winners and Losers. The deception may be mild as with the margin and enthusiasm of a street vendor, or egregious as in the case of multinational corporations avoiding paying tax on their profits. The world is presently dominated by neoliberal capitalism that seeks to maximize the freedom of the rich and advantaged to exploit human and natural resources [for private profit], with the present consequence that the richest One Percent has about 50% of the wealth and 17 million people die avoidably each year from deprivation.

Polya's Third Law of Economics simply states that there is no work, price or profit on a dead planet. Just as the Third Law of Thermodynamics describes a boundary condition of zero motion and zero entropy (zero disorder) at Absolute Zero (zero degrees Kelvin), so the Third Law of Economics describes the terminal state of an Earth in which the Biosphere has been destroyed as a result of unconstrained neoliberal capitalist greed. As the Greens put it: "No jobs on a dead planet".

As well as making a major contribution to physics and thermodynamics, Sir William Thomson (Lord Kelvin, after whom the Kelvin temperature scale is named) famously urged us to express our observations in quantitative terms: "I often say that when you can measure what you are speaking about, and express it in numbers, you know something about it; but when you cannot measure it, when you cannot express it in numbers, your knowledge is of a meagre and unsatisfactory kind; it may be the beginning of knowledge, but you have scarcely, in your thoughts, advanced to the stage of science, whatever the matter may be".

We can take Lord Kelvin's advice and apply some important contemporary economic numbers to Polya's 3 Laws of Economics as outlined below.

1. Multinational corporation tax evasion. In worst case scenario, if the Cost of Production (COP) in a country can be manipulated via a subsidiary in another country such that it equals the Price (P) for Goods and Services, then P = COP and from the First Law of Economics, P – COP = p = 0 i.e. the taxable profit (p) is reduced to zero. It has been estimated that $6 trillion of global wealth is unaccounted for and presumed to be deposited in tax havens like the Cayman Islands, Bermuda, Singapore, Switzerland, Liechtenstein, the Netherlands, Ireland and the UK Channel Islands, and that over 50% of US corporate profit made abroad is now made in countries considered to be tax havens [1, 2]. Thomas Piketty in his book "Capital in the Twenty-First Century" (Figure 12.6, page 466) [3]), has estimated that in the 21st century the foreign assets of rich countries held in tax havens has been about 8% of the world [annual] output of about $78 trillion i.e. $6.2 trillion.

2. World wealth inequality and 17 million avoidable deaths from deprivation annually. The top One Percent have about 50% of the world's wealth and the 85 richest individuals own the same wealth as the bottom 50% [4]. Poverty kills - there is a largely ignored Global Avoidable Mortality Holocaust in which currently 17 million people die avoidably each year from deprivation and deprivation-exacerbated disease in the Developing World (minus China) [5]. This is happening on Spaceship Earth with the flight deck under the control of the 10% richest who have about 90% of the wealth of the World and who in turn are controlled by One Percenters who own about half the wealth of the World. An annual global wealth tax of about 4% would yield US$16 trillion annually and enable raising all countries to annual per capita incomes equivalent to the $6,000 per person per year of China and Cuba, countries for which annual avoidable mortality is zero (0) [6]. This is a feasible option for stopping the Global Avoidable Mortality Holocaust. Indeed a progressive annual wealth tax ranging up to 10% for the richest has been proposed for democracy and economic sustainability reasons by French economist Professor Thomas Piketty in his important book "Capital in the Twenty-First Century" [3, 7].

The world annual output (Price, P) is $78 trillion [3]. If we assume that Third Worlders have the same value as Americans (remember the "all men are created equal" of the American Declaration of Independence), and that the US risk -avoidance-based Value of a Statistical Life (VOSL) is about $7 million per person, then the Cost of Production (COP) in terms of avoidable mortality will be 17 million persons x $7 million per person = $119 trillion. According to the First Law of Economics, P – COP = p = $78 trillion - $119 trillion = - $41 trillion i.e. the world "business" is running at a huge loss [and therefore doesn't have to pay tax].

3. World coal burning and 7 million air pollution deaths annually. World production of coal was 7, 823 million tonnes in 2013 [8]. Assuming the current price of thermal coal of $63 per tonne, the Price (P) = 7,823 million tonnes x $63 per tonne = $0.49 trillion The World Health Organization (WHO) states: "After analysing the risk factors and taking into account revisions in methodology, WHO estimates indoor air pollution was linked to 4.3 million deaths in 2012 in households cooking over coal, wood and biomass stoves. The new estimate is explained by better information about pollution exposures among the estimated 2.9 billion people living in homes using wood, coal or dung as their primary cooking fuel, as well as evidence about air pollution's role in the development of cardiovascular and respiratory diseases, and cancers. In the case of outdoor air pollution, WHO estimates there were 3.7 million deaths in 2012 from urban and rural sources worldwide. Many people are exposed to both indoor and outdoor air pollution. Due to this overlap, mortality attributed to the two sources cannot simply be added together, hence the total estimate of around 7 million deaths in 2012" [9].

If we assume that coal burning is responsible for 25% of the 7 million annual air pollution deaths [10], and that the risk - avoidance-based Value of a Statistical Life (VOSL) is about $7 million per person (see example 2 above), then the Cost of Production (COP) = 0.25 x 7 million x $7 million per person = $12.25 trillion. According to the First Law of Economics, P – COP = p = $0.49 trillion - $12.25 trillion = - $11.76 trillion i.e. properly analyzed, the world coal "business" is running at a huge loss.

4. World greenhouse gas (GHG) production and Carbon Debt. The world annual output (Price of goods and services, P) is $78 trillion [3] and is associated with a greenhouse gas (GHG) production of 64 billion tonnes CO_2-e per year (taking the livestock contribution into proper account) [11]. Climate change economist Dr Chris Hope from 90-Nobel-Laureate University of Cambridge has estimated a damage-related Carbon Price of US$150-$250 per tonne CO_2-e (CO_2-equivalent) depending on location [12]. Based on a Carbon Price of $200 per tonne CO_2-e, the World now has a Carbon Debt from historical CO_2 pollution (mostly from European countries) of $540 trillion that is increasing at 64 billion tonnes CO_2-e per year x $200 per tonne CO_2-e = $12.8 trillion each year [13]. According to the First Law of Economics, P – COP = p = $78 trillion - $12.8 trillion = $65.2 trillion i.e. the world's carbon-based economy involves an annual subsidy to GHG polluters of about $13 trillion, that adds to an inescapable Carbon Debt for future generations of $540 trillion.

5. Coal burning and the cost of removing CO_2 from the atmosphere. It has been estimated that for every $1 for coal it will cost, depending upon location and technology, about $1 to $14 to remove from the atmosphere the CO_2 generated from the subsequent burning of the coal [14]. Noting our First Law of Economics, P - COP = p, this means that the profit, p, from every $1 gained from selling coal for combustion will actually be $0 to minus $13 i.e. coal burning is presently still horrendously subsidized at the expense of future generations.

6. Mainstream lying and the Second Law of Economics. The surprising findings outlined above are of course resolutely hidden from public perception in the Western democracies that have become Murdochracies, Plutocracies, Lobbyocracies, Corporatocracies and Dollarocracies in which Big Money buys people, politicians, parties, policies, public perception of reality and political power – testament to the Second Law of Economics that declares that deception about the Cost of Production (COP) strives to a maximum.

7. Terracide and the Third Law of Economics. The World will exceed its Terminal Carbon Budget for a 75% probability of avoiding plus 2 degrees C in about 3 years, and the world's "goal" of a plus 2 degrees C temperature rise is disastrous for Humanity and the Biosphere, yielding at equilibrium sea levels "at least 6 to 8 metres higher" according Dr James Hansen of NASA and 101-Nobel-Laureate Columbia University [16]. Both Dr James Lovelock FRS (Gaia hypothesis) and Professor Kevin Anderson (Deputy Director, Tyndall Centre for Climate Change Research, University of Manchester, UK) have estimated that only about 0.5 billion people may survive this century due to unaddressed, man-made global warming, this corresponding to a climate genocide involving about 10 billion people dying avoidably this century [17]. The Third Law of Economics states there is no work, price or profit on a dead planet.

Conclusions.

Polya's 3 Laws of Economics mirror the 3 Laws of Thermodynamics and are (1) Price minus COP (Cost of Production) equals profit; (2) Deception about COP strives to a maximum; and (3) No work, price or profit on a dead planet. These 3 Laws of Economics are useful in exposing and assessing massive deceptions in neoliberal capitalism involving the dangerously uncosted exploitation of human and physical resources for private profit.

As exampled above, application of the First Law of Economics enables quantitation of huge hidden subsidies from the remorselesss deception described by the Second Law of Economics. The Third Law of Economics presents a terminal boundary condition of the human condition in which irreversible greenhouse gas (GHG) pollution, ecocide, speciescide and climate genocide have led to omnicide and terracide, the destruction of [most of] Humanity and the Biosphere.

The Paris Climate Change Conference has failed before it has begun – a plus 2C temperature rise is catastrophic and is now inevitable. Humanity is faced with an urgent new task to do everything we can to make the future "less bad" [18]. Dishonest, destructive and deadly neoliberal capitalism that allows the vastly rich to get even vastly richer at the expense of Humanity and the Biosphere must be urgently replaced by an evolving social humanism involving truthful, science-based risk management, and the maximization of human dignity, happiness and opportunity [19].

Peace is the only way but silence kills and silence is complicity – the worsening deception by the ever-self-enriching One Percenters and complicit Mainstream media, politician and academic presstitutes means that 7 million people die annually from air pollution and 17 million people die avoidably each year from deprivation. Decent people must (a) inform everyone they can, and (b) apply Boycotts, Divestment and Sanctions (BDS) against all people, politicians, parties, countries, companies and corporations disproportionately involved in the terracidal rape of the Earth - armed with the 3 Laws of Economics they must ensure Climate Revolution now [20].

References.

[1]. Gabriel Zucman, "The Hidden Wealth of Nations. The scourge of tax havens", University of Chicago Press, 2015.
[2]. Josh Zambrun, "Is wealth inequality hidden in tax havens? Piketty's co-author thinks so", Wall Street Journal, 28 September 2015: http://blogs.wsj.com/economics/2015/09/28/is-wealth-inequality-hidden-in-tax-havens-pikettys-co-author-thinks-so/.
[3]. Thomas Piketty, "Capital in the Twenty-First Century", Belknap, 2014.
[4]. Larry Elliott and Ed Pilkington, "New Oxfam Report says half of global wealth held by the 1% ", The Guardian, 19 January 2015: http://www.theguardian.com/business/2015/jan/19/global-wealth-oxfam-inequality-davos-economic-summit-switzerland
[5]. Gideon Polya, "Body Count. Global avoidable mortality since 1950" that includes an avoidable mortality-related history of every country since Neolithic times and is now available for free perusal on the web: http://globalbodycount.blogspot.com/.
[6]. Gideon Polya, "4 % Annual Global Wealth Tax To Stop The 17 Million Deaths Annually", Countercurrents, 27 June, 2014: http://www.countercurrents.org/polya270614.htm.
[7]. Gideon Polya, "Key Book Review: "Capital In The Twenty-First Century" By Thomas Piketty", Countercurrents, 01 July, 2014: http://www.countercurrents.org/polya010714.htm.
[8]. World Coal Association, "Coal statistics": http://www.worldcoal.org/resources/coal-statistics/.
[9]. World Health Organization (WHO), "7 million premature deaths annually linked to air pollution": http://www.who.int/mediacentre/news/releases/2014/air-pollution/en/.
[10]. "Stop air pollution deaths": https://sites.google.com/site/300orgsite/stop-air-pollution-deaths.
[11]. Robert Goodland and Jeff Anfang. "Livestock and climate change. What if the key actors in climate change are … cows, pigs and chickens?", World Watch, November/December 2009: http://www.worldwatch.org/files/pdf/Livestock%20and%20Climate%20Change.pdf.
[12]. Chris Hope, "How high should climate change taxes be?", Working Paper Series, Judge Business School, University of Cambridge, 9.2011: http://www.jbs.cam.ac.uk/fileadmin/user_upload/research/workingpapers/wp1109.pdf.
[13]. "Carbon Debt Carbon Credit": https://sites.google.com/site/carbondebtcarboncredit/.

[14]. Gideon Polya, "Intergenerational Theft – For Every $1 For Coal Today Future Generations Will Pay $1-$14 To Sequester CO_2", Countercurrents, 8 April, 2015: http://www.countercurrents.org/polya080415.htm.
[15]. Gideon Polya, "G7 Pledge Of Zero Emissions By 2100 Masks Worsening Climate Emergency And Need For Urgent Action", Countercurrents, 10 June, 2015: http://www.countercurrents.org/polya100615.htm.
[16]. Dr James Hansen interviewed by ABC RN's Fran Kelly, "Two degrees of global warming is not "safe": Hansen", ABC RN Breakfast, 5 May 2015: http://www.abc.net.au/radionational/programs/breakfast/two-degrees-of-global-warming-is-not-safe/6444698.
[17]. "Climate Genocide": https://sites.google.com/site/climategenocide/.
[18]. "Too late to avoid global warming catastrophe": https://sites.google.com/site/300orgsite/too-late-to-avoid-global-warming.
[19]. Gideon Polya, "Book Review: "Social Humanism. A New Metaphysics" By Brian Ellis - Last Chance To Save Planet?", Countercurrents, 19 August, 2012: http://www.countercurrents.org/polya190812.htm.
[20]. "Climate Revolution Now": https://sites.google.com/site/300orgsite/climate-revolution.

CHAPTER 17: INESCAPABLE $200-$250 TRILLION GLOBAL CARBON DEBT INCREASING BY $16 TRILLION ANNUALLY

[First published as Gideon Polya, "Inescapable $200-$250 trillion global carbon debt increasing by $16 trillion annually", Countercurrents, 27 April 2019: https://countercurrents.org/2019/04/inescapable-200-250-trillion-global-carbon-debt-increasing-by-16-trillion-annually-gideon-polya.]

Carbon Debt is simply the damage-related cost of greenhouse gas (GHG) pollution that if not addressed now will inescapably have to be paid by future generations. However GHG emissions continue to rise inexorably and there is no global program to draw down CO_2 and other GHGs from the atmosphere. While young people are now vociferously demanding massive climate action, inescapable global Carbon Debt is $200- $250 trillion and increasing by $16 trillion each year.

Unlike Conventional Debt that can be variously expunged by bankruptcy, printing money or default, Carbon Debt is inescapable – thus, for example, national commitments to GHG pollution reduction made to the 2015 Paris Climate Conference amount to a temperature rise of over 3 degrees Centigrade (3C), and unless huge sea walls are built Netherlands-style, coastal cities of the world housing hundreds of millions of people will be submerged by rising sea levels (notably in Asia), mega-delta agricultural lands vital for feeding Humanity will be subject to inundation and salinization, and low-lying Island States will cease to exist [1-3].

While outright, anti-science climate change denialism is politically entrenched in climate criminal Trump America and its climate criminal lackey Australia, most governments around the world are politically committed to effective climate change denialism through climate change inaction. That climate change inaction is most clearly quantitated in terms of Carbon Debt, but the very term has been white-washed out of public perception by US owned or subverted Mainstream media. Thus the Australian ABC (the taxpayer-funded Australian equivalent of the UK BBC) is self-assertedly "progressive" but a Search of the ABC for the term "Climate Debt" reveals zero (0) reportage. A Search of the self-assertedly "ethical" UK BBC for the term "Climate Debt" yields 9 items with none later than 2009, defining the term or quantifying global or national Carbon Debt.

Explanations for this extraordinary Mainstream media lying by omission over Carbon Debt can be variously advanced, ranging from entrenched mendacity by US- and corporate- subverted media to cognitive dissonance in the face of a worsening climate emergency. However I am confident in predicting that if governments do not take action on the world's massive Carbon Debt then intergenerational justice action by the utterly betrayed and robbed young people of the world will make the present Extinction Rebellion climate demonstrations in London look like a proverbial Teddy Bear's Picnic. A young people-led Climate Revolution (non-violent one hopes) is coming [4-8], but the horrendous numbers elicited in this analysis tell us that it is probably much too late.

Set out below are various approaches to estimating the world's horrendous Carbon Debt and the numbingly gigantic annual increase in Carbon Debt. In considering these huge numbers, one must note that the global wealth was $317 trillion in 2018, an estimated $333 trillion in 2019, $350 trillion in 2020 and [a predicted] $399 in 2023 [9]. The aggregate world wealth in the global carbon economy is increasing at about $14 trillion per year [9], but the top 1% (the One Percenters who own half the wealth of the world) have secured 82% of this wealth increase, while the poorest 3.7 billion of the world have secured zero [10].

(1). Requisite atmospheric CO_2 draw-down from the present 407 ppm CO_2 over the next decade will cost up to \$12.7 to \$24.9 trillion annually for various 342 to 280 ppm CO_2 goals.

Professor James Hansen (of 96 Nobel Laureate Columbia University): "One ppm of CO_2 is 2.12 billion tons of carbon or about 7.77 billion tons of CO_2. Recently Keith et al. (2018) achieved a cost breakthrough in carbon capture, demonstrated with a pilot plant in Canada. Cost of carbon capture, not including the cost of transportation and storage of the CO_2, is \$113-232 per ton of CO_2. Thus the cost of extracting 1 ppm of CO_2 from the atmosphere is \$878-1803 billion. In other words, the cost, in a single year, of closing the gap between reality and the IPCC scenario that limits climate change to $+1.5°C$ is already about \$1 trillion. And that is without the cost of transporting and storing the CO_2, or consideration of whether there will be citizen objection to that transportation and storage. This annual cost will rise rapidly, unless there is a rapid slowdown in carbon emissions… cost of CO_2 storage… has been estimated as \$10-20/tCO_2" [11].

Taking Professor Hansen's data, and including his estimates of the cost of transport and storage of CO_2, indicates that this "best so far" cost of atmospheric CO_2 draw-down is \$123-252/tCO_2. Accordingly, the upper estimate of the cost of reducing the atmospheric CO_2 by 1 ppm = 1 ppm x 7.77 Gton CO_2/ppm x \$252/ton CO_2 = \$1,958 billion. Using this upper estimate one can calculate the cost of CO_2 draw-down to various targets.

(a) James Hansen has estimated a required CO_2 draw-down to 342 ppm CO_2 if the Earth's energy imbalance is 1 W/m2 [11]. The upper estimate of the Carbon Debt for a transition from the present 407 ppm CO_2 to 342 ppm CO_2 is 65 ppm CO_2 x \$1,958 billion per ppm CO_2 = \$127.3 trillion or \$12.7 trillion per year if we did this over 1 decade.

(b) The World's top coral scientists have concluded that world coral reefs started dying when the atmospheric CO_2 reached 320 ppm CO_2 and that a CO_2 draw-down to this level is required for sustainability of the world's coral reefs [12, 13]. The upper estimate of the Carbon Debt for a transition from the present 407 ppm CO_2 to 320 ppm CO_2 is 87 ppm CO_2 x $1,958 billion per ppm CO_2 = $170.3 trillion or $17.0 trillion per year if we did this over 1 decade.

(c) Many scientists and science-informed activists demand a reduction of atmospheric CO_2 to a safe and sustainable level of about 300 ppm CO_2 [14, 15]. The upper estimate of the Carbon Debt for a transition from the present 407 ppm CO_2 to 300 ppm CO_2 is 107 ppm CO_2 x $1,958 billion per ppm CO_2 = $209.5 trillion or $21.0 trillion per year if we did this over 1 decade.

(d) Professor Hans Joachim Schellnhuber (director of the Potsdam Institute for Climate Impact Research, Germany) (2008): "Perhaps it will not matter whether we have 270 ppm or 320 ppm [CO_2], but operating well outside the [historic] realm of carbon dioxide concentrations is risky as long as we have not fully understood the relevant feedback mechanisms" [280 ppm is the pre-industrial atmospheric CO_2 concentration] [16]. The upper estimate of the Carbon Debt for a transition from the present 407 ppm CO_2 to 280 ppm CO_2 is 127 ppm CO_2 x $1,958 billion per ppm CO_2 = $248.7.0 trillion or $24.9 trillion per year if we did this over 1 decade.

(2). Upper estimate of $16 trillion for annual increase in Carbon Debt taking all greenhouse gases (GHGs) into account.

Unfortunately there is presently no significant global effort to draw down atmospheric CO_2, and indeed global greenhouse gas (GHG) pollution is remorselessly increasing, and with it a relentlessly increasing Carbon Debt as set out below. In the last year the atmospheric CO_2 as measured by the US National Oceanic and Atmospheric Administration (NOAA) at Mauna Loa, Hawaii, rose by 2.6 ppm CO_2 [17]. The upper estimate of the annual increase in Carbon Debt associated with this annual increase in CO_2 is 2.6 ppm CO_2 per year x $1,958 billion per ppm CO_2 = $5.1 trillion.

However there are GHGs other than CO_2 that contribute to global warming, notably methane (CH_4) and nitrous oxide (N_2O). CH_4 derives from anaerobic bacterial metabolism in the rumens of ruminant livestock (it is voided by eructation or belching) and in anaerobic swamps or urban tips. N_2O derives from oxidation of nitrogenous agricultural fertilizers, and also derives from nitrogen (N_2) oxidation associated with high temperature fossil fuel (coal, gas and oil) combustion. The Global Warming Potential (GWP) of CH_4 relative to the same mass of CO_2 on a 20 year time frame and with aerosol effects considered is 105 times that of CO_2 but is 21 on a 100 year time frame [18-21]. The GWP of N_2O on a 20 year time frame relative to the same mass of CO_2 is 289 [21].

The combined global warming potential of atmospheric GHGs is expressed as CO_2-equivalent (CO_2-e). Thus while the atmospheric CO_2 is presently 407 ppm, the combined GHGs amount to 527 ppm CO_2-e [22]. According to Professor Ronald Prinn (Professor of Atmospheric Science in the 93 Nobel Laureate Massachusetts Institute of Technology (MIT) Department of Earth, Atmospheric and Planetary Sciences), when the CO_2 was 400 ppm in 2013, the CO_2-e was 478 ppm [23], this suggesting a 2019 level of 486 CO_2-e. World Bank analysts have re-assessed the annual global GHG emissions taking livestock, land use and a 20-year time frame GWP of methane of 72 into account – this analysis revised the annual GHG pollution upwards from 41.8 Gt CO_2-e to 63.8 Gt CO_2-e [24].

Assuming that CH_4 and N_2O can be removed from the atmosphere for the same cost as removal of CO_2, then the resultant upper estimated cost of $252 per ton CO_2-e (see section 1 above) translates to an annual increase in Carbon Debt of 63.8 Gt CO_2-e per year x $252 per ton CO_2-e = $16.1 trillion per year.

This upper estimate of the annual cost of $16.1 trillion per year for GHG pollution is similar to the present annual net wealth generation of $14 trillion per year [9]. Indeed this equivalence should not be surprising – in our presently carbon pollution-based economy, productivity can be measured either in the net dollar worth of production or in GHGs produced in the process. One notes that the world's annual GDP (nominal) is about $85 trillion [25, 26]

(3). Using the damage-related cost of carbon pollution of $200 per tonne CO_2 yields a $11.6 trillion annual increase in Carbon Debt.

Dr Chris Hope (Judge Business School, University of Cambridge) has estimated the damage-related cost of carbon pollution (2011): "If the best current scientific and economic evidence is to be believed, and climate change could be a real and serious problem, the appropriate response is to institute today a climate change tax equal to the mean estimate of the damage caused by a tonne of CO_2 emissions. The raw calculations from the default PAGE09 model suggest the tax should be about $100 per tonne of CO_2 in the EU. But correcting for the limited time horizon of the mode, and bringing the calculations forward to 2012, in year 2012 dollars, brings the suggested tax up to about $150 per tonne of CO_2. There are good arguments for setting the initial tax at about $250 per tonne of CO_2 in the US, while starting off at a much lower level, maybe $15 per tonne of CO_2, in the poorest regions of the world, all in the year 2012, in year 2012 dollars" [27].

If we use the $200 per tonne CO_2 average of the damage-related $250 per tonne CO_2 (US) and $150 per tonne CO_2 (EU), we can determine the annual increase in Carbon Debt as 63.8 billion ton CO_2-e per year x $200 per tonne CO_2-e x 0.907 tonne / ton = $11.6 trillion per year, similar to the upper estimate of $16 trillion estimated in section 2 above based on the cost of removing GHGs from the atmosphere [11].

(4). The $16 trillion annual Carbon Debt increase, Value of a Statistical Life (VSL) and the $13.5 trillion annual cost of the Global Avoidable Mortality Holocaust of 15 million deaths from deprivation annually.

There is a largely ignored Global Avoidable Mortality Holocaust in which currently 15 million people die avoidably each year from deprivation and deprivation-exacerbated disease in the Developing World (minus China) [28]. Avoidable mortality from deprivation (avoidable deaths, excess mortality, excess deaths, premature deaths, untimely deaths, deaths that should not have happened) can be simply estimated from UN Population Division demographic data as the difference between the actual deaths in a country and deaths expected for a peaceful, decently-run country with the same demographics. For impoverished, high birth-rate, Developing countries the baseline mortality rate is about 4 deaths per 1,000 of population per year [29]. For such impoverished countries the avoidable mortality is about 1.4 times the under-5 infant deaths, the latter information being readily obtained from the UN Population Division [29].

This carnage is happening on Spaceship Earth with the flight deck under the control of the 10% richest people who have about 90% of the wealth of the World and who in turn are controlled by One Percenters who own about 50% of the wealth of the World. An annual global wealth tax of about 4% would yield $16 trillion annually (as assessed in 2014) [30] and enable raising all countries to an annual per capita GDP equivalent to the presently circa $8,500 per person per year of China and Cuba, countries for which annual avoidable mortality is zero (0) [28, 29, 31, 32]. This is a feasible option for stopping the Global Avoidable Mortality Holocaust. Indeed a progressive annual wealth tax ranging up to 10% for the richest has been proposed for democracy and economic sustainability reasons by French economist Professor Thomas Piketty in his important book "Capital in the Twenty-First Century" [33, 34]. Indeed Islam has mandated a 2.5% annual wealth tax (zakkat) for about 1,400 years [35].

For Americans the risk avoidance-based Value of a Statistical Life (VSL) is about $9 million [36], and for people in the South with GDP per capita values of the order of 10 times lower than for the North [31] we could roughly estimate a 10 times lower VSL of $0.9 million. Accordingly, committing to a 4% annual wealth tax realizing $16 trillion (in 2014) to save 15 million lives annually would be balanced by VSL-based "savings" of 15 million people per year x $0.9 million per person = $13.5 trillion per year.

The profligate $16 trillion annual increase in Carbon Debt (see section 2) is numerically the same as the 4% annual wealth tax realizing $16 trillion (in 2014) that could abolish the Global Avoidable Mortality Holocaust in which currently 15 million people die avoidably each year from deprivation at a VSL-based cost of $13.5 trillion per year.

The top 10 countries in terms of their percentages of total global personal wealth as determined by Allianz, are as follows: (1) United States (41.6%), (2) China (10.5%), (3) Japan (8.9%), (4) U.K. (5.6%), (5) Germany (3.9%), (6) France (3.5%), (7) Canada (3.0%), (8) Italy (2.9%), (9) Australia (2.0%), and (10) South Korea (1.6%) – the rest account for 16.5% of the world's wealth [37]. The total accumulated wealth of the world in 2019 is $333 trillion, and the average [per capita] wealth of Americans in 2018 was $63,100, this being similar to the GDP (nominal) per capita for Americans of $60,000 [26]. A global 4% annual wealth tax would presently garner $13.3 trillion per year, similar to the $15.7 trillion per year for a decade to get back to the coral reef-based "canary in the mine" requirement of 320 ppm CO_2 (see section 1).

However Democratic presidential hopeful Senator Elizabeth Warren's recent "radical" wealth tax proposal – a 2% annual tax on household net worth on all dollars above $50 million and an additional 1% surtax for those with above $1 billion in income – would only garner an utterly paltry $0.275 trillion each year over 10 years from an America with 4.3% of the world's population, 41.6% of the world's wealth [38], and 14.75% of world GHG pollution (minus land use) of 45.367 Gt CO_2-e in 2014 [39] i.e. 6.67 Gt CO_2-e and corresponding to an annual Carbon Debt (minus land use) of at least $1.7 trillion per year. Of course one must recognize that her ignorant opponent, Donald Trump, dismissed climate change as a "hoax" in the 2016 presidential campaign.

(5). The total Carbon Debt of the world is a daunting circa $200-$250 trillion.

The upper estimate of the Carbon Debt for a transition from the present 407 ppm CO_2 back to the pre-Industrial Revolution level of 280 ppm CO_2 is 127 ppm CO_2 x $1,958 billion per ppm CO_2 = $248.7.0 trillion [11], a sum that is 78% of the 2018 accumulated wealth of the world of $317 trillion in 2018 [9], and if paid off over 1 decade would require payment at $24.9 trillion per year.

Alternatively, the historical Carbon Debt of a country can be measured by the amount of greenhouse gas (GHG) it has introduced into the atmosphere since the start of the Industrial Revolution in the mid-18th century. Thus the total Carbon Debt of the world (in Giga tonnes CO_2-e i.e. billions of tonnes CO_2-e) from 1751-2019 (including CO_2 that has gone into the oceans) is about 346 Gt CO_2-e (1750-2008) [40] + 638 Gt CO_2-e (2009-2019) [24] = 984 Gt CO_2-e. Assuming a damage-related Carbon Price of $200 per tonne CO_2-equivalent [27], this corresponds to a Carbon Debt of $197 trillion, about two-thirds of the total wealth of the world ($317 trillion in 2018) and 2.3 times the world's total annual GDP ($85 trillion in 2018). Payment of this debt over 1 decade would require payment at $19.7 trillion per year.

(6). Total and annual Carbon Debt of the US, Australia, China and India.

The US has a huge GHG pollution of 5.14 Gtonnes CO_2-e per year and of about 51.4 Gtonnes CO_2-e (for the period 2009-2019) [44] with an historical GHG pollution of 99.1 Gtonnes CO_2-e (1750-2008) [40, 41], for a total Carbon Debt of 51.4 + 99.1 = 150.5 Gtonnes CO_2-e that corresponds (at $200 per tonne CO_2-e) to a Carbon Debt of $30.1 trillion that is increasing at $1.0 trillion per year (6.3% of the world's $16 trillion per year). However a revised estimate of US GHG pollution taking land use and CH_4 into account [45, 46] is 41 tonnes CO_2-e per person per year for the US, or 13,4 Gt CO_2-e in total (based on a present population of 327 million), and hence an annual Carbon

Debt increase of $2.7 trillion per year (16.9% of the world's $16 trillion per year increase in Carbon Debt). This revised estimate is more consistent with the US (population 327 million, 4.3% of that of the world) being responsible for 24.1 % of the world's GDP in a global carbon economy (world population 7,530 million) [31].

My own country, Australia, that is among world leaders in 14 GHG pollution-related areas [42, 43] has a Carbon Debt of US$5.1 trillion (A$7 trillion) that is increasing at US$400 billion (A$556 billion) per year and at US$40,000 (A$56,000) per head per year for under-30 year old Australians [41]. However a revised estimate of Australian annual GHG pollution taking land use and CH_4 into account [44, 45] is 52.9 tonnes CO_2-e per person per year, or (based on a population of 25 million) 1.32 Gt CO_2-e and an annual increase in Carbon Debt of $0.26 trillion per year (1.7 % of the world's $16 trillion per year Carbon Debt increase). Taking Australia's huge coal and gas exports into account, Australia's annual GHG pollution is 116 tonnes CO_2-e per person per year or (with a population of 25 million) 2.9 Gt CO_2-e and an annual Carbon Debt increase of $0.6 trillion (3.8% of the world's $16 trillion per year increase in Carbon Debt, noting that Australia's population is only 0.3% of the world's population and contributes 1.4 % of world GDP [31]).

The revised annual GHG emission of China (population 1,390 million) taking land use and CH_4 into account, is 7.4 tonnes CO_2-e per person per year (2013), and 10.3 Gt CO_2-e in total per year, with this corresponding to an annual increase in Carbon Debt of $2.1 trillion per year (13.1% of the world's annual Carbon Debt increase, noting that China has 18.5% of the world's population and contributes 12.2% of world GDP).

India does even better than China on a per capita basis. The revised annual GHG emission of India (population 1,340 million) taking land use and CH_4 into account, is 2.1 t CO_2-e per person per year (2013), and 2.8 Gt CO_2-e in total per year, with this corresponding to an annual increase in Carbon Debt of $ 0.56 trillion per year (3.5% of the world's annual Carbon Debt increase, noting that India has 17.8% of the world's population and contributes 2.6% of world GDP).

From the above it is apparent that on a per capita basis the US and Australia contribute disproportionately and vastly more to the annual increase in global Carbon Debt than China that (contributes its fair share) and India that contributes much less to the human world's suicide mission.

(7). VSL and the $7 trillion annual cost of 7 million annual deaths from carbon-based air pollution and 1 million annual deaths from climate change.

About 7 million people die each year from the effects of pollutants from carbon fuel burning [47]. It has been estimated that 0.4 million people die each year from climate change [48, 49], but this is surely an underestimate because each year 15 million people die avoidably from deprivation in the impoverished tropical and sub-tropical South that is disproportionately impacted by climate change [28]. Accordingly, one can reasonably posit that annual climate change deaths are of the order of 1 million. Adopting a risk avoidance-based Value of a Statistical Life (VSL) of $0.9 million per person for impoverished countries, one can see that 8 million annual deaths from carbon fuel burning and climate change correspond to 8 million deaths x $0.9 million per person = $7.2 trillion. Unfortunately the Climate Genocide that presently kills about 1 million people per year is worsening. Several leading climate scientists have estimated that as few as 0.5 billion people will survive this century if man-made climate change is not requisitely addressed, with this translating to about 10 billion climate deaths this century and an extremely academic Carbon Debt of $9,000 trillion [2].

(8). $200 per tonne CO_2-e Carbon Tax can help drive rapid transit to zero emissions and thence massive negative emissions.

If we accept that "all men are created equal" then it is a no-brainer that polluters should fully pay for the social and environmental cost of their pollution. Indeed science-trained Pope Francis has unequivocally stated that the social and environmental cost of pollution should be "fully borne" by the polluters. In stark contrast to the terracidal One Percenters, Pope Francis recognizes a looming climate catastrophe and millions of premature deaths from carbon burning pollutants, demands massive decarbonisation and conversion to renewables starting without delay in the next few years, rejects Carbon Trading as a dishonest ploy, advocates a "fully borne" Carbon Price (Carbon Tax) on polluters, proposes boycotts of polluters, condemns loss of biodiversity, and advises intergenerational solidarity [50, 51].Two expert estimates of a required Carbon Tax (Carbon Price) are considered above, to whit $200 per tonne CO_2-e from climate economist Dr Chris Hope of 118 Nobel Laureate Cambridge University [27], and up to $252 per ton CO_2-e ($229 per tonne CO_2-e) from Professor James Hansen of 96 Nobel Laureate Columbia University [11]. Nevertheless the politically dominant, climate criminal One Percenters who control half the wealth of the world have responded to such expert opinion with massively funded climate change denialism [52], and thence variously with fall-back positions of grossly insufficient, corrupt and counterproductive Carbon Trading schemes to preserve the present suicidal Carbon Economy [53]. A $200 per tonne CO_2-e Carbon Tax can help drive rapid transit to zero emissions and thence massive required negative carbon emissions as set out above.

(9). Any species is priceless – Anthropocene Era mass extinctions imply infinite Carbon Debt.

We must not destroy what we cannot replace – any species is accordingly priceless. Yet mass extinctions are now occurring at such a level to have prompted scientists to call the present era the Anthropocene Era. The extinction rates are now 100-1,000 times above normal. [54-60]. Thus, for example, coral reefs are hugely complex, multi-species ecosystems, and the IPCC has recently projected a further 70-90% decline of coral reefs at global warming of +1.5C versus more than a 99% loss at +2C [56-58]. Anthropocene Era mass extinctions imply infinite Carbon Debt. That said, environmental economists have assessed the economic value of nature as of the same order as the annual world GDP. Dr Andrew Balmford and colleagues have stated (2002): "On the eve of the World Summit on Sustainable Development, it is timely to assess progress over the 10 years since its predecessor in Rio de Janeiro. Loss and degradation of remaining natural habitats has continued largely unabated. However, evidence has been accumulating that such systems generate marked economic benefits, which the available data suggest exceed those obtained from continued habitat conversion. We estimate that the overall benefit:cost ratio of an effective global program for the conservation of remaining wild nature is at least 100:1" [61].

(10). Impossibly high Carbon Debt from unlimited Australian coal and gas exports ($353 trillion) and adumbrated Arctic CH_4 release ($1,050 trillion).

Climate criminal Australia (0.3% of the world's population) has a bipartisan policy – fervently adhered to by the present Liberal Party-National Party Coalition Government and the Labor Party Opposition that are collectively known as the Lib-Labs – of unlimited coal, gas an iron exports. In contrast, the science-informed Australian Greens want to achieve 100% renewable energy and cease all thermal coal exports by 2030 [62]. However it is estimated that feasible exploitation of Australia's presently discovered conventional and unconventional gas resources would generate 61.5 billion tonnes of CO_2-e, and combustion of Australia's huge coal resource potential of 1 trillion tonnes of coal would be an estimated 692.7 Gt CO_2 (from brown coal) plus 1,073.9 Gt CO_2 (from black coal), for a total of 1,766.6 Gt CO_2, with this corresponding to a Carbon Debt of 1.7666 trillion tonnes CO_2 x $200 per tonne CO_2 = $353 trillion (similar to the accumulated global wealth of $350 trillion in 2020 [9]). Australia has always been a deeply racist country [64, 65], but this takes racism and exceptionalism to an extraordinary, genocidal and indeed terracidal level.

It gets worse. Professor Peter Wadhams (professor of Ocean Physics, and Head of the Polar Ocean Physics Group in the Department of Applied Mathematics and Theoretical Physics,

118-Nobel-Laureate University of Cambridge, UK) and colleagues on the threat of a 50Gt methane release from the East Siberian Arctic Shelf (2013): "Economic time bomb. As the amount of Arctic sea ice declines at an unprecedented rate, the thawing of offshore permafrost releases methane. A 50-gigatonne (Gt) reservoir of methane, stored in the form of hydrates, exists on the East Siberian Arctic Shelf. It is likely to be emitted as the seabed warms, either steadily over 50 years or suddenly" [66]. The Global Warming Potential (GWP) of CH_4 on a 20 year time frame and with aerosol impacts included is 105 relative to CO_2 (1), and this 50 Gt CH_4 release corresponds to 50 Gt CH_4 x 105 t CO_2-e/ t CH_4 = 5,250 Gt CO_2-e, which in turn corresponds to a Carbon Debt (at \$200 per tonne CO_2-e) of \$1,050 trillion or 3 times greater than the accumulated wealth of whole world.

Final comments.

The world already has an inescapable Carbon Debt of \$200-250 trillion that – instead of being urgently reduced by massive atmospheric CO_2 draw-down – is relentlessly increasing at about \$16 trillion per year. In a domestic household budget setting this would be utter lunacy – in a global setting it is mass suicide of humanity. This utterly disastrous present scenario means that a catastrophic plus 2C temperature rise is evidently unavoidable. However we are obliged to do everything we can to make the future "less bad" for future generations. A London wall mural attributed to the famed street artist Banksy declares: "From this moment despair ends and tactics begin" [67]. The young have no choice but to revolt (non-violently of course) – as exampled by the present Extinction Rebellion by young people in London [4, 5] – against the destruction of the Biosphere (speciescide, ecocide, omnicide, and terracide).

Decent people must oppose the neoliberal Gadarene rush to global suicide by (a) informing everyone they can, and (b) urging and applying Boycotts, Divestment and Sanctions (BDS) against all people, politicians, parties, collectives, corporations and countries disproportionately involved in the worsening climate emergency, climate genocide and Biosphere destruction.

There is no Planet B and there must be zero tolerance for the neoliberal, genocidal and terracidal climate criminals. In climate criminal Australia that is among world leaders in GHG pollution, Biosphere destruction and climate change inaction, decent people will utterly reject the climate criminal Coalition Government, vote 1 Green and put the Coalition and its extremist associates last.

References.

[1]. Josh Holder, Niko Commenda and Jonathan Watts, "The three-degree world: the cities that will be drowned by global warming", Guardian, 3 November 2017: https://www.theguardian.com/cities/ng-interactive/2017/nov/03/three-degree-world-cities-drowned-global-warming.
[2]. Gideon Polya, editor, "Climate Genocide": https://sites.google.com/site/climategenocide/.
[3. Pacific Islands Development Forum, "Suva Declaration on Climate Change", 4 September 2015: http://pacificidf.org/wp-content/uploads/2013/06/PACIFIC-ISLAND-DEVELOPMENT-FORUM-SUVA-DECLARATION-ON-CLIMATE-CHANGE.v2.pdf.
[4]. "Extinction Rebellion": London climate protest "ending on Thursday"", BBC, 24 April 2019: https://www.bbc.com/news/uk-england-london-48044056.
[5]. Gideon Polya, editor, "Climate Revolution Now": https://sites.google.com/site/300orgsite/climate-revolution.
[6]. Gideon Polya, editor, "Climate Justice & Intergenerational Equity": https://sites.google.com/site/300orgsite/climate-justice.
[7]. Gideon Polya, editor, "Science & economics experts: Carbon Tax needed NOT Carbon Trading": https://sites.google.com/site/300orgsite/sciennce-economics-experts-carbon-tax-needed-not-carbon-trading/.
[8]. Gideon Polya, editor, "Stop climate crime": https://sites.google.com/site/300orgsite/stop-climate-crime.
[9]. Credit Suisse, "Global Wealth Report 2018": https://www.credit-suisse.com/corporate/en/research/research-institute/global-wealth-report.html.
[10]. Oxfam, "Richest 1 percent bagged 82 percent of wealth created last year – poorest half of humanity got nothing", 22 January 2018: https://www.oxfam.org/en/pressroom/pressreleases/2018-01-22/richest-1-percent-bagged-82-percent-wealth-created-last-year.
[11]. James Hansen, "Climate change in a nutshell: the gathering storm", Columbia University, 18 December 2018: http://www.columbia.edu/~jeh1/mailings/2018/20181206_Nutshell.pdf.
[12]. Output of the technical working group meeting, The Royal Society, London, 6th July, 2009, "The Coral Reef Crisis: scientific justification for critical CO_2 threshold levels of less than 350ppm": http://static.zsl.org/files/statement-of-the-coral-reef-crisis-working-group-890.pdf.

[13]. J.E.N. Veron, O. Hoegh-Guldberg, T.M. Lenton, J.M. Lough, D.O. Obura, P. Pearce-Kelly, C.R.C. Sheppard, M. Spalding, M.G. Stafford-Smith and A.D. Rogers, "The coral reef crisis: the critical importance of <350 ppm CO2", Marine Pollution Bulletin, vol. 58, (10), October 2009, 1428-1436:
http://www.sciencedirect.com/science?_ob=ArticleURL&_udi=B6V6N-4X9NKG7-3&_user=10&_rdoc=1&_fmt=&_orig=search&_sort=d&_docanchor=&view=c&_searchStrId=1072337698&_rerunOrigin=google&_acct=C000050221&_version=1&_urlVersion=0&_userid=10&md5=6858c5ff7172f9355068393496a5b35d.
[14]. Gideon Polya, editor, 300.org:
https://sites.google.com/site/300orgsite/300-org.
[15]. Gideon Polya, editor, "300.org – return atmosphere CO_2 to 300 ppm CO_2": https://sites.google.com/site/300orgsite/300-org—return-atmosphere-co2-to-300-ppm.
[16]. Professor Hans Joachim Schellnhuber quoted by David Adam, "Roll back time to safeguard climate, expert warns", Guardian, 15 September 2008:
http://www.guardian.co.uk/environment/2008/sep/15/climatechange.carbonemissions.
[17]. US National Oceanic and Atmospheric Administration (NOAA), "Trends in atmospheric carbon dioxide":
https://www.esrl.noaa.gov/gmd/ccgg/trends/.
[18]. Gideon Polya, editor, "Gas is not clean energy":
https://sites.google.com/site/gasisnotcleanenergy/home.
[19]. Drew T. Shindell, Greg Faluvegi, Dorothy M. Koch, Gavin A. Schmidt, Nadine Unger and Susanne E. Bauer, "Improved Attribution of Climate Forcing to Emissions", Science, 30 October 2009:
Vol. 326 no. 5953 pp. 716-718:
http://www.sciencemag.org/content/326/5953/716.
[20]. Shindell et al. (2009), Fig.2:
http://www.sciencemag.org/content/326/5953/716.figures-only.
[21]. Gideon Polya, "2011 climate change course":
https://sites.google.com/site/300orgsite/2011-climate-change-course.
[22]. "Carbon dioxide equivalent", Wikipedia:
https://en.wikipedia.org/wiki/Carbon_dioxide_equivalent.
[23]. "400 ppm CO_2? Add other GHGs, and its equivalent to 478", Oceans at MIT, 6 June 2013: http://oceans.mit.edu/featured-stories/5-questions-mits-ron-prinn-400-ppm-threshold.html.
[24]. Robert Goodland and Jeff Anfang. "Livestock and climate change. What if the key actors in climate change are … cows, pigs and chickens?", World Watch, November/December 2009:
https://pdfs.semanticscholar.org/6704/c7a0777c82357704d82b9ae800

7c1197cb07.pdf?_ga=2.187734888.1597394103.1556059730-
1006954717.1556059730.
[25]. "The world's top 10 largest economies", Focus Economics,
2019: https://www.focus-economics.com/blog/the-largest-economies-
in-the-world.
[26]. "List of countries by GDP (nominal)", Wikipedia:
https://en.wikipedia.org/wiki/List_of_countries_by_GDP_(nominal).
[27]. Chris Hope, "How high should climate change taxes be?",
Working Paper Series, Judge Business School, University of
Cambridge, 9.2011:
http://www.jbs.cam.ac.uk/fileadmin/user_upload/research/workingpap
ers/wp1109.pdf.
[28]. Gideon Polya, "Body Count. Global avoidable mortality since
1950" that includes an avoidable mortality-related history of every
country since Neolithic times and is now available for free perusal on
the web: http://globalbodycount.blogspot.com/.
[29]. UN Population Division, World Population Prospects, 2013
Revision: http://esa.un.org/unpd/wpp/unpp/panel_population.htm.
[30]. Gideon Polya, "4 % Annual Global Wealth Tax To Stop The 17
Million Deaths Annually", Countercurrents, 27 June, 2014:
https://countercurrents.org/polya270614.htm.
[31]. "List of countries by GDP (nominal) per capita", Wikipedia:
https://en.wikipedia.org/wiki/List_of_countries_by_GDP_(nominal)_
per_capita.
[32]. "List of countries by population", Wikipedia:
http://en.wikipedia.org/wiki/List_of_countries_by_population.
[33]. Thomas Piketty, "Capital in the Twenty-First Century", Harvard
University Press, 2014.
[34]. Gideon Polya, "Key Book Review: "Capital In The Twenty-First
Century" by Thomas Piketty", Countercurrents, 1 July, 2014:
https://www.countercurrents.org/polya010714.htm.
[35]. Gideon Polya, editor, "1% ON 1%: one percent annual wealth
tax on One Percenters" website (
https://sites.google.com/site/300orgsite/1-on-1.
[36]. Timothy Taylor, "Value of a statistical life. 9.1 million?",
Conversable Economist, 22 October 2013:
http://conversableeconomist.blogspot.com.au/2013/10/value-of-
statistical-life-91-million.html.
[37]. Erik Sherman, "America is the richest, and most unequal,
country", Fortune, 30 September 2015:
http://fortune.com/2015/09/30/america-wealth-inequality/.
[38]. David Dayen, "Elizabeth Warren proposes annual wealth tax on
ultra-millionaires", The Intercept, 25 January 2019:

https://theintercept.com/2019/01/24/elizabeth-warren-proposes-annual-wealth-tax-on-ultra-millionaires/.

[39]. "List of countries by greenhouse gas emissions", Wikipedia: https://en.wikipedia.org/wiki/List_of_countries_by_greenhouse_gas_emissions.

[40]. James Hansen, "Letter to PM Kevin Rudd by Dr James Hansen", 2008:
http://www.aussmc.org.au/documents/Hansen2008LetterToKevinRudd_000.pdf

[41]. Gideon Polya, editor, "Carbon Debt Carbon Credit": https://sites.google.com/site/carbondebtcarboncredit/.

[42]. Gideon Polya, "Pro-coal Australia & Trump America Reject Dire IPCC Report & Declare War on Terra", Countercurrents, 17 October 2018: https://countercurrents.org/2018/10/17/pro-coal-australia-trump-america-reject-dire-ipcc-report-declare-war-on-terra/.

[43]. Gideon Polya, "Offences of Pentecostal Christian Scott Morrison, PM after Australia's fourth PM-removing coup in 8 years", Countercurrents, 18 September 2019:
https://countercurrents.org/2018/09/18/offences-of-pentecostal-christian-scott-morrison-pm-after-australias-fourth-pm-removing-coup-in-8-years/.

[44]. "Greenhouse gas emissions by the United States", Wikipedia: https://en.wikipedia.org/wiki/Greenhouse_gas_emissions_by_the_United_States.

[45]. Gideon Polya, "Revised Annual Per Capita Greenhouse Gas Pollution For All Countries – What Is Your Country Doing?", Countercurrents, 6 January, 2016:
http://www.countercurrents.org/polya060116.htm.

[46]. Gideon Polya, "Exposing And Thence Punishing Worst Polluter Nations Via Weighted Annual Per Capita Greenhouse Gas Pollution Scores", Countercurrents, 19 March, 2016:
https://countercurrents.org/polya190316.htm.

[47]. Gideon Polya, editor, "Nuclear weapons ban, end poverty and reverse climate change":
https://sites.google.com/site/300orgsite/nuclear-weapons-ban.

[48]. DARA, "Climate Vulnerability Monitor. A guide to the cold calculus of a hot planet", 2012, Executive Summary pp2-3:
http://daraint.org/climate-vulnerability-monitor/climate-vulnerability-monitor-2012/.

[49]. DARA report quoted by Reuters,"100 mln to die by 2030 if world fails to act on climate", 28 September 2012:
http://in.reuters.com/article/2012/09/26/climate-inaction-idINDEE88P05P20120926.

[50]. Pope Francis, Encyclical Letter "Laudato si", 2015:
http://w2.vatican.va/content/francesco/en/encyclicals/documents/papa
-francesco_20150524_enciclica-laudato-si.html.
[51]. Gideon Polya, "Green Left Pope Francis Demands Climate
Action "Without Delay" To Prevent Climate "Catastrophe"",
Countercurrents, 10 August, 2015:
https://countercurrents.org/polya100815.htm.
[52]. Richard Hil and Gideon Polya, "The silencing of the climate
emergency, and what we should do about it", New Matilda, 5 March
2019: https://newmatilda.com/2019/03/05/silencing-climate-
emergency/.
[53]. Gideon Polya, "Experts: Carbon Tax needed and NOT Cap-and-
Trade Emission Trading Scheme (ETS)", Yarra Valley Climate
Action Group":
https://sites.google.com/site/yarravalleyclimateactiongroup/carbon-
tax-needed-not-cap-and-trade-emission-trading-scheme-ets.
[54]. William J. Ripple et al., 15,364 signatories from 184 countries,
"World scientists' warning to Humanity: a second notice",
Bioscience, 13 November
2017: https://academic.oup.com/bioscience/advance-
article/doi/10.1093/biosci/bix125/4605229.
[55]. Gideon Polya, "Over 15,000 scientists issue dire warning to
humanity on catastrophic climate change and biodiversity loss" ",
Countercurrents, 20 November
2017: https://countercurrents.org/2017/11/20/over-15000-scientists-
issue-dire-warning-to-humanity-on-catastrophic-climate-change-and-
biodiversity-loss/.
[56]. IPCC, "Global warming of 1.5 °C", 8 October
2018: http://www.ipcc.ch/report/sr15/.
[57]. IPCC, "Global warming of 1.5 °C. Summary for Policymakers",
8 October 2018: http://report.ipcc.ch/sr15/pdf/sr15_spm_final.pdf.
[58]. Gideon Polya, "IPCC +1.5C avoidance report – effectively too
late but stop coal burning for "less bad" catastrophes",
Countercurrents, 12 October 2018:
https://countercurrents.org/2018/10/12/ipcc-1-5c-avoidance-report-
effectively-too-late-but-stop-coal-burning-for-less-bad-catastrophes/,
[59]. Andrew Glikson, "The IPCC's final warnings of extreme global
warming", Countercurrents, 10 October
2018: https://countercurrents.org/2018/10/10/the-ipccs-final-
warnings-of-extreme-global-warming/.
[60]. Phillip S. Levin and Donald A. Levin, "The real biodiversity
crisis", Macroscope, January-February 2002:
http://www.soc.duke.edu/~pmorgan/levin&levin.2002.the_real_biodi
versity_crisis.html.

[61]. A. Balmford, A. Bruner, P. Cooper, R. Costanza, S. Farber, R. E. Green, M. Jenkins, P. Jefferiss, V. Jessamy, J. Madden, K. Munro, N. Myers, S. Naeem, J. Paavola, M. Rayment, S. Trumper and R. K. Turner "Economic reasons for conserving wild nature", Science 297: 950-953, 2002: http://www.sciencemag.org/cgi/content/abstract/297/5583/950.
[62]. Katharine Murphy, "Greens set 2030 cut-off for coal exports and coal-fired power stations", Guardian, 28 March 2019: https://www.theguardian.com/australia-news/2019/mar/28/greens-set-2030-cut-off-for-coal-exports-and-coal-fired-power-stations.
[63]. Gideon Polya, "Australia's Huge Coal, Gas & Iron Ore Exports Threaten Planet", Countercurrents, 15 May 2012: http://www.countercurrents.org/polya150512.htm.
[64]. "Aboriginal Genocide": https://sites.google.com/site/aboriginalgenocide/.
[65]. Gideon Polya, "As UK Lackeys Or US Lackeys Australians Have Invaded 85 Countries (British 193, French 80, US 70)", Countercurrents, 9 February, 2015: http://www.countercurrents.org/polya090215.htm.
[66]. Gideon Polya, editor, "Methane Bomb Threat": https://sites.google.com/site/methanebombthreat/.
[67]. Lanre Bakare, "London Extinction Rebellion mural is a Banksy, says expert", Guardian, 26 April 2019: https://www.theguardian.com/artanddesign/2019/apr/26/london-extinction-rebellion-mural-is-a-banksy-says-expert.

CHAPTER 18: CARBON DEBT & DUMPING – CLIMATE CRIMINAL AUSTRALIA HUGELY SUBSIDIZES MEAT, GRAIN & WINE EXPORTS TO CHINA

[First published as Gideon Polya, "Carbon Debt & Dumping – Climate Criminal Australia Hugely Subsidizes Meat, Grain & Wine Exports To China", Countercurrents, 31 August 2020: https://countercurrents.org/2020/08/carbon-debt-dumping-climate-criminal-australia-hugely-subsidizes-meat-grain-wine-exports-to-china/.]

Despite China being Australia's biggest trading partner, US lackey Australia has joined Trump America in its anti-China trade war and military posturing policies. China has been offended by this increasing hostility from US lackey Australia and coincidentally for asserted anti-dumping reasons has applied penalties to Australian beef and barley with threats of extending this to Australian wine. In actuality there is massive dumping by climate criminal Australia through non-application of a Carbon Price.

In 2019 the Australian Gross Domestic Product (GDP) in USD was $1,393 billion [1] for a population of 25.5 million [2] i.e. a per capita GDP of $54,627 (albeit quite unevenly distributed). However rich Australia ignores the huge Carbon Debt associated with that productivity. Australia has a revised annual Greenhouses Gas (GHG) pollution of 1,423 Mt CO_2-e (Domestic) and 3,431 Mt CO_2-e (Domestic plus Exported) [3]. Assuming a damage-related Carbon Price of $200 per tonne of CO_2-equivalent [4, 5] means that Australia has an inescapable Carbon Debt of $5 trillion that is increasing at $686 billion annually and at $70,000 per head per year for under-30 year old Australians (USD)[6].

Thus Australia's annual GDP of $1,393 billion should be inflated to $1,393 billion + $686 billion Carbon Debt = $ 2,079 billion. The ratio of Carbon Debt/GDP = $686 billion/ $1,393 billion = 0.49 i.e. for every $1.0 billion of goods and services generated by Australia each year (or exported to China each year), there is a deliberately hidden but inescapable subsidy of about $0.5 billion to be paid by future generations.

Accordingly every $10 piece of Australian meat, $10 bag of Australian barley or $10 bottle of Australian wine sold to China has been produced through a gigantic and hidden Australian Government subsidy of about $5 to be paid for by future generations because the Australian Coalition Government (with the present support of the Labor Party Opposition) refuses to put a price on Carbon Pollution (Greenhouse Gas or GHG Pollution). The Chinese authorities are presently variously acting on or investigating alleged dumping of Australian beef, barley and wine in China through Australian Government subsidies [7-9], but it is possible that even the Chinese Government is unaware of this massive Australian dumping through non-application of a requisite Carbon Price.

The Australian ABC (Australia's taxpayer-funded equivalent of the UK BBC) reports: "First it was barley, then beef and now wine. China has Australian exporters in a spin over allegations they sold wine below the cost of production… Australian Government, winemakers and farmers deny any wrongdoing" [9]. However it is clear that the circa 49% of the Cost of Production (COP) of these products due to Carbon Debt is deliberately ignored in the pricing. The International Monetary Fund (IMF) argues that introduction of a progressively increasing Carbon Price to about $75 per ton CO_2 is crucial for tackling man-made climate change, and estimates that the global average Carbon Price is presently a mere $2 per ton CO_2. The pro-fossil fuels, climate criminal Australian Coalition Government has bluntly rejected the IMF proposal [10], and the unprincipled, cowardly, spin-driven and electoral popularity-driven Labor Opposition has recently ruled out application of a Carbon Price [11]. The Labor Opposition Leader Anthony Albanese, while advocating a suitably far-off "zero emissions by 2050", has for political advancement reasons stated of a Carbon Price: "That's not necessary … at the moment, what you have is the cheapest forms of energy are renewable" [11].

In the last financial year the value of Australia's annual exports to China totalled about A$151 billion (US$111 billion) of which about A$102 billion (US$75 billion) was due to iron ore [12]. Obviously some exports are "dirtier" GHG-wise than others, but applying the average hidden Carbon Debt as 49% of GDP yields an estimate that Australia's annual exports to China in the last financial year were associated with a hidden, Carbon Price-related and "dumping" subsidy of about $55 billion.

Australia's huge exports to China are at great risk due to US lackey Australia cravenly joining Trump America's anti-China campaign. As they say in Australia, when the US says jump, Australia says "How high?" However the worsening relations between China and Australia have been associated with a re-surfacing of ugly Sinophobia that has been entrenched in a racist White Australia since the gold-rush days of 170 years ago. A US lackey and politicized Australian Intelligence is targeting elected MPs for the asserted "crime" of Chinese associations [13-15], and the Coalition Government is now proposing, with craven Labor Opposition support, to enable the Federal Government to cancel (on "national security" grounds) contracts between foreign (read Chinese) interests and Australian businesses, universities, local government and state governments. A key target in this massive economic sabotage of Australia by the US lackey and Zionist-subverted Coalition Federal Government is the Victorian State Government's association with the Chinese Belt and Road Initiative (BRI) [16]. Of course, no mention is made in this worsening national xenophobia and Sinophobic hysteria [17] of massive subversion of Australia by the US, UK and US- and UK-backed Apartheid Israel [18-20].

Because of the neoliberalism-impelled mendacity of Mainstream media, editor, politician, academic and commentariat presstitutes all of this "Carbon Price-based hidden dumping subsidy" argument may be quite new to the point of implausibility to Western readers and in particular to those in neoliberal, look-the-other-way, and Elephant-in-the-Room-ignoring Australia. Accordingly, set out below are the key elements of the argument that ignoring the price of Carbon Pollution by climate criminal Australia means massive subsidizing of exported goods and hence massive dumping of Australian goods in China (noting that this argument applies to other countries as well as dumpers and dumpees).

(1). Polya's 3 Laws of Economics, economic externalities, damage-related Carbon Price, and inescapable Carbon Debt.

Polya's 3 Laws of Economics are modelled on the 3 Laws of Thermodynamics of chemistry and physics, to whit (1) the energy of a closed system is constant, (2) entropy (S, disorder, chaos, lack of information content) strives to a maximum, and (3) zero molecular mobility at absolute zero (0 degrees Kelvin or minus 273.15 degrees Centigrade). Polya's analogous 3 Laws of Economics are (1) Profit = Price – Cost of Production (COP), (2) deceit about COP strives to a maximum, and (3) zero jobs on a dead planet [21]. The success of the greed-driven denialists and effective climate change denialists in ignoring the cost of carbon pollution is a deadly and disastrous example of Polya's Second Law of Economics, to whit deception about the Cost of Production strives to a maximum.

In stark contrast to the anti-science denialists, science-trained Pope Francis has stated: "Yet only when the economic and social costs of using up shared environmental resources are recognized with transparency and fully borne by those who incur them, not by other peoples or future generations, can those actions be considered ethical" [22, 23]. Eminent UK economist Lord Nicholas Stern has described this massive deceit over these economic "externalities" thus: "The problem of climate change involves a fundamental failure of markets: those who damage others by emitting greenhouse gases generally do not pay. Climate change is a result of the greatest market failure the world has seen. The evidence on the seriousness of the risks from inaction or delayed action is now overwhelming. We risk damages on a scale larger than the two world wars of the last century. The problem is global and the response must be a collaboration on a global scale" [24].

The International Monetary Fund (IMF) has exposed the massive deceit of the neoliberal denialists in stating that while a damage-related Carbon Tax of $75 per tonne CO_2 would be an effective way of addressing the climate threat, the present global average Carbon Price is presently only $2 per ton CO_2, a tiny fraction of what is needed for keeping below the 2°C target [10, 25]. Climate economist Dr Chris Hope (of 120-Nobel-Laureate Cambridge University) and climate scientist Professor James Hansen (of NASA and 96-Nobel-Laureate Columbia University) have independently estimated a damage-related Carbon Price of about $200 per tonne CO_2-equivalent [26-28].

This massive corporate and political deceit in ignoring the gigantic economic externality measured by a damage-related Carbon Price has created a huge, inescapable and assiduously ignored Carbon Debt for future generations of $200-250 trillion that is increasing each year by 63.8 Gt CO_2-e per year x $200 /t CO_2-e = $13 trillion annually [3, 28, 29]. While conventional financial debt can be voided by default, bankruptcy or printing money (quantitative easing), Carbon Debt is inescapable – thus, for example, unless sea walls are built, cities, towns and arable land will be inundated by rising sea levels.

 (2). Methane (CH_4) leakage makes gas worse than coal GHG-wise and makes Australia (population 25.5 million, 0.33% of world population) a world-leading per capita GHG polluter responsible for 5.4% of annual global GHG pollution.

The Australian Government and Australian Mainstream media routinely assert that Australia has a GHG pollution of about 535 million tonnes (Mt) CO_2-equivalent (CO_2-e) year (financial year

2017/18) [30-35] and hence an annual per capita GHG pollution of about 535 Mt CO_2-e per year/25 million persons = 21 tonnes CO_2-e per person per year. However this estimate ignores land use, assumes a Global Warming Potential for methane (CH_4) of 21 relative to the same mass of carbon dioxide (CO_2) on a 100 year time frame, ignores massive gas leakage, and ignores Australia's huge Exported GHG pollution from export of coal, liquefied natural gas (LNG), iron ore and methanogenically-derived meat.

Methane (CH_4) is a gas, is 85% of natural gas, leaks, and has a Global Warming Potential (GWP) 105 times that of the same mass of CO_2 on a 20 year time frame and with aerosol impacts considered [36]. One can readily calculate (assuming natural gas to be 100% CH_4 or CH_4 equivalent) that on this basis a systemic gas leakage of 2.6% would contribute as much GHG pollution as generating the greenhouse gas CO_2 by burning the remaining 97.4% of the gas [37-40]. Thus burning 1 tonne carbon yields 3.67 tonnes CO_2 and combustion of 1 tonne of CH_4 with zero leakage yields 2.75 tonne CO_2. However combustion of 1 tonne CH_4 with 2.6% leakage yields 2.68 tonne CO_2 (from burning 97.4% of the CH_4) plus 2.68 tonne CO_2-equivalent (from the GHG effect of the leaked CH_4) = 5.4 tonnes CO_2-equivalent. One can crudely estimate that with a mere 2.6% of systemic leakage (estimates range from 2.3-5.4% leakage in the US [41-45]), burning 1 tonne of gas generates 1.5 times the CO_2-equivalent produced from burning 1 tonne of coal.

World Bank analysts carefully re-evaluated the contribution of livestock production to world annual GHG pollution, and found that the world's annual total rose from 41.76 billion tonnes CO_2-equivalent (CO_2-e) as estimated by the Food and Agricultural Organisation (FAO) to 63.80 billion tonnes CO_2-e, with livestock production contributing over 51% of the higher figure [46]. A key element of their analysis was to use a Global Warming Potential (GWP) of methane (CH_4) relative to that of carbon dioxide (CO_2) of 72 on a 20-year time frame rather than the 25 on a 100 year time frame used by the FAO [46]. Indeed the World Bank analysis evidently still understates the GHG pollution because NASA scientists have re-evaluated the GWP of CH_4 as 105 on a 20 year time frame with aerosol impacts considered [36].

Accordingly, more properly taking land use into account Australia's revised annual per capita GHG pollution was estimated in 2015 (t CO2-e per person) at 52.9, and at 116 if including its huge GHG-generating exports [47, 48]. Assuming a population of 25 million, this adjusts Australia's annual GHG pollution to 1,323 Mt (Domestic), 1,577 Mt (Exported) and 2,900 Mt (Domestic plus Exported).

Assuming Australian responsibility for gas fugitive emissions both at home and on route to foreign consumers, then Australia's Exported GHG must be updated taking CH_4 emissions and increased LNG and other ultimately GHG-generating exports into account. The land use-accommodating, revised estimate of Australian annual Domestic GHG emissions (1,323 Mt CO_2-e) must be revised upwards by adding the fugitive emissions from Domestic gas exploitation (99.7 Mt CO_2-e) to yield a total of 1,423 Mt CO_2-e (Domestic). The total Exported GHG emissions is by related detailed considerations revised upwards to 2,008 Mt CO_2-e, the breakdown being 997 Mt (coal), 33 Mt (oil), 4 Mt (LPG), 402 Mt (LNG) and 572 Mt (iron ore). Annual Australian Domestic GHG plus Exported GHG = 1,423 + 2,008 = 3,431 Mt CO_2-e, this corresponding to 3,431 x 100/ 63,800 = 5.4% of the global annual GHG pollution total of 63,800 Mt CO_2-e [3].

(3). Australia is among world leaders in 16 areas of climate criminality.

Thanks to the homicidal greed of climate criminal countries such as Australia, the present plus 1.1C temperature rise is already devastating Island Nations, a plus 1.5C is expertly predicted by the IPCC in the coming decade [49, 50], and a catastrophic plus 2C warming is now effectively unavoidable on present trends as Humanity and the Biosphere are existentially threatened by a worsening Climate Emergency, an Anthropocene Era biodiversity collapse, and Climate Genocide [51-64].

Climate criminal Australia is among world leaders for the following 16 climate criminal activities or parameters: (1) annual per capita greenhouse gas pollution, (2) live methanogenic livestock exports, (3) natural gas exports, (4) recoverable shale gas reserves that can be accessed by hydraulic fracturing (fracking), (5) coal exports, (6) land clearing, deforestation and ecocide, (7) speciescide or species extinction, (8) coral reef destruction, (9) whale killing and extinction threat through global warming impacting on krill stocks, (10) terminal carbon pollution budget exceedance, (11) per capita Carbon Debt, (12) ultimately GHG generating iron ore exports, (13) climate change inaction, (14) climate genocide and approach towards omnicide and terracide, (15) increasing Domestic GHG pollution despite Paris commitments to lower GHG pollution, and (16) complicity in 9 million annual air pollution deaths from burning carbon fuels, Australia's share being 75,000 overseas and 10,000 Domestically [64-67]. Australia with a mere 0.33% of the world's population contributes about 5.4% of global GHG pollution (including that due to the burning of Australia's world leading gas and coal exports) [3].

(4). Australia's Carbon Debt is $5 trillion, increasing at $0.7 trillion annually and increasing at $70,000 per head per year for under-30 year old Australians (USD).

Young Australians will have to pay a gigantic Carbon Debt that has been estimated to be increasing at US$70,000 per head per year for under-30 year old Australians [3]. Thus Australia's revised annual Domestic plus Exported GHG pollution is 3,438 Mt CO_2-e that corresponds to 3,431 Mt CO_2-e x $200 /t CO_2-e = $686 billion per year. The Carbon Debt for Australian is thus increasing at $27,520 (A$41,000) per head per year for every Australian, at $70,000 (A$105,000) per head per year for 9.816 million under-30 year old Australians, and at $146,000 (A$218,000) per head per year for 4.7 million 0-14 year old Australian children [29]. The annual increase in Australia's Carbon Debt of $686 billion will ultimately be borne by these 0-14 year old children and is increasing at the rate of $146,000 per head per year (A$218,000) [29] (for relevant demographic data see [68, 69]).

The worsening global Carbon Debt for future generations is $200-250 trillion and is increasing each year by 63.8 Gt CO_2-e per year x $200 /t CO2-e = $13 trillion annually [3, 28, 29]. This represents a massive and continuing outrage against the young – massive Climate Injustice and Intergenerational Inequity that demands Climate Revolution Now (non-violent, of course) [70-72]. Famed physicist Stephen Hawking has stated "We see great peril if governments and societies do not take action now to render nuclear weapons obsolete and to prevent further climate change" [55] – "now" is the operative word here because the World is badly running out of time.

Final comments and conclusions.

Climate criminal Australia is involved in massive dumping of exported goods in China (and elsewhere) through non-application of a required Carbon Price. Every $10 piece of Australian meat, $10 bag of Australian barley or $10 bottle of Australian wine sold to China has been produced through a gigantic and hidden Australian Government subsidy of about $5 of Carbon Debt to be paid for by future generations.

As science-trained Pope Francis has declared, the social and environmental cost of pollution must be "fully borne" by the polluters. This failure of Australia (and indeed the World) to apply a full Carbon Price to economic activities is grand larceny on a gigantic scale that robs the young of their future through imposition of gigantic and inescapable Carbon Debt. Globally this Carbon Debt is $200-250 trillion that is increasing each year by $13 trillion [3, 28, 29]. Australia's Carbon Debt is $5 trillion, increasing at $0.7 trillion annually, and increasing at $70,000 per head per year for under-30 year old Australians [3].

Yet the Silence is Deafening in neoliberal, One Percenter-dominated, look-the-other-way Mainstream Australia [73]. The overwhelming consensus of Australian Mainstream journalist, editor, politician, academic and commentariat presstitutes is that China's imposition of tariffs on Australian beef and barley is "payback" for US lackey Australia's craven support for Trump America's multi-pronged anti-China campaign. For all I know they may right in this instance, but ignoring the Elephant-in-the-Room of remorselessly increasing, inescapable and gigantic Carbon Debt will not make it go away. Eventually the mendacious, ecocidal, anti-science and neoliberal climate criminals running Australia will be brought to account by an indignant World, including the indignant youth of Australia.

References.

[1]. Trading Economics, "Australia GDP", 2019: https://tradingeconomics.com/australia/gdp.
[2]. Australian Bureau of Statistics, "Australia's population grew by 1.4 per cent", 18 June 2020: https://www.abs.gov.au/ausstats/abs@.nsf/mediareleasesbyCatalogue/CA1999BAEAA1A86ACA25765100098A47.
[3]. Gideon Polya, "Methane leakage makes Australia a world-leading per capita Greenhouse Gas polluter", Countercurrents, 18 February 2020: https://countercurrents.org/2020/02/methane-leakage-makes-australia-a-world-leading-per-capita-greenhouse-gas-polluter/.
[4]. Chris Hope, "How high should climate change taxes be?", Working Paper Series, Judge Business School, University of Cambridge, 9, 2011: http://www.jbs.cam.ac.uk/media/assets/wp1109.pdf.
[5]. James Hansen, "Climate change in a nutshell: the gathering storm", Columbia University, 18 December 2018: http://www.columbia.edu/~jeh1/mailings/2018/20181206_Nutshell.pdf.
[6]. "Carbon Debt Carbon Credit": https://sites.google.com/site/carbondebtcarboncredit/.
[7]. Mike Foley and Eryk Bagshaw, "China suspends Australian beef imports from four abattoirs", Sydney Morning Herald, 12 May 2020: https://www.smh.com.au/politics/federal/china-suspends-australian-beef-imports-from-four-abattoirs-20200512-p54s4k.html.
[8]. Dan Conifer, "China imposes 80pc tariff on Australian barley for next five years amid global push for coronavirus investigation", ABC News, 18 May 2020: https://www.abc.net.au/news/2020-05-18/china-to-impose-tariffs-on-australian-barley/12261108.
[9]. Kath Sullivan, "How China hit Australian barley, then beef, and now eyes our wine", ABC News, 19 August 2020: https://www.abc.net.au/news/2020-08-19/china-eyes-australian-wine-export-in-latest-trade-move/12571672.
[10]. Gideon Polya, "Australia rejects IMF Carbon Tax & preventing 4 million pollution deaths by 2030", Countercurrents, 15 October 2019: https://countercurrents.org/2019/10/australia-rejects-imf-carbon-tax-preventing-4-million-pollution-deaths-by-2030/.
[11]. Euan Black, "Climate debate heats up as Anthony Albanese backs away from carbon price", The New Daily, 23 February 2020: https://thenewdaily.com.au/finance/finance-news/2020/02/23/climate-change-insiders-anthony-albanese/.

[12]. Fumi Matsumoto, "Australia's soaring exports to China at risk in diplomatic rift", Nikkei Asian Review, 7 August 2020: https://asia.nikkei.com/Economy/Trade/Australia-s-soaring-exports-to-China-at-risk-in-diplomatic-rift2.

[13]. Gideon Polya, "US lackey Australia attacks free speech of Senator Dastyari, Muslims, Chinese, journalists & truth-tellers", Countercurrents, 10 December 2017: https://countercurrents.org/2017/12/us-lackey-australia-attacks-free-speech-of-senator-dastyari-muslims-chinese-journalists-truth-tellers/.

[14]. Gideon Polya, "Australian Sinophobia & China-bashing from colonial persecution & White Australia to Trump America's Asia Deputy Sheriff", Countercurrents, 26 January 2018: https://countercurrents.org/2018/01/australian-sinophobia-china-bashing-colonial-persecution-white-australia-trump-americas-asia-deputy-sheriff/.

[15]. Anne Davies and Daniel Hurst, "ASIO raids home of NSW Labor MP Shaoquett Moselmane over alleged links to China", Guardian, 26 June 2020: https://www.theguardian.com/australia-news/2020/jun/26/asio-raids-home-nsw-labor-mp-shaoquett-moselmane-links-china.

[16]. Philip Coorey, "Daniel Andrews hits back on China deals", Australian Financial Review, 27 August 2020: https://www.afr.com/politics/federal/daniel-andrews-hits-back-on-china-deals-20200827-p55pq9.

[17]. Gideon Polya, "Review: "Silent Invasion. China's influence in Australia" by Clive Hamilton – feeding Australian Sinophobia", Countercurrents, 6 October 2018: https://countercurrents.org/2018/10/review-silent-invasion-chinas-influence-in-australia-by-clive-hamilton-feeding-australian-sinophobia/.

[18]. "Subversion of Australia": https://sites.google.com/site/subversionofaustralia/.

[19]. Gideon Polya, "Australian xenophobia targets China but ignores huge Israeli subversion of Australia", Countercurrents, 7 July 2018: https://countercurrents.org/2018/07/australian-xenophobia-targets-china-but-ignores-huge-israeli-subversion-of-australia/.

[20]. Gideon Polya, "China's Tibet health success versus passive mass murder of Afghan women and children by US Alliance", Global Research, 7 January 2018: https://www.globalresearch.ca/chinas-tibet-health-success-versus-passive-mass-murder-of-afghan-women-and-children-by-us-alliance/5625151.

[21]. Gideon Polya, "Polya's 3 Laws Of Economics Expose Deadly, Dishonest And Terminal Neoliberal Capitalism", Countercurrents, 17 October, 2015: https://www.countercurrents.org/polya171015.htm.

[22]. Pope Francis, Encyclical Letter "Laudato si'", 2015:
http://w2.vatican.va/content/francesco/en/encyclicals/documents/papa
-francesco_20150524_enciclica-laudato-si.html.
[23]. Gideon Polya, "Green Left Pope Francis Demands Climate
Action "Without Delay" To Prevent Climate "Catastrophe"",
Countercurrents, 10 August, 2015:
https://countercurrents.org/polya100815.htm.
[24]. Alison Benjamin, "Stern: climate change a "market failure"",
Guardian, 29 November 2007:
https://www.theguardian.com/environment/2007/nov/29/climatechang
e.carbonemissions.
[25]. International Monetary Fund (IMF), "Fiscal Monitor: how to
mitigate climate change". Executive Summary", September 2019:
file:///C:/Users/Gideon/AppData/Local/Temp/execsum-6.pdf.
[26]. Chris Hope, "How high should climate change taxes be?",
Working Paper Series, Judge Business School, University of
Cambridge, 9,
2011: http://www.jbs.cam.ac.uk/media/assets/wp1109.pdf.
[27]. James Hansen, "Climate change in a nutshell: the gathering
storm", Columbia University, 18 December 2018:
http://www.columbia.edu/~jeh1/mailings/2018/20181206_Nutshell.pd
f.
[28]. "Carbon Debt Carbon Credit":
https://sites.google.com/site/carbondebtcarboncredit/.
[29]. Gideon Polya, "Inescapable $200-250 trillion global Carbon
Debt increasing by $16 trillion annually", Countercurrents, 27 April
2019: https://countercurrents.org/2020/02/methane-leakage-makes-
australia-a-world-leading-per-capita-greenhouse-gas-polluter/.
[30]. Australian Government, "Australian Energy Update 2018":
https://www.energy.gov.au/sites/default/files/australian_energy_updat
e_2018.pdf.
[31]. Lisa Cox, "Australia's emissions reach the highest on record",
Guardian 9 July 2019: https://www.theguardian.com/australia-
news/2019/jul/09/australias-emissions-reach-the-highest-on-record-
driven-by-electricity-sector.
[32]. Lisa Cox, "Australia's carbon emissions highest on record, data
shows", Guardian, 13 December 2018:
https://www.theguardian.com/australia-news/2018/dec/13/australias-
carbon-emissions-highest-on-record-data-shows.
[33] Penny Timms and Michael Slezak, "Australia's greenhouse gas
emissions rise again, according to delayed Federal Government data",
ABC News, 6 June 2019: https://www.abc.net.au/news/2019-06-
06/australian-emissions-rise-again-delayed-government-data-
shows/11184906.

[34] Michael Slezak, "Australia's greenhouse gas emissions soar in latest figures", Guardian, 4 August 2017: https://www.theguardian.com/australia-news/2017/aug/04/australias-greenhouse-gas-emissions-soar-in-latest-figures ;

[35]. Australian Government, Department of the Environment and Energy, "Quarterly update of Australia's national greenhouse gas inventory: March 2019", March 2019: https://www.environment.gov.au/system/files/resources/6686d48f-3f9c-448d-a1b7-7e410fe4f376/files/nggi-quarterly-update-mar-2019.pdf.

[36]. Drew T. Shindell, Greg Faluvegi, Dorothy M. Koch, Gavin A. Schmidt, Nadine Unger and Susanne E. Bauer, "Improved Attribution of Climate Forcing to Emissions", Science, 30 October 2009: Vol. 326 no. 5953 pp. 716-718: http://www.sciencemag.org/content/326/5953/716.

[37]. "Gas is not clean energy": https://sites.google.com/site/gasisnotcleanenergy/home.

[38]. "2011 climate change course": https://sites.google.com/site/300orgsite/2011-climate-change-course.

[39]. "Methane Bomb Threat": https://sites.google.com/site/methanebombthreat/.

[40]. Gideon Polya, "Australian commitment to unlimited gas exploitation threatens planet & invites global blowback", Countercurrents, 8 April 2017: https://countercurrents.org/2017/04/australian-commitment-to-unlimited-natural-gas-exploitation-threatens-planet-invites-global-blowback.

[41]. Kathryn McKain et al. "Methane emissions from natural gas infrastructure and use in the urban area of Boston, Massachusetts", PNAS, 112 (7) 1941-1946, February 17, 2015: https://www.pnas.org/content/112/7/1941.abstract).

[42]. Megan Geuss, "Study: US oil and gas methane emissions have been dramatically underestimated", Ars Technika, 23 June 2018: https://arstechnica.com/science/2018/06/study-us-oil-and-gas-methane-emissions-have-been-dramatically-underestimated/.

[43]. Robert Howarth, "Methane emissions and climatic warming risk from hydraulic fracturing and shale gas development: implications for policy", Energy & Emission Control Technologies, 8 October 2015: https://www.eeb.cornell.edu/howarth/publications/f_EECT-61539-perspectives-on-air-emissions-of-methane-and-climatic-warmin_100815_27470.pdf.

[44]. "Gas leakage – systemic gas leakage in the US is about 5.8%", Gas is not clean energy: https://sites.google.com/site/gasisnotcleanenergy/gas-leakage.

[45]. Sophie Vorrath, "Australia's new carbon bomb: uncounted coal seam gas emissions", Renew Economy, 26 October 2016: https://reneweconomy.com.au/australias-new-carbon-bomb-uncounted-coal-seam-gas-emissions-14457/.

[46]. Robert Goodland and Jeff Anfang. "Livestock and climate change. What if the key actors in climate change are … cows, pigs and chickens?", World Watch, November/December 2009: https://pdfs.semanticscholar.org/6704/c7a0777c82357704d82b9ae800 7c1197cb07.pdf?_ga=2.187734888.1597394103.1556059730-1006954717.1556059730.

[47]. Gideon Polya, "Revised Annual Per Capita Greenhouse Gas Pollution For All Countries – What Is Your Country Doing?", Countercurrents, 6 January, 2016: https://countercurrents.org/polya060116.htm.

[48]. Gideon Polya, "Exposing And Thence Punishing Worst Polluter Nations Via Weighted Annual Per Capita Greenhouse Gas Pollution Scores", Countercurrents, 19 March, 2016: https://countercurrents.org/polya190316.htm.

[49]. IPCC, "Global warming of 1.5 °C. Summary for Policymakers", 8 October 2018: http://report.ipcc.ch/sr15/pdf/sr15_spm_final.pdf.

[50]. Gideon Polya, "IPCC +1.5C avoidance report – effectively too late, but stop coal burning for "less bad" catastrophes', Countercurrents, 12 October 2018: https://countercurrents.org/2018/10/ipcc-1-5c-avoidance-report-effectively-too-late-but-stop-coal-burning-for-less-bad-catastrophes.

[51]. Andrew Glikson, "Inferno: from climate denial to planetary arson", Countercurrents, 8 September 2019: https://countercurrents.org/2019/09/inferno-from-climate-denial-to-planetary-arson.

[52]. "Are we doomed?": https://sites.google.com/site/300orgsite/are-we-doomed.

[53]. "Nuclear weapons ban, end poverty and reverse climate change": https://sites.google.com/site/drgideonpolya/nuclear-weapons-ban.

[54]. "Too late to avoid global warming catastrophe": https://sites.google.com/site/300orgsite/too-late-to-avoid-global-warming.

[55]. Stephen Hawking, "Brief Answers to the Big Questions", John Murray, 2018, Chapter 7.

[56]. Gideon Polya, "Australian commitment to unlimited gas exploitation threatens planet & invites global blowback", Countercurrents, 8 April 2017: https://countercurrents.org/2017/04/australian-commitment-to-

unlimited-natural-gas-exploitation-threatens-planet-invites-global-blowback.
[57]. "Stop air pollution deaths":
https://sites.google.com/site/300orgsite/stop-air-pollution-deaths.
[58]. Gideon Polya, "Wrong Way Go Back – Global Sectoral Greenhouse Gas Emissions Are All In The Wrong Direction", Countercurrents, 20 March 2020:
https://countercurrents.org/2020/03/wrong-way-go-back-global-sectoral-greenhouse-gas-emissions-are-all-in-the-wrong-direction/.
[59]. William J. Ripple et al., 15,364 signatories from 184 countries, "World scientists' warning to Humanity: a second notice", Bioscience, 13 November
2017: https://academic.oup.com/bioscience/advance-article/doi/10.1093/biosci/bix125/4605229.
[60]. Gideon Polya, "Over 15,000 scientists issue dire warning to humanity on catastrophic climate change and biodiversity loss", Countercurrents, 20 November
2017: https://countercurrents.org/2017/11/20/over-15000-scientists-issue-dire-warning-to-humanity-on-catastrophic-climate-change-and-biodiversity-loss/.
[61]. William Ripple et al., "World scientists' warning of a climate emergency", BioScience, 5 November 2019:
https://academic.oup.com/bioscience/advance-article/doi/10.1093/biosci/biz088/5610806.
[62]. Gideon Polya, "Extrapolating 11,000 scientists' climate emergency warning to 2030 catastrophe", Countercurrents, 14 November 2019: https://countercurrents.org/2019/11/extrapolating-11000-scientists-climate-emergency-warning-to-2030-catastrophe.
[63]. "Climate Genocide":
https://sites.google.com/site/climategenocide/.
[64]. Gideon Polya, "How much negative carbon emissions, negative population growth& negative economic growth is needed to save planet?", Countercurrents, 28 November 2018:
https://countercurrents.org/2018/11/how-much-negative-carbon-emissions-negative-population-growth-negative-economic-growth-is-needed-to-save-planet.
[65]. Gideon Polya, "Australia rejects IMF Carbon Tax & preventing 4 million pollution deaths by 2030", Countercurrents, 15 October 2019: https://countercurrents.org/2019/10/australia-rejects-imf-carbon-tax-preventing-4-million-pollution-deaths-by-2030.
[66]. Gideon Polya, "Latest Lancet data imply Adani Australian coal project will kill 1.4 million Indians", Countercurrents, 21 April 2017: https://countercurrents.org/2017/04/latest-lancet-data-imply-adani-australian-coal-project-will-kill-1-4-million-indians.

[67]. Gideon Polya, "War criminal & climate criminal Australian deception at UN General Assembly", Countercurrents, 29 September 2019: https://countercurrents.org/2019/09/war-criminal-climate-criminal-australian-deception-at-un-general-assembly.

[68]. "UN Population Division World Population Prospects": https://population.un.org/wpp/.

[69]. Australian Bureau of Statistics, "Twenty years of population change", 2019: https://www.abs.gov.au/ausstats/abs@.nsf/0/1CD2B1952AFC5E7AC A257298000F2E76.

[70]. Gideon Polya, editor, "Climate Justice & Intergenerational Equity": https://sites.google.com/site/300orgsite/climate-justice.

[71]. Gideon Polya, editor, "Stop climate crime": https://sites.google.com/site/300orgsite/stop-climate-crime.

[72]. Gideon Polya, editor, "Climate Revolution Now": https://sites.google.com/site/300orgsite/climate-revolution.

[73]. Gideon Polya, "Scores Of Huge Realities Resolutely Ignored By Mendacious, US Lackey Mainstream Australia", Countercurrents, 11 August 2020: https://countercurrents.org/2020/08/scores-of-huge-realities-resolutely-ignored-by-mendacious-us-lackey-mainstream-australia/.

CHAPTER 19: AUSTRALIA REJECTS IMF CARBON TAX & PREVENTING 4 MILLION POLLUTION DEATHS BY 2030

[First published as Gideon Polya, "Australia rejects IMF Carbon Tax & preventing 4 million pollution deaths by 2030", Countercurrents, 15 October 2019: https://countercurrents.org/2019/10/australia-rejects-imf-carbon-tax-preventing-4-million-pollution-deaths-by-2030.]

An International Monetary Fund (IMF) report on climate change mitigation advocates a Carbon Tax of $75 per ton of CO_2 by 2030 that, if progressively implemented in the G20 countries alone, would prevent an estimated 4 million air pollution deaths by 2030. However the effective climate change denialist Coalition Government of Australia, a G20 country that is among world leaders in 15 areas of climate criminality, has flatly rejected a global Carbon Tax. If all G20 countries followed Australia's rejection of a global Carbon Tax then they would be complicit in an unconscionable Climate Genocide and Climate Holocaust killing 4 million people over the next decade.

Here are the key findings of the IMF report on climate change mitigation (September 2019): "Global warming is threatening our planet and living standards around the world, and the window of opportunity for containing climate change to manageable levels is closing rapidly. Carbon dioxide (CO_2) emissions are a key driver of this alarming trend. Fiscal policy has an important role to play...

Action to date has been inadequate. The 2015 Paris Agreement goes in the right direction, but the commitments countries have made fall well short of those needed to limit global warming to the level considered safe by scientists—2°C, at most, above preindustrial temperatures. Furthermore, it remains uncertain whether countries are reducing emissions as agreed. The longer that policy action is delayed, the more emissions will accumulate in the atmosphere and the greater the cost of stabilizing global temperatures—let alone of failing to do so…

Limiting global warming to 2°C or less requires policy measures on an ambitious scale, such as an immediate global carbon tax that will rise rapidly to $75 a ton of CO_2 in 2030. Under such a scenario, over 10 years electricity prices would rise, on average, by 45 percent cumulatively and gasoline prices by 15 percent, for households, compared with the baseline (no policy action). The revenue from such a tax (1.5 percent of GDP in 2030, on average, for the Group of Twenty [G20] countries) could be redistributed, for example, to assist low-income households, support disproportionately affected workers or communities (for example, coal-mining areas), cut other taxes…

The shift from fossil fuels will not only transform an economy but also profoundly change the lives of households, businesses, and communities. Importantly, the shift would generate additional and immediate domestic environmental benefits, such as lower mortality from air pollution (725,000 fewer premature deaths in 2030 for a $75 a ton tax for G20 countries alone). Businesses that deploy new technologies would earn profits and create jobs, which in the renewables sector already reached 11 million globally in 2017…

Some advanced and emerging market economies already use carbon taxes and emission trading systems, but insufficiently. Indeed, the average price on global emissions is currently $2 a ton, a tiny fraction of what is needed for the 2°C target. An early start to reinforce the Paris process could be made through a carbon price floor arrangement among countries with the largest emissions…" [1].

The IMF's assertion of "725,000 fewer premature deaths in 2030 for a $75 a ton tax for G20 countries alone" means (assuming that such a Carbon Tax could be rapidly implemented by all the G20 countries) that lives saved in the period 2020-2030 would average 725,000/2 = 362,500 lives saved per year, and thus would total 362,500 lives per year x 11 years = 4.0 million lives saved (2020-2030).

However in response to the Australian ABC query "Would Australia accept a global Carbon Tax?" the Australian Coalition Government simply and emphatically replied "No" [2, 3].

If all of the G20 followed Australia's example then they would be complicit in a Climate Genocide in which 4 million people would die prematurely from carbon pollution in 2020-2030. Presently each year 9 million people die from air pollution (World Health Organization, WHO) [4, 5], this including about 10,000 Australians and 75,000 dying from the effects of pollutants from the burning of Australia's world leading coal exports [5].

This IMF-adumbrated 4 million Climate Genocide carnage should be considered in relation to other genocidal atrocities – it is of the same order as the carnage of the WW2 Jewish Holocaust (5-6 million deaths from violence or imposed deprivation) [6, 7] and the WW2 Bengali Holocaust (6-7 million Indians starved to death for strategic reasons by the British with Australian complicity) [8-22], but about 10 times smaller than deaths in the WW2 European Holocaust (30 million Slavs, Jews and Roma killed) [12], the WW2 Chinese Holocaust (35 million Chinese killed under the Japanese, 1937-1945) [12, 23], and 40 million Asian deaths from violence or imposed deprivation in Australian Coalition-backed, post-1950 US Asian wars [12].

It gets worse when one considers climate change-related deaths. It is estimated that about 1 million people die from climate change per se each year [24] but this may be an underestimate because 15 million people die avoidably each year from deprivation in the variously tropical and/or sub-tropical Developing World (minus China) that is disproportionately badly impacted by climate change [12]. Further, a variety of scholars from climate scientists to demographers predict a sustainable human population of only 0.5-1.0 billion by 2100, this implying a Climate Genocide involving about 10 billion deaths this century if climate change is not requisitely dealt with [25, 26].

Australia's ongoing commitment to war crimes and climate crimes.

Australia is among leading countries in 15 areas of climate criminal activities: (1) annual per capita greenhouse gas pollution, (2) methanogenic livestock exports, (3) natural gas exports, (4) recoverable shale gas reserves that can be accessed by hydraulic fracturing (fracking), (5) coal exports, (6) land clearing, deforestation and ecocide, (7) speciescide or species extinction, (8) coral reef destruction, (9) whale killing and extinction threat through global warming impacting on krill stocks, (10) terminal carbon pollution budget exceedance, (11) per capita Carbon Debt, (12) ultimately GHG generating iron ore exports, (13) climate change inaction, (14) Climate Genocide (its coal exports ultimately kill 75,000 people per year), and (15) increasing GHG pollution post-Paris (contrary to the Paris demand to decrease GHG pollution) [25].

Thus, for example, in 2018 Australia ranked 57 out of 60 countries for climate change action (only South Korea and Saudi Arabia were worse) [27]. Annual per capita greenhouse gas (GHG) pollution in tonnes CO_2-equivalent per person per year (and taking land use into account) is as follows for Australia and some other countries (2016): Australia (52.9; 116 if including its huge GHG-generating exports), United States (41.0), China (7.4), and India (2.1) [28]. Australia with 0.3% of the world's population is responsible for 4.5% of the world's greenhouse gas emissions (with Australia's exported GHG pollution included).

Australia's "No!" to a global Carbon Tax means an obscene commitment to a 4 million victim Climate Genocide over the next decade. The Australian Coalition has made it absolutely clear that it is opposed to a Carbon Tax. However the Labor Opposition, reeling from an unexpected election defeat in May 2019, remains silent. Indeed the right-wing Labor Shadow

Minister for Agriculture, Joel Fitzgibbon, has put the cat among the pigeons by urging that Labor should pragmatically go soft on climate change action and adopt a pollution reduction position closer to that of the climate criminal Coalition Government in order to increase its electoral appeal [29]. Labor promised before the May 2019 election to cut economy-wide emissions by 45% on 2005 levels by 2030 whereas the Coalition's target is a miserable 26-28% cut, with both involving offsetting emissions by the manipulative trick of purchasing international credits (e.g. from non-destruction of rain forests in the Developing World) [30]. The Australian Coalition Government and Labor Opposition have a common climate criminal policy of indefinite export of coal, gas, iron ore and methanogenically-derived meat.

Australia's being among world leaders in 15 areas of climate criminality make it disproportionately complicit in a worsening Climate Genocide that, in the absence of requisite action, is expected to kill 10 billion people this century en route to a sustainable world population in 2100 of merely 0.5-1.0 billion people [25, 26]. However Australia is notorious for its 2-century Aboriginal Genocide and Aboriginal Ethnocide (2 million Indigenous deaths from violence, 0.1 million, or imposed deprivation; of 350-750 Indigenous languages and dialects only 150 remain and of these all but 20 are endangered) [31]. As UK or US lackeys, Australians have invaded 85 countries (as compared to the British 193 countries, France 82, the US 72 (52 after WW2), Germany 39, Japan 30, Russia 25, Canada 25, Apartheid Israel 12 and China 2) [32, 33], and of these invasions 30 were genocidal as defined by the UN Genocide Convention [34]. Australia as part of a US Alliance continues to occupy Afghanistan in gross violation of the UN Genocide Convention and the Geneva Convention [35]. Through its joint US-Australia Pine Gap electronic spying base US lackey Australia plays a key role in US nuclear terrorism [36] and in the targeting of war criminal US drone strikes in 7 countries [37-39]. The Liberal Party-National Party Coalition Government and the Labor Opposition (aka the Lib-Labs) endless proclaim "Australian values" but it would appear that genocide is a core White Australian value, from the post-1788

British invasion Aboriginal Genocide to the presently worsening Climate Genocide. Australia has been involved in all post-1950 US Asian wars, atrocities associated with 40 million Asian deaths from violence or imposed deprivation [12] – the US lackey Coalition backed all of these vile wars and US lackey Labor backed all but 2 (the Vietnam War and the Iraq War).

About 10,000 Australians die each year from air pollution and 200,000 have died thus this century [5]. However this is just the tip of an iceberg of Australian preventable deaths. The Stupidity, Ignorance and Egregious Greed (SIEG as in Dr Strangelove and "Sieg heil") of successive neoliberal Lib-Lab Australian governments has made them complicit in the preventable deaths of 1.7 million of their fellow Australians this century. Thus each year about 85,000 Australians die preventably from "life style" and "political choice" reasons, the breakdown (including some overlaps) being as follows: (1) 26,000 annual Australian deaths from adverse hospital events, (2) 17,000 obesity-related Australian deaths, (3) 15,500 smoking-related Australian deaths, (4) 10,000 carbon burning pollution-derived Australian deaths, (5) 4,000 avoidable Indigenous Australian deaths, (6) 5,600 Australian alcohol-related deaths, (7) 2,900 Australian suicides (circa100 being veterans), (8) 1,400 Australian road deaths, (9) 630 Australian opiate drug-related deaths with 570 linked to US restoration of the Taliban-destroyed Afghan opium industry, and (10) 300 Australian homicides (80 being of women killed domestically) [40-43].

Final comments.

The IMF has proposed a Carbon Tax rising rapidly to "$75 a ton of CO_2 in 2030" in order to keep global warming to less than a catastrophic plus 2C [1]. Crucially the IMF states that this will "lower mortality from air pollution (725,000 fewer premature deaths in 2030 for a $75 a ton tax for G20 countries alone)". However it must be noted that Dr Chris Hope (of 90-Nobel-Laureate University of Cambridge) has determined a damage–related Carbon Price for Europe of $200 per tonne CO_2 [44].

Presently annual global greenhouse gas (GHG) pollution totals 63.8 billion tonnes (70.3 billion tons) CO_2-equivalent (taking land use into account and a global warming potential for methane based on a 20 year time frame) [45]. Thus the annual value of a Carbon Tax would be about $5.3 trillion (IMF) and $12.8 trillion (Hope). By way of comparison, the world nominal GDP in 2019 was projected to be $88 trillion [46].

By way of further comparison, many scientists consider that we must return the atmospheric CO_2 to about 300 ppm CO_2 for a safe and sustainable planet for all peoples and all species [47, 48]. Lowering the atmospheric CO_2 from the present 410 ppm CO_2 to a requisite 300 ppm CO_2 would mean removing 110 ppm CO_2 x 7.8 billion tonnes CO_2 per ppm CO_2 = 858 billion tonnes CO_2 i.e. a Carbon Debt of 858 billion tonnes CO_2. Assuming a damage-related Carbon Price of US$200 per tonne CO_2-equivalent [44], this corresponds to a Carbon Debt of $172 trillion [49, 50].

The bottom line is that we must pay the true cost of the goods and services we buy through mandatory incorporation of "externalities" into prices. Thus even the social conservatives Pope Francis I and Pope Benedict XVI have clearly stated: "Yet only when the economic and social costs of using up shared environmental resources are recognized with transparency and fully borne by those who incur them, not by other peoples or future generations, can those [economic] actions be considered ethical" [51, 52]. Unfortunately, the mendacious and politically dominant neoliberal One Percenters refuse to pay for these "economic and social costs of using up shared environmental resources" and are happy to exploit and pollute Humanity's common assets of the air, the ocean and the environment for free.

I have translated the 3 Laws of Thermodynamics, to whit, (1) conservation of energy, (2) entropy (disorder, chaos, lack of information content) increases to a maximum, and (3) zero motion at Absolute Zero, to Polya's 3 Laws of Economics, to whit (1) Profit = Price minus Cost of Production, (2) Deceit about Cost of Production increases to a maximum, and (3) no life, work, price or profit on a dead Planet [53]. This massive neoliberal deceit over the true Cost of Production illustrates Polya's Second Law of Economics, and the result is increasingly disastrous for Humanity and the Biosphere. The damage is already so severe as to demand that negative carbon emissions, negative population growth and negative economic growth are needed to save the planet [54].

In response to the worsening Climate Emergency the IMF has issued what is actually a very modest moral challenge to the prosperous countries of the G20 – rapidly and progressively introduce a big Carbon Tax to tackle climate change or become complicit in 4 million pollution deaths in the coming decade. Climate criminal Australia has immediately said "no" – how will the other G20 countries respond? The world is waiting.

What can decent people do in the face of this worsening Climate Emergency? Decent people must (a) inform everyone they can, and (b) urge and apply Boycotts, Divestment and Sanctions (BDS) against all people, politicians, parties, countries, collectives and corporations disproportionately involved in the worsening Climate Emergency that already kills about 1 million people each year.

References.

[1]. International Monetary Fund (IMF), "Fiscal Monitor: how to mitigate climate change". Executive Summary", September 2019: file:///C:/Users/Gideon/AppData/Local/Temp/execsum-6.pdf.
[2]. Paul Karp, "Josh Frydenberg rejects IMF report that Australia will fail to meet Paris target", Guardian, 11 October 2019: https://www.theguardian.com/environment/2019/oct/11/australia-will-fail-to-meet-paris-target-even-with-carbon-price-of-us75-a-ton-imf-says.
[3]. Josh Frydenberg," Interview with Sabra Lane, AM, ABC Melbourne", Josh Frydenberg, 11 October 2019: http://ministers.treasury.gov.au/ministers/josh-frydenberg-2018/transcripts/interview-sabra-lane-am-abc-melbourne.
[4]. World Health Organization (WHO), "Air pollution": https://www.who.int/airpollution/en/.
[5]. "Stop air pollution deaths": https://sites.google.com/site/300orgsite/stop-air-pollution-deaths.
[6]. Martin Gilbert, "Jewish History Atlas", Weidenfeld and Nicolson, London, 1969.
[7]. Martin Gilbert "Atlas of the Holocaust", Michael Joseph, London, 1982.
[8]. Gideon Polya, "Australia And Britain Killed 6-7 Million Indians In WW2 Bengal Famine", Countercurrents, 29 September, 2011: http://www.countercurrents.org/polya290911.htm.
[9]. Gideon Polya, "Jane Austen and the Black Hole of British History. Colonial rapacity, holocaust denial and the crisis in biological sustainability", now available for free perusal on the web: http://janeaustenand.blogspot.com/2008/09/jane-austen-and-black-hole-of-british.html.
[10]. Madhusree Muckerjee, "Churchill's Secret War. The British Empire and the ravaging of India during World War II" (Basic Books, New York, 2010).
[11]. "Bengali Holocaust (WW2 Bengal Famine) writings of Gideon Polya", Gideon Polya: https://sites.google.com/site/drgideonpolya/bengali-holocaust.
[12]. Gideon Polya, "Body Count. Global avoidable mortality since 1950", this including an avoidable mortality-related history of every country since Neolithic times and now available for free perusal on the web: http://globalbodycount.blogspot.com.au/2012/01/body-count-global-avoidable-mortality_05.html.
[13]. Colin Mason, "A Short History of Asia. Stone Age to 2000AD" (Macmillan, 2000).

[14]. Bengal Famine, BBC radio broadcast series "The things we forgot to remember", 2008: http://www.open2.net/thingsweforgot/bengalfamine_programme.html.

[15]. Paul Greenough's "Prosperity and Misery in Modern Bengal: the Famine of 1943-1944" Oxford University Press, 1982.

[16]. Thomas Keneally, "Three Famines", Vintage House, Australia, 2011.

[17]. Gideon Polya, "Economist Mahima Khanna, Cambridge Stevenson Prize And Dire Indian Poverty", Countercurrents, 20 November, 2011: https://countercurrents.org/polya201111.htm.

[18]. Cormac O Grada. "Famine a short history" (Princeton University Press, 2009).

[19]. J. Dreze and Amartya Sen "Hunger and Public Action", Clarendon, Oxford, 1989.

[20]. N.G. Jog, "Churchill's Blind Spot: India", New Book Company, Bombay, 194).

[21]. A. Sen, "Famine Mortality: A Study of the Bengal Famine of 1943" in Hobshawn, E. (1981) (editor), "Peasants In History. Essays in Honour of David Thorner" (Oxford University Press, New Delhi).

[22]. T. Das, "Bengal Famine (1943) as Revealed in a Survey of Destitutes of Calcutta" (University of Calcutta, Calcutta).

[23]. "Backgrounder: China's WWII contributions in figures", New China, 3 September 2015: http://news.xinhuanet.com/english/2015-09/03/c_134582291.htm.

[24]. Damian Carrington, "Save millions of lives by tackling climate change, says WHO", Guardian, 6 December 2018: https://www.theguardian.com/environment/2018/dec/05/save-millions-of-lives-by-tackling-climate-change-says-world-health-organization.

[25]. Gideon Polya, "Millions join Global School Climate Strike – we are running out of time", Countercurrents, 22 September 2019: https://countercurrents.org/2019/09/millions-join-global-school-climate-strike-we-are-running-out-of-time.

[26]. Gideon Polya, editor, "Climate Genocide": https://sites.google.com/site/climategenocide/home.

[27]. Climate Watch Performance Index, "Results 2018": https://www.germanwatch.org/sites/germanwatch.org/files/publication/20504.pdf

[28]. Gideon Polya, "Revised Annual Per Capita Greenhouse Gas Pollution For All Countries – What Is Your Country Doing?", Countercurrents, 6 January, 2016: http://www.countercurrents.org/polya060116.htm.

[29]. Katharine Murphy, "Labor MPs condemn suggestion they adopt Coalition climate change policy", Guardian, 14 October 2019:

https://www.theguardian.com/australia-news/2019/oct/14/labor-mps-condemn-suggestion-they-adopt-coalition-climate-change-policy.
[30]. Phillip Coorey and Ben Potter, "Labor carbon policy relies on global credits: Greens", Australian Financial Review, 8 April 2019: https://www.afr.com/politics/federal/labor-carbon-policy-relies-on-global-credits-greens-20190408-p51bxi.
[31]. Gideon Polya, editor, "Aboriginal Genocide": https://sites.google.com/site/aboriginalgenocide/.
[32]. Gideon Polya, "As UK Lackeys Or US Lackeys Australians Have Invaded 85 Countries (British 193, French 80, US 70)", Countercurrents, 9 February, 2015: http://www.countercurrents.org/polya090215.htm.
[33]. Gideon Polya, editor, "Stop state terrorism": https://sites.google.com/site/stopstateterrorism/.
[34]. Gideon Polya, "Review: "The Cambridge History Of Australia" Ignores Australian Involvement In 30 Genocides", Countercurrents, 14 October, 2013: https://www.countercurrents.org/polya141013.htm.
[35]. Gideon Polya, "China's Tibet health success versus passive mass murder of Afghan women and children by US Alliance", Global Research, 7 January 2018: https://www.globalresearch.ca/chinas-tibet-health-success-versus-passive-mass-murder-of-afghan-women-and-children-by-us-alliance/5625169.
[36]. "Nuclear weapons ban, end poverty and reverse climate change": https://sites.google.com/site/drgideonpolya/nuclear-weapons-ban.
[37]. Philip Dorling, "Australian intelligence "feeding data" for deadly US drone strikes", Sydney Morning Herald, 26 May 2014:http://www.smh.com.au/federal-politics/political-news/australian-intelligence-feeding-data-used-for-deadly-us-drone-strikes-20140526-38ywk.html.
[38]. Mark Corcoran, "Drone strikes based on work at Pine Gap could see Australians charged, Malcolm Fraser says", Sydney Morning Herald, 29 April 2014: http://www.abc.net.au/news/2014-04-28/australians-could-be-charged-over-us-drone-strikes-fraser/5416224.
[39]. John Stapleton, "Australia's dirty secret", UNSW Canberra, 4 December 2015: https://www.unsw.adfa.edu.au/drone-wars-australias-dirty-secret.
[40]. Gideon Polya, "Horrendous Cost For Australia Of US War On Terror", Countercurrents, 14 October, 2012: https://countercurrents.org/polya141012.htm.
[41]. Gideon Polya, "Australian state terrorism (4). Jingoistic, US Lackey Australia's Deadly Betrayal Of Its Traumatized Veterans", Stop state terrorism, 2018:

https://sites.google.com/site/stopstateterrorism/australian-state-terrorism-4.

[42]. Gideon Polya, "Advance Australia Fair" Hides Australian Racism, Theft, Genocide, Ecocide, Speciescide & Terracide", Countercurrents, 1 July 2019: https://countercurrents.org/2019/07/advance-australia-fair-hides-australian-racism-theft-genocide-ecocide-speciescide-terracide.

[43]. "Exposing Australia": https://sites.google.com/site/exposingaustralia/home.

[44]. Chris Hope, "How high should climate change taxes be?", Working Paper Series, Judge Business School, University of Cambridge, 2011: http://www.jbs.cam.ac.uk/fileadmin/user_upload/research/workingpapers/wp1109.pdf.

[45]. Robert Goodland and Jeff Anfang. "Livestock and climate change. What if the key actors in climate change are … cows, pigs and chickens?", World Watch, November/December 2009: http://www.worldwatch.org/files/pdf/Livestock%20and%20Climate%20Change.pdf.

[46]. World Population Review, "GDP ranked by country 2019": http://worldpopulationreview.com/countries/countries-by-gdp/.

[47]. Gideon Polya, editor, 300.org: https://sites.google.com/site/300orgsite/300-org.

[48]. Gideon Polya, editor, "300.org – return atmosphere CO_2 to 300 ppm CO_2": https://sites.google.com/site/300orgsite/300-org—return-atmosphere-co2-to-300-ppm.

[49]. Gideon Polya, editor, "Carbon Debt Carbon Credit": https://sites.google.com/site/carbondebtcarboncredit/.

[50]. Gideon Polya, "Huge Carbon Debt and intergenerational injustice: CO2 drawdown necessity", Global Research, 7 June 2018: https://www.globalresearch.ca/huge-carbon-debt-and-intergenerational-injustice-co2-drawdown-necessity/5643365.

[51]. Gideon Polya, "Green Left Pope Francis Demands Climate Action "Without Delay" To Prevent Climate "Catastrophe"", Countercurrents, 10 August, 2015: https://www.countercurrents.org/polya100815.htm.

[52]. Gideon Polya, "Pope decrees full Carbon Price". MWC News, 28 July 2015: http://www.mwcnews.com/focus/analysis/53226-pope-decree.html?utm_source=twitterfeed&utm_medium=twitter.

[53]. Gideon Polya, "Polya's 3 Laws of Economics expose deadly, dishonest and terminal neoliberal capitalism", Countercurrents, 17 October, 2015: https://www.countercurrents.org/polya171015.htm.

[54]. Gideon Polya, "How much negative carbon emissions, negative population growth & negative economic growth is needed to save

planet?", Countercurrents, 28 November 2018: https://countercurrents.org/2018/11/how-much-negative-carbon-emissions-negative-population-growth-negative-economic-growth-is-needed-to-save-planet.

SECTION F: CASE STUDY # 1 – CLIMATE CRIMINAL AUSTRALIA THREATENS PLANET

"From analysis of published global site biomass data ($n = 136$) from primary forests, we discovered (*i*) the world's highest known total biomass carbon density (living plus dead) of 1,867 tonnes carbon per ha (average value from 13 sites) occurs in Australian temperate moist *Eucalyptus regnans* forest." H. Keith, B.G. Mackey and D. B. Lindenmayer, "Re-evaluation of forest biomass carbon stocks and lessons from the world's most carbon-dense forests", PNAS, 2009.

"Nadi Bay Declaration on the Pacific Islands Climate Change Crisis… The science warns of the real possibility that coral atoll nations could become uninhabitable as early as 2030. By 2100, the coral atoll nations of the Republic of the Marshall Islands, Tuvalu, Kiribati, Tokelau and the Maldives and many SIDS could be submerged." Leaders of the Pacific Small Islands Developing States (PSIDS) and the Small Island States (SIS), 2019.

"What is the relevance of Australia's ongoing protestations in the forum [South Pacific Forum]? If it's going to continue with that line, it poses a danger to the other countries in that forum. How can you justify being part of a family and part of a group which you're trying to destroy?" Anote Tong, former president of Kiribati, 2019.

"Every politician, lobbyist, pundit and journalist who has fought to block serious action on climate change bears responsibility for the increasing risk from a heating planet that is producing these deadly bushfires." Dr Di Natale, former Australian Greens Leader, 2019.

"Australia is among world leaders in 16 areas of climate criminality." Dr Gideon Polya, Countercurrents, 2020.

As an Australian scientist and humanitarian I have been long aware of Australia's disproportionately high contribution to a worsening Climate Crisis, and have written extensively about this glaring reality in books, articles in Countercurrents and in climate-related websites I have created [1-59]. In addition, articles by me have been published in other magazines, notably Media With Conscience News (MWC News), New Matilda, Green Left, Dissident Voice, Information Clearing House, Crime & Power, Arena and Global Research. In the first edition of my book "Jane Austen and the Black Hole of British History" I wrote of Australian climate change inaction (1998): "Australia has been called the Lucky Country because of a variety of historical, geographic, resource, strategic, institutional and social blessings. In short, the Lucky Country is a well-endowed democracy with a well-educated, prosperous and liberal population cognizant of the dark shadows of its past and the problems of the future. If such an "ideal" First World, high technology society is presently unable or unwilling to come to grips with its substantial contribution to global degradation and with the appalling human disaster that is the continuing lot of its own very small residual aboriginal population, will such a society be able to respond adequately to the approach of the global Bergen-Belsen that may be a mere half century away? And if such a resource-rich society is unable to respond in a proper and timely fashion, how then will the world respond? ... From this quick sketch we can see that Australia, a resource-rich, highly-educated, liberal, democratic and orderly nation with an extraordinarily high standard of living, declines to act in a fashion consistent with global environmental responsibility in relation to greenhouse gas emissions. In addition, it is continuing the 200 year process of environmental degradation and biodiversity destruction. If that were not enough, it is continuing the process of destruction of its indigenous cultures and is a passive witness to massive environmental damage and human rights abuse in a swathe of tropical countries to our north" (Chapter 17, pages 174 and 192 [1]). In an epilogue to the 2008 edition of this book I wrote: "Australia has a new Labor Government which is strong on nice-sounding rhetoric (e.g. the farcical 2008 Australia 2020 Summit) but which does not stand up to close scrutiny in a "politically correct racist" (PC racist) country committed to genocidal Bush-ite war policies and in which Coal is King" (page 200 [1]).

I have summarized Australia's disproportionately huge climate criminality (2020): "Thanks to the homicidal greed of climate criminal countries such as Australia, the present plus 1.1C temperature rise is already devastating Island Nations, and a catastrophic plus 2C warming is now effectively unavoidable on present trends. Climate criminal Australia is among world leaders for the following 16 climate criminal activities or parameters: (1) annual per capita greenhouse gas pollution, (2) live methanogenic livestock exports, (3) natural gas exports, (4) recoverable shale gas reserves that can be accessed by hydraulic fracturing (fracking), (5) coal exports, (6) land clearing, deforestation and ecocide, (7) speciescide or species extinction, (8) coral reef destruction, (9) whale killing and extinction threat through global warming impacting on krill stocks, (10) terminal carbon pollution budget exceedance, (11) per capita Carbon Debt, (12) ultimately GHG generating iron ore exports, (13) climate change inaction, (14) climate genocide and approach towards omnicide and terracide, (15) increasing Domestic GHG pollution despite Paris commitments to lower GHG pollution, and (16) complicity in 9 million annual air pollution deaths from burning carbon fuels, Australia's share being 75,000 overseas and 10,000 Domestically… Australia (0.33% of world population) generates 2.5% of upwardly revised global GHG emissions (Australian Domestic use only) and 5.4% (Australian Domestic plus Exported GHG emissions") [33- 44]).

The revised annual per capita greenhouse gas (GHG) pollution for all countries (tonnes CO_2-e per person per year) properly taking land use and methane into account (the world average being 63.80 billion tonnes CO_2-e / 7.137 billion people in 2013 = 8.9 tonnes CO_2-e per person per year): Belize (366.9), Guyana

(203.1), Malaysia (126.0), Papua New Guinea (114.7), Qatar (101.8), Zambia (97.5), Antigua & Barbuda (85.6), United Arab Emirates (82.4), Panama (68.0), Botswana (64.9), Liberia (55.0), Indonesia (53.6), New Zealand (53.2), **Australia (52.9; 116 if including its huge GHG-generating exports),** Nicaragua (51.2), Canada (50.1), Equatorial Guinea (47.5), Venezuela (45.2), Brazil (43.4), Myanmar (41.9), Ireland (41.4), United States (41.0) [22]. On this basis Australia ranks 4th in the world if one includes its huge Exported GHG pollution. However for some very poor countries GHG pollution via fossil fuel exploitation, forestry or other land use is critical for their survival, and hence per capita GDP (nominal) should be taken into account. The "annual per capita GHG pollution" for each country can been multiplied by the ratio of the "annual per capita GDP (nominal)" for each country to the world average (US$10,744) to get a "weighted score" – on this basis Australia (52.9 x 5.80 = 306.8; if including its huge GHG-generating exports, 116 x 5.80 = 672.8) is a world leader with United Arab Emirates (82.4 x 4.09 = 337.0) [23]. Climate criminal Australian politicians of the presently ruling Liberal Party-National Party Coalition or the (presently in Opposition) Labor (aka the Lib-Labs) reject per capita considerations, land use, the GWP of methane on a 20 year time frame, and methane leakage [14, 46] – as Big Brother declared in George Orwell's "1984": "2 plus 2 does not equal 4".

Disproportionately climate criminal countries like Australia (2020 population 25.5 million) and the US (2020 population) are holding the world to ransom. However while the American Empire is so rich and militarily powerful that it can presently largely do as it pleases, small Australia is vastly more vulnerable to global action. Writing on potential "blowback" from an indignant world, I stated (2020): "A damaging plus 1.5C of warming will come in the coming decade, and a catastrophic plus 2C temperature rise is now effectively unavoidable, but we are obliged to do everything we can to make the future "less bad" for future generations. In Australia and other profligately climate criminal countries, decent people will utterly reject the climate criminal climate change deniers and effective climate change deniers at the ballot box. Decent people around the world will subject disproportionately climate criminal people, politicians, parties, collectives, corporations and countries to Boycotts, Divestment and Sanctions (BDS). Decent countries will subject climate criminal people, corporations and countries to legal actions via the International Criminal Court and the International Court of Justice. Time is running out." [37].

Australia's continuing policy of remorseless climate criminality is accompanied by massive actual or effective climate change denialism and mendacity at home and abroad. Thus Australia refused to sign the Kyoto Protocol until 2007 and has consistently opposed effective international action by helping the US sabotage climate-related international meetings such as the 2007 Bali Climate Change Conference [3], the 2008 Poznan Climate Change Conference [3], the 2009 Copenhagen Climate Change Conference [3], the 2015 Paris Climate Conference, the 2019 Pacific Islands Forum [32], and the 2019 UN Climate Action Summit in New York [33]. Thanks to the greed and profound immorality of the US lackey Australia and like countries, the national commitment made at the 2015 Paris Climate Conference actually committed the world to a catastrophic 3.2C temperature rise (UN News, "UN Emissions Report: World on course for more than 3 degree spike, even if commitments are met", 26 November 2019: https://news.un.org/en/story/2019/11/1052171).

Notwithstanding its appalling and continuing record of climate criminality and war criminality [55, 56], Australia disingenuously postures in world forums about diverse matters from biodiversity preservation to human rights. Thus Australia is a world-leader in action against whaling by countries such as Japan. However in reality Australia through its disgraceful position as a world-leading GHG polluter is disproportionately contributing to the successive processes of loss of Antarctic sea ice, loss of phytoplankton, decreased krill and decreased whale nutrition [5, 11]. Australia is among world leaders in speciescide, ecocide and land clearing [33-44]. Australia's climate criminality and war criminality are deadly. Thus each year 9 million people die avoidably from the long-term effects of pollutants from the burning of carbon fuels, and of these 10,000 are Australians and 75,000 are the people who die annually overseas from the burning of Australia's world-leading coal exports [50]. About 1 million people die each year from the effects of climate change but unless requisite action is taken it is predicted that 10 billion people will die thus this century in a worsening Climate Genocide en route to a sustainable human population in 2100 of only circa 1 billion and average annual deaths from climate change of 100 million [51]. Remorselessly climate criminal Australian Governments are guilty of state terrorism, climate terrorism, and carbon terrorism. Similarly, climate criminal Australian citizens and corporations are guilty of egregious non-state terrorism, climate terrorism, and carbon terrorism [50, 52, 55-58]. Australia is a significant biofuel producer (ethanol from sugar and biodiesel fuel from canola), and is thus intimately implicated in the obscene, food-for-fuel Biofuel Genocide [54].

Though not strictly the subject of this book, it must be formally noted that Australia continues to be massively involved in fossil fuels-related state terrorism and war crimes as well as in egregious climate crimes. Thus as a UK or US lackey Australia has invaded 85 countries as compared to the British 193 countries, France 82, the US 72 (52 after WW2), Germany 39, Japan 30, Russia 25, Canada 25, Apartheid Israel 12, China 2, North Korea arguably zero (0), Palestine 0, and US- and Australia-sanctioned Iran zero (0) since the time of Sassanian Empire over 1,300 years ago. Australia is still involved in the illegal and war criminal occupation of Afghanistan, only finally removed forces from US-devastated Iraq after demands by the Iraqi Parliament, and via the joint US-Australia spying base at Pine Gap, Central Australia, Northern Territory, Australia is deeply complicit in US nuclear terrorism and in targeting illegal US drone bombing of 7 countries. Australia has been involved in all post-1950 US Asian wars, atrocities associated with 40 million Asian deaths from violence or imposed deprivation. US lackey Australia has been involved in the ongoing US-imposed post-9-11 Muslim Holocaust and Muslim Genocide (32 million Muslim deaths from violence, 5 million, or from deprivation, 27 million, in 20 Muslim countries invaded by the US Alliance since the US Government's 9-11 false flag atrocity that killed 3,000 people [43, 44]. Mass murder, whether by war or climate change, is the penultimate in racism, and mass murder by genocidal war or by near-terracidal Climate Genocide is the ultimate in racism [43, 44]. One notes that the Iraqi Holocaust and Iraqi Genocide (4.6 million deaths violence or imposed deprivation since imposition of deadly sanctions in 1990 and 9 million since British invasion in 1914) [43, 44] was fundamentally about oil, as explicitly noted by eminent analysts such as Alan Greenspan (long-time former Chairman of the US Federal Reserve) and Professor Noam Chomsky (of 100-Nobel Laureate Massachusetts Institute of Technology) (Peter Beaumont and Joanna Walters, "Greenspan admits Iraq was about oil, as deaths put at 1.2m", The Observer, 16 September 2007: http://www.theguardian.com/world/2007/sep/16/iraq.iraqtimelin e and Noam Chomsky quoted in Sherwood Ross, "Chomsky: Iraq invasion "major crime" designed to control Middle East oil", The Public Record, 3 November 2009: http://pubrecord.org/nation/5953/chomsky-invasion-major-crime/).

Ignorant, stupid, anti-science Trumpist climate change deniers aside, there is now substantial public recognition in the ostensibly democratic West that we are in a worsening Climate Crisis and a Climate Emergency. Like the other Western ostensible democracies Australia is a Plutocracy, Kleptocracy, Murdochracy, Lobbyocracy, Corporatocracy and Dollarocracy in which Big Money purchases people, politicians, parties, policies, public perception of reality, votes and thence more political power and more private profit. Australia is peculiar in that it has a compulsory preferential voting system that means that all adults must vote or face a fine, and for a valid vote people must indicate not just their number 1 preference but also their preferences in descending order for all the other candidates. Thus if a candidate fails to gain 50% of the vote then the second preferences are taken into account so that the electoral officers can find a winner by "2-party preferred vote".

In round figures the presently ruling conservative Liberal Party-ultraconservative National Party Coalition (30% anti-environment climate change deniers, 70% effective climate change deniers through climate change inaction) has about 40% of the primary (first preference) vote and achieves about 50% of the "2-party preferred vote". Labor (100% recognition of the climate crisis and pro-renewable energy but dominated by a pragmatic right wing that is pro-coal and pro-gas) has about 35% of the primary (first preference) vote and achieves about 50% of the "2-party preferred vote". The Australian Greens (science-informed, socially and environmentally progressive) have about 10% of the primary vote and mostly give their second preferences to Labor. And then there are Independent and Minor Parties who get about 15% of the first preference vote. Labor cannot win power without Greens preferences, and has a difficult balancing act involving having some pro-environment policies (e.g. more renewable energy) while supporting the pro-coal and pro-gas policies of the Coalition (notably, unlimited exports of coal, gas, iron ore and methanogenically-derived meat). In my despairing, science-informed, pro-planet view, the 50% of the population voting for the neoliberal Coalition are Stupid, Ignorant and Egregiously Greedy (SIEG as in Dr Strangelove and "Sieg Heil!").

With most Australians recognizing that there is a worsening Climate Crisis (as reflected in catastrophic drought, floods, storms and bushfires), the major parties (the Liberal Party National Party Coalition, the Libs, and the Labor Party, the Labs, and collectively known as the Lib-Labs) have a major problem of fooling as much of the electorate as they can, noting that everyone is compelled to vote. The Libs have little problem garnering the support of the SIEG 50% by supporting fossil fuel exploitation in terms of export earnings and jobs. Labor has a much more difficult job trying to convince the more rational and humane 50% of the electorate that it is "tackling climate" via rhetoric and faster implementation of renewable energy, while simultaneously being in agreement with the Libs that there should be unlimited exports of coal, gas, iron ore and methanogenically-derived meat coal. Labor's cowardice doesn't stop there – since the removal of the progressive Whitlam Labor

Government by a CIA-backed Coup in 1975, Labor has adopted a fervent policy of "all the way with the USA" (see John Pilger, "The British-American Coup that ended Australian independence", Guardian, 23 October 2014: https://www.theguardian.com/commentisfree/2014/oct/23/goug h-whitlam-1975-coup-ended-australian-independence). Labor's US lackey cowardice is in lockstep with that of the Libs, means Australian support for US nuclear terrorism and genocidal militarism, absurd hostility to China (Australia's top trading partner, of which more later), and blind support for ostensibly anti-terrorism laws that have now made Australia a pre-police state (see Brian Toohey, "Secret. The making of Australia's security state", Melbourne University Press, 2019, and George Williams, "Sacrificing Civil Liberties to Counter Terrorism: Where Will it End?", 2018 John Marsden Lecture, NSW Council for Civil Liberties, 22 November 2018: http://www.nswccl.org.au/2018_john_marsden_lecture_sacrifici ng_civil_liberties_to_counter_terrorism_where_will_it_end.)

In its last period of office (2007-2013) Labor under PM Kevin Rudd tried to disguise its failure to "tackle climate change" by proposing an Emissions Trading Scheme (ETS) called the Carbon Pollution Reduction Scheme (CPRS). This was sensibly rejected by the Greens as ineffective and hence counterproductive, but since then the right-wing media have routinely falsely accused the Greens of blocking a Price on Carbon. After PM Kevin Rudd was removed in a pro-Zionist-led Coup in June 2010 his successor PM Julia Gillard notoriously declared "There will be no carbon tax under a government I lead". However when barely returned to power in the 2010 Federal Election she required Greens support and introduced a grossly insufficient "baby steps" price on carbon that the Greens could support as better than nothing [7-17].

The Australian Greens: "We voted against the [Rudd] CPRS [ETS plan] because it was bad policy that would have locked in failure to take action on climate change. According to Treasury modelling, under the CPRS there would have been no reduction in emissions for 25 years. It gave billions in handouts to coal companies and big polluters, while it locked in emissions targets that failed the science… Just months later, we worked with a more collaborative Gillard Labor government and Independent MPs to introduce world-leading climate legislation. We fought for – and achieved – a much better outcome. This Greens-led package included a price on carbon that worked to reduce emissions, the establishment of the Clean Energy Finance Corporation and the Australian Renewable Energy Agency. It drove down pollution and has driven over $20 billion into over 600 energy projects around the country. And what happened? Pollution went down when the carbon price was in force. This was one of the few precious times in our history where emissions actually reduced -- and, in spite of conservative fear-mongering, the sky didn't fall in. It's still regarded as an example of world-leading emissions reduction legislation, and remains the basis of Greens carbon reduction policy today" (Greens, "The Greens and CPRS": https://greens.org.au/cprs). Of course the climate change denialist Coalition Government was returned in 2013 with the slogan "Axe the tax" and did exactly that. Australia's Domestic GHG pollution has increased since the abolition of the price on carbon. Labor's presently pragmatic leader Anthony Albanese has recognized the electoral toxicity of a Price on Carbon and recently announced that it was not necessary (the world's scientists and climate economists think otherwise) [49].

My comment on the Gillard Carbon Price scheme (2011): "The pro-coal, pro-gas Australian Labor Government has proposed a Carbon Tax-ETS-Ignore Agriculture (CTETSIA) policy that means climate change inaction by Australia, a world-leading per capita greenhouse gas (GHG) polluter. This disastrous policy promotes a pointless conversion from coal burning to gas burning (that is just as dirty GHG-wise because of methane leakage); institutes a futile cycle of increased prices coupled with obviating consumer compensation; scuppers science-demanded 100% renewable energy by 2020; institutes a counterproductive and fraudulent carbon emissions trading scheme (ETS); ignores agriculture which is responsible for over 50% of GHG pollution; trashes Australia's international reputation; and has compromised and divided the pro-environment, climate change action movement"[7]. In relation to ETS schemes in general I commented (circa 2013): "Numerous economists, activists, scientists and climate scientists argue that what is needed is a transparent, science-based Carbon Tax rather than an ineffective and fraudulent ETS. The Carbon Trading ETS approach is variously condemned as dodgy because it is empirically ineffective, accordingly counterproductive and is inherently fraudulent because it would involve particular rich countries selling licences to enable corporations to pollute the one common atmosphere and ocean of all countries and all peoples" (for this and numerous other science-informed opinions on the dangerous, counterproductive and fraudulent nature of Emissions Trading Schemes see [49]).

Pro-coal and pro-gas Australia is presently led by a pro-coal and pro-gas Coalition Government licensed by a cowardly, pro-coal and pro-gas Labor Opposition. Australia is trapped in the presently globally dominant neoliberal ideology that demands maximal freedom for the smart and advantaged to exploit natural and human resources for maximal private profit. If that were not bad enough the neoliberal politicians of Stupid, Ignorant and Egregiously Greedy (SIEG) Australia add insult to injury through lying by commission, lying by omission and anti-science spin involving the selective use of asserted facts to support a partisan position. This anti-science, look-the-other-way Australian culture of egregious falsehood has led me to despairingly describe it as Orwellian in articles published in 2011 [14] and 2020 [41]. In a detailed and documented 2020 analysis I wrote: "In George Orwell's prescient, dystopian and frightening novel "1984", Big Brother declared that (A) war is peace, (B) slavery is freedom, (C) ignorance is strength, and (D) 2 plus 2 does not equal 4. Scientists and other humanitarian truth-tellers are alarmed as ostensible democracies as well as authoritarian states head towards this ultimate in comprehensive, state-imposed and blatant falsehood. My country, US lackey Australia, is arguably second only to Trump America for Orwellian government lying in ostensibly democratic societies – but how does your country perform in the mendacity stakes? Science involves the critical testing of potentially falsifiable hypotheses whereas the increasingly prevalent spin that dominates the world involves the uncritical use of asserted facts to support a partisan proposition i.e. the very opposite of the scientific method. However the great irony is that the very success of the scientific method that has now created a highly technological and digital world has also created the electronic vehicles for mass dissemination of lies and spin-based untruths (slies)… Set out below is a succinct account of how Big Brother's declarations that (A) war is peace, (B) slavery is freedom, (C) ignorance is strength, and (D) 2 plus 2 does not equal 4 are rampant in ostensible democracy Australia under a mendacious, racist, pro-war, war criminal, anti-science, anti-environment, climate criminal, human rights-violating, neoliberal, US lackey and Trumpist Liberal Party-National Party Coalition Government)" [41].

As a 50 year career science educator I am profoundly angered by this Australian Lib-Lab lying. While Lib-Lab politicians might think that lying and spin are smart politics, they forget that children are listening to their egregious falsehoods. While a politician might utter falsehoods out of stupidity and ignorance, after receipt of expert scientific advice to the contrary such falsehoods transmute to deliberate lying - and deliberate, anti-science lying to children constitutes egregious and unforgivable intellectual child abuse.

Over the last dozen years I have written detailed and documented excoriations of the resolute dishonesty and mendacity of the Australian Lib-Labs and of the army of lying Mainstream journalist, editor, politician, academic and commentariat presstitutes that support them e.g. "Australia Ignores 25 Huge Climate Change Realities" [11], "Orwellian Australian Labor Government Climate Change Lies" [14], "Australian ABC Lies About Key Climate Emergency Issues" [15], "25 Ways World-Leading Climate Criminal Australia Threatens Planet And Invites Boycotts, Divestment & Sanctions (BDS)" [21], "Advance Australia Fair" hides Australian racism, theft, genocide, ecocide, speciescide & terracide" [33], "Rampant Orwellian falsehood in neoliberal Australia – and in your country too? [41]", and "Scores Of Huge Realities Resolutely Ignored By Mendacious, US Lackey Mainstream

Australia" [46]. This resolute, in-your-face, Trump-like mendacity of the Lib-Labs and their Mainstream presstitute supporters means that Australians are unaware of horrendous, Elephant-in-the-Room realities as exampled below.

(1). Australia's huge coal, gas and iron ore exports threaten planet. In a detailed 2012 analysis I concluded: "In remorseless, greed-driven commitment to climate racism and climate criminality, Australia ignores top German climate change scientists who estimated in 2009 that the world must emit no more than 600 billion tonnes of carbon dioxide (CO_2) before 2050 if it is to have a reasonable chance of avoiding a catastrophic 2 degree Centigrade temperature rise. Australia's annual domestic plus exported greenhouse gas (GHG) pollution is so high that it exceeded its "fair share" of this terminal GHG pollution budget by mid-2011. Worse still, Australia's commitment to unlimited gas, coal and iron ore exports means that it is committed to polluting the atmosphere with over 3 times the world's total terminal GHG pollution budget. Australia's annual domestic plus exported per capita greenhouse gas (GHG) pollution is about 70 times greater than the annual per capita GHG pollution of acutely climate change-threatened Bangladesh" [18].

(2). Australia ranks worst in the world for climate policy and second worst for GDP-weighted annual per capita GHG pollution. Including its Exported as well as Domestic GHG pollution, Australia's revised annual per capita GHG pollution (116 tonnes CO_2-e per person per year) make it 4th worst in the world after Belize (367), Guyana (203), and Malaysia (126) [22]. A revised annual per capita GHG pollution score weighted for GDP places Australia (673) as number 2 in the World for climate criminality after Qatar (924) [23]. Australia (0.33% of world population) generates 2.5% of upwardly revised global GHG emissions (Australian Domestic use only) and 5.4% (Australian Domestic plus Exported GHG emissions) [36]. Australia is among world leaders in 16 areas of climate criminality [33-44]. The "Climate Change Performance Index (CCPI) Results 2020" ranked Australia worst in the world out of 61 advanced countries with a score of 0.0 out of 100 for

"Climate policy". The US ranked second worst with a score of 2.8 out of 100. Both the US and Australia scored "Very low" for "Overall rating", "National climate policy performance" and "International climate policy performance". In terms of overall climate change performance Australia ranked 6th worst in the Developed World and the US ranked worst ("Climate Change Performance Index (CCPI). Results 2020": https://newclimate.org/wp-content/uploads/2019/12/CCPI-2020-Results_Web_Version.pdf).

(3). Climate criminal and climate terrorist Australia kills 85,000 people annually through air pollution. Air pollution kills 9 million people worldwide each year, this including 10,000 Australians and 75,000 people overseas dying from the long-term effects of pollutants from the burning of Australia's world-leading coal exports [50]. The climate criminal Australian Government was returned to power in the 2019 Federal Elections on a pro-coal platform that included fervent support for the proposed Adani coal mine in Queensland that excited opposition from decent people around Australia. However the Coalition's pro-coal stance excited the fervent support of the Stupid, Ignorant and Egregiously Greedy (SIEG) around Australia and especially in the key state of Queensland (named after the world's worst imperialist, genocidal mass murderer and drug pusher in human history, Queen Victoria [1, 2]). In addition to passionate support for neoliberalism, speciescide, ecocide, and climate terrorism, the moronic Queenslander red-necks also passionately support their Rugby League team, appropriately named the Maroons. It is estimated from the latest Lancet data that implementation of the original Adani scheme for exporting Australian coal to India would eventually kill 1.4 million Indians over the proposed life-time of the mine [25, 26, 50].

(4). Biodiversity and environmental devastation from coral to koalas. Australia is a world leader in 16 areas of climate criminality and in particular in land clearing, and destruction of native flora, native fauna, and coral reefs i.e. in speciescide and ecocide en route to omnicide and terracide. Half of Queensland's iconic Great Barrier Reef is now dead through the effects of global warming, ocean acidification and agricultural run-off leading to deadly turbidity, algal blooms and coral-killing Crown of Thorns starfish infestations [48]. Even the whales are threatened by Australian climate criminality as melting Antarctic sea ice decreases the phytoplankton food for krill which in turn is the food of whales [5]. Nothing is sacred in Gadarene Australia that is irrevocably committed to fouling its own nest. Thus, for example, the iconic koala that is beloved of Asian tourists, is under threat of extinction in Queensland and New South Wales due to remorseless urban development. Yet in in September 2020 in the middle of a deadly Covid-19 pandemic the National Party (extreme right-wing farmer's party) part of the New South Wales (NSW) Coalition Government threatened to withdraw from the Coalition and hence bring down the government over legislation that has already been passed in the NSW Parliament and requires landowners not to engage in land use that would threaten koalas. Many koalas perished in the devastating 2019-2020) bushfires in NSW and the koala is predicted to become extinct in NSW by 2050 (Cristina Talacko, "The NSW Nationals' threat to blow up the Coalition over koalas is bizarre as it is misguided ", Guardian, 11 September 2020: https://www.theguardian.com/commentisfree/2020/sep/11/the-nsw-nationals-threat-to-blow-up-the-coalition-over-koalas-is-as-bizarre-as-it-is-misguided#_=_). The farmers of Australia are primarily responsible for over 2 centuries of invasion, dispossession, near-extermination and enslavement of the Indigenous people in an ongoing Australian Aboriginal Genocide and Aboriginal Ethnocide [1, 2, 43, 44], massive deforestation, and world-leading speciescide and ecocide in relation to extinction of native flora and fauna – but for the farmer's party of NSW (the National Party) that is not enough and the iconic koala must also fall victim to their insatiable, near-omnicidal greed.

(5). Terra nullius, aqua nullius, drought, heat stress and Australian Aboriginal Ethnocide - the ongoing Australian Aboriginal Genocide is qualitatively the worst in human history. After the British invasion of Australia in 1788 the Indigenous population dropped from about 1 million to 0.1

million in the first 100 years. There is an ongoing Australian Aboriginal Genocide (0.1 million deaths from violence and about 2 million deaths from dispossession, deprivation and introduced disease). There were an estimated 350-700 Indigenous languages and dialects in 1788 but after 230 years of ongoing Australian Aboriginal Ethnocide there are only 150 languages left and all but 20 are endangered. The Australian Aboriginal Ethnocide continues through remorseless government policies of large-scale removal of Indigenous children from their mothers and communities, limiting bilingual education, and de facto opposition to the viability of remote communities on traditional land [1, 2, 43, 44]. Article 2 of the UN Genocide Convention states: "In the present Convention, genocide means any of the following acts committed with intent to destroy, in whole or in part, a national, ethnical, racial or religious group, as such: (a) Killing members of the group; (b) Causing serious bodily or mental harm to members of the group; (c) Deliberately inflicting on the group conditions of life calculated to bring about its physical destruction in whole or in part; (d) Imposing measures intended to prevent births within the group; (e) Forcibly transferring children of the group to another group" (UN Convention on the Prevention and Punishment of the Crime of Genocide: https://www.un.org/en/genocideprevention/documents/atrocity-crimes/Doc.1_Convention%20on%20the%20Prevention%20and%20Punishment%20of%20the%20Crime%20of%20Genocide.pdf).

The "intent" of the UN Genocide Convention is established by the remorseless continuance of destructive and deadly policies. Thus notwithstanding "politically correct" rhetoric about "closing the gap", the life expectancy gap between Indigenous and non-Indigenous Australians remains at about 10 years due to egregious and sustained deprivation. The ongoing Australian Aboriginal Genocide is qualitatively the worst in human history. The Aboriginal Genocide was enabled by the British lie of "terra nullius" that held that Australia was an empty land – the Indigenous inhabitants were thereby dispossessed, and subject to horrendous avoidable mortality from deprivation, with the survivors often used as effective slave labor. A related de facto policy of "aqua nullius" allocated scarce water resources in "outback" Australia to the European farmers so that in the recent prolonged drought major rivers in western NSW (notably the Darling River of the huge Murray-Darling Basin) dried up and remnant Indigenous populations already largely deprived of language, culture and equitable circumstances were denied their 65,000-year old customary facility of using river water for drinking, swimming, fishing, and washing [31]. Australia disproportionately contributes to worsening global warming which is associated with worsening drought, bushfires, heat stress, high energy tropical storms and floods. I commented (2019): "More than 500 Australians presently die from heat stress each year with this disaster disproportionately impacting the elderly, the impoverished and Indigenous First Nations peoples. Australia is currently beset continent-wide with climate change-exacerbated heat waves with sustained temperatures above 40 degrees Centigrade, extraordinary, nation-wide bush fire emergencies commencing in Winter, and a devastating and widespread drought in Eastern Australia that has lasted for years…. Indeed a racist and climate criminal Australia that has disproportionately contributed to generating over 20 million climate change-displaced persons per each year, is now on the verge of making large numbers of impoverished Indigenous Australians climate refugees within Australia in a foreshadowed jump in the Australian Aboriginal Genocide and Aboriginal Ethnocide. Surely Indigenous Australians have suffered enough in over 230 years of genocidal racism" [38].

(6). Gas is dirty energy and can be dirtier GHG-wise than coal but pro-fossil fuels Australia backs a disastrous coal to gas transition and ignores massive fraud associated with non-accounting of systemic gas leakage. Endlessly greedy, neoliberal, pro-coal and pro-gas Australia is resolutely committed to exploitation of carbon fuels. Even if Australia comes to its senses and kicks out the climate criminal and pro-coal Coalition (COALition) Government at the next election, an incoming Labor Government, while wanting more renewable energy, would reject a Carbon Price and remain in solid agreement with the COALition about unlimited export of coal, gas, iron ore and methanogenically-derived meat. Both the

Coalition and Labor variously accept a disastrous coal-to-gas transition and the falsehood that "gas is cleaner than coal". The Lib-Labs resolutely ignore the realities that gas is dirty energy and that, depending upon the degree of systemic leakage (circa 3% in the US), burning gas can be dirtier GHG-wise than burning coal as set out my 2020 summary: "Methane (CH_4) is 85% of natural gas, leaks, and has a Global Warming Potential (GWP) 105 times that of the same mass of carbon dioxide (CO_2) on a 20 year time frame with aerosol impacts included. At a gas leakage of 2.6% (the average in the US is 3%) the global warming effect of the leaked gas is the same as that from burning the remaining 97.4% of gas. Such considerations reveal that Australia with 0.33 % of the world population has revised annual Domestic greenhouse gas (GHG) emissions that are 2.5% of the world's, and annual Domestic plus Exported GHG emissions that are 5.4% of the world's annual GHG pollution. Claims by pro-coal politicians that "gas is clean" or "gas is cleaner than coal" are utterly false. The "Methane Bomb" in the warming Arctic represents a worsening threat to Humanity. Failure to adequately monitor gas leakage amounts to massive fraud akin to failure to apply a Carbon Price" [46]. The Coalition Government has just announced its intention to build gas-fired power stations notwithstanding outrage from scientists and science-informed energy providers. Thus 25 variously eminent Australian scientists have criticized the Government plan to invest in more gas: "There is no role for an expansion of the gas industry. The combustion of natural gas is now the fastest-growing source of carbon dioxide to the atmosphere, the most important greenhouse gas" (Adam Morton and Katharine Murphy, "Australia's chief scientist rejects experts' letter warning him not to back gas", Guardian, 26 August 2020: https://www.theguardian.com/australia-news/2020/aug/26/australias-chief-scientist-rejects-experts-letter-warning-him-not-to-back-gas). Likewise I wrote in 2020: "Gas is not "relatively low in emissions" as asserted by Dr Finkel [Australia's chief scientist] because (a) combustion of 1 tonne of CH_4 (85% of natural gas) yields 2.75 tonne CO_2 as compared to combustion of 1 tonne of carbon (about 90% of coal) yielding 3.67 tonne CO_2, and (b) depending upon the degree of gas systemic leakage, gas burning can actually be much dirtier GHG-wise than coal burning" [on a mass basis]"[40]. Australia is the world's biggest exporter of Liquid Natural Gas (LNG), and methane leakage makes Australia a world-leading per capita greenhouse gas polluter - Australia has only 0.3% of the world 's population but Australia's Domestic plus Exported GHG emissions represent 5.4% of global GHG pollution [40].

(7). Australia's dishonest commerce, dumping and non-accounting of GHG pollution invite litigation, Carbon Tariffs and Boycotts, Divestment and Sanctions BDS) against a climate criminal Australia. The success of the climate change denialists and the effective climate change denialists is a deadly and disastrous example of Polya's Second Law of Economics, to whit "Deceit about the Cost of Production strives to a maximum". The Second Law of Economics is based on the fundamental Second Law of Thermodynamics that states that entropy (disorder, chaos, lack of information content) strives to a maximum. The International Monetary Fund (IMF) has exposed massive deceit in stating that while a damage-related Carbon Tax of $75 per ton CO_2 would be an effective way of addressing the climate threat, the present global average Carbon Price is only $2 per ton CO_2. However the Australian Coalition Government, backed by a cowardly and irresponsible Labor Opposition, reject any Carbon Price [37]. Australia is a major trading nation and China is now its biggest trading partner. In 2020 I wrote: "Despite China being Australia's biggest trading partner, US lackey Australia has joined Trump America in its anti-China trade war and military posturing policies. China has been offended by this increasing hostility from US lackey Australia and coincidentally for asserted anti-dumping reasons has applied tariffs to Australian beef and barley with threats of extending this to Australian wine. In actuality there is massive dumping by climate criminal Australia through non-application of a Carbon Price… Climate criminal Australia is involved in massive dumping of exported goods in China (and elsewhere) through non-application of a required Carbon Price. For every $10 piece of Australian meat, $10 bag of Australian barley or $10 bottle of Australian wine sold to China, it has been produced through a gigantic and hidden Australian Government subsidy of about $5 of Carbon Debt to be paid for by future generations" [47].

Climate criminal Australia is stealing from the world through non-application of a requisite damage-related Carbon Price ($200 per tonne CO_2-equivalent). Australia has only 0.33% of the world 's population but Australia's Domestic plus Exported GHG emissions represent 5.4% of global GHG pollution i.e. Australia is generating 5.4%/0.33% = 16.3, or about 16 times more than it its "fair share" of global pollution. Further, for every $100 billion of Australian exports there is a massive hidden subsidy (i.e. massive dumping) of about $50 billion annually due to non-application of a damage-related Carbon Price. I concluded (2020): "The overwhelming consensus of Australian Mainstream journalist, editor, politician, academic and commentariat presstitutes is that China's imposition of tariffs on Australian beef and barley is "payback" for US lackey Australia's craven support for Trump America's multi-pronged anti-China campaign. For all I know they may be right in this instance, but ignoring the Elephant-in-the-Room of remorselessly increasing, inescapable and gigantic Carbon Debt will not make it go away. Eventually the mendacious, ecocidal, anti-science and neoliberal climate criminals running Australia will be brought to account by an indignant World, including the indignant youth of Australia" [47]. A remorselessly climate criminal Australia faces the prospect of Green Tariffs, GHG Pollution Tariffs, International Criminal Court (ICC) prosecutions, International Court of Justice (ICJ) litigations, and Boycotts, Divestment and Sanctions (BDS) [21].

The following articles (asterisked in References) are reproduced as chapters in Section F:

Chapter 20. Gideon Polya, "Australia's Huge Coal, Gas & Iron Ore Exports Threaten Planet", Countercurrents, 15 May 2012: https://countercurrents.org/polya150512.htm [18].
Chapter 21. Gideon Polya, "Climate criminal Australia sabotages Pacific Islands Forum & threatens all Island Nations", Countercurrents, 24 August 2019: https://countercurrents.org/2019/08/climate-criminal-australia-

sabotages-pacific-islands-forum-threatens-all-island-nations [34].
Chapter 22. Gideon Polya, "Latest Lancet Data Imply Adani Australian Coal Project Will Kill 1.4 Million Indians", Countercurrents, 21 April 2017: http://www.countercurrents.org/2017/04/21/latest-lancet-data-imply-adani-australian-coal-project-will-kill-1-4-million-indians/.
Chapter 23. Gideon Polya, "Rampant Orwellian falsehood in neoliberal Australia – and in your country too?", Countercurrents, 1 March 2020: https://countercurrents.org/2020/03/rampant-orwellian-falsehood-in-neoliberal-australia-and-in-your-country-too.

References (those marked with an asterisk, *, are reproduced as Chapters in Section F).

[1]. Gideon Polya, "Jane Austen and the Black Hole of British History. Colonial rapacity, holocaust denial and the crisis in biological sustainability", G.M. Polya, 1998 (see: http://janeaustenand.blogspot.com/).

[2]. Gideon Polya, "Jane Austen and the Black Hole of British History. Colonial rapacity, holocaust denial and the crisis in biological sustainability", G.M. Polya, 2008, that is now available for free perusal on the web: http://janeaustenand.blogspot.com/.

[3]. Gideon Polya, "Australia Sabotages Copenhagen By Excluding Huge Agriculture GHG Emissions", Countercurrents, 17 November 2009: https://www.countercurrents.org/polya171109.htm.

[4]. Gideon Polya, "Historic Peak Oil Motion Defeated In Australian Senate", Countercurrents, 22 November 2009: https://www.countercurrents.org/polya221109.htm.

[5]. Gideon Polya, Anti-Whaling Australia Threatens Whales By Global Warming", Countercurrents, 19 January 2010: https://www.countercurrents.org/polya190110.htm.

[6]. Gideon Polya, "Letter To Australia re Climate Denialism, Biofuel Genocide And Climate Genocide", Countercurrents, 3 February 2010: https://countercurrents.org/polya030210.htm.

[7]. Gideon Polya, "Climate Activists Support Australian Government CTETSIA Climate Change Inaction Plan", Countercurrents, 5 March 2011: http://www.countercurrents.org/polya050311.htm.

[8]. Gideon Polya, "Detailed Document Exposing Pro-gas Australia Climate Change Inaction", Countercurrents, 15 March 2011: https://www.countercurrents.org/polya150311.htm.

[9]. Gideon Polya, "Lobbyocracy Australia Opts For Disastrous Coal To Gas Transition", Countercurrents, 16 April 2011: https://www.countercurrents.org/polya160411.htm.

[10]. Gideon Polya, "Australia's Carbon Tax And Coal To Gas Transition Will Double Power Generation Greenhouse Gas Pollution", Countercurrents, 15 May 2011: https://www.countercurrents.org/polya150511.htm.

[11]. Gideon Polya, "Australia Ignores 25 Huge Climate Change Realities", Countercurrents, 6 July 2011: https://www.countercurrents.org/polya060711.htm.

[12]. Gideon Polya, Australian Carbon Price Scheme Is Deceitful Climate Change Inaction", Countercurrents, 18 July 2011: https://www.countercurrents.org/polya090811.htm.

[13]. Gideon Polya, "Anti-Environment Australian Government's Carbon Tax Deceit", Countercurrents, 9 August 2011: https://www.countercurrents.org/polya090811.htm.

[14]. Gideon Polya, "Orwellian Australian Labor Government Climate Change Lies", Countercurrents, 20 August 2011: https://www.countercurrents.org/polya200811.htm.

[15]. Gideon Polya, "Australian ABC Lies About Key Climate Emergency Issues", Countercurrents, 5 September 2011: https://www.countercurrents.org/polya050911.htm.

[16]. Gideon Polya, "Comprehensive Climate Change Inaction By Pro-coal, Pro-gas Gillard Labor Australian Government", Countercurrents, 5 March 2012: https://www.countercurrents.org/polya050312.htm.

[17]. Gideon Polya, "The ABC, Australia's BBC, Turns Climate Inaction Truth Upside-Down Downunder", Countercurrents, 3 April 2012: https://www.countercurrents.org/polya030412.htm.

*[18]. Gideon Polya, "Australia 's Huge Coal, Gas & Iron Ore Exports Threaten Planet", Countercurrents, 15 May 2012: https://countercurrents.org/polya150512.htm.

[19]. Gideon Polya, "Anti-science, Effective Climate Change Denialist Abbott Coalition Australian Government Axes Expert Climate Commission", Countercurrents, 24 September 2013: https://countercurrents.org/polya240913.htm.

[20]. Gideon Polya, "Australian Coalition Government Sacrifices Key Industries While Committing Hundreds Of Billions Of Dollars To Carbon Pollution & War", Countercurrents, 10 February 2014: https://countercurrents.org/polya100214.htm.

[21]. Gideon Polya, "25 Ways World-Leading Climate Criminal Australia Threatens Planet And Invites Boycotts, Divestment & Sanctions (BDS)", Countercurrents, 6 June 2015: http://www.countercurrents.org/polya060615.htm.

[22]. Gideon Polya, "Revised Annual Per Capita Greenhouse Gas Pollution For All Countries – What Is Your Country Doing?", Countercurrents, 6 January, 2016: http://www.countercurrents.org/polya060116.htm.

[23]. Gideon Polya, "Exposing And Thence Punishing Worst Polluter Nations Via Weighted Annual Per Capita Greenhouse Gas Pollution Scores", Countercurrents, 19 March, 2016: https://sites.google.com/site/banyuleclimateactionnow/exposing.

[24]. Gideon Polya, "Australian Commitment To Unlimited Natural Gas Exploitation Threatens Planet & Invites Global Blowback", Countercurrents, 12 April 2017: http://www.countercurrents.org/2017/04/08/australian-commitment-to-unlimited-natural-gas-exploitation-threatens-planet-invites-global-

blowback/.
[25]. Polya, "Pollutants From Adani Coal Mine Will Eventually Kill About 0.5 Million Indians", Countercurrents, 14 April 2017: http://www.countercurrents.org/2017/04/14/pollutants-adani-coal-mine-will-eventually-kill-about-0-5-million-indians/.
*[26]. Gideon Polya, "Latest Lancet Data Imply Adani Australian Coal Project Will Kill 1.4 Million Indians", Countercurrents, 21 April 2017: http://www.countercurrents.org/2017/04/21/latest-lancet-data-imply-adani-australian-coal-project-will-kill-1-4-million-indians/.
[27]. Gideon Polya, "Climate criminal Coalition lying endangers Australia & Planet – Australians must vote 1 Green and put the Coalition last", Countercurrents, 17 June 2017: http://www.countercurrents.org/2016/06/17/climate-criminal-coalition-government-lying-endangers-australia-planet-australians-must-vote-1-green-and-put-the-coalition-last/.
[28]. Gideon Polya, "Australia's Finkel Review ignores 25 key Climate Emergency realities & advocates long term gas & coal use", Countercurrents, 17 June 2017: http://www.countercurrents.org/2017/06/17/australias-finkel-review-ignores-25-key-climate-emergency-realities-advocates-long-term-gas-coal-use/.
[29]. Gideon Polya, "Australian ABC Ignores & Censors Horrendous, Australia-Complicit Air Pollution Deaths", Countercurrents, 15 July 2017: http://www.countercurrents.org/2017/07/15/australian-abc-ignores-censors-horrendous-australia-complicit-air-pollution-deaths/
[30]. Gideon Polya, "Inescapable $200-$250 trillion global carbon debt increasing by $16 trillion annually", Countercurrents, 27 April 2019: https://countercurrents.org/2019/04/inescapable-200-250-trillion-global-carbon-debt-increasing-by-16-trillion-annually-gideon-polya.
[31]. Gideon Polya, "Water Crisis, Global Avoidable Mortality Holocaust, Water Apartheid, Global Warming & Mina Guli", Countercurrents, 17 May 2019: https://countercurrents.org/2019/05/water-crisis-global-avoidable-mortality-holocaust-water-apartheid-global-warming-mina-guli.
[32]. Gideon Polya, "Climate criminal Australia ignores its hugely increasing Carbon Debt & massive subsidies for GHG pollution", Countercurrents, 8 June 2019: https://countercurrents.org/2019/06/climate-criminal-australia-ignores-its-hugely-increasing-carbon-debt-massive-subsidies-for-ghg-pollution.
[33]. Gideon Polya, "Advance Australia Fair" hides Australian racism, theft, genocide, ecocide, speciescide & terracide", Countercurrents, 1 July 2019:

https://countercurrents.org/2019/07/advance-australia-fair-hides-australian-racism-theft-genocide-ecocide-speciescide-terracide.

*[34]. Gideon Polya, "Climate criminal Australia sabotages Pacific Islands Forum & threatens all Island Nations", Countercurrents, 24 August 2019: https://countercurrents.org/2019/08/climate-criminal-australia-sabotages-pacific-islands-forum-threatens-all-island-nations.

[35]. Gideon Polya, "Millions join Global School Climate Strike – we are running out of time", Countercurrents, 22 September 2019: https://countercurrents.org/2019/09/millions-join-global-school-climate-strike-we-are-running-out-of-time.

[36]. Gideon Polya, "War criminal & climate criminal Australian deception at UN General Assembly", Countercurrents, 29 September 2019: https://countercurrents.org/2019/09/war-criminal-climate-criminal-australian-deception-at-un-general-assembly.

[37]. Gideon Polya, "Australia rejects IMF Carbon Tax & preventing 4 million pollution deaths by 2030", Countercurrents, 15 October 2019: https://countercurrents.org/2019/10/australia-rejects-imf-carbon-tax-preventing-4-million-pollution-deaths-by-2030.

[38]. Gideon Polya, "Australian climate criminality, heat stress deaths, & Australian Aboriginal Ethnocide", Countercurrents, 19 December 2019: https://countercurrents.org/2019/12/australian-climate-criminality-heat-stress-deaths-australian-aboriginal-ethnocide.

[39]. Gideon Polya, "Trumpist climate change denial, Australian bushfires, fuel reduction, biochar & Carbon Debt", Countercurrents, 10 January 2020: https://countercurrents.org/2020/01/trumpist-climate-change-denial-australian-bushfires-fuel-reduction-biochar-carbon-debt.

[40]. Gideon Polya "Methane leakage makes Australia a world-leading per capita greenhouse gas polluter", Countercurrents, 18 February 2020: https://countercurrents.org/2020/02/methane-leakage-makes-australia-a-world-leading-per-capita-greenhouse-gas-polluter.

* [41]. Gideon Polya, "Rampant Orwellian falsehood in neoliberal Australia – and in your country too?", Countercurrents, 1 March 2020: https://countercurrents.org/2020/03/rampant-orwellian-falsehood-in-neoliberal-australia-and-in-your-country-too.

[42]. Gideon Polya, "Melbourne's North East Link super-highway project – environmental vandalism & Australian-killing perversion", Countercurrents, 9 March 2020: https://countercurrents.org/2020/03/melbournes-north-east-link-super-highway-project-environmental-vandalism-australian-killing-perversion.

[43]. Gideon Polya, "US-Imposed Post-9/11 Muslim Holocaust & Muslim Genocide", 400 pages, Korsgaard Publishing, Germany, 4

June 2020: https://www.amazon.com/US-Imposed-Post-9-Muslim-Holocaust-Genocide/dp/8793987056.

[44]. Gideon Polya, "Racist Mainstream ignores "US-Imposed Post-9/11 Muslim Holocaust & Muslim Genocide"", Countercurrents, 17 July 2020: https://countercurrents.org/2020/07/racist-mainstream-ignores-us-imposed-post-9-11-muslim-holocaust-muslim-genocide/.

[45]. Gideon Polya, "Covid-19 Pandemic, Climate & Australia: Risky Ignoring Of Science-based Advice", Countercurrents, 28 July 2020: https://countercurrents.org/2020/07/covid-19-pandemic-climate-australia-risky-ignoring-of-science-based-advice/.

[46]. Gideon Polya, "Scores Of Huge Realities Resolutely Ignored By Mendacious, US Lackey Mainstream Australia", Countercurrents, 11 August 2020: https://countercurrents.org/2020/08/scores-of-huge-realities-resolutely-ignored-by-mendacious-us-lackey-mainstream-australia/.

[47]. Gideon Polya, "Carbon Debt & Dumping – Climate Criminal Australia Hugely Subsidizes Meat, Grain & Wine Exports To China", Countercurrents, 31 August 2020: https://countercurrents.org/2020/08/carbon-debt-dumping-climate-criminal-australia-hugely-subsidizes-meat-grain-wine-exports-to-china/.

[48]. Gideon Polya, editor, "2011 climate change course": https://sites.google.com/site/300orgsite/2011-climate-change-course.

[49]. Gideon Polya, editor, "Science & economics experts: Carbon Tax needed NOT Carbon Trading": https://sites.google.com/site/300orgsite/sciennce-economics-experts-carbon-tax-needed-not-carbon-trading/.

[50]. Gideon Polya, editor, "Stop air pollution deaths": https://sites.google.com/site/300orgsite/stop-air-pollution-deaths

[51]. Gideon Polya, editor, "Nuclear weapons ban, end poverty & reverse climate change": https://sites.google.com/site/300orgsite/nuclear-weapons-ban.

[52]. Gideon Polya, editor, "Climate Genocide": https://sites.google.com/site/climategenocide/.

[53]. Gideon Polya, editor, "Gas is not clean energy": https://sites.google.com/site/gasisnotcleanenergy/.

[54]. Gideon Polya, editor, "Biofuel Genocide": https://sites.google.com/site/biofuelgenocide/.

[55]. Gideon Polya, editor, "Stop state terrorism" : https://sites.google.com/site/stopstateterrorism/ [state and corporate complicity in worsening climate genocide and 7 million annual air pollution deaths from carbon fuel burning].

[56]. Gideon Polya, editor, "State crime and non-state terrorism": https://sites.google.com/site/statecrimeandnonstateterrorism/ [state

and corporate complicity in worsening climate genocide and 7 million annual air pollution deaths from carbon fuel burning].
[57]. "Climate terrorism: 400,000 climate change-related deaths globally annually versus an average of 4 US deaths from political terrorism annually since 9-11":
https://sites.google.com/site/statecrimeandnonstateterrorism/climate-terrorism.
[58]. "Carbon terrorism: 3 million US air pollution deaths versus 53 US political terrorism deaths since 9-11 (2001-2015)":
https://sites.google.com/site/statecrimeandnonstateterrorism/carbon-terrorism.
[59]. "Gideon Polya":
https://sites.google.com/site/drgideonpolya/home.

CHAPTER 20: AUSTRALIA'S HUGE COAL, GAS & IRON ORE EXPORTS THREATEN PLANET

[First published as Gideon Polya, "Australia's Huge Coal, Gas & Iron Ore Exports Threaten Planet", Countercurrents, 15 May 2012: https://countercurrents.org/polya150512.htm].

In remorseless, greed-driven commitment to climate racism and climate criminality, Australia ignores top German climate change scientists who estimated in 2009 that the world must emit no more than 600 billion tonnes of carbon dioxide (CO_2) before 2050 if it is to have a reasonable chance of avoiding a catastrophic 2 degree Centigrade temperature rise. Australia's annual domestic plus exported greenhouse gas (GHG) pollution is so high that it exceeded its "fair share" of this terminal GHG pollution budget by mid-2011. Worse still, Australia's commitment to unlimited gas, coal and iron ore exports means that it is committed to polluting the atmosphere with over 3 times the world's total terminal GHG pollution budget. Australia's annual domestic plus exported per capita greenhouse gas (GHG) pollution is about 70 times greater than the annual per capita GHG pollution of acutely climate change-threatened Bangladesh.

In 2009 the German Advisory Council on Climate Change (WBGU) determined that for a 75% chance of avoiding a 2 degree C temperature rise, the World must pollute less than 600 Gt CO_2 (600 billion tonnes CO_2) between 2010 and essentially zero emissions in 2050.

This means that the "fair share" of this global terminal greenhouse gas (GHG) pollution budget of 600 Gt CO_2 (600 billion tonnes CO_2) for Australia (population 22.9 million in 2012 as compared to a World population of 7,010 million) is

600 billion tonnes CO_2 x 22.9 million/7,010 million = 1.960 billion tonnes CO_2 = 1,960 million tonnes CO_2.

Australia's annual domestic greenhouse gas (GHG) pollution was 578 million tonnes CO_2-e in 2010 (this figure taking into account not just CO_2 but other greenhouse gases (GHGs) excluding water (H_2O) that were emitted e.g. methane (CH_4) and nitrous oxide (N_2O) derived from land use) and accordingly Australia had 1,960/578 = 3.4 years at this rate to use up its "fair share" of the world's terminal GHG pollution budget.

However simple "cause and effect" means that we must also consider the GHG impact of Australia's huge coal, natural gas and iron ore exports to the GHG pollution of the one common atmosphere of all countries. If Australia left it all in the ground then this material would not be used to generate GHG pollution of the atmosphere.

Australia's domestic plus exported GHG pollution in 2010 was 1,708 million tonnes CO_2-e, this being made up (in million tonnes of CO_2-e) of 578 (Domestic) + 803 (coal exports) + 34 (LNG exports) + 293 (iron ore exports) (see "2011 Climate Change Course": https://sites.google.com/site/300orgsite/2011-climate-change-course) and hence Australia had 1,960/1,708 = 1.1 years to get to zero emissions i.e. by the middle of 2011).

Australia has thus already used up its "fair share" of the world's terminal GHG pollution budget and is now stealing the entitlement of all of the countries in the world. For a detailed analysis see "Shocking analysis by country of years left to zero emissions", Green Blog, 1 August 2011: http://www.green-blog.org/2011/08/01/shocking-analysis-by-country-of-years-left-to-zero-emissions/ .

However, Australia has huge reserves of black coal, brown coal, natural gas (conventional gas and unconventional gas such as shale gas and coal seam gas or CSG), and iron ore (which contributes to GHG pollution when processed industrially). It is accordingly useful to examine the implications of Australia extracting and exporting all of its reserves of these resources. The answers are quite shocking.

1. Natural gas.

Australian conventional gas resources are 111 Tcf (trillion cubic feet) of Economic Demonstrated Reserves (EDR), 53 Tcf SDR (Sub-economic Reserves (SDR) and 20 Tcf Inferred resources (IR) for a total of 184 Tcf. (see ABARE and Geoscience Australia, "Australian Energy Resource Assessment", Chapter 4 "Gas": http://adl.brs.gov.au/data/warehouse/pe_aera_d9aae_002/aeraCh _04.pdf).

However Australia also has unconventional gas resources, specifically Australian coal seam gas resources are 15.1 Tcf Economic Demonstrated Reserves (EDR), 27.2 Tcf SDR (Sub-economic Reserves (SDR) and 111 Tcf Inferred resources (IR) for a total of 153 Tcf plus tight gas resources (in shale deposits) of 20 Tcf (trillion cubic feet) of Inferred Reserves (IR), for an overall total of unconventional gas reserves of coal seam gas and shale gas variously requiring "fracking" of 173 Tcf.

Now 1 tonne of LNG = 48,700 cubic feet of gas (0.0487 Mcf or million cubic feet of gas) and accordingly Australia's conventional gas reserves = 184 million Mcf x 1 t LNG/0.0487 Mcf = 3,778 million tonnes of LNG.

Australia's unconventional gas reserves = 173 million Mcf x 1 t LNG/0.0487 Mcf = 3,552 million tonnes of LNG.

The major component of LNG is methane (CH_4) which on combustion yields CO_2 and water: $CH_4 + 2O_2 \rightarrow CO_2 + 2 H_2O$. CH_4 has a molecular weight of 16 and CO_2 a molecular weight of 44. Accordingly combustion of 16 tonnes of CH_4 yields 44 tonnes of CO_2 and combustion of 1 tonne of CH_4 yields 2.75 tonnes CO_2.

However CH_4 is a gas, it leaks (an average of 3.3% in the US and up to 7.9% for unconventional gas from fracking shale deposits according to Robert W. Howarth, Renee Santoro and Anthony Ingraffea, "Methane and the greenhouse-gas footprint of natural gas from shale formations", Climatic Change, 2011: http://www.sustainablefuture.cornell.edu/news/attachments/Howarth-EtAl-2011.pdf) and is 105 times worse than CO_2 as a GHG on a 20 year time frame and including aerosol impacts according to Drew T. Shindell, Greg Faluvegi, Dorothy M. Koch, Gavin A. Schmidt, Nadine Unger and Susanne E. Bauer, "Improved Attribution of Climate Forcing to Emissions", Science, 30 October 2009: Vol. 326 no. 5953 pp. 716-718: http://www.sciencemag.org/content/326/5953/716).

For 100 tonnes CH_4 from conventional gas, leakage of 3.3 tonnes CH_4 yields 3.3 tonnes CH_4 x 105 tonnes CO_2-e/ tonne CH_4 = 346.5 tonnes CO_2-e and combustion of the remaining 96.3 tonnes CH_4 yields 96.3 x 2.75 = 264.8 tonnes CO_2 for a total of 611.3 tonnes CO_2-e. Accordingly 1 tonne of CSG-derived CH_4 yields 6.11 tonnes CO_2-e and hence Australia's conventional reserves of 3.778 billion tonnes would yield 3.778 x 6.11 = 23.1 billion tonnes of CO_2-e.

For 100 tonnes CH_4 from unconventional gas, leakage of 7.9 tonnes CH_4 yields 7.9 tonnes CH_4 x 105 tonnes CO_2-e/ tonne CH_4 = 829.5 tonnes CO_2-e and combustion of the remaining 92.1 tonnes CH_4 yields 92.1 x 2.75 = 253. 3 tonnes CO_2 for a total of 1,082.8 tonnes CO_2-e. Accordingly 1 tonne of CSG-derived CH_4 yields 10.83 tonnes CO_2-e and hence Australia's unconventional reserves of 3.552 billion tonnes would yield 3.552 x 10.83 = 38.4 billion tonnes of CO_2-e.

Thus exploitation of Australia's presently discovered conventional and unconventional gas resources would generate 61.5 billion tonnes of CO_2-e or about 10% of the world's terminal GHG pollution budget of 600 billion tonnes CO_2, noting that Australia's "fair share" is only 2.0 billion tonnes CO_2-e.

2. Coal.

Australian black coal resources are 39.2 Gt (billion tonnes) Economic Demonstrated Reserves (EDR), 8.3 Gt Sub-economic Reserves (SDR) and 66.6 Inferred resources (IR) for a total of 114.1 Gt i.e. 114.1 billion tonnes black coal (ABARE and Geoscience Australia, "Australian Energy Resource Assessment", Chapter 5 "Coal": http://adl.brs.gov.au/data/warehouse/pe_aera_d9aae_002/aeraCh _05.pdf).

Australian brown coal resources are 37.2 Gt Economic Demonstrated Reserves (EDR), 55.1 Gt Sub-economic Reserves (SDR) and 101.8 Inferred resources (IR) for a total of 194.0 Gt i.e. 194.0 billion tonnes brown coal.

According to the "Australian Energy Resource Assessment": "Australia ranks fourth in the world in terms of combined recoverable economic coal resources. Australia 's total coal resources are substantially larger than this with total identified resources of black coal being around 114 Gt and brown coal resources of 194 Gt. However, the full extent of Australia's very large coal resource base is not known: potential resources have not been assessed because the existing identified resource base is so large. The resource potential of coal is probably in excess of one trillion tonnes. There are over 25 sedimentary basins with identified resources or coal occurrences and there are areas within these basins that need further exploration. Significant potential also exists in poorly explored basins across the continent" (ABARE and Geoscience Australia, "Australian Energy Resource Assessment", Chapter 5 "Coal": http://adl.brs.gov.au/data/warehouse/pe_aera_d9aae_002/aeraCh _05.pdf).

If the "resource potential" of over "one trillion tonnes" has the same proportion of brown/black as Australia's currently proven and inferred resources (194.0 Gt brown/114 Gt black) then this would involve 194.0 x 1 trillion t/(194.0 + 114.1) = 629.7 Gt brown coal and 370.3 Gt black coal.

Burning 12 t carbon yields 44 t CO_2 i.e. burning 1 t carbon yields 44 /12 = 3.67 t CO_2 Accordingly complete combustion of black coal (about 80% carbon: http://en.wikipedia.org/wiki/Black_coal) would yield 0.8 x 3.67 = 2.9 t CO_2 and complete combustion of 1 t brown coal (about 30% carbon: http://en.wikipedia.org/wiki/Brown_coal) would yield 0.3 x 3.67 = 1.1 t CO_2. Thus, for example, because a lot of brown coal (lignite) is used for power generation in the US, on average burning 1t coal for electricity in the US yields 2.1 t CO_2/t coal (see: http://kilowattcoalco.com/StatsAndLinks.html).

Using these estimates of 2.9 t CO_2/ t black coal and 1.1 t CO_2/ t brown coal we can calculate the CO_2 from combustion of Australian coal resources as set out below.

CO_2 from combustion of established Australian black coal resources would be 113.7 Gt Economic Demonstrated Reserves (EDR), 24.1 Gt Sub-economic Reserves (SDR) and 193.1 Inferred resources (IR) for a total of 330.9 Gt CO_2 i.e. 330.9 billion tonnes CO_2 or 55.2% of the world's terminal pollution budget of 600 billion tonnes of CO_2.

CO_2 from combustion of established Australian brown coal resources are 40.9 Gt Economic Demonstrated Reserves (EDR), 60.6 Gt Sub-economic Reserves (SDR) and 112.0 Inferred resources (IR) for a total of 213.4 Gt CO_2 i.e. 213.4 billion tonnes of CO_2 or 35.6% of the world's terminal pollution budget of 600 billion tonnes of CO_2.

Thus CO_2 from combustion of established Australian black and brown coal resources would total 544.3 billion tonnes CO_2 or 90.7% of the world's terminal pollution budget of 600 billion tonnes of CO_2.

However, CO_2 from combustion of Australia's coal resource potential of 1 trillion tonnes of coal would be 692.7 Gt CO_2 (from brown coal) plus 1,073.9 Gt CO_2 (from black coal) for a total of 1,766.6 Gt CO_2 i.e. 1,766.6 billion tonnes CO_2 or

294.4% (2.9 times) the world's terminal pollution budget of 600 billion tonnes of CO_2.

3. Iron ore.

In 2010 the world produced 2,400 million tonnes of iron ore of which Australia's production was 420 tonnes (see "Iron ore", Wikipedia: http://en.wikipedia.org/wiki/Iron_ore#Production_and_consumption).

According to The Australian: "Australia raised its forecast for 2011/12 iron ore exports to a record 473 million tonnes, from a previous projection of 460 million tonnes, reflecting expectations demand from top buyer China will remain strong…Australia exported 407 million tonnes of ore in 2010/11. Longer term, Australia's exports are expected to rise by an annual rate of around 11 percent to reach 767 million tonnes in 2016/17, the Bureau of Resources and Energy Economics (BREE) predicted" (see "Australia raises iron ore exports forecast to record", The Australian, 20 March 2012: http://www.cnbc.com/id/46802742/Australia_Raises_Iron_Ore_Exports_Forecast_to_Record). At this rate of increase one predicts Australian iron ore exports of 1,048 million tonnes by 2019/2020 and 1,164 million tonnes by 2020/2021 i.e. about 1,106 million tonnes in 2020.

SSAB (the major Swedish steel company) states that: "According to the International Energy Agency (IEA), the iron and steel industry accounts for approximately 4-5% of the global CO_2 emissions. This is due to the fact that steel production requires large amounts of energy, often based on fossil fuels. SSAB's steelworks are among the most efficient in the world" (see SSAB: http://www.ssab.com/en/Investor--Media/Sustainability/32/322/).

In 2010 world CO_2 emissions totalled 33.5 billion tonnes (see "List of countries by carbon dioxide emissions", Wikipedia: http://en.wikipedia.org/wiki/List_of_countries_by_carbon_dioxide_emissions).

Taking the upper estimate of the steel industry accounting for 5% of the global CO_2 emissions (IEA), in 2010 Australia's production of 420 tonnes iron ore was associated with 0.05 x 33,500 million tonnes CO_2 x 420 million tonnes iron ore/ 2,400 million tonnes iron ore = 293.1 million tonnes CO_2. The predicted iron exports of 1,106 million tonnes in 2020 would correspond to 293.1 million tonnes CO_2 x 1,106 million tonnes iron ore /420 million tonnes iron ore = 771.8 million tonnes CO_2.

Australia's "economic demonstrated resources" of iron ore total 24 billion tonnes, this corresponding to 16.7 billion tonnes CO_2 or 2.8% of the world's terminal GHG pollution budget, noting that Australia has only 0.3% of the World's population.

4. Australia's domestic plus exported GHG pollution.

From Treasury, ABARE, US EIA and the above data and assuming an 11% annual growth in iron ore exports, 2.4% annual growth in coal exports and 9% annual growth in gas exports, Australia's Domestic and Exported greenhouse gas (GHG) pollution is as follows (million tonnes CO_2-e; see "2011 Climate Change Course": https://sites.google.com/site/300orgsite/2011-climate-change-course):

2000: 555 (Domestic) + 505 (coal exports) + 17 (LNG exports) + 105 (iron ore exports) = 1,182.

2009: 600 (Domestic) + 784 (coal exports) + 31 (LNG exports) + 97 (iron ore exports) = 1,512.

2010: 578 (Domestic) + 803 (coal exports) + 34 (LNG exports) + 293 (iron ore exports) = 1,708.

2020: 621 (Domestic) + 1,039 (black coal exports) + 80 (LNG exports) + 59 (brown coal exports) + 772 (iron ore exports) = 2,571

In 2009 the German Advisory Council on Climate Change (WBGU) determined that for a 75% chance of avoiding a 2 degree C temperature rise, the World must pollute less than 600 billion tonnes CO_2 between 2010 and essentially zero emissions in 2050. Assuming an Australian population of 22.9 million and a world population of 7,010 million (2012) the Australia's "fair share"of this terminal GHG pollution budget is 600,000 million tonnes CO_2 x 22.9 million /7,010 million = 1,960 million tonnes CO_2 . Accordingly, in 2010 alone Australia used up 1,708 x 100/1,960 = 87% of its "fair share" of the world's terminal 2010-2050 GHG pollution budget. However, on current projections, in 2020 Australia aims to use up 429% (4.3 times) of its "fair share" of the world's terminal 2010-2050 GHG pollution budget.

5. Australian miners are stealing the GHG pollution entitlement of other Australians (notably farmers) as well as of all other people worldwide.

Over 4 decades I have been teaching biochemistry to second year university agricultural science students, many of whom come from farms. In recent years I have drawn their attention to the upwards revision by World Bank analysts Goodland and Anfang of man-made GHG pollution (Robert Goodland and Jeff Anfang. "Livestock and climate change. What if the key actors in climate change are … cows, pigs and chickens?", World Watch, November/December 2009: http://www.worldwatch.org/files/pdf/Livestock%20and%20Cli mate%20Change.pdf .) In short, Goodland and Anfang have estimated that man's GHG pollution is 50% bigger than hitherto thought and that livestock alone contribute over 51% of the bigger estimate. An alternative estimate is given by the Australian Government National Carbon Accounting System (NCAS) that states that "Around 24 per cent of Australia's human-induced greenhouse gas emissions come from activities such as livestock and crop production, land clearing and forestry" (see "National Carbon Accounting System", Australian Government Department of Climate Change and Energy Efficiency:

http://www.climatechange.gov.au/government/initiatives/nation
al-carbon-accounting.aspx).

I have told my students from animal husbandry farms that we
are all in this together, that we all must face up honestly to the
problem facing mankind, and that one group in society cannot
be singled out as we transition to zero emissions and thence the
reduction of atmospheric CO_2 from the present damaging 392
parts per million (ppm) to the circa 300 ppm required for a safe
and sustainable existence for all peoples and all species (see
"300.org – return atmosphere CO_2 to 300 ppm", 300.org: :
https://sites.google.com/site/300orgsite/300-org---return-
atmosphere-co2-to-300-ppm). Put simply, I say "You guys
milk cows or raise sheep, I drive a car".

However Australian farmers have common cause with
environmentalists and scientists over Australia's huge exported
GHG pollution. The iron ore miners, coal miners and coal seam
gas extractors are not merely despoiling the landscape, variously
violating agricultural land (notably through strip mining and
Coal Seam Gas extraction), despoiling aquifers, and generating
a 2-speed economy through revaluation upwards of the
Australian dollar, as shown above they are essentially stealing
the terminal GHG pollution entitlement of other Australians as
well as of all other people around the world.

According to the Australian Greens, 83% of Australia's mining
industry is foreign-owned and it will ship $50 billion in profits
overseas in the next 5 years (see "Greens target billionaires in
tax push:, Australian Financial Review, 29 June 2011:
http://afr.com/p/home/greens_target_mining_billionaires_xDJ3
RMoByzqwCXgGQrvCJL). However, an RMIT University
report has estimated that closure of the coal mining industry
would cost 200,000 jobs and $34 billion pa to the GDP (see
"Greens' plan to shut down coal industry to cost Australia
200,000 jobs, $36b a year in GDP", Courier Mail Business, 4
July 2011: http://www.couriermail.com.au/business/greens-
plan-to-shut-down-coal-industry-to-cost-australia-200000-jobs-
36b-a-year-in-gdp/story-e6freqmx-1226086761615).

The huge economic impact of the mining industry has horribly distorted public discussion in Australia. Thus Australia's huge GHG exports are ignored in the public debate by everyone except the Greens. The neoconservative, pro-coal, pro-gas, pro-iron ore Liberal Party-National Party Coalition Federal Opposition (the Liberals or the Libs) and the neoconservative, pro-coal, pro-gas, pro-iron ore Australian Labor Party Federal Government (the Laborals or the Labs) have a common derisory, Business As Usual (BAU) policy of "5% off 2000 GHG pollution by 2020" coupled with unlimited coal, gas and iron ore exports. A common but utterly false BAU argument is that Australia is so small population-wise (22.9 million out of a global population of 7 billion), and that any GHG pollution constraint by Australia would be negligible – but as shown above, Australia has already exceeded it fair share of the terminal global GHG pollution budget, and exploitation of its likely coal reserves would generate CO_2 pollution about 3 times greater than the total permitted for the whole world before zero emissions in 2050. A difference between the Liberals and Laborals lies in the significant climate change denialism within the Liberals and the fraudulent and ineffective Carbon Tax-ETS policy of the Laborals that pretends to "tackle climate change" for a "clean energy future" while actually doing the opposite (see "Australian PM Julia Gillard's appalling record of climate change inaction", Green Blog, 8 March 2012: http://www.green-blog.org/2012/03/08/australian-pm-julia-gillards-appalling-record-of-climate-change-inaction/). Indeed the Liberals have been aptly described as climate change deniers and the Laborals as climate change pretenders.

The Mining Industry has a malignant effect on public life in Australia. Australia, like the other Western democracies, is a Murdochracy (Big Money buys public perception of reality and votes) and a Lobbyocracy (Big Money buys politicians, policies, parties, public perception of reality and votes). Thus it cost the Mining Industry a mere $22 million in advertising to remove a very popularly elected Australian Prime Minister Kevin Rudd in the June 23/24 2010, overnight Coup. The Mining Lobby is continuing a massive advertising campaign against the current Labor Government's Mining Tax (the Labor voter has shrunk to 30%, one of its lowest levels ever in Australian history). The Australian Mining Australia reports "The nation's leading counter-terrorism organisation is monitoring some environmentalists following Government warnings that protestors pose greater risks than terrorists. According to Fairfax Media Resources and Energy Minister Martin Ferguson has warned coal protests could have "major trade and investment implications" for Australia's economy" (see "Green groups are worse than terrorists: Government", Australian Mining, 12 April 2012: http://www.miningaustralia.com.au/news/green-groups-are-worse-than-terrorists--government).

There is a huge gulf between the perception of reality by scientists and that perceived by the general public. Scientists are intimidated by fear of losing research grants, by codes of conduct restricting comment (see Gideon Polya, "Current academic censorship and self-censorship in Australian universities", Public University Journal, volume 1, Conference Supplement, "Transforming the Australia University", Melbourne, 9-10 December 2001: http://pandora.nla.gov.au/pan/57092/20080218-1150/www.publicuni.org/jrnl/volume/1/jpu_1_s_polya.pdf) and through outrageous censorship by media including taxpayer-funded media (see "Censorship by The Conversation": https://sites.google.com/site/mainstreammediacensorship/censorship-by ; "Censorship by ABC Late Night Live": https://sites.google.com/site/censorshipbyabclatenightlive/ ; "ABC Censorship": https://sites.google.com/site/abccensorship/ ; "Censorship by The Age": https://sites.google.com/site/mainstreammediacensorship/censorship-by-the-age ; and Gideon Polya, "Australian Internet Censorship by Australian Government-funded ABC and Universities-backed The Conversation ", Countercurrents, 10 May 2012: http://www.countercurrents.org/polya100512.htm).

Summary.

In 2009 the German Advisory Council on Climate Change (WBGU) determined that for a 75% chance of avoiding a 2 degree Centigrade temperature rise, the World must pollute less than 600 Gt CO_2 (600 billion tonnes CO_2) between 2010 and essentially zero emissions in 2050. Australia's high domestic greenhouse gas (GHG) pollution means that Australia would use up its "fair share" of this global terminal GHG pollution budget within 3.4 years. However Australia's huge coal, gas and iron ore exports and its consequently high annual domestic plus exported GHG pollution mean that in 2010 alone Australia used up 1,708 x 100/1,960 = 87% of its "fair share" of the world's terminal 2010-2050 GHG pollution budget and, on current projections, in 2020 alone it will use up 4.3 times its "fair share".

However Australia's commitment to climate racism and climate injustice goes even further because the major parties (the Liberal-National Party Coalition and the Labor Party, aka the Lib-Labs or Liberal-Laborals) are committed to the same derisory, Business As Usual (BAU) policy of "5% off 2000 GHG pollution by 2020" coupled with unlimited export of coal, gas and iron ore of which Australia has huge resources.

Australia's "economic demonstrated resources" of iron ore total 24 billion tonnes, this corresponding after industrial steel-making to 16.7 billion tonnes CO_2 or 2.8% of the world's terminal GHG pollution budget, noting that Australia has only 0.3% of the World's population. Exploitation of Australia's presently discovered conventional and unconventional gas resources would generate 61.5 billion tonnes of CO_2-e or about 10% of the world's terminal GHG pollution budget of 600 billion tonnes CO_2, noting that Australia's "fair share" is only 2.0 billion tonnes CO_2-e. However, CO_2 from combustion of Australia's huge coal resource potential of 1 trillion tonnes of coal would be an estimated 692.7 Gt CO_2 (from brown coal) plus 1,073.9 Gt CO_2 (from black coal) for a total of 1,766.6 Gt CO_2 i.e. 1,766.6 billion tonnes CO_2 or 294.4% (2.9 times) the world's terminal pollution budget of 600 billion tonnes of CO_2.

Australia has had an appalling and ongoing secret genocide history that is hidden by a look-the-other-way culture, ignoring, censorship and denial (see "Australia's secret genocide history":

). However Australia's bipartisan commitment to climate racism and climate injustice means that it is committed to hugely disproportionate GHG pollution that is making a disproportionate contribution to a worsening climate genocide that is set to kill 10 billion people this century. Australia's annual domestic plus exported per capita greenhouse gas (GHG) pollution is about 70 times greater than the annual per capita GHG pollution of acutely climate change-threatened Bangladesh. Thus "annual per capita greenhouse gas (GHG) pollution" in units of "tonnes CO_2 - equivalent per person per year" (2005-2008 data) is 0.9 (Bangladesh), 0.9 (Pakistan), 2.2 (India), less than 3 (many African and Island countries), 3.2 (the Developing World), 5.5 (China), 6.7 (the World), 11 (Europe), 16 (the Developed World), 27 (the US) and 30 (Australia; or 64 in 2010 if Australia's huge Exported CO_2 pollution is included) (see "Climate Genocide":
).

Australian politically correct racism (PC racism) asserts love for humanity while being disproportionately involved in war crimes and climate crimes. The world will eventually respond to Australia's dog-in-the-manger climate criminality through Boycott, Divestment, Sanctions, Green Tariffs, Sporting Boycotts, International Court of Justice (ICJ) litigations and International Criminal Court (ICC) prosecutions. Decent people will respond by (a) informing others and (b) urging sanctions against all countries (notably Australia, Canada and the US) involved in disproportionate and deadly pollution of our one common atmosphere.

CHAPTER 21: CLIMATE CRIMINAL AUSTRALIA SABOTAGES PACIFIC ISLANDS FORUM & THREATENS ALL ISLAND NATIONS

[First published as Gideon Polya, "Climate criminal Australia sabotages Pacific Islands Forum & threatens all Island Nations", Countercurrents, 24 August 2019: https://countercurrents.org/2019/08/climate-criminal-australia-sabotages-pacific-islands-forum-threatens-all-island-nations].

The 18 nations of the Pacific Islands Forum recently held their 50th meeting. US lackey and climate criminal Australia brought an anti-China American agenda, offered a $500 million dollar bribe (stolen and rebranded from its miniscule Aid budget), and sabotaged the consensus Forum statement on action over the worsening climate change threat to Island Nations.

Humanity is existentially threatened by nuclear weapons (most of humanity and the Biosphere will be destroyed in a post-nuclear holocaust nuclear winter) [1] and by man-made climate change (10 billion people may perish this century in a worsening Climate Genocide if man-made climate change is not requisitely addressed – indeed various eminent scholars have estimated that the carrying capacity of the Earth for humans in 2100 would be 1 billion or less) [1, 2].

The Small Island States of the Pacific, and indeed all around the world, head the list of countries most immediately threatened by man-made climate change. The nation of Tuvalu hosted the 50th meeting of the Pacific Islands Forum and is acutely threatened by man-made climate change – it may become uninhabitable by as early as 2030 [3].

Australia is a member of the 18-member Pacific Islands Forum but as a rich country occupying all of a 7.7 million square kilometer continent is far less threatened by climate change than tiny, coral atoll-based and low-lying Pacific Islands, most notably Tuvalu, Kiribati, Micronesia, Niue, Tokelau and the Marshall Islands. Indeed all members of the Pacific Islands Forum variously have climate-changed-threatened island and coastal areas. Thus the Pacific Islands Development Forum in its recent "Nadi Bay Declaration on the Pacific Islands Climate Change Crisis" (30 July 2019) stated: "The science warns of the real possibility that coral atoll nations could become uninhabitable as early as 2030. By 2100, the coral atoll nations of the Republic of the Marshall Islands, Tuvalu, Kiribati, Tokelau and the Maldives and many SIDS [Small Island Developing States] could be submerged… Call…on all coal producers to immediately cease any new mining of coal and develop a strategy for a decadal phase-out and closure of all existing coal production" [3].

The 18 members of the Pacific Islands Forum include Australia, Kiribati, New Caledonia, Papua New Guinea, Tuvalu, Cook Islands, Marshall Islands, New Zealand, Samoa, Vanuatu, Fiji, Micronesia, Niue, Solomon Islands, French Polynesia, Palau, Nauru, and Tonga [4].

1. **Australia's horrendous record of colonial abuse, slavery, genocide, subversion and holocaust denial in the Pacific.**

Australia, like other genocidally racist European colonial enterprises (notably the US, the UK, Canada, New Zealand, Argentina, Brazil, Apartheid Israel, the former Apartheid South Africa and the former French Algeria), was founded on invasion, devastation, mass murder, theft and genocide [5-7]. However Australia is among world leaders in climate criminality in 14 key areas and has a disproportionately huge complicity in a worsening Climate Genocide that particularly impacts climate change-threatened tropical Island Nations (the largest being the Philippines and Indonesia) and mega-delta tropical or subtropical nations such as the US, Nigeria, Egypt,

India, Pakistan, Bangladesh, Myanmar, Thailand, Cambodia, Vietnam and China [2].

Before the British Invasion in 1788, Indigenous Australians had been living in Australia for about 60,000 years. There were 350-750 different tribes and a similar number of languages and dialects, of which only 150 survive today and of these all but about 20 are endangered under racist White Australian policies against bilingual education. After the British Invasion, the Aboriginal population dropped from about 1 million in 1788 to about 0.1 million in the first century through introduced disease, deprivation and genocidal violence (an Australian Aboriginal Genocide and Aboriginal Ethnocide). The last officially-sanctioned massacres of Aborigines occurred in Central Australia in 1928 but no Treaty has ever been signed. Indigenous Australians were only counted as citizens after a referendum in 1967, and were finally given some protection by the 1975 Racial Discrimination Act that has been subsequently grossly violated in the ongoing, 21st century Northern Territory Invasion (the 2007 onwards Northern Territory National Emergency Response) [8].

The White Australian-imposed Aboriginal Genocide continues. In 1998 about 4,000 Aborigines out of an Aboriginal population of 500,000 died avoidably every year from deprivation but the avoidable death rate has declined from 0.8% pa in 1998 to 0.6% pa in 2013, with still circa 4,000 annual avoidable deaths out of an Indigenous population of about 670,000. Indigenous Australians are far worse off than White Australians in relation to housing, health, wealth, social conditions, imprisonment, forcible child removal, deaths in custody, avoidable death, and life expectancy [8]. The over 2 century, UK-initiated Aboriginal Genocide and the over 1 century, UK- and US-backed, and Zionist-effected Palestinian Genocide in each case involve about 2 million avoidable deaths from imposed deprivation and about 0.1 million deaths from violence [9]. However present avoidable deaths from deprivation as a percentage of population is about 0.1% for the sorely-oppressed 5 million Occupied Palestinians, whereas it is 0.6% for the 0.7 million sorely oppressed Indigenous Australians in pro-Apartheid Israel and hence pro-Apartheid Australia. Indigenous Palestinians are now 50% of the subjects of Apartheid Israel but the 73% who are Occupied Palestinians cannot vote for the government ruling them [Apartheid] (Indigenous Australians are about 3% of the Australian population and only after a 1967 Referendum were finally counted as citizens with the right and indeed compulsory duty to vote).

However a violent and genocidally racist White Australia did not stop at genocide on the continent of Australia. Racist White Australia (variously as a lackey of the UK or the US) has invaded 85 countries as compared to (in the last millennium) the British 193, France 82, the US 72 (52 after WW2), Germany 39, Japan 30, Russia 25, Canada 25, Apartheid Israel 12, China 2, Korea 0, and Iran 0 [10-15]. 30 of these Australian invasions have been associated with genocide [16]. A racist White Australia has brought invasion, colonialism, slavery ("blackbirding"), and genocide by introduced disease to the Pacific [12, 16-18]. For example and by way of personal disclaimer, all of the grandparents of my dear late wife Zareena *née* Lateef were Indian indentured labourers – 5 year slaves – brought from India to work on British and Australian sugarcane plantations in a Fiji that had been massively depopulated by measles introduced from Australia in 1875 (40,000 Fijians died out of a population of 150,000) [12, 16-18].

War is the penultimate in racism and genocidal war the ultimate in racism. Racist White Australia is America's Deputy Sheriff in the South Pacific and is slavishly and cravenly devoted to a racist and serial war criminal America that has variously brought colonialism, total war, nuclear contamination, war gases, subversion and now Climate Genocide to the remote and innocent Pacific Island Nations. Australia has been involved in all US Asian wars since 1950, atrocities that have been associated with 40 million Asian deaths from violence or war-imposed deprivation [17]. Australia has learned well from its American big brother that subverts every country on earth, and has over 700 military bases in over 70 countries, including Australia [19].

Thus Australian Intelligence was involved in the US-backed overthrow of the democratically-elected Allende government of Chile [20], spying on the East Timor Government [21], and spying on the Indonesian President and his wife [22]. Racist, pro-Apartheid White Australia is second only to Zionist-subverted Trump America as a supporter of Apartheid Israel [23] and there is evidence that Australia, Apartheid Israel and America were variously involved in the overthrow of democratically elected governments in Fiji [24] and Chile. The Australian Government now spies on all Australians, and such bulk crude intelligence is transmitted to Apartheid Israel by the Americans [25]. Through the joint US-Australian Pine Gap electronic spying facility in Central Australia, Australia spies on other countries from the Middle East to the Pacific, targets illegal and war criminal US drone attacks on 7 Muslim nations, and is crucial to US nuclear terrorism [26]. These abuses are on the public record, but now the Australian Government has passed legislation providing for up to 10 years imprisonment for journalists revealing "errors" in operations by Australian Government security agencies agents [27]. Indeed similar legislation provides for up to 2 years' imprisonment for doctors, nurses, teachers and other care providers who disclose abuses in the Nauru and Manus Island (Papua New Guinea) concentration camps for the indefinite and highly abusive Australian imprisonment without charge or trial of innocent refugees [28]. Australian Government agents are very likely committing even worse atrocities throughout the Pacific but it is a serious criminal offence to reveal them [27, 28].

It gets worse. From a qualitative perspective the ongoing Australian Aboriginal Genocide and Aboriginal Ethnocide is the worst genocide in human history (350-750 different languages and dialects originally, of which only 150 survive today and of these all but about 20 are endangered) [8]. However from a quantitative perspective the Australia-complicit, British-imposed WW2 Bengali Holocaust (WW2 Bengal Famine, WW2 Indian Holocaust) was one of the worst genocides and holocausts in human history (6-7 million Indians deliberately starved to death by the British with racist White Australia complicit by withholding food from starving India from the huge wartime Australian grain stores) [16, 17, 29-31]. Indians represent a major group in the South Pacific, specifically in Fiji, Australia and New Zealand.

It gets even worse still. Climate criminal Australia as a world leading exporter of coal and gas, and in being among world leaders in 14 climate-related areas [32], is now a disproportionately high contributor to a worsening Climate Genocide that is predicted to kill about 10 billion people this century if man-made global warming is not requisitely addressed [2]. The damage-related cost of carbon pollution has been variously estimated at US$200 per tonne CO_2, this enabling calculation of the inescapable Carbon Debt we are leaving to be paid by future generations. Australia with a population (25 million) that is only 0.3% of the world's, has an annual increase in GHG pollution (and hence of Carbon Debt) that is 5.4% of that of the world [32]. The worsening Climate Genocide disproportionately impacts the Pacific Island states.

2. **How climate criminal Australia stymied, sabotaged and betrayed the 2019 Pacific Islands Forum.**

The 50th Pacific Islands Forum was held in Tuvalu in August 2019. Meeting the month before the formal Pacific Islands Forum, the leaders of the Pacific Small Islands Developing States (PSIDS) and the Small Island States (SIS) issued a strong "Nadi Bay Declaration on the Pacific Islands Climate Change Crisis" that demanded urgent action: "We the People of the Pacific islands Development Forum, Striving to advance the sustainable and inclusive development of Pacific Island nations … Declare that we … 3. Underscore the serious concerns and stark warnings, documented by the IPCC Special Report on 1.5°C and the Special Report on Oceans and Cryosphere, GHG emissions must be reduced immediately. The science warns of the real possibility that coral atoll nations could become uninhabitable as early as 2030. By 2100, the coral atoll nations of the Republic of the Marshall Islands, Tuvalu, Kiribati, Tokelau and the Maldives and many SIDS could be submerged" [3].

Prior to the 50th Pacific Islands Forum the Pacific Islands states issued a "Tuvalu Declaration on Climate Change for the Survival of Pacific Small Island Developing States (PSIDS)" that critically demanded strong action in a number of key areas [33]. The Tuvalu Declaration supported "the UN Secretary General's call for an immediate global ban on the construction of new coal fired power plants and coal mines and calls on all countries to rapidly phase out their use of coal in the power sector" [33]. However at the Pacific Islands Forum climate criminal Australia, that is the world leader in coal and gas exports, refused to sign up to the final consensus declaration unless it was substantially weakened as set out below [33, 34].

- **On coal use:**

Tuvalu Declaration: "[We support] the UN Secretary General's call for an immediate global ban on the construction of new coal fired power plants and coal mines and calls on all countries to rapidly phase out their use of coal in the power sector".

Australia-subverted consensus Pacific Islands Forum statement: "[We] Invite all parties to the Paris Agreement to reflect [on the UN Secretary-General's remarks on] fossil fuel subsidies and just transition from fossil fuels… [Call on] the members of the G7 and G20 to rapidly implement their commitment to phase out inefficient fossil fuel subsidies". Australia is the world leader in coal exports and managed to remove any mention of "coal" from the final statement.

- **On Green Climate Fund.**

Tuvalu Declaration: "[We support] the significant role that the GCF [Green Climate Fund] plays in supporting developing countries in their efforts to address climate change. We call for a prompt, ambitious and successful replenishment of the GCF".

Australia-subverted consensus Pacific Islands Forum statement: "[The international community] continues efforts towards [meeting international funding commitments] including the replenishment of the Green Climate Fund".

- **On GHG emissions reductions.**

Tuvalu Declaration: "Encourage all countries to revise their nationally determined contributions so as to rapidly reduce greenhouse gas emissions".

Australia-subverted consensus Pacific Islands Forum statement: "Call for… all parties to the Paris Agreement to meet or exceed their nationally determined contributions" [33, 34].

3. **Anti-environment, pro-coal Australian reactions to the Australian sabotage of the 2019 Pacific Islands Forum.**

Not content with Australia sabotaging the 2019 Pacific Islands Forum, various Australian politicians compounded the damage by issuing the following variously ignorant, crass and pro-coal statements. The hard right of the Australian Liberal Party-National Party Coalition government is expressly climate change denialist and the Coalition Government as a whole is effectively climate change denialist by rejecting strong action on climate change. Despite Australia being the world leader in coal and gas exports, third after Russia and Saudi Arabia in fossil fuel exports, and among world leaders in 14 areas relating to climate criminality (notably per capita GHG pollution, fossil fuel exploitation, land clearing and climate change inaction) [32], under a pro-coal Coalition (COALition) Government, Australian greenhouse gas (GHG) emissions are actually increasing but the Coalition offsets this obscenity by using "carry-over credits" and "carbon offsets" carbon accountancy tricks to falsely claim that it is on track to meet its Paris Climate Change Agreement commitments [35-37]. Australia's Exported GHG pollution via its world leading coal and gas exports is about 2 times its Domestic GHG pollution [38]. The Coalition Government and the Labor Opposition differ in rhetoric and in GHG targets (Labor wants much more renewable energy, has better climate rhetoric, and has a better emissions reduction target) but there is a bipartisan Coalition and Labor agreement on unlimited coal and gas exports that are antithetical to the survival of coral atoll Island Nations.

Anthony Albanese (Australian Labor Opposition Leader) asked about Tuvalu and coal exports: "Look, our export of coal will remain an important part of Australia's economy. The fact is the global economy is transitioning. But coal will be a part of the energy mix for a period of time into the future. And so, in terms of the impact as well, to suggest that Australia which has high quality coal compared with what it would be replaced with is not necessarily a productive move forward" [39].

Alan Jones (highly influential, right-wing, pro-Coalition Australian radio "shock jock" commentator) made a highly obnoxious and widely offending statement about New Zealand PM Jacinda Ardern as reported thus by the ABC: "Jones said Mr Morrison should "shove a sock down [Ms Ardern's] throat" after Ms Ardern said "Australia had to answer to the Pacific" on climate change during the Pacific Islands Forum in Tuvalu" [40].

Michael McCormack (pro-coal National Party Deputy PM of Australia): "I also get a little bit annoyed when we have people in those sorts of countries pointing the finger at Australia and say we should be shutting down all our resources sector so that, you know, they will continue to survive. They will continue to survive, there's no question they'll continue to survive and they'll continue to survive on large aid assistance from Australia. They'll continue to survive because many of their workers come here and pick our fruit, pick our fruit grown with hard Australian enterprise and endeavour and we welcome them and we always will. But the fact is we're not going to be hijacked into doing something that will shut down an industry that provides tens of thousands of jobs, that provides two-thirds of our energy needs … and I'm only talking coal, let alone all of our other resources" [41].

Scott Morrison (pro-coal Coalition PM of Australia, who once brought a lump of coal into the Australian Parliament, and who was recently re-elected on a fervently pro-coal stance) on Pacific Islands blowback from his pro-coal position: "I am accountable to the Australia people – that's who I'm accountable for… We want a viable, sustainable, successful, sovereign, independent set of Pacific Island states… and for them to maintain and realise their way of life" [42].

Senator Penny Wong (Labor Leader of the Opposition in the Senate and Labor Shadow Minister for Foreign Affairs): "[Re a coal ban] Of course not. Coal remains an important industry for Australia and it remains part of the global energy mix. But the point is that these negotiations proceeded on the basis of an Australian government which did not understand nor respect the importance of climate to Pacific Island nations and did not bring to the table any realistic policies to reduce emissions, which is the core of what we should be doing as a responsible nation" [43].

4. **Pro-environment and anti-genocide statements re the Pacific Islands Forum by climate activists and Pacific Island leaders.**

Before the Pacific Islands Forum, Marshall Islands President Hilda Heine commented: "We're discouraged and disappointed at the fact that Australia is still actively using coal for their own power generation, and it looks like that is something that is going to continue into the future. That's not helping the issue of emissions and we know they understand that" [33]. Decent people aware of the existential threat of climate change to coral atoll Island Nations were appalled that climate criminal Australia sabotaged the Pacific Islands Forum by insisting on a softened final communique that did not even mention the Heart of Darkness – coal. Below is a selection of views from Pacific Island leaders and some science-informed climate action humanitarians.

Jacinda Ardern ([globally] widely admired and humane PM of New Zealand who attempted to reduce the gulf between

pro-coal, climate criminal Australia and the climate change-threatened Pacific Island Nations): "Like our Pacific Island neighbours, we will continue that international call. We will continue to say that New Zealand will do its bit and we have an expectation that everyone else will as well — we have to. New Zealand, relative to other nations, has a relatively small emissions profile, however, if we all took the perspective that if you're small it doesn't matter, we wouldn't see change. Every single little bit matters" [44].

Frank Bainimarama (forthright PM of Fiji): "I thought [Australian PM] Morrison was a good friend of mine; apparently not. The prime minister at one stage, because he was apparently [backed] into a corner by the leaders, came up with how much money Australia have been giving to the Pacific. He said: 'I want that stated. I want that on the record.' Very insulting… After what we went through with Morrison, nothing can be worse than him. China never insults the Pacific. You say it as if there's a competition between Australia and China. There's no competition, except to say the Chinese don't insult us. They don't go down and tell the world that we've given this much money to the Pacific islands. They don't do that. They're good people, definitely better than Morrison, I can tell you that. The prime minister was very insulting, very condescending, not good for the relationship. We had said 'below 1.5 degrees'. That's what was in our official drafts, but your prime minister didn't want that because it means the Australians will have to come up with a lot of sacrifices. But we're supposed to be here for the Pacific islands, not only for Australia… My sentiments hold true: Australia and New Zealand should not be in the Pacific Islands Forum. [New Zealand PM Jacinda Ardern] was much more compromising. She was really good yesterday. She said the right things about climate change and Morrison did not" [45].

Frank Bainimarama further stated: "We came together in a nation that risks disappearing to the seas, but unfortunately we settled for the status quo in our communique. Watered-down climate language has real consequences — like water-logged homes, schools, communities, and ancestral burial grounds" [46].

Adam Bandt (pro-environment and pro-human rights Australian Greens MP): "The Pacific Islanders desperately want us to phase out coal mining. Instead, Morrison backs it in 100%" [46].

Joseph Moeono-Kolio (head of Greenpeace in the Pacific): "[Australia could become] the pariah of the Pacific… How does Morrison reconcile calling the Pacific family while he persistently ignores our demands for Australia to reduce its emissions?" [47].

Rachel Kyte (Chief Executive Officer of Sustainable Energy for All, and Special Representative of the United Nations Secretary-General for Sustainable Energy for All): "Coal will endanger the environment, threaten jobs and expose even more Australians who are already suffering from the volatility of extreme weather as a result of carbon pollution. This approach to coal is not only short-sighted, it's reckless and cruel" [46].

Patricia Mallam (of the 350 Pacific activist group): "The appalling fact in all this is that Australia is granted a seat at the same PIF [Pacific Island Forum] Meeting table as nations literally struggling to protect the lives and cultural integrity of their people. Australia bullies its way through negotiations, attempting to mask the gravity of the climate crisis on paper — when the visible proof in our lives shows otherwise" [46].

Richie Merzian (Climate and Energy program director at The Australia Institute): "[Morrison's overseas] carbon credits loophole is equivalent to eight years of fossil fuel emissions for the rest of the Pacific and New Zealand — and the Pacific rightly asked Australia to cancel them. The world was watching to see if Australia meets or sinks the hopes of its Pacific Islands neighbours. Australia's Prime Minister waved a lump of coal around Parliament — and now he has put his love of coal ahead of the Pacific's survival" [46].

Ralph Regenvanu (Foreign Minister of Vanuatu and part of the drafting committee of the forum communique): "On emissions reduction and dealing with your own climate, Australia is out there, they're not with us, Pretty much on everything else we're on the same team, in terms of mitigation, adaptation, making sure there's resourcing for all this, it's only in terms of [the coal, gas and land use] domestic emissions strategy that Australia is way out" [48].

Enele Sopoaga (PM of Tuvalu) at a joint press conference with Australian PM Morrison: "You are concerned about saving your economies, your situation in Australia, I'm concerned about saving my people in Tuvalu and likewise other leaders of small island countries… That was the tone of the discussion. Please don't expect that [Australia] comes and we bow down or that … we were exchanging flarey language, not swearing, but of course you know, expressing the concerns of leaders and I was very happy with the exchange of ideas, it was frank. Prime Minister Morrison, of course, stated his position and I stated my position and [that of] other leaders: we need to save these people" [48].

Anote Tong (former president of Kiribati) suggested sanctions against Australia: "What is the relevance of Australia's ongoing protestations in the forum? If it's going to continue with that line, it poses a danger to the other countries in that forum. How can you justify being part of a family and part of a group which you're trying to destroy? … [Australia is increasing coal exploitation] at the same time, it is saying they're a so-called friend. Should there be sanctions imposed? I think that's a question that leaders are no doubt asking themselves. I think these things have got to be expressed. It's not only in my mind, I'm sure it's in everybody else's mind. I'm surprised Australia is not seeing this, or they're ignoring it, which is the height of arrogance" [49].

Malcolm Turnbull (pro-climate action former Coalition PM of Australia, removed in a 2018 Coalition Coup and replaced by pro-coal Morrison): "Climate change and the consequences of it are an existential matter for the Pacific. If

you are a Pacific islander and your home is going to be washed
away by rising sea levels caused by global warming then this is
not a political issue, it's an existential one. So it's critically
important that we show respect to the Pacific islanders and that
we are seen to be helping climate change, both in reducing our
emissions as part of a global effort, and of course as we do
providing them with substantial resources to adapt to climate
change" [41].

**Matthew Wale (Deputy Opposition Leader of the Solomon
Islands):** "What a missed opportunity to really 'step up'.
'Family' has been exploited for domestic Australian politics.
Pacific islanders were hoping for sincerity when we hear 'we're
family'. We were mistaken" [46].

5. **The missing key issue – serial war criminal
   Australia's disproportionate involvement in the
   existential nuclear threat to Island Nations,
   Humanity and the Biosphere.**

Humanity and the Biosphere are existentially threatened not just
by man-made climate change but also by nuclear weapons.
After a major nuclear exchange the consequent nuclear winter
will destroy most of Humanity and the Biosphere [1]. The
people of Hiroshima and Nagasaki aside, the people of the
Pacific Island nations of the Marshall Islands and Kiribati are
the major victims of nuclear weapons, and serial war criminal
Australia has played a major role in UK and US nuclear
terrorism [1].

The US conducted 67 nuclear weapons tests at Bikini and
Enewetak atolls in the Marshall Islands between 1946 and 1958.
The US subsequently established a deadly radioactive waste
dump on Runit Island in Enewetak Atoll. Similarly in Kiribati
there were 9 UK atmospheric nuclear tests on Christmas
(Kiritimati) Island in 1957-58, followed by 24 US nuclear
weapons tests on Kiritimati in 1962 [33]. The UK, US and
France conducted over 300 nuclear detonations in the Pacific
region – in Australia, the Marshall Islands, Kiribati, US-owned
Johnston Atoll and Te Ao Maohi (French Polynesia) [50].

Radioactive contamination from these obscene nuclear tests is a major ongoing issue for the Indigenous inhabitants [33, 50-52], this now being complicated by rising sea levels from climate change and the leaching out of radioactive material from the Enewetak Runit "Dome", albeit into an already badly contaminated area [52].

Australia has played – and continues to play – a major role in nuclear terrorism. The UK used Australia for the testing of nuclear weapons and missile delivery systems. Australia is key part of US nuclear terrorism through the joint US-Australian electronic spying facility at Pine Gap in Central Australia and through the hosting of nuclear armed US naval vessels. Australia is a major uranium producer and has supplied uranium to the nuclear terrorist states of the UK, US and France (and thence to nuclear weapons-armed Apartheid Israel via the US). Further developments include the stationing of up to 2,500 US military near Darwin in the Northern Territory (approved by fervently pro-Zionist and pro-US Labor PM Julia Gillard in 2013), and very recent US Congressional budgetary revelations (Australians being the last to know) that the US is planning to build a US naval base near Darwin, the ABC reporting: "A recent draft US Congressional bill uncovered by the ABC reveals a total of $US211.5 million (A$305.9 million) has been allocated for new "Navy Military Construction" in Darwin, but few other details are available" [53]. Australia is a world leader in laser-based uranium isotope enrichment [54]. One again notes that racist, pro-Apartheid White Australia is second only to Zionist-subverted Trump America as a supporter of Apartheid Israel [23], and there is evidence that Australia, Apartheid Israel and America were variously involved in the overthrow of democratically elected governments in Fiji [24].

It gets worse. Thus journalist Michael Peck has stated (2019): "Australia has a close defense relationship with America. But that doesn't mean Oz [Australia] wants to jeopardize its relations with China by allowing the United States to station missiles on its territory. Both Prime Minister Scott Morrison and Defense Minister Linda Reynolds have quashed the notion that American intermediate-range missiles—which would be aimed at China—could be stationed in northern Australia. Their responses came after U.S. Defense Secretary Mike Esper said that the U.S. hopes to station intermediate-range missiles in Asia, leading to speculation that Australia would be a prime candidate. Esper's statements came after the United States withdrew from the 1987 Intermediate-Range Nuclear Forces (INF) Treaty between the United States and the Soviet Union, which banned land-based ballistic and cruise missiles with a range of 500 to 5,500 kilometers (300 to 3,400 miles)" [55]. But for how long will craven US lackey Australia reject this US request? Indeed a leading Australian security expert Professor Hugh White has recently raised the possibility of Australia acquiring nuclear weapons (in the context of a rising China and declining US), and on a recent "expert" ABC panel discussion of this matter the only person to unequivocally condemn this obscenity was an international human rights lawyer, Diana Sayed [56].

It gets worse still. Craven US lackey Australia has announced that it is joining the nuclear weapons-armed US armada in the Persian Gulf off the coast of Iran, claiming disingenuously that this is simply to support the US and the UK in upholding the "international rules-based order" [56]. US lackey Australia has thus effectively joined the 4-decade US War on Iran in which 4 million Iranians have died from US-backed violence, 1 million (in the 1980-1988 US-backed Iraq-Iran War), or from US-imposed deprivation via sanctions, 3 million [57]. Now UK machinations, US sanctions with threats of "obliteration", and Apartheid Israeli bombing of Iranian facilities in Syria and Iraq threaten a devastating hot war of 3 nuclear powers against a peaceful and non-nuclear-armed Iran [57]. While Iran has no nuclear weapons, 9 other countries do, namely (upper estimates in brackets) the US (7,300), Russia (8,000), Apartheid Israel (400), France (300), UK (250), China (250), Pakistan (120), India (100), and North Korea (circa 10) [1, 57]. US lackey Australia voted against the UN General Assembly resolution to ban nuclear weapons [1].

Serial war criminal and US lackey Australia is already grossly violating the "international rules-based order" by targeting illegal US drone attacks on 7 Muslim countries [26], backed by the UN the Australian Navy interdicts charcoal exports of impoverished Somalians [58], and Australia is part of the US Alliance that is occupying Afghanistan and in doing so grossly violating the Geneva Convention [59]. Thus the most fundamental human right is the right to life. While legitimately criticized for the one party state, the death penalty, censorship, urban air pollution and harsh treatment of dissidents, China has been hugely successful in radically reducing infant mortality and maternal mortality in Tibet and in China as a whole. In stark contrast, the war criminal US Alliance occupation of neighbouring Afghanistan continues to be associated with an under-1 infant mortality and maternal mortality incidence that is 7 times higher and 4-12 times higher, respectively, than that in neighbouring Tibet that is ruled by China (per capita GDP $9,000 as compared to $60,000, $58,000, $40,000 and $600 for the US, Australia, the UK, and Afghanistan respectively – evidence of gross violation by the US Alliance of the UN Genocide Convention and of Articles 55 and 56 of the Geneva Convention relative to the Protection of Civilian Persons in Time of War that both demand that an Occupier must provide its conquered Subjects with life-sustaining food and medical services "to the fullest extent of the means available to it" [59].)

Stephen Hawking (eminent theoretical physicist and cosmologist) addressing the question "Will we survive on Earth?", stated (2018): "In January 2018, the Bulletin of the Atomic Scientists, a journal founded by some of the physicists who had worked on the Manhattan Project to produce the first atomic weapons, moved the Doomsday Clock, their measurement of the imminence if catastrophe – military or environmental – facing our planet, forward to two minutes to midnight… In 1947, the clock was set at seven minutes to midnight. It is now closer to Doomsday than at any time since then, save in the early 1950s at the start of the Cold War… We see great peril if governments and societies do not take action now to render nuclear weapons obsolete and to prevent further climate change" [60]. Who should we believe, one of Humanity's greatest minds, or the blustering, smirking, neoliberal, pro-coal, and anti-science failed advertising executive and now PM of Australia, Scott "Scomo" Morrison? Anti-science neoliberal Scomo (aka Scum-o, and thence also Scam-o, Scheme-o, and Skim-o) absurdly declared "I have always believed in miracles" after being recently re-elected by the variously Stupid, Ignorant and Egregiously Greedy (SIEG as in Dr Strangelove) half of the Australian electorate.

Final comments.

Humanity is existentially threatened by nuclear weapons and man-made climate change. Low-lying Island Nations and major mega-delta countries such as India, Bangladesh, Myanmar, Cambodia, Thailand, Vietnam, and China are particularly threatened by global warming, attendant sea level rise, and storm surges from more intense tropical and sub-tropical storms [1, 2]. While Island Nations beg the world to keep temperature rise under 1.5C, it now appears that 1.5C will be reached in 10 years' time, a catastrophic plus 2C is unavoidable on present trends, and national commitments to the Paris Agreement imply plus 3.2C by 2100 [61-65]. Expert predictions of a sustainable human population of only 0.5-1.0 billion by 2100 [2] imply 10 billion deaths in a worsening Climate Genocide this century if global warming is not requisitely addressed [2].

A climate criminal Australia under PM Scott "Scomo" Morrison is among world leaders in 14 climate-related areas, specifically (1) annual per capita greenhouse gas pollution, (2) live methanogenic livestock exports, (3) natural gas exports, (4) recoverable shale gas reserves that can be accessed by hydraulic fracturing (fracking), (5) coal exports, (6) land clearing, deforestation and ecocide, (7) speciescide – species extinction, (8) coral reef destruction, (9) whale killing and extinction threat through global warming, (10) terminal carbon pollution budget exceedance, (11) per capita carbon debt, (12) GHG generating iron ore exports, (13) climate change inaction, and (14) climate genocide and an approach towards omnicide and terracide [32, 65]. Australia as a UK or US lackey has an appalling history of invasions of other countries, including Island Nations [12, 16], is variously complicit in US, UK and Apartheid Israeli nuclear terrorism, and is opposed to the UN General Assembly-supported a nuclear weapons ban [1].

US lackey Australia has obfuscated the horrendous actuality of its climate criminal threat to the Pacific by beating the drum about a US-asserted threat to the region and the world from an economically burgeoning China, notwithstanding the reality that China is Australia's biggest trading partner. Indeed Australian PM Morrison had the same Trump America-derived Sinophobic message in visiting Vietnam in recent days, ignoring the horrendous reality that the climate criminality of US lackey Australia and Trump America is a huge threat to mega-delta Vietnam. One notes that Australia has been involved in all post-1950 US Asian wars, genocidal atrocities that have been associated with 40 million Asian deaths from violence or from war-imposed deprivation [17]. Now cometh Climate Genocide.

A pro-coal, climate criminal and Trumpist Australia sabotaged the 2019 Pacific Islands Forum by insisting on emasculation of the final communique, and is committed to Climate Genocide of Island Nations as collateral damage to Australian neoliberal greed that indeed threatens Humanity and the Biosphere. Decent Humanity will eventually unite in declaring a Climate Emergency and taking action against disproportionately climate criminal countries like Trumpist Australia and Trump America via International Criminal Court (ICC) prosecutions, International Court of Justice (ICJ) litigations, application of Green Tariffs, and imposition of Boycotts, Divestment and Sanctions (BDS).

References.

[1]. "Nuclear weapons ban, end poverty & reverse climate change": https://sites.google.com/site/300orgsite/nuclear-weapons-ban.
[2]. Gideon Polya, editor, "Climate Genocide": https://sites.google.com/site/climategenocide/.
[3]. Pacific Islands Development Forum, "Nadi Bay Declaration on the Pacific Islands Climate Change Crisis", 30 July 2019: https://drive.google.com/file/d/1cbSloYVSuY4mIZnUUC5ISpZUaV3NTb2y/view.
[4]. "Pacific Islands Forum", Wikipedia: https://en.wikipedia.org/wiki/Pacific_Islands_Forum.
[5]. "Gideon Polya, "Body Count. Global avoidable mortality since 1950", including an avoidable mortality-related history of every country from Neolithic times and is now available for free perusal on the web : http://globalbodycount.blogspot.com.au/.
[6]. Gideon Polya, "Review: "This Land Is Our Land. An Immigrant's Manifest" By Suketu Mehta – Trumping Trumpism", Countercurrents, 26 July 2019: https://countercurrents.org/2019/07/review-this-land-is-our-land-an-immigrants-manifesto-by-suketu-mehta-trumping-trumpism.
[7]. Suketu Mehta, "This Land Is Our Land. An Immigrant's Manifesto", Farrar, Straus and Giroux, New York, 2019.
[8].Gideon Polya, editor, "Aboriginal Genocide": https://sites.google.com/site/aboriginalgenocide/.
[9]. Gideon Polya, "Apartheid Israel's Palestinian Genocide & Australia's Aboriginal Genocide compared", Countercurrents, 20 February 2018: https://countercurrents.org/2018/02/20/apartheid-israels-palestinian-genocide-australias-aboriginal-genocide-compared/.
[10]. Gideon Polya, "The US Has Invaded 70 Nations Since 1776 – Make 4 July Independence From America Day", Countercurrents, 5 July, 2013: http://www.countercurrents.org/polya050713.htm.
[11]. Gideon Polya, "British Have Invaded 193 Countries: Make 26 January (Australia Day, Invasion Day) British Invasion Day", Countercurrents, 23 January, 2015: http://www.countercurrents.org/polya230115.htm.
[12]. Gideon Polya, "As UK Lackeys Or US Lackeys Australians Have Invaded 85 Countries (British 193, French 80, US 70)", Countercurrents, 9 February, 2015: http://www.countercurrents.org/polya090215.htm.
[13]. Gideon Polya, "President Hollande And French Invasion Of Privacy Versus French Invasion Of 80 Countries Since 800 AD",

Countercurrents, 15 January, 2014:
http://www.countercurrents.org/polya150114.htm.
[14]. Gideon Polya, editor, "Stop state terrorism":
https://sites.google.com/site/stopstateterrorism/.
[15]. Gideon Polya, editor, "State crime and non-state terrorism":
https://sites.google.com/site/statecrimeandnonstateterrorism/.
[16]. Gideon Polya, "Review: "The Cambridge History Of Australia"
Ignores Australian Involvement In 30 Genocides", Countercurrents,
14 October, 2013: https://www.countercurrents.org/polya141013.htm.
[17]. "Gideon Polya, "Body Count. Global avoidable mortality since
1950", including an avoidable mortality-related history of every
country from Neolithic times and is now available for free perusal on
the web: http://globalbodycount.blogspot.com.au/.
[18]. Gideon Polya, "Review: "Tears In Paradise. Suffering and
Struggle Of Indians In Fiji 1879-2004" by Rajendra Prasad – Britain's
Indentured Indian "5 Year Slaves"", Countercurrents, 4 March, 2015:
https://www.countercurrents.org/polya040315.htm.
[19]. David Vine, "Where in the world is the US military?", Politico,
July/August 2015:
https://www.politico.com/magazine/story/2015/06/us-military-bases-
around-the-world-119321.
[20]. Peter Boyle interviewing Florencia Melgar, "How Australian
intelligence helped Pinochet dictatorship", Green Left Weekly, 29
November 2013: https://www.greenleft.org.au/content/how-
australian-intelligence-helped-pinochet-dictatorship.
[21]. "Australia-East Timor spying scandal", Wikipedia:
https://en.wikipedia.org/wiki/Australia%E2%80%93East_Timor_spyi
ng_scandal.
[22]. "Australia-Indonesia spying scandal", Wikipedia:
https://en.wikipedia.org/wiki/Australia%E2%80%93Indonesia_spying
_scandal.
[23]. Gideon Polya, "US-, Apartheid Israel- & Zionist-beholden
Australian Government & ABC betray Australia's reputation &
international relations", Countercurrents, 17 November 2018:
https://countercurrents.org/2018/11/us-apartheid-israel-and-zionist-
beholden-australian-government-abc-betray-australias-reputation-
international-relations.
[24]. Gideon Polya, "Anti-Indian subversion of Fiji by Apartheid
Israel, pro-Apartheid Australia & pro-Apartheid America",
Countercurrents, 20 October 2017:
https://countercurrents.org/2017/10/anti-indian-subversion-of-fiji-by-
apartheid-israel-pro-apartheid-australia-pro-apartheid-america.
[25]. Phillip Dorling, "US shares raw intelligence on Australians with
Israel ", Sydney Morning Herald, 12 September 2013:

http://www.smh.com.au/national/us-shares-raw-intelligence-on-australians-with-israel-20130912-2tllm.html.
[26]. Phillip Dorling, "Pine Gap drives US drone kills", Sydney Morning Herald, 21 July 2013: http://www.smh.com.au/national/pine-gap-drives-us-drone-kills-20130720-2qbsa.html.
[27]. "Fact check: journalists face 10 years' jail for exposing security agency bungles", RMIT-ABC Fact Check, 14 October 2014: https://www.abc.net.au/news/2014-10-14/journalists-face-jail-for-exposing-security-agency-bungles/5776504.
[28]. "Doctors challenge border force gag laws", Australian Medical Association, 27 July 2016: https://ama.com.au/ausmed/doctors-challenge-border-force-gag-laws.
[29]. Gideon Polya, "Jane Austen and the Black Hole of British History. Colonial rapacity, holocaust denial and the crisis in biological sustainability", G.M. Polya, Melbourne, 1998, 2008; 2008 edition is now available for free perusal on the web: http://janeaustenand.blogspot.com/.
[30]. Gideon Polya, "Australia And Britain Killed 6-7 Million Indians In WW2 Bengal Famine", Countercurrents, 29 September, 2011: https://countercurrents.org/polya290911.htm.
[31]. "Bengali Holocaust (WW2 Bengal Famine) writings of Gideon Polya", Gideon Polya: https://sites.google.com/site/drgideonpolya/bengali-holocaust.
[32]. Gideon Polya, "Climate criminal Australia ignores its hugely increasing Carbon Debt & massive subsidies for GHG pollution", Countercurrents, 8 June 2019: https://countercurrents.org/2019/06/climate-criminal-australia-ignores-its-hugely-increasing-carbon-debt-massive-subsidies-for-ghg-pollution.
[33]. Nic Maclellan, "Marshall Islands president asks Australia t to reconsider on coal and GCF", Pacific Mining Watch, 14 August 2019: https://mine.onepng.com/2019/08/marshall-islands-president-asks.html.
[34]. Melissa Clarke, "Pacific leaders, Australia agree to disagree on action on climate change", ABC News, 16 August 2019: https://www.abc.net.au/news/2019-08-15/no-endorsements-come-out-of-tuvalu-declaration/11419342.
[35]. "Australia's greenhouse emissions to peak after 2030", Climate Plus, 18 February 2016: http://www.climateplus.info/2016/02/18/australias-greenhouse-emissions-to-peak-after-2030/.
[36]. Alan Pears, "Australia is counting on cooking the books to meet its climate targets", The Conversation, 31 January 2019:

https://theconversation.com/australia-is-counting-on-cooking-the-books-to-meet-its-climate-targets-110768.

[37]. Alan Pears and Tim Baxter, "Carry-over credits and carbon offsets are hot topics this lection – but what are they?", The Conversation, 10 May 2019: https://theconversation.com/carry-over-credits-and-carbon-offsets-are-hot-topics-this-election-but-what-do-they-actually-mean-116748.

[38]. "2011 climate change course": https://sites.google.com/site/300orgsite/2011-climate-change-course.

[39]. Anthony Albanese, "Transcript of doorstop – Marrickville – Thursday 15 August 2019: https://anthonyalbanese.com.au/transcript-of-doorstop-marrickville-thursday-15-august-2019.

[40]. "Alan Jones continues criticism of Jacinda Ardern over climate change comments, despite rebuke from Scott Morrison", ABC News, 16 August 2019: https://www.abc.net.au/news/2019-08-16/alan-jones-doubles-down-on-criticism-of-jacinda-ardern/11420102.

[41]. Ben Smee, "Pacific islands will survive climate crisis because they "pick our fruit", Australia's deputy PM says", Guardian, 16 August 2019: https://www.theguardian.com/australia-news/2019/aug/16/pacific-islands-will-survive-climate-crisis-because-they-can-pick-our-fruit-australias-deputy-pm-says.

[42]. Rob Harrison, ""I'm accountable to the Australian people": Scott Morrison pushes back against Pacific leaders", Sydney Morning Herald 16 August 2019: https://www.smh.com.au/politics/federal/i-m-accountable-to-the-australian-people-scott-morrison-pushes-back-against-pacific-leaders-20190816-p52hoo.html.

[43]. Tom McIlroy, "Labor would have rejected Pacific coal calls", Australian Financial Review, 19 August 2019: https://www.afr.com/politics/federal/labor-would-have-rejected-pacific-coal-calls-wong-20190818-p52i9g.

[44]. Melissa Clarke, "New Zealand caught in the middle of Australia's climate change tussle with the Pacific", ABC News, 14 August 2019: https://www.abc.net.au/news/2019-08-14/australia-new-zealand-climate-change-pacific-island-forum/11414228.

[45]. Kate Lyons, "Fiji PM accuses Scott Morrison of "insulting" and alienating" Pacific leaders", Guardian, 17 August 2019: https://www.theguardian.com/world/2019/aug/16/fiji-pm-frank-bainimarama-insulting-scott-morrison-rift-pacific-countries.

[46]. Erin Handley, "Australia accused of putting coal before Pacific "family" as region calls for climate change action", ABC News, 16 August 2019: https://www.abc.net.au/news/2019-08-16/australia-slammed-watering-down-action-climate-change-pacific/11420986.

[47]. Kate Lyons, "Australia waters down Pacific Islands plea on climate crisis", Guardian, 16 August 2019:

https://www.theguardian.com/world/2019/aug/15/australia-waters-down-pacific-islands-plea-on-climate-crisis.
[48]. Kate Lyons, "Revealed: "fierce" Pacific forum meeting almost collapsed over climate crisis", Guardian, 16 August 2019: https://www.theguardian.com/environment/2019/aug/16/revealed-fierce-pacific-forum-meeting-almost-collapsed-over-climate-crisis.
[49]. Lisa Cox, "Morrison's "arrogance" on climate blasted as Australia accused of "trying to destroy" Pacific Islands", Guardian, 17 August 2019: https://www.theguardian.com/australia-news/2019/aug/17/morrisons-arrogance-on-climate-has-trashed-australias-pacific-standing-labor-says.
[50]. "Don't drink the water: nuclear tests in Kiribati", Abolition 2000, 13 May 2018: http://www.abolition2000.org/en/news/2018/05/13/dont-drink-the-water-nuclear-tests-in-kiribati/.
[51]. Carlisle E. W. Topping et al., "In situ measurements of cesium-137 contamination in fruits from the northern Marshall Islands", PNAS, 30 July 30, 2019, 116 (31) 15414-15419: https://www.pnas.org/content/116/31/15414?fbclid=IwAR16xf-67wTs2BjyuF8Lynbu94jZd1pR1fDiMlO5uzzzKItV32iK_lH3sxg.
[52]. Mark Willacy, "A poison in our island", ABC News, 27 November 2017: https://www.abc.net.au/news/2017-11-27/the-dome-runit-island-nuclear-test-leaking-due-to-climate-change/9161442.
[53]. "America's $300 million push to expand naval facilities in northern Australia", ABC News, 29 July 2019: https://www.abc.net.au/news/2019-07-29/americas-push-to-expand-naval-facilities-in-northern-australia/11354926.
[54]. "Separation of isotopes by laser excitation", Wikipedia: https://en.wikipedia.org/wiki/Separation_of_isotopes_by_laser_excitation.
[55]. "Nukes in an uncertain world", ABC TV Q&A, 8 July 2019: https://www.abc.net.au/qanda/2019-08-07/11262430.
[56]. "Australia to send warship and Defence Force personnel to Middle East to support US", ABC News, 21 August 2019: https://www.abc.net.au/news/2019-08-21/scott-morrison-australia-to-join-us-uk-strait-of-hormuz/11434140.
[57]. Gideon Polya, "Apartheid Israel bombing Syria & Iraq – hotting up deadly 4-decade US war on Iran", Countercurrents, 14 August 2019: https://countercurrents.org/2019/08/apartheid-israel-bombing-syria-iraq-hotting-up-deadly-4-decade-us-war-on-iran.
[58]. Magne Frostad, "United Nations Authorized Embargos and Maritime Interdiction: A Special Focus on Somalia", in G. Andreone (editor), "The Future of the Law of the Sea", Springer, 2017: https://link.springer.com/chapter/10.1007/978-3-319-51274-7_11.

[59]. Gideon Polya, "China's Tibet health success versus passive mass murder of Afghan women and children by US Alliance", Global Research, 7 January 2018: https://www.globalresearch.ca/chinas-tibet-health-success-versus-passive-mass-murder-of-afghan-women-and-children-by-us-alliance/5625151.

 [60]. Stephen Hawking, "Brief Answers to the Big Questions", John Murray, 2018, Chapter 7.

[61]. Gideon Polya, editor, "Are we doomed?": https://sites.google.com/site/300orgsite/are-we-doomed.

[62]. Gideon Polya, editor, "Methane Bomb Threat": https://sites.google.com/site/methanebombthreat/.

[63]. Gideon Polya, editor, "Too late to avoid global warming catastrophe": https://sites.google.com/site/300orgsite/too-late-to-avoid-global-warming.

[64]. Gideon Polya, "The need for global governance" in "The New Enlightenment. On Steven Pinker & beyond", by Brian Ellis (with contributions by Greg Bailey, Tony Lynch and Gideon Polya), Australian Scholarly Publishing, Melbourne, 2019.

[65]. Gideon Polya, "ABC and Mainstream media lying by omission enabled Trumpist & climate criminal Australian Coalition victory", Countercurrents, 28 May 2019: https://countercurrents.org/2019/05/abc-mainstream-media-lying-by-omission-enabled-trumpist-climate-criminal-australian-coalition-victory.

CHAPTER 22: LATEST LANCET DATA IMPLY ADANI AUSTRALIAN COAL PROJECT WILL KILL 1.4 MILLION INDIANS

[First published as Gideon Polya, "Latest Lancet Data Imply Adani Australian Coal Project Will Kill 1.4 Million Indians", Countercurrents, 21 April 2017: http://www.countercurrents.org/2017/04/21/latest-lancet-data-imply-adani-australian-coal-project-will-kill-1-4-million-indians/.]

A multi-author 2017 paper in the prestigious UK medical journal The Lancet estimates 1.1 million Indian deaths annually from ambient (outdoor) air pollution, notably from fine carbon particulates ($PM_{2.5}$). Assuming coal burning causes 50% of ambient air pollution in India, pollutants from the burning of coal exported to India from the giant, Australia-approved Adani coal mine will eventually kill 1.4 million Indians. Racist White Australia has a 2-century history of complicity in mass murder of Indians that is ongoing.

Under a pro-coal, climate criminal Coalition Australian Government and with the support of a compliant, pro-coal, Right-dominated Labor Party Opposition, Australia is heading towards a human and environmental catastrophe represented by the huge, Australian Government-approved Adani Carmichael coal mine in the Galilee Basin of Queensland that is set to be the biggest coal mine in the world, and is adjacent to the seriously ill, acutely industry-threatened, and acutely climate change-threatened Great Barrier Reef.

Coal kills just as surely as bombs, bullets and imposed starvation. The proposed Adani coal mine will be the biggest in the world, and has been approved in the face of expert scientific advice that says we must keep coal and other fossil fuels in the ground to prevent catastrophic global warming. Indeed it is now too late to avoid a catastrophic plus 2 degrees Centigrade global warming. However coal kills and ambient (outside) air pollution from pollutants from burning coal kills about 500,000 Indians each year. On a global comparative basis that ignores national pollution specifics, pollutants from burning Australian coal exported to India from the proposed Adani coal mine are estimated to kill 13,000 Indians annually and 500,000 Indians over the lifetime of the coal mine (most of the coal will go to India and thus most of the victims will be Indians). However based on the latest specific data for India published in The Lancet, pollutants from the burning of Adani Australian coal in India will eventually kill 35,000 Indians annually and 1.4 million Indians over the lifetime of the coal mine.

Properly taking land use into account, Australia has a Domestic annual per capita greenhouse gas (GHG) pollution in "tonnes CO_2-equivalent per person per year" of 52.9 and a Domestic plus Exported annual per capita GHG pollution (i.e. including its huge GHG-generating exports) of 116, as compared to the annual per capita greenhouse gas (GHG) pollution of 8.9 (the world), 7.4 (China), 2.1 (India), 2.5 (Pakistan) and 2.7 (Bangladesh).

Australia with 0.3% of the world's population has a Domestic plus Exported GHG pollution equivalent to 4.4% of the World's total GHG pollution (Gideon Polya, "Revised Annual Per Capita Greenhouse Gas Pollution For All Countries – What Is Your Country Doing?", Countercurrents, 6 January, 2016: https://countercurrents.org/polya060116.htm ; Gideon Polya, "Exposing And Thence Punishing Worst Polluter Nations Via Weighted Annual Per Capita Greenhouse Gas Pollution Scores", Countercurrents, 19 March, 2016: https://countercurrents.org/polya190316.htm). However, the Adani coal mine will lift Australia's outrageously disproportionate percentage to 4.5%.

Pro-coal, neoliberal, US lackey, war criminal, climate criminal and Trumpist Australia ranks second only to war criminal and climate criminal Saudi Arabia for climate change inaction on the German Climate Watch Index (see German Climate Watch Index 2015: https://germanwatch.org/en/download/10407.pdf). The Australia Coalition Government- and Labor Opposition-backed Adani coal mine will eventually kill 1.4 million Indians as argued below.

(A). On a comparative global basis (ignoring country specifics) pollutants from burning Adani coal will eventually kill about 0.5 million Indians.

Below is a letter sent to Australian MPs, media and activists in look-the-other-way Australia on 16 April 2007 re the prediction (based on global comparative data ignoring national pollution specifics) of an eventual 0.5 million Indian deaths due to pollutants from the burning of Adani-extracted Australian coal in India:

LETTER:

"Dear etc,

The anti-science, neoliberal, pro-coal and Trumpist Australian COALition Government and the Labor Opposition both support long-term domestic fossil fuel use and unlimited fossil fuel exports, and ignore the deadly impact of pollutants ($PM_{2.5}$ and NOx) from burning coal. Coal kills as illustrated by the following estimates: (1) 7 million people die globally from the long-term effects of air pollution each year (WHO), including about 1,000,000 Indians (half from coal burning for power), 200,000 Americans, 10,000 Australians and 9,000 Londoners, with several million people dying annually from the long-term effects of coal burning pollutants, (2) 75,000 people die annually from pollutants from the burning of Australian coal exports, and (3) it is estimated that a further 13,000 people will eventually die annually from the burning of proposed Adani Australian coal exports with 500,000 people, mostly Indians, dying thus from the lifetime operation of the proposed Adani coal mine that also threatens Australia's desperately ill Great Barrier Reef.

DARA estimates that 0.4 million people presently die from climate change annually, but this may be a huge under-estimate because 17 million people already die avoidably each year from deprivation in tropical or sub-tropical Developing World countries (minus China) that are already disproportionately impacted by man-made climate change. Presently 20 million people are facing famine and mass starvation in war-, climate change- and drought-impacted northern Nigeria, South Sudan, Somalia and Yemen. Two top climate change scientists have estimated that as few as 0.5 billion people will survive this century if climate change is not requisitely addressed, this translating to a Climate Genocide involving the untimely deaths of about 10 billion people this century. It is now too late to avoid a catastrophic plus 2 degree Centigrade temperature rise but decent people are obliged to do everything they can to make the future "less bad" for future generations e.g. by urging and applying boycotts, divestment and sanctions (BDS) against and voting out ecocidal and homicidal climate criminals.

Yours sincerely, Dr Gideon Polya, Melbourne, Victoria, Australia.

See Gideon Polya, "Pollutants from Adani coal mine will eventually kill 0.5 million Indians", Countercurrents, 14 April 2017: https://countercurrents.org/2017/04/14/pollutants-adani-coal-mine-will-eventually-kill-about-0-5-million-indians/ ; "Stop air pollution deaths": https://sites.google.com/site/300orgsite/stop-air-pollution-deaths ; "Climate Genocide": https://sites.google.com/site/climategenocide/ ; "Too late to avoid global warming catastrophe": https://sites.google.com/site/300orgsite/too-late-to-avoid-global-warming ; Gideon Polya, "Humanity must pledge inescapable dispossession and custodial retribution for climate criminals", Countercurrents, 20 December 2016: https://countercurrents.org/2016/12/20/humanity-must-pledge-inescapable-dispossession-and-custodial-retribution-for-climate-criminals/." END LETTER.

The Silence is Deafening in response to my letter. Silence is Complicity.

(B). Based on the latest, specific national pollution data for India reported in The Lancet, pollutants from burning Adani coal will eventually kill about 1.4 million Indians.

A major problem with ignoring reality is that it keeps on getting worse. A multi-author paper published in the prestigious UK medical journal The Lancet on 10 April 2017 estimates 1,090,400 annual Indian deaths and 1,108,100 annual Chinese deaths from ambient (outdoor) air pollution (Aaron J Cohen, Michael Brauer et al., "Estimates and 25-year trends of the global burden of disease attributable to ambient air pollution: an analysis of data from the Global Burden of Diseases Study 2015", The Lancet, 10 April 2017: http://thelancet.com/journals/lancet/article/PIIS0140-6736(17)30505-6/fulltext).

Presently India burns 924 million tonnes of coal each year. Assuming that all the 60 million tonnes per year of Adani Australian coal will go to India (most will) and that 50% of ambient air pollution in India is due to coal-fired power generation (as asserted by major India media), 0.5 x 1,090,400 Indian deaths per year x 60 million tonnes Adani coal exported per year /924 million tonnes coal burned in India per year = 35,403 or eventually about 35,000 Indian deaths per year, and 0.5 x 1,090,400 Indian deaths per year x 2,300 million tonnes coal exported /924 million tonnes coal burned per year in India = 1,357,100 or eventually about 1.4 million Indian deaths over the lifetime of the Adani project due to pollutants from the combustion of Adani coal from a greed-driven, homicidal, ecocidal, terracidal and genocidally climate criminal Lib-Lab (COALition and Labor Right) Australia (see "Stop air pollution deaths": https://sites.google.com/site/300orgsite/stop-air-pollution-deaths and scroll down to "Australia" and "India").

History ignored yields history repeated – White Australia's 2-century complicity in mass murder of Indians.

History ignored yields history repeated and racist White Australia – politically correct racist or PC racist because overwhelmingly White Australians vigorously and sincerely deny that they are racist – has an appalling past, present and future complicity in the deaths of horrendous numbers of Indians as detailed below (but overwhelmingly missing from Australian historiography, Mainstream media and general public perception):

(1) As part of the British Empire Australia was complicit in the 2-century, British-imposed Indian Holocaust in which 1,800 million Indians died avoidably from extreme deprivation, this regularly punctuated by successive horrendous famines under the British Raj (see Gideon Polya, "Economist Mahima Khanna, Cambridge Stevenson Prize And Dire Indian Poverty", Countercurrents, 20 November, 2011: https://countercurrents.org/polya201111.htm).

(2) White Australians fervently supported the British suppression of the 1857 Indian War of Independence (aka the Indian Rebellion or Indian Mutiny), related captured Indian weaponry were displayed at the Victorian State Library together with Australian Aboriginal weapons, and as British Imperial subjects a number of Australians served in the British forces. Indian historian Amaresh Misra claims in his 2 volume work "War of Civilizations: India AD 1857" that in the decade after the Indian rebellion the British killed 10 million Indians in reprisals for the 2,000 British killed in the 1857 rebellion (Amaresh Misra, "War of Civilisations: India AD 1857"; Randeep Ramesh, "India's secret history: A holocaust, one where millions disappeared", Guardian, 24 August 2007: https://www.theguardian.com/world/2007/aug/24/india.randeepramesh ; Gideon Polya, "Genocidal Racist Charles Dickens (1812-1870), Indian Holocaust And UK – US Muslim Genocide ",Countercurrents, 10 February, 2012: https://countercurrents.org/polya100212.htm).

(3) By withholding grain from its huge wartime granaries, Australia helped Britain deliberately starve 6-7 million Indians to death for strategic reasons in the WW2 Bengali Holocaust

(WW2 Indian Holocaust). This atrocity was associated with horrendous, large-scale, military and civilian sexual abuse of starving women and girls (Gideon Polya, "Australia And Britain Killed 6-7 Million Indians In WW2 Bengal Famine", Countercurrents, 29 September, 2011: https://countercurrents.org/polya290911.htm ; Gideon Polya, "Jane Austen and the Black Hole of British History. Colonial rapacity, holocaust denial and the crisis in biological sustainability", G.M. Polya, Melbourne, 2008 edition that is now available for free perusal on the web: http://janeaustenand.blogspot.com/).

(4) As a disproportionately rich country, Australia is disproportionately complicit in the post-1950 Global Avoidable Mortality Holocaust in which 1,500 million people, including about 400 million Indians, have died avoidably from First World-imposed deprivation since 1950 (see Gideon Polya, "Body Count. Global avoidable mortality since 1950", that includes a succinct history of every country and is now available for free perusal on the web: http://globalbodycount.blogspot.com/).

(5) The 2015 per capita GDP is $51,000 for Australia versus $1,600 for India, $1,400 for Pakistan and $1,200 for Bangladesh. As disproportionately rich people, Australians are complicit in the ongoing Global Avoidable Mortality Holocaust in which 17 million people in the Developing World (minus China) – including 4.7 million Indians, 630,000 Pakistanis and 230,000 Bangladeshis – presently die avoidably from deprivation each year on Spaceship Earth with the First World in charge of the flight deck ("List of countries by GDP (nominal) by capita", Wikipedia: https://en.wikipedia.org/wiki/List_of_countries_by_GDP_(nominal)_per_capita ; Gideon Polya, "Body Count. Global avoidable mortality since 1950"; UN Population Division," 2015 Revision of World Population Prospects": https://esa.un.org/unpd/wpp/).

(6) Australia is a world leader in coal exports and will shortly become the world's top liquefied natural gas (LNG) exporter. According to The Lancet (April 2017) annual deaths from

ambient (outside) pollution total 1,108,100 (China), 1,090,400 (India), 135,100 (Pakistan) and 122,400 (Bangladesh). Coal burning for power is deemed responsible for 50% of ambient pollution in India and one supposes is a major contributor to deadly ambient pollution involving $PM_{2.5}$ carbon particulates in all of these countries. Australia per se kills at least 75,000 people a year, mostly Asians, via pollutants from the burning of its world leading coal exports. The Adani obscenity will increase this annual Indian death toll by about 50%.

(7) Two leading UK climate scientists, Dr James Lovelock FRS (famous for atmospheric gas analysis and his Gaia Hypothesis) and Professor Kevin Anderson (Deputy Director, Tyndall Centre for Climate Change Research, University of Manchester, UK), have separately estimated that only about 0.5 billion people will survive this century if carbon pollution and man-made climate change are not requisitely addressed, this translating to a Climate Genocide of about 10 billion deaths this century, overwhelmingly in Developing countries. Noting that the world population is expected to reach 9.5 billion by 2050 (UN Population Division), this predicted Climate Genocide involving the deaths of 10 billion people this century would involve the premature death of roughly twice the present population of particular mainly non-European groups, specifically 6 billion under-5 year old infants, 3 billion Muslims in a terminal Muslim Holocaust, 2 billion Indians, 1.3 billion non-Arab Africans, 0.5 billion Bengalis, 0.3 billion Pakistanis and 0.3 billion Bangladeshis.

Final comments and conclusions.

As a whole Australians are affable, easy-going, fair, and reject explicit racism within Australia. Indeed these attributes plus Australian prosperity and love of cricket make Australia an attractive destination for Indian university students and skilled Indian immigrants. However there are major exceptions to this idyllic picture. Thus the bigoted and racist Australian One Nation Party now has about 10% of the vote, started life with extremely bigoted and offensive anti-Asian and anti-Indigenous policies, and presently has an extremely bigoted and offensive anti-Muslim position. Further, White Australians have always had a huge fear of their vastly more numerous non-White Asian neighbours that crystallized into the 1901-1974, so-called White Australia Policy that excluded non-Europeans, including Asians and Indians, from Australia (this policy was scrapped by the Whitlam Labor Government before this reformist government was itself removed in a US CIA-backed coup in 1975). However there is a secret New White Australia Policy today that discriminates against non-Europeans wanting a visa to enter Australia.

Australians are politically correct racist or PC racist because, while mostly vehemently rejecting racism at an interpersonal level, they are deeply involved in racist actions at an international level. Thus the ultimate in racism is the invasion and devastation of other countries, and as UK lackeys and thence US lackeys White Australians have invaded 85 countries (with 30 of these invasions involving genocidal atrocities) as compared to the British 193, France 82, the US 72 (52 after WW2), Germany 39, Japan 30, Russia 25, Canada 25, Apartheid Israel 12 and China 2. Australia is presently involved in the ongoing Afghan Holocaust and Afghan Genocide, is engaged in its Third Syrian War and Eighth Iraq War in a century, and via the US-Australian Joint electronic spying facility at Pine Gap in Central Australia, it is involved in US nuclear terrorism and in targeting illegal US drone strikes in Libya, starving Somalia, starving Yemen, Syria, Iraq, Afghanistan and Pakistan… who is next? Australia has been involved in all post-1950 US Asian Wars, atrocities that have been associated with 40 million Asian deaths from violence or war-imposed deprivation. An intrinsically White Supremacist Australia is second only to the US as a blind, racist, White Supremacist supporter of Apartheid Israel, and is a fanatical supporter of the US War on Muslims (aka the US War on Terror) in which 32 million Muslims have died from violence, 5 million, or from imposed deprivation, 27 million, in 20 countries invaded by the US Alliance since the US Government's 9-11 false flag atrocity (Gideon Polya, "As UK Lackeys Or US Lackeys Australians Have Invaded 85 Countries (British 193, French 80, US 70)", Countercurrents, 9 February,

2015: https://countercurrents.org/polya090215.htm ; "Stop state terrorism" : https://sites.google.com/site/stopstateterrorism/ ; "Stop state terrorism" : https://sites.google.com/site/stopstateterrorism/ ; Gideon Polya, "Paris Atrocity Context: 27 Million Muslim Avoidable Deaths From Imposed Deprivation In 20 Countries Violated By US Alliance Since 9-11", Countercurrents, 22 November, 2015: https://countercurrents.org/polya221115.htm ; "Muslim Holocaust Muslim Genocide": https://sites.google.com/site/muslimholocaustmuslimgenocide/). To be fair to the White Australians, Australians Mainstream media fake news through lying by omission means that White Australians are blissfully unaware of the horrendous consequences of their complicities, whether the mass murder of Indians under the British Raj or the global American Empire, the adumbrated eventual killing of 1.4 million Indians by the Australia-approved Adani coal mine or the predicted death this century of 2 billion Indians in a near-terminal Climate Genocide in which a climate criminal Australia is disproportionately complicit. Australian Mainstream media journalist, politician and academic presstitutes ensure that ordinary Australians are like mushrooms – kept in the dark and fed manure (Gideon Polya, "Mainstream media fake news through lying by omission", MWC News, 1 April 2017: https://www.crimeandpower.com/2020/01/18/mainstream-media-fake-news-through-lying-by-omission/).

What can decent people do? Decent people (and especially acutely-threatened Indians) must (a) inform everyone they can (especially Australians) and (b) do everything they can by Boycotts, Divestment and Sanctions (BDS) to ensure that coal and other fossil fuels remain in the ground. Coal kills.

CHAPTER 23: RAMPANT ORWELLIAN FALSEHOOD IN NEOLIBERAL AUSTRALIA – AND IN YOUR COUNTRY TOO?

[First published as Gideon Polya, "Rampant Orwellian falsehood in neoliberal Australia – and in your country too?", Countercurrents, 1 March 2020: https://countercurrents.org/2020/03/rampant-orwellian-falsehood-in-neoliberal-australia-and-in-your-country-too. .]

In George Orwell's prescient, dystopian and frightening novel "1984", Big Brother declared that (A) war is peace, (B) slavery is freedom, (C) ignorance is strength, and (D) 2 plus 2 does not equal 4. Scientists and other humanitarian truth-tellers are alarmed as ostensible democracies as well as authoritarian states head towards this ultimate in comprehensive, state-imposed and blatant falsehood. My country, US lackey Australia, is arguably second only to Trump America for Orwellian government lying in ostensibly democratic societies – but how does your country perform in the mendacity stakes?

Science involves the critical testing of potentially falsifiable hypotheses whereas the increasingly prevalent spin that dominates the world involves the uncritical use of asserted facts to support a partisan proposition i.e. the very opposite of the scientific method. However the great irony is that the very success of the scientific method that has now created a highly technological and digital world has also created the electronic vehicles for mass dissemination of lies and spin-based untruths (slies) [1].

Public recognition of such asserted or actual "fake news" has now provided the excuse for governments and powerful corporations to take action to censor views they consider inimical to their interests. Indeed politically correct, ostensibly "progressive", "liberal" and "pro free speech" Centrists of the Mainstream media commentariat will earnestly discuss the need for censorship in the "public interest" of what "somebody" will find to be "fake news" Thus, for example, Alphabet is the biggest media company in the world and owns Google and YouTube [2]. However the Google Search Robot has been programmed to put progressive websites towards the bottom of Google Searches [3], noting that 90% of people doing Google Searches don't go past the first page of results [4]. Indeed Google omits many results, and after laboriously going through every page of results from an initial Google Search without success, on the last page of Results one is invited to do a more inclusive search: "In order to show you the most relevant results, we have omitted some entries very similar to the [number] already displayed. If you like, you can repeat the search with the omitted results included". By clicking on the link provided one can then laboriously repeat the Google Search (e.g. for the title of an article in inverted commas) and eventually find the thus so exquisitely censored article you have been seeking. In contrast, Searches using Bing (owned by Microsoft) come up with Results on page 1 that are "hidden" by Google – ergo, "Bing it!" [3].

Alphabet's YouTube has boasted of its success in "hiding" stuff it doesn't like. Thus Lesley Stahl (famed anti-racist Jewish American journalist) on YouTube censorship (2019): "She [YouTube's CEO, Susan Wojcicki] told us that earlier this year, YouTube started re-programming its algorithms in the U.S. to recommend questionable videos much less and point users who search for that kind of material to authoritative sources, like news clips. With these changes Wojcicki says they have cut down the amount of time Americans watch controversial content by 70%" [5].

George Orwell in "1984" [6] wrote of a "total control" authoritarian state comprehensively and violently enforcing acceptance of blatant falsehoods using the relatively primitive technological means available in 1948. However the essay "One World Digital Dictatorship" written by Soren Korsgaard in 2019 is a vitally important, must-read and indeed scary account that details the increasingly technologically sophisticated ongoing digital disempowerment and digital imprisonment of Humanity that is pertinent to all societies from the US and UK to India and China. Whereas the dissident in "1984" would be traumatized into submission by exposure to his worst fear in Room 101, in 2020 the dissident faces an accelerating movement by both Western-style democracies and one-party states (notably China and the Gulf States) towards world-wide Digital Dictatorship (Digital Imprisonment) involving mass data collection on everyone, mass surveillance, facial recognition-based tracking, crypotocurrency-based cashless societies, and social credit-based disempowerment [7, 8].

Set out below is a succinct account of how Big Brother's declarations that (A) war is peace, (B) slavery is freedom, (C) ignorance is strength, and (D) 2 plus 2 does not equal 4 are rampant in ostensible democracy Australia under a mendacious, racist, pro-war, war criminal, anti-science, anti-environment, climate criminal, human rights-violating, neoliberal, US lackey and Trumpist Liberal Party-National Party Coalition Government. However, in reading this damning analysis, consider to what extent your own country has joined the Orwellian nightmare.

(A). "War is peace".

Australia has committed a long catalogue of war crimes variously committed in the name of "commerce", "the British Empire" or helping the US to ostensibly bring "peace", "freedom" and "democracy" to numerous distant land, crimes summarized by the Orwellian assertion "war is peace". However it must be stated that while we honor those who serve our country, we offer no such honor to war criminal politicians who put them in harm's way in violation of the Natural Law and International Law.

(1). War is the penultimate in racism and genocide the ultimate in racism. Variously as a UK or US lackey Australians have invaded 85 countries as compared (over the last millennium) to the British 193 countries, France 82, the US 72 (52 after WW2), Germany 39, Japan 30, Russia 25, Canada 25, Apartheid Israel 12, China 2 and Iran zero (0) [9].

(2). US lackey Australia has been involved in all post-9-11 US Asian wars, atrocities associated with 40 million Asian deaths from violence or imposed deprivation [10, 11]. Australian Coalition Governments have supported all these wars but the Labor Opposition opposed the Vietnam War and the 2003-2011 Iraq War.

(3). US lackey Australia presently has armed forces in Iraq but has refused to leave despite the Iraqi Parliament demanding that foreign forces leave Iraq. The US, UK, Australia, Canada and Germany have rejected the Iraqi Parliament's Quit Iraq demand, with the US threatening to instantly collapse the Iraqi economy by a banking freeze if Iraq insists on US Alliance withdrawal from Iraqi territory [12].

(4). US lackey Australia has armed forces of the Australian Defence Force (ADF) in Occupied Afghanistan despite the Indigenous Afghan Taliban controlling 14.5% of the country and 44% being Taliban controlled or contested after 18 years of partial occupation by the world's superpower, the serial invader and serial war criminal US [13].

(5). The Australian Navy is actively involved against Somali sailors in the waters of Occupied Somalia.

(6). Through the joint Australian-US electronic spying base at Pine Gap (Central Australia) Australia actively supports US nuclear terrorism and targets illegal US drone strikes in 7 variously impoverished and devastated countries (Libya, Somalia, Yemen, Iraq, Syria, Afghanistan and Pakistan [14, 15]).

(7). Use of nuclear weapons for mass murder of civilian populations is the worst and most utterly obscene of war crimes. Australia helped UK nuclear terrorism by supplying uranium, nuclear test locations and the Woomera base for testing missiles for nuclear weapons delivery. This UK-Australia program resulted in nuclear contamination of Indigenous lands of the Maralinga Tjarutja (Pitjantjatjara Indigenous Australian people) and of other lands as revealed by Australian journalists Brian Toohey and Ian Anderson [16, 17]. The joint Australia-US Pine Gap electronic spying base is critical for US nuclear terrorism.

(8). Australia hosts nuclear-armed US warships but New Zealand was effectively excluded from US treaty obligations under the ANZUS (Australia, New Zealand and US) Treaty after it refused entry to possibly nuclear armed US warships in 1985 [18].

(9). The UN Charter is very clear about foreign forces being in another country – it can only happen (1) through invitation by the country, (2) in response to invasion of another country by that country, or (3) with the permission of the UN Security Council. None of these criteria were satisfied in the Australian participation in the Anglo-American invasion of Iraq in 2003, and none of the leaders involved – notably President George Bush (US), PM Tony Blair (UK) and PM John Howard (Australia) have been prosecuted for war crimes committed for resources [oil], notwithstanding decisive conclusions to this effect by eminent people [19-22], and calls by eminent people for such prosecutions [19, 20].

(10). Killing people in another country (the penultimate in racism) can happen through invasions and bombing, but Sanctions can be equally deadly. Thus Australia was involved in the UN-imposed 1990-2003 Sanctions against Iraq that were complemented with UK, US and Israeli bombing of Iraqi civil infrastructure, with this resulting in 1.7 million Iraqi deaths from imposed egregious deprivation [23-25]. US-imposed and US-urged sanctions against Iran in a 4-decade US war on Iran have been associated with 3 million Iranian deaths from imposed deprivation in an ongoing Iranian Holocaust and Iranian Genocide [26]. US lackey Australia was involved in such sanctions and more recently has refused to follow the EU example of trying to protect businesses from US retaliation in relation to Sanctions re-imposed on Iran by Trump America [27].

(11). The US subverts all countries and is currently conducting military operations against 7 countries. Crucial to such hostile actions and the global US War on Terror (aka the US War on Muslims) are US military bases around the world, currently estimated at over 700 and variously located in over 70 countries [28, 29]. Australia hosts over 2,000 Marines at a base near Darwin, hosts nuclear-armed US warships, and recently announced an expansion of the Tindal RAAF (Royal Australian Air Force) air base near Katherine in the Northern Territory that PM Scott Morrison has characterized thus: "It will be integral to our alliance with the United States, and [will] increase the reach of Air Force capabilities in the Indo-Pacific" [30].

(12). Back in 1975 the reformist Whitlam Labor Government was removed in a US CIA-backed Coup [31], shortly after Australian Intelligence was covertly involved in the 1973 US-backed overthrow of the democratically-elected Allende Government in Chile [17]. It seems that as America's Deputy Sherriff in the South Pacific, Australia was involved in the 1987 Coup in Fiji (attributed to the US by William Blum [11]), and in the Apartheid Israel-backed George Speight-led Fiji Coup in 2000 [32]. Australia has been up to all kinds of dirty tricks in our region (e.g. spying on the Timor L'Este Cabinet meetings, and on the Indonesian President and his wife). Brian Toohey has revealed Australian urging of the US to back an invasion of President Soekarno-ruled Indonesia in the 1950s and 1960s. However the Americans opposed "overt military intervention in Indonesia". Brian Toohey: "However, the US fostered a rebellion in Sumatra and Sulawesi by giving covert support to Islamist parties…the agency [CIA] managed to arm a rebel group of 42,000 and Australia was expected to supply administrative and medical help if the rebels succeeded" (pages 132-133 [17]). The Aceh rebellion lasted until just after the 2004 Tsunami tragedy (230,000 killed) that shocked the Sumatran rebels and the central government into making peace (dialogue is a humane alternative to war and mass mortality). In 1965 the CIA backed a military coup that put corrupt, right-wing and US-backed General Suharto in power for 31 years, and which was accompanied by horrendous massacres of 0.5-1.0 million Indonesians, notably progressive professionals and Chinese [33]. If slavishly US-linked Australian Intelligence is still up to such dirty tricks against neighbouring countries we will never know because the Australian Government has passed laws (with craven Labor Opposition support) providing for up to 10 years in prison for people exposing such atrocities [34].

(13). Australia gets a lot of its "defence" material from the US, UK, France and Apartheid Israel. These weapons and systems have been developed out of decades of use in US Alliance wars in which millions of civilians have been killed. Thus 32 million Muslims have died from violence, 5 million, or from deprivation, 27 million, in 20 countries invaded by the US Alliance since the US Government's 9-11 false flag atrocity [35, 36]. Australian universities have extensive military research collaborations with the US and Apartheid Israel [37, 38]. Apartheid Israeli merchants of death boast that their weapons have been "pre-tested" on Iraqis, Afghans, Lebanese and Palestinians (most notably on the 2 million inmates of the Gaza Concentration Camp of whom 50% are children and 75% are women and children) [39]. Apartheid Israel and its traitorous Zionist supporters represent a huge threat to Australia in 50 areas [40] but Australian defence has purchased command and control systems and Unmanned Air Vehicles (UAVs) from an Israeli company [41, 42].

(B). "Slavery is freedom".

Australia has a long and evil record of slavery commencing in 1788 with the use of convict labour. Survivors of the violent phase of the Aboriginal Genocide were used as agricultural, domestic and sexual chattels paid with flour, tea and tobacco, with this exploitation lasting into the late 20th century. A massive measles epidemic (sourced in Australia) killed 40,000 of the 150,000 strong Fijian population in 1875. As a consequence, commencing in 1879 the British introduced indentured labour from India into Fiji to work as "5 year slaves" on British and Australian sugar plantations. Outcry over the circumstances of these Girmityas (named thus from mispronunciation of the word "Agreement") resulted in cessation of new imports in 1916 and release of the last "5 year slaves" in 1920. My dear late wife, Zareena *née* Lateef, was Fiji Indian of Bengali and Bihari origin, and came from a large Muslim family. All Zareena's grandparents were "5-year slaves" of the British and Australians in Fiji, and her eminent lawyer father, Abdul Lateef MBE, was a great advocate of inter-ethnic harmony, and helped negotiate independence of Fiji from Britain [43-47]. The Australians sought slave labour throughout the South Pacific in a violent process called "blackbirding" that had the collateral disastrous effect of spreading epidemics of European diseases such as measles with consequent decimation of populations (e.g. 25% in Vanuatu) [10]. Melanesians from countries such as present-day Vanuatu and the Solomon Islands were brought to work on Queensland sugar cane plantations on land ethnically cleansed of Indigenous Australians. On Australian Federation as the nation state of Australia in 1901, racist concerns about a "White Australia" resulted in enforced repatriation of many of these so-called "Kanaks" and their progeny.

(1). The term "The Stolen Generations" refers to the circa 100,000 "mixed race" Indigenous children forcibly removed from their mothers, educated in European ways, and then variously adopted by Europeans or deployed as domestic servants or farm labour on minimal pay [48]. This outrageous crime was justified on the basis of benefit to the children i.e. "slavery is freedom". When Labor PM Kevin Rudd made history by formally offering a national apology to the Stolen Generations on 13 February 2008 at a sitting of the Parliament of Australia, the Aboriginal Genocide-denying former Coalition PM John Howard (yes, the same Australian PM who illegally invaded Iraq in 2003) refused to attend (he was the only living former Australian PM not to attend) [49]. And, of course, both the Labor Government and Coalition Opposition refused to support a Greens motion for proper compensation of the surviving victims [48, 49]. The survivors of the Stolen Generations have not been properly compensated, and indeed removal of Indigenous children from their mothers and their communities is presently continuing at a record rate [50].

(2). In Queensland (and elsewhere in Australia) Aboriginal (Indigenous) workers had their pay withheld by so-called White "Protectors", and former Queensland Premier Peter Beattie estimated that $500 million in wages was stolen from Aboriginal people in Queensland. A class action representing 10,000 Aboriginal workers in Queensland was finally successful in gaining $190 million in compensation for wages stolen between 1939 and 1972 [51]. Living survivors of Aboriginal slave labour remain uncompensated or inadequately compensated around Australia.

(3). In prosperous Australia there is a work participation rate of 66.0%, this referring to the percentage of the working population in the age group of 16-64 who are currently employed or seeking employment (full-time students, housewives and people above 64 years are excluded). However the 34% not participating in the economy as workers are participating as "consumers". Further, of the employed many are the so-called "working poor" who do not earn enough or do not work enough hours to meet modest needs. The fundamental Australian social tenet of "the living wage" has disappeared from neoliberal Australia. 3 million out of 25 million Australians live in poverty, this including 0.8 million children. The social safety net has shrunk so much that the state relief given to the unemployed (the New Start Allowance) condemns these people to dire poverty unless families can give them shelter [this has appallingly worsened in the 2020 Covid-19 pandemic].

(4.) Australia has a large population of temporary migrant agricultural workers from overseas who are vulnerable to exploitation and underpayment. Indeed underpayment of the more poorly paid and vulnerable workers in Australia appears widespread, with guilty parties involved in this "wage theft" ranging from restaurant owners to major corporations. Now this is not slavery, just egregious exploitation, but one is reminded of Winston Churchill's famous description of Chinese indentured labour in South Africa in the House of Commons on 22 February 1906 as Under-Secretary of the Colonial Office: "The conditions of the Transvaal ordinance … cannot in the opinion of His Majesty's Government be classified as slavery; at least, that word in its full sense could not be applied without a risk of terminological inexactitude" [52].

(5). Australia as a major trading nation is a great beneficiary of globalisation. The per capita GDP (nominal) is \$54,000 for Australia as compared to \$1,900 for Bangladesh (2019) [53]. Globalisation means that rich Australians can buy business shirts made by extremely cheap labour in Bangladesh for about \$10. Bangladeshis have to work (slave) for extremely low wages or starve, but the extremely low per capita income in Bangladesh means that 247,000 Bangladeshis die avoidably each year from deprivation [10, 54]. Annual avoidable mortality from deprivation totals 16 million for the Developing World (minus China) and 4 million for India, but is essentially zero (0) for Australia, North America, Western Europe, China, South Korea, Taiwan, Singapore, Apartheid Israel (ignoring the Occupied Palestinians), [Cuba,] and the Gulf States (ignoring South Asian and South East Asian guest workers) [10].

(6). Noting that the world population is 7.6 billion, according to Oxfam: "Economic inequality is out of control. In 2019, the world's billionaires, only 2,153 people, had more wealth than 4.6 billion people… The world's richest 1% have more than twice as much wealth as 6.9 billion people… The combined wealth of the world's 22 richest men equals the wealth of all the women in Africa…New World Bank estimates show that almost half of the world's population lives on less than \$5.50 a day… Billionaires are also able to buy impunity from justice, influence politicians or a pliant media and even rig democratic processes. The use of money to influence elections and public policy is a growing problem all over the world… the United Nations estimates that 820 million people are going hungry, the World Bank estimates that only 735 million are living in extreme poverty… by 2025 up to 2.4 billion people worldwide could be living in areas subject to periods of intense water scarcity… Extreme wealth is a sign of a failing economic system…

Governments must take bold and decisive steps by taxing wealth, high incomes, and cracking down on loopholes and inadequate global tax rules that allow rich corporations and individuals to escape their tax responsibilities" [55]. Each year 16 million people die avoidably from deprivation in the Developing World (minus China) on Spaceship Earth with the rich First World (including Australia) in charge of the flight deck [10]. Slavery implies (i) effectively no means of escape, and (ii) bare subsistence characterized by significant avoidable mortality from deprivation [10]. Australia's annual overseas aid is 0.26% of GNI ($130 per head) as compared to 1.36% for Sweden ($701 per head) [56].

(7). Oxfam: "The Australian Tax Office estimated the tax gap (tax not paid) by large corporations and businesses in 2016-17 was AUD $2 billion. This is the same as the entire National Bushfire Recovery Fund established by the Federal Government in 2020… In Australia the richest 1% (250,000 people) have more than double the wealth of 50% of our population (12.5 million people) and own nearly USD$1.6 trillion. That's 22.2% of all of Australia's wealth… The wealthiest 1 per cent of people in the world have more than double the wealth of 6.9 billion people… It's estimated that tax dodging by multinational companies costs the world's poorest countries at least USD $100 billion every year" [57].

(8). According to Larry Elliott of The Guardian: "The growing concentration of the world's wealth has been highlighted by a report showing that the 26 richest billionaires own as many assets as the 3.8 billion people who make up the poorest half of the planet's population. In an annual wealth check released to mark the start of the World Economic Forum in Davos, the development charity Oxfam said 2018 had been a year in which the rich had grown richer and the poor poorer. It said the widening gap was hindering the fight against poverty, adding that a wealth tax on the 1% would raise an estimated $418bn (£325bn) a year – enough to educate every child not in school and provide healthcare that would prevent 3 million deaths" [58]. French economist Thomas Piketty argues cogently that wealth inequality is bad for the economy because the poor cannot afford to buy the goods and services they produce, and is bad for democracy because Big Money buys public perception of reality and hence votes [59-61]. Piketty advocates a circa 1% annual wealth tax to reduce wealth inequality and indeed France has such a tax that goes up to 1%. For 1,400 years Islam has prescribed an annual wealth tax of 2.5% [61]. It was estimated in 2014 that an annual wealth tax of 4% would bring all countries up to a per capita GDP about that of China and Cuba (countries for which annual avoidable mortality from deprivation is zero) and hence abolish the Global Avoidable Mortality Holocaust that kills 16 million people each year [10, 62]. However at least 50% of Australians would utterly reject such a life-saving tax.

(9). Further to (8), neoliberal Western democracies such as Australia have been variously described as Plutocracies, Kleptocracies, Murdochracies, Lobbyocracies, Corporatocracies and Dollarocracies in which Big Money purchases people, politicians, parties, policies, public perception of reality and hence votes, more political power and more profit in an obscene and deadly positive feedback loop.

(10). In considering "slavery is freedom" note that White Australia appropriated Aboriginal lands subject to Indigenous stewardship for 65,000 years and subjected the Indigenous inhabitants to an Aboriginal Genocide in which the Indigenous population plummeted from 1 million in 1788 to 0.1 million a century later [63]. The White Australians wanted the land but not the Indigenous inhabitants, just as Apartheid Israel wants the land of Palestine but not the Indigenous Palestinian inhabitants [64, 65]. The Australian Aboriginal Genocide (1788 onwards) and the Palestinian Genocide (WW1 onwards) both involved 0.1 million violent deaths and about 2 million avoidable deaths from imposed deprivation. Presently about 5,100 out of 5 million Occupied Palestinians die avoidably annually from violence (500) or from imposed deprivation (4,600) in a process of active and passive mass murder of Indigenous Palestinians in gross violation of the Geneva Convention [66] and thence of the UN Genocide Convention [67]. In contrast, about 5,000 out of 0.9 million Indigenous

Australians die avoidably each year from imposed deprivation – unlike the murderous Israelis, the Australians ceased massacring Indigenous Australians in about 1930. White Australians talk of assimilation and the benefits ("freedom") of modern technology, but many Indigenous Australians are effectively enslaved (unable to escape, barely surviving). Many of the 5 million Occupied Palestinians and 7 million Exiled Palestinians are also effectively enslaved (unable to escape, barely surviving), yet Donald Trump's peace plan is for continued zero human rights and continuing servitude. Australia is second only to Trump America as a supporter of Apartheid Israel and hence of the heinous crime of Apartheid.

(11). Rapacious neoliberalism, that is presently dominant in Australia and globally, seeks to maximize the freedom of the smart and advantaged to exploit human and natural resources for private profit. The system tends to one in which the smart and advantaged 1% increasingly dominate an ever more impoverished 99% who are enslaved (unable to escape, barely surviving) and are convinced by One Percenter-owned media that neoliberalism (unfettered greed and growth) is good for them ("slavery is freedom"). In contrast, social humanism (e.g. socialism, eco-socialism, democratic socialism, Universal Basic Income, the welfare state) seeks to sustainably maximize human happiness, dignity and opportunity for everybody through evolving, pragmatic and culturally-sensitive intra-national and international social contracts [68-73].

(C) "Ignorance is strength".

Australia has a high level of literacy, high school completion and university education. However there is an astonishing amount of ignorance, falsehood, de facto and de jure censorship, and fake news through lying by omission in a look-the-other-way Australia through misplaced patriotism, jingoism, institutional Codes of Conduct, defamation laws, political correctness, intellectual child abuse in religious schools, commercial-in-confidence agreements, non-disclosure agreements, anti-whistleblower laws, anti-leak laws, draconian anti-terrorism laws, intimidation, compulsory metadata acquisition, secret trials, compulsory breakage of encryption, anti-science denialism (notably climate change denialism), sidelining of scientists, rendering truth-tellers "invisible", massive spying on citizens shared with foreign powers, increasing mass surveillance, increasing "digital dictatorship", and massive lying by commission and omission by oligopoly media and by Mainstream editor, politician, academic and commentariat presstitutes. Entrenched active and passive mendacity and Orwellian "ignorance is strength" indeed. There are numerous examples that can be given, but the following are a selection of the more important areas of imposed ignorance in jingoistic, mendacious, look-the-other-way Australia that has an immense capacity for lying and self-deception about Australian atrocities.

(1). Holocaust ignoring and genocide ignoring are far, far worse than repugnant holocaust denial and genocide denial because the latter can at least permit public refutation and public debate. As a UK or US lackey, Australia has invaded 85 countries [74, 75], and of these invasions about 30 have been genocidal [76], with genocide here defined as in Article 2 of the UN Genocide Convention [67].

(2). The ongoing Aboriginal Genocide (about 2 million killed by violence or imposed deprivation and disease since1788) [63] is largely ignored in Australia, with former Coalition PM John Howard dismissing such concerns as "The black arm-band view of history", and former Coalition PM Tony Abbott asserting of pre-invasion Sydney that "back in 1788 it was nothing but bush". A Search of the ABC (Australia's taxpayer-funded equivalent of the UK BBC) for "Aboriginal Genocide" yields 3 results but no mention of the term and no details of the carnage. Before the British invasion in 1788 there were 350-750 distinct Aboriginal languages and dialects but today there are only about 150, and of these all but about 20 are endangered through continuing Aboriginal Ethnocide involving large-scale removal of Indigenous children from their mothers, and government policies inimical to bilingual education and communities living on traditional lands. From a qualitative perspective, the continuing Australian Aboriginal Genocide and Aboriginal Ethnocide have been the worst in human history [63].

(3). "Lest we forget" is the sacred mantra of Anzac Day (25 April) that commemorates the circa 100,000 Australians who have died in wars. However the Australian Government has just spent $500 million on the national war memorial in Canberra that excludes any mention of the Frontier Wars in which about 100,000 Indigenous Australians died violently defending their land from the British invaders [63, 77].

(4). In similar vein, the Australian Coalition Government has allotted $50 million for memorializing Captain James Cook's blatant and criminal 1770 declaration that Australia was British territory, noting that Indigenous Australians had developed numerous unique cultures in Australia over 65,000 years, the survivors of which today represent the oldest living cultures in the world [63].

(5). Australia Day (26 January) commemorates as a national day the violent and genocidal British invasion of Australia on 26 January 1788. The same day is Republic Day in India and commemorates the adoption of an Indigenous Indian constitution after 2 centuries of genocidal British rule. To Indigenous Australians and decent, anti-racist Australians in general, 26 January is "Invasion Day". Some decent local councils decided to move citizenship ceremonies for new citizens to a non-controversial day but were threatened with removal of their right to conduct such ceremonies by the instinctively offensive Australian Coalition Government [78].

(6). During WW2 Australia was complicit in one of the worst genocidal atrocities in living memory, namely the WW2 Bengali Holocaust (WW2 Indian Holocaust, WW2 Bengal Famine) in which the British with Australian complicity deliberately starved 6-7 million Indians to death for strategic reasons in Bengal and the adjoining provinces of Bihar, Assam and Orissa. UK lackey Australia was complicit in this atrocity by withholding wheat in its huge wartime grain stores from its starving ally India. Australia produced 24 million tons of wheat in WW2 but only exported 9 million tons, with as little as 0.3 million tons going to starving India. Huge granaries were constructed to store the unsent grain of which 1 survives as a "Cathedral of the Bush" at Murtoa in Western Victoria. This "forgotten" atrocity was associated with large scale sexual abuse of starving women and girls on a scale commensurate with the "comfort women" abuses of the Japanese Imperial Army. This atrocity has been largely deleted from British and Australian histories by successive generations of racist and mendacious Australian and British journalist, editor, politician, academic and commentariat presstitutes. The WW2 Bengali Holocaust (6-7 million deaths) was indeed the first WW2 atrocity to have been described as a "holocaust" and was commensurate in size with the WW2 Jewish Holocaust (5-6 million deaths) [79-83]. However a Search of the ABC for "Bengali Holocaust" yields zero (0) results whereas a Search for "Jewish Holocaust" yields 20 results.

(7). Racist and US lackey Australian Mainstream journalist, editor, politician, academic and commentariat presstitutes adopt the US position that since WW2 genocides have only been committed by Serbs and non-Europeans the US doesn't like [84, 85]. Thus I have drawn up a list of about 60 significant holocausts and genocides in the last 2 centuries [86-88]. However, by way of example (out of scores of examples that can be given), the 21st century Rohingya Genocide (10,000 killed, 1 million refugees) is rightly described [by Australian and Western Mainstream media] as a "genocide" (the US hates China-supported Myanmar), but the same designation is not given to the 21st century Iraqi Genocide and Iraqi Holocaust (2.7 million Iraqi deaths from violence or deprivation), to the 2001 onwards Afghan Genocide and Afghan Holocaust (7 million Afghan deaths from violence or deprivation) or to the 21st century Muslim Genocide and Muslim Holocaust (32 million deaths from violence, 5 million, or deprivation, 27 million) [10, 23, 35, 89, 90].

(8). Western Mainstream media resolutely accept as an article of faith the "official US version of 9-11" proclaimed by successive mendacious US Administrations, notwithstanding its rejection by numerous science, architecture, engineering, aviation, military and intelligence experts [36]. Indeed the post-9-11 War on Terror subsequent to the US Government's 9-11 false flag atrocity has been associated with huge violations of civil rights, notably in the US via the Patriot Act [91], in Australia by over 70 anti-terrorism laws that have finally elicited Mainstream media opposition when applied to journalists [92-96], and in the UK as exampled by the outrageous persecution of Australian truth-telling journalist and world hero Julian Assange [97, 98].

(9). The UN regards Apartheid as a heinous crime against Humanity [99]. Of the 6.9 million Indigenous Palestinian subjects of Apartheid Israel, who represent about 50% of the population of Palestine, about 72% are the Occupied Palestinians who for 53 years have been excluded from all human rights [100] and cannot vote for the government ruling them i.e. egregious Apartheid that is worse than Apartheid in the former Apartheid South Africa. Under the racist and human rights-violating Australian Coalition Government Australia is second only to racist Trump America as a supporter of Apartheid Israel and hence of Apartheid. There must be zero tolerance for racism. Those who support Apartheid Israel and hence repugnant Apartheid (100% of the Australian Coalition and much of the Labor Opposition) are utterly unfit for public life in a one-person-one-vote, ostensible democracy like Australia. No genuinely anti-racist Australians can support the pro-Apartheid Coalition and will accordingly (a) vote 1 Green and put the Coalition last, and (b) urge and apply Boycotts, Divestment and Sanctions (BDS) [101] against Apartheid Israel and all those supporting this nuclear terrorist, racist Zionist-run, genocidally racist, democracy-by-genocide, Apartheid rogue state that is engaged in an ongoing Palestinian Genocide (2.2 million Palestinian deaths from violence, 0.1 million, or from imposed deprivation, 2.1 million, since the British invasion of the Ottoman Empire in 1914 [9, 64, 65, 102-105].

(10). Extraordinary examples abound of Zionist-backed, jingoistic suppression of free speech and truth-telling in contemporary Australia. The utterly cowardly, timorous, gutless and yellow Australian Lib-Labs (Coalition and Labor) ignore the kidnapping, shooting, tasering, imprisonment, robbing, mangling, killing, torturing, defaming, abusing, deceiving, perverting and subverting of Australians variously by genocidally racist Israeli Zionists or by traitorous Australian Zionists [40, 106]. Human rights lawyer Melissa Parke was forced to remove herself from Labor candidacy in the 2019 Federal elections for being critical of Apartheid Israel [107]. However other decent, anti-racist Australians, including myself, have similarly fallen foul of the Zionist Lobby. Thus Muslim Australian engineer and ABC journalist Yassmin Abdel-Magied made the mistake of simply publishing on her Facebook page the following 7 words: "Lest we forget (Manus, Nauru, Syria, Palestine)". Savaged by public outcry, Ms Abdel-Magied rapidly deleted the post and apologized, the ABC apologized and a month later removed her TV program, and Yassmin Abdel-Magied left for London [108]. The post was correct and her silencing by rabid jingoists was a stain on Australia and an attack on free speech [108]. Other Australian truth-tellers variously subject to attempted ferocious Zionist censorship include author and teacher Paul Gilby (for referring to US and Israeli state terrorism) [109, 110], top columnist Mike Carlton (for criticizing the horrendous 2014 Gaza Massacre by Apartheid Israel) [111], and academic Dr Sandra Nasr (for comparing Biblical and present-day Palestinian Genocides) [112]. A resolutely truth-telling, scientist and anti-racist Jewish Australian humanitarian, I have been rendered "invisible" to Mainstream Australia over the last 10 years, most likely through remorseless, false and indeed anti-Jewish anti-Semitic defamation by Zionists.

(11). Jingoistic censorship continues to be applied to truth tellers in other areas in jingoistic, pro-war, US lackey Australia. Scott McIntyre, a courageous young sports journalist with Australia's substantially taxpayer-funded, multicultural Special Broadcasting Service (SBS), was sacked after he tweeted a succession of truthful anti-war comments on Anzac Day (Australia's [main] war dead remembrance day). Only a few, notably John Pilger, came to his defence, and the rest was silence or strong condemnation of McIntyre [113]. The "One Day of The Year" is an iconic play by Alan Seymour about

Anzac Day, the Australian and New Zealand war dead remembrance day held on the anniversary of the unsuccessful invasion of Turkey at Gallipoli by the Australia and New Zealand Army Corps (ANZAC) on 25 April 1915. The play was banned by the Adelaide Festival in 1960, and in Sydney was delayed through bomb threats [114]. Anti-war sentiment expressed on or about Anzac Day has elicited ferocious responses over the years ranging from death threats to blanket media censorship that hides Australia's appalling history of UK- and US-linked war crimes and genocides. In 1983 about 300 brave and decent women marched on Anzac Day in Sydney with a banner stating: "In memory of all women in all countries raped in all wars". They were abused and many were arrested. In Australia's capital city Canberra, women were prevented from laying a wreath at the War Memorial and a special law was passed making it unlawful to give offence to any ACT Anzac Day march participant [115, 116]. The women have not returned to denounce the obscenities of war. The Armenian Genocide of 1915 is intimately connected with the ANZAC invasion. Months of Anglo-French bombardment of the Dardanelles (linking the Black Sea and the Mediterranean) eventually precipitated violent Turkish xenophobia on the eve of the actual Allied invasion on 25 April 2015. On 24 April 2015 the Armenian Genocide commenced with the rounding up of Armenian community leaders such as councillors, clergy and professionals and expanded to mass killings and the forced expulsion of Armenian communities in horrendous death marches. An estimated 1.5 million Armenians were killed and one of the region's oldest civilizations was decimated. 24 April is commemorated as Armenian Genocide Remembrance Day by Armenians while 25 April is commemorated as Anzac Day by Australians and New Zealanders. Successive Turkish governments have remorselessly rejected the extent and terminology of the Armenian Genocide [117]. In 2013 the Australian New South Wales (NSW) Parliament reiterated its 1997 recognition of the Armenian Genocide and also acknowledged Turkey's Assyrian Genocide and Greek Genocide in the 1920s, but Turkey responded by threatening to block MPs attending the 2015 centenary Anzac Day events at Gallipoli (one notes that Gladys Berejiklian, the newly-elected

Liberal premier of NSW, is of Armenian-origin). However the connection between these 2 events has been scrupulously ignored in look-the-other-way Mainstream Australia, and especially on Anzac Day [118, 119].

(12). Free speech is a core value of decent universities but there is widespread censorship in increasingly corporatized Australian universities through intimidation, fear of losing grants and promotion, through corporate-style Codes of Conduct that confine academic free speech to narrow areas of expertise, and through legislated government constraints on free speech [120-124]. The final conclusion of a detailed analysis of censorship and self-censorship in Australian universities was "Finally, we should publicly insist that universities that constrain free speech are not fit for our children" [121].

(13). On Monday 21 October 2019 Australians around the country picking up their daily newspaper found that page 1 was blacked out in a facsimile of comprehensive redaction by a censor's black pen. The blacking out was a collective protest by Australian Mainstream media (MSM) against appalling press censorship by the mendacious and human rights-violating Australian Coalition Government. The taxpayer–funded ABC strongly supported this protest. However self-censorship by these same MSM means failure to report huge lying by omission by US-, UK- and Zionist-beholden Australian media, politician, academic and commentariat presstitutes i.e. lying by omission about lying by omission about lying by omission… [123, 124].

(14). Corporate constraints on free speech operate via "Codes of Conduct", "Confidentiality Agreements" and "Commercial in Confidence" as well as by draconian defamation and anti-terrorism laws [106, 120-127]. Thus a top Australian rugby player was sacked for publicly expressing his fundamentalist Christian religious views about atheists, homosexuals and other alleged sinners going to "hell". This contentious matter involving violation of a corporate employee "Code of Conduct" was finally settled recently, and he now plays rugby for a

French team under a similar contract forbidding him from making such comments that might deeply offend.

(15). While only about 4 Australians have ever been killed in Australia by jihadi psychopaths (and none before 2014), Australia has enacted draconian anti-terrorism laws that severely impact on civil liberties including free speech. Thus revealing that someone has been arrested without charge or trial for long-term interrogation is punishable by 5 years in prison. Membership, support or advocacy of membership of a proscribed "terrorist" organization is punishable by up to life imprisonment in Australia. Revealing criminal activities by Australian Intelligence is punished by up to 10 years in prison. Australia, while deeply complicit in all post-1950 Asian wars by US state terrorism (with the US lackey Coalition supporting all of these atrocities) has also become a pre-police state in the interests of US state terrorism [123-127].

(16). The asserted justification for the obsessive Australian Intelligence and Australian Government secrecy (lying by omission) is jihadi terrorism, but the reality is that only about 4 Australians have ever been killed in Australia by jihadi terrorists (and none before 2014), whereas 85,000 Australians die preventably each year from fiscally-impacted "lifestyle choices" or "political choices". Indeed 1.5 million Australians have died thus since 9-11. This carnage topically includes about 80 women murdered by their partners or ex-partners each year and about 80 Australian veterans committing suicide each year [128]. The annual breakdown of this carnage (including some overlaps) is as follows: (1) 26,000 deaths from adverse hospital events, (2) 17,000 obesity-related deaths, (3) 15,500 smoking-related deaths, (4) 10,000 carbon burning pollution-derived deaths, (5) 5,600 alcohol-related deaths, (6) 4,000 avoidable Indigenous deaths, (7) 2,900 suicides (circa 80 being veterans), (8) 1,900 opiate drug-related deaths with many linked to US restoration of the Taliban-destroyed Afghan opium industry, (9) 1,400 road deaths, and (10) 300 homicides (80 being of women killed domestically) [128-131]. Successive Australian Governments have been committing to an annual long-term accrual cost of $10 billion for the US War on Terror that is in actuality a genocidal US Alliance War on Muslims [129] – a huge fiscal commitment that could otherwise have been used to help prevent 1.5 million preventable Australian deaths since 9-11.

(17). US lackey Australia is among world leaders in war criminality and is also among world leaders in 16 areas of climate criminality. The massive lying by omission, mendacity, anti-science blustering and gross human rights abuses by the war criminal and climate criminal Australian Governments are self-assertedly justified by "national security", "the national interest" and "jobs and growth". However rational risk management, that is crucial for societal safety and security, successively involves (a) accurate information, (b) scientific analysis involving the critical testing of potentially falsifiable hypotheses, and (c) science-informed systemic change to minimize risk. This rational, science-based protocol is sabotaged by (a) lying by commission, lying by omission, spin, obfuscation, censorship, self-censorship and intimidation, (b) spin-based analysis involving the selective use of asserted facts to support a partisan position, and (c) counterproductive blame and shame with the inhibition of vital reportage, and with the ultimate irrational perversion being war [132]. On this basis, the mendacity of the anti-science, US lackey Australian Government and "sacrosanct" Australian Intelligence is a huge threat to the safety and security of Australians.

(18). In the 19th century Australia was one of the first countries to introduce free, compulsory, secular education. However Christianity still played an important role in social life in terms of public observance (swearing on the Bible in court, public memorial services, Sunday observance, public modesty) and in legislated constraints (notably the criminalization of homosexuality e.g. homosexual sex was illegal in Tasmania until 1997). Two thirds of Australia children attend free, tax-payer-funded public schools (State schools) and one third attend religion-based private schools. For political reasons there is a massive annual tax-payer subsidy for organized religion of about A$31 billion (2013 estimate) [133, 134], noting that the annual Australian defence budget is $39 billion [135]. The

"State Schools" are relatively badly under-funded whereas the top "Private Schools" have lavish facilities. Australia has an Educational Apartheid system that means that State School students are disproportionately excluded from a good education and access to top universities and top courses. However the Private School students variously suffer from gross intellectual child abuse through religious brainwashing that variously teaches them to accept sexism, misogyny, sexual guilt, unsafe sex, homophobia, creationism, intelligent design, death for adultery, blasphemy and apostasy, anti-science miracles (e.g. virgin birth, walking on water, rising from the dead, water into wine, and bread into flesh), and the right to invade, occupy, devastate and ethnically cleanse other countries (there are also heavy penalties in Australia for specifying the religions involved in these particular impositions) [136-138].

(19) The former Gillard Labor Government instituted a Royal Commission into child sexual abuse that was unfortunately confined to investigating horrendous institutional child sexual abuse (up to 40,000 cases over the last 40 years by Catholic Church personnel alone). However both the Coalition and Labor continue to ignore the awful reality that 34% of Australian women and 16% of Australian men – 4.4 million Australians in all – have been subject to child sexual abuse i.e. the Coalition and Labor have both ignored huge non-institutional child sexual abuse [138]. This horrendous child sexual abuse continues unaddressed under the new Coalition Government.

(20). Australia is among world leaders for the following 16 climate criminal activities or parameters: (1) annual per capita greenhouse gas (GHG) pollution, (2) live methanogenic livestock exports, (3) natural gas exports, (4) recoverable shale gas reserves that can be accessed by hydraulic fracturing (fracking), (5) coal exports, (6) land clearing, deforestation and ecocide, (7) speciescide or species extinction], (8) coral reef destruction, (9) whale killing and extinction threat through global warming impacting on krill stocks, (10) terminal carbon pollution budget exceedance, (11) per capita Carbon Debt, (12) ultimately GHG generating iron ore exports, (13) climate change inaction, (14) climate genocide and approach towards omnicide and terracide, (15) increasing Domestic GHG pollution despite Paris commitments to lower GHG pollution, and (16) complicity in 9 million annual air pollution deaths from burning carbon fuels, Australia's share being 75,000 overseas and 10,000 Domestically [139, 140]. Australia with 0.33% of the world's population contributes over 5% of global GHG pollution (including that due to the burning of Australia's world leading gas and coal exports) [141].

(21). There is an enormous amount of outright falsehood from the Right-wing side of Australian public life that is accepted by the "respectful" Mainstream journalist, politician, academic and commentariat presstitutes as "legitimate alternative opinion in a democracy" rather than "utterly unacceptable anti-science lying". Thus the Coalition Government is dominated by right-wing politicians who are either actual anti-science climate change denialist or effective climate change denialist through climate change inaction (see item 20 above). Highly offensive to scientists and science-informed people are the false claims such as "gas is clean energy" (it is dirty energy), "gas is cleaner than coal" (burning gas can be dirtier GHG-wise than coal burning), "clean coal" (coal is dirty energy), and that "the Coalition is tackling climate change" (GHG emissions have been increasing for 10 years under successive Coalition Governments).

(22). The US Murdoch media empire headed by US citizen but Australia-origin Rupert Murdoch has about 70% of the readership of city daily newspapers in Australia. However the Murdoch Yellow Press is variously pro-war and anti-science, climate change denialist or effective climate change denialist [142]. Against this false and anti-humanitarian sludge one can set the world's greatest humanitarians and its greatest scientists e.g. Professor Stephen Hawking from 120-Nobel-Laureate University of Cambridge: "We see great peril if governments and societies do not take action now to render nuclear weapons obsolete and to prevent further climate change" [143]. However we live in a neoliberal world in which Money buys Truth, and mendacious and endlessly greedy One Percenters determine what is "true" and what is "fake news".

(D). "2 plus 2 does not equal 4".

Mainstream media promote sports and gambling as major traditional obsessions in Australia. However while many Australians are familiar with sporting statistics (e.g. cricketer Don Bradman's famous batting average of 99.94) and gambling odds and schemes (e.g. trifectas, quinellas and each way bets), they throw up their hands and profess ignorance when faced with vastly far more important numerical matters such as Australia's contribution to world greenhouse gas (GHG) pollution. Below are some important arithmetical matters that are falsely ignored by Mainstream journalist, politician, academic and commentariat presstitutes in a look-the-other-way Australia that effectively ignores crucial numbers, equations, percentages and inequalities by not insisting on the requisite actions demanded by the numbers.

(1). Taking land use and methane (CH_4) emissions into account, Australia (0.33% of the world's population) contributes 2.5% of annual global GHG emissions (Domestic emissions) and 5.4% (Domestic plus Exported GHG emissions) [141] – but the Australian Government largely ignores CH_4 emissions and Exports, and purveys a figure of about 1% (still 3 times greater than Australia's "fair share" and inconsistent with "all men are created equal").

(2). Taking land use and CH_4 emissions into account, Australia's annual Domestic GHG emissions total 1,423 Mt CO_2-e [141] (ignoring the 2019-20 bushfire contribution of 750 Mt CO_2-e [144]) – but the Australian Government claims annual Domestic emissions of 540 Mt CO_2-e by failing to properly account for CH_4 leakage (underestimating likely fugitive emissions by a factor of 8-54), land use, the CH_4 Global Warming Potential on a 20 year time frame, and ignoring bushfires [141].

(3). Coalition claims that "gas is clean" are false (combustion of 1 tonne of CH_4 (85% of natural gas) yields 2.75 tonne CO_2 as compared to combustion of 1 tonne of carbon (about 90% of coal) yielding 3.67 tonne CO_2) [141].

(4). Claims that "gas is cleaner than coal" and for a "coal to gas transition" are false – while gas burning produces less toxic pollutant than coal burning, massive systemic gas leakage (5.4% in the US) and a Global Warming Potential for CH_4 105 times that of the same mass of CO_2 (on a 20 year time frame and considering aerosol impacts) means that gas burning can actually be dirtier than coal burning GHG-wise [141].

(5). As detailed in item C20 above, Australia is among world leaders in 16 areas of climate criminality and clearly needs to decrease rather than increase its climate criminality in all these areas. However Australia's actions continue be in the wrong direction i.e. "2 plus 2 does not equal 4". Indeed the world as a whole is heading in the wrong direction in over 20 key areas. Thus over 11,000 scientists signed up to a recent World scientists' warning of a Climate Emergency that sets out trends in 24 climate-related areas over the last 40 years. Scientists became aware of the climate change threat from greenhouse gas (GHG) pollution in the 1980s, but in 21 of these 24 areas the trends are (a) huge, (b) in the wrong direction, and (c) linear or quasi-linear functions of time, with this allowing extrapolation from the present climate emergency to a climate catastrophe in 2030 [145, 146].

(6). Further to D5 above, the climate criminal, pro-coal Australian Coalition (COALition) government dishonestly claims that it is meeting its Paris GHG reduction commitments "at a canter". However examination of its own official data shows that the government-asserted Australian Domestic GHG pollution has been steadily increasing over the last decade [147-151].

(7). Fundamental to the worsening Climate Emergency is the refusal of neoliberal Australia to meet the cost of past and present carbon pollution, it preferring to pass these immense and rising costs on to future generations. Eminent economist Lord Nicholas Stern has described this refusal as "the greatest market failure in history" [152]. Pope Francis has demanded that the cost of pollution be "fully borne" by the polluters [153, 154]. Climate economist Dr Chris Hope (120 Nobel Laureate

University of Cambridge) and Professor James Hansen (NASA and 96 Nobel Laureate Columbia University) have estimated a damage-related Carbon Price of $200 per tonne CO_2-e [155, 156]. The International Monetary Fund (IMF) has demanded urgent and progressive application of a Carbon Price of up to $75 per ton CO_2-e, noting that the present world average Carbon Price is a mere $2 per ton CO_2-e [157, 158]. The effective climate change denialist Australian Coalition Government abolished a minimal but modestly effective Carbon Tax introduced by the previous Labor Government, and insists on an utterly dishonest Carbon Price of $0 per tonne CO_2-e. This mendacity and grand larceny is a potent illustration of Polya's Second Law of Economics that states that Deceit about the Cost of Production strives to a maximum (analogous to the fundamental Second Law of Thermodynamics of chemistry and physics that states that entropy, chaos, disorder, or lack of information content strive to a maximum) [159]).

(8). The mendacious Australian Coalition falsely asserts that the annual per capita GHG pollution is 22 tonnes CO_2-e per person per year, but properly taking land use and the CH_4 Global Warming Potential into account, the annual per capita GHG pollution (in tonnes CO_2-e per person per year) is 52.9 (Domestic) and 116 (Domestic plus Exported) as compared to 2.1 for India [160, 161]. However a revised estimate properly taking CH_4 fugitive emissions and Exports into account yields an annual per capita GHG pollution (in tonnes CO_2-e per person per year) for Australia of 56.9 (Domestic) and 137.5 (Domestic plus Exported) – and if we take the 750 Mt CO_2 emissions from the 2019-2020 bushfires into account [144] then Australia's per capita pollution rises to 86.9 (Domestic) and 167.5 (Domestic plus Exported). The racist and climate criminal Australian Coalition Government conveniently rejects consideration of "per capita"(thereby rejecting "all men are created equal") and "Exported GHG" pollution" (thereby rejecting fundamental "cause and effect"). Indeed each year Australia (population 25 million) generates 4.2 Gt CO_2-e (Domestic plus Exported), 50% more than India's total of 2.8 Gt CO_2-e (India's population being 1,339 million, 54 times greater than Australia's).

(9). There is currently a huge spat in Australia because Labor Party Opposition has declared a policy of "zero net emissions by 2050" and the anti-science, neoliberal Coalition Government says "How much will it cost?" Labor replies that not doing anything will cost "lots" but doesn't put a figure on it. According to the Coalition the annual cost of doing nothing is 540Mt CO_2-e per year x \$0 per tonne CO_2-e = \$0. However taking the upper estimate of Australia's Domestic plus Exported annual GHG pollution, the annual cost of doing nothing = 4.2 Gt CO_2-e x US\$200 per tonne CO_2-e = US\$840 billion (to be paid by future generations) [162-167]. One notes that Australia's Gross National Debt is presently A\$700 billion or US\$455 billion. Numbers matter. But the mendacious Coalition plays fast and loose with critical numbers i.e. "2 plus 2 does not equal 4".

(10). Australia reluctantly signed up to the Paris Agreement and a global warming target of ideally no more than plus 1.5C and less than a catastrophic plus 2C. However the temperature rise is already about plus 1.1C, plus 1.5C will be reached within the decade at current rates of pollution [168, 169], plus 2C is already implicit in the system (due to the cooling effect of sulphate aerosols generated by burning coal) [170], a catastrophic plus 2C is effectively unavoidable as CO_2 pollution remorselessly rises above 415 ppm CO_2 at 2-3 ppm CO_2 per year [171], and the Paris Agreement commitments amount to a plus 3.2C by 2100 [172]. And at plus 1.1C immense damage is already being caused by hurricanes and storm surges to tropical Island Nations around the world.

(11). A search of the Australian ABC (Australia's taxpayer equivalent of the UK BBC) yields zero (0) results for the terms "300 ppm CO_2" (to which numerous scientists and science-informed activists say we must return the atmospheric CO_2 [173, 174]), "350 ppm CO_2"(the negative emissions target advocated by the international activist group 350.org [175]), "400 ppm CO_2" (recently exceeded for the first time in 15 million years), "413 ppm CO_2" (the present level) or "415 ppm CO_2" (the seasonal maximum reached in 2019 at Mauna Loa, Hawaii) [172]. Climate criminal Australia is so obsessed with continuing to pollute that the notion of "negative CO_2 emissions" or "CO_2 draw-down" has scarcely registered with Mainstream presstitutes. However the recently announced Labor Party Opposition policy of "zero net emissions by 2050" implicitly recognizes the need for "negative CO_2 emissions" or "CO_2 draw-down" to compensate for certain GHG emissions e.g. CH_4 from cattle and CO_2 from cement manufacture.

(12). Eminent and famed theoretical physicist Professor Stephen Hawking (120 Nobel Laureate University of Cambridge and author of the popular "A Brief History of Time") has clearly stated the key existential threats to Humanity and their solution: "We see great peril if governments and societies do not take action now to render nuclear weapons obsolete and to prevent further climate change" [143]. However this late Time Lord's expert insistence on "Now" is ignored by most of the world except for scientists, climate activists and the children inspired by Greta Thunberg. CO_2 levels continue to rise to record levels at a record rate [172]. In climate criminal Australia there is bipartisan agreement between the Coalition Government and the Labor Opposition for unlimited, GHG-increasing, and world-leading Exports of coal, gas and methanogenically-derived meat.

(13). Deliberately and conveniently ignored by carbon burning Australia are air pollution deaths due to the long-term effects of toxic pollutants from carbon fuel burning e.g. carbon monoxide (CO), sulphur dioxide (SO_2), nitrogen dioxide (NO_2), nitrous oxide (N_2O), radioactivity, heavy metals, and fine carbon particulates (e.g. $PM_{2.5}$). 9 million people die from air pollution each year with 75,000 dying from the effects of pollutants from the burning of Australian coal Exports, and 10,000 being Australians [139]. The recent huge bushfires that destroyed 20% of Mainland Australian forests generated huge amounts of toxic smoke that blanketed 3 capital cities (Sydney, Canberra and Melbourne). Informed Australians await the long-term consequences of this exposure of millions of fellow Australians to toxic smoke [144]. It is estimated that pollutants from the burning of coal exports over the lifetime of the initially proposed Adani coal mine in Queensland would eventually kill 1.4 million Indians. Of course both the Coalition (COALition) Government and the Labor Opposition approved a modified form of this scheme [176].

(14). About 85,000 Australians die preventably each year from a range of 'lifestyle choice" or "political choice" reasons ranging from adverse hospital outcomes, obesity, smoking and air pollution to drugs, suicide and homicide. Yet for all kinds of reasons (notably laziness, unconcern and neoliberal objection to "nanny state" interventions), Australian Governments preside over this carnage that has seen 1.5 million Australians die from such preventable causes since the US Government's 9-11 false flag atrocity that killed 3,000 people. While jihadi psychopaths have only ever killed 4 Australians in Australia (and none before 2014), since 9-11 the US lackey Australian Government has committed to a long-term accrual cost of about A$180 billion for the US War on Terror and helping the US Alliance actively or passively kill tens of millions of Muslims abroad [35, 36] rather than trying to save 1.5 million Australian lives at home [129]. Numbers matter.

(15).The most fundamental human right is the right to life. While legitimately criticized for the one party state, the death penalty, censorship, urban air pollution and harsh treatment of Uyghurs and dissidents, China has been hugely successful in radically reducing infant mortality and maternal mortality in Tibet and in China as a whole. In stark contrast, the war criminal US Alliance occupation of neighbouring Afghanistan continues to be associated with an under-1 infant mortality and maternal mortality incidence that is 7 times higher and 4-12 times higher, respectively, than that in Tibet – evidence of gross violation of the Geneva Convention and the UN Genocide Convention by the US Alliance and by a serial war criminal and US lackey Australia [177]. Killing in war occurs not just through violence (active killing) but also through avoidable death from imposed deprivation (passive mass murder). Mass mortality in an impoverished Subject population occupied by rich Occupiers occurs in gross violation of Articles 55 and 56 of the Geneva Convention relative to the Protection of Civilian Persons in Time of War that unequivocally state that the

Occupier must supply its Subjects with life-sustaining food and medical services "to the fullest extent of the means available to it" [66, 67, 177].

(16). As a deeply racist US lackey, Australia has become second only to Trump America as a supporter of nuclear terrorist, racist Zionist-run, genocidally racist, grossly human rights-abusing, international law-violating, serial invader, serial war criminal, democracy-by-genocide Apartheid Israel. The genocidal maltreatment of Indigenous Palestinians by the numbers: (a) 2.2 million Palestinians have died from violence, 0.1 million, or from imposed deprivation, 2.1 million, since the British invasion of the Ottoman Empire in 1914; (b) of 14 million Indigenous Palestinians, 7 million are Exiled on pain of death from their homeland, 5 million Occupied Palestinians have zero (0) Human Rights as specified by the Universal Declaration of Human Rights and are highly abusively confined to the Gaza Concentration Camp (2 million) or to West Bank ghettoes (3 million), and nearly 2 million "lucky" Palestinian Israelis exist as Third class citizens under over 60 race-based, Nazi-style race laws; (c) each year Apartheid Israel kills about 500 Indigenous Palestinians and passively murders about 4,600 through imposed deprivation; (d) Apartheid Israel has attacked 12 countries and occupied 5 with 1950-2005 avoidable deaths from deprivation in countries neighbouring and variously occupied by Apartheid Israel totalling 24 million; (e) the GDP per capita is $3,000 for Occupied Palestinians as compared to $42,000 for Apartheid Israel; (f) genocidally racist, democracy-by-genocide Apartheid Israel determines that the 72% of its now 50% Indigenous Palestinian subjects who are Occupied Palestinians cannot vote for the government ruling them i.e. egregious Apartheid; (g) 90% of Palestine has now been ethnically cleansed of Indigenous Palestinian inhabitants in an ongoing war criminal ethnic cleansing that has been repeatedly condemned by the UN and most recently by UN Security Council Resolution 2334 that was unanimously supported (with a remarkable Obama US abstention rather than a veto); (h) in its genocidal treatment of the Palestinians, US-, UK-, Canada-, France-, Germany- and Australia-backed Apartheid Israel ignores numerous UN General Assembly Resolutions and UN

Security Council Resolutions, the UN Genocide Convention, the Geneva Convention, the Universal Declaration of Human Rights, the Rights of the Child Convention, the UN Declaration on the Rights of Indigenous Peoples, and many other aspects of International Law [102].

(17). About 1 million people die from climate change each year but climate change disproportionately impacts the Developing World (minus China) in which 16 million people die avoidably from deprivation each year [10]. It is predicted that unless requisite action is taken 10 billion people will die this century in a worsening Climate Genocide en route to a sustainable human population of only 0.5-1.0 billion by 2100 [178]. Australia is disproportionately complicit in this worsening Climate Genocide.

Final comments.

Neoliberal Australia has a greed-driven, look-the-other-way culture characterized by national deception and avoidance of Awful Truths [179, 180]. As detailed above, Big Brother's declarations that (A) "war is peace", (B) "slavery is freedom", (C) "ignorance is strength", and (D) "2 plus 2 does not equal 4" are rampant in ostensible democracy Australia under a mendacious, racist, anti-science, anti-environment, pro-war, US lackey Coalition Government. My country, Australia, is arguably second only to Trump America for Orwellian government lying in ostensibly democratic societies. Poor fellow my country – but how does your country perform in relation to Orwellian falsehood?

References.

[1]. "Gideon Polya: https://sites.google.com/site/drgideonpolya/home
[2]. Gideon Polya, "Zionist subversion, Mainstream media censorship", Countercurrents, 9 March 2018: https://countercurrents.org/2018/03/09/zionist-subversion-mainstream-media-censorship-disproportionate-jewish-board-membership-of-us-media-companies/.
[3]. Gideon Polya, "Do Bing Searches to circumvent mendacious, pro-Zionist Google censorship", Countercurrents, 30 April 2018: https://countercurrents.org/2018/04/do-bing-searches-to-circumvent-mendacious-pro-zionist-google-censorship-bing-it.
[4]. Maryam Mohsin, "10 Google Search statistics you need to know in 2020, Oberlo, 17 November 2019: https://au.oberlo.com/blog/google-search-statistics.
[5]. Lesley Stahl quoted in Ian Schwartz, "YouTube CEO Wojcicki: we've cut the time Americans watch" controversial content by 70%", Real Clear Politics, 2 December 2019: https://www.realclearpolitics.com/video/2019/12/02/youtube_ceo_wojcicki_weve_cut_amount_of_time_americans_watch_controversial_content_by_70.html.
[6]. George Orwell, "1984", 1948.
[7]. Gideon Polya, ""One World Dictatorship": a digital nightmare. Review of Soren Korsgaard's analysis", Global Research, 24 January 2020: https://www.globalresearch.ca/one-world-digital-dictatorship-soren-korsgaard-digital-nightmare/5701587.
[8]. Soren Korsgaard, "One World Digital Dictatorship", Crime & Power, 5 January 2020: https://www.crimeandpower.com/2020/01/05/one-world-digital-dictatorship/.
[9]. "Stop state terrorism": https://sites.google.com/site/stopstateterrorism/.
[10]. Gideon Polya, "Body Count. Global avoidable mortality since 1950", that includes a succinct history of every country and is now available for free perusal on the web: http://globalbodycount.blogspot.com/.
[11]. William Blum, "Rogue State: A Guide to the World's Only Superpower", Common Courage Press, 2000.
[12]. Gideon Polya, "US, UK, Australia, Canada & Germany Reject Iraqi Parliament's Quit Iraq Demand", Countercurrents, 16 January 2020: https://countercurrents.org/2020/01/us-uk-australia-canada-germany-reject-iraqi-parliaments-quit-iraq-demand.

[13]. "Afghanistan: who controls what", Al Jazeera, 23 June 2019: https://www.aljazeera.com/indepth/interactive/2016/08/afghanistan-controls-160823083528213.html.

[14]. Philip Dorling, "Australian intelligence "feeding data" for deadly US drone strikes", Sydney Morning Herald, 26 May 2014:http://www.smh.com.au/federal-politics/political-news/australian-intelligence-feeding-data-used-for-deadly-us-drone-strikes-20140526-38ywk.html.

[15]. Peter Cronau, "Pine Gap plays crucial role in America's wars, leaked documents reveal", ABC News, 21 August 2017: https://www.abc.net.au/news/2017-08-20/leaked-documents-reveal-pine-gaps-crucial-role-in-us-drone-war/8815472.

[16]. "British nuclear tests at Maralinga", Wikipedia: https://en.wikipedia.org/wiki/British_nuclear_tests_at_Maralinga.

[17]. Brian Toohey, "Secret. The making of Australia's security state", Melbourne University Press, 2019.

[18]. "ANZUS", Wikipedia: https://en.wikipedia.org/wiki/ANZUS.

[19]. "Howard is a war criminal, says former colleague", Sydney Morning Herald, 19 July 2004: http://www.smh.com.au/articles/2004/07/18/1090089035899.html.

[20]. Harold Pinter, "Art, Truth and Politics", Countercurrents, 8 December 2005: http://www.countercurrents.org/arts-pinter081205.htm.

[21]. Peter Beaumont and Joanna Walters, "Greenspan admits Iraq was about oil, as deaths put at 1.2m", The Observer, 16 September 2007: http://www.theguardian.com/world/2007/sep/16/iraq.iraqtimeline.

[22]. Noam Chomsky quoted in Sherwood Ross, "Chomsky: Iraq invasion "major crime" designed to control Middle East oil", The Public Record, 3 November 2009: http://pubrecord.org/nation/5953/chomsky-invasion-major-crime/.

[23]. Gideon Polya, editor, "Iraqi Holocaust Iraqi Genocide": https://sites.google.com/site/iraqiholocaustiraqigenocide/.

[24]. Dr Abdul-Haq Al-Ani and Tarik Al-Ani, "Genocide in Iraq Volume I. The case against the UN Security Council and member states", (foreword by Professor Joshua Castellino; Clarity Press, Atlanta, 2012.

[25]. Gideon Polya ""Genocide in Iraq, The Case Against UN Security Council And Member States". Book review", Countercurrents, 8 February, 2013: http://www.countercurrents.org/polya080213.htm.

[26]. Gideon Polya, "Apartheid Israel bombing Syria & Iraq – hotting up 4-decade US War on Iran", Countercurrents, 14 August 2019:

https://countercurrents.org/2019/08/apartheid-israel-bombing-syria-iraq-hotting-up-deadly-4-decade-us-war-on-iran.

[27]. Andrew Tillet, "Government rejects call to protect Australian businesses from Iran sanctions", Financial Review, 7 August 2018: https://www.afr.com/politics/government-rejects-call-to-protect-australian-businesses-from-iran-sanctions-20180807-h13ndo.

[28]. David Vine, "Where in the world is the US military", Politico, July/August 2015: https://www.politico.com/magazine/story/2015/06/us-military-bases-around-the-world-119321.

[29]. Jules Dufour, "The worldwide network of US military bases", Global Research, 1 July 2007: https://www.globalresearch.ca/the-worldwide-network-of-us-military-bases-2/5564.

[30]. Matthew Doran, "Federal Government spends $1.1 billion on Northern Territory air base, expanding reach into the Indo-Pacific", ABC News, 21 February 2020: https://www.abc.net.au/news/2020-02-21/federal-government-spends-1.1-billion-on-top-end-air-base/11986904.

[31]. John Pilger, "The British-American Coup that ended Australian independence", Guardian, 23 October 2014: https://www.theguardian.com/commentisfree/2014/oct/23/gough-whitlam-1975-coup-ended-australian-independence.

[32]. Gideon Polya, "Anti-Indian subversion of Fiji by Apartheid Israel, pro-Apartheid Australia & pro-Apartheid America", Countercurrents, 20 October 2017: https://countercurrents.org/2017/10/anti-indian-subversion-of-fiji-by-apartheid-israel-pro-apartheid-australia-pro-apartheid-america.

[33]. "Indonesian mass killings of 1965-1966", Wikipedia: https://en.wikipedia.org/wiki/Indonesian_mass_killings_of_1965%E2%80%9366.

[34]. "Fact check: Journalists face 10 years' jail for exposing security agency bungles", ABC, 14 October 2014: https://www.abc.net.au/news/2014-10-14/journalists-face-jail-for-exposing-security-agency-bungles/5776504.

[35]. Gideon Polya, "Paris atrocity context: 27 Million Muslim Avoidable Deaths From Imposed Deprivation In 20 Countries Violated By US Alliance Since 9-11", Countercurrents, 22 November, 2015: https://countercurrents.org/polya221115.htm.

[36]. "Experts: US did 9-11": https://sites.google.com/site/expertsusdid911/.

[37]. Gideon Polya, "Australian Universities Complicit With Pro-Zionist Censorship And Genocidal Israeli Militarism", Countercurrents, 24 May, 2012: https://countercurrents.org/polya240512.htm.

[38]. Vacy Vlazna, "Israeli Hawkademia in Australian Universities", Palestinian Chronicle, 5 February 2012: http://palestinechronicle.com/view_article_details.php?id=19268.

[39]. Nabil Salim, "Gaza is the Israeli arms industry's testing ground", Middle East Monitor, 26 September 2018: https://www.middleeastmonitor.com/20180926-gaza-is-the-israeli-arms-industrys-testing-ground/.

[40]. Gideon Polya, "Racist Zionism and Israeli State Terrorism threats to Australia and Humanity", Palestinian Genocide, 2010: https://sites.google.com/site/palestiniangenocide/racist-zionism-and-israeli.

[41]. "Israel's Elbit systems gets $150 mln Australian defence contract", Reuters, 25 January 2018: https://www.reuters.com/article/elbit-systems-australia-contract/israels-elbit-systems-gets-150-mln-australian-defence-contract-idUSL8N1PK1E3.

[42]. "Israel continues to lead TUAV sector", Australian Defence Magazine, 10 January 2008: https://www.australiandefence.com.au/E5ED4F90-F806-11DD-8DFE0050568C22C9.

[43]. Rajendra Prasad, "Tears in Paradise. Suffering and struggle of Indians in Fiji 1879-2004", Glade, Auckland, New Zealand, 200).

[44]. Gideon Polya, "Review: "Tears In Paradise. Suffering and Struggle Of Indians In Fiji 1879-2004" by Rajendra Prasad – Britain's Indentured Indian "5 Year Slaves"", Countercurrents, 4 March, 2015: https://countercurrents.org/polya040315.htm.

[45]. Kavia Ivy Nandan (editor), "Stolen Worlds. FijiIndian Fragments", Ivy Press International, 2005.

[46]. Gideon Polya, "Anti-Indian subversion of Fiji by Apartheid Israel, Pro-Apartheid Australia & pro-Apartheid America", Countercurrents, 20 October 2017: http://www.countercurrents.org/2017/10/20/anti-indian-subversion-of-fiji-by-apartheid-israel-pro-apartheid-australia-pro-apartheid-america/.

[47]. "Abdul Lateef", Wikipedia: https://en.wikipedia.org/wiki/Abdul_Lateef_(Fijian_lawyer).

[48]. "Stolen Generations", Wikipedia: https://en.wikipedia.org/wiki/Stolen_Generations.

[49]. "If you thought sorry was the hardest word, try mentioning compensation", New Matilda, 26 May 2015: https://newmatilda.com/2015/05/26/if-you-thought-sorry-was-hardest-word-try-mentioning-compensation/.

[50]. Paddy Gibson, "Stolen futures", Overland, Spring 2013: http://overland.org.au/previous-issues/issue-212/feature-paddy-gibson/.
[51]. AAP, "Indigenous workers receive $190m stolen wages settlement from Queensland Government", Guardian, 9 July 2019: https://www.theguardian.com/australia-news/2019/jul/09/indigenous-workers-receive-190m-stolen-wages-settlement-from-queensland-government.
[52]. "Terminological inexactitude", Wikipedia: http://en.wikipedia.org/wiki/Terminological_inexactitude.
[53]. "List of countries by GDP (nominal) per capita", Wikipedia: https://en.wikipedia.org/wiki/List_of_countries_by_GDP_(nominal)_per_capita.
[54]." UN World Population Prospects 2019": https://population.un.org/wpp/.
[55]. Oxfam, "Time to care", 2020: https://www.oxfam.org.au/wp-content/uploads/2020/01/Oxfam-report-time-to-care-inequality-200120-.pdf.
[56]. "List of development aid country donors", Wikipedia: https://en.wikipedia.org/wiki/List_of_development_aid_country_donors
[57]. Oxfam, "Fight inequality. Beat poverty", 2020: https://www.oxfam.org.au/what-we-do/inequality/.
[58]. Larry Elliott, "World's 26 richest people own as much as poorest 50%, says Oxfam", Guardian, 21 January 2019: https://www.theguardian.com/business/2019/jan/21/world-26-richest-people-own-as-much-as-poorest-50-per-cent-oxfam-report.
[59]. Thomas Piketty, "Capital in the Twenty-first Century", Harvard, 2014.
[60]. Gideon Polya, "Key Book Review: "Capital In The Twenty-First Century" By Thomas Piketty", Countercurrents, 1 July, 2014: http://www.countercurrents.org/polya010714.htm.
[61]. Gideon Polya, editor, "1 % ON 1%: one percent annual wealth tax on One Percenters": https://sites.google.com/site/300orgsite/1-on-1.
[62]. Gideon Polya, "4 % Annual Global Wealth Tax To Stop The 17 Million Deaths Annually", Countercurrents, 27 June, 2014: http://www.countercurrents.org/polya270614.htm).
[63]. Gideon Polya, editor, "Aboriginal Genocide": https://sites.google.com/site/aboriginalgenocide/.
[64]. Gideon Polya, "Apartheid Israel's Palestinian Genocide & Australia's Aboriginal Genocide compared", Countercurrents, 20 February 2018: https://countercurrents.org/2018/02/apartheid-israels-palestinian-genocide-australias-aboriginal-genocide-compared

[65]. Gideon Polya, editor, "Palestinian Genocide":
https://sites.google.com/site/palestiniangenocide/.
[66]. "Geneva Convention (IV) relative to the Protection of Civilian Persons in Time of War:
https://www.un.org/ruleoflaw/files/Geneva%20Convention%20IV.pdf
.

[67]. "UN Genocide Convention":
http://www.edwebproject.org/sideshow/genocide/convention.html.
[68]. Gideon Polya, "Review: "The New Enlightenment" by Brian Ellis – world government & social humanism to save Humanity", Countercurrents, 7 October 2019:
https://countercurrents.org/2019/10/review-the-new-enlightenment-by-brian-ellis-world-government-social-humanism-to-save-humanity.
[69]. Brian Ellis, "The New Enlightenment. On Steven Pinker & beyond", Australian Scholarly Publishing, Melbourne, 2019.
[70]. Brian Ellis,"Social Humanism. A New Metaphysics", Routledge, UK, 2012.
[71]. Gideon Polya, "Book Review: "Social Humanism. A New Metaphysics" By Brian Ellis – Last Chance To Save Planet?", Countercurrents, 19 August, 2012:
https://countercurrents.org/polya190812.htm.
[72]. Brian Ellis, "Rationalism. A critique of pure theory", Australian Scholarly, Melbourne, 2017.
[73]. Gideon Polya, "Review: "Rationalism" by Brian Ellis, Countercurrents, 14 August 2017:
https://countercurrents.org/2017/08/review-rationalism-by-brian-ellis.
[74]. Gideon Polya, editor, "State crime and non-state terrorism":
https://sites.google.com/site/statecrimeandnonstateterrorism/.
[75]. Gideon Polya, "As UK Lackeys Or US Lackeys Australians Have Invaded 85 Countries (British 193, French 80, US 70)", Countercurrents, 9 February, 2015:
http://www.countercurrents.org/polya090215.htm.
[76]. Gideon Polya, "Review: "The Cambridge History Of Australia" Ignores Australian Involvement In 30 Genocides", Countercurrents, 14 October, 2013: https://www.countercurrents.org/polya141013.htm.
[77]. "Australian frontier wars", Wikipedia:
https://en.wikipedia.org/wiki/Australian_frontier_wars.
[78]. "Gideon Polya, "Comparing India And Australia On 26 January: India's Republic Day And Australia's Australia Day (Invasion Day)", Countercurrents, 26 January, 2012:
https://www.countercurrents.org/polya260112.htm.
[79]. Gideon Polya, "Economist Mahima Khanna, Cambridge Stevenson Prize And Dire Indian Poverty", Countercurrents, 20 November, 2011: https://countercurrents.org/polya201111.htm.

[80]. Gideon Polya, "Australia And Britain Killed 6-7 Million Indians In WW2 Bengal Famine", Countercurrents, 29 September, 2011: http://www.countercurrents.org/polya290911.htm.

[81]. Gideon Polya, "Britain robbed Indian of $45 trillion & thence 1.8 billion Indians died from deprivation", Countercurrents, 18 December 2018: https://countercurrents.org/2018/12/britain-robbed-india-of-45-trillion-thence-1-8-billion-indians-died-from-deprivation.

[82]. "Bengali Holocaust (WW2 Bengal Famine) writings of Gideon Polya": https://sites.google.com/site/drgideonpolya/bengali-holocaust.

[83]. Gideon Polya (2008), "Jane Austen and the Black Hole of British History. Colonial rapacity, holocaust denial and the crisis in biological sustainability" (G.M. Polya, Melbourne, 2008 edition that is now available for free perusal on the web: http://janeaustenand.blogspot.com/.

[84]. Edward S. Herman and David Peterson (foreword by Noam Chomsky), "The Politics of Genocide", Monthly Review Press, New York, 2010.

[85]. Gideon Polya, "Book Review: "The Politics Of Genocide" By Edward Herman And David Peterson", Countercurrents, 5 December, 2011: https://www.countercurrents.org/polya051211.htm.

[86]. Gideon Polya, ""One World Digital Dictatorship" a digital nightmare. Review of Soren Korsgaard's analysis", Global Research, 24 January 2020: https://www.globalresearch.ca/one-world-digital-dictatorship-soren-korsgaard-digital-nightmare/5701587.

[87]. Gideon Polya, "Media Lying, Media Censorship & Australian Federal Police Raids On Media In Pre-Police State Australia", Countercurrents, 15 June 2019: https://countercurrents.org/2019/06/media-lying-media-censorship-australian-federal-police-raids-on-media-in-pre-police-state-australia.

[88]. Gideon Polya, "Review "Enlightenment Now" by Steven Pinker – Climate Genocide & Avoidable Mortality Holocaust ignored ", Countercurrents, 7 September 2019: https://countercurrents.org/2019/09/review-enlightenment-now-by-steven-pinker-climate-genocide-avoidable-mortality-holocaust-ignored.

[89]. Gideon Polya, editor, "Afghan Holocaust Afghan Genocide": https://sites.google.com/site/afghanholocaustafghangenocide/.

[90]. Gideon Polya, editor, "Muslim Holocaust Muslim Genocide": https://sites.google.com/site/muslimholocaustmuslimgenocide/.

[91]. "Patriot Act", Wikipedia: https://en.wikipedia.org/wiki/Patriot_Act.

[92]. Gideon Polya, "Redaction: Mainstream media censorship & self-censorship in pre-police-state Australia", Countercurrents, 24 October

2019: https://countercurrents.org/2019/10/redaction-mainstream-media-censorship-self-censorship-in-pre-police-state-australia.
[93]. George Williams, "Sacrificing Civil Liberties to Counter Terrorism: Where Will it End?", 2018 John Marsden Lecture, NSW Council for Civil Liberties, 22 November 2018: http://www.nswccl.org.au/2018_john_marsden_lecture_sacrificing_civil_liberties_to_counter_terrorism_where_will_it_end.
[94]. Gideon Polya, "50 Ways Australian Intelligence Spies On Australia And The World For UK, Israeli And US State Terrorism", Countercurrents, 11 December, 2013: https://countercurrents.org/polya111213.htm.
[95]. Gideon Polya, "Terror Hysteria – Draconian New Australian Anti-Terrorism Laws Target Journalists, Muslims And Human Rights", Countercurrents, 8 October, 2014: https://countercurrents.org/polya0810114.htm.
[96]. Gideon Polya, "Media lying, media censorship & Australian Federal Police raids on media in pre-police state Australia", Countercurrents, 15 June 2019: https://countercurrents.org/2019/06/media-lying-media-censorship-australian-federal-police-raids-on-media-in-pre-police-state-australia"
[97]. Gideon Polya, "Craven US lackey Australia betrays Australian & world hero Julian Assange & free journalism", Countercurrents, 13 April 2019: https://countercurrents.org/2019/04/craven-us-lackey-australia-betrays-australian-world-hero-julian-assange-free-journalism.
[98]. "Julian Assange", Wikipedia: https://en.wikipedia.org/wiki/Julian_Assange.
[99]. John Dugard, "International Convention on the Suppression and Punishment of the crime of Apartheid", Audiovisual Library of International Law: http://untreaty.un.org/cod/avl/ha/cspca/cspca.html.
[100]. Gideon Polya, "Universal Declaration of Human Rights & Palestinians. Apartheid Israel violates ALL Palestinian Human Rights", Palestine Genocide Essays, 24 January 2009: https://sites.google.com/site/palestinegenocideessays/universal-declaration-of-human-rights-palestinians.
[101]. Gideon Polya, editor, "Boycott Apartheid Israel": https://sites.google.com/site/boycottapartheidisrael/.
[102]. Gideon Polya, "70th anniversary of Apartheid Israel & commencement of large-scale Palestinian Genocide", Countercurrents, 11 May 2018: https://countercurrents.org/2018/05/11/70th-anniversary-of-apartheid-israel-commencement-of-large-scale-palestinian-genocide/.
[103]. Gideon Polya, "Democratic One-State Solution (Unitary State, Bi-National State) for post-Apartheid Palestine", Countercurrents, 22

December 2018: https://countercurrents.org/2018/12/22/democratic-one-state-solution-unitary-state-bi-national-state-for-post-apartheid-palestine/.

[104]. Gideon Polya, "Israeli-Palestinian & Middle East conflict – from oil to climate genocide", Countercurrents, 21 August 2017: https://countercurrents.org/2017/08/21/israeli-palestinian-middle-east-conflict-from-oil-to-climate-genocide/.

[105]. Gideon Polya, "End 50 Years Of Genocidal Occupation & Human Rights Abuse By US-Backed Apartheid Israel", Countercurrents, 9 June 2017: https://countercurrents.org/2017/06/09/end-50-years-of-genocidal-occupation-human-rights-abuse-by-us-backed-apartheid-israel/.

[106]. Gideon Polya, "50 Ways Australian Intelligence Spies On Australia And The World For UK, Israeli And US State Terrorism", Countercurrents, 11 December, 2013: https://countercurrents.org/polya111213.htm.

[107]. Gideon Polya, "Pro-Apartheid Israel Australian Labor Party scraps outstanding anti-Israel Apartheid candidate, Melissa Parke", Global Research, 18 April 2019: https://www.globalresearch.ca/pro-apartheid-israel-australian-labor-party-anti-apartheid-candidate-melissa-parke/5674801.

[108]. Gideon Polya, "Yassmin Abdel-Magied censored on Anzac Day – jingoists trash Australian free speech", Countercurrents, 28 April 2017: https://countercurrents.org/2017/04/28/yassmin-abdel-magied-censored-on-anzac-day-jingoists-trash-australian-free-speech/.

[109]. Letters to The Age newspaper (Melbourne, Australia) re Zionists and politicians demanding the banning of a politics book for high school students and mentioning US and Israeli state terrorism: http://www.theage.com.au/news/letters/he-who-shouts-loudest-not-necessarily-the-best/2006/09/16/1158334730192.html?page=fullpage.

[110]. Paul Heinrichs, "Textbook links US and Israel to "state terrorism"", The Age, 10 September 2006: https://www.theage.com.au/national/textbook-links-us-israel-to-state-terrorism-20060910-ge33s6.html.

[111], Gideon Polya, "Mike Carlton, Top Australian Columnist, Forced From Job For Criticizing Apartheid Israeli Gaza Massacre", Countercurrents, 8 August, 2014: https://www.countercurrents.org/polya080814.htm.

[112]. Gideon Polya, "Academic Free Speech Under Zionist Attack At Notre Dame Australia And LSE, UK", Countercurrents, 16 December, 2015: https://countercurrents.org/polya161215.htm.

[113]. Gideon Polya, "Australia Trashes Free Speech – SBS Sacks Journalist Scott McIntyre Over Anti-War Tweets", Countercurrents, 4 April, 2015: https://www.countercurrents.org/polya040515.htm.

[114]. Gideon Polya, "Review: "The One Day of the Year" – Australian Anzac Day jingoism hides genocidal war crimes", Countercurrents, 30 June 2017: https://countercurrents.org/2017/06/review-the-one-day-of-the-year-australian-anzac-day-jingoism-hides-genocidal-war-crimes.

[115]. Anne Summers, "Lest we forget … the women who marched against war rape", Sydney Morning Herald, 1 May 2015: http://www.smh.com.au/comment/lest-we-forget–the-women-who-marched-against-war-rape-20150430-1mxaxu.html.

[116]. Meredith Burgmann, "The Women Against Rape in War Collective's protests against Anzac Day in Sydney, 1983, and 1984", Cosmopolitan Civil Societies Journal, Vol.6, No.3, 2014: https://epress.lib.uts.edu.au/journals/index.php/mcs/article/view/4222/4498.

[117]. Gideon Polya, "Armenian Genocide And Anglo-French & ANZAC Gallipoli Invasion Centenary – Genocide Ignored Yields Genocide Repeated", Countercurrents, 23 April, 2015: https://countercurrents.org/polya230415.htm.

[118]. Nikki Marczak, "Armenian Genocide forgotten in ANZAC commemorations", The Interpreter, 1 April 2015: https://www.lowyinstitute.org/the-interpreter/armenian-genocide-forgotten-anzac-commemorations

[119]. "20-12-2024 – DFAT releases internal communications on the event in 1915", Australian Turkish Advocacy Alliance (ATA-A), 20 December 2014: https://www.ata-a.org.au/ata-a-media-release-dfat-releases-internal-communications-on-the-event-of-1915/.

[120]. "Impact of the Defence Trade Controls Bill on academic freedom", NTEU: 10 October 2012: http://www.nteu.org.au/article/Impact-of-the-Defence-Trade-Controls-Bill-on-academic-freedom-13461.

[121]. Gideon Polya "Current academic censorship and self-censorship in Australian universities", Public University Journal, volume 1, Conference Supplement, "Transforming the Australia University", Melbourne, 9-10 December 2001; Free University Education: https://sites.google.com/site/freeuniversityeducation/academic-censorship.

[122]. Gideon Polya, "Crisis in our universities", ABC Radio National "Ockham's Razor", 19 August 2001: http://www.abc.net.au/radionational/programs/ockhamsrazor/crisis-in-our-universities/3490214).

[123]. Gideon Polya, "Redaction: Mainstream media censorship & self-censorship in pre-police state Australia", Countercurrents, 24 October 2019: https://countercurrents.org/2019/10/redaction-mainstream-media-censorship-self-censorship-in-pre-police-state-australia.

[124]. Gideon Polya, "Media lying, media censorship & Australian Federal Police raids on media in pre-police state Australia", Countercurrents, 15 June 2015: https://countercurrents.org/2019/06/media-lying-media-censorship-australian-federal-police-raids-on-media-in-pre-police-state-australia.

[125]. Gideon Polya, "Scores of huge realities resolutely ignored by mendacious, US lackey Australia", Countercurrents, 11 August 2020: https://countercurrents.org/2020/08/scores-of-huge-realities-resolutely-ignored-by-mendacious-us-lackey-mainstream-australia/.

[126]. Gideon Polya, "Terror Hysteria –Draconian New Australian Anti-Terrorism Laws Target Journalists, Muslims And Human Rights", Countercurrents, 8 October, 2014: https://countercurrents.org/polya0810114.htm.

[127]. Mark Pearson, "Journalists face jail for reporting intelligence operations – with no public interest defence", Journlaw, 3 October 2014: https://journlaw.com/2014/10/03/journalists-face-jail-for-reporting-intelligence-operations-with-no-public-interest-defence/.

[128]. Gideon Polya, "Australian state terrorism (4). Jingoistic, US Lackey Australia's Deadly Betrayal Of Its Traumatized Veterans", Stop state terrorism, 2018: https://sites.google.com/site/stopstateterrorism/australian-state-terrorism-4.

[129]. Gideon Polya, "Horrendous Cost For Australia Of US War On Terror", Countercurrents, 14 October, 2012: https://countercurrents.org/polya141012.htm.

[130]. Gideon Polya, "Advance Australia Fair" Hides Australian Racism, Theft, Genocide, Ecocide, Speciescide & Terracide", Countercurrents, 1 July 2019: https://countercurrents.org/2019/07/advance-australia-fair-hides-australian-racism-theft-genocide-ecocide-speciescide-terracide.

[131]. "Exposing Australia": https://sites.google.com/site/exposingaustralia/home.

[132]. "Gideon Polya": https://sites.google.com/site/drgideonpolya/home.

[133]. John Perkins and Frank Gomez, "Taxes and Subsidies: The Cost of "Advancing Religion"", The Australian Humanist, No. 93, pages 6-8, Autumn 2009.

[134]. Chris Fotinopoulos, "Religion continues its free ride without our blessing", ABC, 29 August 2013:

https://www.abc.net.au/news/2013-08-29/fotinopoulos-why-does-the-church-still-get-a-free-ride/4918626.
[135]. Australian Government, A safer Australia – Budget 2019-20 – Defence overview", 2 April 2019: https://www.minister.defence.gov.au/minister/cpyne/media-releases/safer-australia-budget-2019-20-defence-overview.
[136]. Gideon Polya, "37 Ways Of Tackling Australian Educational Apartheid And Social Inequity", Countercurrents, 22 May, 2013: https://www.countercurrents.org/polya220513.htm.
[137]. Gideon Polya, editor, "Educational Apartheid": https://sites.google.com/site/educationalapartheid/home.
[138]. Gideon Polya, "Horrendous Child Abuse By Pro-war, Pro-Zionist, Climate Criminal Australian Coalition Governments", Countercurrents, 4 December, 2013: https://countercurrents.org/polya041213.htm.
[139]. Gideon Polya, editor, "Stop air pollution deaths": https://sites.google.com/site/300orgsite/stop-air-pollution-deaths.
[140]. Gideon Polya, "War criminal & climate criminal Australian deception at UN General Assembly", Countercurrents, 29 September 2019: https://countercurrents.org/2019/09/war-criminal-climate-criminal-australian-deception-at-un-general-assembly.
[141]. Gideon Polya, "Methane leakage makes Australia a world leading per capita greenhouse gas polluter", Countercurrents, 18 February 2020: https://countercurrents.org/2020/02/methane-leakage-makes-australia-a-world-leading-per-capita-greenhouse-gas-polluter.
[142]. Gideon Polya, editor, "Boycott Murdoch media": https://sites.google.com/site/boycottmurdochmedia/.
[143]. Stephen Hawking, "Brief Answers to the Big Questions", John Murray, 2018, Chapter 7.
[144]. Gideon Polya, "Trumpist climate change denial, Australian bushfires, fuel reduction, biochar & Carbon Debt", Countercurrents, 10 January 2020: https://countercurrents.org/2020/01/trumpist-climate-change-denial-australian-bushfires-fuel-reduction-biochar-carbon-debt.
[145]. Gideon Polya, "Extrapolating 11,000 scientists' climate emergency to 2030 catastrophe", Countercurrents, 14 November 2019: https://countercurrents.org/2019/11/extrapolating-11000-scientists-climate-emergency-warning-to-2030-catastrophe.
[146]. William Ripple et al., "World scientists' warning of a climate emergency", BioScience, 5 November 2019: https://academic.oup.com/bioscience/advance-article/doi/10.1093/biosci/biz088/5610806.
[147]. Lisa Cox, "Australia's emissions reach the highest on record", Guardian 9 July 2019: https://www.theguardian.com/australia-

news/2019/jul/09/australias-emissions-reach-the-highest-on-record-driven-by-electricity-sector.

[148]. Lisa Cox, "Australia's carbon emissions highest on record, data shows", Guardian, 13 December 2018: https://www.theguardian.com/australia-news/2018/dec/13/australias-carbon-emissions-highest-on-record-data-shows.

[149]. Penny Timms and Michael Slezak, "Australia's greenhouse gas emissions rise again, according to delayed Federal Government data", ABC News, 6 June 2019: https://www.abc.net.au/news/2019-06-06/australian-emissions-rise-again-delayed-government-data-shows/11184906.

[150]. Michael Slezak, "Australia's greenhouse gas emissions soar in latest figures", Guardian, 4 August 2017: https://www.theguardian.com/australia-news/2017/aug/04/australias-greenhouse-gas-emissions-soar-in-latest-figures ;

[151]. Australian Government, Department of the Environment and Energy, "Quarterly update of Australia's national greenhouse gas inventory: March 2019", March 2019: https://www.environment.gov.au/system/files/resources/6686d48f-3f9c-448d-a1b7-7e410fe4f376/files/nggi-quarterly-update-mar-2019.pdf.

[152]. Alison Benjamin, "Stern: climate change a "market failure"", Guardian, 29 November 2007: https://www.theguardian.com/environment/2007/nov/29/climatechange.carbonemissions.

[153]. Pope Francis, Encyclical Letter "Laudato si'", 2015: http://w2.vatican.va/content/francesco/en/encyclicals/documents/papa-francesco_20150524_enciclica-laudato-si.html.

[154]. Gideon Polya, "Green Left Pope Francis Demands Climate Action "Without Delay" To Prevent Climate "Catastrophe"", Countercurrents, 10 August, 2015: https://countercurrents.org/polya100815.htm.

[155]. Chris Hope, "How high should climate change taxes be?", Working Paper Series, Judge Business School, University of Cambridge, 9, 2011: http://www.jbs.cam.ac.uk/media/assets/wp1109.pdf.

[156]. James Hansen, "Climate change in a nutshell: the gathering storm", Columbia University, 18 December 2018: http://www.columbia.edu/~jeh1/mailings/2018/20181206_Nutshell.pdf.

[157]. International Monetary Fund (IMF), "Fiscal Monitor: how to mitigate climate change". Executive Summary", September 2019: file:///C:/Users/Gideon/AppData/Local/Temp/execsum-6.pdf.

[158]. Gideon Polya, "Australia rejects IMF Carbon Tax & preventing 4 million pollution deaths by 2030", Countercurrents, 15 October 2019: https://countercurrents.org/2019/10/australia-rejects-imf-carbon-tax-preventing-4-million-pollution-deaths-by-2030.
[159]. Gideon Polya, "Polya's 3 Laws Of Economics Expose Deadly, Dishonest And Terminal Neoliberal Capitalism", Countercurrents, 17 October, 2015: https://www.countercurrents.org/polya171015.htm.
[160]. Gideon Polya, "Revised Annual Per Capita Greenhouse Gas Pollution For All Countries – What Is Your Country Doing?", Countercurrents, 6 January, 2016: https://countercurrents.org/polya060116.htm.
[161]. Gideon Polya, "Exposing And Thence Punishing Worst Polluter Nations Via Weighted Annual Per Capita Greenhouse Gas Pollution Scores", Countercurrents, 19 March, 2016: https://countercurrents.org/polya190316.htm.
[162]. Gideon Polya, editor, "Carbon Debt Carbon Credit": https://sites.google.com/site/carbondebtcarboncredit/.
[163]. Gideon Polya, editor, "Divest from fossil fuels": https://sites.google.com/site/300orgsite/divest-from-fossil-fuels.
[164]. Gideon Polya, editor, "Climate Justice & Intergenerational Equity": https://sites.google.com/site/300orgsite/climate-justice. Gideon Polya, editor, "Science & economics experts: Carbon Tax needed NOT Carbon Trading": https://sites.google.com/site/300orgsite/sciennce-economics-experts-carbon-tax-needed-not-carbon-trading/.
[166]. Gideon Polya, editor, "Stop climate crime": https://sites.google.com/site/300orgsite/stop-climate-crime.
[167]. Gideon Polya, editor, "Climate Revolution Now": https://sites.google.com/site/300orgsite/climate-revolution.
[168]. IPCC, "Global warming of 1.5 °C. Summary for Policymakers", 8 October 2018: http://report.ipcc.ch/sr15/pdf/sr15_spm_final.pdf.
[169]. Gideon Polya, "IPCC +1.5C avoidance report – effectively too late, but stop coal burning for "less bad" catastrophes', Countercurrents, 12 October 2018: https://countercurrents.org/2018/10/ipcc-1-5c-avoidance-report-effectively-too-late-but-stop-coal-burning-for-less-bad-catastrophes.
[170]. Andrew Glikson, "Inferno: from climate denial to planetary arson", Countercurrents, 8 September 2019: https://countercurrents.org/2019/09/inferno-from-climate-denial-to-planetary-arson.
[171]. "Paris Agreement", Wikipedia: https://www.google.com/search?client=firefox-b-d&q=3.2+paris.

[172]. US National Oceanic and Atmospheric Administration, "Trends in Atmospheric Carbon Dioxide": https://www.esrl.noaa.gov/gmd/ccgg/trends/.

[173]. 300.org:. https://sites.google.com/site/300orgsite/300-org.

[174]. "300.org – return atmosphere CO_2 to 300 ppm CO_2": https://sites.google.com/site/300orgsite/300-org—return-atmosphere-co2-to-300-ppm.

[175]. "350.org": https://350.org/.

[176]. Gideon Polya, "Latest Lancet data imply Adani Australian coal project will kill 1.4 million Indians", Countercurrents, 21 April 2017: https://countercurrents.org/2017/04/latest-lancet-data-imply-adani-australian-coal-project-will-kill-1-4-million-indians.

[177]. Gideon Polya, China's Tibet health success versus passive mass murder of Afghan women and children by US Alliance', Global Research, 7 January 2018: https://www.globalresearch.ca/chinas-tibet-health-success-versus-passive-mass-murder-of-afghan-women-and-children-by-us-alliance/5625169.

[178]. "Climate Genocide": https://sites.google.com/site/climategenocide/.

[179]. Gideon Polya, "War criminal & climate criminal Australian deception at UN General Assembly", Countercurrents, 29 September 2019: https://countercurrents.org/2019/09/war-criminal-climate-criminal-australian-deception-at-un-general-assembly.

[180]. Gideon Polya, "Australia shocked by cricket ball tampering but ignores horrendous Australian crimes from child abuse to genocide", Countercurrents, 24 April 2018: https://countercurrents.org/2018/04/australia-shocked-by-cricket-ball-tampering-but-ignores-horrendous-australian-crimes-from-child-abuse-to-genocide.

SECTION G: CASE STUDY #2 – CLIMATE CRIMINAL & EXCEPTIONALIST AMERICA THREATENS PLANET

No analysis of the worsening Climate Crisis and Climate Genocide would be adequate without consideration of the disproportionately huge and dangerous GHG pollution of the world's atmosphere and oceans by the rich and profligate US. Section G is devoted to my climate change writings relating to America [1-63]. As estimated in 2017, the US (population 325 million) ranked #2 in the World after China in generating 5,270 Mt CO_2 for an annual per capita CO_2 pollution of 16.2 tonnes CO_2 per person. China (population 1,390 million) generated 9,839 Mt CO_2 for an annual per capita CO_2 pollution of 7.0 tonnes CO_2 per person (Thomas Frohlich and Liz Blossom, "These countries produce the most CO_2 emissions", USA Today, 14 July 2019: https://www.usatoday.com/story/money/2019/07/14/china-us-countries-that-produce-the-most-co-2-emissions/39548763/). In contrast, in 2017 India (population 1,339 million) produced 2,410 Mt CO_2 for a per capita of only 1.8 tonnes of CO_2 per person, and the World (population 7,511 million) produced 31,546 Mt CO_2 for a world average of 4.2 tonnes CO_2 per person (Robbie Andrew, "Why India's CO_2 emissions grew strongly in 2017", Carbon Brief, 28 March 2018: https://www.carbonbrief.org/guest-post-why-indias-co2-emissions-grew-strongly-in-2017). Back in 1776 "all men are created equal" did not apply to African slaves or Indigenous Americans, and in 2020 it evidently does not apply to the Developing World (minus China) in relation to per capita utilization of our limited resources on Planet A (there is no Planet B). Poverty kills and each year 15 million people die avoidably from deprivation in the Developing World (minus China) on Spaceship Earth with the remorselessly greedy, neoliberal First World in charge of the flight deck [3].

Back in 2006 it was apparent to me that the rich and egregiously warlike US was involved in a War on Terra i.e. a War on Earth: "Since the 9/11 atrocity (3,000 innocent victims horribly murdered) the world has been brainwashed by the evil Bush Administration into accepting a "War on Terror" that is both semantically absurd (how do you conduct a war on a feeling?) and operationally obscene - the post-invasion excess mortality (avoidable mortality) in Coalition-occupied Iraq and Afghanistan now totals 2.1 million, and the post-invasion under-5 infant mortality 1.7 million (carnage utterly ignored by lying Mainstream media). In actual reality there *is* a War on Terra, a war being waged against Planet Earth (Terra) by the Anglo-American Empire and its SPACEMEN (Super Prosperous Anglo-Celtic Empire Men). This actual War on Terra is the one-sided war of UK-US Empire against defenceless Gaia (planetary homeostasis) and equally defenceless infants and their mothers who are overwhelmingly Muslim i.e. in effect, in its primary deadly consequences this is a War on Muslim Women and Children" [2].

Terrorism exists in 2 utterly repugnant forms, the "state terrorism" of nation states and "non-state terrorism" (non-government terrorism that is variously backed by governments). The US Alliance has used jihadi non-state terrorism as an "excuse" for the deadly and destructive invasion of 20 countries in the post-9-11 US War on Muslims. Indeed jihadi non-state terrorism has been a crucial ally of US and US Alliance state terrorism – every jihadi atrocity provides an excuse for vastly more deadly violence by the US or the US Alliance. The US has a long record of supporting non-state terrorism (e.g. death squads in Latin America, Gladio in post-WW2 Western Europe, Al Qaeda from Kosovo to Afghanistan) (see William Blum, "Rogue State" and Philip Agee, "Inside the Company: CIA Diary"). A subset of "state terrorism" is economic state terrorism through mendacity, subversion and deadly sanctions (e.g. variously deadly US sanctions against Cuba, Venezuela, Iraq, and Iran) (see Johns Perkins, "Confessions of an Economic Hit Man"). Economic terrorism includes "climate terrorism" and "carbon terrorism" by governments and powerful corporations in their remorseless greenhouse gas (GHG) pollution regardless of horrendous consequences for future generations and the Biosphere [15, 31-36].

Professor Laura Westra (Professor Emerita in Philosophy, University of Windsor, Canada) quoted me in her book "Faces of State Terrorism" (2012): "The US has 4.5% of the World's population but consumes 25% of its resources through anti-democratic hegemony maintained by egregious violence (8 million dead in the US War on Terror alone, the breakdown being 2.5 million dead in Iraq, 4.9 million dead in Afghanistan and 0.8 million dead (including 0.1 million Americans) from opiate drug-related causes due to US alliance restoration of the Taliban-destroyed Afghan opium industry). The fundamental message on 1/1/11 is surely 1 (ONE) – 1-man-one-vote, there is only 1 Planet (Spaceship Earth) for Humanity, and equal shares in resources i.e. until we necessarily all cease carbon dioxide (CO_2) pollution in 2050 the ratio of my share to your share should be 1 [Polya, 2011]"(see [5]). This demand for equity going forward is actually quite modest because it ignores historical resource exploitation by rich developed countries, and in particular huge GHG pollution by industrial countries since the mid-18th century.

Professor Dabo Guan (School of International Development, University of East Anglia, UK) (2016) has commented thus on inescapable limits to growth and inequity: "For everyone in the world to have an American lifestyle, we would need seven planets, and three to live as Europeans" (Irene Banos Ruiz, "China's new love affair with dogs – as pets, not food – presents environmental problems", DW, 21 June 2016: https://www.dw.com/en/chinas-new-love-affair-with-dogs-as-pets-not-food-presents-environmental-problems/a-19197523). We have already exceeded the minimal "one planet" requirement. Recent reports collectively endorsed by thousands of expert scientists have warned the world that time is running out to save Humanity and the Biosphere from further catastrophic climate change and further massive biodiversity loss. Massive harm has already occurred due to continuing carbon pollution, population growth and economic growth, and it is clear that zero growth in these areas is insufficient – to halt and reverse the worsening disaster there must be negative carbon pollution (atmospheric CO_2 draw-down to about 300 ppm CO_2 from the present damaging and dangerous 415 ppm CO_2), negative population growth (50% population decline as determined from using coral as a "canary in the mine"), and negative economic growth (50% economic de-growth linked to 50% population decline) [23, 37-54].

America with 4.5% of the world's population presently continues to use about 21% of the world's annually exploited resources i.e. 21%/4.5% = 4.7 or about 5 times more than its fair share. This blatant unsustainability has elicited fervent pleas from scientists and science-informed humanitarians [55, 56] but is enabled by resolute censorship and lying by One Percenter-beholden Mainstream journalist, politician, academic and commentariat presstitutes [57, 58].

An extreme example of this terracidal, neoliberal American greed is the US and Canadian exploitation of the Canadian tars sands in Alberta that Professor James Hansen (NASA and Columbia University) estimated "to contain at least 400 GtC (equivalent to about 200 ppm CO_2 [in the atmosphere])… Phasing out emissions from coal is itself an enormous challenge. However, if the tar sands are thrown into the mix, it is essentially game over". I commented on Dr Hansen's dire "game over" assertion (2011): "400 GtC is equivalent to 400 x 44/12 = 1,467 Gt CO_2. Yet the WBGU (the German Advisory Council on Climate Change), that advises the German Government, in 2009 reported that for a 75% chance of avoiding a disastrous 2C temperature rise the world must emit no more than 600 Gt CO_2 between 2010 and zero emissions in 2050" [3, 55, 56]. Game over, indeed. In 2015 US President Barack Obama refused on environmental grounds to approve the Keystone XL Pipeline from the Canadian tar sands to Texas. However in 2017 climate criminal US President Donald Trump and climate criminal Canadian PM Justin Trudeau approved this "game over for the Planet" project ("Keystone pipeline", Wikipedia: https://en.wikipedia.org/wiki/Keystone_Pipeline).

America disproportionately consumes the world's resources because since WW2 it has dominated the world through force of arms. America has invaded 72 countries (52 since WW2) as compared to British 193 countries, Australia 85, France 82, Germany 39, Japan 30, Russia 25, Canada 25, Apartheid Israel 12, China 2, North Korea arguably zero (0), Palestine 0, and US- and Australia-sanctioned Iran zero (0) since the time of Sassanian Empire over 1,300 years ago [4, 14, 15, 26, 27]. The US today has over 700 military bases located in over 70 countries (David Vine, "Where in the world is the U.S. military?", Politico, July/August, 2015: https://www.politico.com/magazine/story/2015/06/us-military-bases-around-the-world-119321). Indeed back in 2013 I suggested that the World "Make 4 July Independence From America Day" [4]. Post-1950 US Asian wars are atrocities associated with 40 million Asian deaths from violence or imposed deprivation. The ongoing US-imposed post-9-11 Muslim Holocaust and Muslim Genocide has, so far, been associated with 32 million Muslim deaths from violence, 5 million, or from deprivation, 27 million, in 20 Muslim countries invaded by the US Alliance since the US Government's 9-11 false flag atrocity that killed 3,000 people [15, 31, 59-62]. Mass murder, whether by war or climate change, is the penultimate in racism, and mass murder by genocidal war or by near-terracidal Climate Genocide is the ultimate in racism [26, 27]. While the 1776 American Revolution is portrayed as about "freedom, democracy, liberty, and "no taxation without representation", in reality American Independence enabled the American colonies to invade and exterminate Indigenous societies to the West. Indeed in the 1840s the US ethnically cleansed Indigenous Americans from east of the Mississippi [1, 3, 4].

One notes that the Iraqi Holocaust and Iraqi Genocide (4.6 million deaths from violence or imposed deprivation since imposition of deadly sanctions in 1990 and 9 million since British invasion in 1914) [14, 15, 26, 27, 59, 60] was fundamentally about oil, as explicitly noted by eminent analysts such as Alan Greenspan (long-time former Chairman of the US Federal Reserve): "I am saddened that it is politically inconvenient to acknowledge what everyone knows: the Iraq war is largely about oil.'". Likewise Professor Noam Chomsky (of 100-Nobel Laureate Massachusetts Institute of Technology): "There is basically no significant change in the fundamental traditional conception that if we can control Middle East energy resources, then we can control the world" (Peter Beaumont and Joanna Walters, "Greenspan admits Iraq was about oil, as deaths put at 1.2m", The Observer, 16 September 2007: http://www.theguardian.com/world/2007/sep/16/iraq.iraqtimelin e and Noam Chomsky quoted in Sherwood Ross, "Chomsky: Iraq invasion "major crime" designed to control Middle East oil", The Public Record, 3 November 2009: http://pubrecord.org/nation/5953/chomsky-invasion-major-crime/). Indeed Robert F. Kennedy Jr (son of the famed Robert Kennedy) has attributed the US-backed Syrian War to competing plans for gas routing from Iran or Qatar (Robert F. Kennedy Jr "Syria: another pipeline war", EcoWatch, 25 February 2016: https://www.ecowatch.com/syria-another-pipeline-war-1882180532.html). The horrendous deaths in the US War on Muslims (32 million deaths from violence, 5 million, or war-imposed deprivation, 27 million) [26, 27] will be dwarfed by those in the worsening Climate Genocide (unless requisite urgent action is taken, 10 billion people will die this century en route to a sustainable human population in 2100 of as few as 0.5-1.0 billion [45]).

The revised annual per capita greenhouse gas (GHG) pollution (tonnes CO_2-e per person per year) properly taking land use and methane into account (the world average being 63.80 billion tonnes CO_2-e / 7.137 billion people in 2013 = 8.9 tonnes CO_2-e per person per year) is as follows: Belize (366.9), Guyana (203.1), Malaysia (126.0), Papua New Guinea (114.7), Qatar (101.8), Zambia (97.5), Antigua & Barbuda (85.6), United Arab Emirates (82.4), Panama (68.0), Botswana (64.9), Liberia (55.0), Indonesia (53.6), New Zealand (53.2), Australia (52.9; 116 if including its huge GHG-generating exports), Nicaragua (51.2), Canada (50.1), Equatorial Guinea (47.5), Venezuela (45.2), Brazil (43.4), Myanmar (41.9), Ireland (41.4), **United States (41.0)** [16]. On this basis the US ranks 22nd in the world. However for some very poor countries GHG pollution via fossil fuel exploitation, forestry or other land use is critical for their survival, and hence per capita GDP (nominal) should be taken into account. The "annual per capita GHG pollution" for each country can been multiplied by the ratio of the "annual per capita GDP (nominal)" for each country to the world average (US$10,744) to get a "weighted score" – on this basis **the US ranks 7th in the world with a score of 207** after Qatar (924), United Arab Emirates (337), Australia (307 or 673 if including its huge GHG-generating exports), Luxembourg (256), Canada (234), and New Zealand (218) [17].

While the Obama Administration recognized the worsening Climate Crisis and the need for emergency action, it operated in a sloppy, spin-based fashion, and fundamentally ignored the fundamental notions of national Carbon Debt, a damage-related price on GHG pollution, or national per capita-based "fair shares" in a global Terminal Carbon Pollution Budget on the basis that "all men are created equal" [8-13].

In 2009 the German WBGU (that advises the German Government) estimated that the Terminal Carbon Pollution Budget between 2010 and zero emissions in 2050 for a 75% probability of avoiding a catastrophic plus 2C of warning was about 600 billion tonnes of CO_2. I accordingly did a careful, country-by-country analyses of years left at their current rates until national zero greenhouse gas (GHG) emissions because they had used up their "fair share" of this Terminal Carbon Pollution Budget". I concluded that America had to stop GHG pollution in about 2014, having used up its "fair share" [8, 41, 43]. After this America would be polluting at the expense of very low annual per capita pollution countries like India and Bangladesh that had more time left to use up their "fair share". By 2014 the US had used up any time left to use up at current rates of pollution their Terminal Carbon Pollution Budget for a 75% chance of avoiding a catastrophic plus 2C temperature rise. In terms of this Terminal Budget the US went broke in 2014 but since then has been stealing from the rest of the world (of course, the world will eventually say "no more"). However the notion of a Terminal Carbon Pollution Budget has now become academic because the whole world by 2020 has finally exceeded its 600 billion tonnes CO_2 carbon budget. Thus World Bank experts revised global GHG pollution to be 63.8 billion tonnes CO_2-equivalent per year (Robert Goodland and Jeff Anfang. "Livestock and climate change. What if the key actors in climate change are … cows, pigs and chickens?", World Watch, November/December 2009: https://awellfedworld.org/wp-content/uploads/Livestock-Climate-Change-Anhang-Goodland.pdf) i.e. 638 billion tonnes CO_2-equivalent over the last decade.

About 9 million people die each year from the long term effects of air pollution, this including 200,000 Americans [52]. US President Obama recognized both climate change and the massive morbidity and mortality associated with air pollution. However actions speak louder than words, and his actions have been ineffective as set out by me in a careful analysis (2015): "Obama recognizes the seriousness of man-made climate change and pollution-based premature deaths (avoidable deaths) but he deceptively and mendaciously offers too little too late by 2030 – in effect, a derisory 5% reduction in overall greenhouse gas (GHG) pollution, a 9% reduction in premature deaths from coal-fired power plant pollution, and a 1% decrease in child asthma attacks… careful analysis below reveals the comprehensive falsehood of Obama's claimed consequences of a "32% off 2005 coal plant pollution by 2030". Obama's saving grace is that his racist, religious right Republican (R4) opponents represent neoliberalism, corporatism, racism and greed gone utterly mad with a remorseless agenda of ecocide, speciescide, omnicide, climate genocide and terracide for short-term One Percenter profit… However, when a revised natural gas leakage contribution to total GHG pollution is re-assessed, US total annual GHG pollution is revealed as 10,343.3 million tonnes CO_2-e [2], and the Obama Clean Power Plan's 501 million tonnes CO_2-e savings in 2030 will be only a 501 x 100/ 10,343.3 = a 4.8% reduction i.e. about a 5% reduction of what it would otherwise be" [13]. Obama's coal-to-gas transition is a disaster because (a) it involves massive installation and long-term operation of another dirty technology [gas] in the place of coal, and (b) because methane (85% of natural gas) has a Global Warming Potential 105 times greater than that of CO_2 on a 20 year time frame, and that at the level of systemic gas leakage in the US of about 3%, gas burning is actually dirtier GHG-wise than burning coal on a per mass basis [46, 53].

The International Monetary Fund (IMF) has exposed the massive deceit of the neoliberal denialists in stating that while a damage-related Carbon Tax of $75 per tonne CO_2 would be an effective way of addressing the climate threat, the present global average Carbon Price is presently only $2 per tonne CO_2, a tiny fraction of what is needed for keeping below the 2°C target (International Monetary Fund (IMF), "Fiscal Monitor: how to mitigate climate change". Executive Summary", September 2019: file:///C:/Users/Gideon/AppData/Local/Temp/execsum-6.pdf ;.see [36]). Climate economist Dr Chris Hope (of 120-Nobel-Laureate Cambridge University) and climate scientist Professor James Hansen (of NASA and 96-Nobel-Laureate Columbia University) have independently estimated a damage-related Carbon Price of about $200 per tonne CO_2-equivalent (Chris Hope, "How high should climate change taxes be?", Working Paper Series, Judge Business School, University of Cambridge, 9, 2011: http://www.jbs.cam.ac.uk/media/assets/wp1109.pdf ; James Hansen, "Climate change in a nutshell: the gathering storm", Columbia University, 18 December 2018: http://www.columbia.edu/~jeh1/mailings/2018/20181206_Nuts hell.pdf). President Obama failed to put a price on carbon such that, as demanded by Pope Francis, the environmental and social cost of pollution was "fully borne" by the polluters.

While President Barack Obama publicly recognized the serious threat posed by climate change and air pollution, as a politician he was subject to the powerful pressures in Lobbyocracy and Corporatocracy America, and was also constrained by a hostile Congress. He at least was able to exercise presidential power and halt the "game over for the planet" Keystone XL pipeline project and enact environment-protecting regulations. However Obama's anti-environment and climate change denialist successor, President Donald Trump, acted immediately to approve the Keystone XL oil pipeline, and then proceeded to take a wrecking ball to US environmental policies, and in particular to the US Environment Protection Authority (EPA) in the interests of pro-coal and pro-oil corporations and communities. Nadja Popovich et al. (2020: "After three years in office, the Trump administration has dismantled most of the major climate and environmental policies the president promised to undo. Calling the rules unnecessary and burdensome to the fossil fuel industry and other businesses, his administration has weakened Obama-era limits on planet-warming carbon dioxide emissions from power plants and from cars and trucks, and rolled back many more rules governing clean air, water and toxic chemicals. Several major reversals have been finalized in recent months as the country has struggled to contain the spread of the new coronavirus. In all, a New York Times analysis, based on research from Harvard Law School, Columbia Law School and other sources, counts nearly 70 environmental rules and regulations officially reversed, revoked or otherwise rolled back under Mr. Trump. More than 30 additional rollbacks are still in progress" (Nadja Popovich, Livia Albeck-Ripka, and Kendra Pierre-Louis, "The Trump Administration is reversing 100 environmental rules: here's the full list", New York Times, 15 July 2020: https://www.nytimes.com/interactive/2020/climate/trump-environment-rollbacks.html). Sarah Gibbens writing in National Geographic has detailed 15 key anti-environment decisions made by Trump, of which the most internationally recognized and condemned was his decision to pull the US out of the Paris Climate Change Agreement (Sarah Gibbens, "15 ways the Trump administration has changed environmental policies", National Geographic, 1 February 2019: https://www.nationalgeographic.com/environment/2019/02/15-ways-trump-administration-impacted-environment/).

However the greatest crime of Donald Trump is that he not only lies egregiously but is utterly and contemptuously mendacious. The Center for Public Integrity determined that the war criminal Bush Administration told 935 lies about Iraq between 9-11 and the illegal invasion of Iraq in March 2003 (Charles Lewis, "25 years, countless investigations and 935 lies", Center for Public Integrity, 25 June 2014: https://publicintegrity.org/inside-publici/25-years-countless-investigations-and-935-lies/). However the Washington Post has carefully analysed Trump's daily flood of Tweets and other scatter-gun assertions and concluded that he has told over 20,000 lies over about 4 years. Michael Tomasky has analyzed Trump's massive lying (2020): "All right, some still say; Yes, Mr. Trump is worse than normal, but they all lie. What's the big deal, really? Here's the big deal. Mr. Trump's lies are different. Not just in quantity, but also in quality. He lies for a different purpose than every other president… What is that difference? In a nutshell, it is this: Our democracy has, to use a word that former Vice President Joe Biden employed in his powerful June 2 speech in Philadelphia, certain guardrails that, as Mr. Biden put it, "have helped make possible this nation's path to a more perfect union, a union that constantly requires reform and rededication." Every president before Mr. Trump has been mindful of those guardrails. When they lied, they lied out of respect for those guardrails. Mr. Trump lies to crush those guardrails into scrap metal" (Michael Tomasky, "Why does Trump lie?", New York Times, 11 June 2020: https://www.nytimes.com/2020/06/11/opinion/trump-lies.html).

Now science is about the critical testing of potentially falsifiable hypotheses, and is the diametric opposite of politician-favoured spin that involves the selective use of selective facts to support a partisan position. Scientists have zero tolerance for lying because it utterly subverts the scientific process directed at gaining better understanding of the truth and better models of reality [63]. Eminent anti-racist Jewish American journalist I.F. Stone (author of "The Hidden History of the Korean War") famously told journalism students: "Among all the things I'm going to tell you today about being a journalist, all you have to remember is two words: governments lie" [57, 58]. Famed American writer and historian Gore Vidal was even blunter: "Unlike most Americans who lie all the time, I hate lying. And here I am surrounded with these hills [in Hollywood] full of liars — some very talented… Americans are not interested in the truth about anything. They assume everybody is lying because they go out and lie everyday about the automobile they are trying to sell you…This is a country of hoax" [57, 58]. Trump is a neoliberal salesman and anything goes. However there is a paradox here because despite his mendacity, nationalist bellicosity and threats to "obliterate" North Korea and Iran, Trump has got US forces out of Syria and is negotiating to do the same for Afghanistan. In contrast, for all his sophisticated and reasoned dialogue, Obama was awarded the Nobel Peace Prize while being involved in major wars in Libya, Syria, Iraq, and Afghanistan, and despite his war criminal destruction of Libya that was formerly the richest country in impoverished Africa [14, 15, 26, 27, 59].

Lying and remorseless neoliberal greed are at the heart of America's disproportionately huge contribution to the worsening global Climate Emergency, Climate Crisis and Climate Genocide. The deadliness of this lying is illustrated by the Covid-19 pandemic in which Trump's anti-science and neoliberal, pro-business policies have led to over 230,000 American deaths in the first 10 months of the pandemic in America (population 331.6 million with 695 deaths per million of population [as of 26 October 2020]). In Australia (population 25.6 million) ruled by a mendacious, US lackey, Trumpist, anti-science, pro-fossil fuels, neoliberal, pro-business Coalition Government, there have been only 905 deaths (35 deaths per million of population, 20 times less than for the US) because the government chose to listen to science-trained and science-informed medical experts (Worldometer, "Covid-19 coronavirus pandemic": https://www.worldometers.info/coronavirus/ ; [28, 29]). In New Zealand (population 5.0 million) that is culturally similar to the US, there have been only 25 Covid-19 deaths for 5 deaths per million of population. Trump has been responsible for 695 – 5 = 690 avoidable Covid-19 deaths per million of population, and hence for 226,000 avoidable Covid-19 deaths (Gideon Polya, "US Alliance Covid-19 Gerocide – intentional mass killing of elderly by US Alliance countries", Countercurrents, 21 October 2020: https://countercurrents.org/2020/10/us-alliance-covid-19-gerocide-intentional-mass-killing-of-elderly-by-us-alliance-countries/). Trump should be arraigned before the International Criminal Court ICC) but unfortunately the US does not recognize the jurisdiction of the ICC over American war criminals.

Rich, neoliberal America ignores the 15 million people who die preventably from deprivation each year in the Developing World (minus China) on Spaceship Earth with the rich, US One Percenter-dominated First World in charge of the flight deck. In contrast on this global comparative scale there are effectively zero avoidable deaths from deprivation in America, Canada, Western Europe, Australia, New Zealand and East Asia [3]. About 1.7 million Americans die preventably each year from "lifestyle choice" and "political choice" reasons, the annual breakdown (with drug deaths updated from 2017) including (with some overlaps) (1) 443,000 (smoking), (2) 440,000 (adverse events in hospitals), (3) 300,000 (obesity), (4) 200,000 (air pollution), (5) 75,000 (alcohol), (6) 69,000 (drugs), (7) 45,000 (lack of medical insurance), (8) 33,000 (motor vehicles),(9) 31,000 (guns), (10) 30,000 (suicide, with 7,000 being veterans), 21,000 (avoidable under-5 year old US infant deaths), (10) 15,000 (homicide) and (11) 4 (in America by jihadi non-state terrorists) [14, 19, 20]. In the 19 years since 9-11 about 32 million Americans have died preventably from "lifestyle choice" and "political choice" reasons. One notes the successive US administrations have committed $6 trillion to the Zionist-promoted War on Terror in which over 30 million Muslims have died from violence or imposed deprivation [14, 15, 26, 27, 59-63]. Thus successive US administrations have committed $6 trillion to actively or passively killing 30 million Muslims abroad rather than to trying to keep 30 million Americans alive at home.

It is sad that America that leads the World in science, technology and science Nobel laureates is stuck in a rut in which One Percenter neoliberal greed still trumps science-informed policy on climate change, notwithstanding increasingly desperate pleas by leading US and World scientists [56, 57]. There is no move by the religious right-bullied US Establishment towards population control [23], or to atmospheric CO_2 draw-down to a safe and sustainable 300 ppm CO_2 [37-39]. The US remains committed to massive meat eating, exploitation of the warming Arctic, and to the biofuel-based Biofuel Genocide [47, 53]. In addition to being a world leader in state terrorism and backing non-state terrorism in the devastation of other countries, the US is a world leader in climate terrorism and carbon terrorism [32-35].

The consequences of global warming to a present 1.2C above the pre-Industrial Revolution average surface temperature are now impacting the US through high energy tropical storms, storm surges, drought, floods and horrendous forest fires impacting the western half of the country. Yet America is now experiencing a surge of action by the young who will have to bear the cost of the immense Intergenerational Inequity and Intergenerational Injustice imposed by the neoliberal One Percenter Establishment [40, 49, 51].

America's revised annual per capita GHG pollution taking land use into account (2016 estimate) is 41 tonnes CO_2-e per person per year, this making it the 22nd worst in the world after Belize (367), Guyana (203), and Malaysia (126) [17]. A revised annual per capita GHG pollution score weighted for GDP places Australia (673) as number 2 in the World for climate criminality after Qatar (924), with the US (207) ranking 7th [44]. The US generates 13,570 Mt CO_2-e each year with land use considered, this representing 21% of the World's annual total of 63.8 Gt CO_2-e. By way of comparison the figures are 10,648 Mt CO_2-e (16.7% of the World's total) for China, and 429 Mt CO_2-e (4.6% of the World's total) for India [17].

The "Climate Change Performance Index (CCPI) Results 2020" ranked Australia worst in the world out of 61 advanced countries with a score of 0.0 out of 100 for "Climate policy". The US ranked second worst with a score of 2.8 out of 100. Both the US and Australia scored "Very low" for "Overall rating", "National climate policy performance" and "International climate policy performance". In terms of overall climate change performance Australia ranked 6th worst in the Developed World and the US ranked worst ("Climate Change Performance Index (CCPI). Results 2020": https://newclimate.org/wp-content/uploads/2019/12/CCPI-2020-Results_Web_Version.pdf). America will eventually face the ire of the World through Green Tariffs, GHG Pollution Tariffs, International Criminal Court (ICC) prosecutions, International

Court of Justice (ICJ) litigations, and Boycotts, Divestment and Sanctions (BDS). However a war criminal and climate criminal America (population 331 million, 21% of global annual resource utilization) is a law unto itself, and such global punitive actions are accordingly unlikely in the short term, and certainly not from US lackey Western countries. Conversely, America's climate criminal lackey Australia (population 25.5 million) is very vulnerable to such action by an indignant World [40, 43, 49].

The following articles are reproduced as Chapters in Section G:

Chapter 24: Gideon Polya, "Exposing And Thence Punishing Worst Polluter Nations Via Weighted Annual Per Capita Greenhouse Gas Pollution Scores", Countercurrents, 19 March, 2016:
https://sites.google.com/site/banyuleclimateactionnow/exposing [17].
Chapter 25: Gideon Polya, "Obama's Clean Power Plan Will Only Cut 2030 US Greenhouse Gas Pollution By 5%", Countercurrrents, 7 August 2015:
https://www.countercurrents.org/polya070815.htm [13].
Chapter 26: Gideon Polya, "Over 14 million Americans will die preventably under a 2-term Trump administration", Countercurrents, 22 March 2017:
https://countercurrents.org/2017/03/over-14-million-americans-will-die-preventably-under-a-2-term-trump-administration/ [20].

References (those marked with an asterisk * are reproduced as Chapters in Section G).

[1]. Gideon Polya, "Jane Austen and the Black Hole of British History. Colonial rapacity, holocaust denial and the crisis in biological sustainability", G.M. Polya, 1998 (see: http://janeaustenand.blogspot.com/).

[2]. Gideon Polya, "War On Terra: US Empire Versus Gaia And Infants", Countercurrrents, 26 February, 2006: https://www.countercurrents.org/polya260206.htm.

[3]. Gideon Polya, "Body Count. Global avoidable mortality since 1950", G.M. Polya, 2007, that includes an avoidable mortality-related history of every country since Neolithic times and is now available for free perusal on the web: http://globalbodycount.blogspot.com/.

[4]. Gideon Polya, "Jane Austen and the Black Hole of British History. Colonial rapacity, holocaust denial and the crisis in biological sustainability", G.M. Polya, 2008, that is now available for free perusal on the web: http://janeaustenand.blogspot.com/.

[5]. Polya, 2011] Gideon Polya, "New Year's Day 1/1/11 Message: Equal Shares on Our One Planet", Countercurrents, 2 January, 2011: https://www.countercurrents.org/polya020111.htm.

[6]. Gideon Polya, "To Save Planet Stop Canada-US Keystone XL Pipeline", Countercurrrents, 23 August 2011: https://www.countercurrents.org/polya230811.htm.

[7]. Gideon Polya, "The US Has Invaded 70 Nations Since 1776 – Make 4 July Independence From America Day", Countercurrents, 5 July, 2013: http://www.countercurrents.org/polya050713.htm

[8]. Gideon Polya, "Current Carbon Debt Or Carbon Credit For All Countries: Australia, Canada And US Default On Carbon Debt", Countercurrents, 4 October, 2013: https://www.countercurrents.org/polya041013.htm.

[9]. Gideon Polya, "Pro-gas Obama's EPA-based Plan To Reduce Coal-based Pollution Amounts To Climate Change Inaction", Countercurrrents, 7 June, 2014: https://www.countercurrents.org/polya070614.htm.

[10]. Gideon Polya, "Quantitative Analysis Exposes Pro-One Percenter Obama's Pre-G20 Deceit Over Democracy, Climate Change Action & Green Climate Fund", Countercurrents, 17 November 2014: https://www.countercurrents.org/polya171014.htm.

[11]. Gideon Polya, ""Dishonest US GHG Reduction Pledge – US Has Already Exceeded Its Fair Share Of World's Terminal Carbon Budget", Countercurrrents, 5 April 2015: https://www.countercurrents.org/polya050415.htm.

[12]. Gideon Polya, "G7 Pledge Of Zero Emissions By 2100 Masks Worsening Climate Emergency And Need For Urgent Action", Countercurrrents, 10 June 2015: https://www.countercurrents.org/polya100615.htm.

*[13]. Gideon Polya, "Obama's Clean Power Plan Will Only Cut 2030 US Greenhouse Gas Pollution By 5%", Countercurrrents, 7 August 2015: https://www.countercurrents.org/polya070815.htm.

[14]. Gideon Polya, "West Ignores 11 Million Muslim War Deaths & 23 Million Preventable American Deaths Since US Government's False-flag 9-11 Atrocity," Countercurrents, 9 September, 2015: https://www.countercurrents.org/polya090915.htm.

[15]. Gideon Polya, "Paris Atrocity Context: 27 Million Muslim Avoidable Deaths From Imposed Deprivation In 20 Countries Violated By US Alliance Since 9-11", Countercurrents, 22 November, 2015: http://www.countercurrents.org/polya221115.htm).

[16]. Gideon Polya, "Revised Annual Per Capita Greenhouse Gas Pollution For All Countries – What Is Your Country Doing?", Countercurrents, 6 January, 2016: http://www.countercurrents.org/polya060116.htm.

*[17]. Gideon Polya, "Exposing And Thence Punishing Worst Polluter Nations Via Weighted Annual Per Capita Greenhouse Gas Pollution Scores", Countercurrents, 19 March, 2016: https://sites.google.com/site/banyuleclimateactionnow/exposing.

[18]. Gideon Polya, "Terracidal Wealth Transfer By Neoliberal Climate Criminal And Climate Change Denier Donald Trump", Countercurrents, 12 November 2016: http://www.countercurrents.org/2016/11/12/terracidal-wealth-transfer-by-neoliberal-climate-criminal-and-climate-change-denier-donald-trump/.

[19]. Gideon Polya, "Trump's abolition of Obamacare will kill 43,00 Americans over 2 Trump terms", Countercurrents, 16 March 2017: https://countercurrents.org/2017/03/trumps-abolition-of-obamacare-will-kill-43000-americans-over-2-trump-terms.

*[20]. Gideon Polya, "Over 14 million Americans will die preventably under a 2-term Trump administration", Countercurrents, 22 March 2017: https://countercurrents.org/2017/03/over-14-million-americans-will-die-preventably-under-a-2-term-trump-administration/.

[21]. Gideon Polya, "Climate criminal Trump America contributes 20% of world's annual Carbon Debt increase", Countercurrents, 1 June 2017: http://www.countercurrents.org/2017/06/01/climate-criminal-trump-america-contributes-20-of-worlds-annual-carbon-debt-increase/.

[22]. Gideon Polya, "Exceptionalist Trump America exits from Paris Agreement & launches neoliberal War on Terra", Countercurrents, 6 June 2017: http://www.countercurrents.org/2017/06/06/exceptionalist-trump-america-exits-from-paris-agreement-launches-neoliberal-war-on-terra/.

[23]. Gideon Polya, "How much negative carbon emissions, negative population growth & negative economic growth is needed to save planet?", Countercurrents, 28 November 2018: https://countercurrents.org/2018/11/how-much-negative-carbon-emissions-negative-population-growth-negative-economic-growth-is-needed-to-save-planet/.

[24]. Gideon Polya, "Australia rejects IMF Carbon Tax & preventing 4 million pollution deaths by 2030", Countercurrents, 15 October 2019: https://countercurrents.org/2019/10/australia-rejects-imf-carbon-tax-preventing-4-million-pollution-deaths-by-2030.

[25]. Gideon Polya, "Ongoing Global Avoidable Mortality Holocaust and horrendous human carnage from war and hegemony", Crime & Power, 9 December 2019: https://www.crimeandpower.com/2019/12/09/ongoing-global-avoidable-mortality-holocaust-and-horrendous-human-carnage-from-war-and-hegemony/.

[26]. Gideon Polya, "US-Imposed Post-9/11 Muslim Holocaust & Muslim Genocide", 400 pages, Korsgaard Publishing, Germany, 4 June 2020: https://www.amazon.com/US-Imposed-Post-9-Muslim-Holocaust-Genocide/dp/8793987056.

[27]. Gideon Polya, "Racist Mainstream ignores "US-Imposed Post-9/11 Muslim Holocaust & Muslim Genocide"", Countercurrents, 17 July 2020: https://countercurrents.org/2020/07/racist-mainstream-ignores-us-imposed-post-9-11-muslim-holocaust-muslim-genocide/.

[28]. Gideon Polya, "Covid-19 Pandemic, Climate & Australia: Risky Ignoring Of Science-based Advice", Countercurrents, 28 July 2020: https://countercurrents.org/2020/07/covid-19-pandemic-climate-australia-risky-ignoring-of-science-based-advice/.

[29]. Gideon Polya, "Scores Of Huge Realities Resolutely Ignored By Mendacious, US Lackey Mainstream Australia", Countercurrents, 11 August 2020: https://countercurrents.org/2020/08/scores-of-huge-realities-resolutely-ignored-by-mendacious-us-lackey-mainstream-australia/.

[30]. "Carbon Debt & Dumping – Climate Criminal Australia Hugely Subsidizes Meat, Grain & Wine Exports To China", Countercurrents, 31 August 2020: https://countercurrents.org/2020/08/carbon-debt-dumping-climate-criminal-australia-hugely-subsidizes-meat-grain-wine-exports-to-china/.

[31]. Gideon Polya, "Lying Mainstream Media Ignore Expert New 9/11 WTC7 Demolition Report", Countercurrents, 22 August 2020: https://countercurrents.org/2020/08/lying-mainstream-media-ignore-expert-new-9-11-wtc7-demolition-report/.
[32]. Gideon Polya, editor, "Stop state terrorism": https://sites.google.com/site/stopstateterrorism/.
[33]. Gideon Polya, editor, "State crime and non-state terrorism": https://sites.google.com/site/statecrimeandnonstateterrorism/.
[34]. Gideon Polya, editor, "Climate terrorism: 400,000 climate change-related deaths globally annually versus an average of 4 US deaths from political terrorism annually since 9-11": https://sites.google.com/site/statecrimeandnonstateterrorism/climate-terrorism.
[35]. Gideon Polya, editor, "Carbon terrorism: 3 million US air pollution deaths versus 53 US political terrorism deaths since 9-11 (2001-2015)": https://sites.google.com/site/statecrimeandnonstateterrorism/carbon-terrorism.
[36]. Gideon Polya, editor, "Experts: US did 9-11": https://sites.google.com/site/expertsusdid911/.
[37]. Gideon Polya, editor, "2011 climate change course": https://sites.google.com/site/300orgsite/2011-climate-change-course.
[38]. Gideon Polya, editor, 300.org:. https://sites.google.com/site/300orgsite/300-org.
[39]. Gideon Polya, editor, "300.org – return atmosphere CO_2 to 300 ppm CO_2": https://sites.google.com/site/300orgsite/300-org---return-atmosphere-co2-to-300-ppm.
[40]. Gideon Polya, editor, "Climate Revolution Now": https://sites.google.com/site/300orgsite/climate-revolution.
[41]. Gideon Polya, editor, "Cut carbon emissions 80% by 2020": https://sites.google.com/site/cutcarbonemissions80by2020/.
[42]. Gideon Polya, editor, "100% renewable energy by 2020": https://sites.google.com/site/100renewableenergyby2020/.
[43]. Gideon Polya, editor, "Carbon Debt Carbon Credit": https://sites.google.com/site/carbondebtcarboncredit/.
[44]. Gideon Polya, editor, "Banyule Climate Action Now": https://sites.google.com/site/banyuleclimateactionnow/.
[45]. Gideon Polya, editor, "Climate Genocide": https://sites.google.com/site/climategenocide/.
[46]. Gideon Polya, editor, "Gas is not clean energy": https://sites.google.com/site/gasisnotcleanenergy/.

[47]. Gideon Polya, editor, "Biofuel Genocide":
https://sites.google.com/site/biofuelgenocide/.
[48]. Gideon Polya, editor, "Divest from fossil fuels":
https://sites.google.com/site/300orgsite/divest-from-fossil-fuels.
[49]. Gideon Polya, editor, "Climate Justice & Intergenerational
Equity": https://sites.google.com/site/300orgsite/climate-justice.
[50]. Gideon Polya, editor, "Science & economics experts: Carbon
Tax needed NOT Carbon Trading":
https://sites.google.com/site/300orgsite/sciennce-economics-experts-
carbon-tax-needed-not-carbon-trading/.
[51]. Gideon Polya, editor, "Stop climate crime":
https://sites.google.com/site/300orgsite/stop-climate-crime.
[52]. Gideon Polya, editor, "Stop air pollution deaths":
https://sites.google.com/site/300orgsite/stop-air-pollution-deaths.
[53]. Gideon Polya, editor, "Methane Bomb Threat":
https://sites.google.com/site/methanebombthreat/.
[54]. Gideon Polya, editor, "Nuclear weapons ban, end poverty &
reverse climate change":
https://sites.google.com/site/300orgsite/nuclear-weapons-ban.
[55]. Gideon Polya, editor, "Too late to avoid global warming
catastrophe": https://sites.google.com/site/300orgsite/too-late-to-
avoid-global-warming.
[56]. Gideon Polya, editor, "Are we doomed?":
https://sites.google.com/site/300orgsite/are-we-doomed.
[57]. Gideon Polya, editor, "Mainstream media lying:
https://sites.google.com/site/mainstreammedialying/home.
[58]. Gideon Polya, editor, "Mainstream media censorship":
https://sites.google.com/site/mainstreammediacensorship/home.
[59]. Gideon Polya, editor, "Muslim Holocaust Muslim Genocide":
https://sites.google.com/site/muslimholocaustmuslimgenocide/.
[60]. Gideon Polya, editor,"Iraqi Holocaust, Iraqi Genocide":
http://sites.google.com/site/iraqiholocaustiraqigenocide/.
[61]. Gideon Polya, editor, "Palestinian Genocide":
http://sites.google.com/site/palestiniangenocide/.
[62]. Gideon Polya, editor, "Afghan Holocaust, Afghan Genocide":
http://sites.google.com/site/afghanholocaustafghangenocide/.
[63]. "Gideon Polya":
https://sites.google.com/site/drgideonpolya/home.

CHAPTER 24: EXPOSING AND THENCE PUNISHING WORST POLLUTER NATIONS VIA WEIGHTED ANNUAL PER CAPITA GREENHOUSE GAS POLLUTION SCORES

[First published as Gideon Polya, "Exposing And Thence Punishing Worst Polluter Nations Via Weighted Annual Per Capita Greenhouse Gas Pollution Scores", Countercurrents, 19 March, 2016: https://sites.google.com/site/banyuleclimateactionnow/exposing].

The latest shocking global temperature data indicate that it is now too late to avoid a catastrophic plus 2 degrees Centigrade temperature rise. Annual per capita greenhouse gas pollution per se is a flawed measure of national culpability for this because it ignores rich countries outsourcing industrial pollution to China, and impoverished countries compelled to pollute to barely survive. A better measure of culpability is weighted annual per capita greenhouse gas pollution taking relative per capita income into account, this revealing that the worst polluters on this basis include the rich Anglosphere countries of the US, UK, Australia, New Zealand, Canada and Ireland.

Worsening climate emergency.

Meteorologists Dr Jeff Masters (co-founder of Weather Underground) and Bob Henson have recently commented on a shocking spike in global temperature in February 2016: "On Saturday, NASA dropped a bombshell of a climate report.

February 2016 has soared past all rivals as the warmest seasonally adjusted month in more than a century of global recordkeeping. NASA's analysis showed that February ran 1.35°C (2.43°F) above the 1951-1980 global average for the month… if we add 0.2°C as a conservative estimate of the amount of human-produced warming that occurred between the late 1800s and 1951-1980, then the February result winds up at 1.55°C above average. If we use 0.4°C as a higher-end estimate, then February sits at 1.75°C above average… even if we could somehow manage to slash emissions enough to stabilize concentrations of carbon dioxide at their current level, we are still committed to at least 0.5°C of additional atmospheric warming as heat stored in the ocean makes its way into the air. In short, we are now hurtling at a frightening pace toward the globally agreed maximum of 2.0°C warming over pre-industrial levels" [1].

David Spratt (co-author with Phillip Sutton of "Climate Code Red" [2] and author of the Climate Code Red website) has commented: "The 1.35°C figure from NASA is from the norm for February for 1950-1980. The instrumental temperature record goes back to 1880 and early instrumental data and model simulations allow a reasonable estimate for the northern hemisphere for warming since 1750, which is the "pre-industrial" starting point. The record shows 0.3°C warming from ~1900 to the NASA 1950-1980 baseline, and Prof Michael Mann told Climate Code Red that the proxy data shows around another 0.3°C from 1750 to the end of the nineteenth century for the northern hemisphere. "Adding the figures together, we can estimate February 2016 as being around 1.95°C warmer than the 1750 pre-industrial level for the northern hemisphere, and 1.65°C warmer than the beginning of the twentieth century for the planet as a whole", says Mann. "It emphasises the point I have made previously that we have no carbon budget left for the 1.5°C target and the opportunity for holding to 2°C is rapidly fading unless the world starts cutting emissions hard right now" [3]. Dr Stefan Rahmstorf, from Germany's Potsdam Institute of Climate Impact Research commented: "We are in a kind of climate emergency now" [3].

The current El Nino phenomenon has no doubt contributed to this temperature spike but major contributors are the remorseless increase in atmospheric carbon dioxide (CO_2) to the highest level and at the highest rate (3.05 ppm CO_2 per year) observed since records began [4, 5], and the accelerating increase in atmospheric methane (CH_4) [6]. Atmospheric CO_2 was increasing at about 0.9 ppm CO_2 per year in 1959-1970 but the rate has steadily increased and reached 3.05 ppm per year in 2015 [4]. Atmospheric CO_2 (which fluctuates seasonally) reached 400 ppm in 2013 and the peak in 2015 was 400 ppm CO_2. In February 2016 it reached 404 ppm CO_2 [5]. As determined from ice cores, the atmospheric CO_2 concentration has been 180-280 parts per million (ppm) for the last 800,000 years (excluding the last century), during which time *Homo sapiens* finally evolved (glaciation at low CO_2 and inter-glacial periods at high CO_2) [7]. Atmospheric CH_4 increased in 1983-1998 by up to 13 ppb (parts per billion) per year, increased much more slowly in the period 1999-2006 (up to 3 ppb per year, the 2001-2005 average being 0.5 ppb/year), and has increased more rapidly from 2007 onwards, reaching 12.5 ppb per year in 2014 [6]. Atmospheric CH_4 increased to 1,843 ppb CH_4 in December 2015 as compared to a pre-Industrial Revolution level of 700 ppb CH_4 [7].

The Global Warming Potential of CH_4 is 21 times that of CO_2 on a 100 year time frame but is 105 times greater than that of CO_2 on a 20 year time frame and taking atmospheric aerosol impacts into account [8]. The IPCC (the UN Intergovernmental Panel on Climate Change) has released a succession of 5 key reports, the latest being the Fifth Assessment Report (AR5, 2013) which states that "About 450 ppm CO_2-eq [is] likely to limit warming to 2 degrees C above pre-industrial levels" [8]. However in 2013 atmospheric greenhouse gas expert Professor Ron Prinn of 85-Nobel-Laureate MIT stated that we had already reached 478 ppm CO_2-equivalent [10]. On a 20 year time frame and considering aerosol impacts, the present CH_4 contribution to atmospheric CO_2-equivalent (CO_2-e) is 105 x 1.843 ppm = 194 ppm CO_2-e. Accordingly, the current CO_2-e (CH_4 + CO_2) on this 20 year basis = 404 + 194 = 598 ppm CO_2-e.

Urgent international action by legal action, Green Tariffs and Boycotts, Divestment and Sanctions (BDS) to constrain the worst polluters.

As outlined above, it is very unlikely that we will avoid a catastrophic plus 2 degrees Centigrade temperature rise, a position espoused by many scientists [11-13]. Indeed the national proposals to the Paris Climate Change Conference failed Humanity by collectively locking in a catastrophic temperature rise of about plus 2.7 degrees C [14]. However we are obliged to urgently do everything we can to make the future for our children, grandchildren and future generations "less bad". One major course of action by the international community is to identify the worst polluting countries and to constrain their behaviour through litigation via the International Court of Justice, prosecution of climate criminals via the International Criminal Court, Green tariffs and Boycotts, Divestment and Sanctions (BDS).

A first step in identifying the worst national polluters (the worst climate criminal nations) is by determining the "annual per capita greenhouse gas pollution" for all countries. I have made such a detailed determination that takes the huge contribution of livestock and land use into account [14], noting that World Bank analysts carefully re-evaluated the contribution of livestock production to world annual greenhouse gas (GHG) pollution and found that the world's annual total rose from 41.76 billion tonnes CO_2-equivalent (CO_2-e) as estimated by the Food and Agricultural Organisation (FAO) to 63.80 billion tonnes CO_2-e, with livestock production contributing over 51% of the higher figure [15].

However the approach of simply listing all countries in order of "annual per capita greenhouse gas pollution" is flawed because:

(a) it does not fully take into account disproportionate use of limited global resources by rich countries in a present carbon economy;

(b) it ignores the outsourcing by rich countries of high polluting manufacturing to China and other relatively low-wage countries; and

(c) it is unfair to high GHG polluting but impoverished countries for which fossil fuels, cement, timber, deforestation and livestock are variously key components for basic economic subsistence.

A better way of assessing relative national climate criminality is to also take "annual per capita income" into account. As described below, I have done this by multiplying the "annual per capita greenhouse gas pollution" for each country [14] by the ratio of the "annual per capita GDP (nominal)" for each country to the world average (US$10,744) (in nearly all cases UN data, 2014)[16].

Below are listed revised annual per capita greenhouse gas (GHG) pollution for all countries (tonnes CO$_2$-e per person per year), the world average being 63.80 billion tonnes CO$_2$-e / 7.137 billion people in 2013 = 8.9 tonnes CO2-e per person per year. These revised estimates take the impact of methanogenic livestock and land use into account, and the data are grouped into (A) countries above the world average, and (B) countries below the world average [14].

(A) countries above the world average: Belize (366.9), Guyana (203.1), Malaysia (126.0), Papua New Guinea (114.7), Qatar (101.8), Zambia (97.5), Antigua & Barbuda (85.6), United Arab Emirates (82.4), Panama (68.0), Botswana (64.9), Liberia (55.0), Indonesia (53.6), New Zealand (53.2), Australia (52.9; 116 if including its huge GHG-generating exports), Nicaragua (51.2), Canada (50.1), Equatorial Guinea (47.5), Venezuela (45.2), Brazil (43.4), Myanmar (41.9), Ireland (41.4), **United States (41.0),** Cambodia (40.5), Kuwait (37.3), Paraguay (37.2), Central African Republic (35.7); Peru (34.8), Mongolia (32.2), Singapore (31.2), Bahrain (30.5), Trinidad & Tobago (29.8), Cameroon (29.5), Congo, Democratic Republic (formerly Zaire) (29.3), Côte d'Ivoire (29.1), Denmark (27.8), Brunei (27.4), Bolivia (27.3), Guatemala (26.9), Belgium (26.3), Ecuador (26.2), Estonia (25.4), Laos (25.3), Suriname (25.1),

Netherlands (24.9), Libya (24.9), Nepal (24.6), Benin (24.5), Angola (23.8), Madagascar (23.7), Argentina (23.7), Uruguay (23.7)*, Luxembourg (23.6), Turkmenistan (23.5), Czech Republic (23.5), Zimbabwe (23.3), Gabon (23.1), Greece (21.9), United Kingdom of Great Britain and Northern Ireland (21.5), Cyprus (21.4), Congo, Republic (21.0), Spain (20.9), Finland (20.6), Israel (20.2), Norway (20.1), Colombia (19.8), Namibia (19.8), Mauritania (19.7), South Africa (19.4), Ukraine (19.1), Germany (18.6); France (17.7), Italy (17.6), Uzbekistan (17.5), Costa Rica (17.1), Sudan (16.8), Saudi Arabia (16.6), Slovenia (16.5), Azerbaijan (16.4), Russia (16.2), Sierra Leone (16.2), Slovakia (15.9), Honduras (15.8), Hungary (15.5), Kazakhstan (15.4), Portugal (15.0), Sweden (15.0), Iran (14.5), Iceland (14.2), Mexico (13.9), Oman (13.8), Malta (13.3), Austria (13.0), Poland (12.9), Jamaica (12.8), Palau (12.8), South Korea (12.7), Guinea (12.5), North Korea (12.1), Bahamas (12.1), Nigeria (11.7), Nauru (11.7), Malawi (11.7), Mali (11.6), Chad (11.6), Taiwan (11.6), Latvia (11.4), Vanuatu (11.1), Switzerland (11.0), Romania (10.9), Togo (10.9), Japan (10.7), Serbia & Montenegro (10.4), Seychelles (10.2), Bulgaria (10.1), Lebanon (9.8), Syria (9.4), Tanzania (9.3), Turkey (9.2), Barbados (9.1), Jordan (9.1), Occupied State of Palestine (9.1)*, Philippines (9.0), Guinea-Bissau (9.0);

(B) countries below the world average: Ghana (8.9), Thailand (8.7), Chile (8.7), Fiji (8.7), Belarus (8.6), Sri Lanka (8.5), Macedonia (8.5), Tonga (7.4), Croatia (7.4), China (7.4), Burkina Faso (7.3), Bosnia & Herzegovina (7.2), Kenya (7.1), Dominican Republic (7.1), Senegal (7.0), Tunisia (7.0), Algeria (6.6), Grenada (6.4), Samoa (6.2), Rwanda (6.1), El Salvador (6.0), Lithuania (5.9), Mozambique (5.8), Lesotho (5.7), Burundi (5.5), Iraq (5.5), Eritrea (5.3), St Kitts & Nevis (5.1), Uganda (5.1), Haiti (5.0), Mauritius (5.0), Albania (4.3), Dominica (4.2), Bhutan (4.1), Niger (4.1), Ethiopia (4.1), Moldova (4.0), Georgia (4.0), Yemen (3.7), Tajikistan (3.7), Afghanistan (3.6), Swaziland (3.6), Cuba (3.5), Cape Verde (3.5), Kyrgyzstan (3.4), The Gambia (3.0), St Lucia (2.9), Bangladesh (2.7), Egypt (2.6), Niue (2.6), Pakistan (2.5), Morocco (2.5), Djibouti (2.4), St Vincent & Grenadines (2.4), Armenia (2.3), Maldives (2.1), India (2.1), Cook Islands (2.1),

Vietnam (1.9), São Tomé and Príncipe (1.9), Comoros (1.6), Solomon Islands (1.4), Kiribati (1.2), Tuvalu (1.2)* (* estimated from that of a similar, contiguous country) [1].

Below are listed values for each country of "weighted annual per capita greenhouse gas (GHG) pollution" taking per capita GDP (nominal) into account. The "annual per capita GHG pollution" for each country (see above) has been multiplied by the ratio of the "annual per capita GDP (nominal)" for each country to the world average (US$10,744). Thus, for example, relatively poor Belize has a world-leading "annual per capita GHG pollution" of 366.9 tonnes CO_2-e per person per year" which when multiplied by the ratio of the "annual per capita GDP (nominal)" for Belize (US$4,831) to that of the world (US$10,744) (i.e. US$4,831/US$10,744 = 0.450) yields a "weighted annual per capita GHG pollution" of 165.1 tonnes per person per year, with Belize now ranking 10th in the world on this basis.

(A) countries above the world average in annual per capita greenhouse gas (GHG) pollution: Belize (366.9 x 0.450 = 165.1), Guyana (203.1 x 0.376 = 76.4), Malaysia (126.0 x 1.02 = 128.5), Papua New Guinea (114.7 x 0.207 = 23.7), Qatar (101.8 x 9.08 = 924.3), Zambia (97.5 x 0.160 = 15.6), Antigua & Barbuda (85.6 x 1.28 = 109.6), United Arab Emirates (82.4 x 4.09 = 337.0), Panama (68.0 x 1.18 = 80.2), Botswana (64.9 x 0.663 = 43.0), Liberia (55.0 x 0.045 = 2.5), Indonesia (53.6 x 0.325 = 17.4), New Zealand (53.2 x 4.11 = 218.7), Australia (52.9 x 5.80 = 306.8; if including its huge GHG-generating exports, 116 x 5.80 = 672.8), Nicaragua (51.2 x 0.183 = 9.4), Canada (50.1 x 4.67 = 234.0), Equatorial Guinea (47.5 x 1.90 = 90.3), Venezuela (45.2 x 1.55 = 70.1), Brazil (43.4 x 1.06 = 46.0), Myanmar (41.9 x 0.116 = 4.9), Ireland (41.4 x 4.99 = 206.6), **United States (41.0 x 5.05 = 207.1),** Cambodia (40.5 x 0.102 = 4.1), Kuwait (37.3 x 4.06 = 151.4), Paraguay (37.2 x 0.440 = 16.4), Central African Republic (35.7 x 0.036 = 1.3); Peru (34.8 x 0.606 = 21.1), Mongolia (32.2 x 0.386 = 12.4), Singapore (31.2 x 5.20 = 162.2), Bahrain (30.5 x 2.31 = 70.5), Trinidad & Tobago (29.8 x 1.93 = 57.5), Cameroon (29.5 x 0.131 = 3.9), Congo, Democratic Republic (formerly Zaire)

(29.3 x 0.045 = 1.3), Côte d'Ivoire (29.1 x 0.144 = 4.2), Denmark (27.8 x 5.70 = 158.5), Brunei (27.4 x 3.81 = 104.4), Bolivia (27.3 x 0.291 = 7.9), Guatemala (26.9 x 0.342 = 9.2), Belgium (26.3 x 4.41 = 116.0), Ecuador (26.2 x 0.591 = 15.5), Estonia (25.4 x 1.87 = 47.5), Laos (25.3 x 0.163 = 4.1), Suriname (25.1 x 0.901 = 22.6), Netherlands (24.9 x 4.85 = 120.8), Libya (24.9 x 0.614 = 15.3), Nepal (24.6 x 0.064 = 1.6), Benin (24.5 x 0.084 = 2.1), Angola (23.8 x 0.563 = 13.4), Madagascar (23.7 x 0.042 = 1.0), Argentina (23.7 x 1.18 = 28.0), Uruguay (23.7 x 1.56 = 37.0)*, Luxembourg (23.6 x 10.85 = 256.1), Turkmenistan (23.5 x 0.841 = 19.8), Czech Republic (23.5 x 1.81 = 42.5), Zimbabwe (23.3 x 0.090 = 2.1), Gabon (23.1 x 0.96 = 22.2), Greece (21.9 x 1.99 = 43.6), United Kingdom of Great Britain and Northern Ireland (21.5 x 4.32 = 92.9), Cyprus (21.4 x 2.43 = 52.0), Congo, Republic (21.0 x 0.291 = 6.1), Spain (20.9 x 2.78 = 58.1), Finland (20.6 x 4.62 = 95.2), Israel (20.2 x 3.56 = 71.9), Norway (20.1 x 9.05 = 181.9), Colombia (19.8 x 0.736 = 14.6), Namibia (19.8 x 0.520 = 10.3), Mauritania (19.7 x 0.119 = 2.3), South Africa (19.4 x 0.579 = 11.2), Ukraine (19.1 x 0.273 = 5.2), Germany (18.6 x 4.46 = 83.0); France (17.7 x 3.98 = 70.4), Italy (17.6 x 3.33 = 58.6), Uzbekistan (17.5 x 0.199 = 3.5), Costa Rica (17.1 x 0.97 = 16.6), Sudan (16.8 x 0.194 = 3.3), Saudi Arabia (16.6 x 2.27 = 37.7), Slovenia (16.5 x 2.23 = 36.8), Azerbaijan (16.4 x 0.727 = 11.9), Russia (16.2 x 1.21 = 19.6), Sierra Leone (16.2 x 0.072 = 1.2), Slovakia (15.9 x 1.72 = 27.3), Honduras (15.8 x 0.228 = 3.6), Hungary (15.5 x 1.30 = 20.2), Kazakhstan (15.4 x 1.16 = 17.9), Portugal (15.0 x 2.06 = 30.9), Sweden (15.0 x 5.48 = 82.2), Iran (14.5 x 0.507 = 7.4), Iceland (14.2 x 4.84 = 68.7), Mexico (13.9 x 0.96 = 13.3), Oman (13.8 x 1.80 = 24.8), Malta (13.3 x 2.35 = 31.3), Austria (13.0 x 4.77 = 62.0), Poland (12.9 x 1.31 = 16.9), Jamaica (12.8 x 0.466 = 6.0), Palau (12.8 x 1.03 = 13.2), South Korea (12.7 x 2.62 = 33.3), Guinea (12.5 x 0.050 = 0.63), North Korea (12.1 x 0.065 = 0.79), Bahamas (12.1 x 2.07 = 25.0), Nigeria (11.7 x 0.298 = 3.5), Nauru (11.7 x 1.66 = 19.4), Malawi (11.7 x 0.032 = 0.37), Mali (11.6 x 0.065 = 0.75), Chad (11.6 x 0.088 = 1.0), Taiwan (11.6 x 2.06 = 23.9), Latvia (11.4 x 1.46 = 16.6), Vanuatu (11.1 x 0.292 = 3.2), Switzerland (11.0 x 7.95 = 87.5), Romania (10.9 x 0.94 = 10.2), Togo (10.9 x 0.060 = 0.65), Japan (10.7 x 3.38 = 36.2), Serbia &

Montenegro (10.4 x 0.569 = 5.9), Seychelles (10.2 x 1.47 =
15.0), Bulgaria (10.1 x 0.733 = 7.4), Lebanon (9.8 x 0.823 =
8.1), Syria (9.4 x 0.169 = 1.6), Tanzania (9.3 x 0.089 = 0.83),
Turkey (9.2 x 0.96 = 8.8), Barbados (9.1 x 1.43 = 13.0), Jordan
(9.1 x 0.450 = 4.1), Occupied State of Palestine (9.1 x 0.262 =
2.4)*, Philippines (9.0 x 0.267 = 2.4), Guinea-Bissau (9.0 x
0.063 = 0.57), Sweden (15.0 x 5.48 = 82.2);

**(B) countries below the world average in annual per capita
greenhouse gas (GHG) pollution:** Ghana (8.9 x 0.129 = 1.1),
Thailand (8.7 x 0.556 = 4.8), Chile (8.7 x 1.35 = 11.7), Fiji (8.7
x 0.476 = 4.1), Belarus (8.6 x 0.746 = 6.4), Sri Lanka (8.5 x
0.338 = 2.9), Macedonia (8.5 x 0.508 = 4.3), Tonga (7.4 x 0.383
= 2.8), Croatia (7.4 x 1.25 = 9.3), China (7.4 x 0.709 = 5.2),
Burkina Faso (7.3 x 0.067 = 0.49), Bosnia & Herzegovina (7.2 x
0.451 = 3.2), Kenya (7.1 x 0.126 = 0.89), Dominican Republic
(7.1 x 0.572 = 4.1), Senegal (7.0 x 0.099 = 0.69), Tunisia (7.0 x
0.397 = 2.8), Algeria (6.6 x 0.510 = 3.4), Grenada (6.4 x 0.774
= 5.0), Samoa (6.2 x 0.400 = 2.5), Rwanda (6.1 x 0.065 = 0.40),
El Salvador (6.0 x 0.383 = 2.3), Lithuania (5.9 x 1.54 = 9.1),
Mozambique (5.8 x 0.058 = 0.34), Lesotho (5.7 x 0.092 = 0.52),
Burundi (5.5 x 0.026 = 0.14), Iraq (5.5 x 0.595 = 3.3), Eritrea
(5.3 x 0.070 = 0.37), St Kitts & Nevis (5.1 x 1.44 = 7.3),
Uganda (5.1 x 0.068 = 0.35), Haiti (5.0 x 0.076 = 0.38),
Mauritius (5.0 x 0.926 = 4.6), Albania (4.3 x 0.432 = 1.9),
Dominica (4.2 x 0.685 = 2.9), Bhutan (4.1 x 0.239 = 0.98),
Niger (4.1 x 0.040 = 0.16), Ethiopia (4.1 x 0.051 = 0.21),
Moldova (4.0 x 0.182 = 0.73), Georgia (4.0 x 0.381 = 1.5),
Yemen (3.7 x 0.132 = 0.49), Tajikistan (3.7 x 0.104 = 0.38),
Afghanistan (3.6 x 0.062 = 0.22), Swaziland (3.6 x 0.329 = 1.2),
Cuba (3.5 x 0.677 = 2.4), Cape Verde (3.5 x 0.336 = 1.2),
Kyrgyzstan (3.4 x 0.118 = 0.40), The Gambia (3.0 x 0.041 =
0.12), St Lucia (2.9 x 0.712 = 2.1), Bangladesh (2.7 x 0.101 =
0.27), Egypt (2.6 x 0.293 = 0.76), Niue (2.6 x 0.540 = 1.4),
Pakistan (2.5 x 0.126 = 0.32), Morocco (2.5 x 0.302 = 0.76),
Djibouti (2.4 x 0.169 = 0.41), St Vincent & Grenadines (2.4 x
0.621 = 1.5), Armenia (2.3 x 0.337 = 0.78), Maldives (2.1 x
0.790 = 1.7), India (2.1 x 0.148 = 0.31), Cook Islands (2.1 x
1.40 = 2.9), Vietnam (1.9 x 0.188 = 0.36), São Tomé and
Príncipe (1.9 x 0.169 = 0.32), Comoros (1.6 x 0.078 = 0.12),

Solomon Islands (1.4 x 0.179 = 0.25), Kiribati (1.2 x 0.152 = 0.18), Tuvalu (1.2 x 0.353 = 0.42)* (* estimated from that of a similar, contiguous country).

To readily identify the worst climate criminal nations we can now list countries in descending order of "weighted annual per capita greenhouse gas (GHG) pollution" score:

(a) Above 100: Qatar (101.8 x 9.08 = 924.3), United Arab Emirates (82.4 x 4.09 = 337.0), Australia (52.9 x 5.80 = 306.8; if including its huge GHG-generating exports, 116 x 5.80 = 672.8), Luxembourg (23.6 x 10.85 = 256.1), Canada (50.1 x 4.67 = 234.0), New Zealand (53.2 x 4.11 = 218.7), **United States (41.0 x 5.05 = 207.1),** Ireland (41.4 x 4.99 = 206.6), Norway (20.1 x 9.05 = 181.9), Belize (366.9 x 0.450 = 165.1), Singapore (31.2 x 5.20 = 162.2), Denmark (27.8 x 5.70 = 158.5), Kuwait (37.3 x 4.06 = 151.4), Malaysia (126.0 x 1.02 = 128.5), Netherlands (24.9 x 4.85 = 120.8), Belgium (26.3 x 4.41 = 116.0), Antigua & Barbuda (85.6 x 1.28 = 109.6), Brunei (27.4 x 3.81 = 104.4).

(b) 10-100: Finland (20.6 x 4.62 = 95.2), United Kingdom of Great Britain and Northern Ireland (21.5 x 4.32 = 92.9), Equatorial Guinea (47.5 x 1.90 = 90.3), Switzerland (11.0 x 7.95 = 87.5), Germany (18.6 x 4.46 = 83.0); Sweden (15.0 x 5.48 = 82.2), Panama (68.0 x 1.18 = 80.2), Guyana (203.1 x 0.376 = 76.4), Israel (20.2 x 3.56 = 71.9), Bahrain (30.5 x 2.31 = 70.5), France (17.7 x 3.98 = 70.4), Venezuela (45.2 x 1.55 = 70.1), Iceland (14.2 x 4.84 = 68.7), Austria (13.0 x 4.77 = 62.0), Spain (20.9 x 2.78 = 58.1), Trinidad & Tobago (29.8 x 1.93 = 57.5), Italy (17.6 x 3.33 = 58.6), Cyprus (21.4 x 2.43 = 52.0), Estonia (25.4 x 1.87 = 47.5), Brazil (43.4 x 1.06 = 46.0), Greece (21.9 x 1.99 = 43.6), Botswana (64.9 x 0.663 = 43.0), Czech Republic (23.5 x 1.81 = 42.5), Saudi Arabia (16.6 x 2.27 = 37.7), Uruguay (23.7 x 1.56 = 37.0)*, Slovenia (16.5 x 2.23 = 36.8), Japan (10.7 x 3.38 = 36.2), South Korea (12.7 x 2.62 = 33.3), Malta (13.3 x 2.35 = 31.3), Portugal (15.0 x 2.06 = 30.9), Argentina (23.7 x 1.18 = 28.0), Slovakia (15.9 x 1.72 = 27.3), Bahamas (12.1 x 2.07 = 25.0), Oman (13.8 x 1.80 = 24.8), Taiwan (11.6 x 2.06 = 23.9), Papua New Guinea (114.7 x 0.207

= 23.7), Suriname (25.1 x 0.901 = 22.6), Gabon (23.1 x 0.96 =
22.2), Peru (34.8 x 0.606 = 21.1), Hungary (15.5 x 1.30 = 20.2),
Turkmenistan (23.5 x 0.841 = 19.8), Russia (16.2 x 1.21 =
19.6), Nauru (11.7 x 1.66 = 19.4), Kazakhstan (15.4 x 1.16 =
17.9), Indonesia (53.6 x 0.325 = 17.4), Poland (12.9 x 1.31 =
16.9), Costa Rica (17.1 x 0.97 = 16.6), Latvia (11.4 x 1.46 =
16.6), Paraguay (37.2 x 0.440 = 16.4), Zambia (97.5 x 0.160 =
15.6), Ecuador (26.2 x 0.591 = 15.5), Libya (24.9 x 0.614 =
15.3), Seychelles (10.2 x 1.47 = 15.0), Colombia (19.8 x 0.736
= 14.6), Angola (23.8 x 0.563 = 13.4), Mexico (13.9 x 0.96 =
13.3), Palau (12.8 x 1.03 = 13.2), Barbados (9.1 x 1.43 = 13.0),
Mongolia (32.2 x 0.386 = 12.4), Azerbaijan (16.4 x 0.727 =
11.9), Chile (8.7 x 1.35 = 11.7), South Africa (19.4 x 0.579 =
11.2), Namibia (19.8 x 0.520 = 10.3), Romania (10.9 x 0.94 =
10.2).

(c) 1-10: Nicaragua (51.2 x 0.183 = 9.4), Croatia (7.4 x 1.25 =
9.3), Guatemala (26.9 x 0.342 = 9.2), Lithuania (5.9 x 1.54 =
9.1), Turkey (9.2 x 0.96 = 8.8), Lebanon (9.8 x 0.823 = 8.1),
Bolivia (27.3 x 0.291 = 7.9), Iran (14.5 x 0.507 = 7.4), Bulgaria
(10.1 x 0.733 = 7.4), St Kitts & Nevis (5.1 x 1.44 = 7.3),
Belarus (8.6 x 0.746 = 6.4), Congo, Republic (21.0 x 0.291 =
6.1), Jamaica (12.8 x 0.466 = 6.0), Serbia & Montenegro (10.4 x
0.569 = 5.9), China (7.4 x 0.709 = 5.2), Ukraine (19.1 x 0.273 =
5.2), Grenada (6.4 x 0.774 = 5.0), Myanmar (41.9 x 0.116 =
4.9), Thailand (8.7 x 0.556 = 4.8), Mauritius (5.0 x 0.926 = 4.6),
Macedonia (8.5 x 0.508 = 4.3), Côte d'Ivoire (29.1 x 0.144 =
4.2), Dominican Republic (7.1 x 0.572 = 4.1), Cambodia (40.5 x
0.102 = 4.1), Fiji (8.7 x 0.476 = 4.1), Laos (25.3 x 0.163 = 4.1),
Jordan (9.1 x 0.450 = 4.1), Cameroon (29.5 x 0.131 = 3.9),
Honduras (15.8 x 0.228 = 3.6), Uzbekistan (17.5 x 0.199 = 3.5),
Nigeria (11.7 x 0.298 = 3.5), Algeria (6.6 x 0.510 = 3.4), Iraq
(5.5 x 0.595 = 3.3), Sudan (16.8 x 0.194 = 3.3), Bosnia &
Herzegovina (7.2 x 0.451 = 3.2), Vanuatu (11.1 x 0.292 = 3.2),
Sri Lanka (8.5 x 0.338 = 2.9), Cook Islands (2.1 x 1.40 = 2.9),
Dominica (4.2 x 0.685 = 2.9), Tunisia (7.0 x 0.397 = 2.8),
Tonga (7.4 x 0.383 = 2.8), Liberia (55.0 x 0.045 = 2.5), Samoa
(6.2 x 0.400 = 2.5), Cuba (3.5 x 0.677 = 2.4), Occupied State of
Palestine (9.1 x 0.262 = 2.4)*, Philippines (9.0 x 0.267 = 2.4),
El Salvador (6.0 x 0.383 = 2.3), Mauritania (19.7 x 0.119 = 2.3),

Benin (24.5 x 0.084 = 2.1), St Lucia (2.9 x 0.712 = 2.1),
Zimbabwe (23.3 x 0.090 = 2.1), Albania (4.3 x 0.432 = 1.9),
Maldives (2.1 x 0.790 = 1.7), Nepal (24.6 x 0.064 = 1.6), Syria
(9.4 x 0.169 = 1.6), St Vincent & Grenadines (2.4 x 0.621 =
1.5), Georgia (4.0 x 0.381 = 1.5), Niue (2.6 x 0.540 = 1.4),
Congo, Democratic Republic (formerly Zaire) (29.3 x 0.045 =
1.3), Central African Republic (35.7 x 0.036 = 1.3), Swaziland
(3.6 x 0.329 = 1.2), Sierra Leone (16.2 x 0.072 = 1.2), Cape
Verde (3.5 x 0.336 = 1.2), Ghana (8.9 x 0.129 = 1.1),
Madagascar (23.7 x 0.042 = 1.0), Chad (11.6 x 0.088 = 1.0).

(d) Less than 1: Bhutan (4.1 x 0.239 = 0.98), Kenya (7.1 x
0.126 = 0.89), Tanzania (9.3 x 0.089 = 0.83), North Korea (12.1
x 0.065 = 0.79), Armenia (2.3 x 0.337 = 0.78), Morocco (2.5 x
0.302 = 0.76), Egypt (2.6 x 0.293 = 0.76), Mali (11.6 x 0.065 =
0.75), Moldova (4.0 x 0.182 = 0.73), Senegal (7.0 x 0.099 =
0.69), Togo (10.9 x 0.060 = 0.65), Guinea (12.5 x 0.050 = 0.63),
Guinea-Bissau (9.0 x 0.063 = 0.57), Lesotho (5.7 x 0.092 =
0.52), Burkina Faso (7.3 x 0.067 = 0.49), Yemen (3.7 x 0.132 =
0.49), Tuvalu (1.2 x 0.353 = 0.42)* Djibouti (2.4 x 0.169 =
0.41), Kyrgyzstan (3.4 x 0.118 = 0.40), Rwanda (6.1 x 0.065 =
0.40), Haiti (5.0 x 0.076 = 0.38), Tajikistan (3.7 x 0.104 = 0.38),
Malawi (11.7 x 0.032 = 0.37), Eritrea (5.3 x 0.070 = 0.37),
Vietnam (1.9 x 0.188 = 0.36), Uganda (5.1 x 0.068 = 0.35),
Mozambique (5.8 x 0.058 = 0.34), Pakistan (2.5 x 0.126 = 0.32),
São Tomé and Príncipe (1.9 x 0.169 = 0.32), India (2.1 x 0.148
= 0.31), Bangladesh (2.7 x 0.101 = 0.27), Solomon Islands (1.4
x 0.179 = 0.25), Afghanistan (3.6 x 0.062 = 0.22), Ethiopia (4.1
x 0.051 = 0.21), Kiribati (1.2 x 0.152 = 0.18), Niger (4.1 x 0.040
= 0.16), Burundi (5.5 x 0.026 = 0.14), The Gambia (3.0 x 0.041
= 0.12), Comoros (1.6 x 0.078 = 0.12). (* Estimated from data
for a similar, contiguous country. Data were not available for
Timor L'este and the Federated States of Micronesia which are
likely to be in the less than 1 group).

Conclusions

The worst polluters on the basis of "weighted annual capita
GHG pollution" scores greater than 100 include some major
rich fossil fuel producers and users (Qatar, United Arab

Emirates, Australia, Canada, US, Norway, Kuwait and Brunei) and countries with major deforestation (Malaysia and Belize). Denmark, Netherlands, Singapore, Belgium and Ireland score high due to high per capita incomes and relatively high GHG pollution including land use. The surprisingly high "score" for luxury tourist destination Antigua & Barbados, similar to that of the UK, is possibly too high and derives in this analysis from a high 2000 value of "tonnes of CO_2-e with land use change" in the primary data used in estimating "revised annual per capita greenhouse gas (GHG) pollution" [14, 17]. It is surprising that I have only been able to find this one source of information for all countries on GHG pollution with land use considered [17] – is the world serious about tackling climate change or not? New Zealand scores high as a high per capita income country with a large methanogenic livestock population. A notable group of high polluters are Anglosphere countries including the UK itself (21.5 x 4.32 = 92.9), the formerly English-ruled Ireland (41.4 x 4.99 = 206.6), and former European-settled British colonies namely Australia (52.9 x 5.80 = 306.8; if including its huge GHG-generating exports, 116 x 5.80 = 672.8), Canada (50.1 x 4.67 = 234.0), New Zealand (53.2 x 4.11 = 218.7), **United States (41.0 x 5.05 = 207.1),** and UK- and US-backed Israel (20.2 x 3.56 = 71.9) – all very high scores in stark contrast to that of Occupied State of Palestine (9.1 x 0.262 = 2.4).

Countries with "weighted annual capita GHG pollution" scores in the range of 10-100 include most other Western European countries, most Eastern European countries and some other First World countries (notably Japan and South Korea) plus many other countries from around the world variously heavily involved in high polluting industries such as fossil fuels (Angola, Bahrain, Venezuela, Saudi Arabia, Oman, Turkmenistan, Russia, Kazakhstan, Indonesia, Libya, Peru, Azerbaijan, South Africa, and Trinidad & Tobago), deforestation (Panama, Guyana, Brazil, Indonesia, Peru, Papua New Guinea, Suriname, Gabon, Ecuador, and Colombia), livestock (e.g. Botswana, Argentina, Uruguay, and Paraguay), and luxury tourism (e.g. Bahamas, Seychelles, and Barbados). Thus, for example, Botswana (64.9 x 0.663 = 43.0) is a poor, developing country with a big cattle industry.

Countries with "weighted annual capita GHG pollution" scores in the range of 1-10 include the poorest European countries, many Island states and many very poor countries with economies variously heavily involving fossil fuels (Bolivia, Iran, Nigeria, Algeria, Iraq) or significant ongoing deforestation (Nicaragua, Guatemala, Congo, Republic, Myanmar, Thailand, Côte d'Ivoire, Cambodia, Fiji, Laos, Cameroon, Honduras, Nigeria, Vanuatu, Philippines, El Salvador, Benin, Congo, Democratic Republic (formerly Zaire), Central African Republic, Sierra Leone, Ghana, and Madagascar). China (7.4 x 0.709 = 5.2) is notable in this group with a "weighted annual capita GHG pollution" score that is less than the world average value of 8.9 x 1.0 = 8.9.

At the "good" far end of the "weighted annual capita GHG pollution" spectrum are countries with scores of less than 1, this set including many African countries, some global warming-threatened Island Nations (Tuvalu, Haiti, São Tomé and Príncipe, Solomon Islands, Kiribati, and Comoros), some countries devastated by US invasion or US-backed wars (e.g. North Korea, Yemen, Haiti, Eritrea, Vietnam, Afghanistan, Ethiopia), and some countries containing huge river deltas that are accordingly acutely threatened by rising sea levels due to global warming, namely Egypt, Senegal, Vietnam, Mozambique, Pakistan, India, Bangladesh, and the Gambia. Of these acutely threatened, mega-delta countries, Egypt, Vietnam, Pakistan, India, and Bangladesh are extremely populous nations. A sea level rise of 1 metre is predicted by 2100 [18]. David Spratt (co-author of "Climate Code Red") writes: "In Bangladesh, a 1-metre sea level rise would inundate 15-17% of the land and threaten more than a million hectares of agricultural land. The Mekong River Commission warns that a 1-metre sea level would wipe out nearly 40% of the Mekong Delta. A 1-metre rise would flood one-fourth of the Nile Delta, forcing more than 10% of Egypt's population from their homes. Nearly half of Egypt's crops, including wheat, bananas and rice, are grown in the delta" [19]. It is an utterly unacceptable injustice that Island states and mega-delta countries with the lowest "weighted annual capita GHG pollution" scores are so acutely threatened by greedy, heartless and irresponsible countries with the highest scores.

We are badly running out of time to deal with man-made climate change, and a [catastrophic] plus 2C temperature rise appears locked in and inevitable. However atmospheric carbon dioxide (CO_2) is remorselessly increasing to record levels at a record rate. The atmospheric level of methane (CH_4) (which has a global warming potential 105 times that of CO_2 on a 20 year time frame and with aerosol impacts considered), is rising at a much faster rate now (in 2014 at 12.5 parts per billion per year i.e. 12.5 ppb/year) than the average rate of increase in the first decade of the 21st century (0.5 ppb/year). The time for tolerance of climate criminal casuistry and double talk is long gone. The world must adopt a zero tolerance approach, publicly expose the worst polluting countries, and urgently constrain their behaviour through litigation via the International Court of Justice, prosecution of climate criminals via the International Criminal Court, Green Tariffs and Boycotts, Divestment and Sanctions (BDS). Sensible, humane, science-informed people can (a) inform everyone they can about the global warming culprits identified here, and (b) urge and apply Boycotts, Divestment and Sanctions (BDS) against all climate criminal people, politicians, parties, companies, corporations and countries disproportionately involved in the greenhouse gas pollution that so acutely threatens Humanity and the Biosphere.

References.

[1]. Jeff Masters and Bob Henson, "February smashes Earth's all-time global heat record by a jaw-dropping margin", Weather Underground, 13 March 2016: http://www.wunderground.com/blog/JeffMasters/february-smashes-earths-alltime-global-heat-record-by-a-jawdropping.

[2]. David Spratt and Phillip Sutton, "Climate Code Red: the case for emergency action", Scribe, 2008.

[3]. David Spratt, "Mind-blowing February 2016 temperature spike a "climate emergency" says scientist, as extreme events hit Vietnam, Fiji and Zimbabwe", Climate Code Red, 14 March 2016: http://www.climatecodered.org/2016/03/mind-blowing-february-2016-temperature.html.

[4]. National Oceanic and Atmospheric Administration (NOAA), Trends in atmospheric carbon dioxide": http://www.esrl.noaa.gov/gmd/ccgg/trends/gr.html).

[5]. National Oceanic and Atmospheric Administration (NOAA), Trends in atmospheric carbon dioxide": http://www.esrl.noaa.gov/gmd/ccgg/trends/index.html.

[6]. National Oceanic and Atmospheric Administration (NOAA), Trends in atmospheric methane": http://www.esrl.noaa.gov/gmd/ccgg/trends_ch4/index.html.

[7]. Gideon Polya, "2011 climate change course": https://sites.google.com/site/300orgsite/2011-climate-change-course.

[8]. Drew T. Shindell, Greg Faluvegi, Dorothy M. Koch, Gavin A. Schmidt, Nadine Unger and Susanne E. Bauer, "Improved Attribution of Climate Forcing to Emissions", Science, 30 October 2009: Vol. 326 no. 5953 pp. 716-718: http://www.sciencemag.org/content/326/5953/716.

[9]. IPCC, "Climate Change 2014 Synthesis Report, Approved Summary for Policy Makers", 1 November 2014: http://www.ipcc.ch/pdf/assessment-report/ar5/syr/SYR_AR5_SPM.pdf).

[10]. Ron Prinn, "400 ppm CO2? Add other GHGs and its equivalent to 478 ppm", Oceans at MIT, 6 June 2013: http://oceans.mit.edu/featured-stories/5-questions-mits-ron-prinn-400-ppm-threshold.

[11]. "Are we doomed?": https://sites.google.com/site/300orgsite/are-we-doomed.

[12]. "Nuclear weapons ban, end poverty & reverse climate change": https://sites.google.com/site/300orgsite/nuclear-weapons-ban.

[13]. "Too late to avoid global warming catastrophe":

https://sites.google.com/site/300orgsite/too-late-to-avoid-global-warming.

[14]. Gideon Polya, "Revised Annual Per Capita Greenhouse Gas Pollution For All Countries – What Is Your Country Doing?", Countercurrents, 6 January, 2016: http://www.countercurrents.org/polya060116.htm.

[15]. Robert Goodland and Jeff Anfang. "Livestock and climate change. What if the key actors in climate change are … cows, pigs and chickens?", World Watch, November/December 2009: http://www.worldwatch.org/files/pdf/Livestock%20and%20Climate%20Change.pdf.

[16]. "List of countries by GDP (nominal) per capita: Wikipedia: https://en.wikipedia.org/wiki/List_of_countries_by_GDP_%28nominal%29_per_capita.

[17]. "List of countries by greenhouse gas emissions per capita", Wikipedia: https://en.wikipedia.org/wiki/List_of_countries_by_greenhouse_gas_emissions_per_capita.

[18]. "Future sea level", Wikipedia: https://en.wikipedia.org/wiki/Future_sea_level.

[19]. David Spratt, "Climate reality Check. After Paris, counting the cost", March 2016: http://media.wix.com/ugd/148cb0_4868352168ba49d89358a8a01bc5f80f.pdf.

CHAPTER 25: OBAMA'S CLEAN POWER PLAN WILL ONLY CUT 2030 US GREENHOUSE GAS POLLUTION BY 5%

[First published as Gideon Polya, "Obama's Clean Power Plan Will Only Cut 2030 US Greenhouse Gas Pollution By 5%", Countercurrents, 7 August 2015: https://www.countercurrents.org/polya070815.htm].

Obama recognizes the seriousness of man-made climate change and pollution-based premature deaths (avoidable deaths) but he deceptively and mendaciously offers too little too late by 2030 – in effect, a derisory 5% reduction in overall greenhouse gas (GHG) pollution, a 9% reduction in premature deaths from coal-fired power plant pollution, and a 1% decrease in child asthma attacks.

Obama announced his Clean Power Plan on 3 August 2015 thus: "With this Clean Power Plan, by 2030, carbon pollution from our power plants will be 32 percent lower than it was a decade ago. And the nerdier way to say that is that we'll be keeping 870 million tons of carbon dioxide pollution out of our atmosphere… By 2030, we will reduce premature deaths from power plant emissions by nearly 90 percent -- and thanks to this plan, there will be 90,000 fewer asthma attacks among our children each year".

However the careful analysis below reveals the comprehensive falsehood of Obama's claimed consequences of a "32% off 2005 coal plant pollution by 2030". Obama's saving grace is that his racist, religious right Republican (R4) opponents represent neoliberalism, corporatism, racism and greed gone utterly mad with a remorseless agenda of ecocide, speciescide, omnicide, climate genocide and terracide for short-term One Percenter profit.

Obama's announcement of power plant carbon pollution in 2030 that would be 32% lower than in 2005 [1] is a 7% advance on his 2014 announcement of 30% off 2005 greenhouse gas (GHG) pollution of America's dirty power plants by 2030 enforceable by the US Environment Protection Authority (EPA) under the Clean Air Act [2, 3].

This latest policy announcement on climate change by President Obama - 32% off 2005 greenhouse gas (GHG) pollution of America's dirty power plants by 2030 enforceable by the US Environment Protection Authority (EPA) under the Clean Air Act [1] - still represents "too little too late" and falls far short of what is needed to address the worsening climate crisis.

According to the US EPA, in 2012 U.S. greenhouse gas emissions totalled 6,526 million tonnes CO_2-e of which 9% (587 million tonnes CO_2-e) was due to methane (CH_4) and 82% (5,351 million tonnes CO_2-e) was due to CO_2 from coal, gas, oil and wood burning and cement manufacture [4]. Consonant with this estimate, data from US Energy Information Administration (US EIA) indicate that 2012 energy-related CO_2 pollution was about 5,500 million tonnes CO_2 [5]. Past and projected energy-linked CO_2 pollution is in the range of 5,400 -5,600 million tonnes CO_2 annually in the period of 2009-2030 [5]. **The US EPA indicates that in 2012 electricity production accounted for 32% of GHG pollution (2,088 million tonnes CO_2-e),** and about half of this is now gas-fired due to the big shift from coal-fired to gas-fired electricity generation in recent years (the coal-to-gas electricity generation ratio shifting from 2.3 in 2007 to 1.2 in 2012) [6].

The US Energy Information Administration has assessed a remarkably constant US energy-related CO_2 pollution (in millions of tonnes CO_2) as follows: 5,418 (2009), 5,608 (2010), 5,381 (2015), 5,454 (2020), 5,501 (2025), and 5,523 (2030) (average value about 5,500) [5]. If systemic methane leakage is ignored (however see below), the 32% reduction plan would affect only the ostensibly dirtier, coal-based electric power plants and hence encourage a further shift from coal to gas [7]. Ignoring cement production, and assuming that the 32% reduction only affects coal-fired plants, that the 2005 coal/gas electricity generation mix was about 3, and that total GHG pollution was about the same in 2030 as in 2012 (6,526 million tonnes CO_2-e [4]), then the savings in 2030 would be 0.32 x 0.75 x 2,088 million tonnes CO_2-e = 501 million tonnes CO_2-e i.e. 501 x 100/6,526 = 7.7% or a derisory circa 8% decrease in total GHG pollution in 2030.

However, when a revised natural gas leakage contribution to total GHG pollution is re-assessed, US total annual GHG pollution is revealed as 10,343.3 million tonnes CO_2-e [2] and the Obama Clean Power Plan's 501 million tonnes CO_2-e savings in 2030 will be only a 501 x 100/ 10,343.3 = a 4.8% reduction i.e. about a 5% reduction of what it would otherwise be.

Noting that an estimated 52,000 Americans die annually from power generation pollutants [8], making the same assumptions as above, the maximum decrease in energy-linked, carbon burning pollutant-related deaths in 2030 would be 501 x 52,000/ 5,523 = 4,717 or only about 4,700 fewer carbon burning-related deaths per year in 2030 or a reduction of only 9% (rather than Obama's claimed "90%", assuming Obama was referring to annual premature deaths), noting that this is a naïve upper estimate because it may take decades for the pollution exposure to result in mortality.

One notes that the pro-gas Obama plan ignores the deaths from carbon burning pollutants from carbon burning other than coal burning for electricity. Thus MIT researchers have estimated 53,000 untimely deaths annually in the US from vehicle exhaust fine carbon particulates (PM $_{2.5}$) as compared to a total of 200,000 such American deaths annually from PM $_{2.5}$ particulates [8]. One notes that PM 2.5 particulates are a significant pollutant from gas-fired power plants (that are nevertheless much cleaner than coal-fired plants in relation to other toxic pollutants) [9], and that the WHO has found no threshold for adverse effects of such fine carbon particulates [10]. Indeed, a reductio ad absurdum imagines that the "32% off 2005 power plant GHG pollution by 2030" occurs in one go at the end of 2030, in which case there would be insufficient time for the pollutant reduction to have any major effect at all.

Obama's assertion that "Thanks to this plan, there will be 90,000 fewer asthma attacks among our children each year" [1] may also be extremely misleading. Thus the American Academy of Allergy, Asthma and Immunology (AAAAI) reports that "The number of people with asthma continues to grow. One in 12 people (about 25 million, or 8% of the U.S. population) had asthma in 2009, compared with 1 in 14 (about 20 million, or 7%) in 2001. More than half (53%) of people with asthma had an asthma attack in 2008. More children (57%) than adults (51%) had an attack. 185 children and 3,262 adults died from asthma in 2007" [11]. We could estimate from this information that about 10% of a projected circa 360 million Americans in 2030 (i.e. 36 million Americans) would have asthma and that about half of these (i.e. 18 million) would have asthma attacks in 2030, and of these about half would be children (i.e. 9 million). Assuming only 1 asthma attack per child in 2030 and this due solely to power plant pollution, this would indicate at least 9 million child asthma attacks in 2030, and thus Obama is saying that a "32% off 2005 greenhouse gas (GHG) pollution of America's dirty power plants by 2030" would decrease annual asthma attacks by "90,000" or by no more than about 1%.

In summary, in relation to Obama's latest Clean Power Plan, (a) 500 million tonnes CO_2 (rather than the 870 million tonnes CO_2 as claimed) will be kept out of the atmosphere), this corresponding to a derisory 5% reduction in overall annual GHG pollution in 2030; (b) annual deaths from power generation-derived pollutants will decline by 9% in 2030 (rather than by the 90% as claimed); and (c) annual child asthma attacks would decrease by no more than 1%.

The extraordinary insufficiency in Obama's Clean Power Plan is exposed when one compares it with the demands of chemically-trained Pope Francis as set out in his landmark Encyclical

"Laudato si'". Whereas Obama in effect wants a 5-8% reduction of annual GHG pollution by 2030 effected by closing coal-burning power plants, Pope Francis has unequivocally demanded massive decarbonisation and replacement by renewable energy "in the next few years": "There is an urgent need to develop policies so that, in the next few years, the emission of carbon dioxide and other highly polluting gases can be drastically reduced - for example, substituting for fossil fuels and developing sources of renewable energy" [12, 13, 14].

Further, Obama's plan involves a coal-to-gas transition in addition to renewable energy. However this US coal-to-gas transition (that is already massively underway) is flawed because (1) gas is dirty energy and a coal-to-gas policy locks in this dirty energy for decades, and (2) high systemic gas leakage and a 105-times greater Global Warming Potential (GWP) for methane (CH_4) on a 20 year time frame with aerosol impacts considered [15] means that using natural gas to generate electrical power can be dirtier GHG-wise than using coal [on a per mass basis][16].

In contrast to Obama's limited and flawed approach confined to the electricity generation sector, Pope Francis adds a further key dimension to his proposal for tackling climate change, namely a "fully borne" Carbon Price to be applied as a Carbon Tax on all GHG polluters: "Let us hope that the international community and individual governments will succeed in countering harmful ways of treating the environment. It is likewise incumbent upon the competent authorities to make every effort to ensure that the economic and social costs of using up shared environmental resources are recognized with transparency and fully borne by those who incur them, not by other peoples or future generations: the protection of the environment, of resources and of the climate obliges all international leaders to act jointly and to show a readiness to work in good faith, respecting the law and promoting solidarity with the weakest regions of the planet" (Section 195, [12]).

The Obama plan is too little too late, and his recent pronouncements ignore the Elephant in the Room reality that the whole world's 2010-2050 Terminal Carbon Pollution Budget that must not be exceeded if we are to have a 75% chance of avoiding a catastrophic 2C temperature rise is 600 billion tonnes CO_2, and the US finally used up its "fair share" of this Terminal Carbon Pollution Budget before mid- 2014 [17, 18]. The US is already stealing the residual carbon pollution entitlement of all other countries. With an annual per capita GHG pollution 3.6 times greater than the world average and nearly 5 times greater than China's, the US remains remorselessly committed to disproportionate pollution of the one common atmosphere and ocean of all nations.

The war criminal, climate criminal, Neocon American and Zionist Imperialist (NAZI) One Percenters running Obama and running the US are characterized, whether in endless warmongering or in grossly insufficient climate change inaction, by genocidal racism, murderous hegemony, global larceny and utter mendacity.

It is now very unlikely that the World will avoid a catastrophic plus 2 degrees Centigrade temperature rise – it is now effectively too late to prevent a climate catastrophe [19, 20]. Indeed increased temperature has already contributed to increased sea level, storm intensity, flooding and droughts i.e. climate catastrophe has already arrived for vulnerable, impoverished populations of Island Nations, mega-delta nations like Bangladesh and desert countries like those of the Sahel.

However despite his "too-little-too-late", credit must be given to Obama for publicly recognizing the seriousness of man-made climate change and taking steps in the right direction in stark contrast to the racist, religious right Republicans (R4s) who range from actual climate change denialists to effective climate change denialists through corruption- and neoliberal ideology-driven climate change inaction.

Further credit must be given to Obama for radically referring to "premature deaths" that must be avoided by climate change action. Premature death is otherwise referred to as untimely death, avoidable death, avoidable mortality, excess death, excess mortality, preventable death or deaths that could have been prevented [21]. The fundamentally important metric of premature death is resolutely ignored by politicians for the very good reason that it provides a damning measure of the deadly ineffectiveness of greed-perverted public policy. Thus the 3 key existential threats facing Humanity are nuclear weapons, climate change and poverty. Nuclear weapons can destroy all of the 7 billion of Humanity [22]. Already each year about 0.5 million people die from climate change, and 7 million die from air pollution [23, 24]. Poverty kills, and each year about 17 million people (half of them children) die avoidably (prematurely) from deprivation or deprivation–exacerbated disease, and 1.5 billion people have died thus since 1950 on Spaceship Earth with the US in charge of the flight deck. 10 billion people may perish prematurely (avoidably) this century if climate change is not requisitely addressed [21].

However in relation to premature deaths Obama has been further upstaged by Pope Francis who in his landmark Encyclical "Laudato si'" referred to premature deaths occasioned by pollution or poverty: "20. Some forms of pollution are part of people's daily experience. Exposure to atmospheric pollutants produces a broad spectrum of health hazards, especially for the poor, and causes millions of premature deaths ... 48. The human environment and the natural environment deteriorate together; we cannot adequately combat environmental degradation unless we attend to causes related to human and social degradation... The impact of present imbalances is also seen in the premature death of many of the poor, in conflicts sparked by the shortage of resources, and in any number of other problems which are insufficiently represented on global agendas" (Sections 20 and 48, [12]).

Obama's now expressed concern about American "premature deaths from power plant emissions" is certainly welcome but it is unlikely that Neocon American and Zionist Imperialist (NAZI) puppet Obama will extend his concern to the 1.5 million Americans who die preventably (prematurely) each year [25, 26] or the over 20 million Americans who have died thus since the US Government's 9-11 false flag atrocity [27, 28], this post-

9-11 carnage being inescapably linked to the fiscal perversion of killing 10 million Muslims abroad through violence or imposed deprivation in the War on Terror rather than keeping Americans alive at home [26].

Conclusions.

Obama recognizes the seriousness of man-made climate change and pollution-based premature deaths (avoidable deaths) but he deceptively and mendaciously offers too little too late by 2030 – in effect, a 5% reduction in overall greenhouse gas (GHG) pollution, a 9% reduction in premature deaths from coal-fired power plant pollution, and a 1% decrease in child asthma attacks.

In stark contrast, Pope Francis recognizes a looming climate "catastrophe" and "millions of premature deaths" from carbon burning pollutants, demands massive decarbonisation and conversion to renewables starting "in the next few years", rejects Carbon Trading as a dishonest "ploy", advocates a "fully borne" Carbon Price (Carbon Tax) on polluters, and proposes "boycotts" to bring neoliberal polluters to heel [14].

The horrible reality hidden by Obama's smooth rhetoric is that it is now very unlikely that the world will avoid a catastrophic plus 2 degrees Centigrade temperature rise [19, 20]. However this awful reality must invigorate us to do everything we can to make the future "less bad" for our children and future generations [29].

What can decent people – and especially the young - do in the face of worsening climate crisis, climate genocide, climate injustice and intergenerational injustice? Decent people must (a) inform everyone they can, (b) urge and apply Boycotts, Divestment and Sanctions (BDS) against all those climate criminal people, politicians, parties, companies, corporations and countries disproportionately complicit in the worsening climate crisis and climate change inaction, and (c) support pro-environment, pro-Humanity and pro-Biosphere political candidates as exemplified by the European and Australian Greens.

References.

[1]. Barack Obama, "Remarks by the president in announcing the Clean Power Plan", White House, 3 August 2015: https://www.whitehouse.gov/the-press-office/2015/08/03/remarks-president-announcing-clean-power-plan.

[2]. Gideon Polya, "Pro-gas Obama's EPA-based Plan To Reduce Coal-based Pollution Amounts To Climate Change Inaction", Countercurrents, 07 June, 2014: http://www.countercurrents.org/polya070614.htm.

[3]. Suzanne Goldenberg, "Obama unveils historic rules to reduce coal pollution by 30%", Guardian, 3 June 2014: http://www.theguardian.com/environment/2014/jun/02/obama-rules-coal-climate-change.

[4]. US EPA, "National greenhouse gas emissions data", 2014: http://www.epa.gov/climatechange/ghgemissions/usinventoryreport.html.

[5]. US EIA, "Energy-related carbon dioxide emissions", Table A10: http://www.eia.gov/forecasts/ieo/emissions.cfm.

[6]. Steven Macmillan, Alexander Antonyuk and Hannah Schwind, "Gas to coal competition in the U.S. power sector", International Energy Agency, 2013: http://www.iea.org/publications/insights/CoalvsGas_FINAL_WEB.pdf.

[7]. Tim Puko, "How the EPA's plan will impact electricity, coal and natural gas markets", The Wall Street Journal, 2 June 2014: http://blogs.wsj.com/moneybeat/2014/06/02/how-the-epas-plan-will-impact-electricity-coal-and-natural-gas-markets/.

[8]. Caiazzo, Fabio; Ashok, Akshay; Waitz, Ian A.; Yim, Steve H.L.; Barrett, Steven R.H. (November 2013). "Air pollution and early deaths in the United States. Part I: Quantifying the impact of major sectors in 2005", Atmospheric Environment, 79: 198–208: http://www.sciencedirect.com/science/article/pii/S1352231013004548.

[9]. Gideon Polya, "Expert witness testimony to stop gas-fired power plant installation", Countercurrents, 14 June 2013: http://www.countercurrents.org/polya140613.htm.

[10]. WHO, "Air quality and health": http://www.who.int/mediacentre/factsheets/fs313/en/.

[11]. American Academy of Allergy, Asthma and Immunology (AAAAAI), "Asthma attacks": http://www.aaaai.org/about-the-aaaai/newsroom/asthma-statistics.aspx.

[12]. Pope Francis, Encyclical Letter "Laudato si'", 2015: http://w2.vatican.va/content/francesco/en/encyclicals/documents/papa-francesco_20150524_enciclica-laudato-si.html.

[13]. Gideon Polya, "Pope Francis Demands "Fully Borne" Cost of Pollution (Carbon Price) To Prevent "Millions Of Premature Deaths"", Countercurrents, 29 July, 2015: http://www.countercurrents.org/polya290715.htm.

[14]. Gideon Polya, "Papal encyclical to save Planet", MWC News, August 2015: http://mwcnews.net/focus/analysis/53364-papal-encyclical-to-save-planet.html.

[15]. Drew T. Shindell, Greg Faluvegi, Dorothy M. Koch, Gavin A. Schmidt, Nadine Unger and Susanne E. Bauer, "Improved Attribution of Climate Forcing to Emissions", Science, 30 October 2009: Vol. 326 no. 5953 pp. 716-718: http://www.sciencemag.org/content/326/5953/716.

[16]. Gideon Polya, editor, "Gas is not clean energy": https://sites.google.com/site/gasisnotcleanenergy/home.

[17]. Gideon Polya, editor, "Carbon Debt Carbon Credit": https://sites.google.com/site/carbondebtcarboncredit/.

[18]. "Years left for zero greenhouse gas (GHG) pollution relative to 2013": https://sites.google.com/site/carbondebtcarboncredit/years-left-to-zero.

[19]. Gideon Polya, editor, "Are we doomed?": https://sites.google.com/site/300orgsite/are-we-doomed.

[20]. "Too late to avoid global warming catastrophe": https://sites.google.com/site/300orgsite/too-late-to-avoid-global-warming.

[21]. Gideon Polya, "Body Count. Global avoidable mortality since 1950", that includes a history of every country from Neolithic times and is now available for free perusal on the web: http://globalbodycount.blogspot.com/.

[22]. Gideon Polya, editor, "Nuclear weapons ban, end poverty & reverse climate change": https://sites.google.com/site/300orgsite/nuclear-weapons-ban.

[23]. World Health Organization (WHO), "7 million premature deaths annually linked to air pollution": http://www.who.int/mediacentre/news/releases/2014/air-pollution/en/.

[24]. DARA, "Climate Vulnerability Monitor. A guide to the cold calculus of a hot planet", 2012, Executive Summary pp2-3:

http://daraint.org/climate-vulnerability-monitor/climate-vulnerability-monitor-2012/.

[25]. Gideon Polya, "One Percenter Greed & War Means Over 1.5 Million Americans Die Preventably Each Year", Countercurrents, 19 September, 2014: http://www.countercurrents.org/polya190914.htm.

[26]. Gideon Polya, "American Holocaust, millions of untimely American deaths and $40 trillion cost of Israel to Americans", Countercurrents, 27 August, 2013: http://www.countercurrents.org/polya270813.htm.

[27]. Gideon Polya, editor, "Experts; US did 9-11": https://sites.google.com/site/expertsusdid911/.

[28]. Elias Davidsson, "The right to the truth about the mass killings of 11 September 2001", Dissident Voice, 2 August 2015: http://dissidentvoice.org/2015/08/the-right-to-the-truth-about-the-mass-killings-of-11-september-2001/.

[29]. Gideon Polya, "2015 A-to-Z Alphabetical List Of Actions And Advocacies For Climate Change Activists", Countercurrents, 14 January, 2015: http://www.countercurrents.org/polya140115.htm

CHAPTER 26: OVER 14 MILLION AMERICANS WILL DIE PREVENTABLY UNDER A 2-TERM TRUMP ADMINISTRATION

[First published as Gideon Polya, "Over 14 million Americans will die preventably under a 2-term Trump administration", Countercurrents, 22 March 2017: https://countercurrents.org/2017/03/over-14-million-americans-will-die-preventably-under-a-2-term-trump-administration/].

A remorselessly neoliberal, climate change-denying and anti-science Trump aims to slash science, technology, international aid, and medical research funding in favour of extra funding for the military and nuclear terrorism. 30 Americans are predicted to die from terrorism in America in the next 8 years but over 14 million Americans will die preventably under a US- and world-threatening 2-term Trump regime. US aid will be slashed to $20 billion pa for an impoverished world. Meanwhile corporate Mainstream media resolutely ignore the Awful Truth.

It has been estimated that 1.67 million Americans died preventably out of a US population of 319 million in 2014 [1]. In the middle of a 2-term Trump presidency the American population will be about 336 million (UN Population Division, 2015 Revision of World Population Prospects) so we can estimate that about 1.67 million x 336 million/319 million = 1.76 million Americans will die preventably each year under a 2-term Trump presidency, or about 14 million in total. However deadly, anti-science Trump policies, as expertly analysed by the world's leading scientific journal Science [2], are set to increase this carnage i.e. more than 14 million Americans will die preventably under a 2-term American-killing Trump administration.

In stark contrast, 53 Americans were killed by [jihadi] terrorists in America in the 14 years after 9-11, this yielding an estimate of 53 terrorist deaths/14 years = 3.8 terrorism deaths per year [1, 3]. Using this empirical data, we can predict about 3.8 terrorism deaths per year x 8 years = 30 American terrorism deaths in America during a 2-term Trump presidency. Yet Trump has used the prospect of American terrorism deaths as a trumped-up terror hysteria argument for a massive shift of Federal funding and support away from life-saving science (that might lessen the carnage of 14 million preventable American deaths) and instead to the military, advanced weapons, nuclear terrorism, climate criminal fossil fuel exploitation and huge tax breaks for unrestrained, terracidal neoliberal capitalism.

(A). Mainstream lying by omission over American preventable deaths and horrendous Indigenous deaths in US wars.

It is estimated that there will be about 30 American terrorism deaths versus well over 14 million preventable American deaths in a 2-term Trump presidency. However American Mainstream media are endlessly excited by trumped-up terror hysteria over a miniscule 30 terrorism deaths in the next 8 years while resolutely ignoring the prospect of well over 14 million preventable American deaths in a 2-term Trump presidency. This must surely be one of the most outrageous examples of what leading German Nazis described as "the Big Lie" [4].

Thus the democratically-elected Nazi dictator Adolph Hitler (helped to power by Republican war criminal George W. Bush's grandfather, the late US senator Prescott Bush, who was closely linked as a company director to companies backing the rise of Nazism in Germany [5]) is credited with first enunciating the notion of the Big Lie in his book "Main Kampf": "All this was inspired by the principle—which is quite true within itself—that in the big lie there is always a certain force of credibility; because the broad masses of a nation are always more easily corrupted in the deeper strata of their emotional nature than consciously or voluntarily; and thus in the primitive simplicity of their minds they more readily fall victims to the big lie than the small lie, since they themselves often tell small lies in little matters but would be ashamed to resort to large-scale falsehoods. It would never come into their heads to fabricate colossal untruths, and they would not believe that others could have the impudence to distort the truth so infamously. Even though the facts which prove this to be so may be brought clearly to their minds, they will still doubt and waver and will continue to think that there may be some other explanation. For the grossly impudent lie always leaves traces behind it, even after it has been nailed down, a fact which is known to all expert liars in this world and to all who conspire together in the art of lying" [4].

Hitler's propaganda master-mind Joseph Goebbels had a slightly different take on the Big Lie: "The essential English leadership secret does not depend on particular intelligence. Rather, it depends on a remarkably stupid thick-headedness. The English follow the principle that when one lies, one should lie big, and stick to it. They keep up their lies, even at the risk of looking ridiculous" [4].

Napoleon famously declared that "History is a set of *lies* that people have agreed upon" [6], but that exquisitely truthful, Napoleonic era English novelist Jane Austen had evidently swallowed the lies of the English Establishment when in her novel "Northanger Abbey" (unpublished in her lifetime) she has the heroine, Miss Catherine Morland, reproved for her suspicions of wrongdoing by her "lover", Henry Tilney: "If I understand you rightly, you have formed a surmise of such horror as I have hardly words to -. Dear Miss Morland, consider the dreadful nature of the suspicions you have entertained. What have you been judging from? Remember the country and the age in which we live. Remember that we are English, that we are Christians. Consult your own understanding, your own sense of the probable, your own observation of what is passing around you. Does our education prepare us for such atrocities? Do our laws connive at them? Could they be perpetrated without being known, in a country like this, where social and literary intercourse is on such a footing, where every man is surrounded by a neighbourhood of voluntary spies, and where roads and newspapers lay everything open? Dearest Miss Morland, what ideas have you been admitting?" [7]. Appallingly, the answers to Henry Tilney's questions in today's age of lying corporate media are "Yes, yes and yes!" [8-12].

American Mainstream media have an entrenched culture of "manufacturing consent" [13], censoring, lying by commission and, most critically, of egregious lying by omission [8-12]. Lying by omission is far, far worse than lying by commission because at least the latter permits refutation and public debate [12]. Nevertheless, some American and Western journalists – most notably outstanding, anti-racist Jewish journalists – have surfaced from the Mainstream "swamp" to inform the masses and "speak truth to power". Thus anti-racist Jewish American journalist Stone I.F. Stone in "The Hidden History of the Korean War" used Mainstream sources to reveal the horrendous actuality of an American imperialist war [14], a war in which, as revealed decades later by anti-racist Jewish Canadian scholar Michel Chossudovsky, 28% of the North Korean population were killed by US bombing [15].

Anti-racist Jewish American journalist Seymour Hersh won a Pullitzer Prize for exposing the cover-up of the Mylai Massacre (about 500 Vietnamese civilians massacred by US forces in 1968) [16, 17]. Recently, Seymour Hersh condemned the Obama Administration story of having extra-judicially killed Osama bin Laden in Pakistan as being almost comprehensively false [18], but was criticized by outstanding American humanitarian commentator, Dr Paul Craig Roberts, for accepting the Obama claim that Osama bin Laden had actually been killed in the 2011 US operation [19].

Anti-racist Jewish American writer William Blum (author of "Killing Hope: US Military and CIA Interventions Since World War 2", "Rogue State: A Guide to the World's Only Superpower" [20], "West-Bloc Dissident: A Cold War Memoir", "Freeing the World to Death: Essays on the American Empire", and the ongoing series of cogent essays on violent US imperialism called "The Anti-Empire Report") was scathing about Mainstream deception of US voters over endless US war crimes (2012): "As somebody once said, the United States doesn't need a third party. It needs a second party… How many voters does it take to change a light bulb? None. Because voters can't change anything… I can offer no solution to stopping the imperial beast other than this: Educate yourself and as many others as you can, raising their political and ideological consciousness, providing them with the factual ammunition and arguments needed to sway others, increasing the number of those in the opposition until it raises the political price for those in power, until it reaches a critical mass" [21].

Anti-racist Jewish Canadian scholar and editor of Global Research, Professor Michel Chossudovsky, has similar clarity about the serial war criminal American Empire, noting that "Iraq, Afghanistan, Libya and now Syria are but a sequence of stops on a global roadmap of permanent war that also swings through Iran. Russia and China are the terminal targets" [21]. Anti-racist Jewish French writer Emmanuel Todd, who famously predicted the end of the Soviet Union, predicted the end of the American Empire while warning that many would die in the imperial death throes. Emmanuel Todd's "After the Empire. The breakdown of the American Order" (2002) [22] explores American "exceptionalism" that has enabled the brutal creation of a global American Empire. Emmanuel Todd has criticized the neo-fascist [fascoid] Islamophobic terror hysteria sweeping France and the rest of the US Alliance that ignores horrendous crimes of the US Alliance and promotes more such US Alliance crimes in the Muslim world from West Africa to South East Asia [23, 24].

Successive American governments claim to be leading a global fight of Western civilization against Muslim extremists but the Awful Truth is that American governments lie, and that in reality a pro-One Percenter, religious fundamentalist-dominated, Neocon American and Zionist Imperialist (NAZI)-perverted America has an appalling record of supporting and exciting religious fundamentalist rebels (jihadis) against modern, secular regimes in Muslim countries with the collateral trashing of democracy, women's rights, and children's rights [25]. The greatest crime of the jihadis, whether US backed or not, is that their barbarism is used as an excuse for vastly more deadly US Alliance state terrorism. An anti-racist Jewish Hungarian-origin Australian who is haunted by the WW2 Jewish Holocaust that killed all but a dozen of his wider family, I have written endlessly as a humanitarian and out of an inescapable moral obligation about the ever-increasing horrendous death toll in the US Alliance War on Muslims. However Mainstream media, politician and academic presstitutes of the US, US lackey Australia and the US Alliance are almost totally silent about the ongoing Muslim Holocaust and Muslim Genocide [26-29]. Thus for the Sanctions and Occupation period of 1990-2011, under-5 infant deaths in Iraq totalled 2.0 million, avoidable deaths from deprivation totalled 2.9 million, violent deaths totalled 1.7 million, and refugees totalled 5-6 million [27, 29]. Yet Google searches for the terms "Iraqi infant deaths" and "9-11" total 900 and 276 million, respectively – damning evidence of Mainstream lying by omission, genocide ignoring and holocaust ignoring, noting that genocide ignoring and holocaust ignoring are far, far worse than repugnant genocide denial and holocaust denial because the latter at least allow for public debate.

In 1996, well before the post-9-11 invasion of Iraq in 2003, anti-racist Jewish American journalist Lesley Stahl famously asked US Ambassador to the UN (and later US Secretary of State) Madeleine Albright about deadly UN sanctions against Iraq on a "60 Minutes" segment: "We have heard that half a million children have died. I mean, that's more children than died in Hiroshima. And, you know, is the price worth it?" To this humane question, Albright notoriously replied "We think the price is worth it" [30]. Lesley Stahl is a rare example of a US journalist who has actually asked about the mass murder of Iraqis, a subject resolutely avoided by US and indeed Western Mainstream media.

Elias Davidsson is a Palestine-born, anti-racist Jewish writer who has written cogently on starkly evident US involvement in the 9-11 atrocity that killed 3,000 people, mostly Americans [31, 32]. Numerous science, engineering, architecture, aviation, military and intelligence experts have concluded that the US Government must have been responsible for the 9-11 atrocity, with some concluding that America's dirty tricks surrogate, Apartheid Israel, must also have been involved [33]. However the 9-11 atrocity was immediately blamed on "primitive jihadis in caves" and used as an excuse for the US and the US Alliance to launch a racist and genocidal US War on Muslims that now involves Western invasion of a swathe countries from West Africa to South East Asia. The US War on Terror (actually a genocidal US War on Muslims) has been associated with a long-term accrual cost of $6 trillion just for the Iraq War and Afghan War alone, and 32 million Muslim deaths from violence, 5 million, or from imposed deprivation, 27 million, in 20 impoverished countries invaded by the US Alliance since the US Government's 9-11 false flag atrocity [34]. However Google searches for the terms "War on Muslims" and "War on Terror" yield 0.2 million and 11.7 million results, respectively. Google searches for the terms "Muslim Holocaust", "Muslim Genocide", "9-11" and "September 11" yield about 13,000, 53,000, 276 million and 73 million results, respectively.

One might naively think that this huge and totally justified American concern for about 3,000 American lives taken on 9-11 would be reflected in comparable concern for 26 million preventable American deaths since 9-11. However Google searches for the terms "Preventable American deaths" and "American preventable deaths" yield a mere 2,300 and 300 results, respectively, as compared to 276 million and 73 million results, respectively, for "9-11" and "September 11".

American Mainstream media are utterly unconcerned about the preventable deaths of 26 million Americans since 9-11 in which successive pro-One Percenter, Neocon American and Zionist Imperialist (NAZI)-perverted American administrations are inescapably complicit. The projected impact of Trump's policies on this American Holocaust are outlined below.

(B). Over 14 million American preventable deaths predicted under a 2-term Trump Administration.

American and Western Mainstream media won't report the active or passive mass murder of over 30 million Muslims in the

US War on Terror, and mostly won't report the preventable deaths of 1.7 million Americans each year. In stark contrast, American media endlessly report on the "terror threat" and the 60 Americans who have been killed by terrorists in America since the US Government's 9-11 false flag atrocity that killed 3,000 people, mostly Americans, in 2001 [31-33].

As detailed below, an estimated total of about 1.67 million Americans die preventably each year out of a population of 319 million (2014). As indicated above, the UN Population Division data predict a US population of 336 million in the middle of a 2-term Trump presidency [35], this indicating that an average of 1.76 million Americans will die preventably each year under a 2-term Trump presidency or about 14 million in total. One can then briefly assess how retrograde, anti-science Trump policies as set out in his Budget proposals will increase the appalling carnage of preventable American deaths as set out below (circa 2014 estimates; multiply by 336/319 = 1.05 to get estimates for middle of a 2-term Trump presidency):

1. **443,000 Americans die from smoking-related causes annually.** Trump proposes to cut the National Institutes of Health (NIH) budget by 20. This will negatively impact on deaths from smoking-related causes [2].
2. **440,000 Americans die from adverse events in hospitals each year.** One supposes that Trump's proposed 20% cut to the NIH budget [2] will increase hospital deaths from adverse events
3. **300,000 Americans die from obesity-related causes annually. Obesity is a major morbidity and mortality problem in the US.** The US Centers for Disease Control and Prevention (CDC) predict that obesity in the US will to continue to rise and encompass 42% of Americans by 2030 [36]. Trump's proposed 20% cut to the NIH budget [2] will inevitably increase obesity-related deaths.
4. **200,000 Americans die annually from air pollution (e.g. from coal burning, vehicle exhaust, carbon burning in general).** Air pollution from burning carbon fuels kills about 7 million people worldwide each year [37]. Neoliberal and climate change denier Trump's pro-

coal, pro-gas, pro-oil, pro-fossil fuel, anti-science, anti-EPA and anti-NIH policies will inevitably increase American air pollution deaths. Anti-science, climate change denier Trump has appointed non-scientist, anti-science, climate change sceptic, and anti-EPA litigator Scott Pruitt to head the EPA (Environment Protection Agency). Scientifically illiterate politician Scott Pruitt has incorrectly asserted that: "Science tells us that the climate is changing and human activity in some manner impacts that change. The human ability to measure with precision the extent of that impact is subject to continuing debate and dialogue, as well they should be" [38]. The overwhelming scientific consensus is that man-made greenhouse gas (GHG) pollution is warming the planet (as a scientist I was aware of this growing problem way back in the late 1980s). Anti-science, climate change denier Trump's proposed Budget will cut the funding of a whole range of important US scientific bodies that variously relate to the atmosphere, GHG pollution and clean energy, specifically cutting the Advanced Research Projects Agency-Energy (ARPA-E) by 100%, the National Oceanic and Atmosphere Administration (NOAA) research office by 52%, the Environment Protection Agency (EPA) R&D office by 48%, the Department of Energy (DOE) energy program by 44%, the EPA by 30%, the DOE office of science by 17%, NOAA satellites by 16%, the National Institute of Standards and Technology (NIST) by 13%, the United States Geological Survey (USGS) by 10%, the National Aeronautics and Space Administration (NASA) earth science by 5% and NASA by 1% [2].

5. **75,000 American alcohol-related deaths annually.** Trump's ideological deregulation policies and anti-NIH policies [2] will increase American alcohol-related deaths.

6. **45,000 US deaths annually from lack of medical insurance.** The Congressional Budget Office has estimated that the ObamaCare-replacing, Trump Republican American Health Care Act would mean 14

million more uninsured Americans in 2018 and 24 million by 2026. Lack of health insurance kills about 45,000 Americans each year (Harvard Medical School) and Trump's removal of millions of Americans from Health Cover will kill a further 43,000 Americans over 2 Trump terms [39] versus an expected 30 terrorism deaths over the same period [1, 40, 41]. If the terrorist threat is empirically quantitated in terms of the number of deaths expected to be caused, then Trump is by far the worst terrorist threat to America and the world.

7. **38,000 US drug-related deaths annually, this including 21,000 US opiate drug-related deaths annually from US restoration and protection of the Taliban-destroyed Afghan opium industry.** Neofascist militarist Trump has inherited the US War on Muslims, including the ongoing Afghan Genocide [28, 29], from serial war criminal Obama. There is no reason to expect that he will close down the formerly Taliban-destroyed Afghan opium industry. The US Alliance restored the largely Taliban-destroyed Afghan opium industry from 6% of world market share in 2001 to 93% in 2007 [28]. Utterly ignored by Neocon American and Zionist Imperialist (NAZI)-perverted and subverted Western Mainstream media are the 1.2 million people who have died world-wide since 9-11 due to US Alliance restoration of the Taliban-destroyed Afghan opium industry, the breakdown (as of 2015) including 280,000 Americans, 256,000 Indonesians, 68,000 Iranians, 25,000 British, 14,000 Canadians, 10,000 Germans, and 5,000 Australians [28].

8. **33,000 Americans killed by motor vehicles each year.** Trump's pro-coal, pro-gas, pro-oil, pro-fossil fuel, climate change denialist, anti-EPA and anti-NIH policies will increase American motor vehicle deaths. This statistic does not include the tens of thousands of Americans killed by pollutants from vehicle exhaust. Thus a report from 87-Nobel-Laureate Massachusetts Institute of Technology (MIT) (2013): "Researchers from MIT's Laboratory for Aviation and the Environment have come out with some sobering new

data on air pollution's impact on Americans' health. The group tracked ground-level emissions from sources such as industrial smokestacks, vehicle tailpipes, marine and rail operations, and commercial and residential heating throughout the United States, and found that such air pollution causes about 200,000 early deaths each year. Emissions from road transportation are the most significant contributor, causing 53,000 premature deaths, followed closely by power generation, with 52,000" [42]. This may be an under-estimate because scientists from the Environmental Research Group at Kings College, London, UK estimated that annual air pollution deaths in London alone might total about 9,400 [37, 43].

9. **31,000 gun-related US deaths annually.** Trump's pro-gun policies will increase US gun deaths.

10. **30,000 Americans suicide annually with 7,000 being US veterans.** Overblown rhetoric aside, Trump's actual policies may harm veterans and increase an already shocking level of US veteran suicide. Thus CNBC reports: "While the nation's unemployment rate [4.7%] declines, the jobless trend for America's youngest veterans is climbing, jumping to 6.3 percent in January from 4.4 percent in September… President Donald Trump's federal hiring freeze may hit veterans hardest. It's a primary source of civilian employment post-service because of its preferential hiring practices toward veterans. A third of federal workers are veterans" [44]. On the other hand, Veterans Affairs gets a modest 6% budget boost in Trump's proposed Budget. Expert US epidemiological analysis from the US Department of Veterans Affairs reveals that over the last 13.5 years since 9-11 there have been roughly 20 US veteran suicides every day [45] i.e. 365.25 x 20 = 7,305 per year and about 7,305 per year x 15.5 years = 113,000 US veteran suicides since the start of the Zionist-promoted US War on Terror in September 2001. In stark contrast, US military fatalities in Iraq and Afghanistan since 9-11 total about 6,900 [46].

11. **21,000 avoidable under-5 year old US infant deaths annually.** The infant mortality rate is low in the US

(about the same as in impoverished Cuba) but should be much lower given America's high per capita GDP of $56,000 compared to $7,700 for Cuba [47]. Misogynist Trump passionately supports the right to life of unborn infants from the time of ovum fertilization onwards but his anti-NIH and anti-ObamaCare policies will tend to decrease survival of born infants. Trump supports a domestic US abortion ban exception for "rape, incest and the life of the mother" and is set to devastate the Developing World with his primitive, Old Testament views and a policy that requires non-governmental organisations (NGOs) receiving federal funding to agree to "neither perform nor actively promote abortion as a method of family planning in other nations" [48].

12. **15,000 Americans are violently murdered annually.** High murder rate is associated with social disadvantage. Thus, for example, 27.4 % of African Americans were living in poverty in 2011, African American wealth is about 5 times lower than that of Whites, and African Americans are 8 times more likely to murder and 6 times more likely to be murdered than Whites) [49]. Neoliberal billionaire Trump's extremist policies are set to increase social disadvantage. Further, Trump's pro-gun, anti-accessible health and anti-NIH policies will also favour gun deaths. Of course the 15,000 annual American homicide deaths are dwarfed by the 1.7 million annual American preventable deaths that are set to rise under Trump policies.

According to an expert analysis, Trump's plan to slash the corporate tax rate to 15% may cost $1 trillion annually: "The plan would reduce federal revenues by $9.5 trillion over its first decade before accounting for added interest costs or considering macroeconomic feedback effects" [50]. Jason Schenker (president and founder at Prestige Economics LLC) has commented in Bloomberg that Trump may not be able to go that far (2017): "President Donald Trump presented his tax plan to Congress, but there was one notable omission: 15 percent. That's the corporate rate that investors have started pricing into equity markets, and it's the number that Trump has repeatedly stated as his goal. But with almost $20 trillion in U.S. national debt, and $200 trillion in unfunded entitlement obligations, it may too tough to cut taxes that significantly" [51].

Either way, as reflected in his slashing of non-military Federal expenditure in his proposed FY2018 Budget of $1.15 trillion (the discretionary part of the overall $4 trillion US Federal Budget), Trump intends to savagely cut Federal funding that helps keep Americans alive and divert that wealth to climate criminal corporations via deregulation and tax breaks, towards nuclear weapons (an 11% increase in funding for DOE nuclear weapons via the National Nuclear Security Administration (NNSA)) [2], towards national security (7% increase), and to the military (an extra $54 billion or a 10% increase on the huge annual US military budget of about $0.6 trillion) [52, 53].

So much for Trump's promise "to keep America safe" – under Trump's retrograde policies an additional 43,000 Americans will die from lack of Health Cover and a total of well over 14 million Americans will die preventably under a 2-term Trump Administration.

(C). Trump's proposed FY 2018 Budget and a worsening Global Avoidable Mortality Holocaust.

The State Department has summarized Trump's State Department and USAID Budget for Fiscal Year (FY) 2018 thus: "Today the President released the Fiscal Year (FY) 2018 budget blueprint which provides an overview of the Administration's overarching priorities for discretionary spending. It includes $37.6 billion for the Department of State and U.S. Agency for International Development (USAID), of which $12.0 billion is Overseas Contingency Operations funding. The FY 2018 budget advances the national security interests of the United States by focusing on diplomatic efforts and foreign assistance programs that advance the security and prosperity of the American people. The budget blueprint includes $3.1 billion to meet our security assistance commitments to Israel and supports other critical foreign assistance efforts, including global health and humanitarian assistance programs. The budget also supports diplomatic engagement activities, and ensures the safety of our diplomats by applying $2.2 billion towards new embassy construction and maintenance. The budget will support Department of State and USAID efforts to optimize organizational effectiveness, helping us work to efficiently achieve our diplomatic and development goals and objectives. Later this spring the President will release the full FY 2018 budget request with more details on specific funding and programs requested for the Department of State and USAID, along with other Executive Branch agencies" [54].

Trump's FY2018 State Department and USAID Budget is $37.6 billion x 100/ $50.1 = 75% of Obama's proposed FY2017 State Department and USAID Budget (i.e. a 25% reduction) but represents a 28% reduction on the actual FY2017 State Department and USAID Budget [56]. Pro-Zionist Trump's proposed State Department and USAID FY2018 budget of $37.6 billion is almost exactly the same as the $38 billion of military aid pledged by Zionist lackey Obama (with Zionist lackey Trump support) for the next 10 years to nuclear terrorist, racist Zionist-run, genocidally racist, democracy-by-genocide, America-perverting, America-subverting, and American-killing Apartheid Israel.

Massive fiscal diversion from preserving American lives to supporting war, occupation and Apartheid overseas is inextricably linked to 1.7 million preventable American deaths each year and 26 million preventable American deaths since 9-11. A 2013 analysis determined that the horrendous and deadly financial cost of Apartheid Israel to Americans had reached a gigantic $40 trillion in today's dollars. However the human cost since 9-11 has involved the preventable deaths of millions of Americans – passive mass murder of Americans in an American Holocaust inflicted by the fiscal perversion of traitorous Neocon American and Zionist Imperialist (NAZI)-subverted One Percenters committing $8-10 trillion to ethnic cleansing and active and passive mass murder of millions of Muslims abroad in support of Apartheid Israel instead of keeping Americans alive at home. Zionist-subverted American Government support for Apartheid Israel in circa 2008 dollars totals about $40 trillion, the breakdown (as estimated in 2013) being (1) $3 trillion (1948-2003), (2) $4-6 trillion (Zionist-promoted Iraq and Afghan Wars), (3) $0.7 trillion (Value of a Statistical Life- or VSL-based cost of 88,000 US veteran suicides since September 2001), and (4) about $30 trillion (one quarter of the VSL-based cost of 15.6 million preventable American deaths since September 2001). This is an under-estimate because it does not consider the millions of preventable American deaths before 9-11 linked to Zionist subversion and perversion of America [57].

According to the Congressional Research Service (2016): "In FY2015, U.S. foreign assistance, defined broadly, was estimated at $48.57 billion, or 1.3% of total federal budget authority. About 43% of this assistance was for bilateral economic development programs, including political/strategic economic assistance; 35% for military aid and non-military security assistance; 16% for humanitarian activities; and 6% to support the work of multilateral institutions… In FY2015, roughly $26.4 billion – 54% of total estimated U.S. assistance – focused exclusively on mitigating human suffering and poverty and addressing environmental, governance, and other concerns in developing countries" [58].

Trump's $37.6 billion State Department and USAID budget proposal for FY2018 was only 77% of that for FY2015 ($48.57 billion) [58] and only 75% of Obama's proposal for FY2017 ($50.1 billion; i.e. a reduction of 25%) [55]. Assuming that the budget breakdown for FY2018 was the same as for FY2015, then Trump's proposed FY2018 State Department and USAID budget would involve $13.2 billion for military and security aid (of which 23% would go to Apartheid Israel), $4.1 billion for political and strategic purposes (e.g. subverting other governments), and a mere $20.3 billion for humanitarian and development aid.

Trump's proposed FY2018 $20.3 billion State Department and USAID Budget for humanitarian and development aid is only 6.5 times the $3.1 billion military aid component given to Apartheid Israel (Jewish population about 6.1 million versus Developing World population of 6,100 million i.e. $508 per capita for Jewish Israelis with a per capita GDP of $37,000 versus $3 per capita for Developing World people with GDP per capita less than the world average of $10,000) [47].

Trump's proposed $20.3 billion State Department and USAID Budget for humanitarian and development aid in FY2018 is 3% of his proposed circa $650 billion US military budget, 0.5% of the total $4 trillion US Budget, 0.1 % of the 2016 US GDP ($18.56 trillion) and 0.11% of the 2015 US GNI ($18.5 trillion). By way of comparison, aid as a % of GNI in 2015 was 1.4% for Sweden, noting that the UAE, Norway, Luxembourg, Denmark, the Netherlands, and the UK also exceeded the UN Official Development Assistance (ODA) target of 0.7% of GNI [59, 60].

The world is acutely and indeed existentially threatened by (a) nuclear weapons (a nuclear exchange would wipe out humanity and much of the Biosphere) [61], (b) poverty (17 million people die avoidably from deprivation and deprivation-exacerbated disease each year in the Developing World minus China) [61], and (c) by worsening climate change from greenhouse gas (GHG) pollution (presently about 7.5 million people die each year due to carbon burning pollutants, 7 million, or from climate change, 0.5 million; 10 billion deaths are predicted this century if man-made climate change is not requisitely addressed) [63].

Trump acutely threatens the world as (a) a nuclear terrorist and warmonger, (b) a ruthless, merciless, pro-One Percenter, neoliberal capitalist, and (c) as an anti-science, anti-environment, pro-coal, pro-gas, pro-oil, climate change denier committed to unlimited GHG pollution for the private profit of his fellow billionaires.

Trump's actions speak louder than the words of his simplistic and populist rhetoric. Trump's executive orders and proposed FY2018 Budget involve (a) a hugely increased budget for nuclear weapons and the military (while threatening military action against Iran, North Korea and China); (b) slashing the US aid budget to a derisory $20.3 billion or 0.1% of GDP per year, with cessation of US funding of any bodies tackling the core problem of over-population with information on reproductive control (coupled with an extreme neoliberal stance of unrestrained corporate greed for private profit) [64]; and (c) slashing funding for US scientific bodies crucial for tackling man-made climate change and its consequences, approving unlimited GHG pollution, and approving the Keystone XL pipeline to take oil from Canadian oil sands to Texas – "game over for the climate" according Dr James Hansen, former head of the world-leading, "Earth-centric" and Trump-threatened NASA Goddard Institute for Space Research) [65].

Final comments.

In his proposed FY2018 Budget, a remorselessly neoliberal, climate change-denying and anti-science Trump aims to slash science, technology, international aid, and medical research funding in favour of extra funding for the military, war, Apartheid Israel, and nuclear terrorism. 30 Americans are predicted to die from terrorism in America in the next 8 years but well over 14 million Americans will die preventably under a US- and world-threatening 2-term Trump regime.

US aid will be slashed to about $20 billion pa for an impoverished world in which 17 million people already die from deprivation and deprivation-exacerbated disease each year. A racist, pro-war, warmongering, nuclear terrorist, anti-science, and neoliberal extremist Trump significantly adds to the acute, existential threats to the world from nuclear weapons, poverty and climate change.

However corporate US and Western Mainstream media resolutely ignore the hard prediction of over 14 million American preventable deaths in a 2-term Trump presidency, just as they resolutely ignore the active or passive mass murder in the US War on Muslims (US War on Terror) of 32 million Muslims from violence, 5 million, or from imposed deprivation, 27 million, in 20 countries invaded by the US Alliance since the US Government's 9-11 false flag atrocity [34].

What can decent people do as an ignorant, neo-fascist, corporatist, anti-science, neoliberal Trump-led America pushes our nuclear weapons-, poverty- and climate change-threatened world towards the precipice? Decent people everywhere must (a) inform everyone they can, (b) adopt zero tolerance for deadly militarism, neoliberalism and climate criminality [66-68], and (c) urge and apply Boycotts, Divestment and Sanctions (BDS) against a terracidal Trump America and its genocidal lackeys that threaten the whole world with catastrophe.

References.

[1]. Gideon Polya, "West Ignores 11 Million Muslim War Deaths & 23 Million Preventable American Deaths Since US Government's False-flag 9-11 Atrocity", Countercurrents, 9 September, 2015: https://countercurrents.org/polya090915.htm.
[2]. "A grim day for U.S. Science: analysis and reaction to Trump's plan", Science, 16 March 2017: http://www.sciencemag.org/news/2017/03/trumps-first-budget-analysis-and-reaction.
[3]. Ronald Bailey, "How scared of terrorism should you be?", Reason.com, 6 September 2011: http://reason.com/archives/2011/09/06/how-scared-of-terrorism-should.
[4]. "Big lie", Wikipedia: https://en.wikipedia.org/wiki/Big_lie.
[5]. Ben Aris and Duncan Campbell, "How Bush's grandfather help Hitler to power" ", Guardian, 26 September 2004: https://www.theguardian.com/world/2004/sep/25/usa.secondworldwar.
[6]. "Napoleon Bonaparte quotes", Brainy Quotes: https://www.brainyquote.com/quotes/quotes/n/napoleonbo161968.html.
[7]. Jane Austen, "Northanger Abbey", published posthumously, 1818.
[8]. Gideon Polya, editor, "Mainstream media censorship": https://sites.google.com/site/mainstreammediacensorship/home.
[9]. Gideon Polya, editor, "Mainstream media lying": https://sites.google.com/site/mainstreammedialying/.
[10]. Gideon Polya, editor, "Lying by omission": https://sites.google.com/site/mainstreammedialying/lying-by-omission.
[11]. Gideon Polya, "Jane Austen and the Black Hole of British History. Colonial rapacity, holocaust denial and the crisis in biological sustainability", G.M. Polya, Melbourne, 2008 edition that is now available for free perusal on the web: http://janeaustenand.blogspot.com/.
[12]. "Lying by omission is worse than lying by commission because at least the latter permits refutation and public debate": https://sites.google.com/site/mainstreammedialying/lying-by-omission.
[13]. Edward S. Herman and Noam Chomsky, "Manufacturing Consent. The political economy of the mass media", Pantheon, New York, 2002.

[14]. I.F. Stone, "The Hidden History of the Korean War", Monthly Review Press, 1969.

[15]. Michel Chossudovsky, "Know the facts: North Korea lost nearly 30% of its population as a result of US bombings in the 1950s", Global Research, 27 November 2010: http://www.globalresearch.ca/know-the-facts-north-korea-lost-close-to-30-of-its-population-as-a-result-of-us-bombings-in-the-1950s/22131);

[16]. "My Lai Massacre", Wikipedia: https://en.wikipedia.org/wiki/My_Lai_Massacre.

[17]. "Seymour Hersh", Wikipedia : http://en.wikipedia.org/wiki/Seymour_Hersh.

[18]. Seymour M. Hersh, "The killing of Osama bin Laden", London Review of Books, vol. 37, no. 10, 21 May 2015: https://www.lrb.co.uk/v37/n10/seymour-m-hersh/the-killing-of-osama-bin-laden.

[19]. Paul Craig Roberts, "Seymour Hersh Succumbs To Disinformation", Countercurrents, 11 May, 2015: https://countercurrents.org/roberts110515A.htm.

[20]. William Blum, "Rogue State: A Guide to the World's Only Superpower"

[21]. William Blum, "Syria, The Story Thus Far", Countercurrents, 3 October, 2012: https://countercurrents.org/blum031012.htm.

[22]. Emmanuel Todd, "After the Empire: The breakdown of the American Order", Columbia University Pres, 2003.

[23]. Angelique Chrisafis, "Emmanuel Todd: the French thinker who won't toe the Charlie Hebdo line", Guardian, 29 August 2015: https://www.theguardian.com/media/2015/aug/28/emmanuel-todd-the-french-thinker-who-wont-toe-the-charlie-hebdo-line.

[24]. Emmanuel Todd, "Who is Charlie. Xenophobia and the New Middle Class", 2015.

[25]. Gideon Polya, "Fundamentalist America Has Trashed Secular Governance, Modernity, Democracy, Women's Rights And Children's Rights In The Muslim World", Countercurrents, 21 May, 2015: https://countercurrents.org/polya210515.htm.

[26]. "Palestinian Genocide": http://sites.google.com/site/palestiniangenocide/.

[27]. Gideon Polya, editor, "Iraqi Holocaust, Iraqi Genocide": http://sites.google.com/site/iraqiholocaustiraqigenocide/.

[28]. Gideon Polya, editor, "Afghan Holocaust, Afghan Genocide": http://sites.google.com/site/afghanholocaustafghangenocide/.

[29]. Gideon Polya, editor, "Muslim Holocaust Muslim Genocide": https://sites.google.com/site/muslimholocaustmuslimgenocide/.

[30]. Lesley Stahl and Madeleine Albright quoted in "Madeleine Albright", Wikipedia: http://en.wikipedia.org/wiki/Madeleine_Albright.

[31]. Elias Davidsson, "There is no evidence that Muslims committed the crime of 9-11", Op Ed News, 10 January 2008: http://www.opednews.com/articles/1/There-is-no-evidence-that-by-Elias-Davidsson-100811-366.html.

[32]. Elias Davidsson, "Hijacking America 's Mind on 9/11. Counterfeiting Evidence", Algora, New York 2013, 328 pp: http://www.amazon.com/Hijacking-Americas-Mind-11-Counterfeiting/dp/0875869734.

[33]. Gideon Polya, editor, "Experts; US did 9-11": https://sites.google.com/site/expertsusdid911/.

[34]. Gideon Polya, "Paris Atrocity Context: 27 Million Muslim Avoidable Deaths From Imposed Deprivation In 20 Countries Violated By US Alliance Since 9-11", Countercurrents, 22 November, 2015: https://countercurrents.org/polya221115.htm.

[35]. UN Population Division, 2015 Revision of World Population Prospects: https://esa.un.org/unpd/wpp/.

[36]. Melissa Healy, "Obesity in U.S. projected to grow, though pace slows: CDC study", LA Times, 7 May 2012: http://articles.latimes.com/2012/may/07/news/la-heb-obesity-projection-20120507.

[37]. "Stop air pollution deaths": https://sites.google.com/site/300orgsite/stop-air-pollution-deaths.

[38]. Robinson Meyer, "As the planet warms Trump's RPA pick hedges", Atlantic, 19 January 2017: https://www.theatlantic.com/science/archive/2017/01/as-the-planet-warms-senators-shrug/513746/.

[39]. Gideon Polya, "Trump's Abolition Of ObamaCare Will Kill 43,000 Americans Over 2 Trump Terms", Countercurrents, 16 March 2017: https://countercurrents.org/2017/03/16/trumps-abolition-of-obamacare-will-kill-43000-americans-over-2-trump-terms/.

[40]. Gideon Polya, editor, "Stop state terrorism" : https://sites.google.com/site/stopstateterrorism/.

[41]. Ronald Bailey, "How scared of terrorism should you be?", Reason.com, 6 September 2011: http://reason.com/archives/2011/09/06/how-scared-of-terrorism-should.

[42]. "Air pollution causes 200,000 early deaths each year in the U.S.", MIT Laboratory for Aviation and the Environment, 29 August 2013: http://lae.mit.edu/air-pollution-causes-200000-early-deaths-each-year-in-the-u-s/).

[43]. Heather Walton, David Dajnak, Sean Beevers, Martin Williams, Paul Watkiss and Alistair Hunt (King's College London), "Understanding the Health Impacts of Air Pollution in London. For: Transport for London and the Greater London Authority", 14 July 2015: http://www.london.gov.uk/sites/default/files/HIAinLondon_KingsReport_14072015_final.pdf.
[44]. Contessa Brewere, "Trump's hiring freeze hurts young veterans, who were already having trouble getting jobs", CNBC, 14 February 2017: http://www.cnbc.com/2017/02/14/trumps-hiring-freeze-hurts-young-veterans-who-were-already-having-trouble-getting-jobs.html.
[45]. Dr Janet Kemp and Dr Robert Bossarte, "Suicide data report, 2012", Department of Veterans Affairs, Mental Health Services, Suicide Prevention Program, especially Figure 3: http://www.va.gov/opa/docs/Suicide-Data-Report-2012-final.pdf.
[46]. "icasualities": http://www.icasualties.org/.
[47]. "List of countries by GDP (nominal) per capita", Wikipedia: https://en.wikipedia.org/wiki/List_of_countries_by_GDP_(nominal)_per_capita.
[48]. "Trump executive order reverses foreign abortion policy", BBC, 23 January 2017: http://www.bbc.com/news/world-us-canada-38724063.
[49]. Gideon Polya, "Truth & Boycotts, Divestment & Sanctions (BDS) Can Overcome Huge Inequities Suffered By African Americans Under American Apartheid", Countercurrents, 29 September 2014: https://countercurrents.org/polya290914.htm.
[50]. James R. Nunns, Leonard E. Burman, Jeffrey Rohaly, and Joseph Rosenberg. "Analysis of Donald *Trump's* Tax Plan", Tax Policy Center (Urban Institute & Brookings Institution), December 22, 2015: http://www.taxpolicycenter.org/publications/analysis-donald-trumps-tax-plan/full.
[51]. Jason Schenker, "What happened to Trump's 15% corporate tax rate?", Bloomberg, 1 March 2017: https://www.bloomberg.com/view/articles/2017-03-01/what-happened-to-trump-s-15-corporate-tax-rate.
[52]. "Trump budget: Military wins, environment, aid lose", Al Jazeera, 17 Match 2017: http://www.aljazeera.com/news/2017/03/trump-budget-military-wins-environment-aid-lose-big-170316054456207.html.
[53]. Andrew Taylor, "Trump's "America first" budget: More for military, border wall", Boston Globe, 16 March 2017: https://www.bostonglobe.com/news/politics/2017/03/16/trump-unveils-trillion-budget/5Wa4rRxxdgrrBBqrSueUjO/story.html.

[54]. US Department of State, "President's fiscal year 2018 budget outline released today", 16 March 2017: https://www.state.gov/r/pa/prs/ps/2017/03/268480.htm.

[55]. US Department of State, "The Department of State and USAID FY 2017 Budget", 9 February 2016: https://2009-2017.state.gov/r/pa/pl/252281.htm.

[56]. Arshad Mohammed, "Trump plans 28 percent cut in budget for diplomacy, foreign aid", Reuters, 16 March 2017: http://www.reuters.com/article/us-usa-trump-budget-state-idUSKBN16N0DQ.

[57]. Gideon Polya, "American Holocaust, Millions Of Untimely American Deaths And $40 Trillion Cost Of Israel To Americans", Countercurrents, 27 August, 2013: https://countercurrents.org/polya270813.htm.

[58]. Curt Tarnoff and Marian L. Lawson, "Foreign aid: an introduction to U.S. programs and policy", Congressional Research Service, 17 June 2016: https://fas.org/sgp/crs/row/R40213.pdf.

[59]. "Millennium project": http://www.unmillenniumproject.org/press/07.htm.

[60]. "List of development aid country donors", Wikipedia: https://en.wikipedia.org/wiki/List_of_development_aid_country_donors.

[61]. Gideon Polya, editor, "Nuclear weapons ban, end poverty & reverse climate change": https://sites.google.com/site/300orgsite/nuclear-weapons-ban.

[62]. Gideon Polya, "Body Count. Global avoidable mortality since 1950", that includes an avoidable mortality-related history of every country from Neolithic times and is now available for free perusal on the web : http://globalbodycount.blogspot.com.au/.

[63]. "Climate Genocide": https://sites.google.com/site/climategenocide/.

[64]. "Trump's inaugural address: neoliberal anti-Christ Trump's "America first": trumps Jesus' "love thy neighbour"", Countercurrents, 23 January 2017: https://countercurrents.org/2017/01/23/trumps-inaugural-address-neoliberal-anti-christ-donald-trumps-america-first-trumps-jesus-love-thy-neighbour/.

[65]. Andrew Revkin, "Will Trump scarp NASA"s climate research mission", ProPublica, 12 Decemebr 2016: https://www.propublica.org/article/will-trump-scrap-nasas-climate-research-mission.

[66]. Gideon Polya, "Humanity must pledge inescapable dispossession and custodial retribution for climate criminals", Countercurrents, 20 December2016:

https://countercurrents.org/2016/12/20/humanity-must-pledge-inescapable-dispossession-and-custodial-retribution-for-climate-criminals/.
[67]. Gideon Polya, "Exposing And Thence Punishing Worst Polluter Nations Via Weighted Annual Per Capita Greenhouse Gas Pollution Scores", Countercurrents, 19 March, 2016: https://countercurrents.org/polya190316.htm.
[68]. Gideon Polya, editor, "Stop climate crime": https://sites.google.com/site/300orgsite/stop-climate-crime.

SECTION H: SOLUTIONS – TRUTH, SOCIAL HUMANISM, CARBON PRICE, ACCOUNTABILITY, DRAWDOWN TO 300 PPM CO$_2$, RENEWABLE ENERGY, ENERGY EFFICIENCY, RE-AFFORESTATION, POPULATION CONTROL, ECONOMIC DE-GROWTH & BIOLOGICAL SUSTAINABILITY

This final Section H deals with my writings in relation to Solutions, to what actions should be taken to deal with the ever-worsening Climate Crisis and Climate Genocide [1-82]. The major elements of the required solutions are as listed in the title of this Section, namely (1) truthfulness, social humanism and science not spin, (2) adoption of an exactly accounted carbon price, (3) holding climate criminals responsible for their crimes and ballooning Carbon Debt, (4) draw-down of atmospheric CO_2 from the present 415 ppm CO_2 to a safe and sustainable level of circa 300 ppm CO_2, (5) 100% renewable energy with adequate energy storage, (6) energy efficiency, (7) re-afforestation, (8) population control and decrease, (9) economic de-growth, and (10) biodiversity preservation and biological sustainability.

My own country, climate criminal Australia, presently ranks bottom of the list as the world's worst major country for climate change inaction (see Section F). In short, (1) there is extraordinary lying by omission and lying by commission by neoliberal, pro-fossil fuels politicians backed by the mendacious, effective climate change denialist Murdoch media empire as exampled by the false assertions of "clean coal", "gas is clean energy" and "gas is cleaner energy" (coal and gas are dirty energy and depending upon the degree of methane leakage gas burning can be (and very likely is) dirtier GHG-wise than coal burning); (2) both the Coalition Government and the Labor Opposition reject application of a Carbon Price and indeed support a "negative Carbon Price" by hugely subsidizing GHG polluting activities; (3) climate criminals disproportionately involved in Domestic and or Exported GHG pollution are not held responsible for their crimes but conversely are heavily subsidized in their climate criminality; (4) both the Coalition Government and Labor Opposition are committed to endless and unlimited coal, gas and iron ore exports, and while Labor commits to the world-wide near-consensus of "net zero emissions by 2050", the climate criminal Coalition Government refuses to do so, and merely suggests zero net emissions sometime in the latter half of the 21st century i.e. rather than draw-down to 300 ppm CO_2 they are committed to continuing increase in atmospheric CO_2 from the present 415 ppm CO_2; (5) the Labor Opposition wants 100% renewable energy by 2050 but the Coalition still fights implementation of renewable energy tooth and nail to the extent that it now demands that the Labor-installed Australian Renewable Energy Authority (ARENA) and the Clean Energy Finance Corporation (CEFC) must fund dirty energy projects; (6) Australia with 0.33% of the world's population contributes 5.4% of global GHG pollution annually; (7) Australia is among world leaders in land-clearing; (8) neoliberal Australia has a very successful "endless growth" economy based on population increase through immigration; (9) Australia is utterly opposed to economic de-growth and has a speciescidal and ecocidal policy of "jobs and growth" based on endless and massive population growth; and (10) neoliberal and climate criminal Australia rejects sustainability - Australia leads the world in animal species extinction, and recently the National

Party (farmer's party) threatened to bring down the Liberal Party-National Party Coalition Government of New South Wales (NSW) over the asserted right of farmers to drive the threatened and iconic koala to extinction through further clearing of land stolen from the original Indigenous people in the ongoing Australian Aboriginal Genocide (see "Aboriginal Genocide": https://sites.google.com/site/aboriginalgenocide/).

These 10 key areas of Solutions to the worsening Climate Crisis and Climate genocide are detailed below.

(1). Truth and social humanism

Science-based tackling of man-made climate change involves zero tolerance for lying that utterly subverts science-based risk management [80].Yet lying is entrenched in climate criminal Trump America and its fervent climate criminal lackey Australia, the Mainstream media of which is dominated by the mendacious, US-owned Murdoch media empire [17]. Trump lying about everything from Covid-19 to man-made climate change has been quantified by the Washington Post that has determined that Trump has told over 20,000 lies so far in his administration (Michael Tomasky, "Why does Trump lie?", New York Times, 11 June 2020: https://www.nytimes.com/2020/06/11/opinion/trump-lies.html).

As for Australia, in 2015 I published the following thanatological analysis **(carbon-burning-related deaths in bold)**: "Australia has invaded 85 countries, with 30 such invasions associated with genocidal atrocities, and has an appalling record of child abuse both at home and abroad, with as many as 25% of Australians suffering sexual abuse as children. Now on top of draconian anti-terrorism laws grossly violating human rights, Australian doctors, nurses, teachers and care-givers face 2 years in prison for reporting ongoing physical, psychological and sexual abuse of children and other inmates of Australian refugee concentration camps… circa 83,000 Australians die preventably each year, the updated 2015 breakdown being as follows… (1). 26,000 annual Australian deaths from adverse hospital events. (2). 17,000 obesity-related

Australian deaths annually. **(3). 15,500 smoking-related Australian deaths annually. (4). 10,000 carbon burning pollution-derived Australian deaths annually.** (5). 5,600 Australians die alcohol-related deaths annually. (6). 4,000 avoidable Indigenous Australian deaths annually. (7). 2,100 Australian suicides annually. **(8). 1,400 Australian road deaths annually.** (9). 1,400 drug-related deaths annually (this including 630 Australian opiate drug-related deaths annually with 570 annually linked to US restoration of the Taliban-destroyed Afghan opium industry). (10). 300 Australian homicides annually, of which about 100 are domestic homicides and about 75 are women victims (noting that 0.5 million Australian women suffered domestic violence in 2005)… And now this Orwellian, Kafkaesque authoritarianism has spread to criminalizing the fundamental ethical obligation of doctors and other health professionals to report endangerment and abuse of their patients…. Indeed one is reminded of Samuel Butler's novel "Erewhon" (roughly "nowhere" spelled backwards) in which social support is lavished on dangerous sociopaths while the sick are scorned as outcasts" [33]. To this list I could add **500 heat stress deaths annually linked to climate change impacting Australia** - more than 500 Australians presently die from heat stress each year with this disaster disproportionately impacting the elderly, the impoverished and Indigenous First Nations peoples [50]. I have written numerous, detailed accounts of Australian mendacity in relation to climate change impacts see [16, 17, 19, 20, 32, 34, 42, 51].

However it is not just Trump America and Trumpist Australia who are lying. Thus atmospheric CO_2 continues to rise remorselessly (US NOAA, "Monthly average Maua Loa CO_2": https://www.esrl.noaa.gov/gmd/ccgg/trends/) and there is a quasi-linear increase over the last 40 years in over 20 climate-related parameters [48]. Continuing lying by omission, lying by commission, obfuscation and cowardly "softening" of message by Mainstream journalist, editor, politician and commentariat presstitutes is driving Humanity and the Biosphere over the precipice. Indeed while the 2015 Paris Climate Change Conference committed to ideally a 1.5C temperature rise and no more than a catastrophic plus 2C temperature rise, according to the UN the national commitments made correspond to a plus 3.2C temperature rise ("UN emissions report: World on course for more than 3 degree spike, even if climate commitments are met", UN News, 26 November 2019: https://news.un.org/en/story/2019/11/1052171).

Polya's 3 Laws of Economics mirror the 3 Laws of Thermodynamics and are (1) Price minus COP (Cost of Production) equals profit; (2) Deception about COP strives to a maximum; and (3) No work, price or profit on a dead planet. These fundamental laws help expose the failure of neoliberal capitalism in relation to wealth inequality, massive tax evasion by multinational corporations, and horrendous avoidable deaths from poverty and pollution culminating in general ecocide, speciescide, climate genocide, omnicide and terracide [34]. Neoliberalism demands maximal freedom for the smart and advantaged to unsustainably exploit natural and human resources for maximal private profit via private property ownership to obscene degrees and the "hidden hand" of the market. In stark contrast, social humanism (socialism, eco-socialism, democratic socialism, human rights observant communism, the welfare state, Universal Basic Income) strives to maximize happiness, dignity and opportunity for everyone by culturally cognizant and evolving intra-national and international social contracts (see Brian Ellis, "Social Humanism. A New Metaphysics"; [46]).

Neoliberal greed has taken Humanity and the Biosphere to the precipice, and no less than a Climate Revolution now (non-violent, of course) and a rapid transition to social humanism is required to save the Planet, as I have repeatedly advocated in articles and websites [6, 14, 23, 29, 30, 34, 36, 39, 41, 44-46, 63, 72, 78-82].

(2). Carbon Price.

As discussed in Section E, humane, science-informed people from climate scientists to climate economists and science-trained Pope Francis demand that the environmental and social cost of GHG pollution be "fully borne" by the polluters.

However human behaviour operates on a continuum from altruism to neoliberal greed, from "love the stranger" and "love thy neighbour" to deadly inequity and genocidal xenophobia. As detailed by Thomas Piketty in "Capitalism in the Twenty First Century", there is massive and growing inequity in the world, with the One Percenters owning 50% of the world's wealth, inequity being bad for the economy (the poor cannot afford the goods and services they produce), and inequity being bad for democracy (Big Money buys votes). The solution lies in better sharing as for example by a Universal Basic Income (UBI) or a One Percenter-constraining Annual Wealth Tax. The Muslim World has had a 2.5% annual wealth tax (zakat) for 1,400 years, and I estimated in 2015 that an annual wealth tax of 4% would abolish the Global Avoidable Mortality Holocaust in which 15 million people die avoidably from deprivation each year in the Developing World (minus China) [4]. However One Percenter power means that the message is hidden by the One Percenter-beholden Mainstream media and the One Percenter-programmed Google Robot - the deadly inequity remains and is worsening.

Neoliberal deceit about Cost of Production aside (after Polya's Second Law of Economics [34]), it is generally accepted (notably by the Ninety Nine Percenters) that the fully costed, real and damage-related Cost of Production should be incorporated into the price of things. The damage-related Carbon Price is $200 per tonne CO_2-equivalent (CO_2-e) as determined independently in the US and the UK. Applying this to world GHG pollution one find that the world has an inescapable Carbon Debt of $200-$250 trillion that is increasing at about $16 trillion per year [28, 29, 32, 43]. Unlike conventional debt that can be expunged by default, bankruptcy or printing money, Carbon Debt is inescapable because, for example, unless sea walls are built coastal cities and arable land will be inundated [1]. Unfortunately, the global average Carbon Price is only about $2 per tonne CO_2-e [47].

As if this situation were not impossible enough, we have massive biodiversity loss in our present Anthropocene Era, noting that every species is priceless. The animal species extinction rate in the present Anthropocene Era is 100-1,000 times greater than normal (Phillip Levin and Donald Levin, "The real biodiversity crisis", American Scientist, January-February 2002: http://www.americanscientist.org/issues/pub/the-real-biodiversity-crisis). Further, Balmford and colleagues have estimated that wild nature contribute substantially to global annual GDP. The total economic return (TEV) from major biomes (ecological systems) studied can be typically about 50% greater when there is sustainable use, and the economic return from preserving what is left of wild nature is over 100 times the cost of so doing. The total economic value (TEV) of wild nature (e.g. pollination, forestry, fisheries, tourism) was estimated in 1997 as about $18-$60 trillion (average $38 trillion) as compared to a total World GDP (2007) of about $55 trillion (see Andrew Balmford et al., "Economic reasons for conserving wild nature", Science, 9 August 2002: Vol. 297. no. 5583, pp. 950 ñ 953, DOI: 10.1126/science.1073947: http://www.sciencemag.org/cgi/content/abstract/297/5583/950 and http://www.uvm.edu/giee/publications/Balmford_et_al.pdf).

(3). Punishing climate criminals.

Over the years I have written extensively about assessing climate criminality and the judicial punishment of climate criminals [2, 16, 19-21, 34-43, 47, 50, 51, 63, 66-82]. It will be little comfort to Humanity when One Percenter climate criminals disproportionately responsible for the near-terminal catastrophe are brought to account through judicial processes. However that said, the envisaged judicially-mediated punishment of climate criminals would involve (a) exposure of the immense financial cost (Carbon Debt), the environmental damage (speciescide and ecocide), and the immense morbidity (sickness) and mortality (death) associated with their crimes; (b) making them pay financially for their crimes, and (c) judicially-assessed custodial punishment of up to life imprisonment for their climate crimes [41, 72]. One notes that science-trained Pope Francis has declared that the environmental and social cost of pollution should be "fully borne" by the polluters (see Section E).

A dozen years ago I wrote (2007): "It has recently been estimated for Ontario, Canada, that the actual price of coal-based electricity (taking into account the environmental and human cost) can be about 4 times higher than the present commercial cost. However this Canadian estimate was conservative and may have under-estimated the additional environmental cost due to climate change by a factor of ten (10) [Paul Gipe, "Coal's high environmental and social cost in Ontario", Wind Works, 5 July 2005: http://www.wind-works.org/cms/index.php?id=625&tx_ttnews%5Btt_news%5D=523&cHash=828e8d2ac6e89b4224443dd64478a3bc]. Nevertheless this actual cost estimate indicates that a number of renewable options (notably wind power and concentrated solar power, CSP) are already competitive on current costings with the true cost of coal-based electricity i.e. for economic reasons the coal should stay in the ground. Those greedy, corrupt, intrinsically racist climate criminals (climate terrorists) that ignore the science, the technology and economics and remain committed to fossil fuel-based energy to the cost of Humanity will eventually get the bill for the full cost of their racist irresponsibility and the consequent climate genocide. However the Western world is dominated by lying, racist, Bush-ite media who are still giving the climate criminals a free run to pollute the planet at the expense of the Developing World. The words of the world's most eminent scientists, technologists and economists are failing under the weight of corporate Mainstream media lies and spin" [2].

Dr James Hansen (NASA and thence Columbia University) addressing the US Congress in 2008: "CEOs of fossil energy companies know what they are doing. And are aware of long-term consequences of business as usual. In my opinion, these CEOs should be tried for high crimes against humanity and nature. Conviction of ExxonMobil and Peabody Coal CEOs will be no consolation, if we pass on a runaway climate to our children. Humanity would be impoverished by ravages of continually shifting shorelines and intensification of regional climate extremes. Loss of countless species would leave a more desolate planet. If politicians remain at loggerheads, citizens must lead. We must demand a moratorium on new coal-fired power plants. We must block fossil fuel interests who aim to squeeze every last drop of oil from public lands, off-shore and wilderness areas. Those last drops are no solution. They yield continued exorbitant profits for short-sighted self-serving industry, but no alleviation of our addiction or long-term energy source" (James Hansen, "Global warming twenty years later: tipping points near", Address to the US Congress, 2008: http://www.columbia.edu/~jeh1/2008/TwentyYearsLater_20080 623.pdf).

Over the years I have argued that climate criminals – individuals, corporations and nation states - should be punished. Individuals should be punished by exposure, being forced to pay their Carbon Debt, and indeed by judicially-imposed custodial penalty for their complicity in climate genocide. Corporations should be punished by exposure, being forced to pay their dishonestly evaded Carbon Debt, and by Boycotts, Divestment and Sanctions (BDS). Nation states and their responsible leaders and officials should be punished by exposure of their disproportionate GHG pollution as well as by Green Tariffs, International Criminal Court (ICC) prosecutions, International Court of Justice (ICJ) litigations, and by Boycotts, Divestment and Sanctions (BDS) [35, 37, 41, 43].

Prevention of resolute corporate or government climate criminality ultimately requires political change (a peaceful Climate Revolution to establish social humanism and decent, sustainable sharing of the Earth's limited resources) [62], but before this is attained Boycotts, Divestment and Sanctions (BDS) against climate criminal individuals, corporations and countries will be crucial. However a potentially very useful tactic to prevent ongoing climate criminality would be the resolute promise of severe and retrospective punishment in the special circumstance of a worsening Climate Genocide. Thus in 2019 I argued (2019): "There is a real prospect of Climate Justice involving resolutely promised prosecutions of climate criminals and confiscation of their assets (as presently happens to traders in illicit drugs). There is no legislative retrospectivity required in this – climate murder, climate homicide and climate genocide are presently just as deadly as "conventional" murder, homicide and genocide. Something of the order of 1 million people already die climate-related deaths annually, with this set to soar to an average of 100 million such deaths per year this century if man-made climate change is not requisitely addressed. Of course such punishments will be scant comfort to the loved ones of those who have died. However most importantly it is the resolute threat of prosecutions of climate criminals that may finally force urgent climate action. There needs to be a new globally-endorsed social contract to the effect that significant, present-day climate criminals will be inescapably held to account with judicially-imposed penalties ranging up to life imprisonment with hard labour and confiscation of all assets" [41].

(4). Draw-down of atmospheric CO_2 to circa 300 ppm CO_2 for a safe and sustainable planet for all peoples and all species.

While climate scientists say that there must be rapid implementation of "negative CO_2 emissions" (i.e. atmospheric CO_2 draw-down to a safe and sustainable level) [5, 8, 25, 32, 39, 40], in climate criminal Australia the Labor Opposition has a policy of only "net zero emissions by 2050", and the pro-gas, pro-coal Coalition (COALition) Government grudgingly suggests that this limited agenda might happen in the second half of the 21st century. While the excellent 350.org organization demands an atmospheric CO_2 draw-down to no more than 350 parts per million CO_2 (350 ppm CO_2), 300.org reports the view of many climate scientists and science-informed activists that there must be draw-down of atmospheric CO_2 to circa 300 ppm CO_2 for a safe and sustainable planet for all peoples and all species [5, 8, 60, 61].

Wherever I go I wear a big, bold, black on white badge stating "300 ppm CO_2" and people say that it is just not going to happen. But it has to happen for a safe and sustainable environment for all peoples and all species.

"Atmospheric CO_2 draw-down" is very expensive, noting that a large amount of man-generated CO_2 has ended up in the consequently acidifying ocean [58]. Thus in a detailed 2018 analysis I concluded: "Intergenerational Theft – For Every $1 For Coal Today Future Generations Will Pay $1-$14 To Sequester CO_2" [32]. Before entering into gigantic, planet-wide geo-engineering to remove CO_2 from the atmosphere, the sane move would be to cease CO_2 pollution – but it continues remorselessly in all major pollution sectors. One notes that in 2020 the pro-coal Coalition Government of climate criminal Australia wants to invest billions into expensive and presently commercially infeasible Carbon Capture and Storage (CCS) as well as into gas-fired and coal-fired powers stations. Criminal lunacy.

Leading climate scientist Dr James Hansen has quantitated the cost of reducing atmospheric CO_2 by the latest available technology (2018): "Recently Keith et al. (2018) achieved a cost breakthrough in carbon capture, demonstrated with a pilot plant in Canada. Cost of carbon capture, not including the cost of transportation and storage of the CO_2, is $113-232 per ton of CO_2. Thus the cost of extracting 1 ppm of CO_2 from the atmosphere is $878-1803 billion. In other words, the cost, in a single year, of closing the gap between reality and the IPCC scenario that limits climate change to +1.5°C is already about $1 trillion. And that is without the cost of transporting and storing the CO_2, or consideration of whether there will be citizen objection to that transportation and storage. This annual cost will rise rapidly, unless there is a rapid slowdown in carbon emissions. This annual cost is not being paid, and common sense tells us that it will not be paid in the future as the cost rises to astronomical levels. Instead the mess is left for young people. Continued high fossil fuel emissions sentences young people to a massive and likely implausible clean-up and growing deleterious climate impacts. The tragedy of this situation is that it is unnecessary. Honest pricing of energy, economists and common sense concur, would move us toward carbon-free energy" (James Hansen, "Climate change in a nutshell: the gathering storm", Columbia University, 18 December 2018: http://www.columbia.edu/~jeh1/mailings/2018/20181206_Nuts hell.pdf).

A variety of technologies have been proposed for large-scale "atmospheric CO_2 draw-down" including (a) Carbon Capture and Storage, (b) Ocean storage of power plant-derived CO_2, (c) Algal photosynthesis, (d) Ocean algal photosynthesis per fertilization of oceans to promote photosynthetic fixation of CO_2 into biomass, (e) Re-afforestation, (f) Accelerated Weathering of Limestone (AWL) to generate bicarbonate, (g) Biochar (carbon, charcoal) production and storage through anaerobic pyrolysis at 400-700C, (h) Direct Air Capture (DAC) in alkaline solution, and (i) Mineral carbonation of pulverized suitable CO_2-reactive minerals. Such schemes could be complemented by so-called geo-engineering approaches to global warming with variously serious ecological and biological implications including (i) shades in space, (ii) adding sulphate aerosols to the stratosphere or troposphere to promote global dimming, (iii) changing albedo by large-scale plant monocultures or a-biotically (e.g. by painting roofs or roads white) (for details see Section A, [39, 58]). However it is clear that major diminution of GHG pollution can occur through vegetarian diet [38], decrease in population, and economic de-growth [40].

(5). 100% renewable energy ASAP with adequate energy storage and massive reduction of all GHG pollution.

The atmospheric CO_2 is remorselessly increasing notwithstanding increasingly dire warnings from world scientists over 3 decades since the late 1980s (US NOAA, "Monthly average Maua Loa CO_2": https://www.esrl.noaa.gov/gmd/ccgg/trends/). I concluded (2020): "A detailed analysis shows that global greenhouse gas (GHG) emissions in all key economic sectors of energy production and non-energy production are increasing whereas the worsening Climate Emergency demands urgently decreasing GHG emissions and ultimately net zero GHG emissions ASAP. The world is driving headlong in the wrong direction to certain disaster" (Gideon Polya, "Wrong way go back – global sectoral greenhouse gas emissions are all in the wrong direction", Countercurrents, 20 March 2020: https://countercurrents.org/2020/03/wrong-way-go-back-global-

sectoral-greenhouse-gas-emissions-are-all-in-the-wrong-
direction/).

Renewable energy options include solar PV (photovoltaic cells),
floating solar (solar PV floating on hydroelectricity dams), solar
thermal (solar energy collection as molten salts), wind,
geothermal, tidal, wave, and hydroelectricity. Energy storage
backup is required and can be provided by batteries, molten
salts, gas compression and hydrological energy storage in
elevated dams.

The nuclear option is not renewable because required uranium
and thorium ores are non-renewable. Further, excepting the
nuclear fission step, all parts of the nuclear cycle from building
the plant and mining the ore to safely disposing of the waste and
safely retiring the plant, generate large amounts of CO_2 in a
carbon economy. The fast breeder reactor route involves
extremely efficient use of uranium through generation of
plutonium, but a plutonium economy would be a security and
civil rights nightmare. Fusion technology is in its infancy but we
already have a mighty fusion reaction, the Sun, that is 150
million kilometres away, with the Earth being further protected
from solar high energy particles by its magnetosphere.

In 2010 the engineer-based Australian organization Beyond
Zero Emissions (BZE) produced a detailed report showing that
Australia could achieve 100% renewable energy by 2020. Their
plan involved 40% wind energy, 60% concentrated solar
thermal (CST) with molten salts energy storage for 24/7
baseload power, biomass utilization and a hydroelectric backup
(for days of no wind and low sunshine), and a HV DC and HV
AC national power grid. The BZE scheme was costed at $370
billion over 10 years, with roughly half spent on CST, one
quarter on wind and one quarter on the national electricity grid.
I used to give power-point lectures on their behalf around
Greater Melbourne [12, 59].

In 2010 Professor Peter Seligman (a friend of mine) published
an important book entitled "Australian Sustainable Energy by
the Numbers", University of Melbourne, 2010. This book was

Australia-specific and inspired by a book by the former UK Chief Scientific Advisor, Professor David MacKay, entitled "Sustainable Energy-Without the Hot Air", Chicago Review Press, 2009. The Seligman scheme for 100% renewable energy for Australia has been set out by a top electrical engineer (a major player in development of the bionic ear) and involves wind, solar thermal, other energy sources, hydrological energy storage (in dams for pumped sea water on the Nullabor Plain in Southern Australia), an HV AC and HV DC electricity transmission grid, and a cost over 20 years of $253 billion [11, 59].

Ignoring cost-increasing energy storage and transmission grid costs and cost-decreasing economies of scale for a 2- to 10-fold size increase, I made 2 similar cost estimates for installation of wind power for 80% of Australia's projected 325,000 GWh of annual electrical energy by 2020: (1) 90,000 MW capacity, 260,000 GWh/year, $200 billion over 10 years (10-fold scale-up from GL Garrad Hassan) and (2) 96,000 MW, 260,000 GWh/year, $144 billion (2-fold scale up from BZE [57]). To put this into perspective, Australia has spent about $200 billion addressing the Covid-19 pandemic (over 900 Australian Covid-19 deaths), and according to the IMF the World will have spent $9 trillion addressing 1 million global Covid-19 deaths [54, 55].

Energy is a major part of the GHG-generating human economy. However land-use, deforestation and methane generation are major contributors. Taking land-use into account, and adopting a Global Warming Potential (GWP) for methane (CH_4) on a 20 year time frame of about 80 relative to the same mass of CO_2, World Bank analysts upwardly revised annual global GHG pollution from 44 Gt CO_2-e to 64 Gt CO_2-e. Cessation of land clearing, methanogenic livestock production and other methanogenic operations would yield massive reduction of GHG pollution.

(6). Energy efficiency, rail not road, and a vegetarian diet for everybody.

We commonly think of energy efficiency in the context of mechanical efficiency (e.g. of cars and refrigerators), and energy saving from use of devices as in "Turn off the switch!" However the need to reduce atmospheric CO_2 back to the pre-Industrial Revolution level of about 300 ppm CO_2 (see (4) above) exposes the huge energy inefficiency of the "futile cycle" of CO_2 draw-down with simultaneous gross CO_2 emissions [5, 53, 59-61]. This "draw-down with simultaneous GHG pollution" inefficiency is significantly addressed by 100% renewable energy ASAP (Section G; [11, 12, 59].

A major area of energy inefficiency is in fossil fuel-powered private motor transport rather than renewable energy-powered public transport. "Rail not road" is a fundamental mantra of climate activists. A fundamental energy inefficiency is involved in carting circa 80 kilogram humans around in a circa 2,000 kilogram car. The energy inefficiency gets worse when one considers the $6 trillion spent on "oil wars" in the 21st century alone. Thus it is estimated that 32 million Muslims have died from violence, 5 million, or from imposed deprivation, 27 million, in 20 countries invaded by the US Alliance since the US Government's 9/11 false flag atrocity that killed 3,000 people [55, 56]. To this one must add the enormous cost of road transport infrastructure and the human cost of road accidents and air pollution. Thus road accidents kill 1.35 million people each year (WHO, "Road traffic injuries", 7 February 2020: https://www.who.int/news-room/fact-sheets/detail/road-traffic-injuries), and 9 million people die annually from air pollution [73]. Applying a conservative global risk avoidance-based Value of a Statistical Life of (VSL) of $1 million per person puts the annual cost of road deaths and air pollution deaths at $1.35 trillion and $9 trillion, respectively, noting that the pre-Covid-19 World GDP (nominal) was about $80 trillion.

Every species and ecosystem is priceless, and so speciescide and ecocide must be considered somehow in assessing the environmental and social cost of roads in addition to transportation inefficiency, road deaths, and air pollution. Thus in my own neighbourhood in Melbourne the Labor Victorian State Government in collusion with the Coalition Federal Government is about to spend US$12 billion on a huge highway called the North East Link. I protested to all councillors in the various Local Government areas affected, and in a detailed essay stated (2020): "Beautiful Melbourne (capital of Victoria), that vies with Vancouver and Vienna for being the most liveable city in the world, is threatened with a fiscally, morally and environmentally perverse [AUD]$16 billion super-highway North East Link (NEL) project that is set to carve a swathe through its beautiful, tree-rich City of Banyule (aka Heidelberg), the former home of the famed Heidelberg School of Australian Impressionist painters, with 25,000 trees to be felled including an ancient River Red Gum that I adore" [52]. The Green Left Weekly (2020): "Included in this destruction will be the endangered Studley Park Gum, a rare natural hybrid between the River Red Gum and a Swamp Gum… The Banyule municipality and surrounding area provides habitat over winter for the critically endangered swift parrot, a beautiful, bright green bird that spends summer in Tasmania… The area is also home to the powerful owl, another threatened species. The powerful owl is an apex predator that eats possums and requires a large territory" (Mary Merkenish, "North East Link an "ecological misadventure"", Green Left Weekly, 1 September 2020: https://www.greenleft.org.au/content/north-east-link-ecological-misadventure-says-expert).

A major energy efficiency measure would be adoption of a vegetarian diet. Thus I wrote (2017): "Annual global greenhouse gas (GHG) emissions presently total 64 billion tonnes CO_2-equivalent with methanogenic livestock production contributing over 50%. With annual emissions of CO_2 per se from industry at a record high, and with Humanity and the Biosphere existentially threatened by the Methane Bomb of the warming Arctic in coming decades, we must all urgently adopt a vegetarian or vegan diet to help save the Planet. World Bank analysts have carefully re-evaluated the contribution of livestock production to world annual GHG pollution and found that the world's annual total rose from 41.76 billion tonnes CO_2-

equivalent (CO_2-e) as estimated by the Food and Agricultural Organisation (FAO) to 63.80 billion tonnes CO_2-e, with livestock production contributing over 51% of the higher figure… Thus consider the following conversion efficiencies (kg grain to produce 1 kg gain in live weight): herbivorous farmed fish (e.g. carp, tilapia, catfish; less than 2), chicken (2), pork (4), and beef (7) … Decent people must demand cessation of meat eating – in the absence of requisite abolition of about 32 Gt CO_2 annually of industrial pollution (from carbon fuel burning and cement manufacture), a vegetarian or vegan diet for all Humanity is urgently required now to help save the Planet. And it must be done urgently, now, because a predicted release of an estimated 50 Gt methane (50 x 105 = 5,250 Gt CO_2-e) from the steadily warming East Siberian Arctic Shelf in coming decades would mean doom for Humanity and the Biosphere" [38]. The Biofuel Genocide involving food for fuel is not just inefficient but also an obscenity in a hungry and resource-limited world [68].

Resource limitation and energy efficiency as well as social equity considerations compel the need for eco-socialism and just, sustainable sharing. One hopes that the altruism shown in the Covid-19 pandemic will lead to a post-Covid-19 Green New Deal. I commented (2020): "Comprehensive social distancing and economic lockdown measures required to suppress Covid-19 have been associated with rapid socialism-style adoption of a needs-based economy (essential services), a decrease in carbon fuel burning and air pollution, a widened social safety net, and on-line learning. Post-Covid-19 one envisages a socialist needs-based economy, zero emissions, Universal Basic Income (UBI), a Green New Deal (GND), and free university education. Unfortunately neoliberal Australia may well reject this golden opportunity. Every cloud has a silver lining and the deadly Covid-19 pandemic is no exception to this aphorism. Because the coronavirus can infect anyone, from princes to paupers, the Covid-19 pandemic has resulted in a global blooming of altruism, the notion "we are all in this together" and concerted action to save lives – a course of emergency action that in a word is "socialism" but which has not been opposed by the One Percenters who in normal circumstances ruthlessly dominate media and the global economy. The big question is whether this altruism will continue unabated in the post-Covid-19 era or whether the neoliberal One Percenters will restore control with business-as-usual, unsustainable economic growth, merciless austerity and burgeoning inequality" [54]. In Australia that (together with New Zealand and East Asia) has laudably been among world leaders in limiting Covid-19 deaths in the Covid-19 pandemic, we now have the answer: the neoliberal Coalition Government has decided to follow successful suppression of the coronavirus with an appalling post-Covid-19 wealth transfer from the poor to the rich, and Apartheid-style suppression of the unemployed, the young, students, women, older unemployed, children, homeless, disabled, the aged, and extremely vulnerable non-citizen migrant workers, asylum seekers and overseas students.

Acclaimed Indian writer and humanitarian activist Arundhati Roy has urged a post-Covid-19 transformation (2020): "Whatever it is, coronavirus has made the mighty kneel and brought the world to a halt like nothing else could… Nothing could be worse than a return to normality. Historically, pandemics have forced humans to break with the past and imagine their world anew. This one is no different. It is a portal, a gateway between one world and the next. We can choose to walk through it, dragging the carcasses of our prejudice and hatred, our avarice, our data banks and dead ideas, our dead rivers and smoky skies behind us. Or we can walk through lightly, with little luggage, ready to imagine another world. And ready to fight for it" (Arundhati Roy, "The epidemic is a portal", Financial Times, 4 April 2020: https://www.ft.com/content/10d8f5e8-74eb-11ea-95fe-fcd274e920ca).

(7). Re-afforestation and sequestration of forest and agricultural waste as biochar.

There is increasing re-afforestation in Europe and North America but massive and criminal de-forestation continues in the tropics notably in Central America, the Amazon rainforest, West Africa and South East Asia. I have written extensively about re-afforestation and cessation of land clearing as vital for tackling the Climate Crisis [9, 14, 48, 50, 51, 54].

I summarized the overall global de-forestation statistics thus (2019): "Total forest (million hectares): 4.110 (1992) to 3.990 (2016), –2.9% overall, -1.2% per decade, -0.005 per year; 3.92 million hectares by 2030…. Brazilian Amazon forest loss (millions hectares per year) was in the range 1-3 million hectares per year in the period 1988-2008 (very scattered data) and then fell to about 0.7- 0.9 million hectares per year in the period 2009-2018 (- 24.3% /10 years, -2.4% per year)" [48]. However the GHG pollution inherent in this loss is due to (a) CO_2 from combustion, (b) loss of photosynthetic CO_2 sequestration, (c) methane (CH_4) generation from anaerobic bacterial metabolism in resultant swamps, and (d) methane generation from methanogenic livestock raised on pasture replacing forest (as obscenely so in the disappearing Amazon rain forest and land clearing in climate criminal Australia).

Thus I described the 4-decade, quasi-linear increase in ruminant livestock thus (2019): "Ruminant livestock (billion individuals): 2.86 billion (1979) to 3.93 billion (2017) (+ 8.72% per decade, +0.87% per year and +0.028 per year), noting an apparent, short, low-growth period in the middle 1990s. Assuming the 1979-2017 average trend yields a value from extrapolation of 4.29 billion individual ruminants in 2030" [48].

As for solutions, One Tree Planet states: "Scientists have always speculated about how trees can help end the current climate crisis, but the question has always been how many. A new study from the Crowther Lab out of Swiss University ETH Zurich, has finally come up with an answer: 1 trillion more trees on Earth could sequester more than 60% of anthropogenic carbon emissions" (One Tree Planet, "New research: 1 billion hectares of forest could save the Planet ", One Tree Planted: https://onetreeplanted.org/blogs/stories/research-trees-climate). Of course it is the type of tree and ecosystem that is important. Thus in climate criminal Australia that is among world leaders in land-clearing [51], the South East Australian temperate *Eucalyptus regnans* (Mountain Ash) forests are the best forest carbon sinks in the world at about 1,900 tonnes carbon per hectare (H. Keith, B.G. Mackey and D. B. Lindenmayer., "Re-evaluation of forest biomass carbon stocks and lessons from the world's most carbon-dense forests", PNAS, 14 July 2009, 106 (28) 11635-11640: https://www.pnas.org/content/106/28/11635).

The terms "forestry waste" and "agricultural waste" are controversial because such "waste" can contribute to forest floor biology and eventually to soil carbon. Bearing that caveat in mind, forest and agricultural carbon waste can be sequestered as biochar (charcoal, carbon) through the anaerobic pyrolysis of cellulosic material at 400-700C. It is estimated that the amount of biochar that could potentially be generated annually from such waste is 12 Gt i.e. more than the 9 Gt of annual C pollution [9, 48, 51]. The huge forest fires in California, Siberia, Western Europe, Brazil, Indonesia and Australia are linked to climate change through increased temperature, drought and dangerously low humidity. These fires are converting large amounts of ground cover biomass to CO_2 whereas a practical solution is to harvest such forest floor biomass and sequester the carbon safely as biochar [9, 51]. Of course in a hungry world arable land cannot be diverted to generation of biochar, no more than it should be obscenely diverted to generate biofuel i.e. food for fuel Biofuel Genocide [68].

Eminent economist Lord Nicholas Stern (2007): "The evidence on the seriousness of the risks from inaction is now overwhelming. We risk damage on a scale larger than the two world wars of the past century. The problem is global and the response must be collaboration on a global scale. The rich countries must lead the way in taking action… There should be an international programme to combat deforestation, which contributes 15-20% of emissions. For $10bn-$15bn per year, half the deforestation could be stopped" (Alison Benjamin, "Stern: Climate change a "market failure"", Guardian, 29 November 2007: https://www.theguardian.com/environment/2007/nov/29/climate change.carbonemissions).

(8). Population control and population decrease.

Professor Dabo Guan (School of International Development, University of East Anglia, UK) (2016) has commented thus on inescapable limits to growth: "For everyone in the world to have an American lifestyle, we would need seven planets, and three to live as Europeans" (Irene Banos Ruiz, "China's new love affair with dogs – as pets, not food – presents environmental problems", DW, 21 June 2016: https://www.dw.com/en/chinas-new-love-affair-with-dogs-as-pets-not-food-presents-environmental-problems/a-19197523). With a global population in 2020 of 7.6 billion the world is already over-populated by a factor of 2 as determined using coral reefs as a "canary in the mine". Thus I wrote (2018): "Drastic decrease in population is required for cessation of the present appalling level of ecosystem destruction (ecocide) that is associated with catastrophic species loss (speciescide) and leading to omnicide and terracide. If we take world coral as a "canary in the coal mine" then the 320 ppm CO_2 at which coral reefs started to decline was reached in 1965, at which time the world's population was 3.340 billion as compared to the present 7.5 billion. One can therefore plausibly suggest that given the present carbon economy the world's population needs to roughly halve for a safe and sustainable environment for all peoples and all species" [40].

Back in 1998 it was apparent that over-population was a key factor in an emerging Climate Emergency. In Chapter 16, "Global warming and the unthinkable world of 2050", in my book "Jane Austen and the Black Hole of British History. Colonial rapacity, holocaust denial and the crisis in biological sustainability", I wrote (1998): "Global warming and planetary homeostasis. There has been mounting speculation in the last century - and particularly in recent decades - about global warming due to increasing atmospheric carbon dioxide and other "greenhouse" gases due to a combination of industrial activity, fuel burning by internal combustion engines, deforestation and livestock agriculture. The Swedish physical chemist Svante Arrhenius (of thermodynamic Arrhenius Plot fame) suggested one hundred years ago that a doubling of atmospheric carbon dioxide (CO_2) would lead to an increase of average global temperature of about 5° C (degrees Celsius). The current consensus appears to be that a doubling of atmospheric CO_2 in the absence of other mitigating factors will cause an increase of about 3° C in mean global temperature… Of course we can surely all grasp the obvious point that nearly everyone is apparently missing - that there is a limit to how much energy can be released at the earth's surface in the course of our activities without rendering life on earth impossible for all but thermophilic micro-organisms. The dreams of "inexhaustible energy" for terrestrial mankind from nuclear fission or nuclear fusion are hollow dreams indeed - except for those who would go out into the universe, "to go where no man has ever gone before". For the purposes of argument, in relation to biodiversity, quality of life and global energy output we have essentially reached the limits of growth. The vast majority of mankind, who by their actions could be taken to think otherwise, are remorselessly pushing towards the ultimate blasphemy of *terracide* … The world is facing a prospect of a massive gap between available food and population demands in the coming century. Even with a radical change in global attitude to the economic realities (and specifically, the inequalities) of famine, there will be a crisis of moral responsiveness in the face of a situation well beyond normal human experience and conception" [1]).

In my 2007 book "Body Count. Global avoidable mortality since 1950" I concluded: "Global democracy will yield better resource allocation. However population control (advanced by literacy and modest economic security enhanced to annual *per capita* incomes of $1,000- $2,000) will also be required for sustainable use of scarce resources… Notwithstanding better resource allocation, there is a need for fundamental biological sustainability. This is dramatized by considering the global impact if everyone in the growth economies of China and India (over 1/3 of the world population when combined) had an American standard of living and consumption. Greenhouse gas emissions and global warming may well provide the impetus for a general movement to more modest and efficient use of resources… Of course, control of greenhouse gas emissions is crucial to saving the biosphere" (page 186 [4]).

By 2020 there was a wider realization of what a worsening Climate Genocide would mean ([66]; see Section C). Island Nations and low-lying, mega-delta countries like Bangladesh are variously existentially threatened ("in part or in whole" as per the UN Genocide Convention) by global warming-driven sea level rise, greater intensity tropical storms, storm surges, and permanent or disastrously periodic inundation [66]. Drought-ravaged countries of the Sahel, North Africa, the Middle East and South Asia are facing worsening water-deficit circumstances. Presently each year 15 million people die avoidably from deprivation in the tropical to sub-tropical Developing World (minus China) that is threatened by a worsening Climate Crisis. On this account one estimates that about 1 million people presently die annually from climate change, but that this will worsen this century en route to a sustainable human population in 2100 of only about 1 billion [56, 57, 66].

In my 1998 book, "Jane Austen and the Black Hole of British History. Colonial rapacity, holocaust denial and the crisis in biological sustainability", I suggested various ways that population could be humanely decreased, noting that rich First World countries have already reached zero population growth (ZPG). Such suggested mechanisms (some inherently very controversial) include a "one child policy", tax breaks and other financial inducements for people to not have children, protection and indeed encouragement of homosexuality, sociologically-complicated biochemical male child sex selection because only females can bear children (but with the corollary of much greater empowerment of females), celibacy, alternative sexual outlets for men and women, and the "Jane Austen option" (a childless, unmarried state that is pleasant for many women and men) [1].

(9). Economic de-growth.

Unquenchable neoliberal greed means that all countries are committed to endless growth. Indeed the mendacious and terracidal political slogan of the anti-science, anti-environment, neoliberal and climate criminal Australian Coalition Government is "jobs and growth". However as discussed above (part #8, "Population control and population decrease"), the world is over-populated by a factor of 2 [40]. Clearly when population is eventually halved this should be accompanied by a corresponding economic de-growth of at least 50% [40, 54] – but sensible considerations inform that it could go much further. Thus, for example, Cuba under illegal, racist and immoral US-imposed sanctions has a per capita GDP ($8,800) that is 7.4 times lower than that of the US ($65,000) ("List of countries by GDP (nominal) per capita", Wikipedia: https://en.wikipedia.org/wiki/List_of_countries_by_GDP_(nominal)_per_capita). Yet in Cuba "annual avoidable deaths from deprivation" are zero (0) (as for the US) and under-5 infant deaths as a percentage of population are similar in both Cuba and the US [4]. On this basis we could reduce global economic output (after halving the population) by a factor of 15 rather than merely by a factor of 2. Similarly, China has a "per capita GDP"($10,000) that is 6.5 times less than that of the US ($65,000), but its annual avoidable deaths from deprivation" total zero (0) as for the US – on this basis halving the population should be associated with a 13-fold decrease in resource use rather than a mere 2-fold decrease.

I recently concluded (2020): "Comprehensive social distancing and economic lockdown measures required to suppress Covid-19 have been associated with rapid socialism-style adoption of a needs-based economy (essential services), a decrease in carbon fuel burning and air pollution, a widened social safety net, and on-line learning. Post-Covid-19 one envisages a socialist needs-based economy, zero emissions, Universal Basic Income (UBI), a Green New Deal (GND), and free university education" [54]. Once greedy, One Percenter neoliberalism is replaced by all-embracing, "love thy neighbour" social humanism (eco-socialism) then a decent, happy, spiritually rewarding and dignified life for everyone on Earth becomes instantly possible [45, 46].

(10). Biodiversity preservation, sustainability and sharing.

As set out above there is a range of solutions that can and must be applied to save the Planet from a worsening Climate Crisis (presently heading to a catastrophic plus 3.2C this century) and a worsening Climate Genocide (that in the absence of requisite action is predicted to kill 10 billion people this century en route to a sustainable human population in 2100 of only about 1 billion people) [66]. Fundamental to these solutions is sustainability (to preserve what is left of the Biosphere) and sharing (to provide equity and ensure a modest and decent existence for everyone). I have set out what needs to be done in a numerous articles [5, 6, 8, 26, 31, 45, 46, 48, 49].

However we must face up to the reality that at this late stage in Humanity's lemming-like rush to omnicide and terracide it is going to be politically difficult and extremely expensive to save the Planet. However the IMF-estimated $9 trillion cost of addressing the Covid-19 pandemic informs that the world will act in deadly circumstances (albeit woefully insufficiently in the US and UK; see Gideon Polya, "US Alliance Covid-19 Gerocide – intentional mass killing of elderly by US Alliance countries", Countercurrents, 21 October 2020: https://countercurrents.org/2020/10/us-alliance-covid-19-gerocide-intentional-mass-killing-of-elderly-by-us-alliance-countries/). The One Percenters who dominate the world by owning half its wealth will resolutely resist sharing with an increasingly impoverished Ninety Nine Percent of Humanity. As measured by Carbon Debt, the cost of returning to a safe and sustainable 300 ppm CO_2 world [5, 8] is a daunting $200-250 trillion or about 3 times the annual global GDP of $80 trillion. Immense harm to the Biosphere has already occurred, with the animal species extinction rate in the Anthropocene Era now being 100-1,000 times greater than normal [52]. Biosphere damage as a function of time is remorselessly quasi-linear and in the wrong direction [48]. Yet all species and ecosystems are priceless.

Realistically, in the face of resolute inaction [48] it is clear that the Paris Agreement's goal of no more than a plus 1.5C of warming will be exceeded in the coming decade, and that a catastrophic plus 2C is effectively unavoidable. Indeed, according to the UN the national commitments made at the Paris Climate Change Conference amount to an utterly disastrous plus 3.2C. However, that said, decent people are obliged to do everything they can to make the future "less bad" for future generations. However this moral truism can be legitimately criticised as tokenism, and Humanity must engage (non-violently) to actually *win* in this terminal War on Terra, "the war to end all wars".

Fundamental to non-violent Climate Revolution [63] by the Ninety Nine Percenters will be a requisite change in intra-national and international economic philosophy from deadly, speciescidal, ecocidal and One Percenter-advantaging neoliberalism to one of social humanism (socialism, eco-socialism, democratic socialism, human rights-respecting communism, the welfare state, Universal Basic Income) that seeks to sustainably maximize happiness, dignity and opportunity for everyone. A useful suggestion advanced by George Monbiot is that of a one-person-one-vote World Parliament that would have authority in economic matters in particular and break the post-WW2 economic stranglehold of the rich nations (George Monbiot, "Let the People rule the World", Guardian, 17th July 2001: https://www.monbiot.com/2001/07/17/let-the-people-rule-the-world/). I have argued for a one-person-one-vote World Parliament with powers in the areas of economics and sustainability and which initially would act as a third chamber of world government, in addition to the present General Assembly (a one-nation-one-vote chamber of nation states) and the UN Security Council (dominated by world super-powers) [45, 46].

A crucial argument in relation to World Government relates to the fundamental nature of "democracy" and whether One Percenter-dominated Western-style "democracy" is up to the challenge of saving the Planet. Thus Professor Jorgen Randers co-author of The Limits to Growth in 1972, the Report to the Club of Rome) has argued (2012): "I am a climate pessimist. I believe (regrettably) that humanity will not meet the climate challenge with sufficient strength to save our grandchildren from living in a climate-damaged world. Humanity (regrettably) will not make what sacrifice is necessary today in order to ensure a better life for our ancestors forty years hence. The reason is that we are narrowly focused on maximum well-being in the short term. This short-termism is reflected in the systems of governance that we have chosen to dominate our lives: Both democracy and capitalism place more emphasis on costs today that on benefits forty years in the future…In sum, I don't believe that the free market, regulation, political leadership, or public education will solve the climate problem in time. Capitalism is unable to handle this long term challenge, and democratic society is unwilling to modify the market. In my view, we need something stronger, something that can counter the root problem: Man's short-term nature. His tendency to disregard the long term consequences of current action. What can be done? Can democratic society be modified to solve the climate challenge? Eco-dictatorship may be to go too far. But something is needed to temper the short-termism of the nation state, probably something at the supranational level. For example a global central bank for climate gas emissions, introduced through democratic means – like the normal central banks. This is easier said than done. But still necessary. Otherwise, I predict, it will be the Chinese who solve the global climate challenge - singlehandedly. Through a sequence of 5-year plans established with a clear long term vision, and executed without asking regular support from the Chinese. They are already well on the way, for the benefit of our grandchildren" (Jorgen Randers, "Systematic short-termism: Climate, capitalism and democracy", Climate Code Red, 2012: http://www.climatecodered.org/2012/11/systematic-short-termism-climate.html?utm_source=feedburner&utm_medium=email&utm_campaign=Feed%3A+ClimateCodeRed+%28climate+code+red%29).

Notwithstanding the effective dominance of the One Percenters in Western-style "democracies" (Big Money buys votes), this system at least allows for periodic and bloodless transfers of ostensible political power. We love the freedom of Western-style "democracy" notwithstanding One Percenter dominance, the post-9/11 empowerment of the Deep State, and the worrisome, huge and Artificial Intelligence-driven strides in the "surveillance state" both in Western-style democracies and in one-party-ruled China (see Gideon Polya, "Review: "One world digital dictatorship" by Soren Korsgaard – digital nightmare", Countercurrents, 23 January 2020: https://countercurrents.org/2020/01/review-one-world-digital-dictatorship-by-soren-korsgaard-digital-nightmare/). "Democracy" is surely about practical expression of the fundamental wishes of the people e.g. a long, happy and heathy life for them and their children, peace, good governance, economic security and good health and education services – things that are provided by one party state China and which brought 800 million Chinese out of dire and deadly poverty. Thus "annual avoidable deaths from deprivation" total 4 million in ostensible "democracy" India but are zero (0) in altruistic but one party state China [4]. That said, and notwithstanding the adumbrated prospect of a unique, top-down Chinese success in tackling climate change, one must express concern over many matters (mostly not unique to China) including urban air pollution, the death penalty, the one party state, the surveillance state, and human rights abuse in relation to dissidents, Hong Kong and Uighurs. One hopes that just as China has made immense strides in so many areas in its "economic miracle", so it will also advance in relation to human rights. Science that is crucial for societal safety is about rational selection of ever-improving models of reality, and improvement of society similarly requires untrammelled free expression, and selection of useful ideas (memes) for the advancement of Humanity.

The worsening Climate Crisis demands that we must not only sustainably and equitably share the limited resources of the Planet with all members of the Human Family, but we must also sustainably share the Earth with what is left of the Biosphere.

There must be immediate cessation of speciescide, ecocide and Biodiversity loss.

The following articles have been reproduced as Chapters in Section H.

Chapter 27: Gideon Polya, "Worsening Climate Emergency and record CO_2 emissions demand vegetarian diet for all to help save Planet", Countercurrents, 20 June 2017: http://www.countercurrents.org/2016/06/20/worsening-climate-emergency-and-record-co2-emissions-demand-vegetarian-diet-for-all-to-help-save-planet/.
Chapter 28: Gideon Polya, "Resolutely promised prosecutions of climate criminals may force urgent climate action", Countercurrents, 5 January 2019: https://countercurrents.org/2019/01/05/resolutely-promised-prosecutions-of-climate-criminals-may-force-urgent-climate-action/.

References (articles asterisked * have been reproduced as Chapters in Section H).

[1]. Gideon Polya, "Jane Austen and the Black Hole of British History. Colonial rapacity, holocaust denial and the crisis in biological sustainability", G.M. Polya, Melbourne, 1998, 2008, Chapter 17: http://janeaustenand.blogspot.com/.

[2]. Gideon Polya, "Climate Criminals And Climate Genocide ", Countercurrents, 2 July 2007: https://www.countercurrents.org/polya010807.htm.

[3]. Gideon Polya, "Bali Exposes US, Canada And Australian Climate Racism, Climate Terrorism, Climate Criminals And Climate Genocide", Countercurrents, 18 December 2007: https://www.countercurrents.org/polya181207.htm.

[4]. Gideon Polya, "Body Count. Global avoidable mortality since 1950", G.M. Polya, 2007, that includes a n avoidable mortality-related history of every country since Neolithic times and is now available for free perusal on the web: http://globalbodycount.blogspot.com/.

[5]. Gideon Polya, "300.org Urges Reduction Of Atmospheric CO_2 To 300 ppm", Countercurrents, 28 May 2009: https://www.countercurrents.org/polya280509.htm.

[6]. Gideon Polya, "How To SaveThe Planet", Countercurrents, 29 May 2009: https://www.countercurrents.org/polya290509.htm.

[7]. Gideon Polya, "Global Warming - Can Self-interest And Science Save Australia, US And The Planet?", Countercurrents, 17 September 2009: https://www.countercurrents.org/polya170909.htm.

[8]. Gideon Polya, "300.org Message To World – Act Urgently Over Climate Emergency And Climate Genocide", Countercurrents, 23 September, 2009: https://countercurrents.org/polya230909.htm.

[9]. Gideon Polya, "Forest biomass-derived Biochar can profitably reduce global warming and bushfire risk", Yarrra Valley Climate Action group, 2009: https://sites.google.com/site/yarravalleyclimateactiongroup/forest-biomass-derived-biochar-can-profitably-reduce-global-warming-and-bushfire-risk.

[10]. "Submission by Dr Gideon Polya to the Senate Standing Committee on Economics Inquiry into exposure draft of the legislation to implement the Carbon Pollution Reduction Scheme (CPRS)", 2009: file:///C:/Users/Gideon/AppData/Local/Temp/sub05_pdf.pdf.

[11]. Gideon Polya, "Book Review: Australian Sustainable Energy by the Numbers", by Peter Seligman", Open Forum, 17 March 2010: https://sites.google.com/site/bookreviewsbydrgideonpolya/seligman.

[12]. Gideon Polya, "BZE ZCA2020 Plan & Why Developing World Must Go For 100% Renewable Energy by 2020", Countercurrents, 18 July 2010: https://www.countercurrents.org/polya180710.htm.

[13]. Gideon Polya, "New Year's Day 1/1/11 Message: Equal Shares on Our One Planet", Countercurrents, 2 January 2011: https://www.countercurrents.org/polya020111.htm.

[14]. Gideon Polya, "Science And Equality Dictate Eco-Socialism, Climate Socialism And Green Socialism To Save Humanity And Biosphere", Countercurrents, 17 February 2011: https://www.countercurrents.org/polya170211.htm.

[15]. Gideon Polya, "April 22, Earth Day: We Must Stop Climate Catastrophe", Countercurrents, 22 April 2011: https://www.countercurrents.org/polya220411.htm.

[16]. Gideon Polya, "Australian Carbon Price Scheme Is Deceitful Climate Change Inaction", Countercurrents, 18 July, 2011: https://www.countercurrents.org/polya180711.htm.

[17]. Gideon Polya, "Boycott Murdoch Media To Save Planet", Countercurrents, 18 July 2011: https://countercurrents.org/polya180711A.htm.

[18]. Gideon Polya, "World Rapidly Running Out Of Time To Cease Greenhouse Gas Pollution", Countercurrents, 1 August 2011: https://www.countercurrents.org/polya010811.htm.

[19]. Gideon Polya, "Anti-Environment Australian Government's Carbon Tax Deceit", Countercurrents, 9 August 2011: https://www.countercurrents.org/polya090811.htm.

[20]. Gideon Polya, "Orwellian Australian Labor Government Climate Change Lies", Countercurrents, 20 August 2011: https://www.countercurrents.org/polya200811.htm.

[21]. Gideon Polya, "US-Complicit 9-11, War On Terror And Terminal War On Terra", 10 September 2011: https://www.countercurrents.org/polya100911.htm.

[22]. Gideon Polya, "Moving Planet Day 24 September: Stop GHG Pollution", Countercurrents, 24 September 2011: https://www.countercurrents.org/polya240911.htm.

[23]. Gideon Polya, ""Vote 1 Socialist Or Green To Save Humanity And Planet", Countercurrents, 9 October 2011: https://www.countercurrents.org/polya091011.htm.

[24]. Gideon Polya, "CO_2 Rise Drove Global Warming And End Of Ice Age – So Stop Carbon Burning, Return To 300 ppm CO_2", Countercurrents, 7 April 2012: https://www.countercurrents.org/polya070412.htm.

[25]. Gideon Polya, "Expert Witness Testimony To Stop Gas-Fired Power Plant Installation", Countercurrents, 14 June, 2013: https://countercurrents.org/polya140613.htm.

[26]. Gideon Polya, "100 Ideas For Climate Change Activists Trying To Save The Biosphere And Humanity", Countercurrents, 10 August 2013: https://www.countercurrents.org/polya100813.htm.
[27]. Gideon Polya, "Why We Must Divest From Fossil Fuel Corporations To Help Save The Biosphere And Humanity", Countercurrents, 9 October 2013: https://www.countercurrents.org/polya091013.htm.
[28]. Gideon Polya, "$10 Trillion Annual Carbon Debt Increase For Young People On A Threatened Planet", Countercurrents, 8 July 2014: https://www.countercurrents.org/polya080714.htm.
[29]. Gideon Polya, "Letter To Young People Over $220 Trillion Carbon Debt: Revolt (Peacefully)", Countercurrents, 11 July 2014: https://www.countercurrents.org/polya110714.htm.
[30]. Gideon Polya, "This Changes Everything. Capitalism vs. the Climate" by Naomi Klein - Green Socialist Revolution ASAP", Countercurrents, 11 December 2014: https://www.countercurrents.org/polya111214.htm.
[31]. Gideon Polya, "2015 A-to-Z Alphabetical List Of Actions And Advocacies For Climate Change Activists", Countercurrents, 14 January, 2015: https://www.countercurrents.org/polya140115.htm.
[32]. Gideon Polya, "Intergenerational Theft – For Every $1 For Coal Today Future Generations Will Pay $1-$14 To Sequester CO2", Countercurrents, 8 April 2015: https://countercurrents.org/polya080415.htm.
[33]. Gideon Polya, "Australian Coalition Government & Labor Opposition Trash Human Rights, Child Rights, Free Speech & Medical Ethics", Countercurrents, 15 July, 2015: https://countercurrents.org/polya150715.htm.
[34]. Dr Gideon Polya, "Polya's 3 Laws Of Economics Expose Deadly, Dishonest And Terminal Neoliberal capitalism", Countercurrents, 17 October, 2015: https://countercurrents.org/polya171015.htm.
[35]. Gideon Polya, "Exposing And Thence Punishing Worst Polluter Nations Via Weighted Annual Per Capita Greenhouse Gas Pollution Scores", Countercurrents, 6 January 2016: https://sites.google.com/site/banyuleclimateactionnow/revised-annual-per-capita.
[36]. Gideon Polya, "Save Humanity And The Biosphere Through Zero Tolerance For Deadly Neoliberalism And Remorseless Neoliberals", Countercurrents, 13 May, 2016: https://countercurrents.org/polya130516.htm.
[37]. Gideon Polya, "Humanity Must Pledge Inescapable Dispossession And Custodial Retribution For Climate Criminals", Countercurrents, 20 December 2016:

http://www.countercurrents.org/2016/12/20/humanity-must-pledge-inescapable-dispossession-and-custodial-retribution-for-climate-criminals/.

*[38]. Gideon Polya, "Worsening Climate Emergency and record CO_2 emissions demand vegetarian diet for all to help save Planet", Countercurrents, 20 June 2017: http://www.countercurrents.org/2016/06/20/worsening-climate-emergency-and-record-co2-emissions-demand-vegetarian-diet-for-all-to-help-save-planet/.

[39].Gideon Polya, "Huge Carbon Debt & Intergenerational Injustice – CO_2 drawdown necessity, 300.org, & 300 ppm CO_2 target", 6 June 2018: https://countercurrents.org/2018/06/06/huge-carbon-debt-intergenerational-injustice-co2-drawdown-necessity-300-org-300-ppm-co2-target/.

[40]. Gideon Polya, "How much negative carbon emissions, negative population growth & negative economic growth is needed to save Planet?", Countercurrents, 28 November 2018: https://countercurrents.org/2018/11/how-much-negative-carbon-emissions-negative-population-growth-negative-economic-growth-is-needed-to-save-planet/.

*[41]. Gideon Polya, "Resolutely promised prosecutions of climate criminals may force urgent climate action", Countercurrents, 5 January 2019: https://countercurrents.org/2019/01/05/resolutely-promised-prosecutions-of-climate-criminals-may-force-urgent-climate-action/.

[42]. Gideon Polya, "Lying by omission, avoidable mortality from deprivation, holocaust denial & looming terracide", Countercurrents, 4 April 2019: https://countercurrents.org/2019/04/lying-by-omission-avoidable-mortality-from-deprivation-holocaust-denial-looming-terracide.

[43]. Gideon Polya, "Inescapable $200-250 trillion global carbon debt increasing by $16 trillion annually", Countercurrents, 27 April 2019: https://countercurrents.org/2019/04/inescapable-200-250-trillion-global-carbon-debt-increasing-by-16-trillion-annually-gideon-polya.

[44]. Gideon Polya, "Millions join global school climate strike- we are running out of time", Countercurrents, 22 September 2019: https://countercurrents.org/2019/09/millions-join-global-school-climate-strike-we-are-running-out-of-time.

[45]. Gideon Polya, "The need for global governance" in Brian Ellis, "The New Enlightenment. On Steven Pinker & beyond", Australian Scholarly Publishing, Melbourne, 2019, pages 208-222.

[46]. Gideon Polya, "Review: "The New Enlightenment" by Brian Ellis – World Government & Social Humanism to save Humanity", Countercurrents, 7 October 2019:

https://countercurrents.org/2019/10/review-the-new-enlightenment-by-brian-ellis-world-government-social-humanism-to-save-humanity/.

[47]. Gideon Polya, "Australia rejects IMF Carbon Tax & preventing 4 million pollution deaths by 2030", Countercurrents, 15 October 2019: https://countercurrents.org/2019/10/australia-rejects-imf-carbon-tax-preventing-4-million-pollution-deaths-by-2030.

[48]. Gideon Polya, "Extrapolating 11,000 scientists Climate Emergency warning to 2030 catastrophe", Countercurrents, 14 November 2019: https://countercurrents.org/2019/11/extrapolating-11000-scientists-climate-emergency-warning-to-2030-catastrophe.

[49]. Gideon Polya, "Climate scientists: Planetary Emergency, Planet in peril, act now", Countercurrents, 3 December 2019: https://countercurrents.org/2019/12/climate-scientists-planetary-emergency-planet-in-peril-act-now.

[50]. Gideon Polya, "Australian climate criminality, heat stress deaths & Australian Aboriginal Ethnocide", Countercurrents, 19 December 2019: https://countercurrents.org/2019/12/australian-climate-criminality-heat-stress-deaths-australian-aboriginal-ethnocide/.

[51]. Gideon Polya, "Trumpist climate change denial, Australian bushfires, fuel reduction, biochar & carbon debt", Countercurrents, 10 January 2020: https://countercurrents.org/2020/01/trumpist-climate-change-denial-australian-bushfires-fuel-reduction-biochar-carbon-debt/.

[52]. Gideon Polya, "Melbourne's North East Link super-highway project – environmental vandalism & Australian-killing perversion", Countercurrents, 9 March 2020: https://countercurrents.org/2020/03/melbournes-north-east-link-super-highway-project-environmental-vandalism-australian-killing-perversion.

[53]. Gideon Polya, "Post-Covid-19 needs-based economy, zero emissions, UBI, Green New Deal & free university education", Countercurrents, 3 May 2020: https://countercurrents.org/2020/05/post-covid-19-needs-based-economy-zero-emissions-ubi-green-new-deal-free-university-education/.

[54]. Gideon Polya, "Covid-19 & climate: Listen to scientists for a safe & sustainable society & planet", Countercurrents, 28 May 2020: https://countercurrents.org/2020/05/covid-19-climate-listen-to-scientists-for-a-safe-sustainable-society-planet/.

[55]. Gideon Polya, "US-Imposed Post-9/11 Muslim Holocaust & Muslim Genocide", 400 pages, Korsgaard Publishing, Germany, 4 June 2020: https://www.amazon.com/US-Imposed-Post-9-Muslim-Holocaust-Genocide/dp/8793987056.

[56]. Gideon Polya, "Racist Mainstream ignores "US-Imposed Post-9/11 Muslim Holocaust & Muslim Genocide"", Countercurrents, 17 July 2020: https://countercurrents.org/2020/07/racist-mainstream-ignores-us-imposed-post-9-11-muslim-holocaust-muslim-genocide/.

[57]. Gideon Polya, editor, "1% ON 1%: one percent annual wealth tax on One Percenters": https://sites.google.com/site/300orgsite/1-on-1.

[58]. Gideon Polya, editor, "2011 climate change course": https://sites.google.com/site/300orgsite/2011-climate-change-course.

[59]. Gideon Polya, editor, 300.org:. https://sites.google.com/site/300orgsite/300-org.

[60]. Gideon Polya, editor, "300.org – return atmosphere CO_2 to 300 ppm CO_2": https://sites.google.com/site/300orgsite/300-org---return-atmosphere-co2-to-300-ppm.

[61]. Gideon Polya, editor, "Carbon Debt Carbon Credit": https://sites.google.com/site/carbondebtcarboncredit/.

[62]. Gideon Polya, editor, "Climate Revolution Now": https://sites.google.com/site/300orgsite/climate-revolution.

[63]. Gideon Polya, editor, "Cut carbon emissions 80% by 2020": https://sites.google.com/site/cutcarbonemissions80by2020/.

[64]. Gideon Polya, editor, "100% renewable energy by 2020": https://sites.google.com/site/100renewableenergyby2020/.

[65]. Gideon Polya, editor, "Climate Genocide": https://sites.google.com/site/climategenocide/.

[66]. Gideon Polya, editor, "Gas is not clean energy": https://sites.google.com/site/gasisnotcleanenergy/.

[67]. Gideon Polya, editor, "Biofuel Genocide":
https://sites.google.com/site/biofuelgenocide/.
[68]. Gideon Polya, editor, "Divest from fossil fuels":
https://sites.google.com/site/300orgsite/divest-from-fossil-fuels.
[69]. Gideon Polya, editor, "Climate Justice & Intergenerational
Equity": https://sites.google.com/site/300orgsite/climate-justice.
[70]. Gideon Polya, editor, "Science & economics experts: Carbon
Tax needed NOT Carbon Trading":
https://sites.google.com/site/300orgsite/sciennce-economics-experts-
carbon-tax-needed-not-carbon-trading/.
[71]. Gideon Polya, editor, "Stop climate crime":
https://sites.google.com/site/300orgsite/stop-climate-crime.
[72]. Gideon Polya, editor, "Stop air pollution deaths":
https://sites.google.com/site/300orgsite/stop-air-pollution-deaths
[73]. Gideon Polya, editor, "Are we doomed?":
https://sites.google.com/site/300orgsite/are-we-doomed.
[74]. Gideon Polya, editor, "Methane Bomb Threat":
https://sites.google.com/site/methanebombthreat/.
[75]. Gideon Polya, editor, "Nuclear weapons ban, end poverty &
reverse climate change":
https://sites.google.com/site/300orgsite/nuclear-weapons-ban.
[76]. Gideon Polya, editor, "Too late to avoid global warming
catastrophe": https://sites.google.com/site/300orgsite/too-late-to-
avoid-global-warming.
[77]. Gideon Polya, editor, "Stop state terrorism":
https://sites.google.com/site/stopstateterrorism/.
[78].Gideon Polya, editor, "State crime and non-state terrorism":
https://sites.google.com/site/statecrimeandnonstateterrorism/.
[79]. "Climate terrorism: 400,000 climate change-related deaths
globally annually versus an average of 4 US deaths from political
terrorism annually since 9-11":
https://sites.google.com/site/statecrimeandnonstateterrorism/climate-
terrorism.
[80]. "Carbon terrorism: 3 million US air pollution deaths versus 53
US political terrorism deaths since 9-11 (2001-2015)":
https://sites.google.com/site/statecrimeandnonstateterrorism/carbon-
terrorism.
[81]. "Gideon Polya":
https://sites.google.com/site/drgideonpolya/home.
[82]. Gideon Polya, "The lie that gas is cleaner", Green Left, Issue
1284, 7 October 2020: https://www.greenleft.org.au/content/lie-gas-
cleaner.

CHAPTER 27: WORSENING CLIMATE EMERGENCY AND RECORD CO₂ EMISSIONS DEMAND VEGETARIAN DIET FOR ALL TO HELP SAVE PLANET

[First published as Gideon Polya, "Worsening Climate Emergency and record CO_2 emissions demand vegetarian diet for all to help save Planet", Countercurrents, 20 June 2017: http://www.countercurrents.org/2016/06/20/worsening-climate-emergency-and-record-co2-emissions-demand-vegetarian-diet-for-all-to-help-save-planet/.]

Annual global greenhouse gas (GHG) emissions presently total 64 billion tonnes CO_2-equivalent with methanogenic livestock production contributing over 50%. With annual emissions of CO_2 per se from industry at a record high, and with Humanity and the Biosphere existentially threatened by the Methane Bomb of the warming Arctic in coming decades, we must all urgently adopt a vegetarian or vegan diet to help save the Planet.

World Bank analysts have carefully re-evaluated the contribution of livestock production to world annual GHG pollution and found that the world's annual total rose from 41.76 billion tonnes CO_2-equivalent (CO_2-c) as estimated by the Food and Agricultural Organisation (FAO) to 63.80 billion tonnes CO_2-e, with livestock production contributing over 51% of the higher figure [1]. Support for the World Bank analysts' upward revision of agricultural GHG pollution paradoxically comes from a quarter that ostensibly denies it. Thus the Australian Primary Industries Climate Challenges Centre reports that "Methane emissions from agriculture account for 11% of total national greenhouse gas emissions, with the majority coming from dairy, beef and sheep production systems" [2], noting that this is based on the official assumption by the Australian Department of the Environment (formerly the Department of Climate Change and Energy Efficiency) that the Global Warming Potential (GWP) of methane (CH_4) relative to that of carbon dioxide (CO_2) is 21 on a 100 year time frame. However Dr Drew Shindell and colleagues have re-assessed the Global Warming Potential (GWP) of methane at 105 on a 20 year time frame and considering aerosol impacts [3, 4]. Given Australia's annual greenhouse gas (GHG) pollution of 600 Mt CO_2-e (600 million tonnes CO_2-equivalent), the non-methane component is 600 Mt CO_2-e minus 11% = 600 – 66 = 534 Mt CO_2-e. However, assuming a GWP for methane of 105 (5 times bigger than that officially assumed in Australia), the methane component is 66 x 5 = 330 Mt CO_2-e, and the revised total GHG pollution = 534 + 330 = 864 Mt CO_2-e, with the agricultural component representing 38% of the new total.

Taking an upwardly revised estimate of land use (agriculture, deforestation, land-clearing) into account, the revised annual per capita GHG pollution in tonnes per person per year is 8.9 for the world and as follows for the very worst polluters with annual per capita GHG pollution well above the world average: Belize (366.9), Guyana (203.1), Malaysia (126.0), Papua New Guinea (114.7), Qatar (101.8), Zambia (97.5), Antigua & Barbuda (85.6), United Arab Emirates (82.4), Panama (68.0), Botswana (64.9), Liberia (55.0), Indonesia (53.6), New Zealand (53.2), Australia (52.9; 116 if including its huge GHG-generating exports), Nicaragua (51.2), Canada (50.1), Equatorial Guinea (47.5), Venezuela (45.2), Brazil (43.4), Myanmar (41.9), Ireland (41.4), United States (41.0), Cambodia (40.5), Kuwait (37.3), Paraguay (37.2), Central African Republic (35.7) [5].

Revised annual per capita GHG pollution in tonnes per person per year is as follows for lesser polluting countries which nevertheless have per capita pollution above the world average: Peru (34.8), Mongolia (32.2), Singapore (31.2), Bahrain (30.5), Trinidad & Tobago (29.8), Cameroon (29.5), Congo, Democratic Republic (formerly Zaire) (29.3), Côte d'Ivoire (29.1), Denmark (27.8), Brunei (27.4), Bolivia (27.3), Guatemala (26.9), Belgium (26.3), Ecuador (26.2), Estonia (25.4), Laos (25.3), Suriname (25.1), Netherlands (24.9), Libya (24.9), Nepal (24.6), Benin (24.5), Angola (23.8), Madagascar (23.7), Argentina (23.7), Uruguay (23.7)*, Luxembourg (23.6), Turkmenistan (23.5), Czech Republic (23.5), Zimbabwe (23.3), Gabon (23.1), Greece (21.9), United Kingdom of Great Britain and Northern Ireland (21.5), Cyprus (21.4), Congo, Republic (21.0), Spain (20.9), Finland (20.6), Israel (20.2), Norway (20.1), Colombia (19.8), Namibia (19.8), Mauritania (19.7), South Africa (19.4), Ukraine (19.1), Germany (18.6), France (17.7), Italy (17.6), Uzbekistan (17.5), Costa Rica (17.1), Sudan (16.8), Saudi Arabia (16.6), Slovenia (16.5), Azerbaijan (16.4), Russia (16.2), Sierra Leone (16.2), Slovakia (15.9), Honduras (15.8), Hungary (15.5), Kazakhstan (15.4), Portugal (15.0), Sweden (15.0), Iran (14.5), Iceland (14.2), Mexico (13.9), Oman (13.8), Malta (13.3), Austria (13.0), Poland (12.9), Jamaica (12.8), Palau (12.8), South Korea (12.7), Guinea (12.5), North Korea (12.1), Bahamas (12.1), Nigeria (11.7), Nauru (11.7), Malawi (11.7), Mali (11.6), Chad (11.6), Taiwan (11.6), Latvia (11.4), Vanuatu (11.1), Switzerland (11.0), Romania (10.9), Togo (10.9), Japan (10.7), Serbia & Montenegro (10.4), Seychelles (10.2), Bulgaria (10.1), Lebanon (9.8), Syria (9.4), Tanzania (9.3), Turkey (9.2), Barbados (9.1), Jordan (9.1), Occupied State of Palestine (9.1)*, Philippines (9.0), Guinea-Bissau (9.0) (* indicates an estimate based on that for an immediately contiguous, ethnically-related country) [5].

In stark contrast to that of the worst polluting countries, the revised annual per capita GHG pollution in tonnes per person per year for the least polluting countries with annual per capita GHG pollution at or below the world average of 8.9 is as follows: Ghana (8.9), Thailand (8.7), Chile (8.7), Fiji (8.7), Belarus (8.6), Sri Lanka (8.5), Macedonia (8.5), Tonga (7.4), Croatia (7.4), China (7.4), Burkina Faso (7.3), Bosnia & Herzegovina (7.2), Kenya (7.1), Dominican Republic (7.1), Senegal (7.0), Tunisia (7.0), Algeria (6.6), Grenada (6.4), Samoa (6.2), Rwanda (6.1), El Salvador (6.0), Lithuania (5.9), Mozambique (5.8), Lesotho (5.7), Burundi (5.5), Iraq (5.5), Eritrea (5.3), St Kitts & Nevis (5.1), Uganda (5.1), Haiti (5.0),

Mauritius (5.0), Albania (4.3), Dominica (4.2), Bhutan (4.1), Niger (4.1), Ethiopia (4.1), Moldova (4.0), Georgia (4.0), Yemen (3.7), Tajikistan (3.7), Afghanistan (3.6), Swaziland (3.6), Cuba (3.5), Cape Verde (3.5), Kyrgyzstan (3.4), The Gambia (3.0), St Lucia (2.9), Bangladesh (2.7), Egypt (2.6), Niue (2.6), Pakistan (2.5), Morocco (2.5), Djibouti (2.4), St Vincent & Grenadines (2.4), Armenia (2.3), Maldives (2.1), India (2.1), Cook Islands (2.1), Vietnam (1.9), São Tomé and Príncipe (1.9), Comoros (1.6), Solomon Islands (1.4), Kiribati (1.2), Tuvalu (1.2)* (* indicates an estimate based on that for an immediately contiguous, ethnically-related country) [5].

The 2009 Report of the German Scientific Advisory Council on Global Change (WBGU, Wissenshaftlicher Beirat der Bundesregierung Globale Umweltveränderungen) entitled "Solving the climate dilemma: the budget approach" crucially stated: "The budget of CO_2 emissions still available worldwide could be derived from the 2 degree C guard rail. By the middle of the 21st century a maximum of approximately 750 Gt CO_2 (billion metric tons) may be released into the Earth's atmosphere if the guard rail is to be adhered to with a probability of 67%. If we raise the probability to 75%, the cumulative emissions within this period would even have to remain below 600 Gt CO_2. In any case, only a small amount of CO_2 may be emitted worldwide after 2050. Thus, the era of an economy driven by fossil fuels will definitely have to come to an end within the first half of this century" [6].

The consequences of this declaration of less than 600 Gt CO_2 (600 billion tonnes CO_2) in emissions for a 75% chance of avoiding a catastrophic 2 degree C temperature rise are profound. Thus, one can well ask whether a sensible person would board a plane if it had a 25% chance of crashing. Further, the average world population in the period 2010 – 2050 will be 8.321 billion. Accordingly, the per capita share of this Terminal Carbon Pollution Budget is less than 600 billion tonnes CO_2/8.321 billion people = less than 72.1 tonnes CO_2 per person. Thus years left before the world uses up this budget = 72.1 tonnes CO_2 per person/ 8.9 tonnes CO_2-e per person per year = 8.1 years relative to 2010, and thus only about 2 years relative to 2016 [and zero years relative to 2020].

Using the above revised data for the annual per capita greenhouse gas (GHG) emissions that properly account for land use and livestock impacts, one can determine how many years left at current rates of GHG pollution (in units of CO_2-e or CO_2-equivalent i.e. taking other GHGs into account) before a given country uses up its "fair share" of the World's Terminal Carbon Pollution Budget [7]. Thus, for example, for Australia 72.1 tonnes CO_2-e per person / 52.9 tonnes CO_2-e per person per year = 1.4 years left relative to 2010, noting that this analysis does not take into account historical pollution of the atmosphere. Thus Australia used up its "fair share" of the world's Terminal Carbon Pollution Budget in 2011, and since then has been stealing the entitlement of the 50% of other countries which have not yet used up their entitlement [7] – to be fair to Australia, half of the world's nations have also exceeded their share of the world's Terminal Carbon Pollution Budget. Yet in climate criminal Australia there is consensus between the 2 major parties – the Australian Labor Party (the ALP and currently in Opposition) and the Liberal Party-National Party Coalition (aka the pro-coal COALition or KOALition and currently in Government) – for unlimited GHG exports via exports of coal, gas, iron ore and meat. Australia's bipartisan, Lib-Lab (Liberal Party-National Party Coalition Government and Labor Opposition) policy of unlimited gas, coal and iron ore exports means that it is committed to polluting the atmosphere with over 3 times the whole world's total Terminal Carbon Pollution Budget [8], with this estimate ignoring Australia's huge meat exports. Greedy, rich, neoliberal-dominated Australia is a world leader in deforestation [9], species extinction [9], per capita GHG pollution [5], income weighted per capita GHG pollution [10], fossil fuel exports [8, 11] and climate change inaction [10-15].

The world is badly running out of time to deal with the worsening climate emergency due to man-made GHG pollution and consequent global warming and sea level rise. However while it is now too late to avoid a catastrophic plus 2 degrees

Centigrade (plus 2C) temperature rise relative to 1900, we are all obliged to do everything we can to make the future "less bad" for our children and future generations. One way ordinary people (as opposed to climate criminal One Percenters) can collectively have a massive impact on global GHG pollution is to adopt a meat-free diet – if everyone adopted a vegetarian diet, global greenhouse gas pollution would be halved [1]. In between the current profligacy and a universal vegan diet there is a middle way based on the notion that some meats have less environmental and social impact than others. Thus consider the following conversion efficiencies (kg grain to produce 1 kg gain in live weight): herbivorous farmed fish (e.g. carp, tilapia, catfish; less than 2), chicken (2), pork (4), and beef (7) [16].

Annual per capita meat consumption (in kg per person per year) [17] very roughly correlates with annual per capita GDP (in USD per person per year) (UN, 2014) [18] as seen with the following extremes (note the exceptions of livestock-based but impoverished Mongolia and the fishing-rich but impoverished Island States of Saint Lucia, Saint Vincent and the Grenadines and Samoa):

Countries above 80 kg meat per person per year (100 +/- 20): Antigua and Barbuda (84.3 kg per person per year, $13,731 per person per year), Argentina (98.3, $12,645), Australia (111.5, $62,290), Austria (102, $51,296), Bahamas (109.5, $22, 217), Bermuda (101.7, $89,795), Brazil (85.3, $11,387), Canada (94.3, $50,169), Czech Republic (83.4, $19,470), Denmark (95.2, $61,294), French Polynesia (101.9, $20,009), Iceland (86.2, $52,048), Ireland (87.9, $53,648), Israel (96, $38,261), Italy (90.7, $35,812), Kuwait (119.2, $43,600), Luxembourg (107.9, $116, 560), Malta (84.5, $25,222), Mongolia (82.1, $4,147), Netherlands (85.5, $52,129), Netherlands Antilles (91, $18,360), New Caledonia (72.6, $39,392), New Zealand (106.4, $44,189), Portugal (93.4, $22,122), Saint Lucia (93.6, $7,655), Saint Vincent and the Grenadines (91.4, $6,669), Samoa (87.2, $4,294), Slovenia (88.3, $23,954), Spain (97, $29,861), Sweden (80.2, $58,856), United Kingdom (84.2, $46,461), United States of America (120.2, $54,306).

Countries below 20 kg meat per person per year (10 +/- 10):
Algeria (19.5 kg per person per year, $5,484 per person per year), Bangladesh (4, $1,088), Burkina Faso (14.8, $725), Burundi (5.2, $279), Cambodia (16.6, $1,095), Cameroon (12.7, $1,407), Chad (13, $941), Comoros (13.4, $841), Congo (13.4, $480), Eritrea (7.7, $755), Ethiopia (8.5, $553), Gambia (8.1, $441), Ghana (13.9, $1,388), Guinea (8.6, $536), India (4.4, $1,586), Indonesia (11.6, $3,492), Kenya (16.7, $1,358), Lesotho (18.3, $986), Liberia (10.4, $483), Madagascar (14.7, $453), Nigeria (8.8, $3,303), Malawi (8.3, $343), Mozambique (7.8, $628), Nepal (9.9, $692), North Korea (13.4, $696), Pakistan (14.7, $1,561), Rwanda (6.5, $697), Sao Tome and Principe (16.5, $1,811), Senegal (15.8, $1,067) Sierra Leone (7.3, $775), Solomon Islands (11.9, $1,927), Sri Lanka (6.3, $3,635), Sudan (19.1, $2,081), Tanzania (9.6, $952), Tajikistan (14.7, $1,114), Uganda (11, $727), Yemen (17.9, $1,418), Zambia (12.3, $1,715) [17, 18].

Tackling climate change – by the numbers.

For a safe and sustainable environment for all people and all species the atmospheric CO_2 concentration must be reduced from the current dangerous and damaging 400 ppm CO_2 to the pre-Industrial Revolution level of circa 300 ppm CO_2 i.e. when (if) we reach zero carbon emissions we will still need negative CO_2 emissions to get back to 300 ppm CO_2 through removal of about 350 Gt CO_2-e (350 billion tonnes CO_2-equivalent) from the atmosphere. The maximum rate of removal of atmospheric CO_2 by safely sequestering agricultural or forestry cellulosic waste (as waste per se or as biochar from anaerobic pyrolysis of cellulosic material) in combustion-free underground locations is 12 Gt C (44 Gt CO_2-e) per year, and using this mode of negative CO_2 emissions it would take about 30 years if zero emissions (zero GHG pollution) had already been achieved [19].

Rapid achievement of zero emissions will involve a rapid switch to the best non-carbon and renewable energy (solar, wind, geothermal, wave, tide and hydro) options that are currently roughly the same market price as coal burning-based power and 5 times cheaper if the human and environmental cost is considered. Installation of 100% renewable energy is feasible and 100% renewable energy by 2020 is on already track for a number of corporations, towns, cities and countries [20]. Thus 100% renewable energy could be achieved for Australia in 10 years for a cost of A\$350 billion (US\$263 billion) [11, 21], whereas the damage-related Carbon Debt associated with Business As Usual (BAU) GHG pollution would be US\$4,000 billion over 10 years [22].

Achieving zero emissions would also involve energy efficiency, public transport, needs-based production, and re-afforestation coupled with correspondingly rapid cessation of fossil fuel burning, deforestation, population growth and methanogenic livestock production, the latter alone saving about 32 Gt CO_2-e of GHG emissions annually from the 64 Gt CO_2-e generated presently each year [1]. With emissions of CO_2 per se at a record high of over 30 Gt CO_2 annually, and the atmospheric CO_2 concentration of about 400 ppm CO_2 increasing at a record rate of 3 ppm CO_2 per year, something has to go, and it surely cannot be Humanity or the Biosphere.

Decent people must demand cessation of meat eating – in the absence of requisite abolition of about 32 Gt CO_2 annually of industrial pollution (from carbon fuel burning and cement manufacture), a vegetarian or vegan diet for all Humanity is urgently required now to help save the Planet. And it must be done urgently, now, because a predicted release of an estimated 50 Gt methane (50 x 105 = 5,250 Gt CO_2-e) from the steadily warming East Siberian Arctic Shelf in coming decades would mean doom for Humanity and the Biosphere [23-25]. Decent people must (a) inform everyone they can about the worsening climate emergency, (b) urge a vegetarian diet for all Humanity, and (c) urge and apply Boycotts, Divestment and Sanctions (BDS) against all people, politicians, parties, companies, corporations and countries disproportionately involved in the greenhouse gas pollution that so acutely threatens Humanity and the Biosphere.

References.

[1]. Robert Goodland and Jeff Anfang."Livestock and climate change. What if the key actors in climate change are … cows, pigs and chickens?", World Watch, November/December 2009:http://www.worldwatch.org/files/pdf/Livestock%20and%20Climate%20Change.pdf.

[2]. "Reducing methane emissions from livestock", Primary Industries Climate Challenges Centre:http://www.piccc.org.au/research/project/253.

[3]. Drew T. Shindell, Greg Faluvegi, Dorothy M. Koch, Gavin A. Schmidt, Nadine Unger and Susanne E. Bauer, "Improved Attribution of Climate Forcing to Emissions", Science, 30 October 2009: Vol. 326 no. 5953 pp. 716-718: http://www.sciencemag.org/content/326/5953/716.

[4]. D. Shindell et al. (2009), Fig.2: http://www.sciencemag.org/content/326/5953/716.figures-only.

[5]. Gideon Polya, "Revised Annual Per Capita Greenhouse Gas Pollution For All Countries – What Is Your Country Doing?", Countercurrents, 6 January, 2016: https://countercurrents.org/polya060116.htm.

[6]. WBGU, "Solving the climate dilemma: the budget approach", 2009: http://www.wbgu.de/fileadmin/templates/dateien/veroeffentlichungen/sondergutachten/sn2009/wbgu_sn2009_en.pdf.

[7]. Gideon Polya, "Years Left For Zero Emissions By Country – Half Have Exceeded Their Share Of The World's Terminal Carbon Pollution Budget", Countercurrents, 27 February, 2016: https://countercurrents.org/polya270216.htm.

[8]. Gideon Polya, "Australia 's Huge Coal, Gas & Iron Ore Exports Threaten Planet", Countercurrents, 15 May 2012:https://countercurrents.org/polya150512.htm.

[9]. Frances Pike, "Australia: world leader in deforestation and species extinction", Independent Australia, 3 June 2015:https://independentaustralia.net/environment/environment-display/australian-logging-and-the-end-of-species,7788.

[10]. Gideon Polya, "Exposing And Thence Punishing Worst Polluter Nations Via Weighted Annual Per Capita Greenhouse Gas Pollution Scores", Countercurrents, 19 March, 2016: https://countercurrents.org/polya190316.htm.

[11]. Gideon Polya, "2011 climate change course": https://sites.google.com/site/300orgsite/2011-climate-change-course.

[12]. German Climate Watch Index 2015: https://germanwatch.org/en/download/10407.pdf.

[13]. Gideon Polya, "25 Ways World-Leading Climate Criminal Australia Threatens Planet And Invites Boycotts, Divestment & Sanctions (BDS)", Countercurrents, 6 June, 2015: https://countercurrents.org/polya060615.htm.

[14]. Gideon Polya, "Climate Criminal Coalition Government Lying Endangers Australia & Planet – Australians Must Vote 1 Green And Put The Coalition Last", Countercurrents, 17 June, 2016: https://countercurrents.org/polya170616.htm.

[15]. Gideon Polya, "Coalition Climate Crimes & 200 Reasons Why Australia Must Dump Pro-coal, Pro-war Coalition PM Malcolm Turnbull", Countercurrents, 1 November, 2015: https://countercurrents.org/polya011115.htm.

[16]. Gideon Polya, "Biofuel famine, biofuel genocide, meat & global food price crisis", Global avoidable mortality:http://globalavoidablemortality.blogspot.com.au/2008/05/biofuel-famine-biofuel-genocide-meat.html.

[17]. Current worldwide annual meat consumption per capita", ChartsBin: http://chartsbin.com/view/12730.

[18]. "List of countries by GDP (nominal) per capita", Wikipedia:https://en.wikipedia.org/wiki/List_of_countries_by_GDP_(nominal)_per_capita.

[19]. Gideon Polya, "Forest biomass-derived Biochar can profitably reduce global warming and bushfire risk", Yarra Valley Climate Action Group: http://sites.google.com/site/yarravalleyclimateactiongroup/forest-biomass-derived-biochar-can-profitably-reduce-global-warming-and-bushfire-risk.

[20]. Gideon Polya, editor, "100% renewable energy by 2020": https://sites.google.com/site/100renewableenergyby2020/.

[21]. Beyond Zero Emissions Zero (BZE), Zero Carbon Australia by 2020 Report (BZE ZCA2020 Report), 2010:http://www.beyondzeroemissions.org/about/bze-brand.

[22]. Gideon Polya, editor, "Carbon Debt Carbon Credit": https://sites.google.com/site/carbondebtcarboncredit/.

[23]. Gideon Polya, editor, "Methane Bomb Threat": https://sites.google.com/site/methanebombthreat/.

[24]. Gideon Polya, editor, "Too late to avoid global warming catastrophe": https://sites.google.com/site/300orgsite/too-late-to-avoid-global-warming.

[25]. Gideon Polya, editor, "Are we doomed?": https://sites.google.com/site/300orgsite/are-we-doomed.

CHAPTER 28: RESOLUTELY PROMISED PROSECUTIONS OF CLIMATE CRIMINALS MAY FORCE URGENT CLIMATE ACTION

[First published as Gideon Polya, "Resolutely promised prosecutions of climate criminals may force urgent climate action", Countercurrents, 5 January 2019: https://countercurrents.org/2019/01/05/resolutely-promised-prosecutions-of-climate-criminals-may-force-urgent-climate-action/.]

Eminent climate scientist Professor James Hansen has published a must-read, 55-page summary of the worsening climate emergency. In short, to correct the Earth's presently disastrous energy imbalance we must urgently reduce the atmospheric CO_2 to 342-373 ppm CO_2 from the present disastrous 407 ppm CO_2. The cost of extracting 1 ppm of CO_2 from the atmosphere is $878-1803 billion but continuing inaction is not an option – the Paris commitments mean a 3C temperature rise and eventual inundation of coastal areas by a 15-25 meters sea level rise. Hope is not lost – resolutely promised prosecutions of politician, corporate and media climate criminals may finally force urgent climate action.

Professor James Hansen was formerly head of NASA's Goddard Institute for Space Studies and is presently an adjunct professor directing the Program on Climate Science, Awareness and Solutions of the Earth Institute at 96-Nobel Laureate Columbia University [1]. However his latest summary of the science underlying the worsening climate emergency [3] will fall on deaf ears because climate criminal corporate Big Money malignantly dominates, perverts and subverts Mainstream media [4-10], and trumps the voices of Humanity's leading climate scientists such as James Hansen. Indeed while scientists are overwhelmingly concerned about the worsening climate emergency as evidenced by a 2017 letter signed by over 15,000 scientists [11, 12], and the 2018 IPCC Report on plus 1.5C [13, 14], industrial carbon pollution and atmospheric CO_2 continue to rise inexorably [3, 11-20].

Profit-driven, corporate-backed climate change denial has been well documented [7, 8] but Naomi Klein has summarized the fundamental neoliberal capitalist objection to climate change action: "Many deniers are quite open about the fact that their distrust of the science grew out of the powerful fear that if climate change is real, the political implications would be catastrophic … I think these hard-core ideologues understand the real significance of climate change better than most of the "warmists" in the political center, the ones who are still insisting that the response can be gradual and painless and that we don't need to go to war with anybody, including the fossil fuel companies… the real reason we are failing to rise to the climate moment is because the actions required directly challenge our reigning economic paradigm (deregulated capitalism combined with public austerity)… [and] spell extinction for the richest and most powerful industry the world has ever known – the oil and gas industry)" ([5], pages 42, 43 and 63).

While there is "outright climate change denial" by right-wing, anti-science buffoons like US president Donald Trump, there is a dominant global political culture of "effective climate change denial" through climate change inaction that is best illustrated by the woefully insufficient national commitments at the 2015 Paris Climate Change Conference that even if adhered to will amount to a catastrophic 2.9-3.4°C temperature rise by 2100 [21, 22].

Grossly insufficient climate action through "outright climate change denial" and "effective climate change denial" is well illustrated by climate criminal Australia that is among the world leaders in 14 global warming-related areas of climate criminality, specifically (1) annual per capita greenhouse gas (GHG) pollution, (2) live methanogenic livestock exports, (3) natural gas exports, (4) actual or adumbrated exploitation of recoverable shale gas reserves that can be accessed by hydraulic fracturing (fracking), (5) coal exports, (6) land clearing, deforestation and ecocide, (7) speciescide – species extinction, (8) coral reef destruction, (9) whale killing and extinction threat through global warming, (10) terminal carbon pollution budget exceedance, (11) per capita carbon debt, (12) GHG generating iron ore exports, (13) climate change inaction, and (14) climate genocide and approaches towards omnicide and terracide [23]. While about half of the ruling Coalition Australian Government MPs idiotically espouse "outright climate change denial", the whole government is involved in "effective climate change denial" through climate change inaction in these 14 areas. While it pays lip service to the reality of global warming and the need for massive investment in renewable energy, the Labor Opposition is just as "effective climate change denialist" as the Coalition by having a common policy with the Coalition Government of unlimited exports of coal, gas, [iron ore,] and methanogenically-derived meat. A picture says a thousand words, and PM Scott "Scomo" Morrison, leader of the pro-coal Coalition (COALition) Government held up a lump of coal in the Australian Parliament, declaring "This is coal. Don't be afraid, don't be scared" [24].

However Humanity does have serious cause to be afraid and to be scared over coal and greenhouse gas (GHG) pollution. Thus air pollution from carbon fuel burning ultimately kills about 7 million people globally each year [9 million as of 2020], with pollutants from the burning of Australia's world-leading coal exports contributing to about 75,000 of these annual deaths [25]. Global warming from greenhouse gas (GHG) pollution has been estimated to kill about 0.4 million people annually [26], but this is surely an under-estimate because 15 million people die avoidably each year from deprivation in tropical and sub-tropical Developing countries (minus China) [27], countries that are disproportionately impacted by global warming through more severe droughts, forest fires, floods, intense storms, and storm surges. Several leading climate scientists, namely Dr James Lovelock FRS and Professor Kevin Anderson, have suggested that only 0.5 billion may survive this century if global warming is not requisitely tackled, this translating to a worsening Climate Genocide in which 10 billion people will die from climate change this century at an average rate of 100 million per year [26] i.e. 6 times greater than the current 15 million avoidable deaths from deprivation each year [27].

The present and predicted carnage poses the questions of personal responsibility, complicity, and climate criminality. Indeed one must consider individual and collective complicity in manslaughter, murder, genocide and depraved indifference to such deadly events. Briefly, manslaughter is unintentional killing, murder is intentional killing, and genocide is defined by Article 2 of the UN Genocide Convention which states "In the present Convention, genocide means any of the following acts committed with intent to destroy, in whole or in part, a national, ethnic, racial or religious group, as such: a) Killing members of the group; b) Causing serious bodily or mental harm to members of the group; c) Deliberately inflicting on the group conditions of life calculated to bring about its physical destruction in whole or in part; d) Imposing measures intended to prevent births within the group; e) Forcibly transferring children of the group to another group" [28].

Individual or collective complicity is more difficult to define. Thus, for example, I spend a lot of my time researching and writing about the deadly consequences of man-made climate change but I am a citizen of a country, Australia, that is among world leaders in climate criminality and, as a measure of carbon footprint, has a per capita GDP ($56,000) that is 35 times greater than that of acutely global warming-threatened Bangladesh ($1,600) and 28 times greater than that of acutely global warming-threatened India ($2,000) [29]. Setting aside these embarrassing realities, we can nevertheless readily identify the worst individual climate criminals – the directors of fossil fuel companies, "climate change denialist" or "effective climate change denialist" politicians guilty of deadly climate criminality through climate change inaction, and journalists, editors and commentators guilty of depraved indifference through espousing "outright climate change denial" or "effective climate change denial".

There is a compelling case for resolutely promised, global prosecutions of politician, corporate and media climate criminals – a resolute promise by decent Humanity that may finally force urgent climate action. However before considering this promised retribution in detail it is useful to consider the resolute rejection or ignoring by climate criminals of 3 key realities set out by leading climate scientist Professor James Hansen in his recent, must-read analysis of the worsening climate emergency, "The Gathering Storm" [3].

1. **James Hansen on the Earth's energy imbalance and the need for negative CO_2 emissions to 342-373 ppm CO_2:** "Earth is now substantially out of energy balance. The amount of solar energy that Earth absorbs exceeds the energy radiated back to space. The principal manifestations of this energy imbalance are continued global warming on decadal time scales and continued increase in ocean heat content. Quantitative understanding of Earth's energy imbalance has improved over the past decade. The upper two kilometers of the ocean, where most of the excess energy is stored, has been well-monitored by the international Argo floats program since 2005. Over the full solar cycle 2005-2016 Earth's energy imbalance is 0.75 ± 0.25 W/m^2. The range 0.5 to 1 W/m^2 is substantial. For example, in order to restore Earth's energy balance by reducing atmospheric CO_2, which is the principal cause of the imbalance, CO_2 would need to be reduced from its 2018 407 amount to 373 ppm if the imbalance is 0.50 W/m^2, but to 342 ppm if the imbalance is 1 W/m^2. In reality CO_2 is not only continuing to increase, its rate of growth is increasing. The reason is that global population and energy demands continue to increase, and about 85 percent of global energy is provided by fossil fuels" [3].

The climate criminal reality is that atmospheric CO_2 is increasing [3, 18], the rate of growth of atmospheric CO_2 is increasing [3, 18], electricity sector use of coal, gas and oil is remorselessly increasing [17], and annual industrial CO_2

pollution is increasing [19, 20], as is atmospheric pollution by the potent greenhouse gases methane (CH_4) [30, 31] and nitrous oxide (N_2O) [31, 32]. The remorseless increase in carbon pollution and GHG pollution rejects even the notion of zero carbon pollution, and likewise the vital concepts of "negative carbon pollution" and "negative economic growth" are utterly ignored [15]. Using coral reefs as a "canary in the coal mine" one can estimate that coral reefs starting dying when the CO_2 reached 320 ppm at a time (1965) when the world's population was 3.3 billion as compared to the present 7.5 billion i.e. on this basis the Earth is overpopulated by a factor of about 2 [15]. However the presently dominant neoliberal economists argue that present "negative population growth" in some advanced countries is bad for the "economic growth" to which they are ideologically committed.

2. **James Hansen on paleoclimatological evidence for the predicted 3°C of warming meaning a long-term 15-25 meters sea level rise:** "Sea level reached heights as great as 6-9 meters during the prior interglacial period, the Eemian about 120,000 years ago, when global temperature was only about 1°C above the pre-industrial level, i.e., similar to today's global temperature. During the early Pliocene, several million years ago, when global temperature was at most about 3°C warmer than pre-industrial conditions, sea level probably reached as high as 15-25 meters above today's level. In other words, there is plenty of vulnerable ice available to cause eventual sea level rise that would inundate today's coastal cities, in response to a warming level that we could produce this century. Burning all of the readily available fossil fuels would eventually melt almost all the ice on the planet, raising sea level 65-75 meters (more than 300 feet)" [3].

The climate criminal reality is that long-term sea level rise is essentially ignored by the North and the problem, if at all recognized, is left to future generations to deal with in a process of extraordinary intergenerational injustice [3, 33]. The Island Nations and mega-delta countries of the South such as

Bangladesh are acutely aware of the problem that is already impacting them severely, and indeed talk of a worsening Climate Genocide [26, 34].

3. **James Hansen on the \$1-2 trillion cost of lowering atmospheric CO_2 by merely 1 ppm:** "How much CO_2 must be extracted from the air today to offset the excess growth of greenhouse gas forcing in a single year, i.e., to reduce climate forcing by 0.015 W/m^2? Atmospheric CO_2 must be reduced almost exactly 1 ppm CO_2 to increase heat radiation to space by 0.015 W/m^2. [We actually need to suck more than 1 ppm from the air, because the ocean reacts to the reduction of atmospheric CO_2 by increasing the net backflux of CO_2 to the atmosphere. However, we can make our point without including this added difficulty in achieving CO_2 drawdown.] One ppm of CO_2 is 2.12 billion tons of carbon or about 7.77 billion tons of CO_2. Recently Keith et al. (2018) achieved a cost breakthrough in carbon capture, demonstrated with a pilot plant in Canada. Cost of carbon capture, not including the cost of transportation and storage of the CO_2, is \$113-232 per ton of CO_2. Thus the cost of extracting 1 ppm of CO_2 from the atmosphere is \$878-1803 billion" [3].

The climate criminal reality is that in the last decade the atmospheric CO_2 (as measured at Mauna Loa in Hawaii) has been increasing at about 2.5 ppm CO_2 per year [18] and the cost of reducing atmospheric CO_2 by this amount would accordingly be about \$2-4.5 trillion each year, a bill to be eventually and inescapably paid by future generations in a process of gross intergenerational injustice [3, 33]. In his Papal Encyclical "Laudato Si'" Pope Francis cogently argued that the environmental and social costs of pollution must be "fully borne" by those incurring them (i.e. a full Carbon Price) ([35], Section 195 [36]). However the climate criminals insist that the environmental and social costs of pollution must be "fully borne" by the Biosphere, Humanity as a whole, and in particular by the young and future generations i.e. monstrous, "might is right" climate theft, climate larceny, climate murder, and intergenerational injustice. It is notable that climate economist Dr Chris Hope from 90-Nobel laureate Cambridge University has estimated a damage-related carbon price of about $100-$200 per tonne of CO_2 [37, 38].

There still is hope – resolutely promised prosecutions of climate criminals may force urgent climate action

My science-informed and scientist friends and colleagues are very despondent about climate change inaction. As a scientist I am generally optimistic about things but in relation to climate change I have had to concede for several years now that a catastrophic plus 2^oC temperature rise is now essentially inescapable, while nevertheless asserting that we must do everything we can to help make the future "less bad" for our children and future generations. However 2 things augur well for "making the future less bad", specifically Climate Revolution [39] and Climate Justice involving resolutely promised prosecutions of climate criminals [40, 41].

At some point Climate Revolution will happen when the South and the Young revolt (non-violently one hopes) against the rich, neoliberal One Percenters of the North who own half the world's wealth and are hell-bent on pushing Humanity and the Biosphere over the edge of the existential cliff. The key issue that is assiduously hidden by the mendacious, One Percenter-controlled Mainstream media is Humanity's gigantic, ever-increasing and inescapable Carbon Debt [38]. The Historical Carbon Debt (aka Historical Climate Debt) of a country can be measured by the amount of greenhouse gas (GHG) it has introduced into the atmosphere since the start of the Industrial Revolution in the mid-18th century. Thus the total Carbon Debt of the world from 1751-2016 (including CO_2 that has gone into the oceans) is about 1,850 billion tonnes CO_2. Assuming a damage-related Carbon Price of $200 per tonne CO_2-equivalent, this corresponds to a Carbon Debt of $370 trillion, similar to the total wealth of the world and 4.5 times the world's total annual GDP. Using estimates of national contributions to Historical Carbon Debt and assuming a damage-related Carbon Price in

USD of $200 per tonne CO_2, the World has a Carbon Debt of $370 trillion that is increasing at $13 trillion per year.

My own country, Australia, has a Carbon Debt of US$7.5 trillion (A$10.7 trillion) that is increasing at US$400 billion (A$570 billion) per year and at US$40,000 (A$57,000) per head per year for under-30 year old Australians. When young Australians realize the enormity of this imposition they will be enraged and take resolute action – but that day of reckoning has been delayed by the mendacity of largely US-owned Australian Mainstream media. Nevertheless young Australians are well aware that they have been horribly betrayed without realizing the enormity of the betrayal. Already the climate criminal, pro-coal COALition is facing a landslide electoral loss in the mid-year 2019 Federal elections [2020 note: no, unfortunately they won!]. In 2018 thousands of school children across the country skipped school to demonstrate for climate action (after PM Scott Morrison asserted that they should go back to school, the children correctly replied that it is PM Morrison who should go back to school). When under-30 year old young Australians realize their inescapable Carbon Debt is increasing at about A$60,000 per year per person they will utterly reject the climate criminal parties at the ballot box.

There is a real prospect of Climate Justice involving resolutely promised prosecutions of climate criminals and confiscation of their assets (as presently happens to traders in illicit drugs). There is no legislative retrospectivity required in this – climate murder, climate homicide and climate genocide are presently just as deadly as "conventional" murder, homicide and genocide. Something of the order of 1 million people die climate-related deaths annually, with this set to soar to an average of 100 million such deaths per year this century if man-made climate change is not requisitely addressed [26]. Of course such punishments will be of scant comfort to the loved ones of those who have died. However most importantly it is the resolute threat of prosecutions of climate criminals that may finally force urgent climate action. There needs to be a new globally-endorsed social contract to the effect that significant, present-day climate criminals will be inescapably held to account with judicially-imposed penalties ranging up to life imprisonment with confiscation of all assets. Please tell everyone you can – there still is realistic hope that we can stop the present slide to disaster.

References.

[1]. "James Hansen", Wikipedia:
https://en.wikipedia.org/wiki/James_Hansen
[2]. "List of Nobel laureates by university affiliation", Wikipedia:
https://en.wikipedia.org/wiki/List_of_Nobel_laureates_by_university_affiliation.
[3]. James Hansen, "Climate change in a nutshell: the gathering storm", Columbia University, 18 December 2018:
http://www.columbia.edu/~jeh1/mailings/2018/20181206_Nutshell.pdf.
[4]. "Climate change denial", Wikipedia:
https://en.wikipedia.org/wiki/Climate_change_denial.
[5]. Naomi Klein, "This Changes Everything. Capitalism vs. the Climate", Allen Lane, London 2014.
[6]. Gideon Polya, "This Changes Everything. Capitalism vs. the Climate" by Naomi Klein – Green Socialist Revolution ASAP", Countercurrents, 11 December, 2014:
https://www.countercurrents.org/polya111214.htm.
[7]. Robert Kenner, "Merchants of Doubt" movie, 2014.
[8]. Stephen Leahy, "Merchants of Doubt" film exposes slick US industry behind climate denial", Guardian, 20 November 2014:
https://www.theguardian.com/environment/2014/nov/20/merchants-of-doubt-film-exposes-slick-us-industry-behind-climate-denial.
[9]. Andrew Glikson, "The gathering climate storm and the media cover-up", Countercurrents, 2 January 2019:
https://countercurrents.org/2019/01/02/the-gathering-climate-storm-and-the-media-cover-up/.
[10]. Mary Debrett, "The media on climate change: a perfect storm of miscommunication", La Trobe University, 20 December 2018:
https://www.latrobe.edu.au/big-fat-ideas/bold-thinking-social-conscience/the-media-on-climate-change-a-perfect-storm-of-miscommunication.
[11]. William J. Ripple et al., 15,364 signatories from 184 countries, "World scientists' warning to Humanity: a second notice", Bioscience, 13 November 2017:
https://academic.oup.com/bioscience/advance-article/doi/10.1093/biosci/bix125/4605229.
[12]. Gideon Polya, "Over 15,000 scientists issue dire warning to Humanity on catastrophic climate change and biodiversity loss", Countercurrents, 20 November 2017:
https://countercurrents.org/2017/11/20/over-15000-scientists-issue-

dire-warning-to-humanity-on-catastrophic-climate-change-and-biodiversity-loss/.

[13]. IPCC, "Global warming of 1.5 °C. Summary for Policymakers", 8 October 2018: http://report.ipcc.ch/sr15/pdf/sr15_spm_final.pdf.

[14]. Gideon Polya, "IPCC +1.5C avoidance report – effectively too late but stop coal burning for "less bad" catastrophes", Countercurrents, 12 October 2018: https://countercurrents.org/2018/10/12/ipcc-1-5c-avoidance-report-effectively-too-late-but-stop-coal-burning-for-less-bad-catastrophes/.

[15]. Gideon Polya, "How much negative carbon emissions, negative population growth & negative economic growth is needed to save planet?" Countercurrents, 28 November 2018: https://countercurrents.org/2018/11/28/how-much-negative-carbon-emissions-negative-population-growth-negative-economic-growth-is-needed-to-save-planet/.

[16]. Gideon Polya, "Pro-coal Australia & Trump America Reject Dire IPCC Report & Declare War on Terra", Countercurrents, 17 October 2018: https://countercurrents.org/2018/10/17/pro-coal-australia-trump-america-reject-dire-ipcc-report-declare-war-on-terra/.

[17]. David Spratt, "Our energy challenge in 6 eye-popping charts", Climate Code Red, 17 June 2018: http://www.climatecodered.org/2018/06/our-energy-challenge-in-6-eye-popping.html.

[18]. US NOAA, "Full Mauna Loa CO_2 record": https://www.esrl.noaa.gov/gmd/ccgg/trends/full.html.

[19]. Catherine Hours and Marlowe Hood, ""Bad news": CO_2 emissions to rise in 2018, says IAE chief", Phys. Org., 18 October 2018: https://phys.org/news/2018-10-bad-news-co2-emissions-iea.html.

[20]. AP, "Global carbon pollution rose in 2018", New York Post, 5 December 2018: https://nypost.com/2018/12/05/global-carbon-pollution-rose-in-2018/.

[21]. Alex Kirby, "World set for 3.4°C by 2100", Climate News Network, 15 November 2017: https://climatenewsnetwork.net/23422-2/.

[22]. "World is set to warm 3.4°C by 2100 even with Paris climate deal", New Scientist, 3 November 2016: https://www.newscientist.com/article/2111263-world-is-set-to-warm-3-4c-by-2100-even-with-paris-climate-deal/.

[23]. Gideon Polya, "Offences of Pentecostal Christian Scott Morrison, PM after Australia's fourth PM-removing coup in 5 years", Countercurrents, 18 September 2018: https://countercurrents.org/2018/09/18/offences-of-pentecostal-

christian-scott-morrison-pm-after-australias-fourth-pm-removing-coup-in-8-years/.
 [24]. Katharine Murphy, "Scott Morrison brings coal to question time: what fresh idiocy is this?", Guardian, 9 February 2017: https://www.theguardian.com/australia-news/2017/feb/09/scott-morrison-brings-coal-to-question-time-what-fresh-idiocy-is-this.
[25]. "Stop air pollution deaths": https://sites.google.com/site/300orgsite/stop-air-pollution-deaths.
[26]. Gideon Polya, editor, "Climate Genocide": https://sites.google.com/site/climategenocide/.
[27]. Gideon Polya, "Body Count. Global avoidable mortality since 1950", that includes a succinct history of every country and is now available for free perusal on the web: http://globalbodycount.blogspot.com/.
[28]. "UN Genocide Convention": http://www.edwebproject.org/sideshow/genocide/convention.html.
[29]. "List of countries by GDP (nominal) per capita", Wikipedia: https://en.wikipedia.org/wiki/List_of_countries_by_GDP_(nominal)_per_capita.
[30]. Gideon Polya, editor, "Methane Bomb Threat": https://sites.google.com/site/methanebombthreat/.
[31]. Gideon Polya, "2011 climate change course": https://sites.google.com/site/300orgsite/2011-climate-change-course.
[32]. "Global N_2O levels": https://www.n2olevels.org/
[33]. Gideon Polya, editor, "Climate Justice & Intergenerational Equity": https://sites.google.com/site/300orgsite/climate-justice.
[34]. Pacific Islands Development Forum 4 September 2015 "Suva Declaration on Climate Change": http://pacificidf.org/wp-content/uploads/2013/06/PACIFIC-ISLAND-DEVELOPMENT-FORUM-SUVA-DECLARATION-ON-CLIMATE-CHANGE.v2.pdf.
[35]. Gideon Polya, "Green Left Pope Francis Demands Climate Action "Without Delay" To Prevent Climate "Catastrophe"", Countercurrents, 10 August, 2015: https://countercurrents.org/polya100815.htm.
[36]. Pope Francis, Encyclical Letter "Laudato si'", 2015: http://w2.vatican.va/content/francesco/en/encyclicals/documents/papa-francesco_20150524_enciclica-laudato-si.html.
[37]. Dr Chris Hope, "How high should climate change taxes be?", Working Paper Series, Judge Business School, University of Cambridge,
9.2011: http://www.jbs.cam.ac.uk/media/assets/wp1109.pdf.
[38]. Gideon Polya, editor, "Carbon Debt Carbon Credit": https://sites.google.com/site/carbondebtcarboncredit/.

[39]. Gideon Polya, editor, "Climate Revolution Now":
https://sites.google.com/site/300orgsite/climate-revolution.
[40]. Gideon Polya, editor, "Punish climate criminals":
https://sites.google.com/site/punishclimatecriminals/.
[41]. Gideon Polya, editor, "Stop climate crime":
https://sites.google.com/site/300orgsite/stop-climate-crime.

EPILOGUE – SUMMARY OF SOLUTIONS

This concluding epilogue simply, systematically, and with minimal discussion, sets out the things that urgently need to be done as soon as possible (ASAP) in order to mitigate man-made global warming and to minimize harm to Humanity and the Biosphere. As a check list for trying to save the Planet it is broken down into 10 well-defined subsections for clarity.

(1). Science not spin, zero tolerance for lying, and social humanism not neoliberalism.

(a). Science not spin. There must be science and not spin in tackling man-made climate change. Science is about the critical testing of potentially falsifiable hypotheses. In contrast, anti-science spin involves the selective use of asserted facts to support a partisan position. Science-based rational risk management that is crucial for societal safety successively involves (i) accurate information, (ii) science-based analysis of the data, and (iii) science-informed systemic change to minimize harm when things inevitably go wrong. The all too common and dangerous converse of this science-based risk management is the anti-science, spin-based process successively involving (i) censorship, self-censorship, falsehood, lying, obfuscation, intimidation, and spin, (ii) anti-science, spin-based analysis involving the selective use of asserted facts to support a partisan position, and (iii) counterproductive "blame and shame" that militates against the vital primary accurate reportage required for rational risk management, and which in the worst case results in war (or War on Terra in the present context).

(b). Zero tolerance for lying. There must be zero tolerance for lying because it utterly short-circuits the scientific process that is crucial for rational risk management. Lying by omission (e.g. as in spin-based selective use of asserted facts) is far, far worse than outright repugnant lying by commission because the latter at least permits public refutation and public debate. Noting the appalling 9 million annual air pollution deaths and the worsening Climate Genocide that may lead to 10 billion deaths this century, genocide ignoring and holocaust ignoring are far, far worse than repugnant genocide denial and holocaust denial because the latter at least allow for public refutation and discussion. Free expression is vital for democracy, science and scholarly inquiry, and accordingly the only judicially-applied punishment for lying should be the ignominy of judicially-supervised and expertly-effected exposure.

(c). Social humanism not neoliberalism. There must be sustainable and equitable social humanism and not terracidal and inequitable neoliberalism as the World faces serious limitations on sustainable resource exploitation. "All men are created equal and have an inalienable right to life, liberty and the pursuit of happiness". Yet neoliberalism demands maximal freedom for the smart and advantaged to exploit natural and human resources for maximal private profit. Conversely, social humanism (socialism, democratic socialism, eco-socialism, human rights-cognizant communism, the welfare state, Universal Basic Income) seeks to sustainably maximize happiness, opportunity and dignity for everyone through culturally-sensitive and evolving intra-national and international social contracts. Sharing and sustainability are demanded because there is no Planet B, the present exploitation is unsustainable, and each year 15 million people presently die avoidably from deprivation on Spaceship Earth with neoliberal One Percenters in charge of the flight deck.

(2). Exactly accounted Carbon Price and Carbon Tax, species and ecosystems are priceless, and unsustainable human behaviour must be modified.

(a). Exactly accounted, fully borne Carbon Price and Carbon Tax. There must be an exactly accounted, damage-related Carbon Price i.e. a price on carbon pollution. A damage-related Carbon Price is about \$200 per tonne CO_2-equivalent whereas presently the World average Carbon Price is a mere \$2 per tonne CO_2-equivalent. We are familiar in everyday life with the notion of a fair price for things but this 100-fold gap in actually applied Carbon Price is seen as "the greatest market failure in history". It is also grand larceny on a gigantic scale. Thus failure to properly assess methane leakage and to tax it at a correct rate amounts to massive Carbon Tax Evasion. Whereas Conventional Tax Evasion results in poverty and preventable death in rich countries, Carbon Tax Evasion amounts to an existential threat to Humanity and the Biosphere.

(b). Species and ecosystems are priceless. We cannot destroy what we cannot replace. We were horrified by the attack on Michelangelo's "Pieta" statue but enough information exists to replace it by 3-D printing. A multi-billionaire wanted to be buried with one of Van Gogh's $150 million "Irises" but colour reproductions exist. However individual species and complex ecosystems cannot be replaced and are thus priceless. There should be cessation of the ongoing speciescide and ecocide on a remorseless path toward omnicide and terracide.

(c). Mitigate deceitful and greedy human behaviour. It is claimed that while many animals deceive each other, only humans deceive themselves. This is well exampled by the clearly unsustainable and disastrous "economic growth" model accepted by all governments in a World in which de-population and economic de-growth are patently required. Polya's 3 Laws of Economics parallel the 3 Laws of Thermodynamics underlying chemistry and physics, and simply state (1) Profit equals Price minus Cost of Production, (2) Deceit about Cost of Production strives to a maximum, and (3) no price, profit or jobs on a dead Planet. The market's "hidden hand" ignoring the huge Cost of Production externality of a $200 per tonne CO_2-e Carbon Price is massive deceit that now existentially threatens much of Humanity and the Biosphere. It is difficult to modify human behaviour. Thus even a homeless and jobless person selling matches may seek to improve his "margin". Human behaviour is hard-wired through evolution involving selection of mutable genes, but also evolves through societal selection of advantageous "memes" such as 'love thy neighbour as thyself'. Deceitful and greedy human behaviour can be mitigated by such humane memes. Equity and sustainability mean that basic "needs" must be satisfied sustainably for all but "wants" must be mitigated by environmental impact.

(3). Resolute threat to punish climate criminals for speciescide, ecocide, inaction and intergenerational injustice through ignoring inescapable Carbon Debt.

(a). Resolute threat to punish climate criminals. Custodial and cost-recovery-based financial punishment of the worst climate criminals will be of little comfort to wretched future generations in a devastated World. However the resolute promise of severe, retrospective punishment of climate criminal corporate and government leaders may have the effect of constraining their present disastrous conduct. As such, the resolute promise of judicially-determined custodial and financial punishment of climate criminals may be a major item in the armoury of climate change actions. The worst climate criminals are the heads of polluting corporations and of mendacious media corporations spreading a mantra of climate change denialism or of effective climate change denialism through climate change inaction.

(b). Punish climate criminals for speciescide, ecocide and inaction. All species and ecosystems are priceless, but nevertheless some level of monetary accountability for climate criminality can be pragmatically assessed from the economic value of the Biosphere. It has been estimated that the economic value of wild nature is about \$40 trillion annually or about 50% of the annual global GDP, and that the cost of preserving what is left of wild nature is associated with a benefit to cost ratio of 100:1. We are responsible not only for what we do but also for what we fail to do.

(c). Punish climate criminals for intergenerational injustice through their ignoring of inescapable Carbon Debt. Applying a damage-related Carbon Price of \$200 per tonne CO_2-equivalent to an upwardly revised estimate of past and present GHG pollution (properly taking methane and land use into account) yields a global Carbon Debt of \$200-250 trillion that is increasing at \$16 trillion each year. While conventional

debt can be variously evaded by default, bankruptcy or printing
money, Carbon Debt is inescapable. Thus, for example, unless
trillions of dollars are spent on sea walls, coastal cities and
arable land will be inundated.

(d). National and corporate climate criminality can be
punished under International Law by Green Tariffs, Carbon
Pollution Taxes, International Criminal Court (ICC)
prosecutions, International Court of Justice (ICJ) litigations, and
Boycotts, Divestment and Sanctions (BDS).

**(4). Cease CO_2 and CH_4 carbon pollution, gas leaks and is
dirtier than coal GHG-wise, bioengineering, and stop toxic
air pollution, with atmospheric CO_2 draw-down to a safe
and sustainable level of circa 300 ppm CO_2.**

(a). Cease CO_2 pollution. All carbon pollution from burning
carbon fuels – coal, oil, gas and biomass – should cease ASAP.
CO_2 from cement manufacture can be partly mitigated by
improved "green cement" but whatever the mechanism used, all
CO_2 from the high temperature conversion of limestone
($CaCO_3$) to lime (CaO) must be sequestered.

**(b). Cease CH_4 pollution and de-fuse the Arctic Methane
Bomb.** Methane (CH_4) is about 85% of natural gas and has a
Global Warming Potential 105 times greater on a 20 year time
frame than that of CO_2. Atmospheric CH_4 is steadily increasing
due to leakage of CH_4 from gas reticulation systems, from
fracking and conventional gas extraction, from anaerobic
bacterial metabolism in ruminants and swamps, and from
release from huge CH_4 stores as CH_4-water clathrates in the
Arctic tundra and the Arctic Ocean sea bed due to global
warming. It is estimated that 50 billion tonnes of CH_4 will be
released from the Arctic in coming decades, this corresponding
to 50 x 105 = 5,250 billion tonne CO_2-equivalent or about 9
times more than the 600 billion tonne Terminal Carbon
Pollution Budget from 2010 for a 75% chance of avoiding a
catastrophic plus 2C temperature rise. There is a Planet-
threatening positive feedback loop here known as the Arctic

Methane Bomb: global warming -> disproportionate Arctic warming -> methane release -> increased global warming ->....

(c). Gas is leaks and is dirtier than coal GHG-wise, Methane (CH_4) is a gas, leaks, is 85% of natural gas, and has a Global Warming Potential (GWP) of 105 relative to the same mass of CO_2 on a 20 year time frame and with aerosol impacts considered. On this basis and assuming that natural gas is 100% CH_4 or an equivalent hydrocarbon, a gas leakage of 2.6% means that the greenhouse gas (GHG) warming effect from the leaked gas is the same as that of the CO_2 from burning the remaining 97.4% of the gas. With no gas leakage, burning 1 tonne of CH_4 yields 2.75 tonnes CO_2 as compared to burning 1 tonne of carbon (C) yielding 3.67 tonnes CO_2. However with a mere 2.6% of systemic gas leakage (it is about 3% in the US), the burning of 1 tonne gas results in 5.36 tonnes CO_2-equivalent i.e. with systemic gas leakage considered, burning natural gas is about 1.5 times as dirty GHG-wise than burning coal on a per mass basis. Energy from combustion is 50.1 GJ/t (CH_4), 32.6 GJ/t (carbon, C) and 17 GJ/t (lignite or brown coal).The "CO_2-equivalent per GJ" (Giga Joule) of energy produced on burning (e.g. in a 100% efficiency power station) is 5.36/ 50.1 = 0.107 (for CH_4), 3.67/32.6 = 0.113 (for carbon, C) and 0.216 (for brown coal or lignite) i.e. at 2.6% leakage even on a per energy yield basis burning CH_4 is about as dirty GHG-wise as burning carbon (C), but CH_4 becomes dirtier GHG-wise as the degree of leakage increases. Gas is dirty energy, gas is not clean, it is not clean-er than coal GHG-wise, and indeed gas can be dirtier than coal GHG-wise. The present absurd, dangerous and deadly coal-to-gas transition is testament to One Percenter and Mainstream mendacity. While the falsehood of "gas is cleaner than coal GHG-wise" can be attributed to stupidity and ignorance on the part of pro-gas politicians, in the face of expert scientific advice (e.g. from smart, first year high school students) it thence becomes a lie, and utterly unforgivable intellectual child abuse when imposed on children.

(d). Stop toxic air pollution from carbon fuel burning. 9 million people die annually from air pollution, 50% of which is occurring outside the home (e.g. from vehicular exhaust and coal-burning power stations) and half inside the home (e.g. from burning wood biomass and cattle dung in Africa and South Asia). There must be cessation of carbon fuel burning not only to stop this source of GHG emissions but also to stop this huge death toll from the long-term effects of exposure to toxic products from carbon fuel burning - notably fine $PM_{2.5}$ carbon particulates but also N_2O (nitrous oxide), NO_2 (nitrogen dioxide), CO (carbon monoxide), and heavy metals and radioactivity (from burning coal).

(e). Atmospheric CO_2 draw-down to a safe and sustainable level of circa 300 ppm CO_2. Zero net emissions are not enough – there must be atmospheric CO_2 draw-down (negative CO_2 emissions) from the present dangerous and damaging 415 ppm CO_2 to the pre-Industrial Revolution level of circa 300 ppm CO_2 that is safe and sustainable for all peoples and all species. About 320 ppm CO_2 is needed for retention of Arctic sea ice at former levels, and coral reefs started dying around the world at 320 ppm CO_2. The cost of a return to 300 ppm CO_2 is about \$200-250 trillion. Various technologies are available for atmospheric CO_2 draw-down:

(i). Re-afforestation (photosynthetic CO_2 sequestration as wood, tree foliage, undergrowth and ultimately soil carbon);

(ii). Biochar (anaerobic pyrolysis at 400-700C of up to 12 Gt carbon annually of cellulosic forestry or agricultural waste to yield charcoal (biochar, carbon, C) that can be sequestered or added to soil; ecology-cognizant mechanical harvesting of forest undergrowth for biochar would be preferable to CO_2-releasing burning, and would also be useful in reducing fuel load in the context of worsening forest fires world- wide in the Anthropocene and Pyrocene Era).

(iii). Photosynthetic algae in fertilized oceans (a bioengineering proposition involving large scale addition of iron oxide to the oceans or algal biomass production adjacent to fossil fuel power stations).

(iv). Direct Air Capture (DAC) (sequestration of CO_2 as calcium carbonate in an alkaline solution followed by high temperature conversion of the carbonate to CO_2 and alkaline lime (CaO) with sequestration of the CO_2).

(v). Accelerated Weathering of Limestone (AWL)(CO_2 passed through a $CaCO_3$ (limestone, calcium carbonate) suspension is converted to soluble bicarbonate (HCO_3^-) which can be added to the ocean with the additional benefit of lifting the pH of the sea which is presently acidifying due to dissolution of CO_2).

(vi). Mineral carbonation (certain minerals are pulverised and in suspension react with CO_2).

(vii). Carbon Capture and Sequestration (CCS) (removal of 90% of CO_2 from fossil fuel exploitation and sequestration underground in depleted oil and gas fields or in deep saline aquifer formations; it is yet to be applied on a large scale).

(f). Bioengineering to reduce global warming. A variety of bioengineering approaches (variously with attendant risks) have been advanced to reduce global warming:

(i). Re-afforestation (arable land would not be used; there are 1.7 billion hectares of tree-less land suitable for planting 1.2 trillion trees that could remove 200 billion tonnes of carbon over 100 years).

(ii). Biochar from cellulosic biomass from forestry and agricultural "waste" (a maximum of 12 Gt carbon could be sequestered thus annually, noting that arable land would not be involved, and that the richness and biological diversity of both arable land and forests should be maintained).

(iii). Photosynthetic algae in iron-fertilized fertilized oceans to trap CO_2 (downside: massive, global scale ecological disturbance).

(iv). Umbrellas in space to reduce solar radiance hitting the Earth.

(v). Painting of roof tops and roads white to increase surface albedo (reflectivity).

(vi). Changes in vegetation cover to increase surface albedo (downside: severe ecological damage).

(vii). Addition of sulphur to the stratosphere to increase light-reflecting, global dimming sulphate aerosols (there is a huge danger in that cessation of such action in a global emergency and diminution of sulphate particles would result in a sudden spike in temperature).

(5). 100% renewable energy ASAP with adequate energy storage and efficient DC and AC HV electrical transmission grids.

(a). A variety of renewable energy and energy storage technologies are available including (a) wind power (using on-shore or off-shore wind turbine arrays, (b) solar PV using photovoltaic cells, (c) solar thermal (with energy sequestration as molten potassium nitrate or sodium nitrate salts), (d) wave power, (e) tidal power, (f) geothermal power (e.g. as exploited in Iceland), and (g) hydroelectric power. With political will 100% renewable energy with back-up energy storage is possible within 10 years. Back-up energy storage can be provided by batteries, molten salts, gas compression and hydrologically (through hydroelectric dams).

(b). Nuclear energy is often incorrectly cited as a renewable energy source but it is clearly not renewable energy because the required uranium and thorium ores are a limited resource. Further, apart from the actual nuclear fission, the overall nuclear cycle (from building the plant and mining, transporting and processing the ore to eventual safe storage of radioactive waste and de-commissioning the plant) generates a lot of CO_2 in a carbon-based economy. Fast breeder reactors are extremely efficient users of uranium but would generate a plutonium economy that would be a security and human rights nightmare. Nuclear fusion is in its technological infancy, but to state the obvious, in the Sun we already have a massive nuclear fusion reactor 150 million kilometers away from the Earth that is also protected from high energy solar-origin particles by its magnetosphere.

(c). Adequate energy storage is available to supplement wind and solar energy via batteries, molten salts energy storage, pressurizing gas, and hydrological energy storage. Adequate energy storage is required for night-time solar energy operations, to complement wind energy in low wind situations, and for management of renewable energy systems for "dispatchable energy".

(d). High efficiency DC and AC HV electrical transmission is required for extensive power grids so that, for example, high wind and/or solar energy in a particular location can efficiently complement low wind and/or solar energy in a distant location on the grid.

(e). Electrical vehicles to enable renewable energy-powered transport.

(6). Energy efficiency, transport efficiency, rail not road, locally grown food, vegetarian diet, building thermal efficiency, clean steel, and hydrogen economy.

(a). Energy efficiency must be maximized in both electricity usage and in electricity transmission in electricity grids.

(b). Transport efficiency. There must be preference for rail not road and public transport not private transport (renewable energy-powered, of course). Thus a modern tram carrying 100 passengers can have the same energy usage per kilometre as a 4-wheel drive vehicle carrying only 5 passengers. Energy efficiency considerations aside, global vehicle deaths total a horrendous 1.35 million annually.

(c). Eat locally grown food and revel in the seasonal changes as well as the transport efficiency entailed. A reduction ad absurdum would be Arctic musk ox milk transported by jumbo jet to replace dairy cow milk in fertile Tasmania that is contiguous to Antarctica.

(d). Vegetarian diet for all Humanity. There must be a global vegetarian diet and replacement of cow-derived milk with plant substitutes (soybean and almond milk). The deadly food-for-fuel Biofuel Genocide obscenity must be stopped. Food adequacy involves cessation of all of the following: global inequity, population growth, economic growth, GHG pollution, sea level rise, soil salinization, deforestation, damaging land use, unsustainable aquifer depletion and meat consumption.

(e). Building thermal efficiency must be increased through insulation, efficient heat pumps, efficient design and efficient usage.

(f). Clean steel and cleaner metallurgy in general can be achieved by using renewable energy-based production of hydrogen as an energy source and reductant.

(g). A renewable energy-based hydrogen economy would massively sideline dirty energy sources. Hydrogen (H_2) generated by renewable energy-based electrolysis of water (H_2O) would be converted to ammonia (NH_3) for safe bulk transport and thence re-conversion to H_2. A possible downside yet to be thoroughly researched, is that H_2 is a very small molecule, and consequent H_2 leakage into the atmosphere would result in H_2 scavenging of hydroxyl radicals (OH.) otherwise involved in the removal of methane (CH_4), with a consequent increase in atmospheric levels of the potent GHG methane.

(h) Economic efficiency for sustainable provision of basic needs for all would be predicated by population control.

(7). Re-afforestation, increased catastrophic forest fires, and mitigation of forest fire risk.

(a). Re-afforestation on presently tree-less, non-arable land is a major mechanism for reducing atmospheric CO_2 but is inevitably associated with an increased risk of catastrophic forest fires due to increased fuel load, relentless global warming, increased drought, and critically low humidity. The recent catastrophic bush fires in South Eastern Australia more than doubled Australia's already huge annual Domestic GHG pollution.

(b). **Mitigation of forest fire risk.** There is a worsening incidence of unprecedented, catastrophic forest fires in North America, Southern Europe, Siberia, the Amazon, Indonesia and Australia. While much of the criminal damage in Brazil and Indonesia can be attributed to ecocidal greed, high incidence and high energy lightning strike is very important in initiation of forest fires. This huge and worsening risk to species, ecosystems, people and property can be addressed in part by (i) rapid and comprehensive detection systems, (ii) rapid deployment of sophisticated fire suppression operations, and (iii) ecology-cognizant reduction of fuel load, including mechanical harvesting of undergrowth and profitable generation of biochar.

(8). Population control necessity and possible mechanisms (some controversial) for population de-growth.

(a). Population control necessity. Massive GHG pollution, biological diversity loss and ecosystem loss (speciescide and ecocide) demonstrate that the World is already grossly overpopulated. Using coral reefs as a "canary in the mine" one can readily estimate that the world is presently overpopulated by a factor of 2. If everyone had an American lifestyle we would need 7 planets and a European lifestyle would require 3 planets. However there is no Planet B. Numerous scientists (climate scientists and demographers) estimate a human carrying capacity for the Planet of a mere 1 billion by 2100 i.e. a worsening Climate Genocide in which 10 billion people will perish from climate change this century.

(b). Possible mechanisms (some controversial) for population de-growth. Possible mechanisms for population de-growth include (i) Zero Population Growth (ZPG) through contraception (already being achieved in rich nations), (ii) a huge increase in female rights and empowerment throughout the World, (iii) greatly increased female literacy and a modest increase in per capita income leading to lowered infant mortality in the Developing World (minus China); (iv) the "one child policy" as formerly used to great effect in China; (v) humane, pre-conception male sex selection (only females can have children but this controversial measure would have to be accompanied by massive female empowerment and massive social engineering to address fundamental human behaviours), (vi) equal human rights for homosexuals and indeed all LGBTIQ people, (vii) promotion of celibacy, (viii) financial inducements for not having children, (ix) fully-costed debt imposition and long-term debt recovery for having children, and (x) increased adoption of the "Jane Austen option" of more women and men adopting the very satisfying alternative of careers without their own children and with re-directed altruism.

(9). Economic de-growth, equity, a sustainable decent existence for everyone, and a non-violent Climate Revolution now.

(a). Economic de-growth by about 50% follows from the perceived requirement to decrease the human population by about 50%. However much greater de-growth is possible depending on the per capita income – the higher the desired per capita income, the lower the carrying capacity.

(b). A sustainable decent existence for everyone. Cuba, China, the World and the US have per capita GDPs of $8,800, $10,300, $11, 400 and $65,300, respectively. However while 15 million people die avoidably from deprivation in the Developing World (minus China) each year, annual avoidable deaths in Cuba, China and the US are zero (0). If all countries had the same per capita GDP as happy Cuba then the world GDP would be $67 trillion as compared to the present $87 trillion.

(c). Equity and a non-violent Climate Revolution now. The solutions presented here enable further economic de-growth and sustainable preservation of what is left of the Biosphere.

However the fly in the ointment is the neoliberal greed of the One Percent that presently owns about half of the world's wealth and is politically dominant. In the face of intransigent resistance from the presently dominant and neoliberal One Percenters the only course left is a non-violent Climate Revolution now.

(10). Biological sustainability, cessation of speciescide and ecocide, and punishment of the destroyers of the Biosphere.

(a). Biological sustainability and cessation of speciescide and ecocide. In our present Anthropocene Era the animal extinction rate is 100-1,000 times greater than normal. This must stop – there must be absolute cessation of speciescide and ecocide.

(b). Promised punishment of the destroyers of the Biosphere is crucial. My own country Australia ranks worst in the world for climate action policies, and is among world leaders in 16 areas of climate criminality including GHG pollution, speciescide, ecocide, coral reef destruction, whale threatening and land clearing. In Brazil under a neoliberal and fascoid government the vital Amazon rain forest is burning, this prompting governments around the world to call for international action. There must be concerted international action to punish destroyers of the Biosphere through exposure, punitive tariffs, international legal actions and Boycotts, Divestment and Sanctions (BDS). Existing laws protect society from crimes such as deception, embezzlement, theft, corruption and grand larceny. Thus in climate criminal Australia, governments and corporations collude in grossly under-reporting methane fugitive emissions and in not applying a required damage-related Carbon Price of $200 per tonne CO_2-equivalent, this methane leakage-ignoring larceny amounting to an annual defrauding of Australia by an estimated $50 billion.

At the present plus 1.0-1.2 C of warming the World is being savaged by deadly high temperatures and global warming-exacerbated droughts, floods, forest fires, high energy intensity tropical storms, sea level rise, storm surges, massive ecosystem loss and irreversible Biodiversity loss. A plus 1.5C is inevitable in the coming decade and in the face of remorselessly increasing CO_2 and CH_4 in the atmosphere one concludes that a catastrophic plus 2C is effectively unavoidable. Nevertheless we are inescapably obliged to do everything we can to make the future "less bad" for our children, our grandchildren and future generations.

www.ingramcontent.com/pod-product-compliance
Lightning Source LLC
Chambersburg PA
CBHW021951170726
47994CB00020B/1